NITROGEN IN THE MARINE ENVIRONMENT

Academic Press Rapid Manuscript Reproduction

NITROGEN IN THE MARINE ENVIRONMENT

EDITED BY

EDWARD J. CARPENTER
DOUGLAS G. CAPONE
Marine Sciences Research Center
State University of New York
Stony Brook, New York

1983

ACADEMIC PRESS
A Subsidiary of Harcourt Brace Jovanovich, Publishers

New York London
Paris San Diego San Francisco São Paulo Sydney Tokyo Toronto

ACADEMIC PRESS, INC.
111 Fifth Avenue, New York, New York 10003

United Kingdom Edition published by
ACADEMIC PRESS, INC. (LONDON) LTD.
24/28 Oval Road, London NW1 7DX

Library of Congress Cataloging in Publication Data
Main entry under title:

Nitrogen in the marine environment.

Includes index.
1. Nitrogen cycle. 2. Marine biology.
I. Carpenter, Edward J. II. Capone, Douglas G.
QH91.8.N46N57 1983 574.5'2636 83-2829
ISBN 0-12-160280-X

PRINTED IN THE UNITED STATES OF AMERICA

83 84 85 86 9 8 7 6 5 4 3 2 1

To our parents,
Charles and Adelaide Carpenter
and
Louis and Marie Capone

CONTENTS

CONTRIBUTORS

ROBERT R. BIDIGARE, Department of Biochemistry and Biophysics, Texas A&M University, College Station, Texas 77843

DOUGLAS G. CAPONE, Marine Sciences Research Center, State University of New York, Stony Brook, New York, 11974

EDWARD J. CARPENTER, Marine Sciences Research Center, State University of New York, Stony Brook, New York 11974

LOUIS A. CODISPOTI, Bigelow Laboratory for Ocean Sciences, McKown Point, West Boothbay Harbor, Maine 04575

CHRISTOPHER F. D'ELIA, Chesapeake Biological Laboratory, University of Maryland, Center for Environmental and Estuarine Studies, Solomons, Maryland 20688

PAUL G. FALKOWSKI, Oceanographic Sciences Division, Department of Energy and Evironment, Brookhaven National Laboratory, Upton, New York 11973

PATRICIA M. GLIBERT, Woods Hole Oceanographic Institution, Woods Hole, Massachusetts 02543

JOEL C. GOLDMAN, Woods Hole Oceanographic Institution, Woods Hole, Massachusetts 02543

M. DENNIS HANISAK[1], University of Florida, c/o R.R. 1 Box 196, Fort Pierce, Florida 33450

WILLIAM G. HARRISON, Marine Ecology Laboratory, Bedford Institute of Oceanography, Box 1006, Dartmouth, Nova Scotia, Canada B2Y 4A2

AKIHIKO HATTORI, Ocean Research Institute, University of Tokyo, Nakano-ku, Tokyo 164, Japan

WARREN A. KAPLAN, Center for Earth and Planetary Physics, Harvard University, Cambridge, Massachusetts 02138

J. VAL KLUMP, Center for Great Lakes Studies, University of Wisconsin-Milwaukee, Milwaukee, Wisconsin 53201

EDWARD A. LAWS, Department of Oceanography, University of Hawaii, Honolulu, Hawaii 96822

[1]Present address: Center for Marine Biotechnology, Harbor Branch Institution, Fort Pierce, Florida 33450.

CHRISTOPHER S. MARTENS, Marine Sciences Program, University of North Carolina, Chapel Hill, North Carolina 27514

JAMES J. MCCARTHY, Museum of Comparative Zoology, Harvard University, Cambridge, Massachusetts 02138

SCOTT W. NIXON, Graduate School of Oceanography, University of Rhode Island, Narragansett, Rhode Island 02882

JOHN H. PAUL, Department of Marine Science, University of South Florida, St. Petersburg, Florida 33701

MICHAEL E. Q. PILSON, Graduate School of Oceanography, University of Rhode Island, Narragansett, Rhode Island 02882-1197

MICHAEL R. ROMAN, Horn Point Environmental Laboratory, University of Maryland, P. O. Box 775, Cambridge, Maryland 21613

MARY I. SCRANTON, Marine Sciences Research Center, State University of New York, Stony Brook, New York 11974

JONATHAN H. SHARP, College of Marine Studies, University of Delaware, Lewes, Delaware 19958

BARRIE F. TAYLOR, Division of Marine and Atmospheric Chemistry, Rosenstiel School of Marine and Atmospheric Sciences, University of Miami, Miami, Florida 33149

DENNIS L. TAYLOR, Center for Environmental and Estuarine Studies, University of Maryland, Cambridge, Maryland 21613

IVAN VALIELA, Boston University Marine Program, Marine Biological Laboratory, Woods Hole, Massachusetts 02543

RICHARD L. WETZEL, Virginia Institute of Marine Science and School of Marine Science, College of Williams and Mary, Gloucester Point, Virginia 23062

PATRICIA A. WHEELER[2], Museum of Comparative Zoology, Harvard University, Cambridge, Massachusetts 02138

RICHARD G. WIEGERT, Department of Zoology, University of Georgia, Athens, Georgia 30602

[2]Present address: School of Oceanography, Oregon State University, Corvallis, Oregon 97331.

FOREWORD

Investigations of the bases of marine biological productivity logically center on nitrogen, a necessary structural component of living cells, and the element limiting primary production over most of the world ocean. This compilation of the major results and efforts of a decade of intensive investigation into the many aspects of the nitrogen cycle attests to the importance of nitrogen as a central focus of nutrient-based productivity studies. Since the publication of this book represents a milestone in ocean productivity research, it may be helpful to briefly review how this burgeoning field of research arrived at this point, thus providing a guide to future directions.

Brandt's original speculations on nutrient limitation of biological production in the sea, made at the turn of the last century, were based on very few analyses of nutrient concentrations in seawater, the methods available being both tedious and difficult. Phosphorus occupied the center of the nutrient stage for several subsequent decades after relatively simple analytical methods for its determination became available. Nitrogen investigations were held back for lack of suitable analytical procedures, especially for ammonium. Nevertheless, Harvey's "Chemistry and Fertility of Sea Water," published in 1955, presented a useful framework for further studies of the chemical basis of primary production in the sea.

With improved capability to analyze seawater for the primary micronutrients, especially nitrate and ammonium, the realization developed largely through the efforts of John Ryther that inorganic nitrogen limits primary production over much of the world ocean. However, the complexity of the nitrogen cycle with multiple active pathways made it difficult to make further progress with purely chemical, mass balance experiments, especially in the ocean itself. At about that time, in the early 1960s, when John Neess and I began to use the stable isotope of nitrogen, ^{15}N, in the aquatic environment, mass spectrometers were rare and there were few laboratories such as that of Robert Burris at the University of Wisconsin, where this extremely useful tracer was used in biological experiments and where these facilities were made available to aquatic ecologists. In little more than another decade, mass spectrometers have become readily available, and the techniques for sample preparation have been improved and simplified with the result that there has been an explosion of new knowledge as many capable new investigators have directed their attention to marine nitrogen cycling. Their efforts have been based not only upon advances in primary nutrient analyses and stable isotope tracer techniques, but upon a vast array of methodologies that have been developed to probe the activities of living cells. As a result, early hypotheses have been discarded or been modified and new ideas developed to inspire and guide new experimental approaches. Naturally, greater understanding has led to the need for greater accuracies and sensitivities in laboratory instruments and procedures. This volume presents a well-balanced treatment of both the present state of knowledge and techniques that may be used to investigate new and old hypotheses.

What benefits to oceanography as a whole can be envisioned as a result of this newfound wealth of knowledge represented between the covers of this volume and from the research to follow? Beyond the obvious point that marine ecosystem studies generally benefit from advances in the understanding of nitrogen cycling there are some more specific topics that will benefit also. For example, we are entering a period where an understanding of phytoplankton processes and attributes will contribute toward unraveling the nature of some physical processes such as mixed layer dynamics. The intense activity exhibited by bacteria and phytoplankton in eastern boundary regions may conceivably be found to influence world climate through interaction with the carbon dioxide cycle. The local aberrations in sea surface temperature observed in conjunction with intense blooms of migrating phytoplankton populations may be found to have effects that propagate for unexpected distances. Improved understanding of phytoplankton processes may lead also to the development of "phytoplankton engineering" approaches to managing, for example, large anthropogenic introductions of nitrogen into the sea, and to mass culturing of specific phytoplankton species for aquaculture with improved predictability. Reliable simulation models and improved theory are needed if engineering solutions are to be realized. Early attempts to model nitrogen flow suffered from the lack of detailed knowledge of nitrogen processes, some of which is now available from the results of research reported in this volume. The nitrogen story is really just beginning to unfold and with the publication of this volume, we begin a new period of exciting research on ocean production processes.

RICHARD C. DUGDALE

PREFACE

The perception that nitrogen is an important and dynamic ecological factor has roots going back over 200 years. Soon after its discovery in 1772 by Rutherford (or possibly Cavendish or Scheele), in 1784, Berthollet identified "azote" as a component of animal tissue and excreta. Boussingault proposed in 1838 its role as a major controlling factor in plant nutrition and productivity. And although it was long suspected that atmospheric nitrogen was utilized by living organisms, Hillreigel and Wilfarth provided the first definitive evidence in 1880. The pioneering work of Beijernick and Winogradsky in the 1890s confirmed the biological basis of N_2 fixation as well as another suspected bacterial process, nitrification. Thus, by the turn of the century, the rudiments of the nitrogen cycle had been established.

The supposition that nitrogen might also play a crucial role in the ecology of the sea dates back to the early 1900s (see Nixon and Pilson, Chapter 16, for a historical perspective). Research on nitrogen in the sea followed a rather casual course for the first half of the century, a result of minimal scientific interest, limited resources, and inadequate means to investigate the subject.

Today there are active investigations of nitrogen transformations in literally all environments. Increased worldwide demand for nitrogenous fertilizers has paralleled dramatic increases in the cost of their synthesis. Major efforts in the agricultural sectors have been aimed at more efficient fertilizer use, prevention of fertilizer loss, and greater exploitation of biological N_2 fixation to offset requirements for synthetically produced ammonia. An important offshoot of this research has been an elaboration of methods to undertake quantitative studies and a refined understanding of the nitrogen cycle.

The patterns and quantities of nitrogen discharged to the environment have also changed dramatically with increasing population and technological advancement. Along with agricultural leachates, considerable combined nitrogen is released to the environment in industrial and domestic sewage effluents as well as from automobile emissions. Accelerated eutrophication has occurred in some coastal waters as a direct result of nitrogen loading. Furthermore, concern has been expressed over the possible depletion of the atmospheric ozone layer as a result of increased emissions of nitrogen oxides. Hence, there has been considerable stimulus to examine the natural cycling of nitrogen in the biosphere in order to provide a framework to evaluate the interactions and effects of human activities.

Given these factors, the last fifteen years have understandably seen a dramatic surge in research on nitrogen dynamics in the sea. Much of the direct impetus for this interest may be traced to a perceptive monograph written by R. C. Dugdale in 1967 and entitled "Nutrient Limitation in the Sea: Dynamics, Identification, and Significance," (*Limnol. Oceanogr.* **12,** 685–695). Further stimulation and provocation arose from the report of Ryther and Dunstan in 1971 (*Science* **171,** 1008) describing nitrogen as the primary limiting nutrient in coastal waters.

The research interest in nitrogen in the recent past relative to another crucial plant nutrient, phosphorus, is reflected in Figure 1. A substantial divergence in the annual rate of published reports on the two elements in marine ecosystems has occurred over the last decade. An analysis of freshwater ecosystems did not uncover a similar trend.

While nitrogen cycling in the sea is predominantly a biologically oriented subject, the geological, physical, and, in particular, chemical disciplines of the marine sciences have made important contributions to our understanding as well. We have especially profited in oceanic nitrogen cycling studies from the "interdisciplinary" approach.

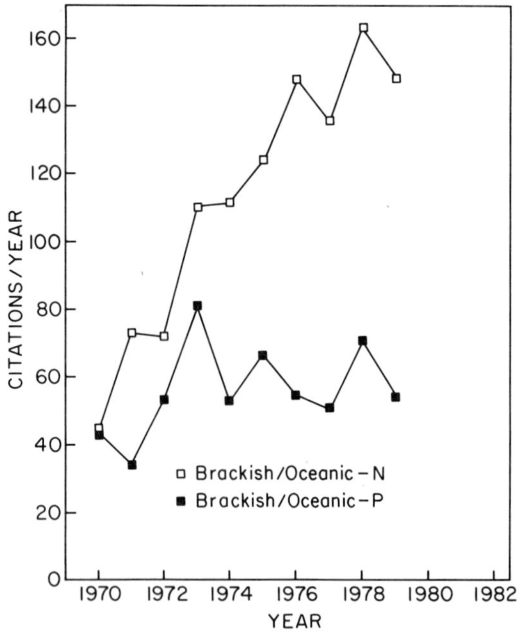

Fig. 1. Comparison of citations in Bibliographic Retrieval Services' Biosis® Data Base for marine-oriented research on nitrogen or phosphorus.

In view of the recent intensity and breadth of research and heightened concern over global nitrogen cycling, we perceived a major need for a synthesis volume on nitrogen cycling in the sea. Given the nature of the topic and its development in the marine sciences, the organization of the volume was straightforward. The volume centers on the role of microbes in nitrogen transformations with excursions to higher trophic levels. We have tried to provide a comparison of the nitrogen cycling of various ecosystems within the marine environment. Furthermore, we have included chapters on chemical distributions and methodology as an aid to those entering the field. We hope this book will provide fresh input and insights by which to align our future directions.

DOUGLAS G. CAPONE
EDWARD J. CARPENTER

Stony Brook, New York
June 1983

ACKNOWLEDGMENTS

The advice, assistance, and service of innumerable people were called upon by the editors in the preparation of this book. We wish to thank them all. Primary credit goes, of course, to our authors for the thoughtful treatment of their subject areas and their (generally) rapid response to our requests. We are particularly indebted to our reviewers who provided us with the ability to critically evaluate manuscripts in the diversity of areas which collectively constitute research in "marine nitrogen cycling." We also specifically thank Ms. Patricia Foster for providing excellent copy editing and Ms. Doris Rompf for secretarial assistance.

While this volume is obviously not a direct outcropping of any one particular research effort, the editors wish to acknowledge the sustained support of the Division of Ocean Sciences of the National Science Foundation in investigations relating to the topic of nitrogen in the marine environment. We specifically acknowledge NSF Grant Nos. OCE-78-25444, OCE 82-000157, and OCE 82-14764 for financial support of our research during the preparation of this work. Financial support of our laboratory is also provided by the National Oceanic and Atmospheric Administration (Grant Nos. NA-80-RAD-00057 and NA-80-RAD-00062), New York State Sea Grants (Grant Nos. 04-715-844-009 and NA-81-AAD-00027), and the Environmental Protection Agency (Grant No. R-809475-01-0).

Chapter 1

THE DISTRIBUTIONS OF INORGANIC NITROGEN AND DISSOLVED AND PARTICULATE ORGANIC NITROGEN IN THE SEA

JONATHAN H. SHARP
College of Marine Studies
University of Delaware
Lewes, Delaware

I. PERSPECTIVE

Nitrogen is found in the sea primarily in five oxidation states: -3 (NH_4^+, NH_2), 0 (N_2), +2 (N_2O), +3 (NO_2^-), and +5 (NO_3^-). The micronutrients have been reviewed by Spencer (1975) and the organic nitrogen compounds by Williams (1975). By far the most abundant species of nitrogen in the sea is N_2, but it is essentially unreactive (see Chapters 3 and 4). Nitrogen gas (N_2) is found throughout most oceanic and coastal waters at very near saturation values (Kester, 1975). It is considered separately (Chapter 2) and is referred to here only for comparative purposes (Table I). The next most abundant species, and a biologically active one, is the nitrate ion (NO_3^-). The other bioactive inorganic ions, nitrite (NO_2^-) and ammonium (NH_4^+) are less abundant overall, but are of local significance. Ammonia gas (NH_3) is lumped with NH_4^+ for discussions in this chapter; NH_3 along with nitrous oxide (N_2O) are

NITROGEN IN THE MARINE ENVIRONMENT

ISBN 0-12-160280-X

treated separately in Chapter 2. Organic nitrogen is usually considered to be found in the amino form (NH_2^-) and values are listed for both dissolved and particulate (Table I).

A note on units is necessary due to differences that have been reported in the literature. For the most part, the basic unit used in this chapter will be microgram-atoms nitrogen per liter (μg·atom N liter^{-1}). Much of the earlier literature used units of micrograms nitrogen per liter (μg N liter^{-1}). One μg·atom N liter^{-1} equals 14 μg·N liter^{-1}. Molar quantities are also often used; one micromolar N (μ*M*) = 1 μg·atom N liter^{-1}. However, these units can be confusing since with dinitrogen species, e.g., N_2 or urea, one micromolar (μ*M*) = 2 μg·atom N liter^{-1}. Geochemists usually use the unit of micromole per kilogram, μ*M* kg^{-1} (*sic*), for nitrate. That unit will be used for the nitrate data in Section II, but only there. One μmol kg^{-1} (=1 μg·atom N kg^{-1}) can be related to volumetric values only by taking density into consideration. As a rough estimate, in oceanic waters, 1 μmol kg^{-1} = 1.03 μg·atom N liter^{-1}. Otherwise, all numbers in this chapter will be given as μg·atom N^{-1}.

TABLE I. Major Nitrogen Species in the Sea[a]

Species	*Surface oceanic (0-100 m)*	*Deep oceanic (>100 m)*	*Coastal*	*Estuarine*
Nitrogen gas (N_2)	800	1150	700-1100	700-1100
Nitrate (NO_3^-)	0.2	35	0-30	0-350
Nitrate (NO_2^-)	0.1	<0.1	0-2	0-30
Ammonium (NH_4^+)	<0.5	<0.1	0-25	0-600
Dissolved organic N	5	3	3-10	5-150
Particulate organic N	0.4	<0.1	0.1-2	1-100

[a]*Approximate average values are given for oceanic waters and appropriate ranges are given for coastal and estuarine waters. All values are listed in microgram-atoms nitrogen per liter (μg·atom N liter*$^{-1}$.

II. OCEANIC NITRATE DISTRIBUTION

A. Distribution with Depth

Nitrate, as a limiting plant nutrient, is almost exhausted in surface waters in the majority of the world's ocean; exceptions at high and low latitudes and in coastal and estuarine waters are discussed below. Beneath the photic zone nitrate increases, usually reaching a maximum in the area of the oxygen minimum layer, and then usually levels off at a lower concentration below there. Typical nitrate profiles found in most text books are attributed to Sverdrup *et al.* (1942). The National Science Foundation GEOSECS program was a 10-year effort in which extensive sampling was done in the Atlantic, Pacific, and Indian Oceans (see Craig and Turekian, 1980). The extensive data sets from GEOSECS cover hundreds of hydrocasts from the three oceans. Nitrate data have been used to construct typical profiles for the three major ocean basins and Antarctic waters (Figs. 1 and 2).

The typical North Atlantic profile shows a pronounced nitrate maximum found near the oxygen minimum layer (Redfield *et al.*, 1963; Broecker, 1974). Below that, a lower, fairly uniform nitrate concentration is found. In the South Atlantic, the nitrate maximum associated with the oxygen minimum is broader and deeper than in the North Atlantic, and a secondary nitrate maximum is found in the bottom water. The South Pacific also has a deeper, broad nitrate maximum, whereas the North Pacific has a shallower maximum. The concentration of nitrate in the maximum is twofold greater in the North Pacific (ca. 45 μg·atom kg^{-1}) than in the North Atlantic (ca. 22 μg·atom kg^{-1}). The Antarctic profile shows much higher surface values than those of lower latitudes and a very shallow nitrate maximum; surface Antarctic values are usually around 25 and a fairly consistent deep-water value of 32 μg·atom N kg^{-1} is characteristic of these waters.

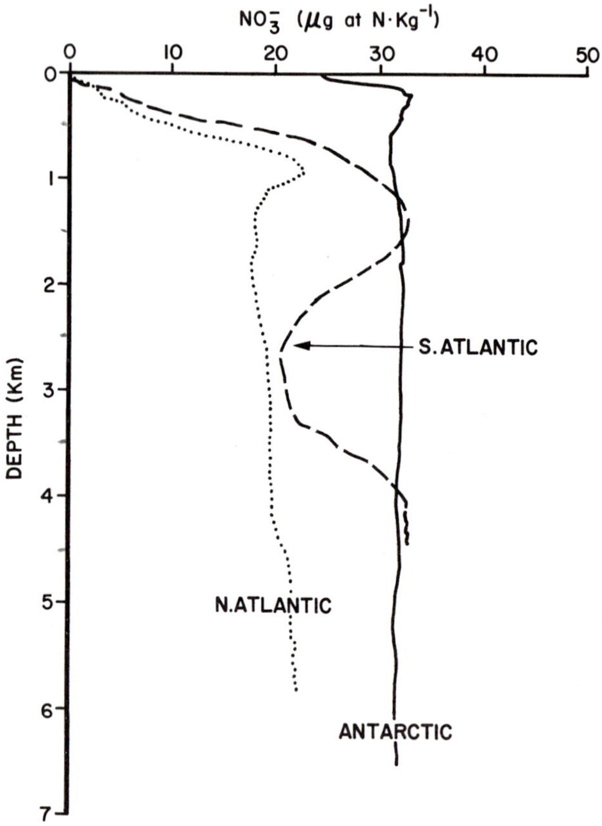

Fig. 1. Typical nitrate profiles from the Atlantic Ocean. These plots were constructed from GEOSECS data. The North Atlantic profile is based on 57 depth samples from station 30, taken 20 September, 1972 at 31° 48'N, 50° 46'W. The South Atlantic profile is based on 47 depth samples from station 60, taken 22 November, 1972 at 32° 58'S, 42° 30'W. The Antarctic profile is based on 43 depth samples from station 82, taken 11 January, 1973 at 56° 16'S, 24° 55'W.

B. Horizontal Distribution

Water masses and deep circulation in the ocean can be illustrated with diagrams of whole ocean basins (Riley, 1951; Reid and Lynn, 1971; Reid et al., 1978). Horizontal nitrate distribution may be best viewed with latitudinal sections through ocean basins. This has been done previously in a number of publications, and re-

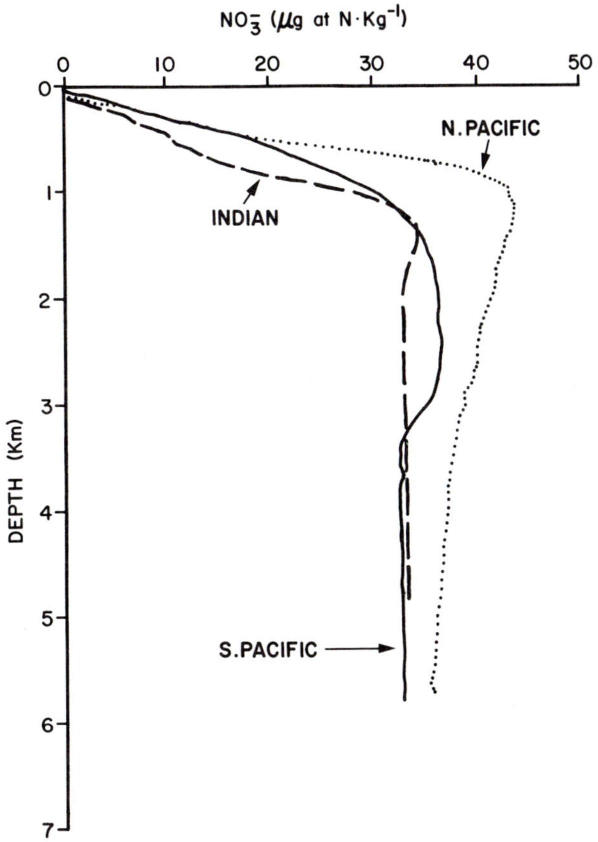

Fig. 2. Typical nitrate profiles from the Pacific and Indian Oceans. These plots were constructed from GEOSECS data. The North Pacific profile is based on 58 depth samples from station 213, taken 22 September, 1973 at 30° 58'N, 168° 28'W. The South Pacific profile is based on 38 depth samples from station 273, taken 22 January, 1974 at 29° 57'S, 175° 44'W. The Indian profile is based on 47 depth samples from station 454, taken 21 April, 1978 at 26° 60'S, 67° 58'E.

cently, sections became available through the GEOSECS program (Bainbridge, 1981; Craig et al., 1981). Rather than using a single ocean basin section, multiple basin sections (Reid and Lynn, 1971) are illustrated here (Figs. 3, 4, and 5). It is recognized now that the deep waters of the Pacific and Indian Oceans are derived partially from North Atlantic and Antarctic surface waters (e.g.,

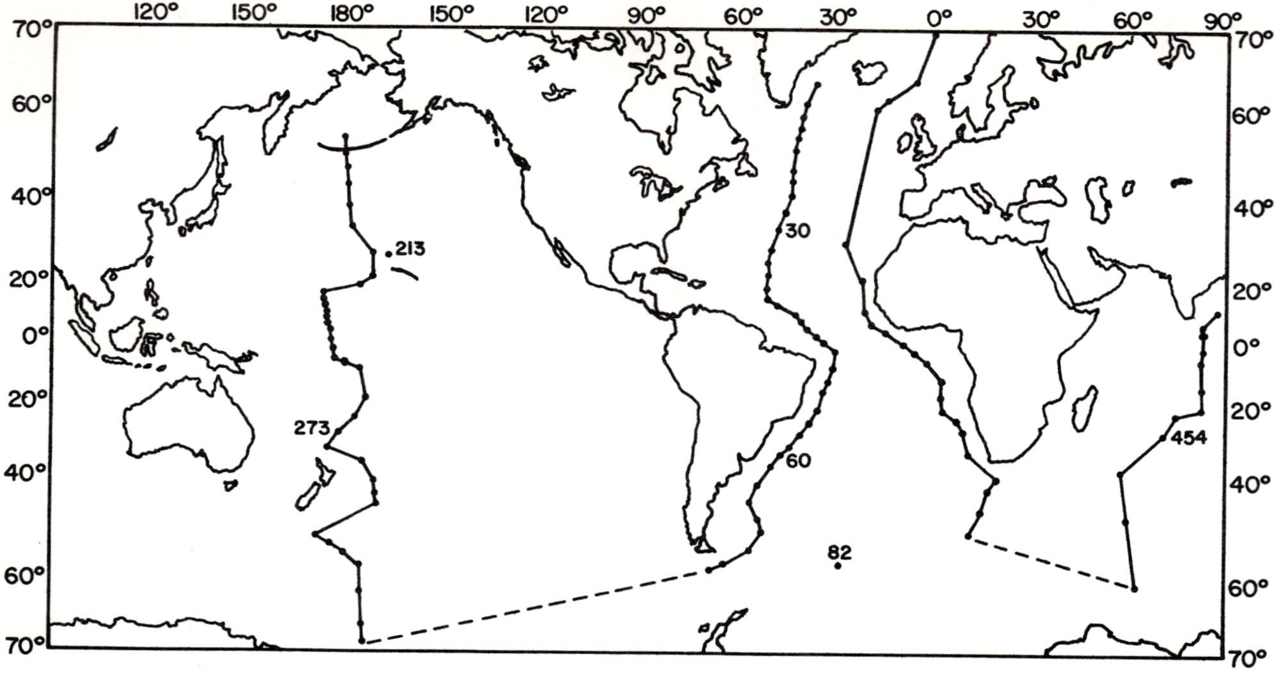

Fig. 3. Location of GEOSECS stations used in Figs. 1, 2, 4, and 5. The six stations indicated are those used for Figs. 1 and 2, other GEOSECS stations are identified in Figs. 4 and 5. The dashed lines in the Antarctic show gaps between stations; nitrate contours on Figs. 4 and 5 are similarly dashed.

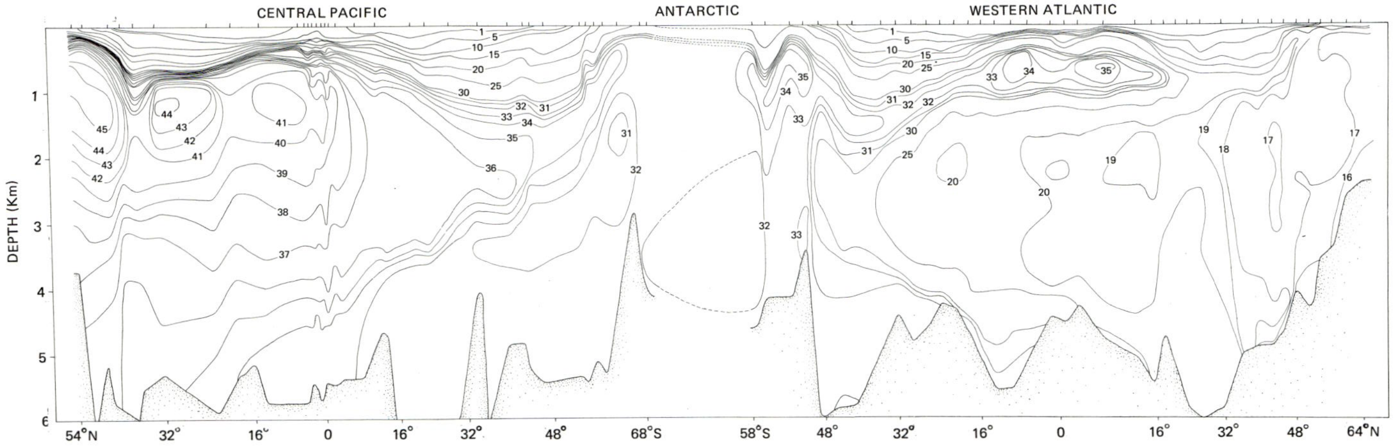

Fig. 4. Ocean section for nitrate, Western Atlantic/Pacific. The horizontal axis shows the GEOSECS stations (indicated by tick marks on axis) spaced proportional to great circle arc distance calculations between adjacent stations. Nitrate values for each station were plotted by depth, and contour lines (as μg·atom N kg^{-1}) were fitted manually. GEOSECS stations (N-S) for the Atlantic are: 11, 8, 5-1, 27-34, 36, 37, 39, 40, 42, 46, 48, 49, 53-61, 66-69, 74-76; for the Pacific (S-N) stations are: 287-293, 298, 300, 301, 303, 278, 273, 269, 268, 254-248, 246, 244-241, 229-227, 214-219.

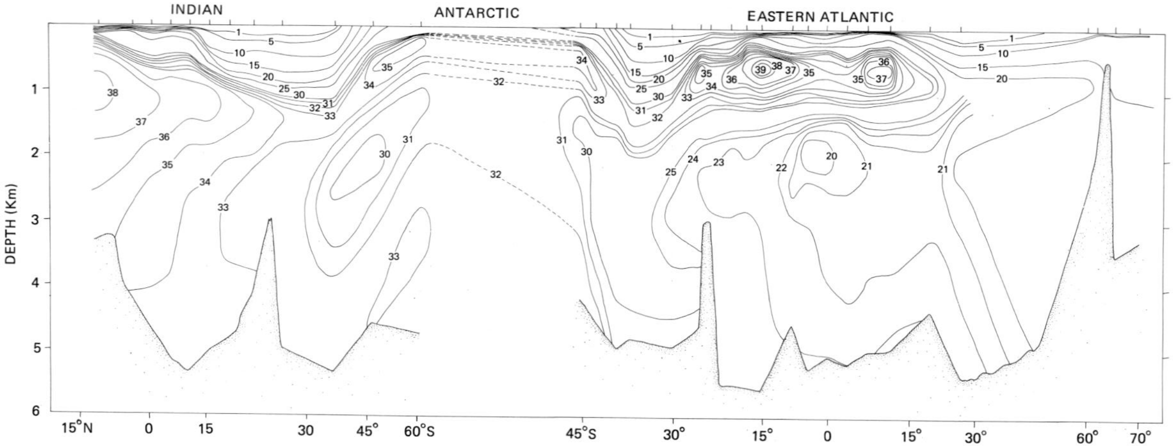

Fig. 5. Ocean section for nitrate, Eastern Atlantic/Indian. See caption Fig. 4. GEOSECS stations (N-S) for the Atlantic are: 18, 19, 22, 23, 115-102, 94-91; for the Indian (S-N) stations are: 430-428, 454-446.

Broecker and Li, 1970; Reid and Lynn, 1971). Figure 3 shows the location of GEOSECS stations that were used for these sections. Consideration of hydrographic and geographic conditions and circulation (Dahm, 1974; Broecker et al., 1976; Fiadeiro, 1980; Broecker et al., 1980; Fine et al., 1981) was made in selecting the stations used for the sections. A section is shown in Fig. 4, starting in the North Atlantic, running south down the western basin, and then from south to north through the center of the Pacific Ocean. North Atlantic Deep Water can be seen between about 1.5- and 4-km depth in the Western Atlantic (isopleths of 16-25 μg·atom N kg^{-1}). The Antarctic waters show little variation with depth and are the obvious source of the Antarctic Bottom Water (32-33 μg·atom kg^{-1}), which is found at the bottom of both basins as far north as the equator. In the Western Atlantic, Antarctic Intermediate Water overlies the North Atlantic Deep Water as a distinct intrusion going as far as 30°N at about 1-km depth. The Antarctic Intermediate Water is less distinct in the Pacific Basin, but it is obvious that the source of subsurface waters is largely Antarctic. Distinct equatorial upwelling can also be seen in both oceans, but it is more dramatic in the Pacific. Figure 5 is a similar two-basin section, running down the eastern basin of the Atlantic and from south to north in the Indian Ocean. Many of the same features as in the other section (Fig. 4) can be seen for the Eastern Atlantic. In the Eastern and Western Atlantic and in the Indian Oceans, the equatorial upwelling is less pronounced than in the Pacific. The subsurface Indian Ocean water clearly shows the influence of the Antarctic source.

C. Seasonal Variations

Normally, when one considers chemical conditions in the majority of the oceanic environment, a general picture is constructed with little thought of seasonal variations. Clearly, this is a partial misconception since there are seasonal cycles in primary productivity in much of the ocean (Menzel and Ryther, 1961; Fogg, 1975; Deuser and Ross, 1980). Since a good seasonal nutrient sequence is not readily demonstrable from the literature, an attempt has been made here to develop one. For this, data are taken from the Bermuda Biological Station hydrostation from the period of 1960 to 1963 (Woods Hole Oceanographic Institution, 1964). The station, which has been sampled regularly since 1956, is at about 3000-m depth and is located at 32° 10'N, 64° 30'W.

To delineate seasonal patterns, temperature, oxygen, and nitrate data were studied. Below 400 m there is no clear seasonal pattern for temperature or for nitrate concentration. In the upper 150 m, typical progressing thermal stratification starts in May, becomes most intense in September-October, decreases in intensity in December-February, and completely mixes in March-April.

There is good evidence for nonseasonal variations in the deep waters of the Sargasso Sea as has been shown for dissolved oxygen (Pocklington, 1972b); this could be caused by cold-core rings (Ring Group, 1981). For the period of time considered here (1960-1963), deep water conditions were probably fairly stable. This is verified with average values for dissolved oxygen in the region of the oxygen minimum layer. Oxygen values for 64 samples taken over 3 years at this 700-900-m level, range from 282 to 330 μg·atom O $liter^{-1}$ (coefficient of variation equal to 3% of the mean).

Therefore, if no physical phenomena have caused major variability during the 3-year period considered, the shallower layers should show *in situ* seasonal patterns. Nitrate values are integrated in the 1- to 150-m range and in the 150- to 400-m range giving mg·atom N m^{-2} in these two depth intervals. It has been shown previously that 150 m is a reasonable approximation of the surface layer for integrated nutrient pictures in these waters (Ryther et al., 1961); this is the maximum extent of the mixed surface layer including the photic zone. Quarterly averages for 3 years are given in Table II. Little seasonal pattern is shown

TABLE II. Average Integrated Nitrate Values for the Surface (1-150 m) and Subsurface (150-400 m) Waters of the Sargasso Sea as a Function of the Season of the Year[a,b]

Season	*mg·atom N m^{-2}* 1-150 m	*mg·atom N m^{-2}* 150-400 m	ΔT(°C)	Sample dates
Fall, 1960	58	741	9.3- 5.3	1-10-60 through 24-11-60
Winter, 1960-1961	54	634	4.0- 1.3	8-12-60 through 8- 2-61
Spring, 1961	40	508	1.2- 5.3	20- 2-61 through 7- 6-61
Summer, 1961	89	799	6.8-10.7	15- 6-61 through 22- 8-61
Fall, 1961	77	771	10.3- 6.2	6- 9-61 through 24-12-61
Winter, 1961-1962	72	824	4.4- 1.6	12-12-61 through 7- 4-62
Spring, 1962	79	541	1.1- 6.2	23- 4-62 through 4- 6-62
Summer, 1962	80	705	6.8-10.3	5- 7-62 through 11- 9-62
Fall, 1962	47	761	9.2- 5.3	26- 9-62 through 22-11-62
Winter, 1962-1963	97	893	4.5- 1.5	11-12-62 through 26- 3-63
Spring, 1963	74	573	1.3- 5.6	16- 4-63 through 4- 6-63
Summer, 1963	54	802	6.6-10.6	21- 6-63 through 7- 8-63

[a]*Data for calculations from Woods Hole Oceanographic Institution, 1964.*

[b]*Seasons were delineated by the temperature difference between the 1 m and 400 m samples (ΔT), with the summer season terminating at maximum ΔT and the spring season beginning at minimum ΔT. The two ΔT values given for each seasonal interval are the beginning and ending values for that interval.*

in the surface layer. Below this layer, a distinct seasonal pattern is demonstrated. With surface water warming and the occurrence of stratification, a buildup of nitrate in these subsurface waters extends through the winter, where it reaches a maximum and then drops abruptly with the spring water turnover. There is a buildup from spring to winter in the 150 to 400-m interval equivalent to 60-65% of the spring value.

III. OTHER INORGANIC IONS IN THE OCEAN ENVIRONMENT

Biogenic nitrogen input to the ocean is in the reduced form, usually as ammonium and amino nitrogen. This is oxidized microbially (nitrification) and in the majority of the world's ocean, almost all of the ionic nitrogen is in the thermodynamically stable form of nitrate. Thus, nitrite and ammonium are generally considered as intermediates in the nitrogen cycle, classically shown by von Brand et al. (1937). As labile compounds, they are usually in comparatively low concentrations and often show discrete maxima near the photic zone. The generalized picture of these two nutrients shows an ammonium maximum near the bottom of the photic zone and a nitrite maximum just below the bottom of the photic zone (Spencer, 1975). A deeper, secondary nitrite maximum is often found in oxygen-depleted waters, and this was demonstrated by Brandhorst (1959) to be due to denitrification (see Chapters 5 and 6).

Figure 6 shows nitrite, ammonium, and nitrate depth profiles for the Sargasso Sea (at the Bermuda hydrostation). The primary nitrite maximum is clearly illustrated, but only a suggestion of the shallower ammonium maximum is seen; both are considerably shallower than the classical nitrate maximum. Additionally, in deeper waters, ammonium values, although low, are distinctly above zero and are varied. Careful examination of other oceanic distributions (McAllister et al., 1960; Hattori and Wada, 1971;

Cline and Richards, 1972; Olson, 1981) also shows that a clear ammonium maximum is only occasionally demonstrable, and that deep water ammonium values are usually above zero and somewhat erratic. The differences between nitrite and ammonium profiles indicate that in oxic waters, nitrite production and destruction are relatively simple and direct, whereas ammonium dynamics are more complicated (see Chapters 10 and 11). Indeed, Beers and Kelly (1965) have shown marked short-term ammonium fluctuations at this same Sargasso Sea station to a depth of 500 m.

IV. COASTAL NITROGEN DISTRIBUTIONS

A. Nonupwelling Regions

The majority of the continental shelf edges of the world's oceans do not have strong, persistent upwelling. These regions are characterized by relatively gentle slopes going out from the shore and terminating in sharp drops at the margin with the open ocean. Such coastal waters can be fairly narrow (a few kilometers) to very broad (100 km or more). The most abundant nitrogenous nutrient in coastal waters is nitrate. Surface nitrate values are usually close to zero in the summer and up to several μg·atom N $liter^{-1}$ in the winter. Deep waters often have nitrate values close to 20 μg·atom N $liter^{-1}$, but usually are below 10 μg·atom N $liter^{-1}$ (McCarthy and Kamykowski, 1972; Dunstan and Atkinson, 1976; Fournier et al., 1977; Postma, 1978; Eppley et al., 1978, 1979a; Butler, 1979; Treguer et al., 1979; Sharp and Church, 1981). Nitrite values are almost always very low and rarely exceed 5% of the nitrate level (McCarthy and Kamykowski, 1972). Ammonium values are variable but usually below 3 μg·atom N $liter^{-1}$ (Postma, 1966; Eppley et al., 1979b; Thomas and Carsola, 1980; Sharp and Church, 1981). With terriginous input, often from sewage, ammonium values can reach 20-25 μg·atom N $liter^{-1}$ in open coastal waters (Atwood et al., 1979; Eppley et al., 1979b; Thomas and Carsola, 1980).

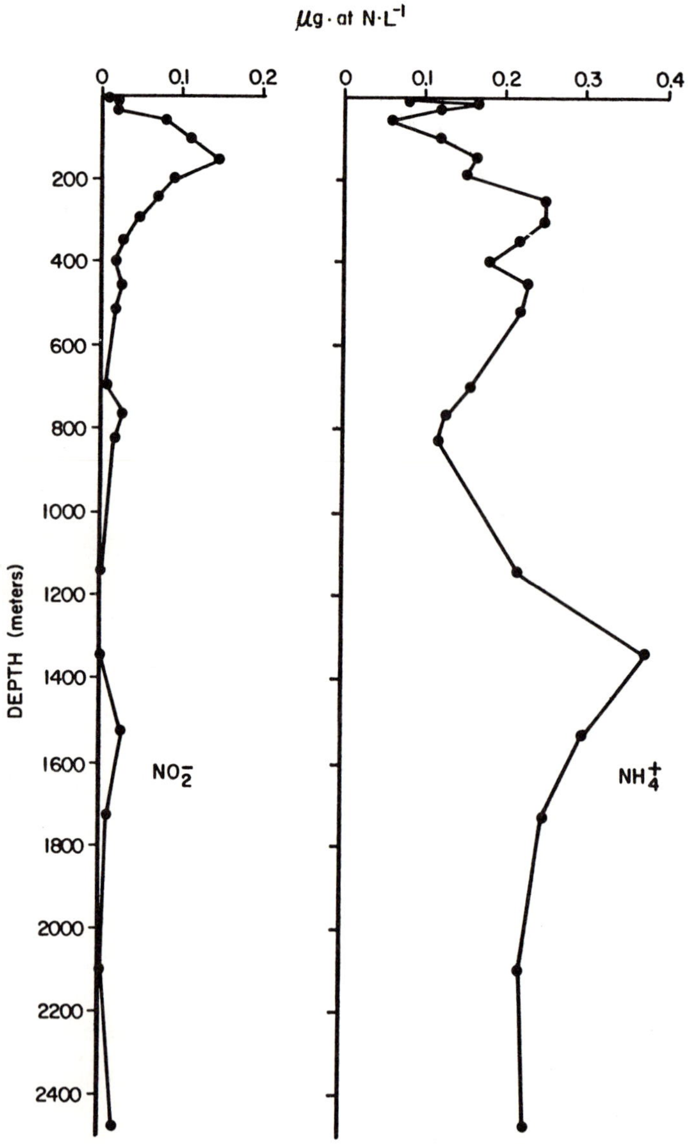

Fig. 6. Nitrite, ammonium, and nitrate in the Sargasso Sea from 18 April, 1981. The errors (±2 standard deviations) calcu-

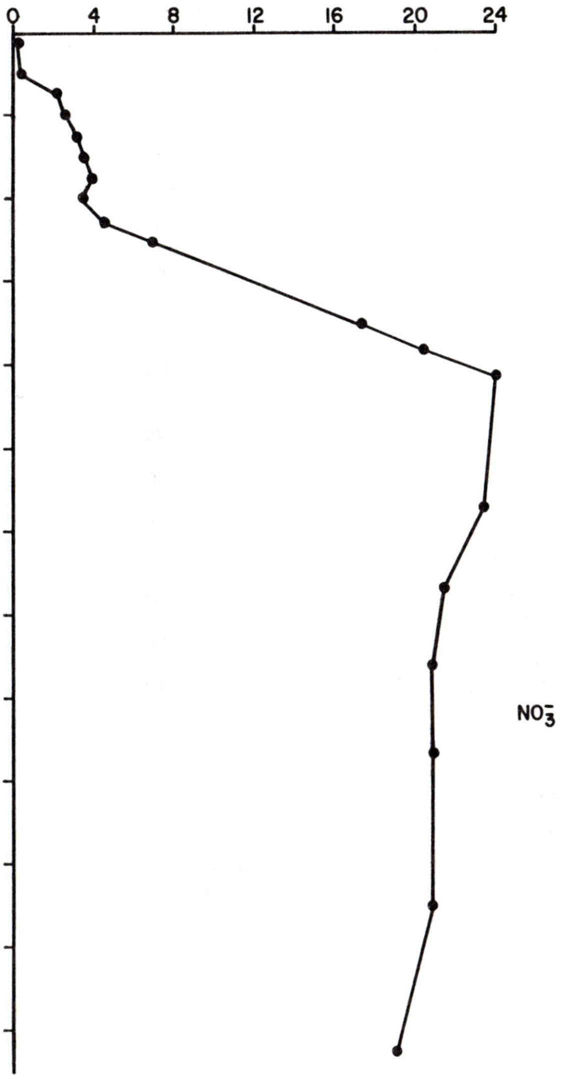

lated for these nitrite and nitrate determinations are each (±) 0.02 µg·atom N liter^{-1} and that for ammonium is (±) 0.09 µg·atom N liter^{-1}. Data from J. H. Sharp (unpublished).

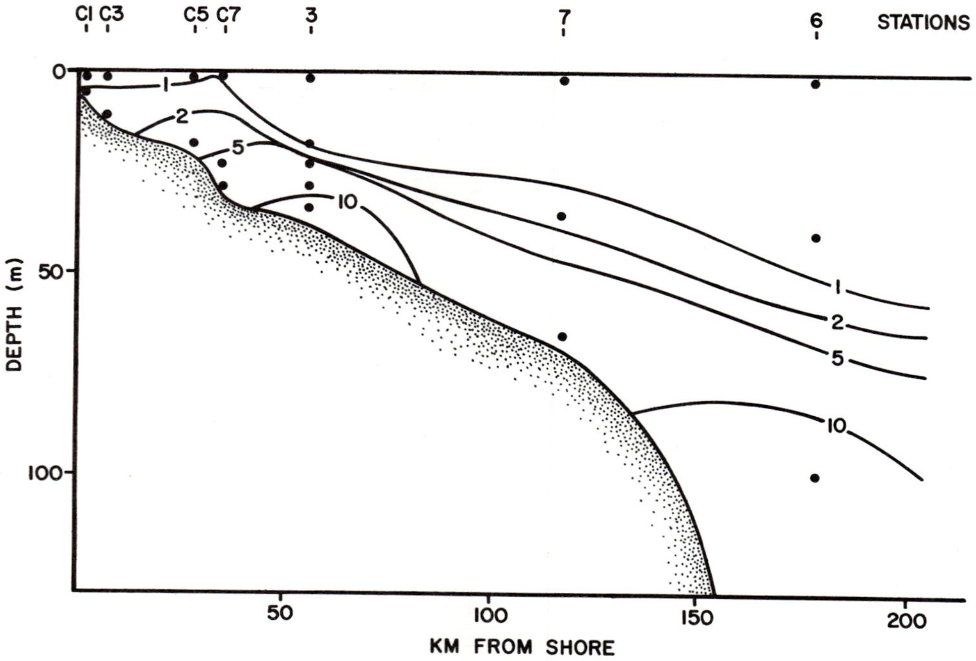

Fig. 7. Coastal nitrogen section. Isopleths indicate total inorganic nitrogen (NO_3^-, NO_2^-, and NH_4^+) in μg·atom N liter^{-1}. Section based on data from middle Atlantic coastal waters (USA) from September, 1977 (Sharp et al., 1980).

Nitrate values in coastal waters usually decrease going seaward in surface waters and increase along the same axis in deep waters. Figure 7 shows an example of cross-shelf inorganic nitrogen concentrations; nitrite and ammonium are included, but constitute less than 20% of the total. Riley (1967) suggested that coastal nitrate distributions were due to diffusive exchange between deep and surface waters (as a function of depth). Depth-dependent nutrient patterns have been attributed by others also to be due to mixing phenomena (Fournier et al., 1977; Eppley et al., 1979a). Appreciable periodic inputs of nitrate occur in coastal waters from short-interval upwelling at the outer shelf (Dunstan and Atkinson, 1976; Fournier et al., 1977; Walsh et al., 1978; see Chapter 15).

A distinct seasonal cycle can be found for dissolved nitrogen in coastal waters, with higher surface values in the winter than the summer and the expected progression from ammonium to nitrite to nitrate from regeneration (Spencer, 1975). A buildup of nitrate below the summer thermocline has been demonstrated for coastal waters (Treguer et al., 1979; Sharp and Church, 1981). This seasonal depth picture is shown in Fig. 8. From analysis of seasonal nitrogen dynamics, it was suggested that much of the coastal nitrogen is recycled on the continental shelf, with only partial exchange with oceanic or estuarine waters (Sharp and Church, 1981).

B. Upwelling Regions

Coastal zones adjacent to eastern boundary currents are often characterized by strong, extended-duration upwelling. Thus, at times, upwelling regions have major inputs of new nitrogen from the subsurface open ocean and are quite different from nonupwelling regions. Subsurface nitrate concentrations in upwelling coastal waters are often near 30 μg·atom N $liter^{-1}$, and surface values of 20 μg·atom N $liter^{-1}$ are not uncommon (Walsh, 1977). Upwelling

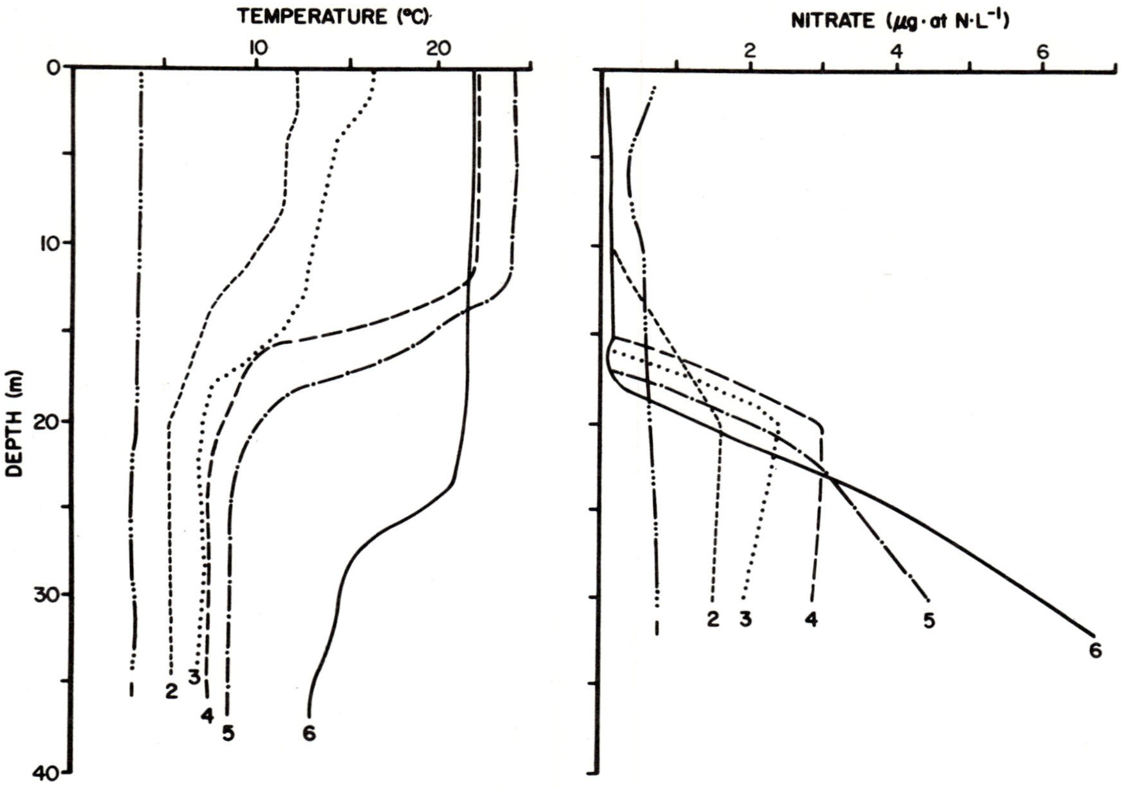

TEMPERATURE (°C)
10
20
NITRATE (μg · at N · L⁻¹)
2
4
6
0
10
20
30
40
DEPTH (m)
1
2
3
4
5
6

regions often have strong oxygen depletion and periodically show high subsurface ammonium and nitrite values. More can be found in chapters on denitrification (Chapter 6) and upwelling (Chapter 15).

V. ESTUARINE NITROGEN DISTRIBUTIONS

One way to describe nitrogenous nutrients in an estuary is with a property-salinity plot. This approach can be used to distinguish between properties without pronounced biological or geochemical activity (conservative behavior) and those that are more reactive (nonconservative behavior). Nutrient-salinity plots of this type have been advocated by Liss (1976) and prove to be very valuable constructs, especially when one is cautious about end-member variability (see Loder and Reichard, 1981). Using nutrient-salinity plots, two types of environments will be considered: low nutrient estuaries that show essentially conservative mixing and high nutrient estuaries that show variations from conservative to highly nonconservative mixing.

A. Low Nutrient Estuaries

Examples of relatively low nutrient loading are found in the tropical estuaries of the Zaire River in Africa (Van Bennekom et al., 1978), the Magdelena River in South America (Fanning and Maynard, 1978), and in the temperate estuaries of the Columbia River (Stefansson and Richards, 1963), St. Lawrence River (Coote and Yeats, 1979), and Sacramento-San Joaquin Rivers (Peterson, 1979), all in North America. In all of these cases, ammonium nitrogen is apparently minor compared to ni-

Fig. 8. Seasonal changes in nitrate in coastal waters. Temperature and nitrate profiles for middle Atlantic coastal waters (USA); bottom depth at the station location was about 40 m. Data are from 1977; sampling times are: 1,9 to 10 March; 2,25 to 26 April; 3,16 to 17 June; 4,26 to 28 July; 5,23 to 25 August; 6,27 to 29 September. (Redrawn from Sharp and Church, 1981.)

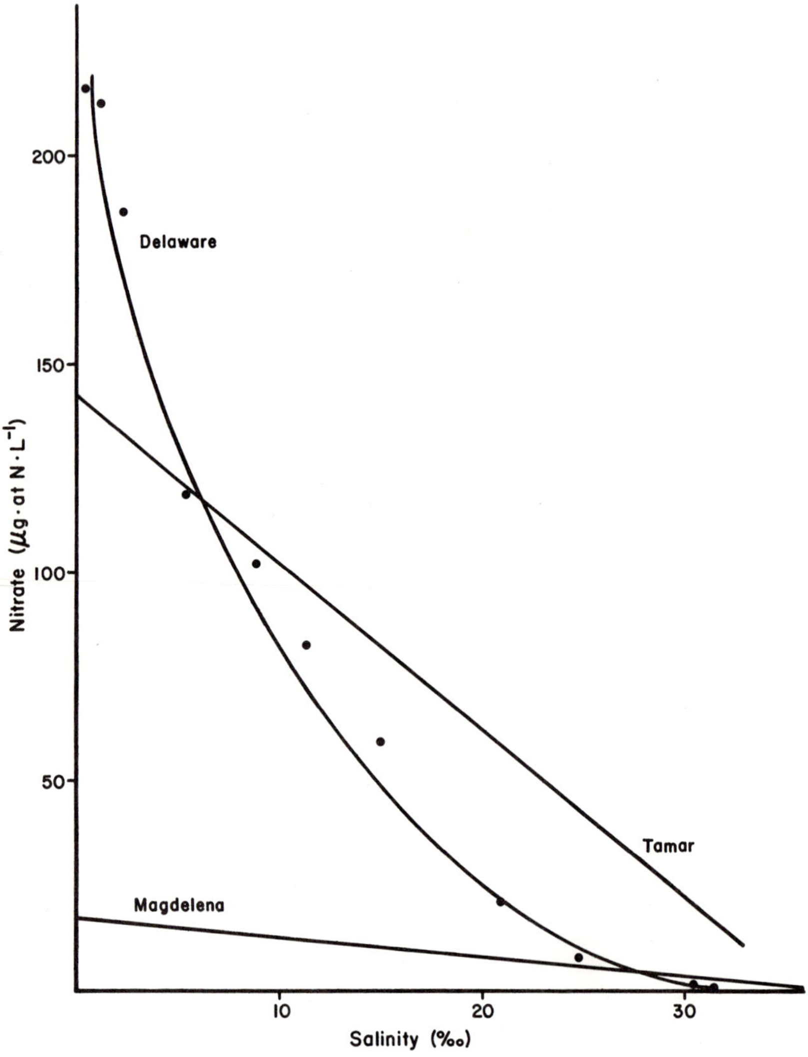

Fig. 9. Estuarine nitrate profiles. For the Tamar and Magdelena estuaries, lines from plots given in the literature were transposed; for the Delaware estuary, data from Culberson et al. (1982) were used from July, 1980.

trate. Also, in all these cases, nitrate values in the fresh water end range from about 10 to 40 µg·atom N liter^{-1} and decrease linearly to near zero at the full salt end of the estuary. An example is given for the Magdelena Estuary in Fig. 9. In all of these cases, the major nitrogen form is nitrate, maximum values are below 50 μg·atom N $liter^{-1}$, and mixing within the estuary is essentially conservative.

B. High Nutrient Estuaries

In contrast to the above, many temperate estuaries, especially those with urban development in their upper reaches, have considerably higher nutrient loading. The Dutch Wadden Sea, which is fed by the Scheldt and Rhine Rivers, shows a range from 0 to 35 parts per thousand salinity. The majority of the inorganic nitrogen is as nitrate with maximum values of 100 μg·atom N $liter^{-1}$; plots against salinity show conservative behavior (Postma, 1966; Helder, 1974). The Tamar Estuary in England also appears to have conservative nutrient mixing with nitrate values at the fresh water end of 100-350 μg·atom N $liter^{-1}$ (Morris et al., 1981). An example from the Tamar is also given in Fig. 9.

Many urbanized estuaries show considerable ammonium input. The Scheldt Estuary in Belgium has values of ammonium in excess of 600 μg·atom N $liter^{-1}$ in its low salinity reaches (Billen, 1975) and, with nitrification, nitrate values down estuary in excess of 300 μg·atom N $liter^{-1}$ (Billen and Smitz, 1978; Wollast, 1978). Similarly in North America, the lower Hudson River (Garside et al., 1976; Duedall et al., 1977) and the Potomac River (Carpenter et al., 1969; Jaworski, 1981) show ammonium values in excess of 100 μg·atom N $liter^{-1}$. In these cases, nonconservative behavior of ammonium is seen usually due to oxygen-demanding nitrification. The Delaware River Estuary in North America also shows some ammonium oxidation, but for much of the salinity gradient, nitrate is the major nitrogen form. Nitrate values of about 200 μg·atom N $liter^{-1}$ decrease nonlinearly almost to zero going down the salinity gradient (Sharp et al., 1982a);

this is also shown in Fig. 9. Maximum values of nitrite, a nitrification intermediate, have been found in the 5-30 μg·atom N $liter^{-1}$ range (Postma, 1966; Culberson et al., 1982).

In summary, in high nutrient estuaries, when ammonium is abundant, it usually displays nonconservative behavior and nitrate, when abundant, can be either conservative or nonconservative.

VI. ORGANIC NITROGEN DISTRIBUTION

A. Dissolved Organic Nitrogen

There are few published values for dissolved organic nitrogen in the marine environment. An extensive sampling was made by Duursma (1961), but more recent analyses make his values appear to be somewhat high. Serious methodological problems have existed in the past (see Chapter 20), which make most of the reported values of the 1960s and 1970s suspect of being low. Additionally, many of the published values include ammonium with the organic nitrogen as a total value (e.g., Holm-Hansen et al., 1966; Butler et al., 1979).

From a few published data (Fraga, 1966) and our own unpublished ones (Culberson et al., 1982), oceanic values for dissolved organic nitrogen usually range from 3 to 7 μg·atom N $liter^{-1}$. A decrease with depth is found for dissolved organic nitrogen in the open ocean (Fraga, 1966; Armstrong et al., 1966), but this decrease is neither as dramatic nor as regular as that seen for dissolved organic carbon (see Williams, 1975).

In coastal waters, a slight increase in maximum values is found, so that a range of 3-20 μg·atom N $liter^{-1}$ is typical (Postma, 1966; Armstrong and Tibbits, 1968; Haines, 1979; Culberson et al., 1982). In estuarine waters, a range of 5-130 μg·atom N $liter^{-1}$ can be found (Postma, 1966; Haines, 1979; Culberson et al., 1982).

In surface oceanic waters and in surface coastal waters (Sharp et al., 1982b), the vast majority of the dissolved nitrogen is thus in the organic rather than the inorganic fraction. Comparison can be made of organic nitrogen to organic carbon and phosphorus. A coastal average atomic C:N ratio for dissolved organic matter is about 8:1 and N:P is about 30:1 (Sharp et al., 1982b). For oceanic samples, a similar average can be calculated from data of Culberson et al. (1982) and the C:N is 13:1. These ratios are quite different from those often discussed in the literature (e.g., Duursma, 1965); values often listed as 2:1 to 8:1. The C:N:P ratios now found show organic nitrogen to be more labile than organic carbon, but probably less labile than organic phosphorus. Overall, organic nitrogen would appear to be less varying than inorganic nitrogen species. Clearly more information is needed.

B. Particulate Organic Nitrogen

Available information on particulate organic nitrogen is better than that for dissolved organic nitrogen. However, certain problems complicate a simple description of particulate distributions. Blank determinations for particulate carbon and nitrogen analyses have been complicated and somewhat controversial (see Gordon and Sutcliffe, 1974; Sharp, 1974); therefore, early publications are *not* considered here. A second problem revolves around sampling. Recent sediment trap work makes it obvious that traditional values for particulate matter from water bottle samples are probably underestimates of real concentrations (see McCave, 1975; Wiebe et al., 1976; Gardner, 1977). The production of particulate matter and the flux to deep water of large particles have spatial heterogeneity, such that the small water collections taken in sample bottles miss most of the particulate matter. Nonetheless, the discussion below is of traditional particulate matter.

In the surface oceanic environment, particulate organic nitrogen values are usually in the range of 0.1-0.5 μg·atom N $liter^{-1}$, while deep water values are usually below 0.07 μg·atom N $liter^{-1}$

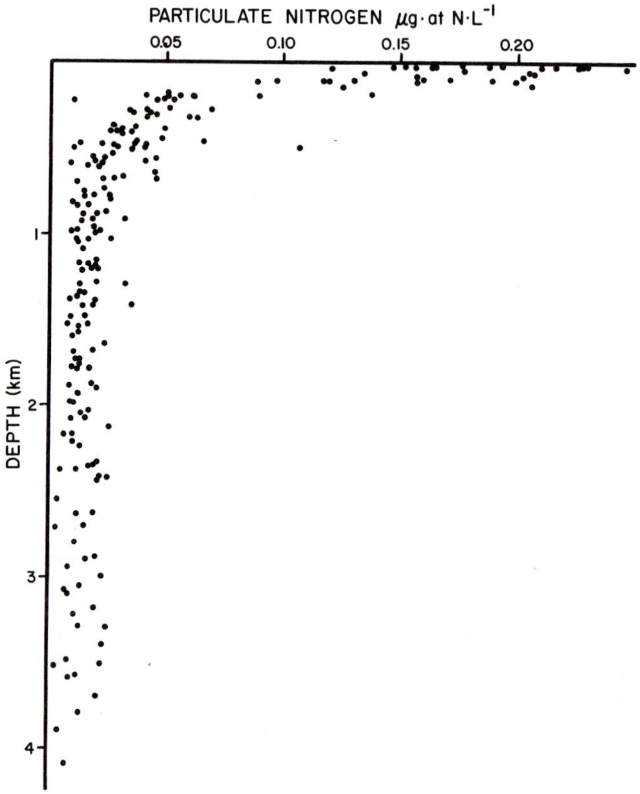

Fig. 10. Particulate organic nitrogen in the Pacific ocean. Samples are from station at 22° 10'N, 158°W and collected over the period from 9 January, 1969 through 4 June, 1970. (Figure redrawn from Gordon, 1971.)

(Gordon, 1971; Ichikawa and Nishizawa, 1975). A fairly regular decrease is seen with depth, so that relatively uniform low values are found below about 1000 m (Gordon, 1971). This picture, illustrated in Fig. 10, may not properly account for horizontal variability due to water masses (Ichikawa and Nishizawa, 1975).

In coastal waters, particulate nitrogen values are found in the range of 0.1-30 μg·atom N liter^{-1} (Postma, 1966; Haines, 1979; Culberson et al., 1982). Estuarine values have been reported in the range of 5-100 μg·atom N $liter^{-1}$ (Postma, 1966; Haines, 1979; Culberson et al., 1982).

For plankton, the atomic ratio of carbon to nitrogen should be about 6.6:1 (Redfield et al., 1963); particulate organic C:N in coastal and surface oceanic waters often range from about 6 to 9:1 (Sharp, 1975; Sharp et al., 1982b). With depth in the open ocean, C:N increases to values around 18:1 (Gordon, 1971). This increase is presumably due to the more labile nature of nitrogen than carbon. A similar phenomenon can be seen with size-class fractionation of surface oceanic organic matter with a C:N of 4:1 for particles larger than 500 μm, increasing to a C:N of 19:1 for organic matter smaller than 1 nm (Sharp, 1975).

C. Specific Compounds

The particulate organic nitrogen pool has not been studied extensively, but protein probably makes up a significant portion, especially in coastal and surface waters. Protein constitutes greater than 70% of the particulate organic nitrogen in surface waters and decreases with depth (Parsons, 1975; Packard and Dortch, 1975; Maita and Yanada, 1978). Chlorophyll, DNA, and ATP are all nitrogenous components of the particulate matter that have been measured with some frequency. Chlorophyll concentrations in the ocean range from essentially zero in deep water to 0.5 μg chlorophyll *a* $liter^{-1}$ as an average for oceanic waters, and 10-40 μg $liter^{-1}$ for fertile coastal waters (Strickland, 1965). DNA in oceanic particulate matter ranges from 0.2 to 30 μg $liter^{-1}$ (Holm-Hansen et al., 1968). ATP in oceanic and coastal particulate matter has been measured in the range of 0.5-1000 ng $liter^{-1}$ (Holm-Hansen and Booth, 1966; Karl and Holm-Hansen, 1978). None of these three components contributes more than about 1% of the particulate nitrogen.

Free amino acids have been measured in seawater fairly often and previous work is summarized in Williams (1975). With the more direct methods available today (see Chapter 14), these older reports are less reliable (Dawson and Liebezeit, 1981). From recent direct analyses, dissolved free amino acids in coastal waters range from 0.04 to 2.2 μg.atom N liter^{-1} (North, 1975; Josefsson et al., 1977; Brockmann et al., 1979; Billen et al., 1980; Mopper and Lindroth, 1982). Few good recent oceanic values are available, but comparison to some of the more exact older works (Riley and Segar, 1970; Pocklington, 1972a; Lee and Bada, 1977) suggests concentrations of dissolved free amino acids in the open ocean in the range of 0.05-0.5 μg·atom N liter^{-1}. Dissolved combined amino acids can be measured by amino acid analysis after hydrolysis. Although there are not many reported values for combined amino acids by the newer methodology, older works (Riley and Segar, 1970; Lee and Bada, 1977) show average concentrations on the order of three times those of dissolved free amino acids. Thus, if 0.5 μg·atom N liter^{-1} is taken as a high value for dissolved free amino acids in surface oceanic waters, 1.5 μg·atom N liter^{-1} as a high value for dissolved combined amino acids, and 7 μg·atom N liter^{-1} as a high value for total dissolved organic nitrogen, dissolved free amino acids constitute 7% and combined amino acids 21% of the dissolved organic nitrogen pool.

Most of the common amino acids have been found in both the dissolved free and combined states. Glycine, ornithine, and serine are often the most prominent, but others show specific patterns over time and depth (Dawson and Liebezeit, 1981; Mopper and Lindroth, 1982).

Urea and creatine are two other organic nitrogen compounds that have been found in seawater and are of importance as zooplankton and nekton excretory products. Creatine has not been measured often, but was shown to be present at 0.05-0.2 μg·atom N liter^{-1} concentrations in coastal waters (Whitledge and Dugdale, 1972). Urea is found in oceanic waters in 0.1-1.0 μg·atom N liter^{-1} concentrations and in higher concentrations in coastal waters (McCarthy, 1970; Remsen, 1971; McCarthy and Kamykowski, 1972; Eppley et al., 1977).

DNA has been measured as a dissolved constituent in seawater (Pillai and Ganguly, 1970), and concentrations given translate into a range of 0.1-0.3 μg·atom N $liter^{-1}$. Vitamins B_{12}, thiamine, and biotin have also been measured in seawater (see Provasoli and Carlucci, 1974); these nitrogenous compounds would contribute a total of less than 0.01 μg·atom N $liter^{-1}$. In pursuing adenylate chemistry, Ammerman and Azam (1981) have recently measured dissolved cyclic adenosine monophosphate in seawater; again, concentrations are less than 0.01 μg·atom N $liter^{-1}$.

An area of considerable importance and neglect is that of amino sugars. These are possibly significant portions of humic substances in seawater (Fox, 1981) and are ubiquitous components of marine biopolymers (Dawson and Liebezeit, 1981). No values are available for free and combined amino sugars dissolved in seawater.

Table III gives a summary of organic nitrogen compounds in seawater as discussed above. As can be seen, about 50% of the dissolved organic nitrogen pool can be accounted for in this table. Recent work in this laboratory helps to further construct a dissolved organic nitrogen inventory. Fox (1981) has shown that free and combined amino acids in a small United States east coast estuary contribute 30-52% of the total dissolved organic nitrogen. Additionally, he demonstrated that humic acids contribute up to 30% of the dissolved organic nitrogen and that a portion of the humic acid extracted was amino acids (Fox, 1981). Looking at the Delaware estuarine gradient, combined and free amino acids plus urea contribute 6-45% of the total dissolved organic nitrogen pool, with the maximum near the mouth of the estuary (Cifuentes, 1982). In the same estuarine gradient, humic acids contributed 2-10% of the total dissolved organic nitrogen. Cifuentes (1982) showed further that coastal waters off the mouth of this estuary had a composition such that 55% of the dissolved organic nitrogen was in the combined and free amino acids plus urea, and 12% was in the humic acid extract. The humic acid extract overlaps partially

TABLE III. Dissolved Organic Nitrogen Values for Oceanic Waters[a]

Compound	*(Range: μg·atom N liter^{-1})*
Total dissolved organic nitrogen	3-7
Free amino acids	0.05-0.5
Combined amino acids	0.15-1.5
Creatine	0.05-0.2
Urea	0.1-1.0
Vitamins	<0.01
Adenosine monophosphate	<0.01
DNA	0.1-0.3

[a]*Reported values are given as ranges, sources are given in the text.*

with the dissolved amino acids analysis. It can be concluded from these works that amino acids, urea, and humic acids may account for up to half of the organic nitrogen in coastal and probably also oceanic waters. It will be interesting in the future to see if amino sugars and fulvic acids can make up much of the difference.

ACKNOWLEDGMENTS

This work, especially preparation of Figs. 1-5, was partially supported by NSF Grant OCE80-09835. The author acknowledges John Casadevall for computer work and Karin Swartz and Lois Butler for graphics work on Figs. 1-5. The data for Fig. 6 were collected with valuable aid from Anthony Knap and Tim Jickells of the Bermuda Biological Station.

REFERENCES

Ammerman, J. W., and Azam, F. (1981). Dissolved cyclic adenosine monophosphate (cAMP) in the sea and uptake of cAMP by marine bacteria. *Mar. Ecol.: Prog. Ser. 5*, 85-89.

Armstrong, F. A. J., and Tibbits, S. (1968). Photochemical combustion of organic matter in seawater, for nitrogen, phosphorous and carbon determinations. *J. Mar. Biol. Assoc. U.K. 48*, 143-152.

Armstrong, F. A. J., Williams, P. M., and Strickland, J. D. H. (1966). Photo-oxidation of organic matter in sea water by ultraviolet radiation, analytical and other applications. *Nature (London) 211*, 481-483.

Atwood, D. K., Whitledge, T. E., Sharp, J. H., Cantillo, A. Y., Berbarian, G. A., Parker, J. M., Hanson, P. F., Thomas, J. P., and O'Reilly, J. E. (1979). Chemical factors. *In* "Oxygen Depletion and Associated Benthic Mortalities in New York Bight, 1976" (R. L. Swanson and C. J. Sindermann, eds.), pp. 79-123. U.S. Natl. Oceanic Atmos. Admin., Washington, D.C.

Bainbridge, A. L. (1981). "GEOSECS Atlantic Expedition," Vol. 2. Nat. Sci. Found., Washington, D.C.

Beers, J. R., and Kelly, A. C. (1965). Short-term variation of ammonia in the Sargasso Sea off Bermuda. *Deep-Sea Res. 12*, 21-25.

Billen, G. (1975). Nitrification in the Scheldt Estuary (Belgium and the Netherlands) estuarine coast. *Mar. Sci. 3*, 79-89.

Billen, G., and Smitz, J. (1978). Mathematical model of water quality in a highly polluted estuary. *In* "Hydrodynamics of Estuaries and Fjords," pp. 55-62. Elsevier, Amsterdam.

Billen, G., Joiris, C., Wijnant, J., and Gillain, G. (1980). Concentration and microbiological utilization of small organic molecules in the Scheldt Estuary, The Belgian Coastal zone of the North Sea and the English Channel. *Estuarine Coastal Mar. Sci. 11*, 279-294.

Brandhorst, W. (1959). Nitrification and denitrification in the eastern tropical North Pacific. *J. Cons., Cons. Int. Explor. Mer 25*, 3-20.

Brockmann, U. H., Eberlein, K., Junge, H. D., Maier-Reimer, E., and Siebers, D. (1979). The development of a natural plankton population in an outdoor tank with nutrient-poor sea water. II. Changes in dissolved carbohydrates and amino acids. *Mar. Ecol.: Prog. Ser. 1*, 283-291.

Broecker, W. S. (1974). "NO," a conservative water-mass tracer. *Earth Planet. Sci. Lett. 23*, 100-107.

Broecker, W. S., and Li, Y.-H. (1970). Interchange of water between the major oceans. *JGR, J. Geophys. Res. 75*, 3545-3557.

Broecker, W. S., Goddard, J., and Sarmiento, J. L. (1976). The distribution of ^{226}Ra in the Atlantic Ocean. *Earth Planet. Sci. Lett. 32*, 220-235.

Broecker, W. S., Toggweiler, J. R., and Takahashi, T. (1980). The Bay of Bengal--A major nutrient source for the deep Indian Ocean. *Earth Planet. Sci. Lett. 49,* 506-512.

Butler, E. I. (1979). Nutrient balance in the western English Channel. *Estuarine Coastal Mar. Sci. 8,* 195-197.

Butler, E. I., Knox, S., and Liddicoat, M. I. (1979). The relationship between inorganic and organic nutrients in sea water. *J. Mar. Biol. Assoc. U.K. 59,* 239-250.

Carpenter, J. H., Pritchard, D. W., and Whaley, R. C. (1969). Observations of eutrophication and nutrient cycles in some coastal plain estuaries. *In* "Eutrophication: Causes, Consequences, Correctives," pp. 210-221. Natl. Acad. Sci., Washington, D.C.

Cifuentes, L. S. (1982). The character and behavior of organic nitrogen in the Delaware Estuary salinity gradient. M.S. Thesis, University of Delaware, Newark.

Cline, J. D., and Richards, F. A. (1972). Oxygen deficient conditions and nitrate reduction in the eastern tropical North Pacific Ocean. *Limnol. Oceanogr. 17,* 885-900.

Coote, A. R., and Yeats, P. A. (1979). Distribution of nutrients in the Gulf of St. Lawrence. *J. Fish. Res. Board Can. 36,* 122-131.

Craig, H., and Turekian, K. K. (1980). The GEOSECS Program: 1976-1979. *Earth Planet. Sci. Lett. 49,* 263-265.

Craig, H., Broecker, W. S., and Spencer, D. (1981). "GEOSECS Pacific Expedition," Vol. 4. Nat. Sci. Found., Washington, D.C.

Culberson, C. H., Sharp, J. H., Church, T. M., and Lee, B. W. (1982). "Data from the Sals X Cruises. May 1978-July 1980," Oceanogr. Data Rep. No. 2. University of Delaware, Newark.

Dahm, C. N. (1974). A study of nutrient dynamics in the Atlantic Ocean. M.A. Thesis, Oregon State University, Corvallis.

Dawson, R., and Liebezeit, G. (1981). The analytical methods for the characterization of organics in seawater. *In* "Marine Organic Chemistry" (E. K. Duursma and R. Dawson, eds.), pp. 445-496. Am. Elsevier, New York.

Deuser, W. G., and Ross, E. H. (1980). Seasonal change in the flux of organic carbon to the deep Sargasso Sea. *Nature (London) 283,* 364-365.

Duedall, I. W., O'Connors, H. B., Parker, J. H., Wilson, R. E., and Robbius, A. S. (1977). The abundances, distribution and flux of nutrients and chlorophyll *a* in the New York Bight Apex. *Estuarine Coastal Mar. Sci. 5,* 81-105.

Dunstan, W. M., and Atkinson, L. P. (1976). Sources of new nitrogen for the South Atlantic Bight. *In* "Estuarine Processes" (M. Wiley, ed.), Vol. 1, pp. 69-78. Academic Press, New York.

Duursma, E. K. (1961). Dissolved organic carbon, nitrogen, and phosphorous in the sea. *Neth. J. Sea Res. 1,* 1-147.

Duursma, E. K. (1965). The dissolved organic constituents of sea water. *In* "Chemical Oceanographer" (J. P. Riley and G. Skirrow, eds.), Vol. 1, pp. 433-475. Academic Press, New York.

Eppley, R. W., Sharp, J. H., Renger, E. H., Perry, M. J., and Harrison, W. G. (1977). Nitrogen assimilation by phytoplankton and other microorganisms in the surface waters of the central north Pacific Ocean. *Mar. Biol. 39,* 111-120.

Eppley, R. W., Sapienza, C., and Renger, E. H. (1978). Gradients in phytoplankton stocks and nutrients off southern California in 1974-1976. *Estuarine Coastal Mar. Sci. 7,* 291-301.

Eppley, R. W., Renger, E. H., and Harrison, W. G. (1979a). Nitrate and phytoplankton production in southern California coastal waters. *Limnol. Oceanogr. 24,* 483-494.

Eppley, R. W., Renger, E. H., Harrison, W. G., and Cullen, J. J. (1979b). Ammonium distribution in southern California coastal waters and its role in the growth of phytoplankton. *Limnol. Oceanogr. 24,* 495-509.

Fanning, K. A., and Maynard, V. I. (1978). Dissolved boron and nutrients in the mixing plumes of major tropical rivers. *Neth. J. Sea Res. 12,* 345-354.

Fiadeiro, M. (1980). The alkalinity of the deep Pacific. *Earth Planet. Sci. Lett. 49,* 499-505.

Fine, R. A., Reid, J. L., and Ostland, H. G. (1981). Circulation of tritium in the Pacific Ocean. *J. Phys. Oceanogr. 11,* 3-14.

Fogg, G. E. (1975). Primary productivity. *In* "Chemical Oceanography" (J. P. Riley and G. Skirrow, eds.), 2nd ed., Vol. 2, pp. 385-453. Academic Press, New York.

Fournier, R. O., Marra, J., Bohrer, R., and Van Det, M. (1977). Plankton dynamics and nutrient enrichment of the Scotian Shelf. *J. Fish. Res. Board Can. 34,* 1004-1018.

Fox, L. E. (1981). The geochemistry of humic acid and iron during estuarine mixing. Ph.D. Dissertation, University of Delaware, Newark.

Fraga, F. (1966). Distribution of particulate and dissolved nitrogen in the Western Indian Ocean. *Deep-Sea Res. 13,* 413-425.

Gardner, W. D. (1977). Incomplete extraction of rapidly settling particles from water samplers. *Limnol. Oceanogr. 22,* 764-768.

Garside, C., Malone, T. C., Roels, O. A., and Sharfstein, B. A. (1976). An evaluation of sewage-derived nutrients and their influence on the Hudson Estuary and New York Bight. *Estuarine Coastal Mar. Sci. 4,* 281-289.

Gordon, D. C., Jr. (1971). Distribution of particulate organic carbon and nitrogen at an oceanic station in the central Pacific. *Deep-Sea Res. 18,* 1127-1134.

Gordon, D. C., Jr., and Sutcliffe, W. H., Jr. (1974). Filtration of seawater using silver filters for particulate nitrogen and carbon analysis. *Limnol. Oceanogr. 19,* 989-993.

Haines, E. B. (1979). Nitrogen pools in Georgia coastal waters. *Estuaries 2,* 34-39.

Hattori, A., and Wada, E. (1971). Nitrite distribution and its regulating processes in the equatorial Pacific Ocean. *Deep-Sea Res. 18,* 557-568.

Helder, W. (1974). The cycle of dissolved inorganic nitrogen compounds in the Dutch Wadden Sea. *Neth. J. Sea Res. 8,* 154-173.

Holm-Hansen, O., and Booth, C. R. (1966). The measurement of adenosine triphosphate in the ocean and its ecological significance. *Limnol. Oceanogr. 11,* 510-519.

Holm-Hansen, O., Strickland, J. D. H., and Williams, P. M. (1966). A detailed analysis of biological important substances in a profile of southern Calif. *Limnol. Oceanogr. 11,* 548-561.

Holm-Hansen, O., Sutcliffe, W. H., and Sharp, J. (1968). Measurement of deoxyribonucleic acid in the ocean and its ecological significance. *Limnol. Oceanogr. 13,* 507-514.

Ichikawa, T., and Nishizawa, S. (1975). Particulate organic carbon and nitrogen in the eastern Pacific Ocean. *Mar. Biol. 29,* 129-138.

Jaworski, N. A. (1981). Sources of nutrients and the scale of eutrophication problems in estuaries. *In* "Estuaries and Nutrients" (B. J. Neilson and L. E. Cronin, eds.), pp. 83-110. Humana Press, Clifton, New Jersey.

Josefsson, B. O., Lindroth, P., and Ostling, G. (1977). An automated fluorescence method for the determination of total amino acids in natural waters. *Anal. Chim. Acta 89,* 21-28.

Karl, D. M., and Holm-Hansen, O. (1978). Methodology and measurement of adenylate energy charge ratios in environmental samples. *Mar. Biol. 48,* 185-197.

Kester, D. R. (1975). Dissolved gases other than CO_2. *In* "Chemical Oceanography" (J. P. Riley and G. Skirrow, eds.), 2nd ed., Vol. 1, pp. 497-556. Academic Press, New York.

Lee, C., and Bada, J. L. (1977). Dissolved amino acids in the equatorial Pacific, the Sargasso Sea, and Biscayne Bay. *Limnol. Oceanogr. 22,* 502-510.

Liss, P. S. (1976). Conservative and non-conservative behaviour of dissolved constituents during estuarine mixing. *In* "Estuarine Chemistry" (J. D. Burton and P. S. Liss, eds.), pp. 93-130. Academic Press.

Loder, T. C., and Reichard, R. P. (1981). The dynamics of conservative mixing in estuaries. *Estuaries 4,* 64-69.

McAllister, C. D., Parsons, T. R., and Strickland, J. D. H. (1960). Primary productivity and fertility a station "P" in the North-East Pacific Ocean. *J. Cons., Cons. Int. Explor. Mer 25,* 240-259.

McCarthy, J. J. (1970). A urease method for urea in seawater. *Limnol. Oceanogr. 15,* 303-313.

McCarthy, J. J., and Kamykowski, D. (1972). Urea and other nitrogenous nutrients in La Jolla Bay during February, March, and April 1970. *Fish. Bull. 70,* 1261-1274.

McCave, I. N. (1975). Vertical flux of particles in the ocean. *Deep-Sea Res. 22,* 491-502.

Maita, Y., and Yanada, M. (1978). Particulate protein in coastal waters, with special reference to seasonal variation. *Mar. Biol. 44,* 329-336.

Menzel, D. W., and Ryther, J. H. (1961). Annual variations in primary production of the Sargasso Sea off Bermuda. *Deep-Sea Res. 7,* 282-288.

Mopper, K., and Lindroth, P. (1982). Diel and depth variations in dissolved free amino acids and ammonium in the Baltic Sea determined by shipboard HPLC analysis. *Limnol. Oceanogr. 27,* 336-347.

* Morris, A. W., Bale, A. J., and Howland, R. J. M. (1981). Nutrient distributions in an estuary: Evidence of chemical precipitation of dissolved silicate and phosphate. *Estuarine Coastal Mar. Sci. 12,* 205-216.

North, B. B. (1975). Primary amines in California coastal waters: Utilization by phytoplankton. *Limnol. Oceanogr. 20,* 20-27.

Olson, R. J. (1981). ^{15}N tracer studies of the primary nitrite maximum. *J. Mar. Res. 39,* 203-226.

Packard, T. T., and Dortch, Q. (1975). Particulate protein-nitrogen in North Atlantic surface waters. *Mar. Biol. 33,* 347-354.

Parsons, T. R. (1975). Particulate organic carbon in the sea. *In* "Chemical Oceanography" (J. P. Riley and G. Skirrow, eds.), 2nd ed., Vol. 2, pp. 365-383. Academic Press, New York.

Peterson, H. (1979). Sources and sinks of biologically reactive oxygen, carbon, nitrogen, and silica in Northern San Francisco Bay. *In* "San Francisco Bay, The Urbanized Estuary" (T. J. Conomos, ed.), pp. 175-193. Am. Assoc. Adv. Sci., Washington, D.C.

Pillai, T. N. V., and Ganguly, A. K. (1970). Nucleic acids in the dissolved constituents of sea water. *Curr. Sci. 39,* 501-504.

Pocklington, R. (1972a). Determination of nanomolar quantities of free amino acids dissolved in North Atlantic Ocean waters. *Anal. Biochem. 45,* 403-421.

Pocklington, R. (1972b). Secular changes in the ocean off Bermuda. *JGR, J. Geophys. Res. 77,* 6604-6607.

Postma, H. (1966). The cycle of nitrogen in the Wadden Sea and adjacent areas. *Neth. J. Sea Res. 3,* 186-221.

Postma, H. (1978). The nutrient contents of North Sea water: Changes in recent years, particularly in the southern bight. *Rapp. P.-V. Reun., Cons. Int. Explor. Mer 172,* 350-357.

Provasoli, L, and Carlucci, A. R. (1974). Vitamins and growth regulars. *In* "Algal Physiology and Biochemistry" (W. D. P. Stewart, ed.), pp. 741-787. Univ. of California Press, Berkeley.

Redfield, A. C., Ketchum, B. H., and Richards, F. A. (1963). The influence of organisms on the composition of sea water. *In* "The Sea" (M. N. Hill, ed.), Vol. 2, pp. 26-77. Wiley (Interscience), New York.

Reid, J. L., and Lynn, R. J. (1971). On the influence of the Norwegian-Greenland and Weddell Seas upon the bottom waters of the Indian and Pacific oceans. *Deep-Sea Res. 18,* 1063-1088.

Reid, J. L., Brinton, E., Fleminger, A., Venrick, E. L., and McGowan, J. A. (1978). Ocean circulation and marine life. *In* "Advances in Oceanography" (H. Charnock and G. Deacon, eds.), pp. 65-130. Plenum, New York.

Remsen, C. C. (1971). The distribution of urea in coastal and oceanic waters. *Limnol. Oceanogr. 16,* 732-740.

Riley, G. A. (1951). Oxygen, phosphate, and nitrate in the Atlantic Ocean. *Bull. Bingham Oceanogr. Collect. 13,* 1-126.

Riley, G. A. (1967). Mathematical model of nutrient conditions in coastal waters. *Bull. Bingham Oceanogr. Collect. 19,* 72-80.

Riley, J. P., and Segar, D. A. (1970). The seasonal variation of the free and combined dissolved amino acids in the Irish Sea. *J. Mar. Biol. Assoc. U.K. 50,* 713-720.

Ring Group (1981). Gulf Stream cold-core rings: Their physics, chemistry, and biology. *Science 212,* 1091-1100.

Ryther, J. H., Menzel, D. W., and Vaccaro, R. F. (1961). Diurnal variations in some chemical and biological properties of the Sargasso Sea. *Limnol. Oceanogr. 6,* 149-153.

Sharp, J. H. (1974). Improved analysis for "particulate" organic carbon and nitrogen from sea water. *Limnol. Oceanogr. 19,* 984-989.

Sharp, J. H. (1975). Gross analyses of organic matter in sea water: Why, how, and from where? *ACS Symp. Ser. 18,* 682-696.

Sharp, J. H., and Church, T. M. (1981). Biochemical dynamics in coastal waters of the Middle Atlantic states. *Limnol. Oceanogr. 26,* 843-854.

Sharp, J. H., Church, T. M., and Culberson, C. H. (1980). "Data from the 1977 Trans X Cruises," Oceanogr. Data Rep. No. 1. University of Delaware, Newark.

Sharp, J. H., Culberson, C. H., and Church, T. M. (1982a). The chemistry of the Delaware Estuary. General considerations. *Limnol. Oceanogr. 27,* 1015-1028.

Sharp, J. H., Frake, A. C., Hillier, G. B., and Underhill, P. A. (1982b). Modeling nutrient regeneration in the ocean with an aquarium system. *Mar. Ecol.: Prog. Ser. 8,* 15-23.

Spencer, C. P. (1975). The micronutrient elements. *In* "Chemical Oceanography" (J. P. Riley and G. Skirrow, eds.), 2nd ed., Vol. 2, pp. 245-300. Academic Press, New York.

Stefansson, V., and Richards, F. A. (1963). Processes contributing to the nutrient distributions off the Columbia River and Strait of Juan de Fuca. *Limnol. Oceanogr. 8,* 394-410.

Strickland, J. D. H. (1965). Production of organic matter in the primary stages of the marine food chain. *In* "Chemical Oceanography" (J. P. Riley and G. Skirrow, eds.), pp. 477-610. Academic Press, New York.

Sverdrup, H. U., Johnson, M. W., and Fleming, R. H. (1942). "The Oceans." Prentice-Hall, Englewood Cliffs, New Jersey.

Thomas, W. H., and Carsola, A. J. (1980). Ammonium input to the sea via large sewage outfall. Part 1. Tracing sewage in southern California waters. *Mar. Environ. Res. 3,* 277-289.

Treguer, P., LeCorre, P., and Grall, J. R. (1979). The seasonal variations of nutrients in the upper waters of the Bay of Biscay region and their relation to phytoplankton growth. *Deep-Sea Res*. *26*, 1121-1152.

Van Bennekom, A. J., Berger, G. W., Helder, W., and De Vries, R. T. P. (1978). Nutrient Distribution in the Zaire Estuary and river plume. *Neth. J. Sea Res*. *12*, 296-323.

von Brand, T., Rakestraw, N. W., and Renn, C. E. (1937). The experimental decomposition and regeneration of nitrogenous organic matter in seawater. *Biol. Bull. (Woods Hole, Mass.) 72*, 165-175.

Walsh, J. J. (1977). A biological sketchbook for an eastern boundary current. *In* "The Sea" (E. D. Goldberg, I. H. McCave, J. J. O'Brien, and J. H. Steele, eds.), Vol. 6, pp. 923-968. Wiley (Interscience), New York.

Walsh, J. J., Whitledge, T. E., Barvenik, F. W., Wirick, C. D., Howe, S. O., Esaias, W. E., and Scott, J. T. (1978). Wind events and food chain dynamics within the N.Y. Bight. *Limnol. Oceanogr*. *23*, 659-683.

Whitledge, T. E., and Dugdale, R. C. (1972). Creatine in seawater. *Limnol. Oceanogr*. *17*, 309-354.

Wiebe, P. H., Boyd, S. H., and Winget, C. (1976). Particulate matter sinking to the deep-sea floor at 2000 m in the Tongue of the Ocean, Bahamas with a description of a new sedimentation trap. *J. Mar. Res*. *34*, 341-354.

Williams, P. J. L. (1975). Biological and chemical aspects of dissolved organic material in sea water. *In* "Chemical Oceanography" (J. P. Riley and G. Skirrow, eds.), 2nd ed., Vol. 2, pp. 301-362. Academic Press, New York.

Wollast, R. (1978). Modeling of biological and chemical processes in the Scheldt Estuary. *In* "Hydrodynamics of Estuaries and Fjords" (J. C. J. Nihoul, ed.), pp. 53-77. Elsevier, Amsterdam.

Woods Hole Oceanographic Institution. "Physical, Chemical and Biological Observations in the Sargasso Sea off Bermuda 1960-1963." WHOI Ref. No. 64-8, Appendix II. WHOI, Woods Hole, Massachusetts.

Chapter 2

Gaseous Nitrogen Compounds in the Marine Environment

MARY I. SCRANTON
Marine Sciences Research Center
State University of New York
Stony Brook, New York

I. INTRODUCTION

Dissolved gaseous nitrogen compounds, although mentioned only briefly in most discussions of dissolved inorganic nitrogen in the ocean, play important roles in the oceanic nitrogen cycle. As gases, these compounds can be supplied to or lost from the ocean across the air-sea interface. They also may be produced or consumed within the ocean and its sediments by various biological and chemical mechanisms. Therefore, an understanding of their distributions and controlling processes are important if a full understanding of the marine nitrogen cycle is to be obtained. The dissolved gaseous nitrogen compounds discussed in this chapter are dissolved ammonia

ISBN 0-12-160280-X

gas (NH_3), dinitrogen gas (N_2),* nitrous oxide (N_2O), and nitric oxide (NO). Other nitrogen gases, such as nitrogen dioxide (NO_2) and nitrous acid (HONO), are known in the environment, but have never been studied in marine systems.

Of the nitrogen compounds found in the environment, a gaseous species (dinitrogen) is by far the most abundant. The predominance of N_2 over nitrate (the thermodynamically stable form of nitrogen in the presence of oxygen) demonstrates the global importance of denitrification in controlling nitrogen speciation. However, the oceanic N_2 distribution is largely controlled by physical parameters, with biological perturbations only significant in specialized environments, such as reducing sediments and the Cariaco Trench. Similarly, the oceanic distribution of dissolved gaseous ammonia $(NH_3)_{aq}$ is controlled by chemical as well as biological phenomena. The total oceanic content of ammonia ($NH_3 + NH_4^+$) is due to biological activity, but the NH_3 content is controlled by the pH of sea water and factors affecting the dissociation of NH_4^+. Nitrous oxide distributions are controlled by biological production and removal during nitrification and denitrification. In contrast, although an intermediate in denitrification, free NO (nitric oxide) seems to be an abiotic product of photochemical reactions of nitrite in surface seawater. Clearly the nitrogen gases represent a fascinating suite of compounds with a diverse set of controlling parameters.

II. PHYSICAL PROCESSES

Before it is possible to identify the influence of biological or chemical factors on any dissolved gas, it is necessary to understand the various physical processes that influence the concen-

**Throughout this review "dinitrogen" refers to N_2 gas, while "nitrogen" refers to the element.*

trations of gases in the oceans. These include processes influencing the solubility of gases in seawater; the air-sea exchange mechanism; processes that may cause surface seawater to be either super- or undersaturated with gas before it is isolated from the atmosphere; and processes that may affect the apparent degree of saturation after the water is isolated from the surface. These topics have been covered in detail by Kester (1975), Broecker and Peng (1971), and Craig and Weiss (1971) among others, but will be summarized briefly here to provide a background for the sections on specific gases to follow.

A. Solubility

The solubility of a gas in seawater and the concentration of that gas in the atmosphere determine the amount of gas present in surface seawater at a given salinity, temperature, and atmospheric pressure, unless some other process is occurring. Thus, most discussions of dissolved gas distributions relate observed gas concentrations to the value predicted from solubility considerations.

Solubilities are determined from Henry's law, which states that the partial pressure of a gas in solution (P_g) is directly proportional to the mole fraction of the gas in solution c_g, so

$$P_g = K_H c_g \tag{1}$$

where K_H is a form of the Henry's law constant with dimensions of atm·liter·mol^{-1}. At equilibrium between the atmosphere and seawater, the partial pressure of the gas in the atmosphere p_g equals the partial pressure of the gas in solution P_g, so the solubility of the gas c_g^* can be represented as

$$c_g^* = \frac{1}{K_H} p_g \tag{2}$$

Except for the units used, the Bunsen coefficient β_g, which is defined as the volume of dry gas at standard temperature and pressure that can be absorbed by a unit volume of solution when the partial pressure of the gas is one atmosphere, is conceptually similar to an inverse Henry's law constant, so

$$c_g^* = \beta_g \, p_g \tag{3}$$

Because Bunsen coefficients are calculated for dry air, a correction for water vapor pressure must be made before tabulated Bunsen coefficients can be used to calculate solubilities for oceanic gases. If P equals the total atmospheric pressure and p^* is the water vapor pressure at the temperature of the sample, assuming 100% humidity

$$c_g^* = \beta_g p_g \frac{(P - p^*)}{P} \tag{4}$$

For very precise work, corrections must be made for the temperature and pressure dependence of the volume of seawater, and therefore it is preferable to use units of moles gas per kilogram of seawater instead of moles per liter seawater, since the former unit is independent of temperature and pressure. At present the oceanographic community uses a variety of units, and while attempts will be made to provide appropriate conversion factors in particularly unconventional cases, in general the units originally reported will be used in this chapter.

High-quality solubility data for the temperature and salinity ranges found in seawater are available for N_2 (Weiss, 1970) and for N_2O (Weiss and Price, 1980). Seawater solubility coefficients for ammonia are not available, but Hales and Drewes (1979) present data for pure water and for rain water. For NO, Zafiriou and McFarland (1980) have presented a solubility coefficient for seawater. Values of the solubility coefficients for these gases at a salinity

TABLE I. Solubility Coefficients for Gaseous Nitrogen Compounds

Gas	*Solubility coefficient*	*Value*[a]	*Units*	*Reference*
N_2	Bunsen coefficient (β)[b]	11.61	$m\ell\ N_2 \cdot \ell^{-1} \cdot atm^{-1}$	Weiss (1970)
		(5.18×10^{-4}	$mol\ N_2 \cdot \ell^{-1} \cdot atm^{-1}$)	
N_2O	Equilibrium constant (K_o)[c]	2.077×10^{-2}	$mol\ N_2O \cdot \ell^{-1} \cdot atm^{-1}$	Weiss and Price (1980)
NO	Henry's law constant (K_H)[d]	26		Zafiriou and McFarland (1980)
NH_3	Henry's law constant (K_H)[d]	5.4×10^{-4}		Hales and Drewes (1979)

[a] *Values given are constants for 25°C and 35°/oo salinity except for ammonia. For ammonia, the solubility coefficient is for 25°C in distilled water.*

[b] $c_g = \beta p_g$ *where* c_g *= concentration gas in solution;* β *= Bunsen coefficient;* p_g *= partial pressure gas in dry atmosphere at one atmosphere total pressure.*

[c] $K_o = \frac{\beta}{V^+}$ *where* V^+ *= volume of one mole pure gas at standard conditions (STP).*

[d] $K_H = \frac{X_g}{c_g}$ *where* X_g *= moles gas per liter air;* c_g *= moles gas per liter solution (*K_H *is therefore dimensionless).*

of 35.0×10^{-3} (except for ammonia) and at a temperature of 25°C are given in Table I.

B. Air-Sea Fluxes

In the event that a gas in the oceanic mixed layer is significantly super- or undersaturated with respect to solubility equilibrium with the atmosphere, transport of the gas across the air-sea interface will occur until equilibrium is reached. This fact will be important, for example, in seasonal studies of gas concentrations (e.g., Pytkowicz, 1964). The process by which air-sea exchange occurs has been successfully studied using a simple laminar layer or thin film model (Danckwerts, 1970). The model outlined in Fig. 1 describes the air-sea interface as consisting of three zones: a well-mixed atmospheric layer, a well-mixed fluid layer, and at the interface, a thin film in which transport of gas is by molecular diffusion (i.e., there is no turbulence). The model also assumes that the surface of the water is at equilibrium with the atmosphere, and thus the gas concentration at the interface is c_g^*. In the bulk liquid, the gas concentration is c_g and in the thin film near the interface, there is a linear concentration gradient since molecular diffusion is controlling the gas flux across the layer. Assuming a thin film thickness (z), a diffusion coefficient for the gas in seawater (D), and Fickian diffusion, the flux of gas across the air-sea interface will be

$$F = \frac{D}{z} (c_g - c_g^*) \tag{5}$$

Although it is unlikely that this model accurately represents the physics of the ocean (e.g., consider the difficulty of maintaining a uniform thin film thickness in the presence of breaking waves), extensive research has shown that the model gives a realistic representation of flux (see Broecker and Peng, 1974).

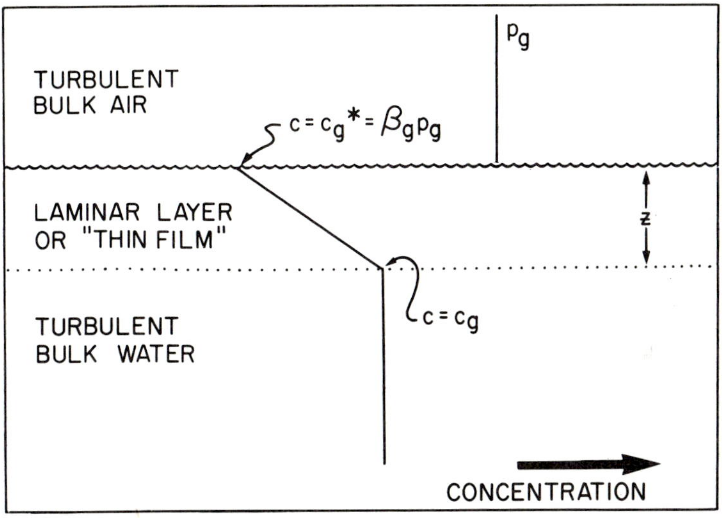

Fig. 1. Schematic diagram for the "thin film" model of gas exchange. All symbols are defined in the text.

Broecker and Peng (1971), Peng et al. (1974), and others have determined the "thin film" thickness from radon data and obtained values of from 20 μm to 126 μm for typical oceanic conditions. Kanwisher (1963) and others have shown that the thin film thickness is a function of the wind velocity, so gas exchange is more rapid at high wind speeds. A plot of thin film thickness versus wind speed taken from Emerson (1975) is shown in Fig. 2.

Values for diffusion coefficients of N_2 and N_2O as a function of temperature are presented by Broecker and Peng (1974) in Table II. For NO, Zafiriou and McFarland (1981) assumed a diffusion coefficient of twice that for radon (Peng et al., 1979) (Table II).

C. Abiotic Causes for Dissolved Gas Saturation Anomalies

Extensive work with inert gases (He, Ar, Ne) has indicated the presence of several physical processes that can create significant gaseous saturation anomalies in the absence of biological and chemical activity (Carritt, 1954; Benson and Parker, 1961; Craig and

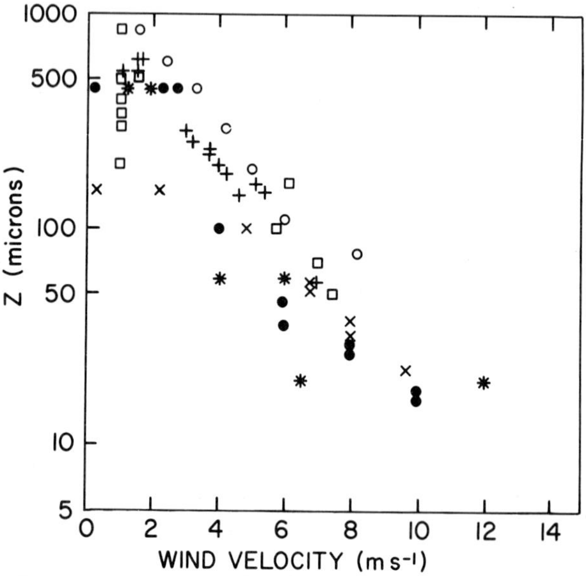

Fig. 2. Thin film thickness as a function of wind speed. Data plotted are from a number of different environments (published with permission from Emerson, 1975).

TABLE II. Diffusivities of Gaseous Nitrogen Compounds[a]

Temperature (°C)	*0*	*5*	*10*	*15*	*20*	*25*	*30*
N_2	0.95	1.11	1.29	1.49	1.71	1.96	2.23
N_2O	1.16	1.35	1.56	1.80	2.07	2.36	2.69
(Rn)[b]	0.67	0.79	0.91	1.05	1.20	1.37	1.56

[a] *Values (in $cm^2 \cdot s^{-1}$) from Broecker and Peng (1974).*

[b] *Included because Zafiriou and McFarland (1981) use Rn diffusivities as a basis for estimating NO diffusivities.*

Weiss, 1971). The relative importance of these processes for a specific gas depends on the solubility and atmospheric abundance of the gas.

Three general mechanisms for producing anomalies are discussed by Craig and Weiss (1971) as being the most important for He, Ne, and Ar, and thus by inference for other gases as well. The first type is known as the pressure component, which includes saturation anomalies caused by atmospheric pressure changes or by *partial* dissolution of bubbles forced down into the ocean by breaking waves. The second type of anomaly is caused by temperature changes after gas equilibration once the parcel has become isolated from the atmosphere. This second type of anomaly includes anything causing the potential temperature of the parcel to be different from the temperature at the time of water-mass sinking (e.g., geothermal heating). It also includes anomalies due to the nonlinear nature of mixing phenomena in the ocean [e.g., the apparent gas concentration in a 50/50 mixture of gas-equilibrated $0^{\circ}C$ water and $30^{\circ}C$ water is not equal to the equilibrium gas concentration of $15^{\circ}C$ water, but in fact is substantially higher (Kester, 1975)]. The third type of anomaly is known as air injection and is caused by the total dissolution of bubbles forced below the air-sea interface.

The pressure component will affect all gas concentrations by a constant percentage as can be seen by reference to Eq. (4). On the other hand, total dissolution of bubbles (air injection) will create anomalies of different sizes for different gases. Given two atmospheric gases with different solubilities, total dissolution of a bubble will yield water with a higher percent saturation anomaly[+] for the less soluble gas. For example, air injection of 1 mℓ of air containing 300 ppbv (parts ber billion by volume) N_2O and 78% N_2 will produce saturation anomalies of 0.1% for N_2O and 4.4% for N_2. Temperature anomalies, causing apparent gas saturation anomalies, will also vary from gas to gas, depending on solubility be-

[+]*Percent saturation anomaly is defined as the measured gas concentration divided by the equilibrium solubility concentration times 100.*

havior as a function of temperature. A more detailed discussion can be found in Kester (1975).

For most gases in the environment, the effect of the various physical processes discussed above is to yield super- or under-saturations of only a few percent. Without simultaneous data from truly inert gases such as argon or neon, it is unwise to attribute concentration anomalies of this size (95-105% of saturation) in field studies to processes other than air injection, temperature anomalies, or a pressure component. Unfortunately, the analytical procedures used for noble gases are sufficiently difficult that few studies of physical anomalies beyond that of Craig and Weiss (1971) have been carried out.

III. DISSOLVED GASEOUS NITROGEN COMPOUNDS

Except for nitrous oxide (N_2O), gaseous nitrogen compounds have not received a great deal of attention by oceanographers. This is due largely to the low concentrations of most gases. In the case of dinitrogen (N_2), however, studies have been inhibited by the large and relatively constant background of N_2 present due to gaseous equilibrium with the atmosphere. Current evidence suggests that gaseous nitrogen compounds should not be overlooked in discussions of the marine nitrogen cycle.

A. Aqueous Gaseous Ammonia [$(NH_3)_{aq}$]

Ammonia is present in very low concentrations in most of the world's oceans, and therefore analytical problems have limited the extent to which this fraction of dissolved inorganic nitrogen has been studied. Even less attention has been paid to that fraction of the ammonia in seawater present as aqueous ammonia gas, because the predominant form in seawater is the protonated NH_4^+. Nevertheless, a small fraction (2-6%) of total ammonia is present as dissolved ammonia gas, and this component may play an important role as a source of gaseous ammonia to the atmosphere in remote oceanic locations (Ayers and Gras, 1980; Georgii and Gravenhorst, 1977). It is also possible that, since terrestrial atmospheres may contain elevated ammonia concentrations (Tsunogai, 1971; Georgii and Müller, 1974), the transport of gaseous ammonia across the air-sea interface may provide a small source of the compound to the mixed layer in areas near the coast.

The fraction of ammonia present in seawater as aqueous ammonia gas $(NH_3)_{aq}$ is determined predominately by two factors: the total ammonia concentration and the pH of the water. Ammonia is a weak base, and in aqueous solution the reaction

$$NH_3 + H_2O \rightleftarrows NH_4^+ + OH^- \tag{I}$$

is important. The equilibrium constant for this reaction is K_b

$$K_b = \frac{[NH_4^+]\ [OH^-]}{[H_2O]\ [NH_3]} \tag{6}$$

where brackets represent chemical activities. Since the analytically determined ammonia concentration a is

$$a = [NH_4^+] + [NH_3] \tag{7}$$

and the activity of the solvent, water, can be taken to be 1, Eq. (6) can be written

$$K_b = \frac{[NH_4^+][OH^-]}{a - [NH_4^+]} \tag{8}$$

which, on rearrangement, becomes

$$[NH_4^+] = \frac{K_b}{[OH^-] + K_b} \tag{9}$$

Thus, the fraction of ammonia present as ammonium ion $[NH_4^+]$ will range from 94 to 98% over the oceanic pH range of 8.2 - 7.6, using a value of 7.9×10^{-5} at 25°C for K_b [data from Sillen and Martell (1971) for constants in 1 *N* $NaClO_4$] and the pK_w^{sw} reported by Culberson and Pytkowicz (1973).

Although the fraction of total ammonia present as $(NH_3)_{aq}$ is small, Ayers and Gras (1980) and Georgii and Gravenhorst (1977) suggest that it may play an important role in controlling gaseous ammonia concentrations in the atmosphere over the ocean. From Henry's law,

$$p_{NH_3} = [NH_3]_{aq} K_H \tag{10}$$

where p_{NH_3} is the atmospheric partial pressure of the gas and K_H is the Henry's law constant for ammonia in seawater [about 7.3×10^{-3} atm·liter·mol^{-1} at 25°C (Hales and Drewes, 1979)]. Substituting Eq. (10) in Eq. (6)

$$K_b = \frac{[NH_4^+][OH^-] K_H}{p_{NH_3}} \tag{11}$$

or

$$p_{NH_3} = \frac{[NH_4^+][OH^-] K_H}{K_b} \tag{12}$$

Further substituting Eq. (9) into Eq. (12)

$$p_{NH_3} = \frac{a K_H [OH^-]}{(OH^-) + K_b} \tag{13}$$

which relates the equilibrium partial pressure of ammonia gas in the atmosphere with the pH and total ammonia concentration. Both K_H and K_b are functions of temperature. Total ammonia in the oceans ranges from less than 0.05 μmol liter^{-1} to 1 or 2 μmol liter^{-1}, with higher values occasionally occurring in coastal regions (see Chapter 1; Lenhard and Georgii, 1980). Based on Eq.(13), the equilibrium concentration of ammonia gas in the atmosphere would then range from less than 0.94 nmol NH_3 m^{-3} (2.2 x 10^{-11} atm) to 36 nmol NH_3 m^{-3} (8.6 x 10^{-10}atm). Measured values for atmospheric ammonia range from 3.5 nmol NH_3 m^{-3} (8.4 x 10^{-11} atm) south of Australia (Ayers and Gras, 1980) to 0.12 - 0.59 μmol m^{-3} (14 x 10^{-9} atm) over the continental United States and Central Europe (Lau and Charlson, 1977; Georgii and Müller, 1974). The most common atmospheric values over the oceans are around 5.9 - 17.6 nmole m^{-3} (1.4 x 10^{-10} - 4.2 x 10^{-10} atm) (Tsunogai, 1971; Georgii and Gravenhorst, 1977), although the latter group found values as high as 0.47 μmol m^{-3} (11.2 x 10^{-10}atm) over the Sargasso Sea and Caribbean. Thus, especially in truly oceanic locations, the surface ocean appears to be at or near equilibrium with ammonia gas with respect to the atmosphere. Research still remains to be done to clarify the relative importance of interactions of the coastal ocean and atmosphere in the oceanic gaseous ammonia cycle.

B. Dinitrogen (N_2)

Dinitrogen is the most abundant form of nitrogen on earth in spite of the fact that the thermodynamically stable form for nitrogen in the presence of oxygen is nitrate. The principal reason for the maintenance of the disequilibrium is that biological denitrification continuously recycles nitrate to N_2 (Delwiche, 1956, 1970). On the other hand, the coupled activities of N_2-fixing and nitrifying bacteria provide a biological route of conversion of N_2 back to NO_3^-. Because of the potential of these processes to perturb an oceanic dinitrogen distribution caused by gas equilibrium with the atmosphere, substantial efforts have been made to determine the abundance of N_2 throughout the ocean.

The earliest measurements of N_2 in the ocean, such as those by Buch (1929), Rakestraw and Emmel (1938), and Hamm and Thompson (1941) indicated that N_2 concentrations were between 93 and 111% of saturation. However, analytical precision was inadequate to determine whether the observed variations represented true deviations from equilibrium. Benson and Parker (1961) noted that, by assuming argon to be truly inert, mass spectrometric measurements of the N_2:Ar ratio would improve the likelihood of identifying biological perturbations in N_2 concentrations. They recognized that undersaturations and supersaturations could be caused by pressure changes, bubble dissolution, and so on. Since these processes should also affect argon concentrations, by using the N_2:Ar ratio, they were able to correct for these effects. Their data suggest that N_2 behaved conservatively (concentrations unperturbed after isolation from the atmosphere) in the open ocean to within (±)1%. These results were checked using ($^{14}N^{15}N:^{14}N^{14}N$) ratio measurements. Since it was expected that biological processes should yield N_2 with a significantly different isotopic signal than that present in air, the fact that dissolved N_2 had an isotopic concentration indistinguishable from air suggests the dissolved gas presumably derived from the atmosphere and was not produced biologically. Their data confirm the original idea that N_2 was conservative in the open ocean within available analytical precision. Similar results were obtained by Linnenbom et al. (1966) and Swinnerton and Sullivan (1962) who measured N_2 saturation symmetrically distributed around a mean of 99.4% and ranging from 92 to 108% in the Greenland Sea and Andaman Sea, and by Craig et al. (1967) who found N_2 saturations averaging 101.9 (±) 1.1% in the South Pacific. Within oxygenated waters, biological processes do not create detectable change in N_2 concentrations.

Nevertheless, it appears that biological N_2 production does occur in the ocean in areas of denitrification. Inorganic nitrogen mass balances have been made in upwelling areas such as the eastern tropical North Pacific (Cline and Richards, 1972). Even when N_2O

is included as part of the mass balance between inorganic compounds calculated to have been present initially in the water (preformed nitrate plus nitrogen regenerated from the decay of organic matter) and reduced nitrogen compounds (nitrite plus ammonia plus nitrous oxide), a nitrogen imbalance still may occur (Cline and Richards, 1972; Cohen and Gordon, 1978). In these cases the "missing" nitrogen is assumed to be N_2 produced by the denitrifying bacteria. Goering (1968), Cline and Kaplan (1975), and others (see Hattori, Chapter 6, this volume; Codispoti, Chapter 15, this volume) have observed denitrification in these environments using ^{15}N but no direct observations of N_2 production have been made.

In permanently anoxic systems, the amounts of nitrate reduced by denitrifying bacteria and the amounts of organic material decomposed may be larger, and nitrate reduction may go completely to N_2 (Cohen, 1978). Richards and Benson (1961) made mass spectrometric measurements of dinitrogen and argon in the Cariaco Trench, a large marine anoxic basin off Venezuela, and in the Dramsfjord, Norway. Again, by assuming argon concentrations to be unaffected by biological activity, ratio determinations permit calculation of the amount of N_2 produced bacterially. Richards and Benson (1961) found that about 20 μmol liter^{-1} of excess or biogenic N_2 were present in the Cariaco Trench (102% saturation), and about 40 μmol liter^{-1} (104% saturation) were found in the Dramsfjord. These values agree well with the amount of N_2 expected to be generated from decaying organic matter as calculated from phosphate concentrations and using the Redfield ratios (16 N: 1 P) of Redfield et al. (1963). Similar results were obtained by Cline (1973) using N_2/Ar ratios measured by differential thermal analysis. Linnenbom and Swinnerton (1971) measured N_2 in the Cariaco Trench using a gas chromatographic method and found saturations of about 104%.

From these results, it is obvious why N_2 anomalies have not been detected in more oceanic regimes. The background N_2 content (about 900 μmol liter^{-1}) is so large that the biogenic contribution (20 μmol liter^{-1}) is barely detectable. In sediments, much larger anomalies have been observed. Barnes et al. (1975) found N_2 saturations of up to 119.5% at depth in Southern California borderland basin sediments (with one value at 393 cm in the Santa Cruz Basin reaching 159% of equilibrium). Wilson (1978) also reported elevated N_2 levels in pore waters (100-120% saturation) and concluded that pore water denitrification may be a source of N_2 to oceanic bottom watter. More recently Seitzinger et al. (1980) measured large N_2 fluxes (50 μmol $N_2 \cdot m^{-2} \cdot h^{-1}$) from coastal sediments in summer. Fluxes of this size might provide a significant source of excess nitrogen in the bottom waters overlying these sediments. No data are available to describe N_2 gradients in coastal waters.

C. Nitrous Oxide (N_2O)

While not the best understood of the oceanic gaseous nitrogen compounds, nitrous oxide is perhaps the most studied. This is due to interest generated by the major role the gas appears to play in controlling the abundance of stratospheric ozone (Crutzen, 1970, 1981). To provide a baseline from which the importance of various anthropogenic sources of nitrous oxide can be viewed, a major effort has been made to describe the natural (including the oceanic) N_2O cycle. Hahn (1981) has recently published a review of oceanic N_2O distributions, so these comments will be relatively brief.

It must be noted here that instrumental calibration has been a major problem in N_2O work over the past 15-20 years. Different groups have obtained substantially different atmospheric nitrous oxide concentrations, even though within each researcher's data set the values are quite uniform and precisions are very high. Therefore, from the point of view of considering the oceanic N_2O distribution in the following pages, the relative concentration variations and the saturation anomalies calculated from water and air measurements published by the same group rather than absolute values are noted. Most recent data indicate that an atmospheric N_2O concentration of very near 300 ppbv is accurate (Weiss, 1981; Goldan et al., 1981).

Early work (Craig and Gordon, 1963) indicated that the South Pacific was generally undersaturated in N_2O with respect to solubility equilibrium. Due to analytical sensitivity, however, samples from different locations and depths had been combined for analysis. Therefore, the results are not comparable with more recent work.

Junge et al. (1971) and Junge and Hahn (1971) presented the next studies of oceanic N_2O for the North and equatorial Atlantic. Their data suggested that the tropical ocean, in particular, was a major source of N_2O to the atmosphere (i.e., the mixed layer had N_2O saturations of up to 230%) and that the degree of saturation decreased northward. Deep waters (below about 1000 m) were undersaturated with respect to solubility equilibrium with the atmosphere. Further work (Hahn, 1974, 1975) apparently confirmed the observation of significantly supersaturated surface water in the North Atlantic, although the observed saturations were closer to 120-125% than the 230% observed during the earliest cruise in 1969. Subsequent work has shown that even those supersaturations were probably slightly too high. Yoshinari (1976) reported N_2O values for the surface waters of the North Atlantic and Caribbean, and found that, in contrast to the work by Hahn and co-workers, most of the surface waters sampled were at saturation or, at most, slightly supersaturated. No large saturation anomalies were found. Cohen and Gordon (1978) sampled the surface waters of the east tropical North Pacific, a region of active upwelling in which N_2O would be expected to be high, and found surface waters to average about 110% of saturation with N_2O. Pierotti and Rasmussen (1980) also investigated the east tropical North Pacific, from 4° to $22^\circ N$, although at stations closer to the coast than Cohen and Gordon, finding saturations averaging 123%. Singh et al. (1979) reported N_2O data from $46^\circ N$ to $40^\circ S$ that showed an average saturation of 133% with maximum values of 190%. Weiss (1978), in extensive surface sampling in both upwelling and nonupwelling regimes, has found

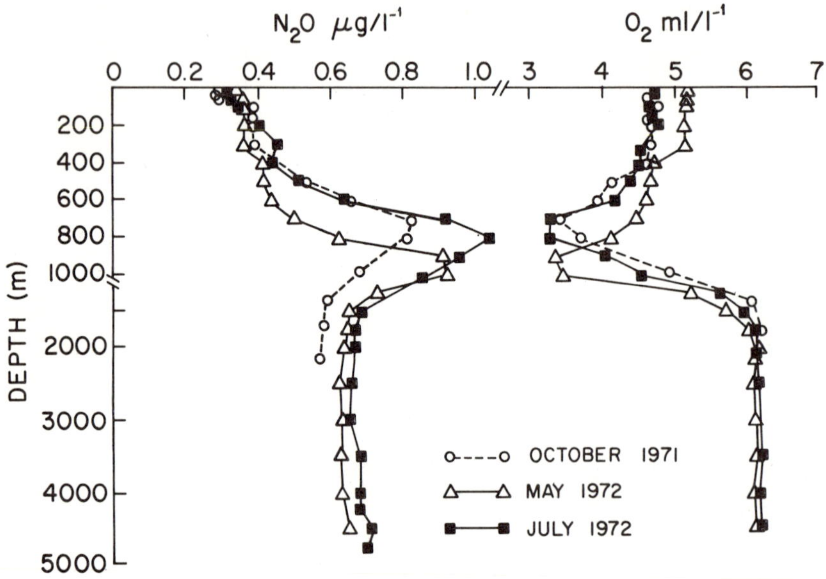

Fig. 3. Vertical profiles for nitrous oxide and oxygen from the Sargasso Sea (published with permission from Yoshinari, 1976).

that the surface oceans are, in general, nearly at equilibrium with the atmosphere, but that in areas where upwelling occurs (east tropical North Pacific, equatorial zones in both oceans, near coasts), N_2O levels can become substantially elevated and the ocean can act as a locally important N_2O source.

The cause for the occurrence of elevated surface layer N_2O concentration in upwelling regions can best be understood in the context of processes causing deep-water N_2O variations. Hahn (1974) first pointed out the inverse correlation present between N_2O and O_2 in the deep ocean. Figure 3 from Yoshinari (1976) shows typical oxygen and nitrous oxide profiles from the Atlantic, demonstrating the simultaneous N_2O maximum and O_2 minimum. Yoshinari (1976) and Cohen and Gordon (1978) confirmed that, in areas of the ocean where oxygen concentrations were greater than 17 μmol liter^{-1}, a strong inverse correlation exists between the two gases (see Fig. 4). Although Hahn (1974) had interpreted this to mean that denitrification

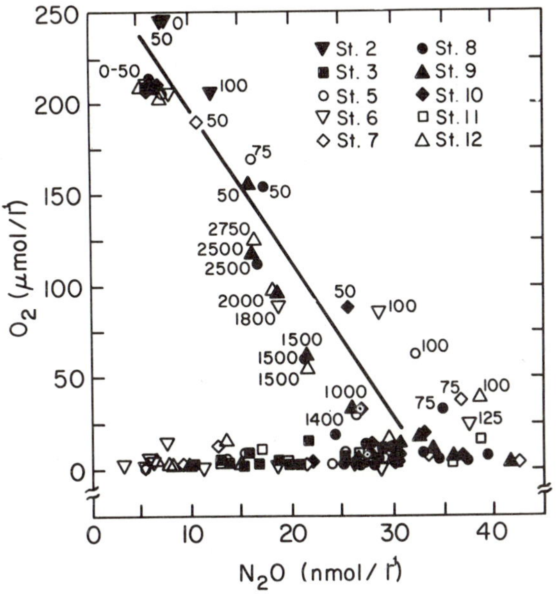

Fig. 4. Correlation between oxygen and nitrous oxide from the east tropical North Pacific. The numbers next to the symbols are sampling depths (published with permission from Cohen and Gordon, 1978).

(reduction of nitrate to N_2 via N_2O) was the major source of N_2O in the ocean, Cohen and Gordon (1978, 1979) and Cohen (1978) suggested that nitrification (oxidative regeneration of nitrate from organic nitrogen or ammonia) was a more likely source. In particular, they note that denitrification generally does not begin until O_2 levels are very low [less than 5 μmol liter^{-1}, according to Goering (1968)]. In addition, they observed that in the east tropical North Pacific, N_2O concentrations actually went through a minimum coincident with the secondary nitrite maximum at the core of the O_2 minimum. This feature is also obvious in the plot of N_2O versus O_2, in which the linear trend between the gases disappears at O_2 concentrations less than 17 μmol liter^{-1} (Fig. 4).

Data from Saanich Inlet, an anoxic fjord in British Columbia (Cohen, 1978), also demonstrate that N_2O is consumed in anoxic environments as N_2O disappears within the interface layer between oxic and anoxic zones. Above the interface, in the region where oxygen decreases and nitrification might be expected, N_2O increases to a maximum of about 180-200% of saturation. Below the interface, where oxygen and nitrate go to zero, N_2O also disappears, apparently being consumed by denitrifying bacteria. Thus it appears that aerobic nitrate regeneration (see Kaplan, Chapter 5 this volume) and microaerophilic denitrification (Firestone et al., 1980) is accompanied by N_2O production, whereas anaerobic dentrification processes results in N_2O consumption.

Other possible explanations for the observed N_2O distributions have been proposed. Pierotti and Rasmussen (1980) suggest that assimilatory nitrate reduction (nitrate uptake) results in N_2O production. Pierotti and Rasmussen (1980), Elkins et al. (1978), and Weiss (1978) all report the highest surface concentrations of N_2O in regions of upwelling (where nitrate would be expected to be elevated). In regions of upwelling and the associated high primary productivity, nitrate reduction will be rapid (Dugdale and Goering, 1967), and this may be accompanied by N_2O production. It is also possible that N_2O found in the surface waters of upwelling regions may be produced at greater depths in the water column during nitrification, or as suggested by Pierotti and Rasmussen (1980), during denitrification in zones of intermediate (0.1-0.3 $m\ell$ $liter^{-1}$) oxygen concentration. The latter possibility is also suggested by the observations of Firestone et al. (1980) that $^{13}NO_3^-$ was converted to N_2O in soils in the presence of low levels of oxygen. At present, it is not possible to resolve the controversy between the different possibilities without studies combining microbiological and other oceanographic data.

D. Nitric Oxide (NO)

Only in recent years has information become available on an additional and more "exotic" gaseous nitrogen species found in the ocean. Nitric oxide (NO) is a highly reactive chemical species (a free radical) that is not thermodynamically stable under the conditions found either in the lower troposhere or upper ocean (Cotton and Wilkinson, 1966).

Zafiriou (1974) pointed out, however, that among the photochemical reactions which should occur in the surface waters of the ocean are the photolysis of aqueous nitrite and nitrate, both reactions which require light of the wavelength range available in seawater. The reactions are

$$NO_2^- + h\nu \leftrightarrow NO + O^-$$

(II)

and

$$NO_3^- + h\nu \leftrightarrow NO_2 + O^-$$

(III)

These reactions convert nitrite and nitrate to highly unstable nitrogen oxides, which may react further with organic matter or other components of seawater. The detailed consequences of the further reactions of NO or NO_2 with other oceanic species have not been studied yet under natural conditions.

In studies over the past 6 or 7 years, Zafiriou and co-workers have shown that reaction III is not very significant in seawater (Zafiriou and True, (1979b), but that II, the conversion of nitrite to NO, is very important (Zafiriou and True, 1979a). Studies have shown both the disappearance of nitrite on irradiation of sterilized seawater and the appearance of NO (Zafiriou and True, 1979a,c; Zafiriou and McFarland, 1981). Nitrite in illuminated samples consistently showed losses estimated to be about 10% per day, a rate significant for the nitrite pool (Zafiriou and True, 1979a).

More recently, using both on-deck incubations and a floating *in situ* stripping device that bubbled gas through surface seawater and then transferred the partially equilibrated gas to an NO detec-

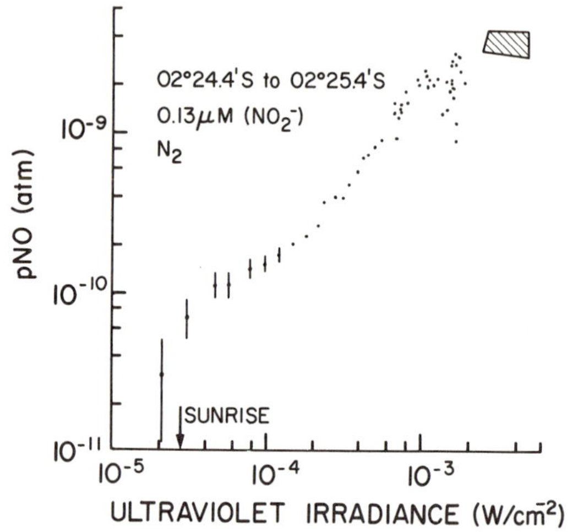

Fig. 5. The atmospheric partial pressure of nitric oxide as a function of UV insolation as measured in an in situ stripper. The hatched area represents data taken during the middle of the day.

tor on deck, Zafiriou and McFarland (1981) obtained values for the partial pressure of NO in equilibrium with seawater (p_{NO}) under different light regimes. NO generated photochemically on deck in the central Equatorial Pacific had a 20- to 150-s half-life in darkness. Figure 5 from Zafiriou et al. (1980), for data obtained in the Pacific, shows the relationship between UV irradiance and the p_{NO} obtained by *in situ* stripper. With exposure to sunlight, surface seawater NO levels rise by at least two orders of magnitude. McFarland et al. (1979) found that the partial pressure of NO in the air was less than 8×10^{-12} atm compared with p_{NO} calculated for surface seawater of 7×10^{-8} atm. Thus, the surface ocean is markedly supersaturated with NO during the day. At night, due to the rapid decay of NO, p_{NO} is very low [near the limit of detection for Zafiriou and McFarland (1981); Zafiriou et al. (1980)]. Zafiriou and co-workers argue convincingly that although NO can be produced biologically (see, e.g., Chapter 6 this volume; Firestone et al., 1979; Sorenson, 1978), the NO observed in the surface ocean is a photochemical product because NO can be generated in significant quantities in illuminated sterilized seawater. In addition, the rate of NO production agrees with the rate calculated from nitrite

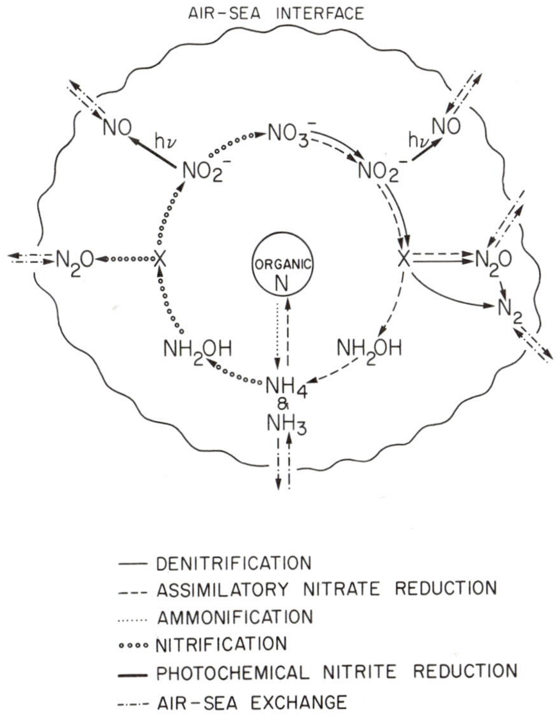

Fig. 6. Portions of the marine nitrogen cycle influencing dissolved gaseous nitrogen compounds. (Redrawn with permission from Cohen and Gordon, 1978). X may represent NOH. The NO present as an intermediate in the denitrification pathway may be enzyme bound. ——, Denitrification; ---, Assimilatory nitrate reduction; ..., Ammonification; ooo, Nitrification; ▬, Photochemical nitrite reduction; —·—·—, Air-Sea exchange.

disappearance (Zafiriou et al., 1980). Although still new, this type of research may turn out to be very significant, as the NO_2^- to NO reaction is certainly not the only important photochemical reaction producing free radicals and other reactive species in the irradiated mixed layer.

IV. CONCLUSION

Dissolved gaseous nitrogen compounds are involved in the complex series of reactions known as the marine nitrogen cycle (Fig. 6).

In the case of N_2O, the distribution of the compound is dominated by biologically mediated reactions. In the case of others (N_2, NH_3), biological processes play an important role in controlling their distributions but physical and chemical processes are also of major significance. For still others (NO), the dominant source is abiotic (photochemical dissociation of nitrite) and the main sink potentially involves reaction with any of a large number of compounds found in the mixed layer.

If substantial progress is to be made in understanding the roles of the gaseous nitrogen compounds in the ocean, innovative methods will be required. Studies of nitrous oxide should include specific microbiological assays (preferably using isotopes) of the roles of nitrification and denitrification. A new, more sensitive method for N_2, which can identify concentration variations of less than 1%, will be needed if direct confirmation of N_2 production in the ocean is to be made.

For dissolved ammonia gas, whose importance lies in its ability to move across the air-sea interface, studies investigating atmospheric ammonia concentrations should be made in conjunction with surface water total ammonia measurements, especially in coastal zones. In the case of nitric oxide, research is still at a very early stage, but the most interesting questions are related to the fate of the reactive NO species and how it (and other photochemical products) interact with other constituents of seawater. Further study of gaseous nitrogen compounds seems certain to yield significant insights into many oceanic processes.

ACKNOWLEDGMENTS

This publication is contribution number 311 from the Marine Sciences Research Center. Support for preparation of the manuscript was provided by ONR under Contract N00014-80-C0771 to M. I. Scranton.

REFERENCES

Ayers, G. P., and Gras, J. L. (1980). Ammonia gas concentrations over the Southern ocean. *Nature (London) 284*, 539-540.

Barnes, R. O., Bertine, K. K., and Goldberg, E. D. (1975). N_2:Ar, nitrification and denitrification in Southern California borderland basin sediments. *Limnol. Oceanogr. 20*, 962-970.

Benson, B. B., and Parker, P. D. M. (1961). Nitrogen/argon and nitrogen isotope ratios in aerobic seawater. *Deep-Sea Res. 7*, 237-253.

Broecker, W. S., and Peng, T.-H. (1971). The vertical distribution of radon in the BOMEX area. *Earth Planet. Sci. Lett. 11*, 99-108.

Broecker, W. S., and Peng, T.-H.(1974). Gas exchange rates between air and sea. *Tellus 26*, 21-35.

Buch, K. (1929). Die Verwendung von Stickstoff- und Sauerstoffanalysen in der Meeresforschung. *J. Cons. Cons. Int. Explor. Mer. 4*, 162-191.

Carritt, D. E. (1954). Atmospheric pressure changes and gas solubility. *Deep-Sea Res. 2*, 59-62.

Cline, J. D. (1973). Denitrification and isotopic fractionation in two contrasting marine environments: The eastern tropical North Pacific and the Cariaco Trench. Ph.D. Thesis, University of California, Los Angeles.

Cline, J. D., and Kaplan, I. R. (1975). Isotopic fractionation of dissolved nitrate during denitrification in the eastern tropical North Pacific ocean. *Mar. Chem. 3*, 271-299.

Cline, J. D., and Richards, F. A. (1972). Oxygen deficient conditions and nitrate reduction in the eastern tropical North Pacific ocean. *Limnol. Oceanogr. 17*, 885-900.

Cohen, Y. (1978). Consumption of dissolved nitrous oxide in an anoxic basin, Saanich Inlet, British Columbia. *Nature (London) 272*, 235-237.

Cohen, Y., and Gordon, L. I. (1978). Nitrous oxide in the oxygen minimum of the eastern tropical North Pacific: Evidence for its consumption during denitrification and possible mechanisms for its production. *Deep-Sea Res. 25*, 509-524.

Cohen, Y., and Gordon, L. I. (1979). Nitrous oxide production in the ocean. *JGR, J. Geophys. Res. 84*, 347-353.

Cotton, F. A., and Wilkinson, G. (1966). "Advanced Inorganic Chemistry." Wiley, New York.

Craig, H., and Gordon, L. I. (1963). Nitrous oxide in the ocean and the marine atmosphere. *Geochim. Cosmochim. Acta 27*, 949-955.

Craig, H., and Weiss, R. F. (1971). Dissolved gas saturation anomalies and excess helium in the ocean. *Earth Planet. Sci. Lett. 10*, 289-296.

Craig, H., Weiss, R. F., and Clarke, W. B. (1967). Dissolved gases in the equatorial and South Pacific ocean. *JGR, J. Geophys. Res. 72*, 6165-6181.

Crutzen, P. J. (1970). The influence of nitrogen oxides on the atmospheric ozone content. *Q. J. R. Meteorol. Soc. 96,* 320-325.
Crutzen, P. J. (1981). Atmospheric chemical processes of the oxides of nitrogen, including nitrous oxide. *In* "Denitrification, Nitrification and Atmospheric Nitrous Oxide" (C. C. Delwiche, ed.), pp. 17-44. Wiley (Interscience), New York.
Culberson, C. H., and Pytkowicz, R. M. (1973). Ionization of water in seawater. *Mar. Chem. 1,* 309-316.
Danckwerts, P. V. (1970). "Gas-liquid Reactions." McGraw-Hill, New York.
Delwiche, C. C. (1956). Denitrification. *In* "Inorganic Nitrogen Metabolism" (W. D. McElroy and B. Glass, eds.), pp. 233-256. Johns Hopkins Press, Baltimore, Maryland.
Delwiche, C. C. (1970). The nitrogen cycle. *Sci. Am. 223,* 136-147.
Dugdale, R. C., and Goering, J. J. (1967). Uptake of new and regenerated forms of nitrogen in primary productivity. *Limnol. Oceanogr. 12,* 196-206.
Elkins, J. W., Wofsy, S. C., McElroy, M. B., Kolb, C. E., and Kaplan, W. A. (1978). Aquatic sources and sinks for nitrous oxide. *Nature (London) 275,* 602-606.
Emerson, S. (1975). Gas exchange in small Canadian Shield lakes. *Limnol. Oceanogr. 20,* 754-761.
Firestone, M. K., Firestone, R. B., and Tiedje, J. M. (1979). Nitric oxide as an intermediate in denitrification: Evidence from nitrogen-13 isotope exchange. *Biochem. Biophys. Res. Commun. 91,* 10-16.
Firestone, M. K., Firestone, R. B., and Tiedje, J. M. (1980). Nitrous oxide from soil denitrification: Factors controlling its biological production. *Science 208,* 749-751.
Georgii, H. W., and Gravenhorst, G. (1977). The ocean as a source or sink of reactive trace gases. *Pure Appl. Geophys. 115,* 503-511.
Georgii, H. W., and Müller, W. J. (1974). On the distribution of ammonia in the middle and lower troposphere. *Tellus 26,* 180-184.
Goering, J. J. (1968). Denitrification in the oxygen minimum layer of the eastern tropical Pacific ocean. *Deep-Sea Res. 15,* 157-164.
Goldan, P. D., Kuster, W. C., Schmeltekopf, A. L., Fehsenfeld, F. C., and Albritton, D. L. (1981). Correction of atmospheric N_2O mixing ratio data. *JGR, J. Geophys. Res. 86,* 5385-5386.
Hahn, J. (1974). The North Atlantic ocean as a source of atmospheric N_2O. *Tellus 26,* 160-168.
Hahn, J. (1975). N_2O measurements in the Northeast Atlantic Ocean. *"Meteor" Forschungsergeb. Reihe A 16,* 1-14.
Hahn, J. (1981). Nitrous oxide in the oceans. *In* "Denitrification, Nitrification and Atmospheric Nitrous Oxide" (C. C. Delwiche, ed.), pp. 191-277. Wiley (Interscience), New York.
Hales, J. M., and Drewes, D. R. (1979). Solubility of ammonia in water at low concentrations. *Atmos. Environ. 13,* 1133-1147.

Hamm, R. E., and Thompson, T. G. (1941). Dissolved nitrogen in the seawater of the northeast Pacific with notes on the total carbon dioxide and the dissolved oxygen. *J. Mar. Res. 4,* 11-27.

Junge, C., and Hahn, J. (1971). N_2O measurements in the North Atlantic. *JGR, J. Geophys. Res. 76,* 8143-8146.

Junge, C., Bockholt, B., Schütz, K., and Beck, R. (1971). N_2O measurements in air and seawater over the Atlantic. *"Meteor" Forschungsergeb., Reihe B 6,* 1-11.

Kanwisher, J. (1963). On the exchange of gases between the atmosphere and the sea. *Deep-Sea Res. 10,* 195-207.

Kester, D. R. (1975). Dissolved gases other than CO_2. *In* "Chemical Oceanography" (J. P. Riley and G. Skirrow, eds.), 2nd ed., Vol. 1, pp. 497-556. Academic Press, New York.

Lau, N.-C., and Charlson, R. J. (1977). On the discrepancy between background atmospheric ammonia gas measurements and the existence of acid sulfate as a dominant atmospheric aerosol. *Atmos. Environ. 11,* 475-478.

Lenhard, U., and Georgii, H. W. (1980). Der Ozean als Quelle reaktiver Stickstoffverbindungen. *Pure Appl. Geophys. 118,* 1145-1154.

Linnenbom, V. J., and Swinnerton, J. W. (1971). Distribution of low molecular weight hydrocarbons and excess molecular nitrogen in the Cariaco Trench. *Symp. Invest. Resour. Caribb. Sea Adjacent Reg. Pap., 1968* FAO Fish. Rep. No., 71.1.

Linnenbom, V. J., Swinnerton, J. W., and Cheek, C. H. (1966). Statistical evaluation of gas chromatography for the determination of dissolved gases in seawater. *NRL Rep. 6344,* 1-16.

McFarland, M., Kley, D., Drummond, J. W., Schmeltekopf, A. L., and Winkler, R. M. (1979). Nitric oxide measurements in the equatorial Pacific region. *Geophys. Res. Lett. 6,* 605-609.

Peng, T. H., Takahashi, T., and Broecker, W. S. (1974). Surface radon measurements in the North Pacific, Ocean Station PAPA. *JGR, J. Geophys. Res. 79,* 1772-1780.

Peng, T. H., Broecker, W. S., Mathieu, G. G., and Li, Y. H. (1979). Radon evasion rates in the Atlantic and Pacific oceans as determined during the GEOSECS Program. *JGR, J. Geophys. Res. 84,* 2471-2486.

Pierotti, D., and Rasmussen, R. A. (1980). Nitrous oxide measurements in the eastern tropical Pacific Ocean. *Tellus 32,* 56-72.

Pytkowicz, R. M. (1964). Oxygen exchange rates off the Oregon coast. *Deep-Sea Res. 11,* 381-389.

Rakestraw, N., and Emmel, V. M. (1938). The relation of dissolved oxygen to nitrogen in some Atlantic waters. *J. Mar. Res. 1,* 207-216.

Redfield, A. C., Ketchum, B. H., and Richards, F. A. (1963). The influence of organisms on the composition of seawater. *In* "The Sea" (M. N. Hill, ed.), Vol. 2, pp. 65-77. Wiley (Interscience), New York.

Richards, F. A., and Benson, B. B. (1961). Nitrogen/argon and nitrogen isotope ratios in two anaerobic environments, the Cariaco Trench in the Caribbean Sea and Dramsfjord, Norway. *Deep-Sea Res. 7,* 254-264.

Seitzinger, S., Nixon, S., Pilson, M. E. Q., and Burke, S. (1980). Denitrification and N_2O production in near shore marine sediments. *Geochim. Cosmochim. Acta 44*, 1853-1860.

Sillen, L. G., and Martell, A. E. (1971). Stability constants of metal ion complexes. *Spec. Publ. Chem. Soc. 25*, Suppl. No. 1 to Spec. Publ. No. 17.

Singh, H. B., Salas, L. J., and Shigeishi, H. (1979). The distribution of nitrous oxide (N_2O) in the global atmosphere and the Pacific Ocean. *Tellus 31*, 313-320.

Sorenson, J. (1978). Occurrence of nitric and nitrous oxides in a coastal marine sediment. *Appl. Environ. Microbiol. 36*, 809-813.

Swinnerton, J. W., and Sullivan, J. P. (1962). Shipboard determination of dissolved gases in seawater by gas chromatography. *NRL Rep. 5806*, 1-13.

Tsunogai, S. (1971). Ammonia in the oceanic atmosphere and the cycle of nitrogen compounds through the atmosphere and the hydrosphere. *Geochem. J. 5*, 57-67.

Weiss, R. F. (1970). The solubility of nitrogen, oxygen and argon in water and seawater. *Deep-Sea Res. 17*, 721-735.

Weiss, R. F. (1978). Nitrous oxide in the surface water and marine atmosphere of the North Atlantic and Indian oceans. *Trans. Am. Geophys. Union 59*, 1101-1102.

Weiss, R. F. (1981). The temporal and spatial distribution of tropospheric nitrous oxide. *JGR, J. Geophys. Res. 86*, 7185-7195.

Weiss, R. F., and Price, B. A. (1980). Nitrous oxide solubility in water and seawater. *Mar. Chem. 8*, 347-359.

Wilson, T. R. S. (1978). Evidence for denitrification in aerobic pelagic sediments. *Nature (London) 274*, 354-356.

Yoshinari, T. (1976). Nitrous oxide in the sea. *Mar. Chem. 4*, 189-202.

Zafiriou, O. C. (1974). Sources and reactions of OH and daughter radicals in seawater. *JGR, J. Geophys. Res. 79*, 4491-4497.

Zafiriou, O. C., and McFarland, M. (1980). Determination of trace levels of nitric oxide in aqueous solution. *Anal. Chem. 52*, 1662-1667.

Zafiriou, O. C. and McFarland, M. (1981). Nitric oxide from nitrite photolysis in the central equatorial Pacific. *JGR, J. Geophys. Res. 86*, 3173-3182.

Zafiriou, O. C., and True, M. B. (1979a). Nitrite photolysis in seawater by sunlight. *Mar. Chem. 8*, 9-32.

Zafiriou, O. C., and True, M. B. (1979b). Nitrate photolysis in seawater by sunlight. *Mar. Chem. 8*, 33-42.

Zafiriou, O. C., and True, M. B. (1979c). Nitrite photolysis as a source of free radicals in productive surface waters. *Geophys. Res. Lett. 6*, 81-84.

Zafiriou, O. C., McFarland, M., and Bromund, R. H. (1980). Nitric oxide in seawater. *Science 207*, 637-639.

Chapter 3

NITROGEN FIXATION BY MARINE *OSCILLATORIA* (TRICHODESMIUM) IN THE WORLD'S OCEANS

EDWARD J. CARPENTER
Marine Sciences Research Center
State University of New York
Stony Brook, New York

I. INTRODUCTION

In the open ocean there are at least 26 genera of planktonic cyanobacteria, many of which are known to fix nitrogen (Sournia, 1970). One relatively abundant N_2-fixing species is the endosymbiont *Richelia intracellularis* which is clearly present in five species of the diatom *Rhizosolenia* and possibly five others, as well as being epiphytic on the diatom *Chaetoceros compressus* (Sournia, 1970). This cyanobacterium has also been observed in the diatom *Hemiaulus membranaceus* (Kimor et al., 1978). Its abundance in *Rhizosolenia* in the central north Pacific Ocean has been noted by Mague et al. (1974), and rate of N_2 fixation at 31°N, 155°W in June was approximately 800 μg N m^{-2} day^{-1}. Unfortunately, very little information exists on the contribution of fixed nitrogen to the ocean by this species. One major problem in assessing the contri-

ISBN 0-12-160280-X

bution is that in many species, *Richelia* is difficult to identify clearly in the host cytoplasm.

In the pelagic *Sargassum* community, the epiphyte *Dichothrix fucicola* has been shown to fix a significant amount of nitrogen relative to community needs (Carpenter, 1972). However, in relation to the overall nitrogen budget of the Sargasso Sea, the introduced nitrogen is virtually insignificant, amounting to about 0.1-0.2 μg N m^{-2} day^{-1}.

In the Indian Ocean, Bernard and Lecal (1960) have noted the presence of three species of *Nostoc,* which at times may be very abundant. Highest concentrations were at 5°S, 75°E at a depth of 1000 m. To date, there is no information on N_2-fixing properties of these species. Taylor (1966), in a study off southwest Africa, also observed high populations of *Nostoc* and states that at times they were "surprisingly abundant." Clearly more research is needed to assess the importance of nitrogen fixation by *Nostoc* in the Indian Ocean.

Recently, Johnson and Sieburth (1979) observed the presence of minute chroococcoid cyanobacteria in the open sea. These organisms are predominantly of the genus *Synechococcus,* and usually present in concentrations ranging from 10^3 to 10^4 cells per ml. There has been confirmation of N_2 fixation under aerobic conditions in some marine isolates of *Synechococcus* (Duerr, 1981). However, their importance as N_2 fixers in the sea also remains to be quantified.

In the Baltic Sea, blooms of N_2-fixing cyanobacteria *Aphanizomenon flos-aquae, Nodularia spumigena, Anabaena lemmermannii,* and *A. baltica* occur during the summer (Nemi, 1979); however, these blooms appear to be confined to this region and have little impact as oceanic N_2 fixers.

In the sea, some pelagic bacteria have been shown to be oceanic N_2 fixers. Kawai and Sugahara (1971) observed a small population (maximum 200 cells $liter^{-1}$) of aerobic N_2 fixers in the mid-Pacific Ocean and a slightly larger population (maximum 400 cells $liter^{-1}$)

in the Japan Sea. Similarly, Maruyama et al. (1970) in the central Pacific found populations of N_2-fixing bacteria ranging from nil to 10^4 cells·100 mℓ$^{-1}$. They isolated N_2 fixers at depths from surface down to 3000 m. Nitrogen-fixing bacteria within *Rhizosolenia* mats have recently been observed by L. Martinez et al. (unpublished) in the North Pacific gyre. In a 3-year study off Wales, Wynne-Williams and Rhodes (1974) were able to isolate a "few" aerobic or facultatively anaerobic N_2-fixing bacteria. They were unable to isolate any species of *Azotobacter,* in spite of the fact that in the Black Sea, Pshenen (1963) has obtained numerous isolates of *Azotobacter.* Wynn-Williams and Rhodes (1974) believe that Pshenen's work should be reappraised since he isolated the *Azotobacter* in 50%, rather than full strength, seawater. At this time, it appears that rates of bacterial N_2 fixation in the open ocean are low relative to cyanobacterial rates.

Evidence to date suggests that the major N_2-fixing organisms in the marine plankton are in the genus *Oscillatoria.* There are four pelagic species in this genus, all formerly in the genus *Trichodesmium* (Sournia, 1968). While the genus name *Oscillatoria* has priority, for convenience in this report the archaic genus *Trichodesmium* will be used to refer collectively to the four marine species and to distinguish them from other species of *Oscillatoria* that are occasionally present in coastal waters.

This report is an attempt to calculate the annual rate of N_2 fixation by *Trichodesmium* in the pelagic realm of the sea. Nitrogen fixation by other organisms such as *Richelia intracelluaris,* chroococcoid cyanobacteria, and heterotrophic bacteria remains to be assessed accurately; however, at this time their input appears to be considerably less than that of *Trichodesmium.* Fortunately, since *Trichodesmium* colonies are large enough to be seen by the unaided eye from shipboard and often form noticeable blooms, a large number of observations have accumulated in the literature. These observations plus recent quantitative measurements are used here to map the distribution and density of this genus.

II. MINIMUM TEMPERATURE REQUIREMENT

While *Trichodesmium* is usually considered a tropical species, there are numerous reports of its occurrence in cold water. For example, Farran (1932) frequently collected it in the winter immediately south of Ireland in water of temperatures between 9 and 12°C. He observed it in 75% of his plankton samples between October and March, but found it to be scarce in the summer. Its most northerly record appears to be the observation at 64°N, immediately south and west of Iceland, by the Edinburgh Oceanographic Laboratory (1973). Other observations in cold water are given in Section III. However, while it can be present in relatively cold water, it would appear that 20°C is the thermal minimum for physiologically active populations. There are extensive temperature records and physiological measurements made by Marumo and Nagasawa (1976) in the Kuroshio Current, and 20°C appears to be an appropriate thermal minimum. Over a 3-year period in the 1930s, Marumo and Nagasawa (1976) made extensive observations on *Trichodesmium* and found that it did not appear in Kuroshio water until it warmed to 20°C (Fig. 1). Similarly, it disappeared from Sagami Bay in the 1970s when water cooled to 20°C (Fig. 2). Measurements of photosynthesis on field-collected material show a maximum at 30°C and a major decrease below 20°C (Fig. 3). While populations may have some activity below 20°C [nitrogen-fixing *Trichodesmium* were observed at 18.3°C southwest of Spain by McCarthy and Carpenter (1979)], these extensive measurements support the conservative use of 20°C as a minimum for active N_2-fixing populations.

III. DISTRIBUTION

A. Assumptions

Concentrations of *Trichodesmium* are expressed as cells, trichomes, or colonies by various investigators. In this report, data

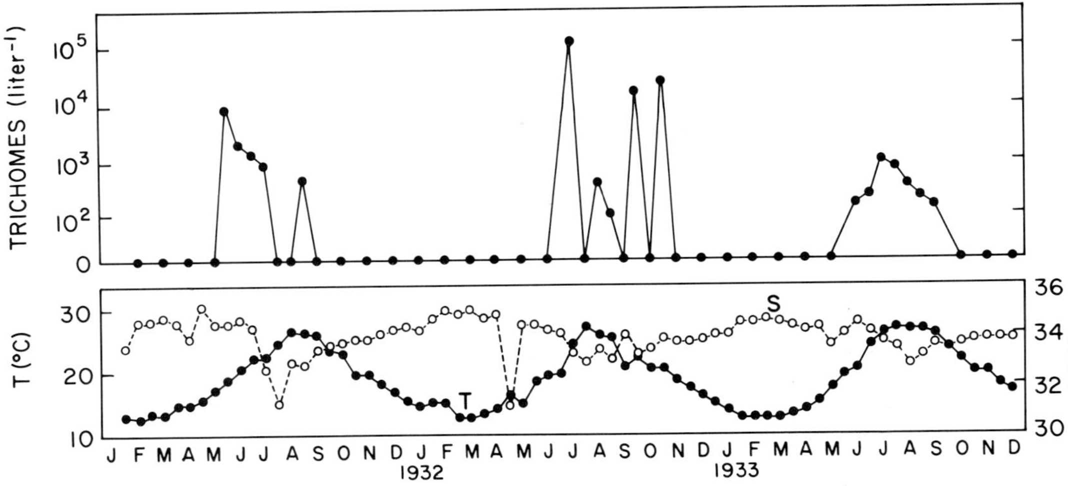

Fig. 1. Trichodesmium concentrations, salinity, and temperature of surface water in the Kuroshio Current from 1931 to 1933 (from Marumo and Nagasawa, 1976).

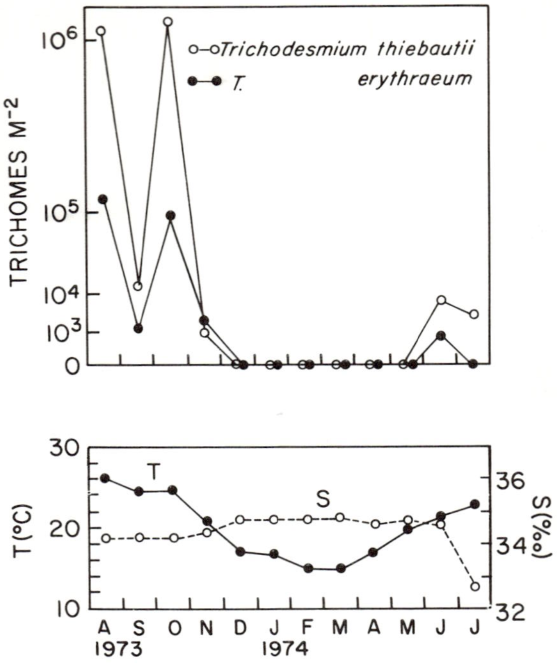

Fig. 2. Water column density of Trichodesmium per m^2 and surface salinity and temperature in Sagami Bay, Japan (from Marumo and Nagasawa, 1976).

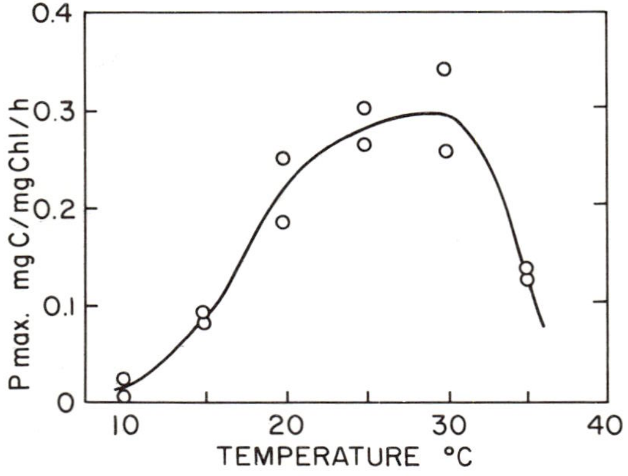

Fig. 3. Rate of photosynthesis by Trichodesmium versus seawater temperature (from Aruga et al., 1975).

are given as trichomes m^{-3}, and in converting data from other investigations, it is assumed that a trichome contains 100 cells and a colony has 200 trichomes. This yields a colony slightly smaller than the mean of 22,000 cells for a tuft (fusiform)-shaped colony and 38,700 for a puff (spherical) colony observed by McCarthy and Carpenter (1979) in the central North Atlantic Ocean. Mean combined carbon content of McCarthy and Carpenter's puff and tuft colonies in the Atlantic was 10.9 and 11.6 μg, respectively, similar to the mean of 9.7 μg obtained by Mague et al. (1977) in the Pacific.

From collections by Carpenter and Price (1977) in the North Atlantic Ocean and Caribbean Sea, it is apparent that most (over 85%) of the *Trichodesmium* in a water column is present in the upper 50 m, with peak density between about 15 and 20 m and the surface. Some investigators present cell densities on a square meter basis. In the conversion of their data to an m^3 basis, it is assumed that the bulk of cells were encountered in the upper 50 m, and calculations have been made to an m^3 on this assumption. In the text of this report we present all data as trichomes m^{-3}, and concentrations given are those that are present between the surface and a 15-m depth.

B. Red Sea

To estimate oceanic N_2 fixation by *Trichodesmium,* it is necessary first to know the density and range of this genus. *Trichodesmium* was first described by Ehrenberg (1830) from the Bay of Tor in the Red Sea. Blooms of *T. erythraea* were abundant in this sea in the 1800s and were thought to be the origin of its red coloration (Carter, 1863). Ehrenberg (1830) observed blooms in the northern Red Sea in December, 1823 and January, 1824, whereas Mobius (1880) observed a bloom in August, Montagne (1844) in July, and Carter (1863) in June.

More recently, Kimor and Goldansky (1977) in the Gulf of Elat followed *Trichodesmium* populations through an annual cycle. It was present in water 22°C or warmer and had a maximum abundance in November, 1974 and May, 1975. Lower concentrations were observed in December, 1974 and April, June, and July, 1975. Concentrations of about 10^5 trichomes were present in autumn and 2.5×10^5 trichomes per plankton tow in the spring in a 100-m deep water column. Samples were taken with vertical tows using a 0.25-m mouth diameter net. Concentrations per m^3 would then be about 4×10^3 trichomes m^{-3} in autumn and 1×10^4 trichomes m^{-3} in the spring. Since there appear to be no other quantitative data for the Red Sea, the Gulf of Elat data were used in estimating seasonal densities. Thus it is assumed from these data that the fall and spring populations are 10^4 trichomes m^{-3}, and in summer and winter it is 10^2 trichomes m^{-3} (Fig. 4).

C. Mediterranean Sea

In the Aegean Sea and Sarconicos Gulf, *Trichodesmium* is characterized as occurring occasionally and "without any definite areal or seasonal pattern" (L. Ignatiades, personal communication). In the western Mediterranean Sea, Feldman (1932) saw a bloom near Banyuls, France in September, 1931 and Voltolina (1975) collected it in the Lagoon of Venice, Italy in winter. Vatova (1928) also observed it nearby in the northern Adriatic Sea in winter.

In the western Mediterranean Sea, Margalef (1969) states that its density near southern Spain was 2×10^3 trichomes m^{-3} in the upper 75 m between November and May. He did not note its presence in the summer, nor was it observed when sampling near Barcelona in northern Spain. It would appear that overall, the concentration of

---

Fig. 4. Distribution of Trichodesmium in summer. Shown are estimated concentrations (trichomes m^{-3}) in the interval between surface and 15-m depth. For the Marumo and Nagasawa (1974) transect in the Pacific Ocean, Trichodesmium was not observed at the three northernmost stations shown.

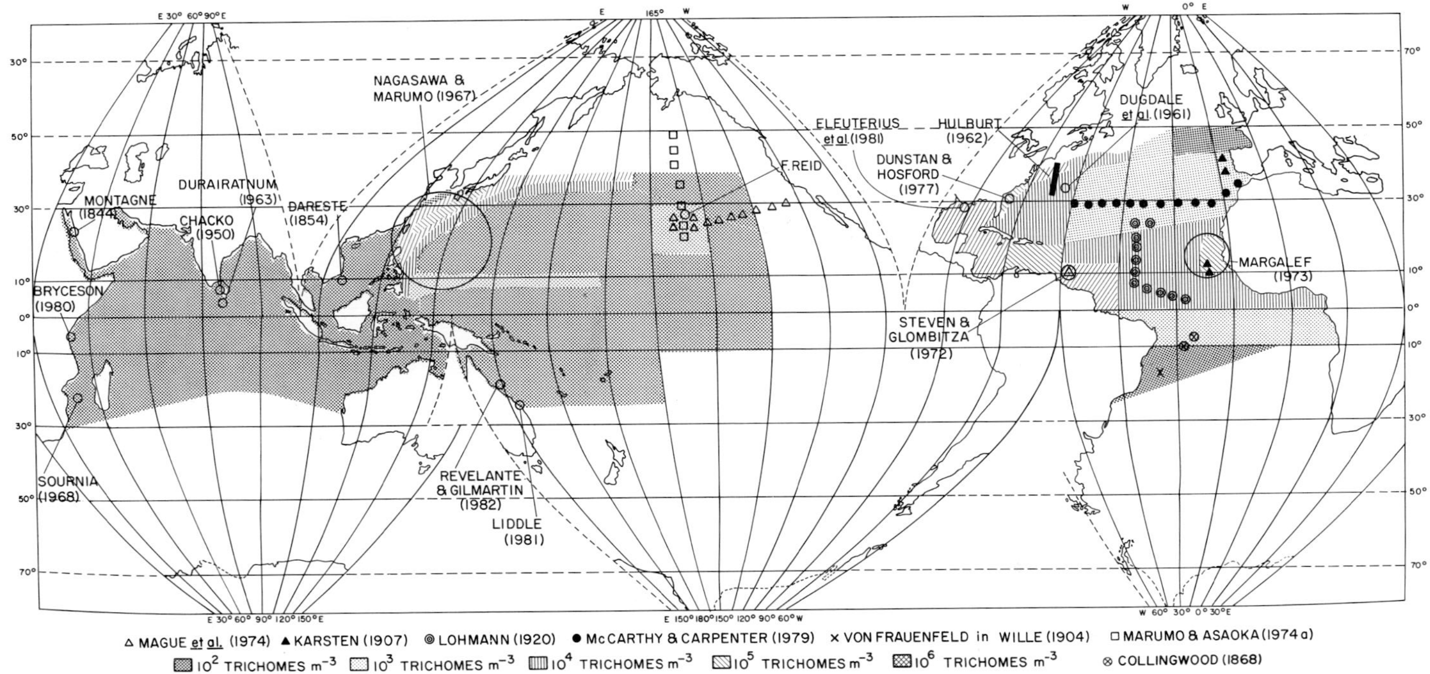
NAGASAWA & MARUMO (1967)
DURAIRATNUM (1963)
MONTAGNE (1844)
CHACKO (1950)
DARESTE (1854)
BRYCESON (1980)
SOURNIA (1968)
REVELANTE & GILMARTIN (1982)
LIDDLE (1981)
F. REID
ELEUTERIUS et al. (1981)
HULBURT (1962)
DUNSTAN & HOSFORD (1977)
DUGDALE et al. (1961)
STEVEN & GLOMBITZA (1972)
MARGALEF (1973)
△ MAGUE et al. (1974) ▲ KARSTEN (1907) ◎ LOHMANN (1920) ● McCARTHY & CARPENTER (1979) × VON FRAUENFELD in WILLE (1904) □ MARUMO & ASAOKA (1974a)
10^2 TRICHOMES m^{-3}
10^3 TRICHOMES m^{-3}
10^4 TRICHOMES m^{-3}
10^5 TRICHOMES m^{-3}
10^6 TRICHOMES m^{-3}
⊗ COLLINGWOOD (1868)

Trichodesmium in the Mediterranean is low. Surface water temperature in winter is between 13 and 15°C, below the 20°C level of assumed viability. In the summer, surface water temperatures are generally 25°C and the existing population would be active. We assume no N_2 fixation occurs in winter and spring, even though a population is present. During summer there do not appear to be enough data to estimate population size, hence we cannot calculate a rate of N_2 fixation for this area.

D. Atlantic Ocean

1. *Summer*

In June, McCarthy and Carpenter (1979) found a mean concentration of 2.8×10^3 trichomes m^{-3} on a transect between Spain and Bermuda (Fig. 4). In the summer near Bermuda, Dugdale et al. (1961) observed a mean of 0.5×10^3 trichomes m^{-3}. Between Cape Cod and Bermuda in August, Hulburt (1962) noted a somewhat greater concentration of 1.4×10^5 trichomes m^{-3}. In the western Sargasso Sea in late summer, Carpenter and McCarthy (1975) calculate a density of about 1×10^3 trichomes m^{-3}. Hence for most of the open North Atlantic in summer, trichome concentration is about 10^3 m^{-3}.

Higher concentrations are found outside the Sargasso gyre. Near the coast of Georgia in summer, Dunstan and Hosford (1977) measured 1.9×10^6 trichomes m^{-3}. Off the northwest coast of Africa, Margalef (1973) observed 4.2×10^6 trichomes m^{-3} in the upper 50 m. There appear to be few data for the Caribbean Sea in summer; however, Steven and Glombitza (1972) measured *Trichodesmium* concentration over a period of 3 years near Barbados (on 66 dates), and obtained a mean concentration of 3.65×10^5 trichomes m^{-3}. There was no measurable seasonal change in concentration; however, the population exhibited a 120-day oscillation in density. Based on these data and current flow patterns into the Caribbean, the summer mean for the

Carribean Sea is placed at 10^5 trichomes m^{-3} for this area and the Gulf of Mexico. According to King (1950), *Trichodesmium* occurs along the west coast of Florida in the Gulf of Mexico through the summer, and a bloom of *Trichodesmium* was reported to have occurred along the Mississippi Gulf coast by Eleuterius et al. (1981) during this season.

For the North Atlantic Ocean lying between 20°N and the equator, the concentration of *Trichodesmium* in summer is about 10^4 trichomes m^{-3}. This is based on data collected by Lohmann (1920) in the Deutschland Expedition in 1911. Between June 26 and July 19 from 25° to 5°N at 11 stations, the concentration was 1.0×10^4 trichomes m^{-3} at surface and 2.7×10^3 trichomes m^{-3} at 50-m depth. Lohmann observed *Trichodesmium* only sporadically south of 5°N; however, because he collected discrete water samples of small volume, this method would not be suitable for estimates of *Trichodesmium* density at low concentrations.

Wille (1904) also observed high concentrations of *Trichodesmium* in the North Equatorial Current. For the open Atlantic Ocean, it appears that this is one of the areas of highest concentration. On the Humbolt Institute Plankton Expedition, Wille (1904) observed the greatest concentration of his entire voyage in the North Equatorial Current. Similarly, during the Meteor Expedition, transects made in autumn, winter, and spring between the equator and 20°N, concentrations were very high, averaging 4.7×10^4 trichomes m^{-3} (Hentschel, 1932).

It is clear that *Trichodesmium* is present in the South Atlantic Ocean in summer (austral winter). For example, von Frauenfeld (in Wille, 1904) observed a bloom at $19^{\circ}24'$S, $38^{\circ}19'$W in August, 1857. Similarly, Collingwood (1868) collected *Trichodesmium* in a bloom in June at $8^{\circ}28'$S, $28^{\circ}32'$W. It was observed on the Humbolt Expedition in 1889 at 2°S, 49°W. In the Deutschland Expedition, it was not observed in east-west profiles made between South America and Africa at 22°-24°S in July, 1926 and 27°-30°S in August, 1925.

These observations indicate that it was not present south of 22°S in summer in this area except in the southward flowing Brazil Current at about 25°S. The southerly limit of its distribution is approximated by the 20°C isotherm, which is about 8°S on the African coast and 25°S off Brazil. While its presence has been substantiated, there are virtually no density data for the central South Atlantic Ocean in summer. In this area it is assumed that the density is low, about 10^3 trichomes m^{-3} south to 10°S and beyond this, 10^2 trichomes m^{-3}.

2. *Autumn*

In the North Atlantic Ocean in autumn, *Trichodesmium's* range is extreme, with citations as far north as 59°N (Ostenfeld, 1898). The Edinburgh Oceanographic Laboratory (1973) collected it with continuous plankton recorders at 64°N near Iceland, but no sample collection dates were given. Farran (1932) reports it from the south coast of Ireland, and Establier and Margalef (1964) found it in September and October in 20°-21°C water near Cadiz, Spain (Fig. 5). Cleve (1901a), using samples collected on a steamer in October, states it was present between 25°N, 22°W and 12°N, 26°W. In the western North Atlantic at Bermuda, Dugdale et al. (1961) calculated a mean of 2.1×10^3 trichomes m^{-3} in September and October, and Carpenter and McCarthy (1975) noted a mean of 10×10^3 trichomes m^{-3} at 11 stations in the western Sargasso Sea. In the western North Atlantic, H. G. Marshall (personal communication) commonly collects it in samples as far north as Nova Scotia in autumn. From the information given above, the northern limit of active populations in autumn is 45°N, with about 10×10^2 trichomes m^{-3} occurring south of this, to 40°N where the population averages 10×10^3 m^{-3}. Along the coast of Georgia, Dunstan and Hosford (1977) es-

---

Fig. 5. Distribution of Trichodesmium in autumn.

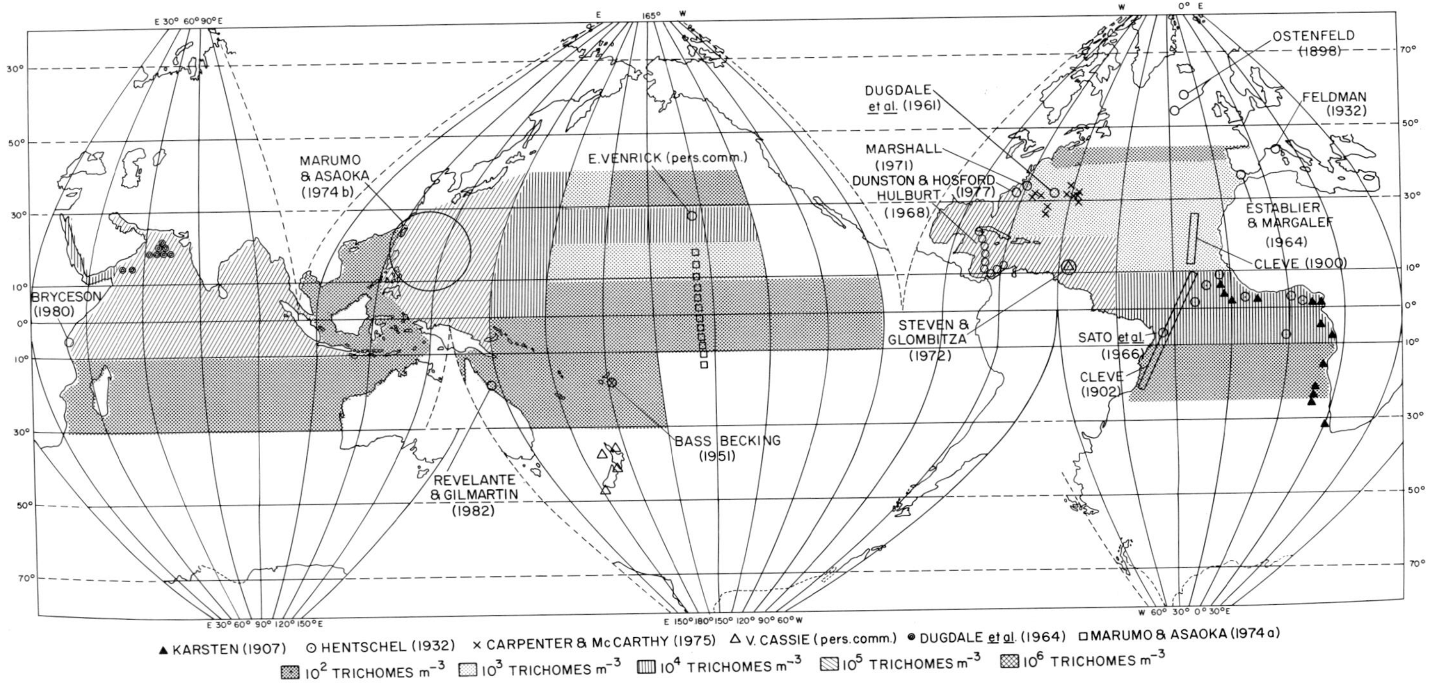
OSTENFELD (1898)
FELDMAN (1932)
DUGDALE et al. (1961)
MARSHALL (1971)
DUNSTON & HOSFORD (1977)
HULBURT (1968)
ESTABLIER & MARGALEF (1964)
CLEVE (1900)
STEVEN & GLOMBITZA (1972)
SATO et al. (1966)
CLEVE (1902)
E. VENRICK (pers. comm.)
BASS BECKING (1951)
MARUMO & ASAOKA (1974 b)
REVELANTE & GILMARTIN (1982)
BRYCESON (1980)
▲ KARSTEN (1907) ⊙ HENTSCHEL (1932) × CARPENTER & McCARTHY (1975) △ V. CASSIE (pers. comm.) ⊚ DUGDALE et al. (1964) □ MARUMO & ASAOKA (1974 a)
10² TRICHOMES m⁻³ 10³ TRICHOMES m⁻³ 10⁴ TRICHOMES m⁻³ 10⁵ TRICHOMES m⁻³ 10⁶ TRICHOMES m⁻³

timated a mean density of 2.9×10^5 trichomes m^{-3} in autumn, and Marshall (1971), also working along the southeast coast of the United States, observed up to 1.28×10^5 trichomes m^{-3}. Using these data, the density in this area is estimated to be 10^5 trichomes m^{-3}.

In the Gulf of Mexico, Hulburt (1968) examined 26 stations between Panama and the western tip of Cuba in November and obtained a mean of 1.67×10^5 trichomes m^{-3}. Farther north, in the northeastern Gulf of Mexico, Curl (1959) encountered a bloom in November but did not give densities. King (1950) states that it is present along the west coast of Florida during this season.

Farther south, the annual means from Barbados (Steven and Glombitza, 1972) are used to estimate a concentration of about 10^5 trichomes m^{-3} for this area. Also, near Recife, Brazil, Sato et al. (1966) investigated a *Trichodesmium* bloom, and concentrations of *Trichodesmium* at 5-m depth on two dates in October after the bloom had subsided averaged 1.25×10^4 m^{-3} and 5.2×10^4 m^{-3}. Also in the autumn, Cleve (1902) observed it between 10°N, 25°W and 21°S, 38°W. Along the African coast, Karsten (1907c) collected it as far south as Cape Town. However, during the Meteor Expedition, Karsten (1907a) did not observe *Trichodesmium* in east-west profile III in September at 48°S or in profile IV in November. Based on these observations, it would appear that Cleve's collection at 21°S is probably the southwestern limit of the species in autumn. In the southeastern region, Karsten's collections at three stations between 20° and 30°S firmly establish it in this area, so the eastern and western boundaries are set at 25°S. Density in the southern area is assumed to be low, 10^2 trichomes m^{-3}.

3. *Winter*

In the northeastern Atlantic Ocean, *Trichodesmium* was found in December and January near the coast of Cadiz, Spain at 17°C (Fig. 6); however, in February, with lower water temperatures of 15°C, it disappeared and was not present again until summer (Establier and Margalef, 1964). It was not present at Bermuda in winter (Dugdale et al., 1961) in 18°C water; however, it was found in the Florida Current near Miami (Taylor et al., 1973) in January. Off the Georgia coast, Dunstan and Hosford (1977) calculated a mean concentration of 1.4×10^5 trichomes m^{-3}. Farther south in the North Equatorial Current and southern Sargasso Sea, Hulburt (1962) sampled 19 stations on a transect between Trinidad and 30°N, 48°W. *Trichodesmium* was not observed north of 17°N and was only present in the North Equatorial Current. In this current, trichome concentrations averaged 1.1×10^6 m^{-3}. This is a relatively high density, and in all probability, *Trichodesmium* was present north of 17°N; however, Hulburt only collected and examined 50 mℓ of water at each site, so at northern stations a colonial organism like *Trichodesmium* could easily have been missed using such a small sample size. It is clear that it occurs north of 17° since Hargraves et al. (1970) collected *Trichodesmium* at 19°N, 64°W north of Puerto Rico.

Based on the observations given above and on surface water isotherms, the northern limit of active populations is set at 30°N for the open Atlantic Ocean. In the western Atlantic, based on Dunstan and Hosford's (1977) observations, it extends to 35°N in the Gulf Stream, and in the eastern Atlantic it is assumed not to be active in cold water (below 20°C), east of 20°W and north of 20°N. In the western North Atlantic between 20°N and 30°N, the density is conservatively estimated at 10^3 trichomes m^{-3} based on quantitative measurements of densities of 10^5 trichomes m^{-3} in peripheral areas and on current directions. In the eastern North Atlantic, because of the presence of colder water in the Canaries Current, the density of 10^3 trichomes m^{-3} extends to 10°N. The northern boundary of active *Trichodesmium* populations is approximately at the 20°C isotherm in winter.

South of this, Hulburt's (1962) estimate of 10^6 trichomes m^{-3} is somewhat higher than observed at Barbados (Sander and Steven, 1973; Steven and Glombitza, 1972). At Barbados the means observed

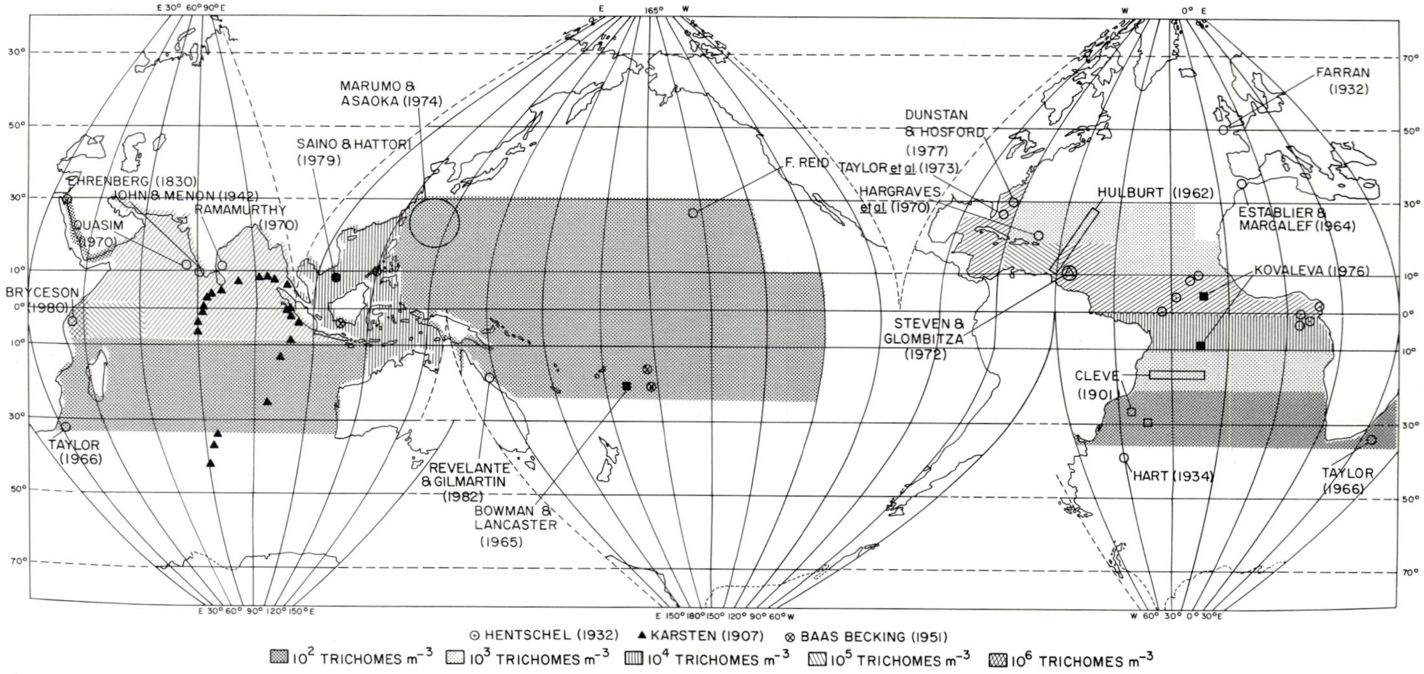

MARUMO & ASAOKA (1974)
SAINO & HATTORI (1979)
EHRENBERG (1830)
JOHN & MENON (1942)
RAMAMURTHY (1970)
QUASIM (1970)
BRYCESON (1980)
TAYLOR (1966)
F. REID
DUNSTAN & HOSFORD (1977)
TAYLOR et al (1973)
HARGRAVES et al (1970)
HULBURT (1962)
FARRAN (1932)
ESTABLIER & MARGALEF (1964)
KOVALEVA (1976)
STEVEN & GLOMBITZA (1972)
CLEVE (1901)
HART (1934)
TAYLOR (1966)
REVELANTE & GILMARTIN (1982)
BOWMAN & LANCASTER (1965)
HENTSCHEL (1932)
KARSTEN (1907)
BAAS BECKING (1951)
10^2 TRICHOMES m^{-3}
10^3 TRICHOMES m^{-3}
10^4 TRICHOMES m^{-3}
10^5 TRICHOMES m^{-3}
10^6 TRICHOMES m^{-3}

in these two studies were 5.4×10^5 and 3.65×10^5 trichomes m^{-3}, respectively. The mean concentration in equatorial water is thus assumed to be 5×10^5 trichomes m^{-3}.

In the North Equatorial Current in winter, Hentschel (1932) made eight observations of the organism's presence, ranging through the whole equatorial region from southwestern Africa to eastern South America. Also, off southwest Africa, Kovaleva (1976) reported a mean of 5×10^5 trichomes m^{-3} in the upper 100 m at 6°N 24°W in February. These two sample sites were in the North and South Equatorial Currents, respectively.

There are several other reports of the presence of *Trichodesmium* in the South Atlantic in winter. Montagne (1844) observed it off the eastern shoulder of South America at 8°S in February, 1836. Cleve (1902) collected it in a zone between 13°-15°S and 25°-37°W in January, as well as along the coast of Brazil from 14° to 26°S in December. Its density in the South Atlantic Ocean was noted by Karsten (1907c) in the Valdivia Expedition. Stations where present ranged from just north of the equator between 18° and 8°E along the Ivory Coast, then southerly along the African west coast to 33°S. It was not present south of 33°S. However, in November Karsten collected it south of Cape Town at 40°S, 15°E, but the Agulhas Current probably carried it there from the Indian Ocean. There is a second observation of *Trichodesmium* at 40°S reported by Hart (1934). He found "vast swarms" of *Trichodesmium* at 40°S, 45°W between Montevideo and South Georgia Island. This would appear to be the approximate position of the Subtropical Convergence.

On the Deutsche Atlantic Expedition, Hentschel (1932) did not observe *Trichodesmium* on profile V at 55°S in January, 1926 nor in profile IV at 35°S in November, 1925. Since the 20°C isotherm lies at about 33°S in winter and is approximately the division between

---

Fig. 6. Distribution of Trichodesmium in winter.

the West Wind Drift and South Atlantic Gyre, the probable southerly extent of *Trichodesmium* is 33°S (as also observed by Karsten, 1907c)

Information on density of *Trichodesmium* in the South Atlantic in winter is nonexistent, even though its presence has been established to 33°S and along the east coast of South America to 28°S. Since the density from the equator north to 10°N is 5×10^5 trichomes m^{-3}, it would seem reasonable that south of the equator to 10°S its density would be at least 10^3 trichomes m^{-3} and south of this to 33°S it would average 10^2 m^{-3}.

4. *Spring*

In the spring, *Trichodesmium* is found in the Gulf Stream near Miami (Taylor et al., 1973) and in coastal water near Georgia (Dunstan and Hosford, 1977). Its mean concentration near Georgia was high, 1.44×10^6 trichomes m^{-2} (Fig. 7). Water temperature at Georgia ranged from 19 to 23°C in the spring. The 20°C isotherm reaches the latitude of Bermuda (ca. 32°N) by April (Hulburt, 1964), and the northern limit of viable *Trichodesmium* is placed at this latitude. In the eastern Atlantic Ocean, because of cooler water in the Canaries Current, the eastern border is placed at 25°N, the approximate location of the 20°C isotherm. Concentration in the northern Sargasso Sea between 25° and 32°N is assumed to be low, 10^2 trichomes m^{-3}. The Southern Sargasso Sea is assumed to have a density of 10^3 trichomes m^{-3}.

Trichodesmium is known to be present in the North Equatorial Current in the spring, as Hentschel (1932) collected it frequently in a region approximately between 34° and 50°W longitude and 4° and 18°N latitude. West of this area at Barbados, Steven and Glombitza (1972) and Sander and Steven (1973) report annual means of about 3 to 5×10^5 trichomes m^{-3}.

Fig. 7. Distribution of Trichodesmium in spring.

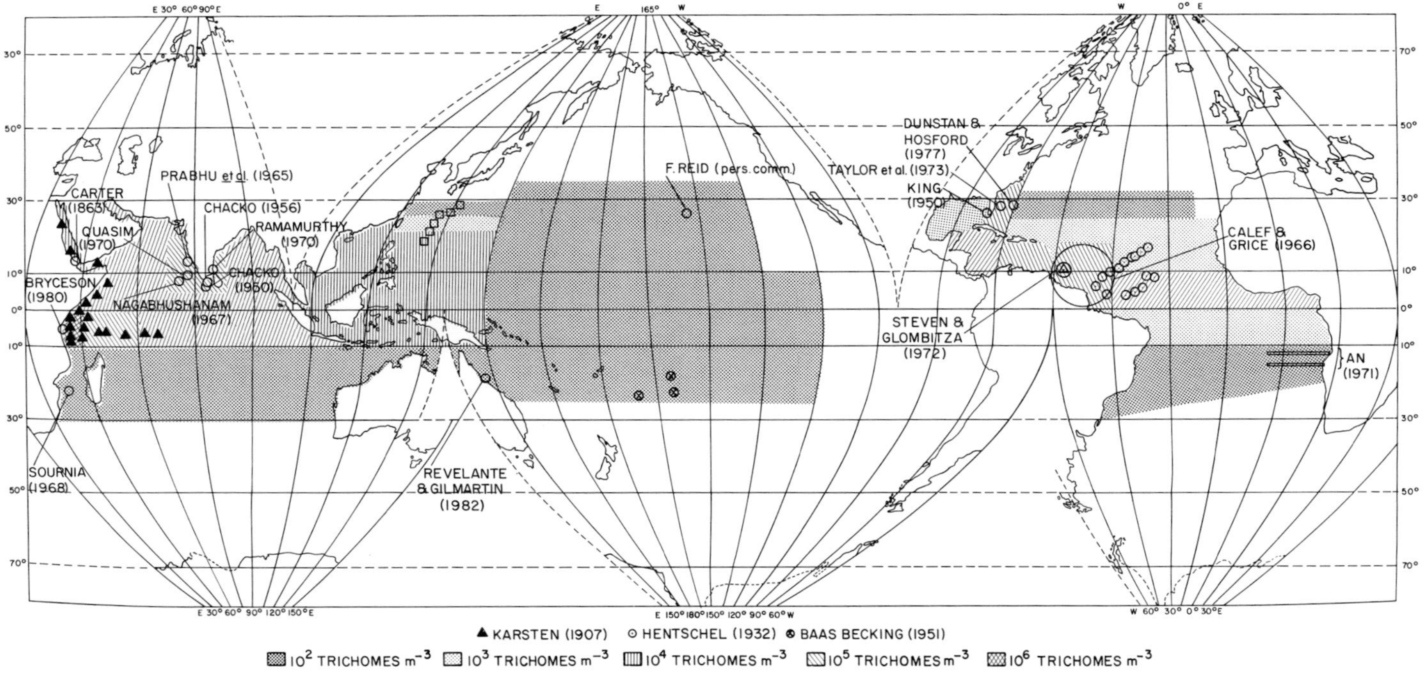
E 30° 60° 90° E
E 165° W
W 0° E
30°
50°
70°
10°
0°
CARTER (1863)
QUASIM (1970)
PRABHU et al. (1965)
CHACKO (1956)
RAMAMURTHY (1970)
CHACKO (1950)
BRYCESON (1980)
NAGABHUSHANAM (1967)
SOURNIA (1968)
REVELANTE & GILMARTIN (1982)
F. REID (pers. comm.)
DUNSTAN & HOSFORD (1977)
TAYLOR et al. (1973)
KING (1950)
CALEF & GRICE (1966)
STEVEN & GLOMBITZA (1972)
AN (1971)
E 30° 60° 90° 120° 150° E
E 150° 180° 150° 120° 90° 60° W
W 60° 30° 0° 30° E
KARSTEN (1907)
HENTSCHEL (1932)
BAAS BECKING (1951)
10^2 TRICHOMES m^-3
10^3 TRICHOMES m^-3
10^4 TRICHOMES m^-3
10^5 TRICHOMES m^-3
10^6 TRICHOMES m^-3

In a large area between 50 and 60°W, from the northeastern coast of South America north to 20°M, Calef and Grice (1966) quantitatively counted the density of *Trichodesmium* while noting its relationship to the copepod *Macrosetella gracilis*. Converting their colony ("bundle") counts at 35 stations to trichomes, a mean of 2.61 x 10^5 m^{-3} is obtained for this area in the spring. Counts in the study were done by R. R. L. Guillard. R. R. L. Guillard (personal communication) states that the term "bundles" refers to the entire *Trichodesmium* colony rather than to individual trichomes.

Along the west coast of Florida, King (1950) states that blooms are common in the spring, and C. VanBaalen (personal communication) reports that it normally appears in coastal waters of the western Gulf near Port Aransas, Texas in mid-May.

In the South Atlantic Ocean in April, An (1971) found *Trichodesmium* to be very abundant in transects at 11°S and 14°S along the African Coast. Concentrations were as high as 1.8 x 10^5 trichomes m^{-3} at one station. In the northerly transect, *Trichodesmium* averaged about 3 x 10^3 trichomes m^{-3} (62% of all cells). In going farther south its density decreased, so at 14°S it constituted only 3.8% of all cells, suggesting that the southerly limit in spring was being approached. Based on this, plus interpolating between winter and summer southerly distribution, its spring limit is placed at 20°S near Africa. Due to the southerly flow of the Brazil Current, its limit is at 30°S on the South American coast.

E. Pacific Ocean

1. *Summer*

In a north-south transect through the Central Pacific Ocean (at longitude 155°W), Marumo and Asaoka (1974b) first encountered *Trichodesmium* at 35°N in late August at water temperatures of 22°C (Fig. 4).

Concentrations over most of Marumo and Asaoka's (1974b) transect were about 10^4 trichomes m^{-2} except for the stations immediately north and south of the Hawaiian Islands where concentrations approached 10^5 m^{-2}. On a cubic meter basis, for the upper 50 m this yields concentrations of about 2 x 10^2 trichomes m^{-3} with 10^3 m^{-3} near the Hawaiian Islands.

Similarly, on a number of Scripps Institution of Oceanography cruises in 1973 and 1974, Freida Reid (personal communication) observed trichome concentrations much like those noted by Marumo and Asaoka (1974b). For June and August, Reid observed mean concentrations at 20m of about 2.5 x 10^3 trichomes m^{-3}, and at 40-m depth it was 4.0 x 10^2 m^{-3}. These stations were clustered around 28°N, 155°W, due north of Hawaii.

The easterly range of *Trichodesmium* in the North Pacific in summer is about 130°W since this is the approximate location of the westerly influence of the cold California Current. Furthermore, Mague et al. (1974) did not observe *Trichodesmium* east of this longitude on cruise Cato I in July, 1972. The cruise was approximately along latitude 30°N. In sailing west from San Diego, he first encountered *Trichodesmium* at about 132°W.

In the western North Pacific in the summer of 1966, Nagasawa and Marumo (1967) found the concentration was about 10^5-10^6 trichomes m^{-3} in the west Equatorial Drift (ca. 135°E) as this current passed by the Philippine Islands (Figs. 8 and 9). Mean concentration for stations at 130-135°E, between 5° and 25°N was 10^4 trichomes m^{-3} in a 100-m water column. In 1966, the authors sampled stations in a north-south transect at 130°E between 7° and 21°N, and concentration ranged from nil to 10^6 trichomes m^{-2}, or for a 100-m water column, about 10^3 trichomes m^{-3}. Thus, for the West Equatorial Drift, concentration of *Trichodesmium* was about one to two orders of magnitude greater than that in the North Pacific gyre. This may be due to an island mass effect from the current passing among the numerous islands located between about 5° and 15°N in the western Pacific. The density of *Trichodesmium* similarly increases

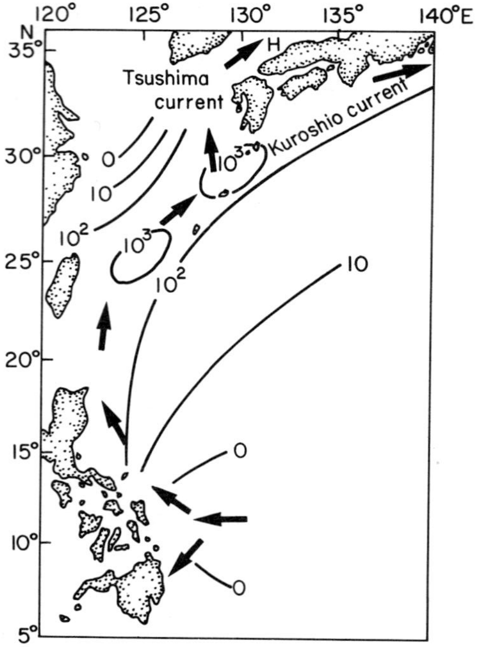

Fig. 8. Distribution of Trichodesmium as trichomes per liter in the Kuroshio region in summer (from Marumo and Asaoka, 1974b).

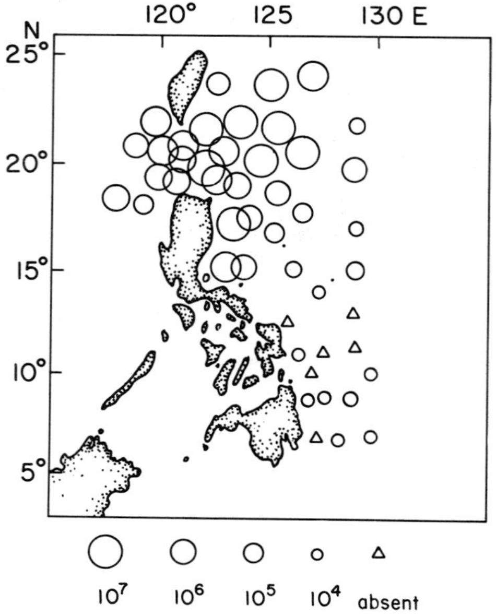

Fig. 9. Distribution of Trichodesmium per m^2 of sea surface near the Philippine Islands between July and September, 1966 (from Nagasawa and Marumo, 1967).

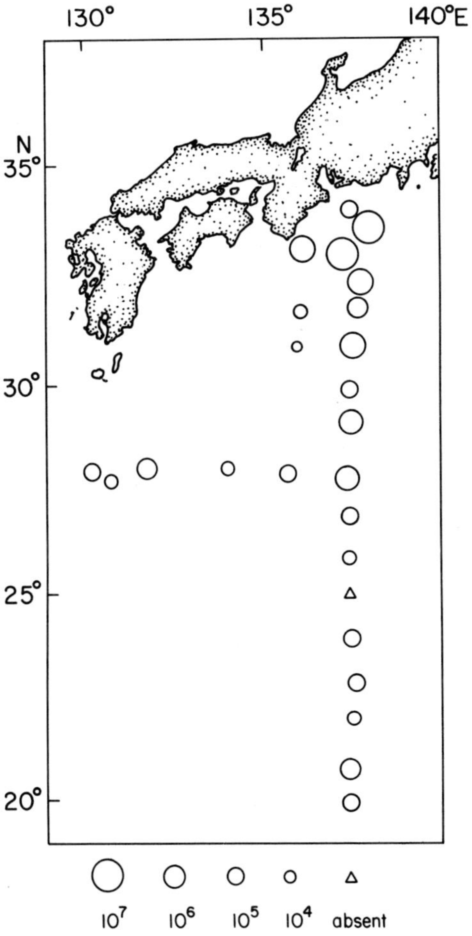

Fig. 10. Water column density (per m^2) of Trichodesmium south of Japan in June, July, and August, 1966 (from Nagasawa and Marumo, 1967).

as currents bring populations near the mainland of Japan (Fig. 10). Carpenter and Price (1977) similarly saw an increase in *Trichodesmium* in the Caribbean Sea after seawater passed among the lesser Antilles.

In the Kuroshio Current in summer, concentrations north of 25°N were typically 10^6 trichomes m^{-3} (Marumo and Asaoka, 1974a). Marumo and Asaoka state "*Trichodesmium thiebautii*, thus propagates flourishingly when Kuroshio water takes in the character of coastal water by running along lands such as the Philippine Islands, Taiwan, the Okinawa Islands and the Main Land of Japan." The alga is most abundant in the Kuroshio Current from July to September, while it is almost absent in the winter (Aikawa, 1936).

For the south Central Pacific it is assumed that the southerly range of this species in summer is 15°S. In August, E. Venrick (personal communication) sampled at 25°S, 155°W in the south Pacific, and she did not observe any *Trichodesmium*. However, near the Great Barrier Reef around Heron Island, Liddle (1981) saw an extensive bloom of *Trichodesmium* while traveling in a helicopter. On the Great Barrier Reef near Townsville (ca. 19°S), Revelante and Gilmartin (1982) found *Trichodesmium* to be abundant in autumn, with a mean of about 5×10^{5} trichomes m^{-3} in 1977, but much lower concentrations in 1976. Thus, for the south Pacific Ocean the southerly range is 25°S.

2. Autumn

In the open Pacific in September, E. Venrick (personal communication observed a mean of 1.78×10^{4} trichomes m^{-3} at 28°N, 155°W, north of Hawaii. On a transect south of Hawaii in September and October, Marumo and Asaoka (1974b) sampled *Trichodesmium* at stations as far south as 15°S. In equatorial water, *Trichodesmium* concentration was from 10^{3} to 10^{4} trichomes m^{-2}. Assuming that most were present in the upper 50 m, this is a density of about 10^{2} trichomes m^{-3}. Concentrations were 10-fold higher at the northern stations.

In the western Pacific, the range of *Trichodesmium* extends farther south than it does in the central region. Baas-Becking (1951) reports blooms at Viti Levu (18°S, 190°E) in October. Near New Zealand in September, V. Cassie (personal communication) collected *Trichodesmium* at Hawke Bay (east coast of North Island), the Tasman Sea (20-60 miles west of North Island) in October, at Foveau Strait in southern New Zealand, and in Cook Strait in November. While the species is recorded from New Zealand waters, it is probably a moribund population brought there in the East Australian Current. Surface water temperatures near New Zealand are virtually always below 20°C, except in the northerly areas in the southern hemisphere summer. Its southerly extent in autumn is assumed to be 30°S.

In the western North Pacific, Marumo and Asaoka (1974a) report that by October and November it disappears from the inner East China Sea, but it was present in the Kuroshio Current. Concentrations in this area are typically 10^5-10^6 trichomes m^{-3} in summer, and for the autumn 10^5 m^{-3} is assumed.

3. *Winter*

In January and February, F. Reid (personal communication) calculated a density of 2.5 x 10^2 trichomes m^{-3} in the top 20 m, and at 60 m it was 1.3 x 10^2 m^{-3}. In the Tonga Islands in January, 1963 a bloom was reported (Bowman and Lancaster, 1965). Other blooms occurred in February, 1936 at 15°S, 110°E and at New Caledonia in February, 1950 (Baas-Becking, 1951). In the Kuroshio Current it is usually very rare in winter; however, high densities were found at 32°N, 128-129°W in February, 1963 (Marumo and Asaoka, 1974a). This was considered an exceptional occurrence and probably due to an abnormal northward movement of warm water.

In the Pacific, the southern record for distribution of *Trichodesmium* in winter is assumed to be 25°S, based on the observations of Baas-Becking (1951). On the Great Barrier Reef (19°S), Revelante and Gilmartin (1982) observed high *Trichodesmium* concentrations in summer, reaching 1.5 x 10^7 trichomes m^{-3} in August. Cell density data for the Pacific Ocean in winter are sparse. Based on F. Reid's (personal communication) collections north of Hawaii, we assume that the population density is low, and for the entire Pacific it is 10^2 trichomes m^{-3}.

4. *Spring*

In the Pacific in spring, the concentration of *Trichodesmium* north of Hawaii in March was 8.9×10^2 m^{-3} in the upper 20 m and 2×10^2 at 60 m. In May it averaged 3.5×10^3 m^{-3} in the upper 20 m and 9.3×10^2 at 60 m (F. Reid, personal communication). For the North Pacific Gyre, the concentration is approximated at 10×20^2 trichomes m^{-3} for the spring season. The northern boundary is placed between that of winter and summer at 35°N.

In the South Pacific, Baas-Becking (1951) reports blooms of *Trichodesmium* at 24°S, 170°E in March and also in the Society Islands at 18°S, 160°W in April and at the Cook Archipelago in March (ca. 22°S). In the spring on the Great Barrier Reef, *Trichodesmium* density reached 1.5×10^7 trichomes m^{-3} in 1976, but there were virtually none present in 1977 (Revelante and Gilmartin, 1982). The southerly distribution in the central Pacific is placed near the sightings of blooms by Baas-Becking (1951) at 25°S.

In the western Pacific Ocean in spring, Marumo et al. (1975), on a transect south of Japan (cruise KH-72-1), measured *Trichodesmium* density. At stations north of 30°N it was not present. However, the concentration at about 28°N was about 5×10^2 trichomes m^{-3}. Density increased to 10^3 m^{-3} at stations near 25°N and at 20°N it was 10^4 m^{-3}.

F. Indian Ocean

1. *Winter and Spring*

There are numerous reports of blooms in the Indian Ocean during the northern monsoon. This monsoon lasts for about 6 months, beginning approximately in late autumn or early winter, and extending through spring. At this time, winds are calm and conditions are favorable for blooms (Carpenter and Price, 1976; Bryceson and Fay, 1981). For example, in the Arabian Sea, Nagabhushanam (1967) observed a bloom near Minicoy Island (ca. 8° 15'N, 73°E), which began in early May and lasted until late June. A bloom in March was noted off Ullal (Mangalore) by Prabhu et al. (1965). Similarly, Cleve (1901b) observed a *Trichodesmium* bloom in the Maldiva Islands 8°N, 75°E at the turn of the century that "colored the sea for great distances." It was commonly observed in a large lagoon in Sri Lanka through the spring, with concentrations ranging from ca. 25 to 200 "individual" (trichomes?) per milliter (Durairatnam, 1963). If an individual is a trichome, then density during this period was 2.5 to 20 x 10^{7} trichomes m^{-3}. If an individual is a cell, then density would be 2.5 to 20 x 10^{5} trichomes m^{-3}.

On the west coast of India near Goa, blooms occur continuously from February to May, and mean concentration is about 10^{7} trichomes m^{-3}. On one occasion, the near-surface layer concentration reached 39,000 trichomes $m\ell^{-1}$ on the east coast of India (Devassy et al., 1978). Another bloom was seen by Chidambaram and Unny (1944) in May near the southern coast of Pamban.

Ramamurthy (1970) states that a bloom began near Porto Novo (ca. 11°30'N on the southeast Indian Coast) in mid-February and lasted through mid-March. In an extensive area covering several thousand square miles between 8°-110°N and 73°-75°E in the Laccadive Islands, Quasim (1970) determined trichome density in an April bloom. For three stations mean density was 2.67 x 10^{6} trichomes m^{-3} at 5 m and 3.2 x 10^{4} at 10 m, for a mean of 9.3 x 10^{5} trichomes m^{-3} for the upper 10 m. Because of the numerous blooms reported near India, it is assumed that mean concentration in spring is 10^{6} trichomes m^{-3}. However, very high concentrations were observed by Bryceson (1980) near Dar es Salamm (ca. 6°S) on the African coast. His mean for the whole northern monsoon was 1.2 x 10^{7} trichomes m^{-3} for an offshore station. Thus, the area west of 45°E between the equator and 10°S is assumed to have a concentration of 10^{7} trichomes m^{-3}, and east of this, in the easterly flowing counter current, it is 10^{6} m^{-3}.

Trichodesmium is present over an extensive area of the Indian Ocean in winter and spring (Karsten, 1907b). Karsten (1907b) collected it as far south as 41°5'S, 76°23'E in January. However, it was not observed in 25 stations on a north-south transect between 43° and 62°S taken along longitude 67°E in March (Jacques et al., 1979). Taylor (1966) collected it working off southern Africa in January at about 35°S, 30°E. He did not give density but stated it was one of the more abundant species. We have conservatively placed the southern distribution of *Trichodesmium* at 35°S for the winter and 30°S for the spring. Concentration in both seasons in the Indian Ocean south of 10°S is assumed to be low, 10^2 trichomes m^{-3}.

2. *Summer*

In the summer, the Southern Monsoon reduces the concentration of *Trichodesmium* and it drops to very low levels. Bryceson (1980) did not observe it in 125-mℓ samples collected monthly during this period. However, since the sample size was small, it could have been missed if present at low density. For example, if there was one colony per liter, trichome concentration would be 2×10^3 m^{-3}, a significant concentration. Also, Sournia (1968) found *Trichodesmium* to be present in the Mozambique Channel near South Africa throughout the year; however, he reports that red water blooms only occur in the summer. Despite the fact that *Trichodesmium* density is low during the southern monsoon, there are still records of its blooming at this time. Chacko (1950) cites blooms ("swarms") having over 200 individuals $m\ell^{-1}$ near southern India in July. Assuming that an "individual" is a trichome, density in "swarms" would be 20×10^7 trichomes m^{-3}. In July and August it was commong (ca. 25-75 individuals $m\ell^{-1}$) in a Sri Lanka lagoon (Durairatnam, 1963). Unfortunately, Durairatnam does not state whether an individual is a cell or trichome. Based on these observations, the southerly distribution of *Trichodesmium* is assumed to approximate the 20°C isotherm, and overall for the Indian Ocean the population density appears low, with an estimated density of 10^2 trichomes m^{-3}.

3. Autumn

In autumn, numerous blooms are reported between Sri Lanka and India (Durairatnam, 1963; Chacko, 1950). In the northern Arabian Sea, Dugdale et al. (1964) collected *Trichodesmium* and measured rates of N_2 fixation at nine stations in October and November. Farther south at Dar es Salaam, *Trichodesmium* appears with the onset of the northern monsoon, which begins in October or November. When the monsoon occurs, the concentration is about 10^6 trichomes m^{-3} (Bryceson, 1980).

There are a few data available on the density of *Trichodesmium* in the Indian Ocean for autumn; therefore, the concentration and distribution is estimated by interpolating between summer and winter distributions.

G. South China Sea and Arafura Sea

In the South China Sea in February, *Trichodesmium* averaged 4×10^4 trichomes m^{-3} in the upper 50 m at $5^{\circ}02'N$, $107^{\circ}29'W$ (Saino and Hattori, 1979). Blooms are reported in this region in autumn at $5^{\circ}S$, $110^{\circ}E$ and in winter at $10^{\circ}N$, $120^{\circ}E$, so this organism would appear to be present throughout the year. Dareste (1854, in Wille, 1904) came across a bloom in September at approximately $10^{\circ}N$, $112^{\circ}W$ in the South China Sea. It is also present throughout the Arafura Sea north of Australia as Moseley (1879) on the Challenger Expedition reports, "We met with this alga in greatest abundance in the Arafura Sea between Torres Straits and the Aru Islands. Here it was at first encountered discoloring the sea-surface in bands and streaks; as the ship moved farther on, it became thicker, and at length the whole sea, far and wide, was discolored with it."

In the South China Sea a bloom was reported for April (von Frauenfeld, 1862; in Wille, 1904). Von Frauenfeld climbed the ship's topsails and with a telescope saw the bloom extending from horizon to horizon or a radius of "12 to 15 miles." We assume that for summer and autumn the concentration in this area is low, only 10^2 trichomes m^{-3}. For winter and spring, it increases somewhat to 10^4 trichomes m^{-3}.

IV. CALCULATION OF N_2 FIXATION RATE

Considering the differences in measurement techniques, incubation conditions, and the capacity of *Trichodesmium* to bloom, the rates reported in Table I are surprisingly similar, having a range within an order of magnitude. The results from nine of the studies shown in Table I were used to calculate a mean water column N_2 fixation rate. Measurements made under bloom conditions (Saino and Hattori, 1979) or those where only surface incubations were done (Bryceson, 1980) were not used in this calculation. The remaining nine studies average 7.23 pg N•trichome^{-1}•h^{-1}. Assuming that over a 12-h period of daylight the mean rate would be about half of this, we assume a mean pg N•trichome^{-1}•h^{-1} of 3.75 ± 0.8(S.E.).

The standing crop of *Trichodesmium* was calculated by assuming that trichome concentration is uniform between the surface and 15 m. Below this, it is assumed to decrease linearly to 10% of the surface value at 50 m. According to Carpenter and McCarthy (1975), virtually all *Trichodesmium* N_2 fixation occurs in the upper 50 m, so only this layer is considered. This profile approximates typical water column concentrations observed by others (Carpenter and Price, 1976; Saino, 1977). For a water column having near-surface (upper 15 m) concentrations of 1 x 10^3 trichomes m^{-3}, we thus obtain a total of 3.38 x 10^4 trichomes m^{-2}. For other near surface concentrations the total is scaled up accordingly. Applying the rate of 3.75 pg N•trichome^{-1}h^{-1} over a 12-h period, the summed N_2 fixation

TABLE I. Rate of N_2 Fixation by *Trichodesmium*

Reference	*Location*	*Stations*	*pg N•trichome•h^{-1}*
Goering et al., 1966	Tropical Atlantic, surface only	6	7.8
Mague et al., 1977	Central N. Pacific, maximum rate	1	28.5
Carpenter and Price, 1977	Caribbean Sea (mpz)[a]	12	7.7
Carpenter and Price, 1977	Western Sargasso (mpz)[a]	16	3.3
McCarthy and Carpenter, 1979	Central N. Atlantic (mpz)[a]	5	1.9
Saino and Hattori, 1979	E. China Sea		
	T. erythraea (bloom)	1	33.0
	T. thiebautii (bloom)	1	81.0
D. G. Capone, unpublished	St. Croix, W. I. (23 samples)	1	1.5
Bryceson, 1980	Tanzania, Indian Ocean (mean 5 surface samples)	1	14.8
Saino, 1977	S. China Sea (mpz)[a]	1	14.9
	Kuroshio Water (mpz)[a]	1[b]	6.5
	Kuroshio Water (mpz)[a]	1[c]	6.5

[a] *mpz, Mean photic zone rate.*

[b] *20 measurements in 4 profiles.*

[c] *24 measurements in 6 profiles.*

for a trichome concentration of $3.38 \times 10^4\ m^{-2}$ would be 1.52 μg $N \cdot m^2 \cdot day^{-1}$. The oceanic areas of various concentrations were calculated by planimitry from Figs. 4-7 and then multiplied by appropriate scaled-up rates.

Total N_2 fixation by *Trichodesmium* is conservatively estimated to be 4.79×10^{12} g N annually (Table II). The greatest annual rate occurs in the Indian Ocean, having 3.12×10^{12} g N $year^{-1}$, followed by the Atlantic Ocean with 1.32×10^{12} g N and the Pacific Ocean with 0.337×10^{12} g N. On a global basis, the highest rate was in the spring with 2.0×10^{12} g N fixed, and lowest was in the autumn with only 0.56×10^{12} g N. The cause of the high spring rate is the favorable conditions for N_2 fixation provided by the Northern Monsoon in the Indian Ocean. High rates were also observed in the winter in this area as a result of the monsoon.

The estimate of 4.79×10^{12} gN fixed annually by *Trichodesmium* is within the range of pelagic N_2 fixation presented by others. Delwich (1981) calculated a rate of 1.0×10^{12} g N. Soderlund and Svensson (1976) estimated a pelagic rate of 20 to 120×10^{12} g N annually, and Saino (1977) calculated 11×10^{12} g N. Saino's (1977) calculation primarily involved N_2 fixation by *Trichodesmium* in the South China Sea, northern Indian Ocean, and Caribbean Sea; it did not include fixation by *Trichodesmium* in extensive open areas of

TABLE II. Estimated Annual Rate of N_2 Fixation by *Trichodesmium* ($\times 10^9$ g N)

Ocean	*Spring*	*Summer*	*Autumn*	*Winter*	*Total*
Pacific	11.4	162.8	162.1	0.854	337.1
Atlantic	100.9	473.6	132.5	614.1	1321.1
Indian	1889.0	0.45	266.7	965.9	3122.1
South China and Arafura Sea	7.0	0.88	0.094	10.2	18.2
	2008.3	637.7	561.4	1591.1	4798.5

of the world's oceans. Soderlund and Svensson's (1976) N_2 fixation rate was calculated by extrapolating measurements made by Goering et al. (1966) in the tropical Atlantic Ocean under non-bloom conditions (0.13 and 0.66 g $N \cdot m^{-2} \cdot year^{-1}$) and by Gundersen (1974) near Hawaii (0.69 g $N \cdot m^{-2} \cdot year^{-1}$). Soderlund and Svensson (1976) assumed that one-third of the world's oceans were tropical and had the above rates. A lower aerial rate was assumed for colder waters. These rates for tropical waters are about the same as the estimated N_2 fixation (0.55 g $N \cdot m^{-2} \cdot year^{-1}$), which is calculated for a *Trichodesmium* concentration of 10^4 trichomes m^{-3} in this report.

It is evident that on a global basis when pelagic N_2 fixation is summed with that in marine sediments (Capone, Chapter 4, this volume), the oceans contribute 20 Tg N annually, somewhat more nitrogen than fixed in terrestrial ecosystems. If N_2 fixation by the endosymbiont *Richelia intracellularis*, as well as possible aerobic N_2 fixation by chroococcoid cyanobacteria (Duerr, 1981) and pelagic bacterial N_2 fixation are included, oceanic N_2 fixation could be considerably greater.

There are several deficiencies in this estimate of N_2 fixation by *Trichodesmium*. First, for some large oceanic areas (i.e., Pacific Ocean) there are few data points. Second, the data on frequency and extent of blooms are scant. Rate of N_2 fixation per trichome can be an order of magnitude, or more, greater in blooms, and these blooms may contribute substantially to N_2 fixation in the sea. One major deficiency in this compilation is the inability to assess accurately the frequency and extent of N_2 fixation in blooms. Because of the inability to quantify N_2 fixation in blooms accurately, this estimate of pelagic N_2 fixation is certainly an underestimate of the actual rate. The use of satellites to map the occurrence of near-surface accumulations of phycobiliproteins would be valuable in assessing bloom occurrence. More information on viability of N_2-fixing *Trichodesmium* in cold water is needed. Last, more data on N_2 fixation by other pelagic organisms is required before an accurate total N_2 fixation estimate can be made.

REFERENCES

Aikawa, H. (1936). On the diatom communities in the waters surrounding Japan. *Rec. Oceanogr. Works Jpn. 3,* 1-159.

An, C. N. (1971). Atlantic Ocean phytoplankton south of the Gulf of Guinea on profiles along 11° and 14° S. *Oceanology 11,* 896-901.

Aruga, Y., Ichimura, S., Fujita, Y., Shimura, S., and Yamaguchi, S. (1975). Characteristics of photosynthesis of planktonic marine blue-green algae, *Trichodesmium. In* "Studies on the Community of Marine Pelagic Blue-green Algae" (R. Marumo, ed.) pp. 48-55, Ocean Research Institute, Univ. of Tokyo.

Baas-Becking, L. G. M. (1951). Notes on some cyanophyceae of the Pacific region. *Proc. K. Ned. Akad. Wet., Ser. C 54,* 213-225.

Bernard, F., and Lecal, J. (1960). Plancton unicellulaire reculte dans l'ocean Indian par le *Charcot* (1950) et le *Norsel* (1955-56). *Bull. Inst. Oceanogr. 5,* 1-59.

Bowman, T. E., and Lancaster, L. J. (1965). A bloom of the planktonic blue-green alga *Trichodesmium erythraeum* in the Tonga Islands. *Limnol. Oceanogr. 10,* 291-293.

Bryceson, I. (1980). Nitrogen fixation and the autecology of *Oscillatoria erythraea* (Ehrenberg) Kuetzing, a planktonic cyanophyte from the coastal waters of Tanzania: A preliminary investigation. *Pap. Symp. Coastal Mar. Environ. Red Sea. Gulf of Aden Trop. West. Indian Ocean, 1980.* 1-20.

Bryceson, I., and Fay, P. (1981). Nitrogen fixation in *Oscillatoria (Trichodesmium) erythraea* in relation to bundle formation and trichome differentiation. *Mar. Biol. 61,* 159-166.

Burns, R. C., and Hardy. R. W. F. (1975). Nitrogen fixation in bacteria and higher plants. *Mol. Biol. Biochem. Biophys. 21,* 1-189.

Calef, G. W., and Grice, G. D. (1966). Relationship between the blue-green alga *Trichodesmium thiebautii* and the copepod *Macrosetella gracilis* in the plankton off northeastern South America. *Ecology 47,* 855-856.

Carpenter, E. J. (1972). Nitrogen fixation by a blue-green epiphyte on pelagic *Sargassum. Science 178,* 1207-1208.

Carpenter, E. J., and McCarthy, J. J. (1975). Nitrogen fixation and uptake of combined nitrogenous nutrients by *Oscillatoria (Trichodesmium) thiebantii* in the western Sargasso Sea. *Limnol. Oceangor. 20,* 389-401.

Carpenter, E. J., and Price, C. C., IV. (1976). Marine *Oscillatoria (Trichodesmium):* An explanation for aerobic nitrogen fixation without heterocysts. *Science 191,* 1278-1280.

Carpenter, E. J., and Price, C. C., IV (1977). Nitrogen fixation, distribution, and production of *Oscillatoria (Trichodesmium)* spp. in the western Sargasso and Caribbean Seas. *Limnol. Oceanogr. 22,* 60-72.

Carter, H. J. (1863). Notes on the coloring matter of the Red Sea. *Ann. Mag. Nat. Hist.* [3] 182-188.

Chacko, P. I. (1950). Marine phytoplankton from the waters around the Krusadai Island. *Proc. Indian Acad. Sci. 31,* 162-174.

Chacko, P. I., and Mahadevan, S. (1956). Swarming of *Trichodesmium erythraeum* in Waters around Krusadai Island, Gulf of Mannar. Fish St. Rep. Yearb. April 1954 to 1955. Government of Madras.

Chidambaram, K., and Unny, M. (1944). Note on the swarming of the planktonic algae *Trichodesmium* in the Pamban area and its effects on the fauna. *Curr. Sci. 13,* 263.

Cleve, P. T. (1901a). The seasonal distribution of Atlantic plankton-organisms. *Goteborgs Vetensk. Handl. 4,* 369.

Cleve, P. T. (1901b). Plankton from the Indian Ocean and the Malay Archipelago. *K. Sven. Vetenskaps-akad. Handl. 35,* 1-58.

Cleve, P. T. (1902). Additional notes on the seasonal distribution of Atlantic plankton-organisms. *Goteborgs K. Vetensk.-Vitterhets-Samh. Handl., Ser.* 4. *4.* 1-41.

Collingwood, C. (1868). Observations on the microscopic alga which causes the discoloration of the sea in various parts of the world. *Trans. R. Microsc. Soc. 16.*

Curl, H. (1959). The phytoplankton of Apalachee Bay and the northeastern Gulf of Mexico. *Inst. Mar. Sci. 6,* 277-320.

Dareste, C. (1854). Note sur la coloration des eaux de la mer de Chine. *C. R. Hebd. Seances Acad. Sci. 38.*

Delwiche, C. C. (1981). The nitrogen cycle and nitrous oxide. *In* "Denitrification, Nitrification and Atmospheric Nitrous Oxide" (C. C. Delwich, ed.), pp. 1-15. Wiley, New York.

Devassy, V. P., Bhattathiri, P. M. A., and Quasim, S. Z. (1978). *Trichodesmium* phenomenon. *Indian J. Mar. Sci. 7,* 168-186.

Duerr, E. O. (1981). Aerobic nitrogen fixation by two unicellular marine cyanobacteria (*Synechococcus* spp.) Ph. D. Thesis, University of Miami, Coral Gables, Florida.

Dugdale, R. C., Menzel, D. W., and Ryther, J. H. (1961). Nitrogen fixation in the Sargasso Sea. *Deep-Sea Res. 7,* 298-300.

Dugdale, R. C., Goering, J. J., and Ryther, J. H. (1964). High nitrogen fixation rates in the Sargasso Sea and the Arabian Sea. *Limnol. Oceanogr. 9,* 507-510.

Dunstan, W. M., and Hosford, J. (1977). The distribution of planktonic blue-green algae related to the hydrography of the Georgia Bight. *Bull. Mar. Sci. 27,* 824-829.

Durairatnam, M. (1963). Studies on the seasonal cycle of sea surface temperature, salinities and phytoplankton in Puttalam Lagoon, Dutch Bay and Portugal Bay, along the west coast of Ceylon. *Bull. Fish. Res. Stn., Ceylon 16,* 9-24.

Edinburg Oceanographic Laboratory (1973). Continuous plankton records: A plankton atlas of the North Atlantic and the North Sea. *Bull. Mar. Ecol. 7,* 1-174.

Ehrenberg, C. G. (1930). Neue Beobachtungen uber bluartige Erbscheinungen in Aegypten Arabien und Siberien nebst einer Uebersicht und Kritik der fruher bekannnten. *Ann. Phys. Chem. 18,* 477-514.

Eleuterius, L., Perry, H., Eleuterius, C., Warren, J., and Caldwell J. (1981). Causative analysis on a nearshore bloom of *Oscillatoria erythraea (Trichodesmium)* in the northern Gulf of Mexico. *Northeast Gulf Sci. 5,* 1-12.

Establier, R., and Margalef, R. (1964). Fitoplancton e hidrografi de las costas de Cadiz (Barbate), dejunio de 1961 a agosto de 1962. *Invest. Pesq. 25,* 5-31.

Farran, G. P. (1932). The occurrence of *Trichodesmium thiebautii* off the south coast of Ireland. *Rapp. P.-V. Reun. Cons. Perm. Int. Explor. Mer 77,* 60-64.

Feldman, J. (1932). Sur la biologie des *Trichodesmium* Ehrenberg. *Rev. Algo. 6,* 357-358.

Goering, J. J., Dugdale, R. C., and Menzel, D. W. (1966). Estimates of *in situ* rates of nitrogen uptake by *Trichodesmium spp* in the tropical Atlantic Ocean. *Limnol. Oceanogr. 11,* 614-620.

Gundersen, K. (1974). A study of biological nitrogen transformations in the water masses of the north central Pacific Ocean. *HIG-74-12, Hawaii Inst. Geophys.* [Tech. Rep.] 1-33.

Hargraves, P. E., Brody, R. W., and Burkholder, P. R. (1970). A study of phytoplankton in the lesser Antilles Region. *Bull. Mar. Sci. 20,* 331-349.

Hart, T. J. (1934). On the phytoplankton of the south-west Atlantic and the Bellingshausen sea. *"Discovery" Rep. 8,* 3-268.

Hentschel, E. (1932). Uberden Bewuchs auf den Treibenden Tanger der Sargassosee (Ergebnisse von der Ausreise der "Deutschland" 1911) *Beih. Jahrb. Hamb. Wissenschaftl. Anst. 38.*

Hulburt, E. M. (1962). Phytoplankton in the southwestern Sargasso sea and north equatorial current, February 1961. *Limnol. Oceanogr. 7,* 307-315.

Hulburt, E. M. (1968). Phytoplankton observations in the western Caribbean Sea. *Bull. Mar. Sci. 18,* 388-399.

Jacques, G., Descolas-Gros, C., Grall, J., and Sournia, A. (1979). Distribution in phytoplancton dans la partie Antarctique de l'Ocean Indian enfin d'ete. *Int. Rev. Gesamten Hydrobiol. 64,* 609-628.

Johnson, P. W., and Sieburth, J., McN. (1979). Chroococcoid cyanobacteria in the sea: A ubiquitous and diverse phototrophic biomass. *Limnol. Oceanogr. 24,* 928-935.

Karsten, G. (1907a). Das Phytoplankton des Antaktischen Meers nach dem Material der deutschen Tiefsee-Expedition 1898-1899. *Wiss. Ergeb. Dsch. Tiefsee-Exped. 'Valdivia' 2,* 1-136.

Karsten, G. (1907b). Das Indische phytoplankton nach dem material Tiefsee-Expedition 1898-1899. *Wiss Ergeb. Dsch. Tiefsee-Exped. 'Valdivia' 2,* 138-221.

Karsten, G. (1907c). Das phytoplankton des Atlantischen Oceans nach dem material der dentschen Tiefsee-Expedition 1898-1899. *Wiss.*

Ergeb. Dsch. Tiefsee-Exped. 'Valdivia' 2, 221-548.
Kawai, A., and Sugahara, I. (1971). Microbiological studies on nitrogen fixation in aquatic environments III. On the nitrogen fixing bacteria in offshore regions. *Bull. Jpn. Soc. Sci. Fish. 37,* 981-985.
Kimor, B., and Golandsky, B. (1977). Microplankton of the Gulf of Elat: Aspects of seasonal and bathymetric distribution. *Mar. Biol. 42,* 55-67.
Kimor, B., Reid, F. M. H., and Jordan, J. B. (1978). An unusual occurrence of *Hemiaulus membranaceus* Cleve (Bacillariophyceae) with *Richelia intracellularis* Schmidt (Cyanophyceae) off the coast of southern California in October 1976. *Phycologia 17,* 162-166.
King, J. E. (1950). A preliminary report on the plankton of the west coast of Florida. *Q. J. Fla. Acad. Sci. 12,* 109-137.
Kovaleva, T. M. (1976). Vertical distribution of phytoplankton in the tropical region of the Atlantic Ocean. *Hydrobiol. J. (Engl. Transl.) 12,* 1-6.
Liddle, L. (1981). Report on XIIIth International Botanical Congress. *Phycol. Newsl. 17,* 3-4.
Lohmann, H. (1920). Die Bevolkerung dez Ozeans mit plankton nach den Ergebnissen der Zentrifugenfange wahrend der Ausreise der "Deutschland," 1911. *Arch. Biontol. Berlin 4,* 1-617.
McCarthy, J. J. and Carpenter, E. J. (1979). *Oscillatoria (Trichodesmium) thiebautii* (Cyanophyta) in the central north Atlantic Ocean. *J. Phycol. 15,* 75-82.
Mague, T. H., Weare, M. M., and Holm-Hansen, O. (1974). Nitrogen fixation in the north Pacific Ocean. *Mar. Biol. 24,* 109-119.
Mague, T. H., Mague, F. C., and Holm-Hansen, O. (1977). Physiology and chemical composition of nitrogen fixing phytoplankton in the central North Pacific Ocean. *Mar. Biol. 41,* 213-227.
Margalef, R. (1969). Composition especifica del fitoplancton de la costa catalano-levantina (Mediterraneo occidental) en 1962-1967. (1973). Fitoplancton marino de la region de afloramiento del NW de Africa. II. Composicion y distribucion del fitoplancton (Campana Sahara II del Cornide de Saavedra). *Res. Exp. Cient. B/O Cornide 2,* 65-94.
Marshall, H. G. (1971). Composition of phytoplankton off the southeastern coast of the United States. *Bull. Mar. Sci. 21,* 806-825.
Martinez, L., Silver, M. W., King, J. M., and Alldredge, A. L. (1982). Nitrogen fixation by floating diatom mats: A source of new nitrogen to oligotrophic waters (unpublished manuscript).
Marumo, R., and Asaoka, O. (1974a). *Trichodesmium* in the east China Sea. 1. Distribution of *Trichodesmium Thiebautii* Gomont during 1961-1967. *J. Oceanogr. Soc. Jpn. 30,* 48-53.
Marumo, R., and Asaoka, O. (1974b). Distribution of pelagic blue-green algae in the north Pacific Ocean. *J. Oceanogr. Soc. Jpn. 30,* 77-85.

Marumo, R., and Nagasawa, S. (1976). Seasonal variation of the standing crop of a pelagic blue-green alga, *Trichodesmium* in the Kuroshio water. *Bull. Plankton Soc. Jpn. 23,* 19-25.

Marumo, R., Murano, M., and Aizawa, Y. (1975). Distribution, seasonal variation and red tide of *Trichodesmium. In* "Studies on the Community of Marine Pelagic Blue-Green Algae. 1972-1974" (R. Marumo, ed.), pp. 17-27. Ocean Res. Inst., University of Tokyo, Tokyo.

Maruyama, Y., Taga, N., and Matsuda, O. (1970). Distribution of nitrogen-fixing bacteria in the central Pacific Ocean. *J. Oceanogr. Soc. Jpn. 26,* 360-366.

Mobius, K. (1880). "Beitrage zur Meeresfauna der Insel Mauritius und der Seychellen." Berlin.

Montagne, C. (1844). Memoire sur le phenomene de la coloration des eaux de la mer rouge. *Ann. Sc. Nat. Bot. Biol. Veg.* [3] 2, 332-362.

Moseley, H. N. (1879). "Notes by a Naturalist on the 'Challenger'". Macmillan & Co., London.

Nagabhushanam, A. K. (1967). On an unusually dense phytoplankton 'bloom' around Minicoy island (Arabian Sea), and its effect on the local tuna fisheries. *Curr. Sci. 36,* 611-612.

Nagasawa, S. and Marumo, R. (1967). Taxonomy and distribution of *Trichodesmium* in Kuroshio water. *Inf. Bull. Planktonol. Jpn.* pp. 139-144.

Nemi, A. (1979). Blue-green algal blooms and N:P ratio in the Baltic Sea. *Acta Bot. Fenn. 110,* 57-61.

Ostenfeld, C. (1898). Plankton. Jagttagelser over Overfladevandets Temperatur, Saltholdighed og Plankton paa islandske og gronlandske Skibsrouter i 1898 foretagne under Led else af C. F. Wanderl Kjobenhaven.

Prabhu, M. S., Ramamurthy, S., Kuthalingham, M. D. K., and Dhulkhed, M. H. (1965). On an unusual swarming of the planktonic blue-green algae, *Trichodesmium spp.* off Mangalore. *Curr. Sci. 34,* 95.

Pshenen, L. N. (1963). Distribution and ecology of *Azotobacter* in the Black Sea. *In* "Symposium on Marine Microbiology" (C. H. Oppenheimer, ed.), pp. 383-391. Thomas, Springfield, Illinois.

Quasim, S. Z. (1970). Some characteristics of a *Trichodesmium* bloom in the Laccadives. *Deep-Sea Res. 17,* 655-660.

Ramamurthy, V. D. (1970). Studies on red water phenomenon in Portonovo waters (11 29'N-79 49'E, S. India) caused by *Trichodesmium erythraeum* (marine blue-green alga). *Ocean World-J. Oceanogr. Assembly, 1970,* pp. 13-25. Contrib. Biol. Oceanogr.

Revelante, N., and Gilmartin, M. (1982). Dynamics of phytoplankton in the Great Barrier Reef lagoon. *J. Plankton Res. 4,* 47-76.

Saino, T. (1977). Biological nitrogen fixation in the ocean with emphasis on the nitrogen fixing blue-green alga *Trichodesmium* and its significance in the nitrogen cycling in the low latitude sea areas. Ph.D. Thesis, University of Tokyo.

Saino, T., and Hattori, A. (1979). Nitrogen fixation by *Trichodesmium* and its significance in nitrogen cycling in the Kuroshio area and adjacent waters. *Proc. CSK Symp., 4th, 19* , pp. 1-13.

Sander, F., and Steven, D. M. (1973). Organic productivity of inshore and offshore waters of Barbados: A study of the island mass effect. *Bull. Mar. Sci. 23,* 771-792.

Sato, S., Paranagua, M. N., and Eskinazi, E. (1966). On the mechanism of red tide of *Trichodesmium* in Recife, northeastern Brazil, with some considerations of the relation to the human disease "Tamandare Fever." *Trab. Inst. Oceanogr. Univ. Recife 5,* 7-49.

Soderlund, R., and Svensson, B. H., (1976). The global nitrogen cycle. *Ecol. Bull. 22,* 23-73.

Sournia, A. (1968). La Cyanophyceae *Oscillatoria (Trichodesmium)* dans le plancton marin: Taxanomic, et observations dans le canae de Mozambique. *Nova Hedwigia 15,* 1-12.

Sournia, A. (1979). La cyanophycee *Oscillatoria (Trichodesmium)* dans la plancton marin. *Nova Hedwigia 15,* 1-12.

Steven, D. M., and Glombitza, R. (1972). Oscillatory variation of a phytoplankton population in a tropical ocean. *Nature (London) 237,* 105-107.

Taylor, B. F., Lee, C. C., and Bunt, J. S. (1973). Nitrogen-fixation associated with the marine blue-green alga, *Trichodesmium* as measured by the acetylene reduction technique. *Arch. Mikrobiol. 88,* 205-212.

Taylor, F. J. R. (1966). Phytoplankton of the southwestern Indian Ocean. *Nova Hedwigia 12,* 433-476.

Vatova, A. (1928). Compendio della flora e della fauna del mare Adriatico presso Rovigno. *Mem. Com. Talass. Ital. 143,* 1-614.

Voltolina, D. (1975). The phytoplankton of the lagoon of Venice. November 1971-November 1972. *Publ. Stn. Zool. Napoli 39,* 206-340.

Wille, N. (1904). Die Schizophyceen der Plankton Expedition. *Ergeb. Humbolt-Stift. 4,* 1-88.

Wynn-Williams, D. D., and Rhodes, M. E. (1974). Nitrogen fixation in seawater. *J. Appl. Bacteriol. 37,* 203-216.

Chapter 4

BENTHIC NITROGEN FIXATION

DOUGLAS G. CAPONE
Marine Sciences Research Center
State University of New York
Stony Brook, New York

I. INTRODUCTION

A. Physiological and Environmental Considerations

It has long been suspected that benthic processes complement, contribute to, and possibly control those biological processes occurring in the water column (Waksman *et al.*, 1933; Oppenheimer, 1960). The implication of nitrogen as a primary regulator of oceanic productivity (Ryther and Dunstan, 1971) has induced researchers to examine the nature and importance of nitrogen metabolism in benthic marine communities, and also to determine the extent of coupling of these activities with pelagic ecosystems (Rowe *et al.*, 1975). Logically, a portion of this attention has been focused on the primary input of combined nitrogen through biological reduction of dinitrogen (i.e., nitrogen fixation).

ISBN 0-12-160280-X

Historically, N_2 fixation has been invoked as a likely source of requisite nutrients in marine communities that display high apparent productivities despite meager concentrations of available nitrogenous nutrients (Odum and Odum, 1955; Patriquin, 1972). The occurrence of known N_2-fixing organisms in such systems provided circumstantial evidence for this suggestion (Keutner, 1904; Waksman *et al.*, 1933; Pshenin, 1963). Over the last decade many of these suspicions have been directly confirmed in a variety of benthic environments. Quantitative estimates from recent studies indicate that N_2 fixation may indeed provide substantial nutrient resources for many ecosystems. These assessments must be viewed, at present, with some caution because of the methodological limitations and discrepancies among the results of various investigators working in similar environments.

Research into all aspects of N_2 fixation, including its biochemistry, genetics, physiology, and ecology, has intensified over the last several decades in response to the need for alternatives to synthetic fertilizers for agriculture (Hardy and Havelka, 1975). Therefore, considerable advances in our knowledge of biological N_2 fixation and its physiological and environmental controlling factors have been made.

The common denominator in all N_2-fixing systems is the enzyme nitrogenase, which occurs in a diverse group of prokaryotes. Nitrogenase has been highly conserved during evolution (Burns and Hardy, 1975), being strikingly similar in both structure and regulation in all N_2-fixing organisms (Burns and Hardy, 1975). Characteristics common to most nitrogenase systems studied include a large demand for ATP and reducing equivalents, a high degree of sensitivity to O_2, and similar forms of transcriptional regulation (Shanmugam *et al.*, 1978; Brill, 1980). ATP-dependent hydrogen evolution is also directly associated with active N_2 fixation (Schubert and Evans, 1976).

In nature, diazotrophic organisms have employed a diversity of metabolic and ecologic strategies to meet the energetic demands of N_2 fixation. The most efficient heterotrophic N_2 fixers known, the rhizobia, have evolved intimate symbiotic associations with higher plants (Evans and Barber, 1977). Photoautrophic N_2 fixers, including the heterocystous cyanobacteria and members of the anoxygenic photobacteria, obtain energy and reducing power for nitrogenase activity from photosynthesis, and therefore, as a group, probably exhibit the greatest degree of nutritional independence in the biosphere (Stewart, 1971).

The benthic marine environment has representatives from the spectrum of N_2-fixing organisms. Free-living heterotrophs occur at and below the sediment surface. In those benthic areas within the euphotic zone, heterotrophic N_2 fixers are associated with many of the major plant communities. Photosynthetic diazotrophs also occur as both free-living and epiphytic forms (Peters, 1978). One may speculate that chemolithotrophs capable of N_2 fixation (e.g., members of the thiobacilli and hydrogen-oxidizing bacteria) may also provide important primary inputs of nitrogen in those ecosystems based on chemoautotrophy such as the deep sea hydrothermal vents (Rau, 1981a,b).

While nitrogenase is a highly O_2-sensitive enzyme, N_2 fixation occurs under fully oxic to anoxic conditions because N_2-fixing organisms have evolved various strategies to protect the enzyme from O_2 inactivation. The benthic interface is generally accompanied by a sharp O_2 gradient, from fully aerobic at or near the interface to highly reduced conditions within millimeters to centimeters below the interface (Revsbech *et al.*, 1980). Hence, we also find the range of physiological types (with respect to O_2 tolerance) of N_2 fixers in the benthic environment. These extend from strict aerobes such as *Azotobacter* near the sediment surface (Dicker and Smith, 1980a) to strict anaerobes including the clostridia (Waksman *et al.*, 1933) and SO_4^{-2}-respiring bacteria (Patriquin and Knowles, 1972; Dicker and Smith, 1980a; Patriquin and McClung, 1978).

Sulfate-respiring bacteria, as a group, are important in oxidizing much of the organic material in many anoxic environments (Howarth and Teal, 1979; Sorensen *et al.*, 1979; Fenchel and Blackburn, 1979; Jorgensen, 1982). Postgate (1979) pointed out that they may also be the dominant N_2-fixing flora of marine sediments. Recent results appear to confirm this supposition (Dicker and Smith, 1980b; Nedwell and Aziz, 1980; Capone, 1982; Capone and Taylor, unpublished).

Nitrogenase synthesis is often regulated by the enzyme glutamine synthetase (Striecher *et al.*, 1974; Evans and Barber, 1977). High intracellular levels of NH_4^+ promote adenylation of glutamine synthetase, which in turn acts as a repressor of the nitrogen fixation (*nif*) genes (Evans and Barber, 1977; Brill, 1980). High concentrations of NH_4^+ generally occur in marine sediments and may play a role in modulating *in situ* N_2 fixation in these environments.

Nitrate can also affect N_2 fixation. Dicker and Smith (1980b) proposed that NO_3^- inhibition of salt marsh sediment N_2 fixation was a result of competition for reducing power by the assimilatory nitrate reductase pathway. Substantial denitrification and dissimilatory NO_3^- reduction occurs in marine sediments (Sorensen, 1978; Koike and Hattori, 1978), which may also compete for reducing power. The concomitant accumulation of NH_4^+ through dissimilatory reduction also suggests that direct NH_4^+ inhibition might result from high NO_3^- concentrations. In this context, it is interesting to note that some free-living diazotrophic bacteria couple the dissimilation of NO_3^- to N_2 (denitrification) to N_2 fixation (Neyra *et al.*, 1977; Bothe *et al.*, 1981).

B. Methodological Limitations

The recent expansion in our realization of the extent of N_2 fixation in marine ecosystems has come primarily as a result of the introduction of facile assay methods. Specifically, these include adaptation of ^{15}N isotopic methods (see Harrison,

Chapter 21) and the C_2H_2 reduction technique (see Taylor, Chapter 22) for aquatic studies. The application of these assays has, however, outstripped a full understanding of their limitations. This, coupled with the different procedures employed by various workers, makes quantitative comparisons difficult.

Numerous caveats have been issued with regard to the use of the C_2H_2 reduction method. Acetylene, besides acting as a substrate analog for nitrogenase in the C_2H_2 reduction assay, directly inhibits a variety of microorganisms that occur in sediment environments. These include denitrifying (Yoshinari and Knowles, 1976; Balderston *et al.*, 1976), methanogenic (Oremland and Taylor, 1975), methane-oxidizing (DeBont and Mulder, 1974; Dalton and Whittenburg, 1976), sulfate-respiring (Payne and Grant, 1982), and nitrifying bacteria (Hynes and Knowles, 1978; Bremner and Blackmer, 1979) as well as N_2-fixing bacteria themselves (Brouzes and Knowles, 1971). The inhibitory effect of C_2H_2 on N_2-fixing bacteria most likely results from the inhibition of N_2 fixation by C_2H_2 (Rivera-Ortiz and Burris, 1975), with concomitant acceleration of nitrogenase synthesis in response to nitrogen starvation. Indeed, David and Fay (1977) showed that N_2-fixing cyanobacteria exposed to C_2H_2 for extended periods exhibited stimulated nitrogenase activity and recommended brief exposure periods to avoid this problem.

Ethylene, the product of acetylene reduction, may be aerobically degraded by methane-oxidizing bacteria (Flett *et al.*, 1975). However, this may be of little consequence since C_2H_2 appears to inhibit this activity (DeBont, 1976). Acetylene may itself be metabolized aerobically or anaerobically other than via nitrogenase (DeBont and Peck, 1980; Culbertson *et al.*, 1981).

For assessing *in situ* activities in benthic samples, and in particular sediments, a number of precautions should be observed. First, sample disturbance and the disruption of "microsites" within the sample should be minimized. Second, for assays requiring the introduction of a substrate (e.g., $^{15}N_2$, C_2H_2) into the sample, rapid dissemination of the addition to sites of activity is important. Unfortunately, effectively achieving the second goal may comprise efforts toward the first. Finally, and particularly where the substrate for the assay is also a metabolic inhibitor as in C_2H_2 reduction, assay periods must be brief to minimize artifacts produced by the inhibition.

In situ approaches (e.g., bell jar) rely most heavily on natural diffusion of substances and likely reflect activity at or near the sediment surface. Long lag periods and loss of reactants are associated with such systems, and these aspects have been discussed in detail by Patriquin and Denike (1978).

Because of these constraints, small samples of sediment are often assayed in a container, thereby entailing some physical disturbance. Sediment samples are commonly collected by coring, and assay methods described range from introduction of the substrate or inhibitor directly into the collection device, to extrusion of the sediment sample into an assay vessel with (slurry) or without a liquid phase. Assays within a core minimize physical disturbance of the sample but generally require extended incubation to overcome diffusional limitations. Conversely, slurries assure rapid dispersion of introduced compounds to target sites but disrupt microzones and gradients originally present. In both cases, the isolation of the sample may conceivably result in the buildup of inhibitory metabolites or in the depletion of substrates.

The effect of sample size generally has not been considered. Smith (1980) recently compared the effects of sample size on apparent rates of nitrogenase activity and found a negative correlation between the specific rate of C_2H_4 production (per gram dry sediment) and total sample size. He ascribed this to decreased penetration of C_2H_2 with increasing size. Small samples would minimize the time lag in C_2H_2 penetration through the sample while maximizing the effects of spatial heterogenity. The issue of sample-size effects certainly needs closer scrutiny, particularly

TABLE I. Summary of Studies Comparing $^{15}N_2$ Fixation and C_2H_2 Reduction Using Samples from Benthic Marine Environments

Sample	Mole ratio[a] ($C_2H_2:N_2$)	Reference
Zostera marina rhizosphere sediments	2.6 (0.5-6.2)	Patriquin and Knowles (1972)
Thalassia testudinum rhizosphere sediments	2.6 (2.1-3.3)	Patriquin and Knowles (1972)
Syringodium filiforme rhizosphere sediments	4.6 (2.6-15.4)	Patriquin and Knowles (1972)
washed roots and rhizomes	2.8 (0.8-10.5)	Patriquin and Knowles (1972)
Suspensions of cyanobacteria (*Scytonema* sp.) from coral reef environment	1.9 (1.4-2.5)	Burris (1976)
Cyanobacteria on dead coral	3.3	Hanson and Gundersen (1976)
Cyanobacterial mat from salt marsh environment	3.0	Carpenter *et al.* (1978)
Cyanobacteria on surface sediments, intertidal	5.4 (4.7-5.7)	Potts *et al.* (1978)
Zostera marina roots and rhizomes	2.6 (1.7-4.1)	Capone and Budin (1982)

[a] *Values reported are means with ranges in parentheses.*

considering the range in sample sizes now used by various investigators.

The use of the C_2H_2 reduction method (Stewart *et al.*, 1967; Hardy *et al.*, 1968) has far exceeded that of $^{15}N_2$ reduction. This is understandable because of the greater sensitivity, economy, and simplicity of C_2H_2 reduction as compared to the isotopic method. However, the underlying assumption generally employed, that 3 moles of C_2H_2 are reduced per mole of N_2 which would have been fixed, has often times been shown to be invalid (Burris, 1974). Several reasons may account for the empirically observed deviations from the theoretical 3:1 ratio. These include differential affinity of nitrogenase for N_2 and C_2H_2 (Rivera-Ortiz and Burris, 1975), substantially reduced leakage of H_2 from nitrogenase under C_2H_2-reducing conditions (Hardy *et al.*, 1968; Rivera-Ortiz and Burris, 1975), and possibly, inhibition of conventional uptake hydrogenases (which may be coupled to N_2 fixation) by C_2H_2 (Smith *et al.*, 1976).

Direct calibration of the C_2H_2 reduction method is strongly recommended for each system (Burris, 1974) but has been performed in only a limited number of C_2H_2 reduction studies of benthic (or, for that matter, pelagic) marine N_2 fixation (Table I). Relatively good agreement with a 3:1 ratio has been noted but a wider comparison of various samples and environments is needed in order to substantiate our extrapolations of C_2H_2 reduction.

II. SEDIMENT SURFACE

Our knowledge of N_2 fixation in the oceans started at the edge, in the intertidal zone. Stewart (1965, 1967) undertook one of the earliest studies, focusing on the cyanobacterial (*Calothrix scopulorum*) mats of the rocky shores of Scotland. He clearly established the quantitative importance of N_2 fixation within these communities, as well as identifying the potential fate of recently fixed nitrogen (Jones and Stewart, 1969). Since then, cyanobacterial mats have also been found to be highly active sites of N_2 in other environments (Table II).

For salt marsh ecosystems, several distinct surface environments have been identified. Highest surface activity in salt marshes also appears to be associated with cyanobacterial mats, whereas generally lower nitrogenase activity occurs in bare mud or at the base of *Spartina* stands (Table II).

Coral reef environments typically are characterized as highly productive ecosystems existing in low nutrient environments (Odum and Odum, 1955). This enigma has been explained partially by the common occurrence of cyanobacterial mats capable of intensive N_2 fixation on many reef flats (Table II). Reported areal rates ap-

TABLE II. Sediment Surface Nitrogen Fixation

Sample	N_2 fixation[a] (mg N $m^{-2} \cdot day^{-1}$)	Methods[b]	References
Intertidal/subtidal			
Calothrix mat, rocky shore	7 (15)	1,3,8	Stewart (1965)
Calothrix mat, rocky shore	3 (11)	2,3,5,8	Warmling (1973)
Calothrix mat, coral reef	180	2,3,5,7	Wiebe *et al.* (1975)
Calothrix mat, coral reef	6	2,3,5	Mague and Holm-Hansen (1975)
Calothrix mat, coral reef	43	2,3,5	Potts and Whitton (1977)
Scytonema mat, coral reef	5.6	1,2,3,5	Burris (1976
Scytonema mat, coral reef	168	2,3,5	Potts and Whitton (1977)
Nostoc mat, coral reef	40	2,3,5	Mague and Holm-Hansen (1975)
Oscillatoria, photosynthetic bacteria	22	2,4,6,10	Bohlool and Wiebe (1978)
Algal mats	11	2,3,5,8	Gotto *et al.* (1981)
Salt marsh			
Bare mud	1.2 (6)	2,3,6	Jones (1974)
Algal mat	55 (293)	2,3,6	Jones (1974)
Algal mat	28 (89)	2,3,5,10	Whitney *et al.* (1975)
Algal mat	6.3 (70)	2,3,5,8	Carpenter *et al.* (1978)
Base of *Spartina*	48 (304)	2,3,6	Jones (1974)
Base of *Spartina*	1.8 (7)	2,4,5,7,8	Hanson (1977)
Base of *Spartina*	12 (135)	2,4,6,7,9	Patriquin and Denike (1978)
Base of *Spartina*	2.3 (25)	2,3,5,8	Carpenter *et al.* (1978)

[a] *Values reported are averages over indicated periods with maximum (if given) in parentheses.*

[b] *Key: 1. $^{15}N_2$ uptake measurements; 2. C_2H_2 reduction determinations; 3. small vial assay; 4. bell jar assay, in situ; 5. short-term assay, 1-5 h; 6. long-term assay, >5 h; 7. variable lag period observed; 8. average from annual survey; 9. average from survey during growing season; 10. average from survey during summer only.*

proach values similar to the most intense agricultural N_2-fixing systems, such as the legumes (Burns and Hardy, 1975; Evans and Barber, 1977).

As may be evident from Table II, a considerable divergence in results does exist among the various studies employing similar samples. While this may be partially attributed to environmental variations between sampling sites, the accompanying differences in methodology and data interpretation must also contribute to the observed disparities.

III. PLANT ASSOCIATIONS

For heterotrophic diazotrophs, association with a plant insures a source of substrate for the energetically demanding process of N_2 fixation. The benefit to the plant host is a supply of nitrogenous nutrients. Numerous examples of such bacterial-plant consortia are found in terrestrial systems and range from the highly integrated endosymbiosis of rhizobia with leguminous crops (Burns and Hardy, 1975) to the looser symbioses of *Azospirillum* spp. and tropical grasses (Van Berkum and Bohlool, 1980). Nitrogen-fixing cyanobacteria also associate with plants (Peters, 1978).

Diazotrophic bacteria are found on a variety of higher marine plants (Table III). Benthic macroalgae from both tropical and temperate environments host epiphytic assemblages capable of high rates of nitrogenase activity. These include representatives of both the Rhodophyta (*Laurencia* sp.) and Chlorophyta (*Codium* spp., *Enteromorpha* sp., *Microdictyon* sp.). Screenings of benthic phaeophytes have generally proved negative (Capone *et al.*, 1977; E. J. Carpenter, personal communication). Cyanobacterial epiphytes are the active agents for the majority of recognized associations (Mague and Holm-Hansen, 1975; Capone, 1977; Capone *et al.*, 1977; Penhale and Capone, 1981; Rosenberg and

TABLE III. Nitrogen Fixation Associated with Benthic Marine Plants

Plant association	N_2 Fixation[a]		Reference
	mol $C_2H_4 \cdot g^{-1} \cdot h^{-1}$	mg $N \cdot m^{-2} \cdot day^{-1}$	
Macroalgal epiphytes			
Codium/heterotrophic bacteria	780	1.3	Head and Carpenter (1975)
Jania/*Hormothamnion*	300-700	31	Mague and Holm-Hansen (1975)
Microdictyon/cyanobacteria	250	4.3	Capone (1977)
Laurencia/cyanobacteria	2250	64	Capone (1977)
Enteromorpha/cyanobacteria	--	1.1	Bohlool and Wiebe (1978)
Codium/cyanobacteria	129	--	Rosenberg and Paerl (1981)
Higher plant epiphytes			
Thalassia/cyanobacteria	11,000	150	Goering and Parker (1972)
Thalassia/cyanobacteria	0-0.5	--	McRoy *et al.* (1973)
Zostera/epiphytes	0	--	McRoy *et al.* (1973)
Spartina/epiphytes	--	0.6	Hanson (1977)
Thalassia/cyanobacteria	80-900	3.7	Capone and Taylor (1980a)
Zostera/epiphytes	0-20	<0.02	Capone (1982)
Root and rhizome associations			
Spartina roots	70-538	--	Patriquin (1978a)
Spartina sods (washed)	--	35	Patriquin and Denike (1978)
Spartina roots	2-75	--	Van Berkum and Sloger (1979)
Mangrove roots	1200-1600	--	Zuberer and Silver (1978)
Thalassia roots and rhizomes	300-400	--	Capone and Taylor (1980b)
Zostera roots and rhizomes	100	1.8	Capone and Budin (1982)
Plant detritus			
Mangrove leaves/photosynthetic bacteria	1200	--	Gotto and Taylor (1976)
Thalassia leaves	100-300	0.8	Capone and Taylor (1980a)
Spartina leaves and shoots	--	0.6	DeLaune and Patrick (1980)

[a]*Weight specific rates are reported ranges or means (as indicated), whereas areal rates are reported averages.*

Paerl, 1981). For *Codium fragile*, however, heterotrophic bacteria are the predominant N_2 fixers (Head and Carpenter, 1975).

The leaf surfaces of several marine grasses, such as *Thalassia testudinum* (Goering and Parker, 1972) and *Spartina alterniflora* (Hanson, 1977), may also support a vigorous N_2-fixing flora. For *T. testudinum*, cyanobacteria were identified as the principal diazotrophs (Goering and Parker, 1972). Generalizations about leaf-associated N_2 fixation in seagrasses should be avoided, since such associations were found to be highly variable or even absent in further studies of *T. testudinum* or *Zostera marina* (McRoy *et al.*, 1973; Capone and Taylor, 1977; Smith and Hayasaka, 1982; Capone, 1982).

Nitrogenase activity is also associated with the excised roots and rhizomes of seagrasses, marsh grass, as well as with mangroves (Table III). Root and rhizome surfaces presumably provide a highly suitable habitat for heterotrophic N_2 fixers, given the potential for exudation of labile organic compounds by the roots (Wetzel and Penhale, 1979) and the varying redox environment in the rhizosphere of these marine plants (Patriquin and Knowles, 1975). Patriquin (1978b; Boyle and Patriquin, 1980) found some evidence of diazotrophic bacteria within the cortical tissues of *S. alterniflora* roots. McClung and Patriquin (1980) recently isolated a N_2-fixing *Campylobacter* from *Spartina alterniflora* roots, whereas diazotrophic spirilla have been isolated from the root tissue of *Potamogeton filiforme* (Sylvester-Bradley, 1976) and *Zostera marina* (Budin and Capone, unpublished).

Plant detritus may also support substantial populations of N_2-fixing organisms (Table III). Senescing plant tissue often has a much higher C:N ratio than healthy material (Thayer *et al.*, 1977; Harrison and Mann, 1975; Capone and Taylor, 1977), and therefore represents a substrate particularly suited to diazotrophs.

Nitrogen-fixing bacteria have also been found in association with several marine animals, including sea urchins (Guerniot and Patriquin, 1981), sponges (Wilkinson and Fay, 1979), and shipworms (Carpenter and Culliney, 1975) (see D. Taylor, Chapter 18).

IV. RHIZOSPHERE AND NONRHIZOSPHERE SEDIMENTS

A prime environmental constraint of the nitrogenase enzyme is its O_2 sensitivity. In this regard, marine sediments by their generally anoxic nature are a likely site of N_2 fixation and, for benthic environments, the bulk of observations have in fact been directed to the sediments (Table IV).

Little quantitative information is available on N_2 fixation in deep-sea sediments. Hartwig and Stanley (1978) found no C_2H_2 reduction in sediments retrieved from 4800 m in the Atlantic Ocean (Table V), while Soderlund and Svensson (1976) cite a similar result from an unpublished study. Detectable C_2H_2 reduction was noted by Hartwig and Stanley (1978) in sediments from a shallower station (2800 m). In both cases, assays were conducted onboard ship after recovery of sediments from depth. Thus, one cannot rule out *in situ* N_2 fixation by bacteria adapted to abyssal environments (Yayanos *et al.*, 1979).

A greater number of studies have been conducted on samples obtained within the 200 m isobath (Tables IV and V). Hartwig and Stanley (1978) obtained relatively high rates of activity in coastal shelf sediments, in comparison with their deeper stations (Table IV). Their values were somewhat lower than those reported by other investigators for samples from similar depths on the shelf.

Despite a diversity of approaches in applying the C_2H_2 reduction technique to nonrhizosphere estuarine environments, results have been generally consistent, varying over only a twofold range (Tables IV and V). However, the same has not held for rhizosphere sediments as typified by studies of seagrass (subtidal) or salt marsh (intertidal) systems. For *Thalassia testudinum* communities, Patriquin and Knowles (1972) reported extraordinarily high rates of N_2 fixation, sufficient to account for the entire nitrogen demand of these very productive communities. Their estimates were based on long-term assays (2-3 days)

TABLE IV. Nitrogen Fixation in Marine Sediments

Environment	Comment[b]	N_2 Fixation[a]: nmol $C_2H_4 \cdot g^{-1} \cdot h^{-1}$	N_2 Fixation[a]: mg $N \cdot m^{-2} \cdot day^{-1}$	Reference
Offshore				
Beaufort Sea (25-200 m)	1,3,5,9,11	0.004-0.013	0.07	Knowles and Wishart (1977)
Atlantic Shelf (150 m)	1,2,8	0.0005	--	Hartwig and Stanley (1978)
Atlantic (2800 m)	1,2,7	0.0013	--	Hartwig and Stanley (1978)
Norton Sound (27 m)	1,3,7,11	0.02	0.54	Haines *et al*. (1981)
Beaufort Sea (19 m)	1,3,7,11	0.02	0.48	Haines *et al*. (1981)
Shallow subtidal, nonrhizosphere				
Waccasassa Estuary, FL	1,3,7,11	0.06-0.54	1.0	Brooks *et al*. (1971)
Nearshore, Barbados	1,6,11	0.02-0.33	0.9	Patriquin and Knowles (1972)
Nearshore, Nova Scotia	1,6,11	0.14-0.21	0.9	Patriquin and Knowles (1972)
Rhode River Estuary, MD	1,3,5,8,11,12	0.26	0.7	Marsho *et al*. (1975)
Tay Estuary, Scotland	2,4,7,10	0.19	0.2	Herbert (1975)
Kaneohe Bay, HA	1,3,8,10	0.2-1.1	1.7	Hanson and Gundersen (1977)
Scottish Loch	1,2,8	0.08-0.50	--	Hartwig and Stanley (1978)
Shelikot Strait, AL	1,3,7,11	0.07	1.6	Haines *et al*. (1981)
Elson Lagoon, AL	1,3,7,11	0.07	1.5	Haines *et al*. (1981)
Subtidal, rhizosphere				
Thalassia, Barbados	1,3,6,11	0.14-27	82 (137)	Patriquin and Knowles (1972)
Zostera, Nova Scotia	1,6,10	0.16-3.1	--	Patriquin and Knowles (1972)
Thalassia, FL	1,3,10	0-0.01	--	McRoy *et al*. (1973)

Mangrove, FL	2,4,9,11	0.06-1.3	7.8	Zuberer and Silver (1978)
Thalassia, FL	1,3,4,10,12	0.06-0.7	21 (38)	Capone and Taylor (1980b)
Zostera, NY	1,4,10,12	0.04-0.2	5.5 (7)	Capone (1982)
Salt marsh, nonrhizosphere				
Panne, Flax Pond, NY	1,3,5,10,11	--	1.5 (6.6)	Whitney *et al.* (1975)
Bare mud, Flax Pond, NY	1,3,5,10,11	--	3.3 (19)	Whitney *et al.* (1975)
Bare mud, Sippewissett Marsh, MA	1,9,11,12	--	4.8 (29)	Teal *et al.* (1979)
Sand, Sippewissett Marsh, MA	1,9,11,12	--	1.8 (7)	Teal *et al.* (1979)
Salt marsh, rhizosphere				
Tall *Spartina*, Flax Pond, NY	1,3,5,10,13	--	2.8 (6.8)	Whitney *et al.* (1975)
Short *Spartina*, Flax Pond, NY	1,3,5,10,13	--	1.6 (2.4)	Whitney *et al.* (1975)
Salt Marsh, Sapelo Is. GA	1,9,11,12	0.22	142 (270)	Hanson (1977)
Spartina soil, Nova Scotia	1,3,6,7,11	--	82 (158)	Patriquin and Denike (1978)
Tall *Spartina*, Sippewissett, MA	1,3,9,11,12	--	23 (132)	Teal *et al.* (1979)
Short *Spartina*, Sippewissett, MA	1,3,9,11,12	--	33 (132)	Teal *et al.* (1979)
Salt Marsh, DE	1,7,11,12	0.2-10 (28)	--	Dicker and Smith (1980c)
Salt Marsh, LA	14		43	DeLaune and Patrick (1980)

[a]*Areal values are averages for indicated periods with reported maxima in parentheses.*

[b]*Key: 1. Bottle assays; 2. in situ cores; 3. slurries; 4. stimulated by anoxic conditions; 5. unaffected by anoxic conditions; 6. variable effect of oxygen; 7. assays of 0-5 cm sediment depth; 8. assays of 0-10 cm sediment depth; 9. assays > 10 cm; 10. short-term assays (<10 h); 11. long-term assays (>10 h); 12. average from annual survey; 13. average from survey during growing season only; 14. information not available.*

TABLE V. Annual Areal Estimates of Nitrogen Fixation in Various Marine Communities

Environment	N_2 Fixation		Reference
	mg $N \cdot m^{-2} \cdot day^{-1}$	g $N \cdot m^{-2} \cdot year^{-1}$	
>200-m Sediments			
Atlantic (2800 m)	(0.002)	(0.0007)	Hartwig and Stanley (1978)
Atlantic (4800 m)	0	0	Hartwig and Stanley (1978)
<200-m Sediments			
Beaufort Sea (25-200 m)	0.07	0.03	Knowles and Wishart (1977)
Atlantic Shelf (150 m)	(0.05)[a]	(0.02)	Hartwig and Stanley (1978)
Norton Sound (27 m)	0.54	0.20	Haines *et al.* (1981)
Beaufort Sea (19 m)	0.48	0.17	Haines *et al.* (1981)
Cook Inlet (1-206 m)	0.2	0.07	Haines *et al.* (1981)
Georgia shelf (13-68 m)	0.28	0.10	Hanson *et al.* (1981)
Average		0.10 ± 0.03	
Estuarine sediments			
Shelikot Strait, AL	0.80	0.58	Haines *et al.* (1981)
Elson Lagoon, AL	1.5	0.54	Haines *et al.* (1981)
Waccasassa Estuary, FL	1.0	0.36	Brooks *et al.* (1971)
Rhode River Estuary, MD	0.7	0.25	Marsho *et al.* (1975)
Kaneohe Bay, HA	1.7	0.62	Hanson and Gundersen (1977)
Average		0.40 ± 0.07	
Seagrass communities			
Thalassia epiphytes and seds, FL	25	9.0	Capone and Taylor (1980a,b)
Zostera sediments, NY	5.5	2.0	Capone (1982)
Average		5.5	

Coral reef environments			
Reef macroalgae, Eniwetok	31	11	Mague and Holm-Hansen (1975)
Reef macroalgae, 3-30 m, Bahamas	5	1.8	Capone (1977)
Reef macroalgae, reef flats, Bahamas	64	23	Capone (1977)
Calothrix mat, Eniwetok	180	66	Wiebe *et al.* (1975)
Calothrix mat, Aldebra	43	16	Potts and Whitton (1977)
Calothrix mat, Eniwetok	6	2.0	Mague and Holm-Hansen (1975)
Nostoc mat, Eniwetok	110	40	Mague and Holm-Hansen (1975)
Scytonema mats, Lizard Island	5	1.9	Burris (1976)
Scytonema mats, Aldabra	168	61	Potts and Whitton (1977)
Average		25 + 8.4	
Mangrove communities			
Rhizosphere			
Mangrove sediments, FL	7.8	2.8	Zuberer and Silver (1978)
Mats			
Cyanobacterial mats, Sinai	23.2	8.4	Potts (1980)
Mangrove total		11.2	
Salt marsh communities			
Rhizosphere			
Salt Marsh, Sapelo Island, GA	142	51	Hanson (1977)
Spartina soil, Nova Scotia	82	15	Patriquin and Denike (1978)
Spartina, Sippewissett, MA	28	10	Teal *et al.* (1979)
Salt Marsh, LA	43	15	DeLaune and Patrick (1980)
Average		23 ± 10	

TABLE V. Annual Areal Estimates of Nitrogen Fixation in Various Marine Communities

Environment	N_2 Fixation		Reference
	mg $N \cdot m^{-2} \cdot day^{-1}$	g $N \cdot m^{-2} \cdot year^{-1}$	
Epiphytes			
Salt marsh, Sapelo Island, GA	0.6	0.2	Hanson (1977)
Surface			
Salt marsh, Sapelo Island, GA	2	0.66	Hanson (1977)
Spartina soil, Nova Scotia	12	2.2	Patriquin and Denike (1978)
Spartina, Sippewissett, MA	2	0.84	Carpenter *et al.* (1978)
Average		1.2 ± 0.5	
Salt marsh total		24.4	

[a] *Areal values calculated assuming sediment density of 0.45 g* cm^{-3}, *a 24-h fixation period, and a 3:1 ratio of* C_2H_2 *reduced to* N_2 *fixed.*

of samples amended with glucose. Subsequent studies, however, using briefer assay periods, have found either negligible (McRoy *et al.*, 1973) or considerably lower (Capone *et al.*, 1979; Capone and Taylor, 1980a,b; Capone, 1982) input of nitrogen through N_2 fixation in seagrass ecosystems. Capone and Taylor (1980a) estimated that about 25-50% of the nitrogen demand of *Thalassia testudinum* may be met by rhizosphere N_2 fixation when recycling is considered. More recently, Capone (1982) found that N_2 fixation in the *Zostera marina* rhizosphere, while detectable, was less important than in *T. testudinum* communities.

Similarly, in salt marsh ecosystems, a wide range of estimates for sediment N_2 fixation has been reported. Using a short-term (2 h) assay, Whitney *et al.* (1975) found N_2 fixation in *Spartina alterniflora* sediments from Flax Pond to be about two- to threefold greater than reported for bare estuarine areas (Brooks *et al.*, 1971; Marsho *et al.*, 1975). In many subsequent studies of salt marsh sediments, workers have often been confronted with extended (10-48 h) lag periods before the onset of measurable nitrogenase activity (Hanson, 1977; Patriquin and Denike, 1978; Teal *et al.*, 1979; Dicker and Smith, 1980c). The *in situ* estimates derived from activity after these lag periods are considerably greater than those of Whitney *et al.* (1975) (Table IV). Based on their studies, Valiela and Teal (1979; Valiela, Chapter 17) have suggested that bacterial N_2 fixation may account for 9-20% of the total nitrogen input in the Great Sippewissett Salt Marsh. Both Haines *et al.* (1976) and Patriquin and McClung (1978) estimate that inputs of nitrogen through N_2 fixation at their study sites (Sapelo Island, Georgia and Nova Scotia, respectively) are roughly equivalent to the accretion rate of nitrogen in plant material. It should be noted that in salt marsh environments, denitrification may considerably exceed N_2 fixation (Valiela and Teal, 1979; Valiela, Chapter 17).

V. ESTIMATE OF TOTAL BENTHIC NITROGEN FIXATION

Several previous efforts aimed at estimating oceanic N_2 fixation (e.g., Burns and Hardy, 1975; Soderlund and Svensson, 1976; Delwiche, 1981; Fogg, 1982) in order to place the marine contribution in a global perspective. In these studies, only limited attempts were made to differentiate contributions in the water column from the sediment environment.

Any extrapolation is limited by the accuracy of the assays of N_2 fixation, as well as the estimation of the areal (or volumetric) extent of each ecosystem component. Areal estimates for the seafloor of the coastal shelf and beyond are generally agreed upon (Reid, 1974). Ironically, there is less precise information on the areal extent of shallower marine regions. This is a result of the dynamic nature of the biota that define many shallow water communities, as well as the physical processes that occur in shallow environments (e.g., sea-level changes, shoreline erosion and modification). Woodwell *et al.* (1973) surveyed the extent of estuaries ($1.36 \times 10^6 km^2$) and intertidal marsh ($0.4 \times 10^6 km^2$) and suggested that their estimates were accurate within 50%. Their calculation of intertidal marsh area combined tropical mangroves with temperate *Spartina* communities. DeVooys (1979) recently estimated the area of coral reefs at $0.11 \times 10^6 km^2$. No estimates of the extent of seagrass communities were found.

Soderlund and Svensson (1976) used the areal estimates of Whittaker and Likens (1973) (apparently derived from the analyses of Woodwell *et al.* 1973) in their approximation of sediment N_2 fixation. They assumed three major regions of active N_2 fixation: coral reefs, estuaries, and the continental shelf. Data on sediment N_2 fixation were limited at the time of their compilation, which estimated benthic N_2 fixation on individual studies of a coral reef, an intertidal, and a subtidal sediment. Using the reported area of coral reefs and (tropical and temperate ?) algal beds given by Whittaker and Likens (1973) of 0.6 × $10^6 km^2$, they scaled up the maximum reported activity (40 g $N \cdot m^{-2} \cdot year^{-1}$) in the report of Mague and Holm-Hanson (1975). For estuaries, they assumed half the area to be comparable in activity to the intertidal zone (35 g $N \cdot m^{-2} \cdot year^{-1}$, calculated from Herbert, 1975) and the remainder of the estuary and the area of the entire continental shelf to have activities similar to those determined by Brooks *et al.* (1971) for estuarine sediments (0.4 g $N \cdot m^{-2} \cdot year^{-1}$). They thus arrived at an upper limit of 60 Tg N $year^{-1}$. Using a more conservative approach by scaling up all estuarine areas based on the value for subtidal sediments, they calculated a lower limit of 10 Tg N $year^{-1}$.

In a recent workshop (Paul, 1978), the annual contribution of N_2 fixation by salt marsh ecosystems was estimated using values of activity from several geographically distinct areas and the estimate of areal coverage of salt marshes (and mangroves) from Woodwell *et al.* (1973): Annual N_2 fixation was calculated to be about 2 Tg.

For the present calculation, I divided the benthic environment into several zones (Table V and VI). Five depth ranges in the open ocean were recognized, from 3000 m to coastal waters shallower than 200 m. The areal estimate of Woodwell *et al.* (1973) for subtidal estuaries was subdivided into vegetated (seagrass) and nonvegetated areas in a 1:4 ratio. Also, I subdivided their areal estimates of intertidal marsh (0.39 × $10^6 km^2$) into a salt marsh and mangrove component at a 2:1 ratio, based on their description of the latitudinal distribution of "marshes." Finally, I used deVooys's (1979) calculation of the global extent of coral reefs.

The areal estimates for annual N_2 fixation in shallow benthic environments were derived by averaging those studies which incorporated a seasonal or annual component (Table V). For benthic communities with several distinct segments, total N_2 fixation was

TABLE VI. Estimate of the Total Annual Contribution of Combined Nitrogen to the Global Nitrogen Cycle by Nitrogen Fixation in the Benthic Environments of the Ocean[a]

Environment	Area ($km^2 \times 10^6$)	N_2 Fixation $g\ N \cdot m^{-2} \cdot year^{-1}$	N_2 Fixation $Tg\ N \cdot year^{-1}$
>3000 m	272	0	0
2000-3000 m	31	0.0007	0.022
1000-2000 m	16	0.001	0.016
200-1000 m	16	0.01	0.16
0-200 m	27	0.1	2.7
Bare estuary	1.08[b]	0.4	0.43
Seagrass	0.28[b]	5.5	1.5
Coral reefs	0.11[c]	25	2.8
Salt marsh	0.26[d]	24	6.3
Mangroves	0.13[d]	11	1.5
Total	363		15.4

[a] *Adapted with permission from Capone and Carpenter (1982b).*

[b] *Areal estimate of estuaries worldwide from Woodwell et al. (1973) and assuming 20% coverage by seagrasses.*

[c] *From DeVooys (1979).*

[d] *Areal estimate of intertidal marshes from Woodwell et al. (1973) and assuming a ratio of mangroves to temperate salt marshes of 1:2.*

derived by summing the contributions of each component (mangrove and salt marsh).

Coral reefs and temperate salt marshes, for which a number of studies have been conducted, had similarly high annual areal rates of N_2 fixation (Table V). Estimates for tropical mangroves, for which there was a limited data base, were also high at about 11 g N $m^{-2} \cdot year^{-1}$. Seagrass communities, both temperate and tropical, averaged about 6 g $N \cdot m^{-2} \cdot year^{-1}$. Annual estimates for estuarine sediments were available from tropical, temperate, and boreal areas and varied over about a twofold range, averaging 0.4 g $N \cdot m^{-2} \cdot year^{-1}$, whereas N_2 fixation in coastal sediments (to 200 m) averaged about 0.1 g $N \cdot m^{-2} \cdot year^{-1}$.

For shallow water ecosystems, I scaled up the calculated averages (Table V) using the appropriate areal values (Table VI). Over the 200 to 1000 m depth range, I assumed a 10-fold lower activity than that in coastal sediments, while for 1000-2000 m, a 100-fold decrease, relative to coastal sediments, was assumed. For 2000-3000 m, a rate was calculated from the work of Hartwig and Stanley (1978) and it was assumed that no substantial N_2 fixation occurred below 3000 m.

Given the above-mentioned procedures, total annual N_2 fixation in the marine benthic environment was calculated at about 15 Tg. This is similar to the lower estimate of Soderlund and Svensson (1976) and is about threefold greater than the estimate for pelagic fixation made by Carpenter (Chapter 3). By the present calculation the largest contribution is made by salt marsh ecosystems (6.3 Tg N·year^{-1}), followed by the coral reefs (2.8 Tg N·year^{-1}) and coastal sediments (2.7 Tg N·year^{-1}). Mangrove and seagrass systems contributed similar amounts of (1.5 Tg N·year^{-1} each), whereas the nonvegetated portions of estuaries and deeper (>200) sediments are responsible for only minor inputs. Summing the estimated inputs of salt marsh, coral reef, seagrass and mangrove ecosystems, it appears that a major fraction (80%) of the N_2 fixation in the oceans occurs in shallow littoral and sublittoral areas that cumulatively account for only about 0.2% of the total benthic area of the seas.

Because of the larger data set available, these estimates are more substantive than previous extrapolations. However, it should be reemphasized that they are constrained by the accuracy of the estimate of N_2 fixation (see Section I,B) and uncertainties in the areal extent of each subdivision of the benthic zone, especially those in shallow water ecosystems. One should hope for, and certainly expect, a refinement of this approximation in the future.

VI. CONCLUSIONS AND RECOMMENDATIONS

A much better understanding of benthic N_2 fixation has been gained as a result of the intensified efforts over the last decade. Several ecosystems have been identified in which this activity is substantial and probably contributes a major fraction of the nitrogen demand. For salt marsh, coral reef, mangrove, and seagrass ecosystems, further work is in order to define more precisely the regional generality and quantitative magnitude of N_2 fixation. This is particularly important for seagrass and mangrove communities for which there is a paucity of information. Complementary efforts aimed at identifying other major transformational pathways of nitrogen should be encouraged.

For the largest areas of the seas, only minor inputs are suggested in the present approximation. These are based, however, on a very limited number of observations and hence require further work to confirm or refute this notion. The potential for significant input through N_2 fixation in the deep sea in association with the indigenous fauna and in the vicinity of hydrothermal vents should be further explored.

On the horizon, work is proceeding in a number of laboratories exploiting the ability to discern small natural differences in $^{14}N:^{15}N$ ratios within various ecosystems. This approach potentially may provide independent information on environmental N_2 fixation (Delwiche *et al.*, 1979). Insights gained using natural isotope distributions should complement ongoing field efforts.

A more critical application of existing techniques is in order for all future studies. Because of the diversity of sample types encountered in benthic environments, a true standardization of methodologies (e.g., C_2H_2 reduction) would be difficult to achieve. However, general criteria for determining *in situ* nitrogenase activity (as well as other activities in the sediments) need be considered by workers in the field, with the aim of more rigorously defining and establishing a basis for comparison of existing and developing data sets.

ACKNOWLEDGMENTS

The author kindly acknowledges the valuable comments of E. J. Carpenter, R. Hanson, M. Scranton, D. Smith, and B. F. Taylor on this chapter. Financial assistance was provided by NSF Grants OCE 78-25444 and 82-00157 from Biological Oceanography, NOAA Grants NA-80-RAD-00057 from the Office of Marine Pollution Assessment, Grant 04715844009 from N.Y. State Sea Grant, and EPA Grant R-809475-01-0.

REFERENCES

Balderston, W., Sherr, B., and Payne, W. (1976). Blockage by acetylene of nitrous oxide reduction in *Pseudomonas perfectomarinus*. *Appl. Environ. Microbiol. 31,* 504-508.

Bohlool, B. B., and Wiebe, W. J. (1978). Nitrogen-fixing communities in an intertidal ecosystem. *Can. J. Microbiol. 24,* 932-938.

Bothe, H., Klein, B., Stephan, M. P., and Dobereiner, J. (1981). Transformations of inorganic nitrogen by *Azospirillum* spp. *Arch. Microbiol. 130,* 96-100.

Boyle, C. D., and Patriquin, D. G. (1980). Endorhizal and exorhizal acetylene-reducing activity in a grass (*Spartina alterniflora* Loisel) diazotroph association. *Plant Physiol. 66,* 267-280.

Bremner, J., and Blackmer, A. (1979). Effects of acetylene and soil water content on emission of nitrous oxide from soils. *Nature (London) 280,* 330-331.

Brill, W. J. (1980). Biochemical genetics of nitrogen fixation. *Microbiol. Rev. 44,* 449-467.

Brooks, R. H., Brezonik, P. L., Putnam, H. D., and Keirn, M. A. (1971). Nitrogen fixation in an estuarine environment: The Waccasassa on the Florida coast. *Limnol. Oceanogr. 16,* 701-710.

Brouzes, R., and Knowles, R. (1971). Inhibition of growth of *Clostridium pasteurianum* by acetylene: Implication for nitrogen fixing assay. *Can. J. Microbiol. 17,* 1483-1489.

Burns, R., and Hardy, R. (1975). "Nitrogen Fixation in Bacteria and Higher Plants." Springer-Verlag, Berlin and New York.

Burris, R. H. (1974). Methodology. *In* "The Biology of Nitrogen Fixation" (A. Quispel, ed.), pp. 9-33. North-Holland Publ., Amsterdam.

Burris, R. H. (1976). Nitrogen fixation by blue-green algae of the Lizard Island area of the Great Barrier Reef. *Aust. J. Plant Physiol. 3,* 41-51.

Capone, D. G. (1977). N_2 (C_2H_2) fixation by macroalgal epiphytes. *Proc. Int. Coral Reef Symp. 3rd, 1970,* Vol. 1, pp. 337-342.

Capone, D. G. (1982). Nitrogen fixation (acetylene reduction) by rhizosphere sediments of the eelgrass, *Zostera marina* L. *Mar. Ecol.: Prog. Ser. 10,* 67-75.

Capone, D. G., and Budin, J. (1982). Nitrogen fixation associated with rinsed roots and rhizomes of the eelgrass, *Zostera marina*. *Plant Physiol. 70,* 1601-1604.

Capone, D. G., and Carpenter, E. J. (1982a). Perfusion method for assaying microbial activities in sediments: Applicability to studies of N_2 fixation by C_2H_2 reduction. *Appl. Environ. Microbiol. 43,* 1400-1405.

Capone, D. G., and Carpenter, E. J. (1982b). Nitrogen fixation in the marine environment. *Science 217,* 1140-1142.

Capone, D. G., and Taylor, B. F. (1977). Nitrogen fixation (acetylene reduction) in the phyllosphere of *Thalassia testudinum*. *Mar. Biol. (Berlin) 40,* 19-28.

Capone, D. G., and Taylor, B. F. (1980a). N_2 fixation in the rhizosphere of *Thalassia testudinum*. *Can. J. Microbiol. 26,* 998-1005.

Capone, D. G., and Taylor, B. F. (1980b). Microbial nitrogen cycling in a seagrass community. *In* "Estuarine Perspectives" (V. S. Kennedy, ed.), pp. 153-161. Academic Press, New York.

Capone, D. G., Taylor, D. L., and Taylor, B. F. (1977). Nitrogen-fixation (acetylene-reduction) associated with macroalgae in a coral reef community in the Bahamas. *Mar. Biol. 40,* 29-32.

Capone, D. G., Penhale, P., Oremland, R., and Taylor, B. F. (1979). Relationship between productivity and N_2 (C_2H_2) fixation in *Thalassia testudinum* community. *Limnol. Oceanogr. 24,* 117-125.

Carpenter, E. J., and Culliney, J. (1975). Nitrogen fixation in marine shipworms. *Science 187,* 551-552.

Carpenter, E. J., VanRaalte, C. D., and Valiela, I. (1978). Nitrogen fixation by algae in a Massachusetts salt marsh. *Limnol. Oceanogr. 23,* 318-327.

Culbertson, C., Zehnder, A., and Oremland, R. (1981). Anaerobic oxidation of acetylene by estuarine sediments and enrichment cultures. *Appl. Environ. Microbiol. 41,* 396-403.

Dalton, H., and Whittenbury, R. (1976). The acetylene reduction technique as an assay for nitrogenase activity in the methane oxidizing bacterium *Methylococcus capsulatus* strain Bath. *Arch. Microbiol. 109,* 147-151.

David, K., and Fay, P. (1977). Effects of long-term treatment with acetylene on nitrogen-fixing microorganisms. *Appl. Environ. Microbiol. 34,* 640-646.

DeBont, J. A. M. (1976). Bacterial degradation of ethylene and the acetylene reduction test. *Can. J. Microbiol. 22,* 1060-1062.

DeBont, J. A. M., and Mulder, E. G. (1974). Nitrogen fixation and co-oxidation of ethylene by a methane-utilizing bacterium. *J. Gen. Microbiol. 83,* 113-121.

DeBont, J. A. M., and Peck, M. (1980). Metabolism of acetylene by *Rhodococcus* Al. *Arch. Microbiol. 127,* 99-104.

DeLaune, R., and Patrick, W. (1980). Nitrogen and phosphorus cycling in a Gulf coast salt marsh. *In* "Estuarine Perspectives" (V. Kennedy, ed.), pp. 143-151. Academic Press, New York.

Delwiche, C. C. (1981). The nitrogen cycle and nitrous oxide. *In* "Denitrification, Nitrification and Atmospheric Nitrous Oxide" (C. C. Delwiche, ed.), pp. 1-15. Wiley (Interscience), New York.

Delwiche, C. C., Zinke, P. J., Johnson, C. M., and Virginia, R. A. (1979). Nitrogen isotope distribution as a presumptive indicator of nitrogen fixation. *Bot. Gaz. (Chicago) 140,* S65-S69.

DeVooys, C. G. (1979). Primary production in aquatic environments. *In* "The Global Carbon Cycle" (B. Bolin, E. Degans, S. Kempe, and P. Ketner, eds.), pp. 259-292. Wiley, New York.

Dicker, H. J., and Smith, D. W. (1980a). Enumeration and relative importance of acetylene-reducing (nitrogen-fixing) bacteria in a Delaware salt marsh. *Appl. Environ. Microbiol. 39,* 1019-1025.

Dicker, H. J., and Smith, D. W. (1980b). Physiological ecology of acetylene reduction (nitrogen fixation) in a Delaware salt marsh. *Microb. Ecol. 6,* 161-171.

Dicker, H. J., and Smith, D. W. (1980c). Acetylene reduction (nitrogen fixation) in a Delaware, USA salt marsh. *Mar. Biol. (Berlin) 57,* 241-250.

Evans, H., and Barber, L. (1977). Biological nitrogen fixation for food and fiber production. *Science 197,* 332-339.

Fenchel, T., and Blackburn, T. (1979). "Bacteria and Mineral Cycling." Academic Press, New York.

Flett, R. J., Rudd, J. W. M., and Hamilton, R. D. (1975). Acetylene reduction assays for nitrogen fixation in freshwaters: A note of caution. *Appl. Microbiol. 29,* 580-583.

Flett, R., Hamilton, R., and Campbell, N. (1976). Aquatic acetylene-reduction techniques: Solutions to several problems. *Can. J. Microbiol. 22,* 43-51.

Fogg, G. E. (1982). Nitrogen cycling in sea water. *Philos. Trans. R. Soc. London Ser. B, 296,* 511-520.

Goering, J. J., and Parker, P. L. (1972). Nitrogen fixation by epiphytes on sea grasses. *Limnol. Oceanogr. 17,* 320-323.

Gotto, J. W., and Taylor, B. F. (1976). N_2-fixation associated with decaying leaves of the red mangrove (*Rhizophora mangle*). *Appl. Environ. Microbiol. 31,* 781-783.

Gotto, J. W., Tabita, F. R., and vanBaalen, C. (1981). Nitrogen fixation in intertidal environments of the Texas gulf coast. *Estuarine Coastal Shelf Sci. 12,* 231-235.

Guérinot, M., and Patriquin, D. (1981). N_2-fixing vibrios isolated from the gastrointestinal tract of sea urchins. *Can. J. Microbiol. 27,* 311-317.

Haines, E., Chalmers, A., Hanson, R., and Sherr, B. (1976). Nitrogen pools and fluxes in a Georgia salt marsh. *In* "Estuarine Processes" (M. Wiley, ed.), Vol. 2, pp. 241-254. Academic Press, New York.

Haines, J. R., Atlas, R., Griffiths, R., and Morita, R. (1981). Denitrification and nitrogen fixation in Alaskan continental shelf sediments. *Appl. Environ. Microbiol. 41,* 412-421.

Hanson, R. B. (1977). Nitrogen fixation (acetylene reduction) in a salt marsh amended with sewage sludge and organic carbon and nitrogen compounds. *Appl. Environ. Microbiol. 33,* 846-852.

Hanson, R. B., and Gundersen, K. (1976). Influence of sewage discharge on nitrogen fixation and nitrogen flux from coral reefs in Kaneohe Bay, Hawaii. *Appl. Environ. Microbiol. 31,* 942-948.

Hanson, R. B., and Gundersen, K. (1977). Relationship between nitrogen fixation (acetylene reduction) and the C:N ratio in a polluted coral reef ecosystem, Kaneohe Bay, Hawaii. *Estuarine Coastal Mar. Sci. 5,* 437-444.

Hanson, R. B., Tenore, K. R., Bishop, S., Chamberlain, C., Pamatmat, M. M., and Tietjen, J. (1981). Benthic enrichment in the Georgia Bight related to Gulf Stream intrusions and estuarine outwelling. *J. Mar. Res. 39,* 417-441.

Hardy, R. W. F., and Havelka, U. D. (1975). Nitrogen fixation research: A key to world food? *Science 188,* 633-643.

Hardy, R. W. F., Holsten, R. D., Jackson, E. K., and Burns, R. C. (1968). The acetylene-ethylene assay for N_2 fixation: Laboratory and field evaluation. *Plant Physiol. 43,* 1185-1207.

Harrison, P. G., and Mann, K. H. (1975). Chemical changes during the seasonal cycle of growth and decay in eelgrass (*Zostera marina*) on the Atlantic coast of Canada. *J. Fish. Res. Board Can. 32,* 615-621.

Hartwig, E. O., and Stanley, S. O. (1978). Nitrogen fixation in Atlantic deep-sea and coastal sediments. *Deep-Sea Res. 25,* 411-417.

Head, W. D., and Carpenter, E. J. (1975). Nitrogen fixation associated with the marine macroalga *Codium fragile*. *Limnol. Oceanogr. 20,* 815-823.

Herbert, R. A. (1975). Heterotrophic nitrogen fixation in shallow estuarine sediments. *J. Exp. Mar. Biol. Ecol. 18,* 215-225.

Howarth, R. W., and Teal, J. M. (1979). Sulfate reduction in a New England salt marsh. *Limnol. Oceanogr. 24,* 999-1013.

Hynes, R., and Knowles, R. (1978). Inhibition by acetylene of ammonia oxidation in *Nitrosomonas europaea*. *FEMS Microbiol. Lett. 4,* 319-321.

Jones, K. (1974). Nitrogen fixation in a salt marsh. *J. Ecol. 62,* 553-565.

Jones, K., and Stewart, W. D. P. (1969). Nitrogen turnover in marine and brackish habitats. IV. Uptake of the extracellular products of the nitrogen-fixing alga *Calothrix scopularum*. *J. Mar. Biol. Assoc. U.K. 49,* 701-716.

Jorgensen, B. B. (1982). Mineralization of organic matter in the sea bed--The role of sulphate reduction. *Nature (London) 296,* 643-645.

Keutner, J. (1904). Uber das Vorkommen und die Verbreitung stickstoffbendender Bakterien im Meere. *Wiss. Meeresunters. Abt. Kiel* [N.S.] *8,* 1-29.

Knowles, R., and Wishart, C. (1977). Nitrogen fixation in arctic marine sediments: Effect of oil and hydrocarbon fractions. *Environ. Pollut. 13,* 133-149.

Koike, I., and Hattori, A. (1978). Denitrification and ammonia formation in anaerobic coastal sediments. *Appl. Environ. Microbiol. 35,* 278-282.

McClung, C. R., and Patriquin, D. G. (1980). Isolation of a nitrogen-fixing *Campylobacter* species from the roots of *Spartina alterniflora* Loisel. *Can. J. Microbiol. 26,* 881-886.

McRoy, C. P., Goering, J. J., and Chaney, B. (1973). Nitrogen fixation associated with sea grasses. *Limnol. Oceanogr. 18,* 998-1002.

Mague, T., and Holm-Hansen, O. (1975). Nitrogen fixation on a coral reef. *Phycologia 14,* 87-92.

Marsho, T. V., Burchard, R. P., and Fleming, R. (1975). Nitrogen fixation in Rhode River estuary of Chesapeake Bay. *Can. J. Microbiol. 21,* 1348-1356.

Nedwell, D., and Aziz, S. (1980). Heterotrophic nitrogen fixation in an intertidal salt marsh sediment. *Estuarine Coastal Mar. Sci. 10,* 699-702.

Neyra, C. A., Dobereiner, J., Lalande, R., and Knowles, R. (1977). Denitrification by N_2 fixing *Spirillum lipoferum. Can. J. Microbiol. 23,* 300-305.

Odum, H., and Odum, E. (1955). Trophic structure and productivity of a windward coral reef community of Eniwetok Atoll. *Ecol. Monogr. 25,* 291-320.

Oppenheimer, C. H. (1960). Bacterial activity in sediments of shallow marine bays. *Geochim. Cosmochim. Acta 19,* 244-260.

Oremland, R. S., and Taylor, B. F. (1975). Inhibition of methanogenesis in marine sediments by acetylene and ethylene: validity of the acetylene reduction assay for anaerobic microcosms. *Appl. Microbiol. 30,* 707-709.

Patriquin, D. G. (1972). The origin of nitrogen and phosphorus for growth of the marine angiosperm *Thalassia testudinum. Mar. Biol. (Berlin) 15,* 35-46.

Patriquin, D. G. (1978a). Factors affecting nitrogenase activity (acetylene reducing activity) associated with excised roots of the emergent halophyte *Spartina alterniflora* Loisel. *Aquat. Bot. 4,* 193-210.

Patriquin, D. G. (1978b). Nitrogen fixation (acetylene reduction) associated with cord grass, *Spartina alterniflora* Loisel. *Ecol. Bull. 26,* 20-27.

Patriquin, D. G., and Denike, D. (1978). *In situ* acetylene reduction assays of nitrogenase activity associated with the emergent halophyte *Spartina alterniflora* Loisel: Methodological problems. *Aquat. Bot. 4*, 211-226.

Patriquin, D. G., and Knowles, R. (1972). Nitrogen fixation in the rhizosphere of marine angiosperms. *Mar. Biol. (Berlin) 16*, 49-58.

Patriquin, D. G., and Knowles, R. (1975). Effects of oxygen, mannitol and ammonium concentrations on nitrogenase (C_2H_2) activity in a marine skeletal carbonate sand. *Mar. Biol. (Berlin) 32*, 49-62.

Patriquin, D. G., and McClung, C. (1978). Nitrogen accretion, and the nature and possible significance of N_2 fixation (acetylene reduction) in a Nova Scotian *Spartina alterniflora* stand. *Mar. Biol. (Berlin) 47*, 227-242.

Paul, E. A. (1978). Contribution of nitrogen fixation to ecosystem functioning and nitrogen fluxes on a global basis. *Ecol. Bull. 26*, 282-293.

Payne, W. J., and Grant, M. A. (1982). Influence of acetylene on growth of sulfate-respiring bacteria. *Appl. Environ. Microbiol. 43*, 727-730.

Penhale, P., and Capone, D. G. (1981). Primary productivity and nitrogen fixation in two macroalgae-cyanobacteria associations. *Bull. Mar. Sci. 31*, 164-169.

Peters, G. (1978). Blue-green algae and algal associations. *BioScience 28*, 580-585.

Postgate, J. R. (1979). "The Sulfate-Reducing Bacteria." Cambridge Univ. Press, London and New York.

Potts, M. (1980). Blue-green algae (Cyanophyta) in marine coastal environments of the Sinai Peninsula; distribution, zonation, stratification and taxonomic diversity. *Phycologia 19*, 60-73.

Potts, M., and Whitton, B. A. (1977). Nitrogen fixation by blue-green algal communities in the intertidal zone of the lagoon of Aldabra Atoll. *Oecologia 27*, 275-283.

Potts, M., Krumbein, W. E., and Metzger, J. (1978). Nitrogen fixation rates in anaerobic sediments determined by acetylene reduction, a new ^{15}N field assay and simultaneous total N determination. *In* "Environmental Biogeochemistry and Geomicrobiology" (W. Krumbein, ed.), Vol. 3, pp. 753-769. Ann Arbor Sci. Publ., Ann Arbor, Michigan.

Pshenin, L. N. (1963). Distribution and ecology of *Azotobacter* in the Black Sea. *In* "Marine Microbiology" (C. H. Oppenheimer, ed.), pp. 383-391. Thomas, Springfield, Illinois.

Rau, G. H. (1981a). Low $^{15}N/^{14}N$ in hydrothermal vent animals: Ecological implications. *Nature (London) 289*, 484-485.

Rau, G. H. (1981b). Hydrothermal vent clam and tube worm $^{13}C/^{12}C$: Further evidence of nonphotosynthetic food source. *Science 213*, 338-340.

Reid, J. L. (1974). Physical oceanography. *In* "Handbook of Marine Science" (F. G. Walton Smith, ed.), pp. 71-174. CRC Press, Cleveland, Ohio.

Revsbech, N. P., Sorensen, J., Blackburn, T., and Lomholt, J. (1980). Distribution of oxygen in marine sediments measured with microelectrodes. *Limnol. Oceanogr. 25*, 403-411.

Rivera-Ortiz, J. M., and Burris, R. H. (1975). Interactions among substrates and inhibitors of nitrogenase. *J. Bacteriol. 123*, 537-545.

Rosenberg, G., and Paerl, H. (1981). Nitrogen fixation by blue-green algae associated with the siphonous green seaweed *Codium decorticatum*: Effects on ammonium uptake. *Mar. Biol. (Berlin) 61*, 151-158.

Rowe, G. T., Clifford, C. H., and Smith, K. L., Jr. (1975). Benthic nutrient regeneration and its coupling to primary productivity in coastal waters. *Nature (London) 255*, 215-217.

Ryther, J. H., and Dunstan, W. M. (1971). Nitrogen, phosphorus and eutrophication in the coastal marine environment. *Science 171*, 1008-1012.

Schubert, K. R., and Evans, H. J. (1976). Hydrogen evolution a major factor affecting the efficiency of nitrogen fixation in nodulated symbionts. *Proc. Natl. Acad. Sci. USA 73*, 1207-1211.

Shanmugam, K. T., O'Gara, F., Anderson, K., and Valentine, R. C. (1978). Biological nitrogen fixation. *Annu. Rev. Plant Physiol. 29*, 263-276.

Smith, D. W. (1980). An evaluation of marsh nitrogen-fixation. *In* "Estuarine Perspectives" (V. S. Kennedy, ed.), pp. 135-142. Academic Press, New York.

Smith, G. W., and Hayasaka, S. S. (1982). Nitrogenase activity associated with *Zostera marina* from a North Carolina estuary. *Can. J. Microbiol. 28*, 448-451.

Smith, L. A., Hill, S., and Yates, M. (1976). Inhibition by acetylene of conventional hydrogenase in nitrogen-fixing bacteria. *Nature (London) 262*, 209-210.

R Soderlund, R., and Svensson, B. H. (1976). The global nitrogen cycle. *Ecol. Bull. 22*, 23-74.

Sorensen, J. (1978). Capacity for denitrification and reduction of nitrate to ammonia in a coastal marine sediment. *Appl. Environ. Microbiol. 35*, 301-305.

Sorensen, J., Jorgensen, B. B., and Revsbech, N. P. (1979). A comparison of oxygen, nitrate and sulfate respiration in coastal marine sediments. *Microb. Ecol. 5*, 105-115.

R Stewart, W. D. P. (1965). Nitrogen turnover in marine and brackish habitats. I. Nitrogen fixation. *Ann. Bot. (London)* [N.S.] *29*, 229-239.

Stewart, W. D. P. (1967). Nitrogen turnover in marine and brackish habitats. II. Use of ^{15}N in measuring nitrogen fixation in the field. *Ann. Bot. (London)* [N.S.] *31*, 385-407.

Stewart, W. D. P. (1971). Nitrogen fixation in the sea. *In* "Fertility of the Sea" (J. D. Costlow, Jr., ed.), Vol. 2, pp. 537-564. Gordon Breach, New York.

Stewart, W. D. P., Fitzgerald, G. P., and Burris, R. H. (1967). *In situ* studies on N_2 fixation using the acetylene reduction technique. *Proc. Natl. Acad. Sci. USA 58*, 2071-2078.

Striecher, S. L., Shanmugan, K. T., Ausubel, F., Morandi, C., and Goldberg, R. B. (1974). Regulation of nitrogen fixation in *Klebsiella pneumoniae*: Evidence for a role of glutamine synthesis as a regulator of nitrogenase synthesis. *J. Bacteriol. 120,* 815-821.

Sylvester-Bradley, R. (1976). Isolation of acetylene-reducing spirilla from the roots of *Potamogeton filiformis* from Loch Leven (Kinross). *J. Gen. Microbiol. 97,* 129-132.

Teal, J. M., Valiela, I., and Berlo, D. (1979). Nitrogen fixation by rhizosphere and free-living bacteria in salt marsh sediments. *Limnol. Oceanogr. 24,* 126-132.

Thayer, G. W., Engel, D. W., and LaCroix, M. W. (1977). Seasonal distribution and changes in the nutritive quality of living, dead and detrital fractions of *Zostera marina* L. *J. Exp. Mar. Biol. Ecol. 30,* 109-127.

Valiela, I., and Teal, J. M. (1979). The nitrogen budget of a salt marsh ecosystem. *Nature (London) 280,* 652-656.

Van Berkum, P., and Bohlool, B. (1980). Evaluation of nitrogen fixation by bacteria in association with roots of tropical grasses. *Microbiol. Rev. 44,* 491-517.

Van Berkum, P., and Sloger, C. (1979). Immediate acetylene reduction by excised grass roots not previously incubated at low oxygen tension. *Plant Physiol. 64,* 739-743.

Waksman, S. A., Hotchkiss, M., and Carey, C. L. (1933). Marine bacteria and their role in the cycle of life in the sea. II. Bacteria concerned in the cycle of nitrogen in the sea. *Biol. Bull. (Woods Hole, Mass.) 65,* 137-167.

Warmling, P. (1973). Nitrogen fixation on rocks in Oslofjord. *Bot. Mar. 16,* 237-240.

Wetzel, R. G., and Penhale, P. A. (1979). Transport of carbon and excretion of dissolved organic carbon by leaves and root/rhizomes in seagrass and their epiphytes. *Aquat. Bot. 6,* 149-158.

Whitney, D., Woodwell, G., and Howarth, R. (1975). Nitrogen fixation in Flax Pond: a Long Island salt marsh. *Limnol. Oceanogr. 20,* 640-643.

Whittaker, R. H., and Likens, G. E. (1973). Carbon in the biota. *In* "Carbon and the Biosphere" (G. Woodwell and E. Pecan, eds.), Natl. Tech. Inf. Serv. CONF-720510, pp. 281-302. U.S. Govt. Printing Office, Washington, D.C.

Wiebe, W. J., Johannes, R. E., and Webb, K. L. (1975). Nitrogen fixation in a coral reef community. *Science 188,* 257-259.

Wilkinson, C., and Fay, P. (1979). Nitrogen fixation in coral reef sponges with symbiotic cyanobacteria. *Nature (London) 279,* 527-529.

Woodwell, G., Rich, P. H., and Hall, C. (1973). Carbon in estuaries. "Carbon and the Biosphere" (G. Woodwell and E. Pecan, eds.), Natl. Tech. Inf. Serv. CONF-720510, pp. 221-240. U.S. Govt. Printing Office, Washington, D.C.

Yayanos, A. A., Dietz, A. S., and vanBoxtel, R. (1979). Isolation of a deep-sea barophilic bacterium and some of its growth characteristics. *Science 205,* 808-810.

Yoshinari, T., and Knowles, R. (1976). Acetylene inhibition of reduction of nitrous oxide by denitrifying bacteria. *Biochem. Biophys. Res. Commun.* *69*, 705-710.

Zuberer, D. A., and Silver, W. S. (1978). Biological dinitrogen fixation (acetylene reduction) associated with Florida mangroves. *Appl. Environ. Microbiol.* *35*, 567-575.

Chapter 5

Nitrification

WARREN A. KAPLAN
Center for Earth and Planetary Physics
Harvard University
Cambridge, Massachusetts

Nothing of him that doth fade
But doth suffer a sea-change
Into something rich and strange...
The Tempest, William Shakespeare

I. INTRODUCTION

In the marine environment it has been repeatedly demonstrated that nitrogen is a limiting resource for the photosynthetic production of organic carbon (Eppley et al., 1979; Caperon and Meyer, 1972; McCarthy and Goldman, 1979). Nitrogen is an essential component of all living materials, and to a large extent its availability, through transfer between various N reservoirs, can regulate the cycle of organic matter in the world's oceans and estuaries.

ISBN 0-12-160280-X

The most prevalent form of nitrogen among the global nitrogen reservoirs is gaseous nitrogen (McElroy, 1975). Dissolved nitrogen gas in the oceans is about 30 times more abundant than the sum of its inorganic forms (ammonium, nitrite, or nitrate). However, gaseous dinitrogen is relatively inert and must be converted into more readily available forms (NH_4^+, NO_2^-, NO_3^-) before it can be used by organisms. Dugdale and Goering (1967) separated nitrogen limitation of phytoplankton in the photic zone into those fractions supported by "regenerated" forms such as ammonium or dissolved organic nitrogen and "new" forms such as gaseous nitrogen or nitrate. In their model, nitrate is produced in the deep water reservoir by microbial transformation of organic matter and ultimately mixed back into the photic zone. Inspection of the vertical distribution of nitrate in the sea (Vaccaro, 1965; Sharp, Chapter 1, this volume) supports the validity of this two-layer system. The source of this "new" nitrate-nitrogen (NO_3^--N) is the sequential oxidation of reduced inorganic nitrogen to nitrite and nitrate. This nitrate-based new production accounts for 10-20% of the total global marine carbon production (Eppley and Peterson, 1979). Other inputs of new nitrogen are N_2 fixation and riverine NO_3^- fluxes. Marine N_2 fixation is not well quantified and although it has been considered to be small compared to N_2 fixation in terrestrial environments (Carpenter and McCarthy, 1975; Delwiche, 1970; Mague et al., 1979), recent estimates (Carpenter, Chapter 3, Capone, Chapter 4, this volume) suggest that this may not be so.

Our knowledge of river-borne transport of nitrogenous nutrients is even less complete. A recent estimate of global NO_3^- runoff (Walsh et al., 1981) of 2×10^{13} g NO_3^- - N year^{-1} suggests, given the marine NO_3^- reservoir of 6×10^{17} g NO_3^- - N (McElroy, 1975), a "residence time" of the order 3×10^4 years, which is 30 times longer than the mean residence time of deep water (Broecker, 1974). This calculation suggests that *in situ* generative processes are responsible for the appearance of oxidized nitrogen in the sea. The mechanism, site(s), and rates of its formation are clearly of great biological and geochemical interest.

By the late 1920s the basic transformations of the marine N cycle were fairly well described (Harvey, 1927), although the microbial populations responsible for them and their locations were a matter of some dispute. Pioneering work by Waksman and co-workers (1933a,b; Carey and Waksman, 1934) suggested that marine nitrifying activity was primarily confined to the sediments rather than the water column. Carey (1937) found that raw seawater "apparently cannot produce nitrite or nitrate" during 52-day incubations. On the other hand, Rakestraw (1936) recognized the intermediate position of nitrite in the nitrification reaction and, in his important study of nitrite distributions in the Gulf of Marine, suggested that nitrification in the water column was responsible for the observed NO_2^- and NO_3^-. Although he was unclear about the exact processes leading to its formation and destruction, he also investigated the now well-known primary nitrite maximum (see Section V, this chapter) lying at the base of the photic zone. The mechanism(s) for its formation have been the subject of research for many years (Vaccaro and Ryther, 1960; Carlucci et al., 1970; Hattori and Wada, 1971; Olson, 1981b). The significance of nitrification in marine sediments is largely unknown (see Section V).

During the past 10-15 years, several apparently contradictory facts have emerged regarding nitrification in the marine environment. Laboratory experiments (Carlucci and Strickland, 1968; Watson, 1965; Yoshida, 1967) have shown that marine nitrifying bacteria grow and oxidize their substrates very slowly. While pure cultures of marine nitrifiers show a fairly low affinity for their substrates (see Section II,A), these levels are often environmentally unrealistic. It is equally clear that, given the preponderance of NO_3^- relative to NH_4^+ and NO_2^- in oceanic waters below the thermocline, nitrification must be complete and very efficient indeed, even at the low levels of NH_4^+ and NO_2^- found in the sea.

This chapter will, in part, be an attempt to focus on this apparent paradox of nitrification in the sea by summarizing recent relevant information concerning the environmental physiology of marine nitrifiers and the rates of activities of nitrification in the water column and sediments of marine and estuarine ecosystems.

II. PHYSIOLOGICAL ECOLOGY OF MARINE NITRIFYING BACTERIA

A. Growth and Substrates

The biochemistry and physiology of nitrification has been covered in a number of reviews (Painter, 1970; Aleem, 1970; Kelly, 1971; Suzuki, 1974; Hooper, 1978). Most of the information regarding environmental effects comes from the sewage and/or soils literature (see above reviews, plus Focht and Chang, 1975; Focht and Verstraete, 1977; Belser, 1979), and only a few reports have dealt specifically with marine nitrifiers.

Nitrifying bacteria are chemolithotrophs (Kelly, 1971) and can obtain all the energy necessary for growth and carbon assimilation from the aerobic oxidation of ammonium to nitrite or nitrite to nitrate. In the marine environment, two genera of bacteria, *Nitrosomonas* and *Nitrosococcus*, mediate the first step in nitrification ($NH_4^+ \rightarrow NO_2^-$) while a second set, the *Nitrobacter*, *Nitrospira*, and *Nitrococcus* group, oxidize NO_2^- to NO_3^- (Table I, adapted from Schmidt, 1978). Attempts to isolate ammonium-oxidizing bacteria from a variety of marine environments have resulted in mostly *Nitrosomonas*-like strains, with *Nitrosococcus* types much less prevalent (Ward et al., 1982). The extent to which these two genera represent the entire suite of marine ammonium oxidizers is unknown. Watson and Waterbury (1971) obtained over 200 enrichment cultures of marine nitrite oxidizers and found that *Nitrobacter* species predominated. In terrestrial environments, *Nitrobacter* also appears to be the principal genus detected. As Belser and Schmidt (1978) point out, "virtually nothing is

TABLE I. Marine Nitrifying Bacteria

Genus	*Species*	*Comments*
		Ammonium to nitrite
Nitrosomonas[a]		Straight rods; cytomembranes on periphery of cell; 5°-30°C growth range; 2°-30°C optimal; 47.4-51% G + C ratio; when motile, one or two subpolar flagella.
Nitrosococcus	*oceanus*[b]	Spherical; cytomembranes in central region of cell; 2°-30°C growth range; 25°-30°C optimal; 50.5-51% G + C ratio; when motile, single or tuft of peritrichous flagella; obligate Na requirement.
	mobilis[c]	Spherical; cytomembranes on periphery of cell; 5°-40°C growth range; 25°-30°C optimal; polar tuft of flagella; obligate Na requirement.
		Nitrite to Nitrate
Nitrobacter	*winogradskyi*[d]	Short rods; cytomembranes on polar cap; 5°-40°C growth range; 25°-30°C optimal; 60.7-61% G+C ratio.
Nitrospina	*gracilis*[e]	Straight rods; no cytomembrane system; 14°-40°C growth range; 25°-30°C optimal; 57.7% G + C ratio; nonmotile; grows in 70-100% seawater.
Nitrococcus	*mobilis*[e]	Spherical; cytomembranes as branched tubes in cytoplasm; 14°-40°C growth range; 25°-30°C optimal; 61.2% G + C ratio; grows in 70-100% seawater.

[a]*The species of marine Nitrosomonas has not yet been identified.* [b]*Watson (1965), formerly called Nitrosocystis oceanus.* [c]*Koops et al. (1976).* [d]*N. winogradskyi contains at least two serotypes (Schmidt, 1968).* [e]*Watson and Waterbury (1971).*

known of the genera that comprise a given nitrifying population." They were speaking of soil systems, but this applies to the marine environment as well. In view of the fact that marine nitrification supports such a large percentage of global primary production (Eppley and Peterson, 1979), the brevity of Table I is striking.

During growth by nitrifying bacteria at seawater pH, the reactions for the first step in nitrification are

$$NH_4^+ + 1.5\ O_2 \xrightarrow{6e^-} NO_2^- + H_2O + 2\ H^+$$

$$\Delta G = -\ 65\ \text{kcal mol N}^{-1} \qquad (1)$$

which occurs in at least two steps:

$$NH_4^+ + 0.5\ O_2 \xrightarrow{2e^-} NH_2OH + H^+$$

$$\Delta G = +\ 4.0\ \text{kcal mol N}^{-1} \qquad (2)$$

$$NH_2OH + O_2 \xrightarrow{4e^-} NO_2^- + H_2O + H^+$$

$$\Delta G = -\ 69\ \text{kcal mol N}^{-1} \qquad (3)$$

Hydroxylamine does not usually accumulate in nature. It is unstable in aqueous solutions and rapidly decomposes to N_2, N_2O, or NH_4^+ (von Breymann et al., 1982; Rajendran and Venugopalan, 1976). The oxidation of nitrite to nitrate is a simple two-electron transfer, with molecular oxygen as the terminal electron acceptor.

$$NO_2^- + 0.5\ O_2 \xrightarrow{2e^-} NO_3^-$$

$$\Delta G = -\ 18.18\ \text{kcal mol N}^{-1} \qquad (4)$$

Differences in ΔG by various authors are based on differing assumptions in these calculations (Painter, 1970).

During nitrification, carbon dioxide is reduced via the reductive pentose phosphate (Calvin) cycle (Gottschalk, 1979). The electron potentials of the inorganic compounds oxidized by nitrifiers, NH_2OH^+/NO_2^- (+ 0.45 V) and NO_2^-/NO_3^- (+ 0.42 V), are much more positive than the NAD/NADH couple (- 0.32 V), and so the generation of reduced pyridine nucleotides coupled to the oxidation of NH_4^+, NH_2OH, or NO_2^- is thermodynamically improbable (Wallace and Nicholas, 1969). There is an *energy-requiring* reversal of electron transfer from the respiratory electron transport chain to NAD, as demonstrated in *Nitrosomonas* and *Nitrobacter* (Aleem, 1965; Aleem et al., 1963). This energy-demanding reduction, in which ATP is required, is a major factor in the slow growth of nitrifiers (Table II).

In general, oxidation of substrate by nitrifying bacteria follows saturation-type rate kinetics. The Michaelis-Menten constant (K_m), the substrate concentration at which the initial reaction velocity is half-maximal, has been determined from a wide variety of nitrification experiments. Values of K_m for ammonium oxidation, primarily characterized on pure cultures or in sewage-treatment processes, range from 70 to 700 μM NH_4^+ for temperatures between 20 and 30°C (Painter, 1970). Nitrite oxidation shows K_m values of 350-600 μM NO_2^- over the same temperature range (Focht and Verstraete, 1977).

The dilemma in these measurements is that, given the slow growth and oxidation rates of marine nitrifiers, the appearance of nitrate in the sea would require bacterial nitrification to be several orders of magnitude more efficient than that postulated from laboratory work.

Kinetic studies on nitrification in the marine environment are only now being performed and it appears that marine nitrifiers in nature can utilize their substrate at much lower levels than those observed in culture studies (Table III). Olson (1980) found no

TABLE II. Growth Constants of Nitrifying Bacteria

Organism	Substrate (μM N)	Generation time (h)	Activity (pmol N· cell^{-1}·h^{-1})	Yield (cells μg N^{-1})	Reference
NH_4^+ Oxidizers					
N. oceanus	10	500	--	--	Carlucci and Strickland (1968)
	42	552	--	--	Carlucci and Strickland (1968)
	50	331	--	--	Carlucci and Strickland (1968)
	5×10^3	21	0.058	1×10^6	Carlucci and Strickland (1968)
	25×10^3	--	0.0012-0.083	0.5×10^6	Watson (1965)
	1×10^5	24	--	--	Carlucci and Strickland (1968)
N. mobilis	4×10^3	12-13	--	--	Koops et al. (1976)
Nitrosomonas sp.[a]	25×10^3	57-83	0.002-0.014	0.6×10^6	Goreau et al. (1980)
Nitrosomonas sp.[b]		34.5			Belser and Mays (1980)
Mud isolate	292	40			Yoshida (1967)
Water isolate	1.5	500			Carlucci and Strickland (1968)
	5×10^3	16			Carlucci and Strickland (1968)
NO_2^- Oxidizers					
N. gracilis	1×10^3	24-48	--	--	Watson and Waterbury (1971)
N. mobilis	1×10^3	10-24	--	--	Watson and Waterbury (1971)
Nitrobacter sp[c]	--	--	0.009-0.042	--	Belser (1979)
	--	--	--	2×10^6	Bock (1978)
Nitrosospira sp	--	20	.004	8×10^6	Belser (1979)

[a]Range of values given for 0.5-20% O_2 levels. *[b]Marine strain D-41.* *[c]Range of activity values among five serotypes.*

TABLE III. Half-Saturation Constants (K_m) for Nitrifying Bacteria

Organism/Observation	Substrate	K_m [a]	Reference
Nitrosomonas sp.[b]	NH_4^+	$1\text{-}2 \times 10^3$	Belser and Schmidt (1980)
Nitrosomonas sp.[c]	NH_4^+	71-714	Focht and Verstraete (1977)
N. europaea[d]	NH_4^+	$0.3\text{-}1.6 \times 10^3$	Suzuki et al. (1974)
Nitrosomonas sp.[e]	NH_4^+	14.2	Knowles et al. (1965)
Nitrosomonas sp.[f]	NH_4^+	2×10^3	Knowles et al. (1965)
N. oceanus[g]	NH_4^+	1×10^3 (0.44)	Watson (1965)
		500	Carlucci and Strickland (1968)
		2-10 ($0.89\text{-}4 \times 10^{-3}$)	Carlucci and Strickland (1968)
Unknown isolate	NH_4^+	7.5-75 ($3.3\text{-}33 \times 10^{-3}$)	Yoshida (1967)
Natural populations	NH_4^+	0.10 (0.04×10^{-3})	Olson (1980)
		0.10 (0.04×10^{-3})	Hashimoto et al. (1983)
		5.0 (2.2×10^{-3})	Wada and Hattori (1971)
	NO_2^-	0.02	Olson (1980)

[a] *K_m values in μM N. Values in parentheses are for NH_3.*
[b] *pH 7.5-8.0.*
[c] *Pure cultures, sewage; 25°-30°C.*
[d] *pH 7.5-8.5.*
[e] *8°C.*
[f] *46°C.*
[g] *Pure cultures.*

consistent increase in the rate of ammonium oxidation in natural samples from coastal California when substrate levels ranged from 0.1 to 20 μ*M* NH_4^+. The organisms responsible for the activity apparently have a very high affinity for ammonium (K_m < 0.1 μ*M* NH_4^+). Wada and Hattori (1971) reported a kinetic constant of 5 μ*M* NH_4^+ for nitrite production in water samples collected from the central North Pacific Ocean. A kinetic experiment on nitrite production in waters of the Cariaco Trench (Hashimoto et al., 1983) was performed with substrate levels ranging from 0.1 to 5 μ*M* NH_4^+. The data (Fig. 1a,b) suggest that the sharpest increase in rate takes place as substrate concentration is increased from zero to about 0.1 μ*M* NH_4^+. Olson (1980) also estimated the K_m for nitrite oxidation to nitrate. He found, in contrast to the situation for ammonium oxidation, a marked dependence of rate on substrate level with an apparent K_m of 0.07 μ*M* NO_2^-. Discrepancies between uptake kinetics in batch culture and rates of nitrification inferred from the concentration of nitrate (Watson, 1965; Carlucci and Strickland, 1968) can now be explained by adaptation of marine nitrifiers to low levels of NH_4^+ and NO_2^-.

Interestingly, the exact form of the substrate during the first step in nitrification is not known. Suzuki et al. (1974) studied the effect of pH on the K_m values for ammonium oxidation by *Nitrosomonas* cells. The K_m decreased with increasing pH, and they suggested that undissociated NH_3 rather than NH_4^+ was the substrate. The ionization of ammonia

$$NH_3 + H_2O \rightleftarrows NH_4^+ + OH^- \qquad (5)$$

has a p*K* value of 8.25, although some small corrections need to be applied for pH, temperature, and salt. Georgii and Gravenhorst (1977) calculated that at a seawater pH of 8.3 and a total $NH_3 + NH_4^+$ concentration of 0.3 μ*M*, levels of NH_3 in equilibrium with the atmosphere would be 0.029 n*M*.

The K_m values for undissociated NH_3 have been recalculated for marine strains at pH 8.3 (Table III). The resultant K_m data for

TABLE IV. Oxygen Effects on Nitrification

Organism/Observation	Oxygen mg $liter^{-1}$	Oxygen μM	Comments	Reference
Nitrosomonas sp.	0.15	9.3	K_m for substrate oxidation; growth rate slower at high oxygen	Goreau et al. (1980)
N. oceanus	0.08	5.0	Minimum level at which nitrification occurs on solid media	Gundersen (1966)
	0.10	6.2	Minimum level at which nitrification occurs on liquid media	Carlucci and McNally (1969)
Nitrosomonas sp. *Nitrobacter* sp.	0.3-1	18.7-62.5	K_m for pure cultures	Focht and Verstraete (1977)
Marine aquarium	0.6	37.5	K_m	Forster (1974)
	0.07	5.0	Minimum level at which nitrification occurs; ^{15}N assay	Suguhara et al. (1974)

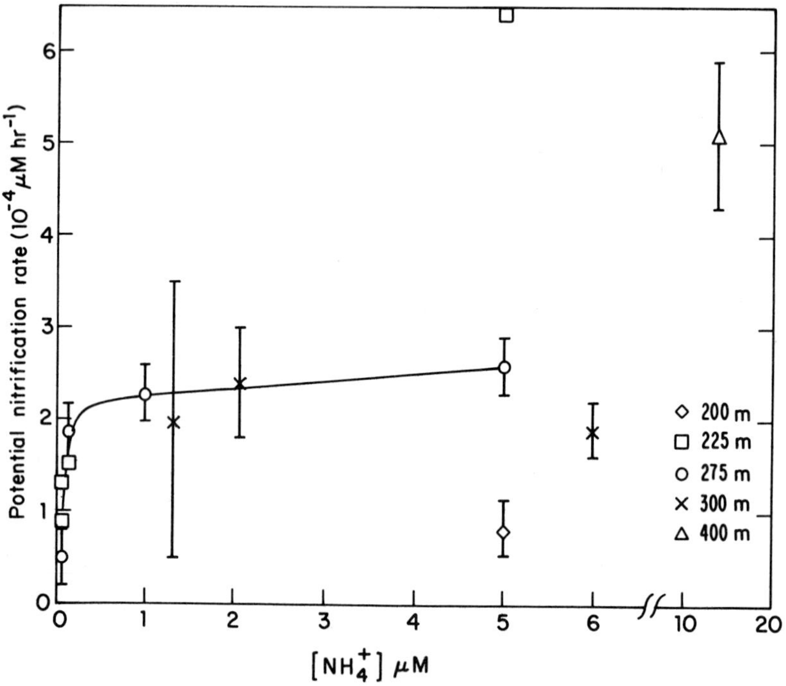

Fig. 1a. The potential rate of nitrification plotted against the concentration of NH_4^+ (including label) for samples from various depths in the Cariaco Trench.

natural samples are low enough (<0.04 n*M* NH_3) such that dissolved NH_3 gas may be used as the substrate. Clearly, then, future work should be designed to look at *in situ* rate kinetics for substrate oxidation and/or any pH effects on substrate affinities in pure cultures of marine nitrifiers.

B. Oxygen

In the oxidation of ammonium to nitrite by *Nitrosomonas*, molecular oxygen serves as the terminal electron acceptor and is incorporated directly into the substrate (Rees and Nason, 1966). Oxygen's role in nitrite oxidation is that of an electron acceptor because the oxygen in nitrate is derived solely from water, not molecular oxygen (Aleem et al., 1965).

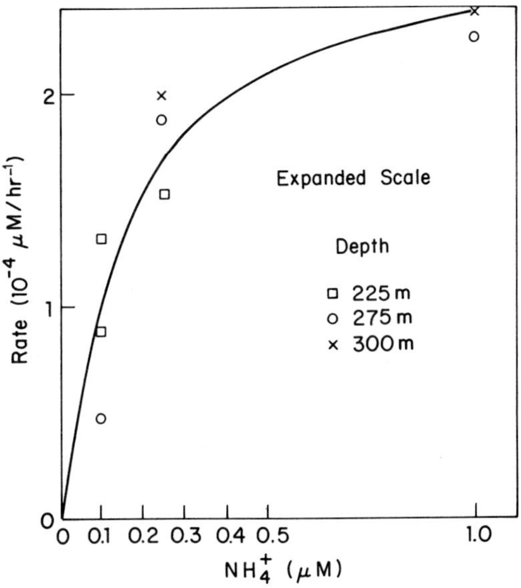

Fig. 1b. The same data as in Fig. 1a on an expanded scale, displaying the drop-off in nitrification rate at NH_4^+ concentrations below 0.5 μM.

Much of the nitrifying activity in aquatic and terrestrial environments occurs in sediments and biological films attached to detrital particles. These sites almost invariably contain low concentrations of dissolved oxygen. The response of nonmarine nitrifiers to oxygen concentration has also been investigated in pure cultures or in sewage facilities (Painter, 1970; Focht and Verstraete, 1977). Although the K_m for oxygen (Focht and Verstraete, 1977) is smaller for *Nitrosomonas* (16 μ*M* O_2) than for *Nitrobacter* (62 μ*M* O_2), they might still be expected to be outcompeted for oxygen by heterotrophs at low O_2 levels (Belser, 1979). Focht and Verstraete (1977) pointed out that heterotrophic microbial respiration is relatively independent of oxygen tension because of the extremely high affinity of the respiratory electron transport chain for oxygen (K_m = 1 μ*M* O_2). However, marine nitrifiers can grow and oxidize their substrate at very low oxygen tensions (Table IV). No experimental work has been performed to measure the respiratory K_m for marine nitrifiers.

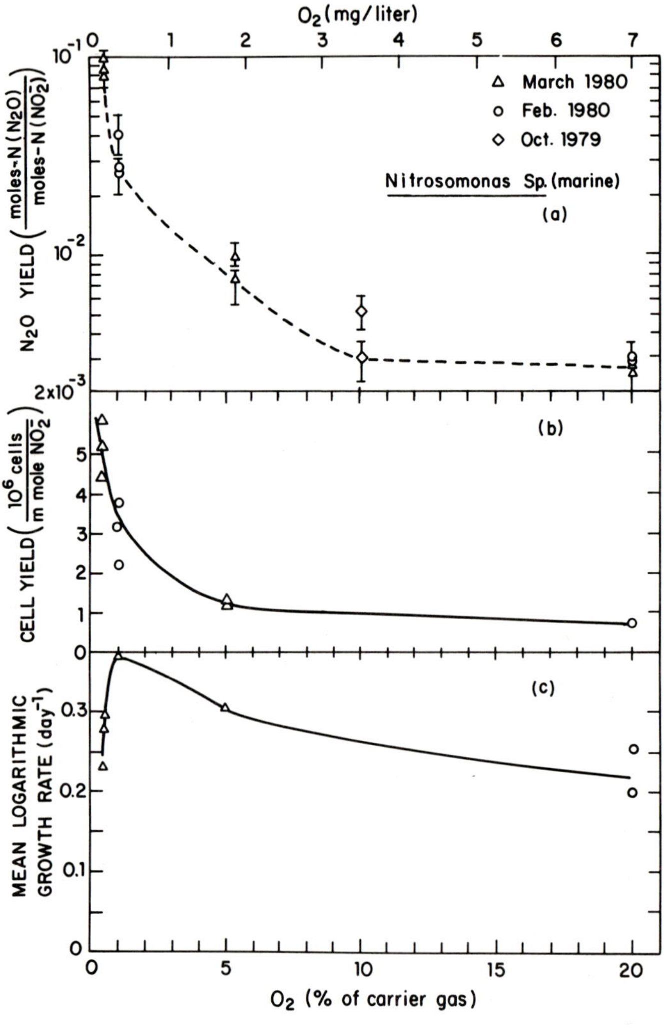

Fig. 2. Summary of results of replicate flasks of Nitrosomonas sp. (marine) at different oxygen tensions. (a) The N_2O yield increases sharply at low oxygen concentrations. (b) At lower oxygen levels more cells were produced for a given quantity of nitrite, as compared to higher oxygen treatments. (c) The generation time for cells changed little over the range of oxygen tensions, with a suggestion of a slight optimum near $p_{O_2} = 0.01$ atm (1% O_2 in carrier gas is approximately 0.37 mg O_2 liter^{-1}).

Gundersen (1966) found that *N. oceanus* could oxidize NH_4^+ at oxygen levels as low as 5 μ*M* O_2. Carlucci and McNally (1969) found total $^{14}CO_2$ uptake rates consistently faster at low oxygen tensions, whereas nitrite production was most rapid at O_2 levels near saturation. This differential effect on the ratio of CO_2 fixed to nitrogen oxidized has important consequences for the use of *in situ* activity assays that depend on empirical C:N ratios (Section IV,A). Marine nitrifying bacteria produce more nitrous oxide (N_2O) at low oxygen concentrations (Goreau et al., 1980). Low oxygen conditions (down to 10 μ*M* O_2), induced a marked decrease in the rate for production of nitrite, from 3.6×10^{-10} to 0.5×10^{-10} m*M* $NO_2^- \cdot cell^{-1} \cdot day^{-1}$. In contrast, N_2O evolution increased as did the yield of N_2O relative to NO_2^- (Fig. 2). Poth and Focht (1982) found that the nitrifier *N. europaea* was capable of denitrification and N_2O formation. Ritchie and Nicholas (1974) isolated a denitrifying nitrite reductase from *N. europaea* that produced N_2O from nitrite under anaerobic conditions in the presence of nitrite. This enzyme may serve to remove nitrite that would otherwise become toxic. The role of these reductive pathways in nitrifying bacteria under situations such as in the deep secondary nitrite maxima of upwelling regions (Section V) is unknown. A detailed study of the oxygen relationships of marine nitrifiers is long overdue.

C. Temperature

Most of the water in the ocean is at fairly low temperatures. Almost half is below 2°C (McLellan, 1965). Nitrification may proceed at temperatures much lower than this. Horrigan (1981), on the basis of dark ^{14}C-bicarbonate incorporation and the presence of nitrate, suggested that nitrification occurs under the Ross Ice Shelf in waters of <-2°C.

In waters of the main oceanic thermocline, the oxygen minimum and nitrate maximum correspond to temperatures between 8° and 11°C in the Atlantic (Wüst, 1978) and 3 and 4°C for the Pacific (Chung, 1976). Although the standard Arrenhius equation can be used to model the effect of temperature on nitrification with temperature limits of 15°-35°C, temperatures below 12°-15°C, which represent the bulk of the normal marine thermal regime, do not conform to the exponential Arrhenius temperature model. Thus, it may not be surprising to find that in pure cultures and in sewage studies, the optimal temperature for nitrification ranges from 25°-35°C (Focht and Verstraete, 1977). Similarly Watson and Waterbury (1971) noted that the nitrite oxidizers *Nitrospina* and *Nitrococcus* both grew optimally at 25°-30°C, but failed to grow below 14°C.

Carlucci and Strickland (1968) isolated several nitrifiers from the northern Pacific Ocean. The optimum temperature for nitrification by all cultures was 28°C. As the incubation temperature decreased, the cultures exhibited longer lag times prior to nitrification. None of the cultures, including one isolated from 9°C subarctic waters, oxidized the substrate at 5°C even after 3-4 months. Values of Q_{10} for the oxidation of substrate varied between 1.3 and 3.0 over a temperature range of 18°-28°C.

A review of most of the available literature gives the impression that below 4°-5°C, nitrifiers do not grow (but see Horrigan, 1981). Given that the bulk of the ocean lies at temperature near or below this value, the effect of low temperatures on the activity and growth of marine nitrifiers should be rigorously examined, especially in chemostat studies.

The relative extent of the effect of temperature on ammonium and nitrite oxidation has not been investigated for marine nitrifiers and may depend on the pH of the medium. In sewage experiments, Wong-Chong and Loehr (1975) found that at optimal pH (7-8), changes in temperature would be more pronounced for ammonium oxidizers. At acid or alkaline pH, the reverse would be true. Given the rather constant pH of seawater, these kinds of effects would only be significant in low pH sediments or in shallow, highly productive systems.

Seasonal changes in temperature often play a significant role in selecting for different genera of heterotrophic bacteria

(Sieburth, 1967). In thermally stable environments, temperature growth optima usually reflect the mean environmental temperature (Brock and Brock, 1968). In thermally fluctuating regimes, bacteria must have their optimal growth and/or activity near the maximum environmental temperature encountered for the season when they are active, rather than near the mean temperature. No information on seasonal temperature selection is yet available for the activities of nitrifiers in the marine environment. Seasonal temperature selection may be more likely, as in the case of heterotrophic bacteria (Sieburth, 1967), where there is a wide diversity of microbial genera.

Jones and Hood (1980b) showed that a *Nitrosomonas* sp. isolated from a Florida estuary had an optimum activity temperature of nearly 40°C, with sharp decreases on either side of this. Therefore, whether these same bacteria are responsible for nitrification in the winter months is not clear.

D. Organic Matter

Recent information on the organic nutrition of autotropic bacteria has been summarized by Matin (1978). A large number of organic compounds in the concentration range 10^2-10^3 μM inhibit the growth of autotrophs. The concentration of dissolved organic carbon in the oceans (1-2 mg C liter^{-1} or 80-160 μM) is on the lower end of this range, and there may not necessarily be any *a priori* reason to expect ambient levels of dissolved organic matter to be toxic to marine nitrifiers.

The biochemical basis for an apparent lack of heterotrophic ability in nitrifying bacteria is due, in part, to the lack of Krebs cycle enzymes, particularly α-ketoglutarate dehydrogenase. Although the data have been judged to be equivocal by Matin (1978), there have been several claims of limited heterotrophic potential by cultures of *N. agilis* and *N. europaea* (Smith and Hoare, 1968; Pan and Umbreit, 1972; Bock, 1976). The strain of *N. agilis* used by Smith and Hoare (1968) did, in fact, contain this enzyme, although other workers have been unable to cultivate different strains of *N. agilis* heterotrophically (Matin, 1978).

Watson and Waterbury (1971) attempted to grow the marine NO_2^- oxidizers *N. gracilis* and *N. mobilis* on a variety of organic media. No growth was detected. While *N. mobilis* could assimilate "small amounts" (15 μM) of acetate, neither organism could oxidize significant quantities of organic matter.

Matin fails to mention the finding by Williams and Watson (1968) of a complete Krebs cycle in *N. oceanus*. *Nitrosomonas oceanus* (Williams and Watson, 1968) could incorporate ^{14}C-labeled glucose, glutamate, pyruvate, and methionine into cellular material, although rates were too low to sustain the energy and carbon needs of the cell. There has been no attempt to look for Krebs cycle enzymes in other marine nitrifiers.

The presence of Krebs cycle enzymes in *N. oceanus* (Williams and Watson, 1968) again raises the question of whether, in certain environments, nitrifiers may act as mixotrophs and combine the utilization of inorganic carbon/energy sources with that of organic matter. Certainly in coastal sediments, salt marshes, or wetlands and plankton blooms, these nitrifying organisms presumably encounter both types of substrate.

It is also known that autotrophic organisms, including cyanobacteria, algae, and nitrifiers will excrete organic compounds into their growth medium (Cohen et al., 1979). These excreted organics may serve as substrates for the growth of heterotrophs. Jones and Hood (1980a) showed that an estuarine *Nitrosomonas* sp. increased its ability to oxidize ammonium by almost a factor of 2 when grown in the presence of heterotrophs isolated from the same environment. Furthermore, they detected no heterotrophic nitrification. Steinmüller and Bock (1976) stimulated nitrite oxidation and growth of *N. agilis* with culture filtrates of *Pseudomonas fluorescens*.

The interactions between nitrifiers and heterotrophs and possible mixotrophic nutrition of marine nitrifiers may be of ecological importance and should be investigated further. Carlucci et al. (1970) noted an increase in nitrite production by *N. oceanus* when added to a suspension of decomposing phytoplankton, and attributed this to the effect of increased surface area (see Section II,E). An alternative explanation could be the stimulatory influence of the heterotrophic bacteria.

E. Salt, Surfaces, and Sulfides

Early ideas about nitrification in the sea summarized by Waksman et al. (1933b) centered around the idea that nitrifying bacteria were essentially terrestrial forms washed into the sea. In coastal rivers, this is almost certainly true. Chen et al. (1976) studied salinity effects on nitrification in New York City's East River. Little quantitative information is available, particularly since their samples were collected during the winter but incubated at 30°C. Qualitatively, the rate of nitrification was inversely proportional to salinity, suggesting that most of the nitrifying activity they observed could be attributed to terrestrial or sewage-derived forms. A similar result was found for the Scheldt estuary (Billen, 1975).

However, ever since Carey and Waksman (1934) were able to demonstrate nitrification in sediments taken from a depth of 4742 m, the evidence argues that the bacteria responsible for marine nitrification are indigenous and not land derived.

Salinity effects on marine nitrifiers are at best, contradictory. Srna and Baggaley (1975) found no change in nitrification rate of an aquarium filter bed over a salinity change of 13 parts per thousand. Yoshida (1967) noted a rather complicated inhibition and stimulation of unknown marine isolates to Na^+, K^+, and Ca^{2+}, Mg^{2+}, respectively, although his experiments were not performed on pure cultures. In a culture of an ammonium-oxidizer isolated from sediments of Maizuru Bay, Japan, Yoshida found optimal growth at half-strength seawater but optimal nitrite production at full-strength seawater or at 30 ppt NaCl. In a strain isolated from the water column of this bay, optimal nitrite production occurred at half-strength seawater.

Unknown factors present in seawater may modulate the growth and activities of nitrifiers in pure cultures. Watson (1965) found *N. oceanus* to be an obligate halophile whose cells lyse in distilled water. After repeated transfers, the bacterium would only grow in natural, not artificial seawater. Nitrite production could occur in distilled water with added 0.5 *M* sodium.

Finstein and Bitsky (1972) found that in environments with fluctuating salinity regimes, the environment of origin determined to a large extent the salt tolerance. Freshwater strains were inhibited by salts, estuarine strains required them. However, a continuum also existed in the relationship with dissolved salts, such that enrichments of marine samples eventually led to strains of nitrifiers that were relatively independent of salt concentrations.

Jones and Hood (1980b) isolated *Nitrosomonas* sp. from a freshwater marsh and estuarine bay. Their estuarine isolate lost all activity at 0 ppt Na^+, with a requirement of 7.5-10 ppt Na^+ for optimal activity. Cultures incubated in natural seawater showed an optimal activity spread over a similar range, 8-12 ppt. To date, there have been no unambiguous studies performed on the relative effect of salts on ammonium or nitrite oxidation.

The effect of solid surfaces on the metabolism of bacteria has been studied since the work of Zobell (1943). In nutrient-poor systems, characteristic of most oceanic environments, surfaces enhance the establishment of microbial populations by concentrating nutrients and providing attachment sites.

Lees and Quastel (1946) suggested that nitrifiers grow on the surface of particles because they utilize adsorbed NH_4^+ ions. On the other hand, Kholdebarin and Oertli (1977) found that particles merely served as a physical support rather than a site for base exchange. They noted that the stimulatory effect of particles on nitrification became less as the surface area to volume ratio decreased, contradicting the Lees and Quastel (1946) NH_4^+ adsorption hypothesis.

Surfaces may play an important role in nitrification in rivers and estuaries (Focht and Verstraete, 1977), although not enough information is available to draw any generalizations. It seems obvious, however, that the water residence time and the surface area-to-volume ratio of the entire sediment/water interface are major factors in determining whether sedimentary/particulate or planktonic/free-living nitrification is more important. Olson (1980) performed a simple, size-fractionation experiment on NH_4^+-oxidizing activity in seawater and found 13-49% of the NH_4^+-oxidizing activity did not pass a 0.6 μm filter. More work of this type needs to be done, especially in estuarine areas where the flux of particulates may be high.

Brandhorst (1959), on observational evidence alone, concluded that nitrifiers "do not live freely suspended in the open ocean but are attached to plankton organisms." Carlucci et al. (1970) determined NH_4^+ oxidation rates in decomposing plankton with and without enrichment of *N. oceanus*. The rate of NO_2^- formation (0.2 μ*M* NO_2^- $week^{-1}$) was doubled in the presence of *N. oceanus*. They suggested that nitrification may occur in microenvironments where nutrients and bacteria are concentrated, i.e., decomposing detrital matter. Paerl et al. (1975) thought the same mechanism responsible for nitrification in Lake Tahoe. Webb and Wiebe (1975) showed that significant nitrification could occur in algal flat regions of coral reefs.

In the marine environment, the three most important oxidizing agents for the decomposition of organic matter are oxygen, nitrate, and sulfate. In areas where water circulation is inadequate to meet the biochemical oxygen demand, nitrate and then sulfate are used as terminal respiratory electron acceptors (Deuser, 1975). In such environments (Black Sea, Cariaco Trench; see Broenkow, 1969; Richards, 1975) and in all marine sediments, there is invariably a transition zone where oxygen and nitrate are depleted and sulfide begins to accumulate as the result of sulfate reduction (Richards, 1975). This "transition zone" would be an ideal environment for chemolithotrophic bacteria capable of oxidizing reduced inorganic ions (NH_4^+, S^{2-}) diffusing upwards.

Vertical profiles of nonconservative properties in the water column of these regions show negligible amounts of NH_4^+ and S^{2-} until conditions become anoxic. The concomitant buildup of NH_4^+ and S^{2-} below the oxic/anoxic interface (Broenkow, 1969) is due to the following scheme (Redfield et al., 1963):

$$(CH_2O_{106}\ (NH_3)_{16}H_3PO_4 + 53\ H_2SO_4 \rightarrow 106\ CO_2 + 53\ H_2S$$

$$+ 16\ NH_3 + H_3PO_4 + 106\ H_2O$$

Nitrification can occur at low oxygen levels, but dissolved sulfides and volatile sulfur compounds inhibit nitrification. The sudden increase in NH_4^+ observed at the interface of these isolated basins and fjords may actually be caused by a mechanism that suppresses nitrification.

Carbon disulfide (CS_2), found in high concentrations in stagnant waters and anoxic muds (Lovelock, 1974), inhibits ammonium oxidation (Powlson and Jenkinson, 1971; Bremner and Bundy, 1974). Yoshida (1967) found that a sulfide concentration of 40 μM S^{2-} inhibited NO_2^- production of a marine nitrifier isolated from sediments. Srna and Baggaley (1975) showed that a sulfide concentration of only 0.9 μM S^{2-} almost completely inhibited ammonium oxidation in an aquarium; oxidation was suppressed until sulfide fell below 0.4 μM S^{2-}. Interestingly, nitrite oxidation in the same system was only slightly inhibited, even when sulfide levels were as high as 3 μM S^{2-}.

The effect of dissolved sulfur compounds on nitrification in marine systems has not been critically examined and may be important when measuring activities in anoxic basins and marine sediments.

III. ALTERNATE MECHANISMS FOR NITRIFICATION

A. Heterotrophic Nitrification

A large number of bacteria and fungi can convert N-containing organic compounds into nitrite or nitrate (Painter, 1970; Focht and Verstraete, 1977). Products excreted during the biochemical reaction sequences of heterotrophic nitrification include hydroxamic acids, nitroso compounds, and hydroxylamine. Verstraete and Alexander (1973) showed that natural samples of river and lake water when amended with large amounts of acetate (3 mg C liter^{-1}) and NH_4^+ (1 mg ml^{-1}) produced detectable quantities of hydroxylamine, nitrosoethanol, nitrite, and nitrate. In this context, the appearance of hydroxylamine during the decomposition of marine plankton (Rajendran and Venugopalan, 1976) is of some interest.

However, the ecological importance of heterotrophic nitrification has been difficult to assess because of the low level of products. Under axenic conditions, the rate of heterotrophic $NO_2^- + NO_3^-$ formation is 10^3-10^4 times slower than in autotrophs (Focht and Verstraete, 1977).

Convincing evidence for the participation of heterotrophic nitrifiers in marine systems is not yet available. Castellvi and O'Shanahan (1977) isolated a heterotrophic bacteria from the Peruvian upwelling system, capable of growing on sodium citrate and ammonium sulfate, which liberated nitrate and nitrite. It is not clear if the culture was pure or whether intermediate forms of N (oximes, hydroxylamine) were found. Carlucci et al. (1970) were unable to demonstrate nitrification in seven isolates of open-ocean heterotrophic bacteria.

However, the number of active heterotrophic bacterial cells in seawater may approach 10^8 liter^{-1} (Fuhrman and Azam, 1982), while recent estimates of direct counts of nitrifiers (living and dead) are on the order of only 10^4 liter^{-1} (Ward et al., 1982). On the basis of this 10^4 difference in standing crop, the presence and significance of heterotrophic nitrifiers should be pursued further.

A first step in determining if heterotrophs are at all important in the production of oxidized nitrogen would be to identify any N-containing intermediates dissolved in seawater or bound to particulate matter.

B. Methane Oxidation

The microbial oxidation of methane and other C_1 compounds as an energy-yielding mechanism is widespread. Methane-oxidizing bacteria have a worldwide distribution in freshwater and marine systems and have been the subject of several reviews (Rudd and Taylor, 1980; Hanson, 1980; Higgins et al., 1981; Quayle, 1976). Methane oxidizers participate in the first step of nitrification by oxidizing ammonium to nitrite (Rudd and Taylor, 1980). This involves the ability to co-oxidize but not use the NH_4^+. The mechanism for this process is somewhat speculative but may be related to free radical formation (Hutchinson et al., 1976).

There is a striking similarity between methane oxidizers and nitrifying bacteria (Dalton, 1977; Hutton and ZoBell, 1949; O'Neill and Wilkinson, 1977; Whittenbury et al., 1975):

1. Both have extensive and complex cytomembranes which contain the CH_4- and NH_4^+-oxidizing enzymes.
2. Acetylene inhibits both CH_4 and NH_4^+ oxidation.
3. Ammonium inhibits CH_4 oxidation in cell-free extracts of *Methylomonas methanica* and whole cells of *Methylosinus trichosporium*
4. Methane inhibits NH_4^+ oxidation in *Nitrosomonas*.
5. Other inhibitors of CH_4 monooxygenases also inhibit NH_4^+ oxidation by *Nitrosomonas* cells. Both oxygenases contain copper.
6. Hydroxylamine is an intermediate in NH_4^+ oxidation by *Methylococcus capsulatus*.
7. *Methylosinus trichosporium* oxidizes NH_4^+ according to the same stoichiometric relationship as autotrophic nitrifiers.
8. Both "type I" methane oxidizers and nitrifiers (except *N. oceanus*) lack a complete Krebs cycle.

Ammonium oxidation by methylotrophs may be significant in freshwater lakes where high rates of nitrification and methane oxidation occur at the oxic/anoxic interface of sediments and stratified water columns. Aerobic CH_4-oxidation mechanisms are much less significant in the marine environment where sulfate reduction exerts powerful controls on CH_4 recycling (Rudd and Taylor, 1980). Aerobic CH_4 oxidation may, however, take place at the sediment/water interface of organic-rich muds where sulfate levels are depleted (Sansone and Martens, 1978), or perhaps at the oxic/anoxic transition zones of fjords and basins (see Section II,E). In the oceans, the significance of methane oxidation as a source of oxidized nitrogen is not known but may be trivial.

Washed suspensions of *M. trichosporium*-oxidized NH_4^+ at the rate of 6.2×10^{-3} μmol NH_4^+ per mg cell per min (O'Neill and Wilkinson, 1977). One mg of bacterial biomass could contain, based on an average size and weight of marine bacteria (Watson et al., 1977), on the order of 10^9 cells. The daily yield per cell would then be 8.7×10^{-9} μmol $NO_2^- \cdot cell \cdot ^{-1} \cdot day^{-1}$. Preliminary *in situ* assays of ammonia oxidation rate per cell (Ward et al., 1982) in the photic zone are on the order 5×10^{-7} μmol NO_2^- $cell^{-1} \cdot day^{-1}$. This discrepancy, even at laboratory rates, would suggest that *in situ* CH_4 oxidation could probably not account for the nitrification observed in the open ocean.

IV. MEASUREMENT TECHNIQUES

Analytical methods for assaying *in situ* nitrification rates and numbers of microorganisms have been reviewed in this and other volumes (Schell, 1978). Here, several specific new approaches will be briefly considered.

A. N-Serve and the Problem of Dark $^{14}CO_2$ Fixation

In the environment, nitrite is oxidized to nitrate as fast as it is being produced and it rarely, if ever, accumulates. Thus, the rate of NH_4^+ oxidation is equal to that at which NO_2^- + NO_3^- accumulates. If ammonium oxidation is blocked, the difference in nitrate production rates with and without the blockage would be an estimate of the nitrification rate. Likewise, if nitrite oxidation alone could be inhibited, the rate of NO_2^- accumulation would equal the nitrification rate.

Chlorate specifically inhibits NO_2^- oxidation by autotrophic nitrifiers (Lees and Simpson, 1957) and nitrapyrin [N-serve or 2-chloro-6-(trichloromethylpyridine] specifically blocks ammonium oxidation (Goring, 1962; Campbell and Aleem, 1965). N-serve, however, also inhibits sulfate reduction (Somville, 1978), methanogenesis (Salvas and Taylor, 1980), and phytoplankton photosynthesis (R. Olson, personal communication).

Henricksen (1980) found that 5-20 ppm N-serve injected into sediment cores did not affect the rate of ammonification or NH_4^+ incorporation, and up to 10 ppm N-serve had no measurable influence on benthic photosynthetic O_2 production. He also found that N-serve was adsorbed to sedimentary organic matter. In mud with 12% organic matter, only half of the added N-serve was recovered in the pore water. Measurements of dark $^{14}CO_2$ fixation with and without N-serve have been used to determine the rate of CO_2 fixation by nitrifiers (Billen, 1976; Somville, 1978). Other attempts to use this method in aquatic environments have not always been successful (H. Ducklow, unpublished; C. D'Elia, and A. F. Carlucci, personal communication). A severe problem occurs when one attempts to convert the resultant C fixation rate to a rate for nitrification. The ratio nitrogen oxidized/carbon fixed varies considerably (Table V). S. G. Horrigan (personal communication) observed N:C ratios of 40 during exponential growth of marine nitrifiers and only 4 during stationary phase. Data kindly provided by Dr. H. Glover show that, as the specific growth rate of chemostat-grown

N. oceanus decreases five-fold, there is nearly an eightfold decrease in N:C ratio. *Nitrosomonas* and *Nitrobacter*, respectively, oxidize about 35 and 100 atoms of N for every molecule of CO_2 fixed (Alexander, 1965). On thermodynamic principles alone, this is to be expected, since the free energy change for nitrite oxidation (Section II,A) is about three to four times less than that for ammonium oxidation. The data presented in Table V do not support either this theoretical argument or Alexander's (1965) results.

The effect of environmental factors on the N:C ratio in both marine and terrestrial nitrifiers is well worth studying and unless simultaneous estimates of *in situ* N:C ratios are made, the use of the N-serve, dark $^{14}CO_2$ fixation method is unwarranted.

B. Assays of Carboxylating Enzymes--Nitrifiers as Primary Producers

All autotrophic organisms contain ribulose bisphosphate carboxylase (RUBPCase), the primary carboxylating enzyme of the Calvin cycle. The presence of RUBPCase and the ratio RUBPCase:CO_2 fixed might allow for an estimation of the potential for autotrophic CO_2 fixation. The potential for measuring nitrification in this way would only be realized in areas below the photic zone where the fraction of phytoplankton RUBPCase is minimal, or can be estimated where nitrifiers are thought to comprise a substantial part of the microbial population.

In pure cultures of nitrifying bacteria (H. Glover, personal communication), the cellular rates of chemoautotrophic CO_2 fixation and RUBPCase activities had a relatively constant ratio when grown under different experimental conditions including growth in batch culture and nitrogen-limited chemostats. Under the appropriate conditions, it may be possible to compute *in situ* rates of autotrophic CO_2 assimilation by nitrifiers from enzyme assays.

TABLE V. Ratio of Nitrogen Oxidized to Carbon Assimilated by Nitrifying Bacteria

Organism	*Substrate* (μ*M N*)	*Oxygen* (*mg* O_2 *liter*$^{-1}$)	*N/C*	*Comments*	*Reference*
N. oceanus	25×10^3	8	12:1	$^{14}CO_2$ uptake, NO_2^- production	W. A. Kaplan (unpublished data)
	$0.05–5 \times 10^3$	8	6–17:1	$^{14}CO_2$ uptake, NO_2^- production	Carlucci (1965)
	--	8	13±3:1[a]	$^{14}CO_2$ uptake, NO_2^- production	Carlucci and Strickland (1968)
	--	8	7:1	O_2 uptake, NO_2^- production	Gundersen and Mountain (1973)
	--	--	7–14:1	O_2 uptake, NO_2^- production	Gundersen (1966)
Nitrosomonas sp.	666	8	11:1	O_2 uptake, NO_2^- production	Wezernak and Gannon (1967)
Unknown isolate	5	0.008	5:1	$^{14}CO_2$ uptake, NO_2^- production	Carlucci and McNally (1969)
Unknown isolate	1×10^3	8	11:1	NO_2^- production	Billen (1976)
			$\bar{x}$ = 10.8±3.6:1		
Nitrobacter agilis	533	8	83:1	O_2 uptake, N oxidized	Wezernak and Gannon (1967)

Nitrobacter sp.	50	8	25:1	O_2 uptake, N oxidized	Gundersen and Mountain (1973)
N. gracilis	1×10^3	--	132:1	$^{14}CO_2$ uptake, N oxidized	Watson and Waterbury (1971)
N. mobilis	1×10^3	--	44:1	N oxidized	Watson and Waterbury (1971)
Nitrobacter winogradskyi	--		71–77:1 $\bar{x}$ = 68.8±34:1		Schön (1965)

[a] *Seventeen separate experiments; ratio varied slightly with substrate concentration.*

C. Production of Gaseous N-Oxides

Gaseous nitrous oxide (N_2O) and nitric oxide (NO) are released during the oxidation of ammonium (Goreau et al., 1980; Lipshultz et al., 1981). Field observations suggest that a relatively constant proportion (0.1-0.3%) of the NO_2^- produced is released as N_2O (Elkins et al., 1978; Cohen and Gordon, 1979). Laboratory experiments support this view and this yield ratio is strongly dependent on oxygen (see Fig. 2). If the yield is fairly constant at a given oxygen tension, one might be able to estimate rates of nitrification by measuring increases in N_2O during short-term incubations. Like the other indirect N-serve assay, the "yield ratio" factor must be calibrated in the laboratory and/or measured in the field. Experiments were performed in Chesapeake Bay (McCarthy et al., 1983) to measure N_2O changes and $^{15}NH_4^+$ oxidation in water samples simultaneously. Yields ranged from 0.3 to 0.6%, similar to laboratory results. This kind of assay may be useful in productive estuaries but its application may be limited in the open ocean where rates of nitrification and gas metabolism are too slow.

V. SYSTEMS

A. Pelagic Systems

Until recently, most of the direct evidence that nitrification was occurring in the open ocean centered around the relationship, however dubious, between rates of biological activity and concentrations of putative reaction products (NO_2^- and NO_3^-). Vertical profiles of nitrate in the sea show an increasing gradient below the thermocline reaching maxima of 20-40 μ*M* (Vaccaro, 1965). Watson (1965) suggested that a possible site for nitrification may be in the upper 500 m along these nitrate gradients.

On the other hand, maxima of nitrite are associated with depths at the base of the photic zone (Venrick et al., 1973; Kiefer et al., 1976) and were originally studied by Rakestraw (1936). Several mechanisms have been proposed to account for this "primary" nitrite maximum. Vaccaro and Ryther (1960) suggested that nitrite is produced by photosynthetic reduction of nitrate by phytoplankton and laboratory experiments of Carlucci et al. (1970) supported this idea. Hattori and Wada (1971), on the basis of an indirect cell-counting method, concluded that nitrifiers were not abundant enough to account for the nitrite in the Central Pacific, south of the equator. They suggested that nitrate-reducing bacteria were important, although this reductive process (Payne, 1973) would have to occur under aerobic conditions, a fairly unlikely event.

In the central North Pacific, observations on the rate of nitrite production after ammonium enrichment in water samples from this maxima (Wada and Hattori, 1971; Miyazaki et al., 1973) suggest that the nitrite originates from ammonium oxidation. Kiefer et al. (1976) argued that this "primary" nitrite peak was not the result of nitrification, since the ambient NH_4^+ levels (1 μM) were much lower than the K_m values reported in the literature at that time (Carlucci and Strickland, 1968). It is apparent from the data collected in the Cariaco Trench and from the California coast (Olson, 1981a) that the K_m for NH_4^+ is less than 0.2 μM NH_4^+, and one can no longer use this "kinetic" argument.

The most up-to-date conceptual model of the biological processes leading to the formation of the primary nitrite maximum has been proposed by Olson (1980). There exists a "continuum" of nitrite production potential with depth, that gradually shifts from nitrite excretion by phytoplankton to ammonium oxidation. The K_m for nitrifiers in natural samples appears to be low enough so that nitrifiers could actually compete with phytoplankton for available NH_4^+ in surface waters. However, Miyazaki et al. (1973), Wada and Hattori (1971), and Olson (1980) all showed that the oxidation of ammonium was negligible above the depth of the NO_2^- peak.

The master variable controlling the onset of nitrification may be the light intensity. Photoinhibition of cultures of *N. winogradskyi* and *N. europaea* has been reported (Müller-Neuglück and

Engel, 1961; Schön and Engel, 1962). Inhibition was caused by wavelengths below 480 nm and light intensities as low as 0.3-1% of full sunlight were effective. Horrigan et al. (1981) showed that the high nitrate concentrations in seasurface films could not be the result of nitrite oxidation, since light at approximately 0.1% full sunlight inhibited nitrite oxidation.

Olson (1980) found that nitrite oxidizers are more sensitive to light than ammonium oxidizers, and he hypothesized that this differential photoinhibition, i.e., NH_4^+-oxidizing activity occurring higher up in the water column than NO_2^--oxidizing activity, will lead to the formation of the primary nitrite maximum. There have been no *in situ* measurements of nitrification taken over a diel cycle, and light-dark effects may be significant but should be examined further. Horrigan et al. (1981) found that NH_4^+ oxidizers showed higher activity on an 8-h light/16-h dark than on a 16/8 light-dark (L/D) cycle. A culture on a 16/8 L/D cycle, when placed in continuous darkness, recovered to show nitrifying activity similar to cultures that had never been exposed to light, although the lag period was close to 1 month. It is important to note that nitrite oxidizers only showed activity in the dark.

Deep "secondary" nitrite maxima below the thermocline associated with low-oxygen zones have also been observed (Spencer, 1975; Brandhorst, 1959). Goering (1968) provided direct evidence for nitrate reduction to dinitrogen gas at the secondary nitrite maxima of the Eastern Tropical North Pacific. Carlucci and Schubert (1969) isolated nitrate-reducing bacteria from the low-oxygen, high-nitrite zones off the Peruvian coast. Wada and Hattori (1971), using a sensitive chemical method, found nitrite production in enclosed water samples from deep Pacific waters (800-3900 m) during prolonged incubations (3-15 days). Added nitrate (110 μM) consistently stimulated nitrite production, and these authors suggested that nitrate reduction to nitrite was primarily responsible for the deep NO_2^- maximum. This conclusion is not entirely convincing. Given the low K_m for nitrite oxidizers and the long incubation times, the addition of so much nitrate probably selected for a population of nitrate-reducing bacteria.

Packard et al. (1978) came to a similar conclusion in a less equivocal way by measuring nitrate reductase activity in samples from the core of the secondary nitrite maximum off the Peruvian coast. At this time, it is not known to what extent nitrification contributes to the nitrite in these low oxygen areas. Nitrifiers can metabolize at very low oxygen levels (section II,B). Nitrous oxide has been measured in low-oxygen regions (Cohen and Gordon, 1979; Elkins et al., 1978). This gas, once believed to be primarily a product of denitrification, is also released by marine nitrifiers (Section IV,C; Goreau et al., 1980).

Rates of nitrification that have been measured in aquatic marine environments (Table VI) range from 10^{-8} to 10^{-9} *M* N day^{-1} over a wide range of NH_4^+ concentrations, locations, and depths. Note the relative lack of data available on nitrite oxidation. It is difficult, if not foolhardy, to generalize from such few data, although the more productive estuarine and coastal areas do show higher rates of activity than open ocean areas. One might expect some relationship with organic production, since the ultimate source of regenerated nitrate is decomposing organic matter. The data in Table VI represent nitrification rates in enriched samples. In addition, both ammonium and nitrite may be regenerated during incubations. These considerations, plus those experimental caveats introduced in sampling, filtration, and incubation steps, should make it clear that the data in Table VI represent potential rates of activity.

Accurate methods for counting the numbers of nitrifying bacteria responsible for these rates have not been available until recently. Most probable number (MPN) assays, the commonly used method, are fairly insensitive and inaccurate (Belser, 1979, Matulewich et al., 1975). Estimates of MPN abundances (Watson, 1965; Ezura et al., 1974) would suggest that nitrifying bacteria in the water column of oceans and bays are low, on the order of

TABLE VI. Nitrification Rates in Marine Systems: The Water Column

Location	*Rate*[a] *(mol N·liter^{-1}·day^{-1})× 10^{-8}*	*Comments*	*Reference*
Estuaries, coastal bays			
Alaska, Skan Bay	1.6-15	^{15}N assay: 10 μ*M* NH_4^+	Hattori et al. (1978)
	2-22	Chemical assay	Hattori et al. (1978)
Japan, Sagami Bay	1.5-2.7	^{15}N assay: 5-10 μ*M* NH_4^+	Miyazaki et al. (1973)
York River, Virginia	24-96	^{15}N assay: 0.05-5 μ*M* NH_4^+	McCarthy et al. (1983)
Chesapeake Bay	72-240	^{15}N assay: 0.05-5 μ*M* NH_4^+	McCarthy et al. (1983)
Offshore, open ocean			
Alaska, shelf waters	7-13	Chemical assay (3°-8°C)	Schell (1974)
Southern California	0.45-6	^{15}N assay	Olson (1981a)
	1.9	^{15}N assay	Ward et al. (1982)
	1.5-7.5[b]	^{15}N assay	Olson (1981a)
Cariaco Trench	0.21-1.5	^{15}N assay: 0.1-10 μ*M* NH_4^+	Hashimoto et al. (1983)
Sargasso Sea	0.024-2.4[c]	Chemical assay: 10 μ*M* NH_4^+ enrichment	Vaccaro (1962)

East China Sea	0.38-0.64	^{15}N assay: 10 μ*M* NH_4^+	Miyazaki (1975)
Phillipine Sea	0.5-0.72	^{15}N assay: 10 μ*M* NH_4^+	Miyazaki (1975)
North Pacific	6.8	Chemical assay	Hattori and Wada (1971)
North Pacific	0.9-1.6	Chemical assay: 5°C	Wada and Hattori (1971)
West Pacific	0.88-2.8	^{15}N assay: 10 μ*M* NH_4^+	Miyazaki et al. (1975)

[a] *Production of nitrite from ammonium unless noted otherwise.*
[b] *Nitrite oxidation.*
[c] *Production of nitrite plus nitrate.*
[d] *Samples from nitrite maximum.*

TABLE VII. Nitrification Rates in Marine Systems: Sediments

Location	Rate[a] $(mol\ N \cdot cm^{-3} \cdot h^{-1}) \times 10^8$	Rate[a] $(mg\ N \cdot m^{-2} \cdot day^{-1}$	Reference
Inshore			
Alaska, eelgrass bed	0.27	--	Iizumi et al. (1980)
Massachusetts, salt marsh	0-2	0-2.1[b]	Kaplan et al. (1979)
Japan, coastal bays	2-5	130-330[c]	Koike and Hattori (1978)
North Sea	0.36-2.8	60-480[d]	Vanderborght and Billen (1975)
	0.03-21	3-240[d]	Billen (1978)
	0-2	0-134[d]	Billen (1976)
Danish Coast	0-2.5	0-42[e]	Henriksen (1980)
		16.2[f]	Henricksen et al. (1980)
		28-35[g]	Henricksen et al. (1980)
New Zealand, intertidal	2.2		Belser and Mays (1980)
Coral Reef, Enewetok		4.6	Webb and Wiebe (1975)
Freshwater sediment		29[f]	Chaterpaul et al. (1980)
		69[g]	Chaterpaul et al. (1980)
Offshore			
North Sea		2.1-48.2[d]	Billen (1978)
Denmark		4.2-18.2[e]	Henricksen et al. (1981)
Santa Barbara Basin, USA		0.60[h]	Rittenberg et al. (1955)
Atlantic Ocean		0.95[i]	Bender et al. (1977)

$1\text{-}10^3$ liter^{-1}. Watson (1965) calculated population densities from observed NO_2^- production rates in enriched samples and got approximately the same result. Watson then estimated that *in situ* ammonium oxidation rates approach 3.8×10^{-11} M NH_4^+ day^{-1}, slower than the potential activities measured in Table VI. Indrebo et al. (1979) used N-serve inhibition/$^{14}CO_2$ assays to measure nitrification in a stratified estuary. Rates appeared much too high (14×10^{-6} M NH_4^+ day^{-1}) for the observed MPN number (30 cells liter^{-1}).

A major breakthrough occurred when Ward and Perry (1980) developed the immunofluorescent antibody assay (FA) for *N. marinus* and *N. oceanus*. The IFA assay has the capability of detecting and enumerating specific bacteria directly (Schmidt, 1974). A major practical limitation is that the IFA reagents are so specific that a considerable number of IFAs may be needed for a given population. Using the FA assay, Ward et al. (1982) found cell abundances of the order of 1×10^4 cells liter^{-1} in Southern California surface waters, at least an order of magnitude larger than all previous PMN and culture estimates. They found no obvious vertical population structure even though rates of ammonium oxidation were up to 10 times faster at depth. The efficiency of nitrification per cell varies with depth and the reasons for this are not well understood, although the influence of light intensity alluded to previously may be one factor.

Footnotes to Table VII.

[a]*Assuming density of sediment: 1 g cm*$^{-3}$.
[b]*Chemical assays: 18-day incubation.*
[c]*Depth-integrated rates:* ^{15}N *dilution assay, 36- to 48-h incubation.*
[d]*Depth-integrated rates: N-serve* ^{14}C *assay; N:C = 8:1*
[e]*Depth-integrated rates: N-serve, chemical assay.*
[f]*Sediment without infauna.*
[g]*Sediment with infauna.*
[h]*Mass balance.*
[i]*Calculated from nutrient profiles in pore waters.*

B. Sediments

Chemical analyses of nitrogenous nutrients in marine and coastal sediments (Barnes et al., 1975; Bender et al., 1977; Rittenberg et al., 1955) often show the presence of nitrate values far in excess of those in the overlying water. Various theoretical models have been proposed to account for this (Jahnke et al., 1982; Goloway and Bender, 1982; Vanderborght and Billen, 1975). In sediments, nitrification may occur primarily at the oxygenated sediment/water interface. The nitrate diffuses away from its zone of production to be released into the water or reduced to nitrogenous gases or ammonium in deeper sediments. Grundmanis and Murray (1977) found nitrate at intermediate depths in sediments of Puget Sound, Washington. They postulated that oxygen was supplied from surface waters by irrigation through burrows of benthic organisms, thus enhancing the nitrification rate. In an excellent paper, Chaterpaul et al. (1980) determined nitrification and denitrification rates in stream sediments with and without the presence of tubificid worms. The presence of the worms doubled the nitrification rate. The authors suggested that the worms accelerated the upward flux of ammonium into the aerobic surface layers; a quite different mechanism than that of Grundmanis and Murray (1977). Henriksen et al. (1980) found that the presence of benthic infauna (*Corophium volutator, Nereis virens*) also increased nitrification rates in marine sediments. Bioturbation is therefore an important sedimentary process that may have profound effects on the nitrification in sediments. There is almost no discussion of these effects in any of the literature devoted to assays of sedimentary nitrification (Table VII).

Interestingly, rates vary over a narrow range, 0-5 × 10^{-8} mol $N \cdot cm^{-3} \cdot h^{-1}$, regardless of assay method. There are few data taken over a seasonal cycle. Billen's (1978) results, when recalculated, show a reasonable relationship to seasonal *in situ* temperature (r = 0.74) with an apparent Q_{10} of 2.

Whenever you can, count.
Sir Francis Galton

VI. CONCLUSIONS

There are several ways in which to estimate global rates of nitrification. Given a yearly production rate for newly synthesized marine organic matter of 2-3 × 10^{16} g C and a C:N ratio of 15:1, McElroy (1975) estimated that the *net* rate at which oceanic nitrogen is transformed by nitrification would be 3 × 10^{15} g N $year^{-1}$. It is also possible to estimate independently nitrification rates from the flux of nitrous oxide across the sea/air interface. Cohen and Gordon (1979) assumed a constant fractional yield of N_2O during nitrification and calculated the annual N_2O flux from the sea surface. A nitrification rate of 2 × 10^{15} g N $year^{-1}$ would be needed to support this flux, in good agreement with McElroy's (1975) estimate.

King and Devol (1979) calculated a diffusive flux of nitrate across the thermocline of the Eastern Tropical North Pacific. By assuming steady-state with respect to nitrogen cycling over time scales much greater than mixing in the photic zone and little or no N_2 fixation and nitrification in the photic zone, they suggested that the upward flux of deep nitrate is equal to the integrated rate of nitrate uptake by phytoplankton with nitrate uptake by plankton controlling the nitrate gradient of the deep water. Although processes that control advection of nutrients across the thermocline are not well quantified and may be episodic (storms, internal waves and tides, spin-off eddies), at steady state the rate of diffusive nitrate flux from deep water into the photic zone must also equal the rate of nitrification.

King and Devol (1979) used a Fickian diffusive model to estimate an average nitrate flux of 48.9 mol $N{\cdot}m^{-2}{\cdot}h^{-1}$ across the thermocline. While this does not take into account horizontal transport, multiplying this flux by the surface area of the open ocean

(332×10^{12} m^2; Whittaker and Likens, 1972) we can estimate a possible upper limit for steady-state nitrate production of 1.9×10^{15} g N $year^{-1}$, again similar to previous calculations. Bender et al. (1977) calculated a nitrate release rate from sediments in the equatorial Atlantic of 3.3 mol $N \cdot m^{-2} \cdot h^{-1}$, an order of magnitude less than King and Devol's (1979) water column flux.

There are so few measurements of nitrification in marine sediments that it is not known, at this time, where the bulk of the nitrate in the ocean is produced. However, the efficient utilization of NH_4^+ (see Section II,A) may provide some clues. Given a nitrate concentration in the deep ocean of about 30 μmol N $liter^{-1}$ and a water residence time of 1600 years (Broecker, 1974), a mean oceanic nitrification rate would be 5×10^{-11} mol $N \cdot liter^{-1} \cdot day^{-1}$. Hashimoto et al. (1983) used the following equation for nitrification by active bacteria (Carlucci and Strickland, 1968) to calculate R, an *in situ* nitrification rate.

$$R = aN\, K_{max}\, [NH_4^+]/(K_m + [NH_4^+])$$

$[NH_4^+]$ is the ambient ammonium (about 0.1 μM in deep water), a is the amount of NH_4^+ used per cell per doubling (1.5×10^{-12} mol; Goreau et al., 1980), N is the average number of cells (10^3 $liter^{-1}$) in deep waters (Watson, 1965), K_{max} is the growth rate at high NH_4^+ ($10\text{-}100^{-3}$ h^{-1}; Watson, 1965; Goreau et al., 1980), and K_m is the half-saturation constant (0.1-0.2 μM, see Section II). Using these values, estimates for R range from $18\text{-}180 \times 10^{-11}$ M N day^{-1} close to this mean global rate on the order of 10^{-11} M N day^{-1}. If the efficiency for ammonium utilization is a few tenths micromolar (Fig. 1 and Olson, 1981a), then nitrification in the water column alone could account for the bulk of nitrate in the deep sea.

It is also instructive to ask how similar the geochemical flux calculations are to actually measured rates. King and Devol's (1979) average flux of nitrate into the mixed layer can be integrated over a 50-m depth range (see their Fig. 2) to give a steady-state nitrate production rate needed to support this flux of 2.5×10^{-8} *M* N day^{-1}, in excellent agreement with most of the field data in Table VI.

It also may be possible to estimate a generation time for nitrifying bacteria in the mixed layer by multiplying an *in situ* nitrification rate (mol N·liter^{-1} day^{-1}) by an empirical cell yield (cells mol N^{-1}) to give a production rate (cells·liter^{-1}·day^{-1}). Given a measured nitrification rate of the order 10^{-8} mol N day^{-1} (Table VI) and a laboratory yield of the order 10^{6} cells·μmol N^{-1} (Goreau et al., 1980), we obtain a production rate of 1×10^{4} cells·liter^{-1}·day^{-1}. To calculate a generation time (day^{-1}), we divide the production rate by a standing crop (cells liter^{-1}). Using the estimate of Ward et al. (1982) of $1\text{-}5 \times 10^{4}$ cells liter^{-1}, we derive a generation time of 1-0.2 day^{-1} or 16-83 h. Generation times for laboratory cultures of marine nitrifiers grown at near *in situ* substrate levels are 5 or 6 times slower than this (Table II). The discrepancy may be an artifact of work in laboratory batch cultures that have become adapted to high levels of ammonium.

At the present time we have only a limited understanding of the factors regulating nitrification in the oceans. Since we are already conducting, however inadvertently, several global perturbation experiments, our knowledge of "undisturbed" global nutrient cycling is becoming increasingly difficult to obtain. The anthropogenic fixation of nitrogen gas as fertilizer or by cultivation of legumes may eventually find its way into coastal regions (Walsh et al., 1981), perhaps increasing phytoplankton production, decay, and nitrate regeneration.

Perhaps the next review on this subject will tell us how far we have advanced in our understanding of the global N cycle and how much of the promise of future work mentioned here was, in fact, realized.

ACKNOWLEDGMENTS

Financial support for this work has been provided by NSF (DEB79-20282) and NASA (NSG1-55). I wish to thank Drs. Michael McElroy and Steven Wofsy for support during the writing of this article.

Dr. I. Foster Brown supplied some of the epigrams.

REFERENCES

Aleem, M. I. H. (1965). Path of carbon and assimilation power in chemosynthetic bacteria. 1. *Nitrobacter agilis*. *Biochim. Biophys. Acta 107*, 14-28.

Aleem, M. I. H. (1970). Oxidation of inorganic nitrogen compounds. *Annu. Rev. Plant Physiol. 21*, 67-90.

Aleem, M. I. H., Lees, H., and Nicholas, D. J. D. (1963). Adenosine triphosphate dependent reduction of nicotinamide adenine dinucleotide by ferro-cytochrome *C* in chemoautotrophic bacteria. *Nature (London) 200*, 759-761.

Aleem, M. I. H., Hoch, G. E., and Varner, J. E. (1965). Water as the source of oxidant and reductant in bacterial chemosynthesis. *Proc. Natl. Acad. Sci. USA 54*, 869-873.

Alexander, M. (1965). Nitrification. *In* "Soil Nitrogen" (W. V. Bartholomew and F. E. Clark, eds.), pp. 307-343. Am. Soc. Agron., Madison, Wisconsin.

Barnes, R., Bertine, K. K., and Goldberg, E. D. (1975). N_2:Ar, nitrification and denitrification in Southern California borderland basin sediments. *Limnol. Oceanogr. 20*, 962-970.

Belser, L. W. (1979). Population ecology of nitrifying bacteria. *Annu. Rev. Microbiol. 33*, 309-333.

Belser, L. W., and Mays, E. L. (1980). Specific inhibition of nitrite oxidation by chlorate and its use in assessing nitrification in soils and sediments. *Appl. Environ. Microbiol. 39*, 505-510.

Belser, L. W., and Schmidt, E. L. (1978). Diversity in the ammonia-oxidizing nitrifier population of a soil. *Appl. Environ. Microbiol. 36*, 584-588.

Belser, L. W., and Schmidt, E. L. (1980). Growth and oxidation kinetics of three genera of ammonia-oxidizing nitrifiers. *FEMS Microbiol. Lett. 7*, 213-216.

Bender, M. L., Fanning, K. A., Froelich, P. N., Heath, G. R., and Maynard, V. (1977). Interstitial nitrate profiles and oxidation of sedimentary organic matter. *Science 198*, 605-609.

Billen, G. (1975). Nitrification in the Scheldt estuary (Belgium and The Netherlands). *Estuarine Coastal Mar. Sci. 3*, 79-89.

Billen, G. (1976). A method for evaluating nitrifying activity in sediments by dark ^{14}C-bicarbonate incorporation. *Water Res. 10,* 51-57.

Billen, G. (1978). A budget of nitrogen recycling in North Sea sediments off the Belgian coast. *Estuarine Coastal Mar. Sci. 7,* 127-146.

Bock, E. (1976). Growth of *Nitrobacter* in the presence of organic matter. II. Chemo-organic growth of *Nitrobacter agilis. Arch. Microbiol. 108,* 305-312.

Bock, E. (1978). Lithoautotrophic and chemo-organotrophic growth of nitrifying bacteria. *In* "Nitrification and Reduction of Nitrogen Oxides" (D. Schlessinger, ed.), pp. 310-314. Am. Soc. Microbiol., Washington, D.C.

Brandhorst, W. (1959). Nitrification and denitrification in the Eastern Tropical North Pacific. *J. Cons. Int. Explor. Mer 25,* 3-20.

Bremner, J. M., and Bundy, L. G. (1974). Inhibition of nitrification in soils by volatile sulfur compounds. *Soil Biol. Biochem. 6,* 161-165.

Brock, T. D., and Brock, M. L. (1968). Relationship between environmental temperature and optimum temperature of bacteria along a hot spring thermal temperature. *J. Appl. Bacteriol. 31,* 54-58.

Broecker, W. S. (1974). "Chemical Oceanography." Harcourt Brace Jovanovich, Inc., New York.

Broenkow, W. W. (1969). The distributions of nonconservative solutes related to the decomposition of organic material in anoxic marine basins. Ph.D. Thesis, University of Washington, Seattle.

Campbell, N. E. R., and Aleem, M. I. H. (1965). The effect of 2-chloro,6 (trichloromethyl) pyridine on the chemoautotrophic metabolism of nitrifying bacteria. *Antonie van Leeuwenhoek 31,* 124-144.

Caperon, J., and Meyer, J. (1972). Nitrogen-limited growth of marine phytoplankton. *Deep-Sea Res. 19,* 601-632.

Carey, C. L. (1937). The occurrence and distribution of nitrifying bacteria in the sea. *J. Mar. Res. 1,* 291-304.

Carey, C. L., and Waksman, S. A. (1934). The presence of nitrifying bacteria in deep seas. *Science 79,* 349-350.

Carlucci, A. F. (1965). "Nitrification in the Sea," Res. Mar. Food Chain Prog. Dep., pp. 50-53. Scrips Inst. Oceanogr., La Jolla, California.

Carlucci, A. F., and McNally, P. M. (1969). Nitrification by marine bacteria in low concentrations of substrate and oxygen. *Limnol. Oceanogr. 14,* 736-739.

Carlucci, A. F., and Schubert, H. R. (1969). Nitrate reduction in seawater of the deep nitrite maximum off Peru. *Limnol. Oceanogr. 14,* 187-193.

Carlucci, A. F., and Strickland, J. D. H. (1968). The isolation, purification and some kinetic studies of marine nitrifying bacteria. *J. Exp. Mar. Biol. Ecol. 2,* 156-166.

Carlucci, A. F., Hartwig, E. O., and Bowes, P. M. (1970). Biological production of nitrite in seawater. *Mar. Biol. (Berlin) 7,* 161-166.

Carpenter, E. J., and McCarthy, J. J. (1975). Nitrogen fixation and uptake of combined nitrogenous compounds by *Oscillatoria (Trichodesmium) thiebauti* in the Western Sargasso Sea. *Limnol. Oceanogr. 20,* 389-401.

Castellvi, J., and O'Shanahan, L. (1977). Nitrificatión heterotrofica par bacterias marinas. *Invest. Pesq. 41,* 501-507.

Chaterpaul, L., Robinson, J. B., and Kaushik, N. K. (1980). Effects of tubificid worms on denitrification and nitrification in stream sediments. *Can. J. Fish. Aquat. Sci. 37,* 650-663.

Chen, M., Canelli, E., and Fuhs, G. W. (1976). Effects of salinity on nitrification in the East River. *J. Water Pollut. Control Fed. 47,* 2474-2481.

Chung, Y.-C. (1976). A deep ^{226}Ra maximum in the North Pacific. *Earth Planet. Sci. Lett. 32,* 249-257.

Cohen, Y., and Gordon, L. I. (1979). Nitrous oxide production in the ocean. *JGR, J. Geophys. Res. 84,* 347-353.

Cohen, Y., de Jonge, I., and Kuenen, J. G. (1979). Excretion of glycolate by *Thiobacillus neapolitanus* grown in continuous culture. *Arch. Microbiol. 122,* 189-194.

Dalton, H. (1977). Ammonia oxidation by the methane-oxidizing bacterium *Methylococcus capsulatus* strain Bath. *Arch. Microbiol. 114,* 273-279.

Delwiche, C. C. (1970). The nitrogen cycle. *Sci. Am. 223,* 136-146.

Deuser, W. G. (1975). Reducing environments. *In* "Chemical Oceanography" (J. P. Riley and G. Skirrow, 2nd ed., eds.), Vol. 3, pp. 1-37. Academic Press, New York.

Dugdale, R. C., and Goering, J. J. (1967). Uptake of new and regenerated forms of nitrogen in primary production. *Limnol. Oceanogr. 12,* 196-206.

Elkins, J. W., Wofsy, S. C., McElroy, M. B., Kolb, C. E., and Kaplan, W. A. (1978). Aquatic sources and sinks for nitrous oxide. *Nature (London) 275,* 602-606.

Eppley, R. W., and Peterson, B. J. (1979). Particulate organic carbon flux and planktonic new production in the deep ocean. *Nature (London) 282,* 677-680.

Eppley, R. W., Renger, E. H., and Harrison, W. G. (1979). Nitrate and phytoplankton production in Southern California coastal waters. *Limnol. Oceanogr. 24,* 483-494.

Ezura, Y., Daiku, K., Tajma, K., Kimur, T., and Sakai, M. (1974). Seasonal differences in bacterial counts and heterotrophic bacterial flora in Ahkeshi Bay. *In* "Effect of the Ocean Environment on Microbial Activities" (R. Colwell and R. Y. Morita, eds.), pp. 113-123. Univ. Park Press, Baltimore, Maryland.

Finstein, M. S., and Bitsky, M. R. (1972). Relationship of autotrophic ammonium-oxidizing bacteria to marine salts. *Water Res. 6,* 31-40.

Focht, D. D., and Chang, A. C. (1975). Nitrification and denitrification processes related to waste water treatment. *Adv. Appl. Microbiol. 19,* 153-186.

Focht, D. D., and Verstrate, W. (1977). Biochemical ecology of nitrification and denitrification. *Adv. Microbiol. Ecol. 1,* 135-214.

Forster, J. R. M. (1974). Studies on nitrification in marine biological filters. *Aquaculture 4,* 387-397.

Fuhrman, J. A., and Azam, F. (1982). Thymidine incorporation as a measure of heterotrophic bacterioplankton production in marine surface waters: Evaluation and field results. *Mar. Biol. (Berlin) 66,* 109-120.

Georgii, H. W., and Gravenhorst, G. (1977). The ocean as a source or sink of reactive trace gases. *Pure Appl. Geophys. 115,* 503-511.

Goering, J. J. (1968). Denitrification in the oxygen minimum layer of the Eastern Tropical Pacific Ocean. *Deep-Sea Res. 15,* 157-164.

Goloway, F., and Bender, M. (1982). Diagenetic models of interstitial nitrate profiles in deep sea suboxic sediments. *Limnol. Oceanogr. 27,* 624-638.

Goreau, T. J., Kaplan, W. A., Wofsy, S. C., McElroy, M. B., Valois, F. A., and Watson, S. W. (1980). Production of NO_2^- and N_2O by nitrifying bacteria at reduced concentrations of oxygen. *Appl. Environ. Microbiol. 40,* 526-532.

Goring, C. A. I. (1962). Control of nitrification by 2-chloro-6-(trichloromethyl) pyridine. *Soil Sci. 93,* 211-218.

Gottschalk, G. (1979). "Bacterial Metabolism." Springer-Verlag, Berlin and New York.

Grundmanis, V., and Murray, J. W. (1977). Nitrification and denitrification in marine sediments from Puget Sound. *Limnol. Oceanogr. 22,* 804-813.

Gundersen, K. (1966). The growth and respiration of *Nitrosocystis oceanus* at different partial pressures of oxygen. *J. Gen. Microbiol. 42,* 387-396.

Gundersen, K., and Mountain, C. W. (1973). Oxygen utilization and pH change in the ocean resulting from biological nitrate formation. *Deep-Sea Res. 20,* 1083-1091.

Hanson, R. S. (1980). Ecology and diversity of methylotrophic organisms. *Adv. Appl. Microbiol. 20,* 3-39.

Harvey, H. W. (1927). Nitrate in the Sea. *J. Mar. Biol. Assoc. U.K. 14,* 71-88.

Hashimoto, L. K., Kaplan, W. A., Wofsy, S. C., and McElroy, M. B. (1983). Transformations and fixed nitrogen and N_2O in the Cariaco Trench. *Deep-Sea Res. 30,* 575-590.

Hattori, A., and Wada, E. (1971). Nitrite distribution and its regulating processes in the equatorial Pacific Ocean. *Deep-Sea Res. 18,* 557-568.

Hattori, A., Goering, J. J., and Boisseau, D. B. (1978). Ammonium oxidation and its significance in the summer cycling of nitrogen in oxygen depleted Skan Bay, Unalaska Island, Alaska. *Mar. Sci. Commun. 4,* 139-151.

Henriksen, K. (1980). Measurement of *in situ* rates of nitrification in sediment. *Microb. Ecol. 6,* 329-337.

Henriksen, K., Hansen, J. I., and Blackburn, T. H. (1980). The influence of benthic infauna on exchange rates of inorganic nitrogen between sediment and water. *Ophelia, Suppl. 1,* 249-256.

Henriksen, K., Hansen, J. I., and Blackburn, T. H. (1981). Rates of nitrification, distribution of nitrifying bacteria and nitrate fluxes in different types of sediment from Danish waters. *Mar. Biol. (Berlin) 61,* 299-304.

Higgins, I. J., Best, D. J., Hammond, R. C., and Scott, D. (1981). Methane-oxidizing microorganisms. *Microbiol. Rev. 45,* 556-590.

Hooper, A. B. (1978). Nitrogen oxidation and electron transport in ammonia-oxidizing bacteria. *In* "Nitrification and Reduction of Nitrogen Oxides" (D. Schlessinger, ed.), pp. 299-304. Am. Soc. Microbiol., Washington, D.C.

Horrigan, S. G. (1981). Primary production under the Ross Ice Shelf, Antarctica. *Limnol. Oceanogr. 26,* 378-382.

Horrigan, S. G., Carlucci, A. F., and Williams, P. M. (1981). Light inhibition of nitrification in sea-surface films. *J. Mar. Res. 39,* 557-565.

Hutchinson, D. W., Whittenbury, R. W., and Dalton, H. (1976). A possible role of free radicals in the oxidation of methane by *Methylococcus capsulatus. J. Theor. Biol. 58,* 325-335.

Hutton, W. E., and Zobell, C. E. (1949). Production of nitrite from ammonia by methane-oxidizing bacteria. *J. Bacteriol. 65,* 216-219.

Iizumi, H., Hattori, A., and McRoy, C. P. (1980). Nitrate and nitrite in interstitial waters of eelgrass beds in relation to the rhizosphere. *J. Exp. Mar. Biol. Ecol. 47,* 191-201.

Indrebo, B., Pengerud, B., and Dundas, I. (1979). Microbial activities in a permanently stratified estuary. II. Microbial activities at the oxic-anoxic interface. *Mar. Biol. (Berlin) 51,* 305-309.

Jahnke, R., Emerson, S., and Murray, J. W. (1982). A model of oxygen reduction, denitrification, and organic matter mineralization in marine sediments. *Limnol. Oceanogr. 27,* 610-623.

Jones, R. D., and Hood, M. A. (1980a). Interaction between an ammonium oxidizer, *Nitrosomonas* sp. and two heterotrophic bacteria, *Nocardia atlantica* and *Pseudomonas* sp.: A note. *Microb. Ecol. 6,* 271-276.

Jones, R. D., and Hood, M. A. (1980b). Effects of temperature, pH salinity and inorganic nitrogen on the rate of ammonium oxidation by nitrifiers isolated from wetland environments. *Microb. Ecol. 6,* 339-347.

Kaplan, W. A., Valiela, I., and Teal, J. M. (1979). Denitrification in a salt marsh ecosystem. *Limnol. Oceanogr. 24,* 726-734.

Kelly, D. P. (1971). Autotrophy: Concepts of lithotrophic bacteria and their organic metabolism. *Annu. Rev. Microbiol. 25,* 177-209.

Kholdebarin, B., and Oertli, J. J. (1977). Effect of suspended particles and their sizes on nitrification in surface water. *J. Water Pollut. Control. Fed. 49,* 1693-1697.

Kiefer, D. A., Olson, R. J., and Holm-Hansen, O. (1976). Another look at the nitrite and chlorophyll maxima in the Central North Pacific. *Deep-Sea Res. 23,* 1199-1208.

King, F. D., and Devol, A. H. (1979). Estimates of vertical eddy diffusion through the thermocline from phytoplankton nitrate uptake rates in the mixed layer of the Eastern Tropical Pacific. *Limnol. Oceanogr. 24,* 645-651.

Knowles, G., Downing, A. L., and Barrett, M. J. (1965). Determination of kinetic constants for nitrifying bacteria in mixed culture, with the aid of an electronic computer. *J. Gen. Microbiol. 38,* 263-278.

Koike, I., and Hattori, A. (1978). Simultaneous determination of nitrification and nitrate reduction in coastal sediments by an ^{15}N dilution technique. *Appl. Environ. Microbiol. 35,* 853-857.

Koops, H. P., Harms, H., and Wehrman, H. (1976). Isolation of a moderate halophilic ammonia-oxidizing bacterium, *Nitrosococcus mobilis* nov. sp. *Arch. Microbiol. 108,* 277-282.

Lees, H., and Quastel, J. H. (1946). Biochemistry of nitrificaion in soil. 2. The site of soil nitrification. *Biochem. J. 40,* 815-823.

Lees, H., and Simpson, J. R. (1957). The biochemistry of the nitrifying organisms. 5. Nitrite oxidation by *Nitrobacter. Biochem. J. 65,* 297-305.

Lipshultz, F., Zafiriou, O. C., Wofsy, S. C., McElroy, M. B., Valois, F. W., and Watson, S. W. (1981). Production of NO and N_2O by soil-nitrifying bacteria. *Nature (London) 294,* 641-643.

Lovelock, J. E. (1974). CS_2 and the natural sulphur cycle. *Nature (London) 248,* 625-626.

McCarthy, J. J., and Goldman, J. C. (1979). Nitrogenous nutrition of marine phytoplankton in nutrient-depleted waters. *Science 203,* 670-672.

McCarthy, J. J., Kaplan, W. A., and Nevins, J. L. (1983). Sources and sinks of nitrite in the York River and Chesapeake Bay. *Limnol. Oceanogr.* (in press).

McElroy, M. B. (1975). Chemical processes in the solar system: A kinetic perspective. *Int. Rev. Sci., Phys. Chem., Ser. Two 9,* 127-211.

McLellan, H. J. (1965). "Elements of Physical Oceanography." Pergamon, Oxford.

Mague, T. H., Mague, F. C., and Holm-Hansen, O. (1979). Physiology and chemical composition of nitrogen-fixing phytoplankton in the Central North Pacific Ocean. *Mar. Biol. (Berlin) 41,* 213-227.

Matin, A. (1978). Organic nutrition of chemolithotrophic bacteria. *Annu. Rev. Microbiol. 32,* 433-468.

Matulewich, U. A., Strom, P. F., and Finstein, M. S. (1975). Length of incubation for enumerating nitrifying bacteria present in various environments. *Appl. Microbiol. 29,* 265-268.

Miyazaki, T., Wada, E., and Hattori, A. (1973). Capacities of shallow waters of Sagami Bay for oxidation and reduction of inorganic nitrogen. *Deep-Sea Res. 20,* 571-577.

Miyazaki, T., Wada, E., and Hattori, A. (1975). Nitrite production from ammonia and nitrate in the euphotic layer of the Western North Pacific Ocean. *Mar. Sci. Commun. 1,* 381-394.

Müller-Neuglück, M., and Engel, H. (1961). Photoinaktivierung von *Nitrobacter winogradskyi Buch. Arch. Microbiol. 39,* 130-138.

Olson, R. J. (1980). Studies of biological nitrogen cycle processes in the upper waters of the ocean, with special reference to the primary nitrite maximum. Ph.D. Thesis, University of California, San Diego.

Olson, R. J. (1981a). ^{15}N tracer studies of the primary nitrite maximum. *J. Mar. Res. 39,* 203-226.

Olson, R. J. (1981b). Differential photoinhibition of marine nitrifying bacteria: A possible mechanism for the formation of the primary nitrite maximum. *J. Mar. Res. 39,* 227-238.

O'Neill, J. G., and Wilkinson, J. F. (1977). Oxidation of ammonia by methane-oxidizing bacteria and the effects of ammonia on methane oxidation. *J. Gen. Microbiol. 100,* 407-412.

Packard, T. T., Dugdale, R. C., Goering, J. J., and Barber, R. T. (1978). Nitrate reductase activity in the subsurface waters of the Peru Current. *J. Mar. Res. 36,* 59-76.

Paerl, H. W., Richards, R. C., Leonard, R. L., and Goldman, C. R. (1975). Seasonal nitrate cycling as evidence for complete vertical mixing in Lake Tahoe, California-Nevada. *Limnol. Oceanogr. 20,* 1-8.

R Painter, H. A. (1970). A review of the literature on inorganic nitrogen metabolism in microorganisms. *Water Res. 4,* 393-450.

Pan, P. C., and Umbreit, W. W. (1972). Growth of obligate autotrophic bacteria on glucose in a continuous flow-through apparatus. *J. Bacteriol. 109,* 1149-1155.

Payne, W. J. (1973). Reduction of nitrogenous oxides by microorganisms. *Bacteriol. Rev. 37,* 409-452.

Poth, M. A., and Focht, D. D. (1982). Production of nitrous oxide during denitrification by *Nitrosomonas europaea* ATCC 19718. *Abstr., 82nd Meet. Am. Soc. Microbiol.,* p. 187.

Powlson, D. S., and Jenkinson, D. S. (1971). Inhibition of nitrification in soil by carbon disulfide from rubber bungs. *Soil Biol.-Biochem. 3,* 267-269.

Quayle, J. R. (1976). Mechanisms of C_1-oxidation by methane utilizers and their correlation with growth yields. *In* "Microbial Production and Utilization of Gases" (H. G. Schlegel, G. Gottschalk, and N. Pfennig, eds.), pp. 353-357. E. Goltze, KG, Göttingen.

Rajendran, A., and Venugopalan, V. K. (1976). Hydroxylamine formation in laboratory experiments on marine nitrification. *Mar. Chem. 4,* 93-98.

Rakestraw, N. W. (1936). The occurrence and significance of nitrite in the sea. *Biol. Bull. (Woods Hole, Mass.) 71,* 133-167.

Redfield, A. C., Ketchum, B. H., and Richards, F. A. (1963). The influence of organisms on the composition of sea-water. *In* "The Sea" (N. H. Hill, E. D. Goldberg, C. O. D. Iselin, and W. H. Munk, eds.), pp. 26-77. Wiley, New York.

Rees, M., and Nason, A. (1966). Incorporation of atmospheric oxygen into nitrite formed during ammonia oxidation by *Nitrosomonas europaea*. *Biochim. Biophys. Acta 113,* 398-402.

Richards, F. A. (1975). The Cariaco Basin (Trench). *Oceanogr. Mar. Biol. 13,* 11-67.

Ritchie, G. A. F., and Nicholas, D. J. D. (1974). The partial characterization of purified nitrite reductase and hydroxylamine oxidase from *Nitrosomonas europaea*. *Biochem. J. 138,* 471-480.

Rittenberg, S. C., Energy, K. O., and Orr, W. L. (1955). Regeneration of nutrients in sediments of marine basins. *Deep-Sea Res. 3,* 23-45.

Rudd, J. W. M., and Taylor, C. D. (1980). Methane cycling in aquatic environments. *Adv. Aquat. Microbiol. 2,* 77-150.

Salvas, P. L., and Taylor, B. F. (1980). Blockage of methanogenesis in marine sediments by the nitrification inhibitor 2-chloro-6(trichloromethyl)pyridine (nitrapyn or N-serve). *Curr. Microbiol. 4,* 305-308.

Sansone, F. J., and Martens, C. S. (1978). Methane oxidation in Cape Lookout Bight, North Carolina. *Limnol. Oceanogr. 23,* 349-355.

Schell, D. M. (1974). Regeneration of nitrogenous nutrients in arctic Alaskan estuarine waters. *Symp. Beaufort Sea Coastal Shelf Res., Arct. Inst. North Am.,* p. 649-663.

Schell, D. M. (1978). Chemical and isotropic methods in nitrification studies. *In* "Nitrification and Reduction of Nitrogen Oxides" (D. Schlessinger, ed.), pp. 292-295. Am. Soc. Microbiol., Washington, D.C.

Schmidt, E. L. (1974). Quantitative autoecological study of microorganisms in soil by immuno-fluorescence. *Soil Sci. 118,* 141-149.

Schmidt, E. L. (1978). Nitrifying organisms and their methodology. *In* "Nitrification and Reduction of Nitrogen Oxides" (D. Schlessinger, ed.), pp. 288-291. Am. Soc. Microbiol., Washington, D.C.

Schön, G. (1965). Untersuchungen über den Nutzeffekt von *Nitrobacter winogrodskyi Buch.* *Arch. Mikrobiol. 50,* 111-132.

Schön, G., and Engel, H. (1962). Den Einfluss des Lichtes of *Nitrosomonas europaea* Win. *Arch. Microbiol. 42,* 415-428.

Sieburth, J. McN. (1967). Seasonal selection of estuarine bacteria by water temperature. *J. Exp. Mar. Biol. Ecol. 1,* 98-121.

Smith, A. J., and Hoare, D. S. (1968). Acetate assimilation by *Nitrobacter agilis* in relation to its "obligate autotrophy." *J. Bacteriol. 95,* 844-855.

Somville, M. (1978). A method for the measurement of nitrification rates in water. *Water Res. 12*, 843-848.

Spencer, C. P. (1975). The micronutrient elements. *In* "Chemical Oceanography" (J. P. Riley and G. Skirrow, eds.), 2nd ed., Vol. 2, pp. 245-295. Academic Press, New York.

Srna, R. F., and Baggaley, A. (1975). Kinetic responses of perturbed marine nitrification systems. *J. Water Pollut. Control Fed. 47*, 472-486.

Steinmüller, W., and Bock, E. (1976). Growth of *Nitrobacter* in the presence of organic matter. I. Mixotrophic growth. *Arch. Microbiol. 108*, 299-304.

Sugahara, I., Sugiyama, M., and Kawai, A. (1974). Distribution and activity of nitrogen cycle bacteria in water-sediment systems with different concentrations of oxygen. *In* "Effect of the Ocean Environment on Microbial Activities" (R. Colwell and R. Y. Morita, eds.), pp. 327-340. Univ. Park Press, Baltimore, Maryland.

Suzuki, I. (1974). Mechanisms of inorganic oxidation and energy coupling. *Annu. Rev. Microbiol. 28*, 85-102.

Suzuki, I., Dular, V., and Kuok, S. C. (1974). Ammonia or ammonium as substrate for oxidation by *Nitrosomonas europaea* cells and extracts. *J. Bacteriol. 120*, 556-558.

Vaccaro, R. F. (1962). The oxidation of ammonia in sea water. *J. Cons. Explor. Mer. 27*, 3-14.

Vaccaro, R. F. (1965). Inorganic nitrogen in seawater. *In* "Chemical Oceanography" (J. P. Riley and G. Skirrow, eds.), pp. 365-408. Academic Press, London.

Vaccaro, R. F., and Ryther, J. H. (1960). Marine phytoplankton and the distribution of nitrite in the sea. *J. Cons. Cons. Int. Explor. Mer. 25*, 260-271.

Vanderborght, J. P., and Billen, G. (1975). Vertical distribution of nitrate concentration in interstitial water of marine sediments with nitrification and denitrification. *Limnol. Oceanogr. 20*, 953-961.

Venrick, E. L., McGowan, J. A., and Mantlya, A. W. (1973). Deep chlorophyll maxima in the oceanic Pacific. *Fish. Bull. 71*, 41-52.

Verstraete, W., and Alexander, M. (1973). Heterotrophic nitrification in samples of natural ecosystems. *Environ. Sci. Technol. 7*, 39-42.

von Breymann, M. T., de Angelis, M. A., and Gordon, L. I. (1982). Gas chromatography with electron capture detection for determination of hydroxylamine in seawater. *Anal. Chem. 54*, 1209-1210.

Wada, E., and Hattori, A. (1971). Nitrite metabolism in the euphotic layer of the Central North Pacific Ocean. *Limnol. Oceanogr. 16*, 766-772.

Waksman, S. A., Reuzer, H. W., Carey, C. L., Hotchkiss, M., and Renn, C. E. (1933a). Studies on the biology and chemistry of the Gulf of Maine. *Biol. Bull. (Woods Hole, Mass.) 64,* 183-205.

Waksman, S. A., Hotchkiss, M., and Carey, C. L. (1933b). Marine bacteria and their role in the cycle of life in the sea. *Biol. Bull. (Woods Hole, Mass.) 65,* 137-167.

Wallace, W., and Nicholas, D. J. D. (1969). The biochemistry of nitrifying microorganisms. *Biol. Rev. Cambridge Philos. Soc. 44,* 359-391.

Walsh, J. J., Rowe, G. T., Iverson, R. L., and McElroy, C. P. (1981). Biological export of shelf carbon is a sink of the global CO_2 cycle. *Nature (London) 291,* 196-201.

Ward, B. B., and Perry, M. J. (1980). Immunofluorescent assay for the marine ammonium-oxidizing bacterium *Nitrosococcus oceanus. Appl. Environ. Microbiol. 39,* 913-918.

Ward, B. B., Olson, R. J., and Perry, M. J. (1982). Microbial nitrification rates in the primary nitrite maximum off Southern California. *Deep-Sea Res. 29,* 247-255.

Watson, S. W. (1965). Characteristics of a marine nitrifying bacterium *Nitrosocystis oceanus* sp.n. *Limnol. Oceanogr. 10,* Suppl., 274-289.

Watson, S. W., and Waterbury, J. B. (1971). Characteristics of two marine nitrite oxidizing bacteria, *Nitrospina gracilis nov. gen. nov.* sp. and *Nitrococcus mobilis nov. gen. nov.* sp. *Arch. Microbiol. 77,* 203-230.

Watson, S. W., Novitsky, T. J., Quinby, H. C., and Valois, F. W. (1977). Determination of bacterial number and biomass in the marine environment. *Appl. Environ. Microbiol. 33,* 940-946.

Webb, K. L., and Wiebe, W. J. (1975). Nitrification on a coral reef. *Can. J. Microbiol. 21,* 1427-1431.

Wezernak, C. T., and Gannon, J. J. (1967). Oxygen-nitrogen relationships in autotrophic nitrification. *Appl. Microbiol. 15,* 1211-1214.

Whittaker, R. H., and Likens, G. E. (1972). Carbon in the biota. *In* "Carbon in the Biosphere" (G. M. Woodwell and E. V. Pecan, eds.), pp. 281-302. U.S. At. Energy Comm., Washington, D.C.

Whittenbury, R., Dalton, H., Eccleston, M., and Reed, R. L. (1975). The different types of methane-oxidizing bacteria and some of their more unusual properties. *In* "Microbial Growth on C_1 Compounds," pp. 1-9. Soc. Ferment. Technol., Tokyo, Japan.

Williams, P. J. Le B., and Watson, S. C. (1968). Autotrophy in *Nitrosocystis oceanus. J. Bacteriol. 96,* 1640-1648.

Wong-Chong, G. M., and Loehr, R. C. (1975). The kinetics of microbial nitrification. *Water Res. 9,* 1099-1106.

Wüst, G. (1978). Stratosphere of the Atlantic Ocean. *In* "Scientific Results of the German Atlantic Expedition of the Research Vessel Meteor" (W. J. Emery, ed.), 112 pp. U.S. Dept. of Commerce, Washington, D.C.

Yoshida, Y. (1967). Studies on the marine nitrifying bacteria: With special reference to characteristics and nitrite formation of marine nitrite formers. *Bull. Misaki Kenkyu Hokoku Maizuru 11*, 2-58.

ZoBell, C. E. (1943). The effect of solid surfaces upon bacterial activity. *J. Bacteriol. 46*, 39-56.

Note Added in Proof

Jones and Morita (*Appl. Environ. Microbiol. 45*, 401-410) were able to demonstrate substantial methane oxidation by cultures of *N. oceanus* and *N. europaea*. Hynes and Knowles (*Appl. Environ. Microbiol. 45*, 1179-1182) criticized the chlorate-inhibition assay for nitrite oxidation. The chlorite (ClO_2^-) produced by the cellular reduction of ClO_3^- will, in turn, inhibit NH_4^+ oxidation. Care should be exercized when using this assay in mixed populations of primary and secondary nitrifiers. French et al. (*Deep-Sea Res. 30*, 707-722) studied diel changes in the primary NO_2^- maximum. Using a ID-vertical model, they calculated nitrification rates 2-5 orders of magnitude faster than other estimates.

Chapter 6

Denitrification and Dissimilatory Nitrate Reduction

AKIHIKO HATTORI
Ocean Research Institute
University of Tokyo
Nakano-ku, Tokyo 164, Japan

I. INTRODUCTION

Although nitrate and nitrite are reduced photochemically, these processes are restricted to the top thin layer of surface water (Hamilton, 1964). Only biological processes are of quantitative importance in the transformation of nitrate in marine environments.

Two types of biological nitrate reduction can be distinguished: (1) assimilatory nitrate reduction, and (2) dissimilatory nitrate reduction (Verhoeven, 1956; Takahashi et al., 1963). In the first type, nitrate is reduced to ammonium, which is assimilated as a source of nitrogen for the synthesis of cellular materials. Plants including phytoplankton and a number of aerobic bacteria and fungi share this capacity. This process proceeds at the expense of light energy (photosynthetic organisms) or energy reserved in organic

ISBN 0-12-160280-X

compounds (heterotrophic organisms). In the second type, nitrate serves primarily as terminal electron acceptor of respiration in place of oxygen, and energy yielded in association with nitrate reduction is utilized for biosynthetic and other cellular endergonic reactions. This capacity is found in a variety of facultatively anaerobic bacteria and in a limited number of strictly anaerobic bacteria. Nitrate is not used as a nitrogen source, and products of nitrate reduction are excreted from the cells. When the end products of nitrate reduction are gases, e.g., N_2 and N_2O, the process is called denitrification, because they are ultimately lost to the atmosphere.

This chapter is primarily concerned with processes of dissimilatory nitrate reduction occurring in marine environments. Various aspects of assimilatory nitrate reduction are dealt with by Goldman (Chapter 7), Wheeler (Chapter 9), and Falkowski (Chapter 23) in this volume. Estimates of oceanic nitrogen budget (Emery et al., 1955; Tsunogai, 1971; Holland, 1973) indicate that the annual supply of combined nitrogen to the oceans by river runoff and by precipitation on the sea surface exceeds the loss by burial in sediments by $10\text{-}70 \times 10^{12}$ g of nitrogen. If a steady state is to be maintained in the sea, this excess nitrogen must be released to the atmosphere. Denitrification appears to be the mechanism by which this is accomplished. In the above estimates, input of combined nitrogen by biological nitrogen fixation is not included. However, oceanic nitrogen fixation of the order of 10^{13} g N year^{-1} is suggested by Söderlund and Svensson (1976) and Saino (1977). Carpenter (Chapter 3, this volume) estimates an annual nitrogen fixation by *Trichodesmium* populations of 5×10^{12} g N and Capone (Chapter 4, this volume) estimates a benthic fixation of 15×10^{12} g N. Thus, dissimilatory nitrate reduction not only supports marine life under anoxic or oxygen-depleted conditions, but also plays an important role for the balance of combined nitrogen in the sea. Denitrification in coastal and estuarine areas is also the subject of active investigation because of increased eutrophication caused by increase in anthropogenic input.

II. TYPES OF BACTERIAL DISSIMILATORY NITRATE REDUCTION

This section summarizes the types and processes of dissimilatory nitrate reduction by various groups of bacteria. Information has been obtained mainly from bacteria of terrestrial origin, but discussion of them is useful in order to understand events occurring in the sea.

Products of dissimilatory nitrate reduction vary depending on the species or genera of bacteria. Under oxygen-depleted conditions, many facultative anaerobes reduce nitrate to nitrite, but no further. Relatively limited numbers of bacteria can reduce nitrate to N_2 through a series of consecutive reduction steps, and N_2O is occasionally accumulated. According to ZoBell (1946), about 40% of the marine bacteria he had isolated could reduce nitrate to nitrite in the presence of sufficient organic matter, but fewer than 5% could reduce nitrate to N_2.

Terminology in this field is somewhat imprecise, because the terms nitrate respiration, dissimilatory nitrate reduction, respiratory nitrate reduction, and denitrification encompass each other to a varying extent. Zumft and Cárdenas (1979) proposed to specify a reaction by identifying substrate and product. Since the nitrate reductions to nitrite by *Escherichia coli*, *Micrococcus denitrificans*, *Pseudomonas aeruginosa*, and *P. denitrificans* are coupled to electron transport phosphorylation (Haddock and Jones, 1977; Thauer et al., 1977), they can be termed nitrate respiration to nitrite, or simply nitrate respiration. Some strains of *Pseudomonas* are unable to reduce nitrate, but they can use nitrite as a terminal electron acceptor; nitrite is reduced to N_2O and then to N_2 (Vangnai and Klein, 1974). The first and second steps of this reaction can be designated nitrite respiration to nitrous oxide and nitrous oxide respiration to dinitrogen, respectively, or simply nitrite respiration and nitrous oxide respiration, respectively. However, enzymes responsible for individual steps of nitrite reduction and associated electron transport components are poorly characterized and vary depending on bacterial species or genera. We, therefore, prefer to reserve the conventional term denitrification to situations when the end products are gaseous, irrespective of the electron acceptor, nitrate or nitrite. Respiratory nitrate reduction can be used as a general term in which nitrate, nitrite, nitrous oxide respirations and denitrification are included.

The pathway of denitrification as suggested by Payne (1973) is as follows:

$$NO_3^- \rightarrow NO_2^- \rightarrow NO \rightarrow N_2O \rightarrow N_2 \quad (1)$$

Nitrite and N_2O are generally accepted as the obligatory and stable intermediates of denitrification. Evidence for this was provided by transient accumulation of nitrite during nitrate reduction (Payne, 1973), accumulation of N_2O in the presence of acetylene (Balderston et al., 1976; Yoshinari and Knowles, 1976), and ^{15}N-labeling of N_2O and N_2 from $^{15}NO_2^-$ (St. John and Hollocher, 1977). Some strains of *Pseudomonas fluorescens* and *P. chlororaphis* (Greenberg and Becker, 1977) and *Corynebacterium nephridii* (Hart et al., 1965; Renner and Becker, 1970) reduce nitrate only to N_2O.

Although NO is identified as reaction product of the nitrite reductases from *Alcaligenes faecalis* (Matsubara and Iwasaki, 1971), *Pseudomonas perfectomarinus* (Cox and Payne, 1973), and *P. aeruginosa* (Wharton and Weintraub, 1980) and NO reductases are also found in these bacteria, evidence for the positioning of NO in a linear reduction sequence as an obligatory intermediate is not adequate. St. John and Hollocher (1977) have shown that only trace amounts of ^{15}NO and $^{14}N^{15}N$ are produced when the actively denitrifying cells of *P. aeruginosa* are incubated in the presence of $^{15}NO_2^-$ and ^{14}NO. On the other hand, Firestone et al. (1979) demonstrated ^{13}N exchange from $^{13}NO_2^-$ to the NO pool in *Pseudomonas aureofaciens* and *P. chlororaphis*. The exchange was significant only when sufficient amounts of unlabeled NO were added. The reason for the disparity in results of the two experiments is not clear, but this is probably due to the difference in bacterial species. Recent extensive ^{15}N tracer studies on *P. denitrificans, P. aureofaciens, P. stutzeri,* and *Paracoccus denitrificans* (Garber and Hollocher, 1981) support this inference. Alternative pathways for denitrification Eqs. (2) and (3) have been proposed by Zumft and Cárdenas (1979) and Firestone et al. (1979):

(b) NO_{free} (formed from NO_2^- and reduced to N_2O)

$$\text{(a)}\quad NO_2^- \rightarrow [X_{bound}] \rightarrow N_2O \rightarrow N_2 \qquad (2)$$

$$NO_2^- \rightarrow X \rightarrow N_2O \rightarrow N_2 \qquad (3)$$
$$X \rightleftarrows NO$$

In Eq. (2), nitrite is reduced either directly to N_2O (reaction a) or indirectly by way of NO (reaction b). Equation (3) includes an unknown intermediate "X" with which NO is in equilibrium. A more sophisticated scheme, in which a nitrosyl complex of ferrous porphyrin is postulated as an intermediate, was recently presented by Averill and Tiedje (1982) for conversion of nitrite to N_2O, and its plausibility was discussed in connection with the known inorganic chemistry of related systems.

Cox and Payne (1973) found that nitrite and NO reductases of *P. perfectomarinus* are soluble, but nitrate and N_2O reductases are membrane-bound. They inferred that throughout the whole process of denitrification, only two steps, the reduction of nitrate to nitrite and the reduction of N_2O to N_2, function as energy-yielding systems. Phosphorylation is presumably coupled to N_2O reduction, because many denitrifiers and nitrate respirers can grow with N_2O as a sole electron acceptor of respiration (Barbaree and Payne, 1967; Matsubara, 1971; Koike and Hattori, 1975; Yoshinari, 1980). The energy-yield data of *P. denitrificans* (Koike and Hattori, 1975) suggest that phosphorylations are coupled to each reduction step of denitrification.

A comprehensive list of the variety of species involved in respiratory nitrate reduction and its products is given by Payne (1973). Supplementary information can be found in the review of Focht and Verstraete (1977).

Denitrifiers are, in general, facultative anaerobes, but they cannot grow anaerobically by fermentation. In other words, they need substitute electron acceptors such as nitrate, nitrite, and N_2O for their anaerobic growth. Exceptions are found in a limited number of propionic acid bacteria (Van Gent-Ruijters et al., 1975; Kaneko and Ishimoto, 1978). They are strict anaerobes, but contain cytochromes *a* and *b* and cytochrome oxidase and produce N_2 and N_2O from nitrate. Other strict anaerobes, e.g., *Veillonella alcalescens* and *Selenomas ruminatum*, reduce nitrate to nitrite, but no further (de Vries et al., 1974). They can grow using energy yielded by fermentation.

Most of the bacteria that are capable of reducing nitrate and other nitrogenous oxides are heterotrophs. Organic compounds, e.g., formate, lactate, and glucose, usually serve as reductants. *Thiobacilus denitrificans*, a lithotrophic autotroph, is also able to utilize reduced sulfur compounds (sulfide and thiosulfate) to reduce nitrate (Aminuddin and Nicholas, 1973). Isolates from anoxic waters and sediments of the Black Sea and Cariaco Trench exhibited this activity (Tuttle and Jannasch, 1972, 1973).

Clostridium perfringens and *C. tertitium*, strict anaerobes, contain no respiratory cytochrome components, but they reduce nitrate in an apparently dissimilatory way to ammonium as an end product. Nitrate serves as an electron sink rather than as an electron acceptor of respiration, and contributes to reoxidation of reduced NAD which is produced in fermentation. Ferredoxin is an immediate electron donor for the nitrate reduction (Chiba and Ishimoto, 1973; Seki et al., 1979). The term fermentative nitrate reduction is used to distinguish this reaction from the respiratory nitrate reduction (Fenchel and Blackburn, 1979). Fermentation balance studies in the presence and absence of nitrate indicate that ATP is formed from acetyl coenzyme A in association with the nitrate reduction (Hasan and Hall, 1975, 1977; Ishimoto et al., 1974). Therefore, this process should be included in the category of dissimilatory nitrate reduction. Although ammonium can also be produced as an end product of respiratory nitrate reduction by *Aerobacter aerogenes* (Hadjipetrou and Stouthamer, 1965), *Achromobacter fischeri* (Prakash and Sanada, 1973), and *Escherichia coli* (Cole, 1978), ammonium production from nitrate can, in general, be taken as an indication of a fermentative pathway (Fenchel and Blackburn, 1979).

Dissimilatory nitrate reduction exhibits a high diversity among bacteria, but it is common in the following respects and distinct from the assimilatory nitrate reduction: (1) The reaction is inhibited by oxygen but not by ammonium; (2) the energy for synthesis of cellular materials and proper functioning of cell machinery is generated; (3) the end products are liberated from cells; (4) nitrate reductase is membrane-bound and reduces chlorate; and (5) syntheses of the responsible reductases and associated cytochromes (except for *Clostridium*) are repressed by oxygen, induced (or derepressed) by nitrate and other nitrogenous oxides, but not affected by ammonium. Comprehensive references on biochemical and bioenergetic aspects of dissimilatory nitrate reduction can be found in the reviews of Payne (1973), Focht and Verstraete (1977), Haddock and Jones (1977), Thauer et al. (1977), Zumft and Cárdenas (1979), Whatley (1981), Payne (1981), and Knowles (1982). Focht and Verstraete (1977) and Knowles (1982) also deal with ecological aspects of dissimilatory nitrate reduction. Earlier literature is compiled in the review of Painter (1970).

III. DISSIMILATORY NITRATE REDUCTION IN MARINE AQUATIC SYSTEMS

A. Estimates of Denitrification from Oxygen, Nitrate, Nitrite, and Phosphate Distributions

Bacterial denitrification takes place in restricted sea areas where oxygen concentrations are sufficiently low. Since dissolved N_2 is abundant in seawater, it is difficult, though not impossible (cf. Section III,B), to assess the extent of denitrification from direct measurement of small amounts of N_2 produced by denitrification. Among dissolved oxygen, phosphate, and nitrate concentrations in subsurface water, there exist, in general, definite relationships that can be well described by the stoichiometric model of Redfield et al. (1963). When nitrate is lost by denitrification as N_2 and N_2O, significant anomaly appears on an oxygen-versus-nitrate diagram but not on an oxygen-versus-phosphate diagram (cf. Fiadeiro and Strickland, 1968; Deuser et al., 1978). Anomalies are invariably found in water masses in which oxygen is extremely depleted, and are associated with high concentrations of nitrite, the first product of dissimilatory nitrate reduction. The presence of a secondary nitrite maximum in the low-oxygen subsurface waters of the Eastern Tropical North Pacific was first described by Brandhorst (1959), and associated nitrate deficiencies later by Thomas (1966). Using water samples collected from the oxygen minimum layer of the same area, Goering and Cline (1970) demonstrated experimentally that nitrate is reduced and nitrite is accumulated in nearly a 1:1 ratio before the onset of the further reduction of nitrite to gaseous nitrogen. A similar relationship was also found by Richards and Broenkow (1971) in their observations in Darwin Bay, an oxygen-depleted embayment in the Galapagos Islands.

If the water mass under consideration is *closed* or if material exchange between this water mass and that outside is negligibly small compared to the biological change within this water mass, one can estimate the extent of denitrification, or nitrate deficit, from nitrogen mass balance on the basis of the stoichiometric model. The nitrate deficit can be calculated by the equation

$$\Delta N = [NO_3^-]_{exp} - [NO_3^-]_{obs} - [NO_2^-]_{obs}$$

where $[NO_3^-]_{exp}$ is the nitrate concentration expected before the onset of denitrification, and $[NO_3^-]_{obs}$ and $[NO_2^-]_{obs}$ are observed nitrate and nitrite concentrations, respectively. Data on distributions of nutrients and other environmental variables can also be used to identify the extent of water mass in which denitrification proceeds intensively. Many investigations have been conducted along this line over extended sea areas where oxygen data suggest the active occurrence of denitrification. The closed system assumption is not always valid. Corrections and modifications have been made in several ways. To estimate the rate of oceanic denitrification, additional parameters are also needed that contain information on rate processes. These include residence time of water, horizontal and vertical advective transport of water, diffusivities of solutes, and biological activities such as electron transport system (ETS) activities. Basic assumptions and techniques used are concisely described by Goering (1978). Details can be found in the papers of Cline (1973), Codispoti and Richards (1976), and Codispoti and Packard (1980).

Table I summarizes estimates of oceanic denitrification presented by various authors. With a few exceptions, the areal estimates of denitrification obtained by different methods are in good agreement with each other, and so probably of the right order of magnitude. The estimates suggest that denitrification in sulfide-bearing waters, as found in the Black Sea and the Cariaco Trench in the Caribbean Sea, is not quantitatively important for maintenance of the nitrogen balance in the sea. The low-oxygen intermediate waters of the Eastern Tropical Pacific Ocean are the major sites for denitrification. Denitrification in the low-oxygen intermediate water of the Arabian Sea and the bottom water of the

TABLE I. Summary of Water Column Denitrification Rates

Site	Area[a] ($m^2 \times 10^{12}$)	Volume[b] ($m^3 \times 10^{13}$)	Annual denitrification (g N·$year^{-1} \times 10^{12}$)	Method[c]	Reference
Eastern Tropical North Pacific			10	(1)	Goering et al. (1973)
	6.7	140	10	(2)	Goering et al. (1973)
	3.1	110	230	(3)	Cline and Richards (1972)
	2.9	140	16	(4)	Cline (1973)
			23	(5)	Codispoti and Richards (1976)
	1.9	86	19	(2)	Codispoti and Richards (1976)
Eastern Tropical South Pacific	1.1	14	25	(2)	Codispoti and Packard (1980)
			19	(5)	Codispoti and Packard (1980)
Arabian Sea	0.65	9.8	0.1-1	(6)	Deuser et al. (1978)
Eastern Tropical Atlantic	0.01	0.1	5	(1)	I. R. Kaplan, quoted in Goering (1978)

Black Sea			0.007	(7)	Goering et al. (1973)
		0.3	0.2	(8)	Goering et al. (1973)
Cariaco Trench	0.01	0.52	0.011	(6)	Goering et al. (1973)
			0.01	(4)	Cline (1973)

[a]*Estimated area of low-oxygen water.*

[b]*Estimated volume of low-oxygen water.*

[c]*(1) Estimated from nitrate deficit and volume transport of low-oxygen intermediate water; (2) estimated from volume of low-oxygen water and electron transport system (ETS) activity; (3) estimated from volume of low-oxygen water and denitrification rate observed in Darwin Bay (Richards and Broenkow, 1971); (4) estimated based on a one-dimensional diffusion-advection model; (5) estimated in the same way as (1), but corrected for horizontal and vertical diffusion losses; (6) estimated from volume of low-oxygen water and its residence time; (7) estimated from influx of outside water; and (8) estimated on assumed yearly renewal of anoxic water.*

southwest African shelf are also substantial. The Bay of Bengal in the Indian Ocean may contribute to oceanic denitrification, as oxygen concentrations of less than 0.2 ml/liter and high nitrite concentrations have been observed (Wyrtki, 1971). The volume of low-oxygen water in this bay has not been computed, and no estimates of its yearly denitrification activity have been made.

The volume of the low-oxygen water with elevated nitrite concentrations in the Eastern Tropical South Pacific is only one-tenth the volume of the Eastern Tropical North Pacific. Surprisingly, the estimated rates of annual denitrification are similar (Table I). Among various environmental factors, availability of organic matter as an electron donor is probably of major importance in regulating the rates of nitrate reduction and denitrification in marine systems. Evidence supporting this inference is provided by studies of ^{15}N enrichment in nitrate (Section III,D). Organic matter produced photosynthetically in the euphotic layer settles down and serves as electron donor for nitrate reduction in subsurface waters. Coastal upwelling prevails in the area of the Eastern Tropical South Pacific in question. Consequently, primary productivity there is maintained at a higher level than that in the northern zone. This might explain comparable denitrification on other side of the equator. The high denitrification rate in the area off the southwest coast of Africa might also result from coastal upwelling.

Nitrite concentrations in the oxygen minimum layer of the Eastern Tropical South Pacific are consistently higher than those in the Eastern Tropical North Pacific by a factor of 3-5 (Wooster et al., 1965; Fiadeiro and Strickland, 1968; Brandhorst, 1959; Thomas, 1966; Goering, 1968). A positive correlation between nitrite concentrations and denitrification rates has been shown (Codispoti and Packard, 1980). Approximately 10% of primary production is utilized as fuel for denitrification in the Eastern Tropical South Pacific. This value is a little higher than, but falls within, the range estimated in the Eastern Tropical North

Pacific. Codispoti and Packard (1980) suggest a possible extra supply of organic matter that may be carried into the nitrite maximum layer by the net offshore flow that exists over the Peruvian shelf. However, it should be noted that the high denitrification rate in the Eastern Tropical South Pacific might be merely a reflection of natural and sporadic events, e.g., *El Niño* (Wyrtki, 1977) and *Aguaje* (Dugdale et al., 1977), which occurred before and/or during the period when the data used for the denitrification estimates were collected (Codispoti and Packard, 1980).

Complete denitrification and subsequent sulfide production was observed by Dugdale et al. (1977) at depths from 135 to 175 m in the upwelling region off Peru. High ammonium concentrations of up to 6 μ*M* were also found at this depth. They suggested that an unusually massive production of the dinoflagellate *Gymnodinium splendens* (*Aguaje*) and a subsequent supply of large amounts of organic matter to depths were responsible for this phenomenon. Some sulfur bacteria reduce nitrate using sulfide as an electron donor (Section II). However, this activity in marine aquatic systems is probably insignificant. Although Ivanenkov and Rozanov (1961) reported the occurrence of H_2S in the Arabian Sea where nitrate and nitrite were also present, their observations were recently questioned by Deuser et al. (1978). The Peru upwelling region data (Dugdale et al., 1977) and the observations in Golfo Dulce, Costa Rica (Richards et al., 1971) and the Cariaco Trench (Cline, 1973) commonly show a clear spatial separation of distributions of H_2S and NO_3^-. It is more likely, as generally accepted, that reduction of sulfate is initiated after the exhaustion of both nitrate and nitrite.

From his closed-system model calculation using data sets of nutrients, oxygen, and ^{14}C abundance in total carbonate, Tsunogai (1972) suggested that denitrification might take place, at a rate of 10-20 n*M* year^{-1}, in the deep water of the South Pacific with salinity of >34.7 o/oo. The occurrence of denitrification in this water body seems unlikely from a biochemical point of view because of oxygen concentrations excessive for denitrification and of low nitrite concentrations, but its possibility cannot be ruled out. If it occurs, denitrification would likely proceed within suspended organic particles which may produce, in oxic environments, microsites of anoxic condition (Gundersen, 1977).

B. Denitrification and Excess Dissolved N_2 and N_2O

Richards and Benson (1961) earlier noted the occurrence of denitrification in anoxic waters of the Cariaco Trench by mass spectrometric determination of the N_2:Ar ratio. Using a similar technique, the presence of excess N_2 associated with denitrification in a brackish lake was demonstrated by Hattori et al. (1970). A gas chromatographic technique for precise determination of the dissolved N_2:Ar ratio has been developed by Cline and Ben-Yaakov (1973). This technique has been successfully used by Cline (1973) and Cline and Ben-Yaakov (1973) for determination of excess N_2 in the low-oxygen waters of the Eastern Tropical North Pacific and Cariaco Trench. The excess N_2 in the Cariaco Trench was in good agreement with the nitrate deficit calculated from the mass balance of combined nitrogen, evidence for a complete reduction of nitrate to N_2.

Nitrous oxide is produced by some denitrifiers (Section II) and N_2O supersaturation has been observed in the water columns of the North Atlantic, Caribbean Sea, and Eastern Tropical Pacific (Hahn, 1974, 1975; Yoshinari, 1976; Cohen and Gordon, 1978, 1979; Elkins et al., 1978). Nitrous oxide is evidently released to the atmosphere, and this process undoubtedly contributes to the maintenance of nitrogen balance in the sea. However, N_2O concentrations are negatively correlated with oxygen concentrations (Yoshinari, 1976; Cohen and Gordon, 1979). It is more likely that N_2O in the open oceans is produced in association with nitrification (see Kaplan, Chapter 5, this volume). Nitrous oxide production by pure cultures of marine nitrifying bacteria have been found by Goreau et al.

(1980). Cohen and Gordon (1978) suggested that N_2O is reduced in the oxygen minimum layer of the Eastern Tropical North Pacific. If so, the low-oxygen intermediate waters may act as a sink for N_2O.

C. Direct Determination of Denitrification and Nitrate Reduction

Goering and Dugdale (1966) were the first to directly determine denitrification rates in marine aquatic systems by measuring the production of $^{15}N_2$ from $^{15}NO_3^-$ added as a tracer. The technique they used is essentially the same as that developed by Hauck et al. (1958) for denitrification studies in soil systems. Because of the relatively low sensitivity of ^{15}N detection, the addition of large quantities of $^{15}NO_3^-$ and a long incubation time were required. Therefore, it is difficult to assess how the denitrification rates obtained in this way reflect *in situ* activities. Actually, the denitrification rate measured with the water samples from Darwin Bay (0.9-1.3 μM day^{-1}, Goering and Dugdale, 1966) is one order of magnitude higher than the value derived from time series observations of nitrate and nitrite in the same bay (0.16 μM day^{-1}, Richards and Broenkow, 1971). A much greater difference is found, in the Eastern Tropical North Pacific, between the denitrification rates measured by the ^{15}N technique (Goering, 1968) and those estimated by a one-dimensional diffusion-advection model (Cline, 1973). The directly measured denitrification rates undoubtedly give overestimates, and should be regarded, at best, as potential capacity for denitrification. The solution to the problem may be to use a much more sensitive mass spectrometric technique for ^{15}N determination, such as differential ratiometry of $^{15}N:^{14}N$. This technique has been used for the determination of denitrification in a brackish lake (Koike et al., 1972) but not in any major oceanic denitrification sites. The ^{13}N technique of Gersberg et al. (1976) is very sensitive, but the short lifetime of ^{13}N (10 min) and the requirement of cyclotron facilities restrict its application to the field studies.

Nitrate reduction to nitrite can also be determined by monitoring ^{15}N incorporation from $^{15}NO_3^-$ into nitrite. Although investigations have so far been confined mainly to nitrite production in or near the primary nitrite maximum layer (Miyazaki et al., 1973, 1975; Olson, 1981), some limited data suggest that nitrate is reduced to nitrite and then to ammonium in the deep waters of Suruga Bay, Japan, and in the bottom waters of the Philippine Sea (Wada and Hattori, 1972; Wada et al., 1975b). It is not known whether this reduction proceeds in a dissimilatory way by the action of nitrate respirers or in assimilatory way by the action of aerobic bacteria. Data on nitrate reductase activity in deep waters have been presented by Packard et al. (1978). Since respiration is, in general, regulated by the energy charge, it is difficult to unequivocally determine *in situ* rates of nitrate reduction from the activity of nitrate reductase.

The acetylene inhibition technique is a very promising alternative for determining denitrification rates in the field (see B. F. Taylor, Chapter 22, this volume). Acetylene blocks the terminal step of N_2O reduction to N_2. In the presence of 10^{-2} atm of C_2H_2, nitrate nitrogen is quantitatively recovered as N_2O, which can be easily measured by gas chromatography with very high sensitivity (Balderson et al., 1976; Yoshinari and Knowles, 1976). To my knowledge, there have been no studies of oceanic denitrification using C_2H_2 blockage, although this technique is now widely used in studies of denitrification in soils, inland waters, and coastal and lacustrine sediments.

D. Nitrogen Isotope Fractionation in Denitrification and Nitrate Respiration

Nitrogen isotope fractionation in association with bacterial denitrification and nitrate respiration has been documented (Table II). Cline and Kaplan (1975) disclosed ^{15}N enrichment in nitrate in the oxygen minimum layer of the Eastern Tropical North Pacific. Using a one-dimensional diffusion-advection model, they estimated

TABLE II. Nitrogen Isotope Fractionation for Nitrate Reduction

Reaction	*Fractionation factor*	*Reference*
Denitrification		
Pseudomonas stutzeri	1.02, 1.03	Wellman et al. (1968)
Pseudomonas denitrificans	1.017	Delwiche and Steyn (1970)
Unidentified denitrifier	1.02	Miyake and Wada (1971)
Soil	1.019, 1.0065	Chien et al. (1977)
Soil	1.011-1.017	Blackmer and Bremner (1977)
Eastern Tropical North Pacific	1.03-1.04	Cline and Kaplan (1975)
Nitrate respiration		
Serratia marinorubra	1.039	Miyazaki et al. (1980)
Assimilatory nitrate reduction		
Phaeodactylum tricornutum	1.023	Wada and Hattori (1978)
Nitrate reduction		
(inorganic reaction)	1.075	Brown and Drury (1967)
N—O bond rupture (theoretical)	1.09	Wellman et al. (1968)

the fractionation factor for nitrate loss and suggested that $^{14}NO_3^-$ is consumed 3-4% faster than $^{15}NO_3^-$. This value is somewhat greater than those reported for biological nitrate reduction in laboratory experiments.

In their study, Miyazaki et al. (1980) showed that the nitrate reduction system of the marine nitrate respirer *Serratia marinorubra* consists of the two reaction steps that produce different kinetic isotope effects: active transport of nitrate across the cell membrane and reduction of nitrate to nitrite. Highest fractionation of 3.9% is obtained when nitrate respiration is limited at the latter step. The extent of isotope fractionation is almost identical with that estimated by Cline and Kaplan (1975), and is the highest ever reported for biological nitrate reduction. The rate of nitrate reduction mediated by nitrate reductase is limited either by nitrate concentration within the cells or by supply of electrons from substrate of respiration. Chien et al. (1977) found that the nitrogen isotope fractionation associated with denitrification in soils is lowered when glucose is added. The studies of Wada and Hattori (1978) and Wada (1980) showed that nitrogen isotope fractionation in nitrate reduction by *Phaeodactylum tricornutum* is inversely correlated with the supply of reductant. The high fractionation in the low-oxygen water of the Eastern Tropical North Pacific (Cline and Kaplan, 1975) thus suggests that the nitrate respiration there is limited by the supply of electron donor or organic matter.

Using a simplistic model based on nitrogen isotope mass balance, available data on ^{15}N abundance in nitrogenous components and on isotope fractionation in biochemical reactions, Wada et al. (1975a) assessed the relative contributions of four types of denitrifications in the sea: (1) estuarine and coastal denitrification, (2) oceanic denitrification in the low-oxygen intermediate waters, (3) microsite denitrification in oxygenated waters, and (4) denitrification in anoxic fjords and trenchs. No consideration has been given in their model to denitrification in sediment systems. Their estimates suggest that types 1 and 4 of denitrification are insignificant as expected from Table I, but that the microsite denitrification (type 3) is approximately seven times the oceanic denitrification (type 2). Their conclusion apparently conflicts with the widely accepted view that the low-oxygen intermediate waters of the Eastern Tropical Pacific are major sites of oceanic denitrification, but supports Tsunogai's (1972) hypothesis of denitrification in the deep water of the South Pacific.

IV. DISSIMILATORY NITRATE REDUCTION IN MARINE SEDIMENT SYSTEMS

Because of the lack of suitable techniques and practical difficulties, denitrification in marine sediments has long been left unexplored. Its importance has been increasingly understood in recent years. The contribution to the global biogeochemical cycling of nitrogen is probably equal to that of denitrification in the low-oxygen waters of the Eastern Tropical Pacific Ocean and Arabian Sea or even more so (Goering, 1978). In this section, recent advances in this study area are discussed.

A. Dissimilatory Nitrate Reduction in Coastal and Estuarine Sediments

Table III summarizes the experimentally determined rates of denitrification and nitrate reduction in coastal and estuarine sediments. Discrete sediment samples from various depths or undisturbed sediment cores as such were used in these experiments. For reference, brief notes on assay techniques used are also given. *In situ* denitrification rate is affected by many factors, e.g., temperature, pH, Eh (oxidation-reduction potential), and concentrations of oxygen, nitrate, and organic matter. These factors are not always independent of each other, and are, at times, difficult to control or to simulate throughout assaying processes. Sediment structure consisting of solid and liquid phases introduces another complexity. Although almost all of the experiments quoted in Table III were carried out under *in situ* or simulated *in situ* conditions, it is still uncertain how closely the values obtained reflect *in situ* denitrification rates, because the sediment structure is inevitably disturbed during the experiments and because bottling may affect microbial activity owing to the solid surface effects of the container. However, we can use these results to illustrate the characteristics of denitrification in various marine sediments.

In embayments where the terrigenous input is abundant and exchange of water with the outside is restricted, concentrations of nitrate and nitrite in overlying waters are, in general, higher than those in sedimentary pore waters. Primary productivity is high, and large amounts of particulate organic matter settle down. Sediments below the top few millimeters become anoxic because bacterial and chemical oxygen consumptions exceed the diffusive supply

TABLE III. Summary of Rates of Nitrate Reduction and

Site	Date	Nitrate reduction[a] $\mu mol\ N \cdot m^{-2} \cdot h^{-1}$	$nmol\ N \cdot g^{-1} \cdot h^{-1}$
Belgian coast	March, 1974–June, 1976	50(0–86)	
Limfjord, Denmark	April, 1977	4200(1070)[f]	
Randers Fjord, Denmark	January, 1978		
	June, 1978		
Kysing Fjord, Denmark	October, 1977		
	January, 1978		
	June, 1978		
New England coast[g]	January, 1975–March, 1976		
Izembek Lagoon, Alaska[h]	July, 1977		0.3(0.07–0.7)
Mangoku-Ura, Japan[h]	March, 1975–September, 1976		4.4(1.2–8.9)
Narragansett Bay, Rhode Island	July, 1978		
Tokyo Bay, Japan	September, 1980		11[d](10–12)
	September, 1980	51(32–70)	
Odawa Bay, Japan	April, 1980–June, 1981	25(0.3–90)	
Tama Estuary, Japan	May, 1980–June, 1981	1100(280–1900)	
	December, 1978		150[d]
	March, 1981		22[d]

[a]*Ranges are given in parentheses.*

[b]*Key: A, denitrification rates estimated from N_2O production in the presence of acetylene; B1, denitrification rates estimated, from decrease in nitrate concentration during bottle incubation, on the basis of one-dimensional diffusion model and assumed complete nitrate reduction to N_2; B2, denitrification rates estimated from N_2 production in bottle incubation experiment; BI, denitrification rates estimated from $^{15}N_2$ production in bottle incubation experiments; BJ, denitrification rates estimated from N_2 production measured by setting a bell jar on sediments; C, denitrification and nitrate reduction rates determined simultaneously in a continuous flow sediment-water system in combination with ^{15}N tracer technique.*

Denitrification in Coastal and Estuarine Sediments

Denitrification[a]			
nmol $N \cdot cm^{-3} \cdot h^{-1}$	$\mu mol\ N \cdot m^{-2} \cdot h^{-1}$	*Method*[b]	*Reference*
	50(0-86)	Bl	Billen (1978)
36[c] (7.5)[f]	1800(560)[f]	BI & A	Sørensen (1978a)
1.5[c]	41	A	Sørensen (1978b)
	5.8	A	Sørensen et al. (1979)
0.5[c]	6.9	BI	Oren and Blackburn (1979)
	130	A	Sørensen et al. (1979)
	0.8	A	Sørensen et al. (1979)
	0-360	BI & BJ	Kaplan et al. (1979)
		BI	Iizumi et al. (1980)
	240	BI	Koike and Hattori (1978a,b); Hattori et al. (1977, 1979)
	100	B2	Seitzinger et al. (1980)
1.9[d] (1.7-2.0)	97[e] (74-120)	BI	Nishio (1982)
	25(16-33)	C	Nishio et al. (1982)
	13(0.2-39)	C	Nishio (1982); Nishio et al. (1982)
	490(130-790)	C	Nishio et al. (1982); Nishio (1982)
56[d]	610[e]	BI	Nishio et al. (1981)
8.4[d]	87[e]	BI	Nishio (1982)

[c]*Maximum values observed.*

[d]*Maximum values estimated in this paper using data from the given sources.*

[e]*Integrated rates estimated on assumed exponential decrease in denitrification with depth.*

[f]*Denitrification rates in the presence of 1.8 mM of nitrate; numbers in parentheses refer to in situ rates calculated using a* K_m *value of 0.6 mM (Nedwell, 1975) and given in situ nitrate concentrations.*

[g]*Salt marsh sediments.*

[h]*Seagrass bed sediments.*

TABLE IV. Half-Saturation Constants (K_m) for Denitrification and Nitrate Reduction

System	Reaction	K_m (μ*M*)	Reference
Water column			
Bottom water of a brackish lake, Hamana, Japan	$NO_3^- \rightarrow NO_2^-$	4	Koike et al. (1972)
	$NO_3^- \rightarrow N_2$	13	Koike et al. (1972)
Bottom water of CEPEX enclosure, Saanich Inlet, B.C.	$NO_3^- \rightarrow N_2$	31	Koike et al. (1978)
Coastal and estuarine sediment			
Vacuwaqua Estuary, Fiji low organic	NO_3^- loss	180	Nedwell (1975)
high organic	NO_3^- loss	600	Nedwell (1975)
Mangoku-Ura, Japan	$NO_3^- \rightarrow N_2$	27-42	Koike et al. (1978)
Tokyo Bay, Japan	$NO_3^- \rightarrow N_2$	24	Koike et al. (1978)
Kysing Fjord, Denmark	$NO_3^- \rightarrow N_2$	344	Oren and Blackburn (1979)
Izembek Lagoon, Alaska	$NO_3^- \rightarrow N_2$	53	Izumi et al. (1980)
Offshore sediment			
Bering Sea shelf	$NO_3^- \rightarrow N_2$	7	Koike and Hattori (1979)
Land soil			
St. Bernard Sandy loam	$NO_3^- \rightarrow N_2O$	130-1200	Yoshinari et al. (1977)
Coachella sand, glucose amended	NO_3^- loss	12,000	Bowman and Focht (1974)
Central Illinois cultivated soils	NO_3^- loss	170-2100	Kohl et al. (1976)

of oxygen from the overlying water. The K_m values for nitrate reduction in soils vary depending on the availability of the organic matter (Kohl et al., 1976); the lower the concentration of organic matter, the lower the apparent K_m value. A similar trend can be seen in coastal and estuarine sediments, although the values are much smaller than those reported on soils (Table IV). The observed K_m values probably have little quantitative meaning on their own, but suggest that rates of denitrification and nitrate reduction in coastal and estuarine sediments are mainly controlled by *in situ* concentrations of nitrate and nitrite in pore waters. Actually, the denitrification activities in Tokyo Bay sediments, as expressed on sediment dry weight basis, were not altered when the amount of sedimentary organic matter was reduced up to one-tenth; the addition of either glucose or peptone did not accelerate the rates (I. Koike and A. Hattori, unpublished data, quoted in Hattori et al., 1977). In other words, the rate of denitrification in such sediments is not limited by electron donors. Denitrification rates in the sediments of Randers and Kysing fjords, Denmark, were 7-100 times higher during winter (3°C) than during summer (18°C) (Table III), although summer rates of oxygen consumption and sulfate reduction were 3-5 times the winter rates (Sørensen et al., 1979). Maximum nitrate concentrations in the sedimentary pore water were ca. 150 μM during winter and ca. 15 μM during summer. Denitrification was apparently limited by the availability of nitrate, although it was not known why the nitrate concentration was lower during summer. First-order kinetics with respect to nitrate is usually assumed in model calculations of denitrification and nitrate reduction rates in coastal and estuarine sediments.

In principle, areal denitrification activity can be computed by numerical integration (Oren and Blackburn, 1979) or by applying a one-dimensional diffusion-advection model (Vanderborght and Billen, 1975; Billen, 1978; Koike and Hattori, 1979), if depth profiles of concentrations of nitrate and nitrite and of specific activity of denitrification (the amount of N_2 produced per unit volume of sediment per unit time) are known. However, precise determination of nitrate and nitrite concentrations in sedimentary pore waters is difficult because of sharp decline of nitrate and nitrite concentrations with depth. Contamination of small amounts of overlying water rich in nitrate may also introduce a large error. Since oxygen is present in the near-surface layer of sediment, denitrification is undoubtedly suppressed but its effect is rarely considered. Actually, the amount of excess N_2, as estimated from N_2:Ar ratio in pore waters of an estuarine sediment, was far smaller than the amount computed based on the experimentally determined denitrification rates (Nishio et al., 1981). The discrepancy can be attributed to error associated with *in situ* rate estimation, or to rapid loss of excess N_2 caused by physical and biological disturbances of the near surface sediment, or both. Nishio et al. (1981) have shown that if diffusivity of N_2 in the surface layer of sediment is two orders of magnitude greater than the molecular diffusivity as suggested by Vanderborght et al. (1977), observed vertical profiles of excess N_2 can be described on the basis of bacterial denitrification rates estimated by bottle incubation experiments. They emphasize the importance of bioturbation in early diagenetic processes in sediments as discussed by Martens and Berner (1977).

To avoid this problem, Nishio et al. (1982) used a continuous-flow system consisting of undisturbed sediment and overlying water. By combining this technique with a ^{15}N tracer technique, they determined, simultaneously, denitrification, nitrate reduction, and oxygen consumption in coastal and estuarine sediments. The rates of denitrification and nitrate reduction measured in this way tend to be somewhat smaller than those estimated based on a bottle incubation experiment. When the inhibitory effect of oxygen is included in the model calculation, the two estimates were in agreement with each other with a deviation of less than 40% (Nishio, 1982).

Payne and Riley (1969) showed that the reduction step leading to N_2 evolution in *P. perfectomarinus* denitrification is inhibited by high concentrations of nitrate. However, this inhibitory effect of nitrate is probably not significant in the range of nitrate concentrations in sedimentary pore waters (Blackmer and Bremner, 1978). Koike and Hattori (1978a) found that the rate of N_2 production from nitrite is almost identical with that from nitrate. They concluded that the step of nitrite reduction is rate limiting for the overall process of nitrate reduction in the coastal sediments (Tokyo and Shimoda bays, Japan). The presence of NO and N_2O in the anoxic coastal marine sediments of Kysing Fjord has been demonstrated by Sørensen (1978c). Maximum concentrations of NO and N_2O as high as 200 μM and 50 μM, respectively, were found at 4- and 6-cm depths where nitrate concentration was ca. 10 μM. This depth was located at the lower edge of the activity profile of denitrification, and near the redox transition zone below which H_2S was accumulated. The nitrogen oxides were probably produced in association with denitrification. Nitrous oxide and NO can also be produced from nitrite by chemical reactions (Nelson and Bremner, 1970; Stevenson et al., 1970), but the active occurrence of these reactions is unlikely because of neutral alkaline pH and low nitrite concentrations in the marine sediments. In the light of the recent finding by Sørensen et al. (1980) that H_2S partially inhibits NO reduction and strongly inhibits N_2O reduction, the inhibition by H_2S diffused from deeper sediments is probably the mechanism responsible for the accumulation of NO and N_2O.

In some coastal marine sediments, nitrate and nitrite maxima appear below the sediment surface. In such systems, nitrate and nitrite must be produced from ammonium within the sediment by the action of nitrifiers. The possibility cannot be discounted that nitrate and nitrite are supplied from ground water, but experimental evidence supporting this inference has not been presented. Billen (1978) postulated that nitrification and denitrification proceed in separate layers, oxic upper layer and anoxic deeper layer, and discussed the importance of nitrification as a source of nitrate for denitrification in sediments off the Belgian coast. He estimated that nitrate produced by *in situ* nitrification is three to four times the amount required for sedimentary denitrification. His hypothesis cannot be widely applied, since there is now evidence that shows concurrent occurrence of nitrification and denitrification at the same depth (Koike and Hattori, 1978b; Kaplan et al., 1979; Iizumi et al., 1980; Nishio, 1982). The presence of anoxic microsites in oxic sediments and of oxic microsites in anoxic sediments seems responsible for this phenomenon. The contribution of *in situ* nitrification to denitrification varies depending upon the nitrate concentration in the overlying water. According to the estimates of Nishio (1982), nitrate produced by *in situ* nitrification in Tama Estuary, where the concentrations of nitrate and nitrite in overlying water were 144 μM, supported only 7% of denitrification (39 $\mu mol\ N \cdot m^{-2} \cdot h^{-1}$ against a total of 593, 20 May, 1981), whereas more than 70% of denitrification was supported by nitrate produced within the sediment in Odawa Bay where nitrate concentration in overlying water was 5.3 μM (27 $\mu mol\ N \cdot m^{-2} \cdot h^{-1}$ against a total of 37, 8 May, 1981).

Experimental evidence for nitrate reduction to ammonium in coastal marine sediments has been presented by Koike and Hattori (1978a) and Sørensen (1978a). This process might be mediated by strictly anaerobic bacteria such as *Clostridium* (Section II) or facultative anaerobes (Cole, 1978). A portion of the ammonium yielded is further utilized as a nitrogen source for the growth of bacteria. Nitrate reduction exceeds denitrification, and more than 50% of nitrate reduced is sometimes recovered as ammonium or in organic nitrogenous compounds (Koike and Hattori, 1978a; Sørensen, 1978a; see also Table III). The occurrence of nitrate reduction to ammonium in a mangrove swamp has also been suggested by Nedwell (1975) based on nitrogen mass balance calculation.

Based on the calculation of free energy change associated with nitrate reduction with H_2 (Table V), Delwiche (1978) concluded that the most efficient reaction under electron donor-limited conditions is denitrification or production of N_2. However, when

TABLE V. Free Energy Changes in Nitrate Reduction[a]

Reaction	$-\Delta G'_{298}$ (kcal) at pH 7 (per H_2)	(per NO_3^-)
$NO_3^- + 2\ H_2 + H^+ \rightarrow \frac{1}{2}\ N_2O + 2\frac{1}{2}\ H_2O$	46.67	93.35
$NO_3^- + 2\frac{1}{2}\ H_2 + H^+ \rightarrow \frac{1}{2}\ N_2 + 3\ H_2O$	53.62	134.07
$NO_3^- + 4\ H_2 + 2\ H^+ \rightarrow NH_4^+ + 3\ H_2O$	37.25	149.00

[a]*After Delwiche (1978).*

nitrate is limiting and organic matter (electron donor) is abundant, the reduction of nitrate to ammonium would be more advantageous for anaerobic bacteria. Eight electrons consumed in the reduction of nitrate to ammonium might provide an effective electron sink and stimulate fermentation, as discussed by Caskey and Tiedje (1979). The nitrate reduction to ammonium might also act as a short circuit in the biogeochemical nitrogen cycle, and its ecological significance has been discussed by Cole and Brown (1980).

B. Dissimilatory Nitrate Reduction in Pelagic Sediments

Barnes et al. (1975) were probably the first to provide convincing evidence for the occurrence of denitrification in pelagic sediments. They demonstrated in the sedimentary pore waters of the southern California borderland basin the presence of excessive N_2, up to 17% over adjacent bottom water. Data on nitrate distribution in deep sediments of the eastern equatorial Atlantic investigated by Bender et al. (1977) also suggest that nitrate reduction takes place at depths shallower than 40 cm. Based on a one-dimensional diffusion model and the stoichiometric model for organic diagenesis with nitrate (Richards, 1965), they estimated denitrification rate in the sediments. The presence of excess N_2 was later confirmed in deep sediments of northeastern North

TABLE VI. Summary of Denitrification Rates in Pelagic Sediments

Site	Denitrification		Reference
	μmol $N \cdot m^{-2} \cdot day^{-1}$	*nmol* $N \cdot g^{-1} \cdot h^{-1}$	
Eastern equatorial Atlantic	3.6		Bender et al. (1977)
Northeast Atlantic	10		Wilson (1978)
Eastern North Pacific	3.9-16		See text[a]
Central North Pacific	1.0-5.8		See text[a]
Bering Sea Basin	54		Tsunogai et al. (1979)
Santa Barbara Basin	4.1		See text[b]
Bering Sea shelf	220	1.2	Koike and Hattori (1979)
	430	0.9	Haines et al. (1981)

[a]*Estimated in this chapter based on the nitrate data of Pamatmat (1973) and assumed diffusion coefficient of* $4 \times 10^{-6} cm^2\ s^{-1}$ *(Bender et al., 1977).*

[b]*Estimated in this chapter based on the excess* N_2 *data of Barnes et al. (1975) and assumed diffusion coefficient of* $4 \times 10^{-6} cm^2\ s^{-1}$ *(Wilson, 1978).*

Atlantic by Wilson (1978). At some locations, amounts of the excess N_2 are much larger in the topmost oxygenated layers of the sediments. He suggested that denitrification probably occurs within temporarily isolated anoxic microsites.

Direct measurement of denitrification in pelagic sediments was first made by Goering and Pamatmat (1970) on the continental shelf off Peru. A value of 0.3 g $N \cdot m^{-2} \cdot day^{-1}$, calculated by Goering (1978), is probably too high because of prolonged incubation. Denitrification rates in the sediments on the Bering Sea shelf have been reported by Koike and Hattori (1979) and Haines et al. (1981). The two sets of data were very similar, although these authors used different techniques. Koike and Hattori (1979) further showed that denitrification rate estimated, based on a one-dimensional diffusion-advection model, is almost identical with the rate directly determined. Available values for denitrification rates in pelagic sediments are summarized in Table VI. These estimates suggest that denitrification proceeds in shelf sediments at a rate one order of magnitude higher than in the Bering Sea basin sediments, and two orders of magnitude higher than in open ocean sediments.

C. Global Denitrification in Marine Sediments

Although variations of the denitrification rates in marine sediments summarized in Tables III and VI are rather substantial, some general trends can be seen. For the purposes of illustration, the sea floor is divided into the four regions: (1) open ocean areas deeper than 200 m, (2) continental shelf areas, (3) estuarine and embayment areas, and (4) marginal basin areas. The averages of reported values of denitrification rates are taken as representative of the respective regions. The areas of these regions given by Svedrup et al. (1942) and Whittaker (1975) are used for the first and second regions and for the third region, respectively. The Bering Sea basin sediments exhibited denitrification rates one order of magnitude greater than open ocean sediments. The extent

TABLE VII. Global Denitrification in Marine Sediments

Site	Denitrification rate ($g\ N \cdot m^{-2} \cdot day^{-1}$)	Area ($10^{12}\ m^2$)		Annual denitrification ($10^{12}\ g\ N\ year^{-1}$)
		Total	Denitrification site[a]	
Estuaries and embayments	4×10^{-2}	2	1	15
Continental shelfs	5×10^{-3}	27	14	25
Marginal basins	8×10^{-4}	?	3	0.9
Open oceans	8×10^{-5}	331	110	3
Total				44

[a] *Estimated as described in text.*

of these types of basins is unknown, but it is probably not so large. A total of twice the Bering Sea basin area is assumed. Out of 15 deep stations in the central North Pacific surveyed by Pamatmat (1973), nitrate concentrations in the sedimentary pore water were lower than those in the bottom water at five stations. Therefore, one-third of the open ocean area is assumed to have sediments where denitrification takes place. According to Koike and Hattori (1979), half of the shelf areas is considered to be sites of denitrification. The same assumption is applied to estuarine and embayment areas. The estimate of global denitrification rate in marine sediments obtained in this way is 44×10^{12} g N year^{-1} (Table VII). This value compares with the annual denitrification in the low-oxygen intermediate waters of the world oceans (Table I) and is 3 times that for total benthic nitrogen fixation (Capone, Chapter 4, this volume).

V. CONCLUDING REMARKS

Data reviewed in this chapter indicate that water column denitrification and sedimentary denitrification contribute equally to the nitrogen budget in the world ocean. The estimates presented by various authors, however, are not always consistent with each other. This is not surprising because all of these estimates are based on many assumptions and contain methodological and conceptual uncertainties. Turbulence caused by burrowing of benthic animals, commonly called bioturbation, obstructs precise determination of *in situ* denitrification rate in coastal and estuarine sediments. Bioturbation is not uncommon even in deeper sediments, as shown by Grundmanis and Murray (1977). In model calculation, we must postulate certain relationships among denitrification rate, nitrate and nitrite concentrations, and organic matter (electron donor) concentrations. Limited data suggest that denitrification is limited by the availability of nitrate or nitrite in sediment systems and by the availability of organic matter in the water column, but the validity of this assumption has not been confirmed. Denitrification in soil systems has been shown to be controlled largely by the supply of readily decomposable organic matter (Burford and Bremner, 1975; Stanford et al., 1975). Observations by Billen (1978) suggest that denitrification in offshore sediments is limited by the supply of organic matter. In some coastal anoxic sediments, nitrate may also be reduced to ammonium. The occurrence of this type of reaction in deep sediments and low-oxygen intermediate waters is unknown. If this reaction also takes place at these sites, denitrification rates calculated based on nitrate data would lead to overestimates. Many techniques have been developed for determination of *in situ* denitrification rate in recent years. An extensive survey using these techniques, either individually or in combination, should provide important information needed to assess, with greater certainty, the extent of marine denitrification.

Some data on N_2:Ar ratio and nitrate concentrations are available which evidently show the occurrence of denitrification and nitrate reduction in pelagic sediments, but data are too scanty to make an accurate estimate over all oceanic areas. Although denitrification appears to proceed less rapidly in deep oceanic sediments than in the shelf and coastal sediments, the collection of detailed information on excess N_2 and nitrate in oceanic sediments is of critical importance, because sediments cover extensive areas.

The mode and type of denitrification and nitrate reduction vary depending on bacterial species. In soil systems, many investigations have been carried out to identify bacterial species' composition, which is fundamental to characterizing the types of reactions, but little is known with respect to marine systems. Increased effort should be made on this subject. Data on the ^{15}N abundance in nitrate and N_2 in seawater and sedimentary pore water would undoubtedly provide additional information in understanding *in situ* processes.

Existing data suggest that N_2O is lost from the sea to the atmosphere, and that this process may act as an important sink of combined nitrogen in marine systems. It is uncertain, however, whether N_2O production results from denitrification or nitrification. The use of ^{15}N to identify the responsible process should provide the information to solve this problem.

ACKNOWLEDGMENTS

The author expresses his sincere gratitude to Drs. D. G. Capone, E. J. Carpenter, E. Wada, and Miss M. Ohtsu for their suggestions and assistance in the preparation of this manuscript.

REFERENCES

Aminuddin, M., and Nicholas, D. J. D. (1973). Sulphide oxidation linked to the reduction of nitrate and nitrite in *Thiobacillus denitriticans*. *Biochim. Biophys. Acta 325,* 81-93.

Averill, B. A., and Tiedje, J. M. (1982). The chemical mechanism of microbial denitrification. *FEBS Lett. 138,* 8-12.

Balderston, W. L., Sherr, B., and Payne, W. J. (1976). Blockage by acetylene of nitrous oxide reduction in *Pseudomonas perfectomarinus*. *Appl. Environ. Microbiol. 31,* 504-508.

Barbaree, J. M., and Payne, W. J. (1967). Products of denitrification by a marine bacterium as revealed by gas chromatography. *Mar. Biol. (Berlin) 1,* 136-139.

Barnes, R. O., Bertine, K. K., and Goldberg, E. D. (1975). N_2:Ar, nitrification and denitrification in Southern California borderland basin sediments. *Limnol. Oceanogr. 20,* 962-970.

Bender, M. L., Ganning, K. A., Froelich, P. N., Heath, G. R., and Maynard, V. (1977). Interstitial nitrate profiles and oxidation of sedimentary organic matter in the eastern equatorial Atlantic. *Science 198,* 605-609.

Billen, G. (1978). A budget of nitrogen recycling in North Sea sediments off the Belgian coast. *Estuarine Coastal Mar. Sci. 7,* 127-246.

Blackmer, A. M., and Bremner, J. M. (1977). Nitrogen isotope discrimination in denitrification of nitrate in soils. *Soil Biol. Biochem. 9,* 73-77.

Blackmer, A. M., and Bremner, J. M. (1978). Inhibitory effect of nitrate on reduction of N_2O to N_2 by soil microorganisms. *Soil Biol. Biochem. 10*, 187-191.

Bowman, R. A., and Focht, D. D. (1974). The influence of glucose and nitrate concentrations upon denitrification rates in sandy soils. *Soil Biol. Biochem. 6*, 297-301.

Brandhorst, W. (1959). Nitrification and denitrification in the Eastern Tropical North Pacific. *J. Cons., Int. Explor. Mer 25*, 3-20.

Brown, I. L., and Drury, J. S. (1967). Nitrogen-isotope effects in the reduction of nitrate, nitrite, and hydroxylamine to ammonia. I. In sodium hydroxide solution with Fe(II). *J. Chem. Phys. 46*, 2833-2837.

Burford, J. R., and Bremner, J. M. (1975). Relationships between the denitrification capacities of soils and total water-soluble and readily decomposable soil organic matter. *Soil Biol. Biochem. 7*, 389-394.

Caskey, W. H., and Tiedje, J. M. (1979). Evidence for Clostridia as agents of dissimilatory reduction of nitrate to ammonium in soils. *Soil Sci. Soc. Am. J. 43*, 931-936.

Chiba, S., and Ishimoto, M. (1973). Ferredoxin-linked nitrate reductase from *Clostridium perfringens*. *J. Biochem. (Tokyo) 73*, 1315-1318.

Chien, S. H., Shearer, G. B., and Kohl, D. H. (1977). The nitrogen isotope effect associated with nitrate loss from waterlogged soils. *Soil Sci. Soc. Am. J. 41*, 63-69.

Cline, J. D. (1973). Denitrification and isotopic fractionation in two contrasting marine environments: The Eastern Tropical North Pacific Ocean and the Cariaco Trench. Doctoral Dissertation, University of California, Los Angeles.

Cline, J. D., and Ben-Yaakov, S. (1973). Nitrogen/argon ratios by difference thermal conductivity. *Deep-Sea Res. 20*, 763-768.

Cline, J. D., and Kaplan, I. R. (1975). Isotope fractionation of dissolved nitrate during denitrification in the Eastern Tropical North Pacific Ocean. *Mar. Chem. 3*, 271-299.

Cline, J. D., and Richards, F. A. (1972). Oxygen-deficient conditions and nitrate reduction in the Eastern Tropical North Pacific Ocean. *Limnol. Oceanogr. 17*, 885-900.

Codispoti, L. A., and Packard, T. T. (1980). Denitrification rates in the Eastern Tropical South Pacific. *J. Mar. Res. 38*, 453-477.

Codispoti, L. A., and Richards, F. A. (1976). An analysis of the horizontal sequence of denitrification in the Eastern Tropical North Pacific. *Limnol. Oceanogr. 21*, 379-388.

Cohen, Y., and Gordon, L. I. (1978). Nitrous oxide in the oxygen minimum of the Eastern Tropical North Pacific: Evidence for its consumption during denitrification and possible mechanisms for its production. *Deep-Sea Res. 25*, 509-524.

Cohen, Y., and Gordon, L. I. (1979). Nitrous oxide production in the ocean. *JGR, J. Geophys. Res. 84*, 347-354.

Cole, J. A. (1978). The rapid accumulation of large quantities of ammonium during nitrite reduction by *Escherichia coli*. *FEMS Microbiol. Lett. 4*, 327-329.

Cole, J. A., and Brown, C. M. (1980). Nitrite reduction to ammonia by fermentative bacteria: A short circuit in the biological nitrogen cycle. *FEMS Microbiol. Lett. 7*, 65-72.

Cox, C. D., Jr., and Payne, W. J. (1973). Separation of soluble denitrifying enzymes and cytochromes from *Pseudomonas perfectomarinus*. *Can. J. Microbiol. 19*, 861-872.

Delwiche, C. C. (1978). Biological production and utilization of N_2O. *Pure Appl. Geophys. 116*, 414-422.

Delwiche, C. C., and Steyn, P. L. (1970). Nitrogen isotope fractionation in soils and microbial reactions. *Environ. Sci. Technol. 4*, 929-939.

Deuser, W. G., Ross, E. H., and Mlodzinska, Z. J. (1978). Evidence for and rate of denitrification in the Arabian Sea. *Deep-Sea Res. 25*, 431-445.

de Vries, W., Van Wijck-Kapteyn, W. M. C., and Oosterhuis, C. K. H. (1974). The presence and function of cytochromes in *Selenomonas ruminantum*, *Anaerovibrio lipolytica* and *Veillonella alcalescens*. *J. Gen. Microbiol. 81*, 69-78.

Dugdale, R. C., Goering, J. J., Barber, R. T., Smith, R. L., and Packard, T. T. (1977). Denitrification and hydrogen sulfide in the Peru region during 1976. *Deep-Sea Res. 24*, 601-608.

Elkins, J. W., Wofsy, S. C., McElroy, M. B., Kolb, C. E., and Kaplan, W. A. (1978). Aquatic sources and sinks for nitrous oxide. *Nature (London) 275*, 602-606.

Emery, K. O., Orr, W. L., and Rittenberg, S. C. (1955). Nutrient budgets in the ocean. *In* "Essays in the Natural Sciences in Honor of Captain Allen Hancock," pp. 299-310. Univ. of Southern California Press, Los Angeles.

Fenchel, T., and Blackburn, T. H. (1979). "Bacteria and Mineral Cycling." Academic Press, New York.

Fiadeiro, M., and Strickland, J. D. H. (1968). Nitrate reduction and the occurrence of a deep nitrite maximum in the ocean off the west coast of South America. *J. Mar. Res. 26*, 187-201.

Firestone, M. K., Firestone, R. B., and Tiedje, J. M. (1979). Nitric oxide as an intermediate in denitrification: Evidence from nitrate-13 isotope exchange. *Biochem. Biophys. Res. Commun. 91*, 10-16.

Focht, D. D., and Verstraete, W. (1977). Biochemical ecology of nitrification and denitrification. *Adv. Microb. Ecol. 1*, 135-214.

Garber, E. A. E., and Hollocher, T. C. (1981). ^{15}N tracer studies on the role of NO in denitrification. *J. Biol. Chem. 256*, 5459-5465.

Gersberg, R., Krohn, K., Peek, N., and Goldman, C. R. (1976). Denitrification studies with ^{13}N-labelled nitrate. *Science 192*, 1229-1231.

Goering, J. J. (1968). Denitrification in the oxygen minimum layer of the Eastern Tropical Pacific Ocean. *Deep-Sea Res. 15,* 156-164.

Goering, J. J. (1978). Denitrification in marine systems. *In* "Microbiology--1978" (D. Schlessinger, ed.), pp. 357-361. Am. Soc. Microbiol., Washington, D.C.

Goering, J. J., and Cline, J. D. (1970). A note on denitrification in seawater. *Limnol. Oceanogr. 15,* 306-309.

R Goering, J. J., and Dugdale, R. C. (1966). Denitrification rates in an island bay in the equatorial Pacific Ocean. *Science 154,* 505-506.

Goering, J. J., and Pamatmat, M. M. (1970). Denitrification in sediments of the sea off Peru. *Invest. Pesq. 35,* 233-242.

Goering, J. J., Richards, F. A., Codispoti, I. A., and Dugdale, R. C. (1973). Nitrogen fixation and denitrification in the ocean: Biogeochemical budgets. *Proc. Symp. Hyrogeochem. Biogeochem. 1970,* Vol. 2, pp. 12-27.

Goreau, T. J., Kaplan, W. A., Wofsy, S. C., McElroy, M. B., Valois, F. W., and Watson, S. W. (1980). Production of NO_2^- and N_2O by nitrifying bacteria at reduced concentrations of oxygen. *Appl. Environ. Microbiol. 40,* 526-532.

Greenberg, E. P., and Becker, G. E. (1977). Nitrous oxide as end product of denitrification by strains of fluorescent *Pseudomonas. Can. J. Microbiol. 23,* 903-907.

Grundmanis, V., and Murray, J. W. (1977). Nitrification and denitrification in marine sediments from Puget Sound. *Limnol. Oceanogr. 22,* 804-813.

Gundersen, K. (1977). Biological Nitrogen Transformations in the Upper Water Column of the Central North Pacific Ocean, Part I, Report. Department of Microbiology, University of Hawaii, Honolulu.

Haddock, B. A., and Jones, C. W. (1977). Bacterial respiration. *Bacteriol. Rev. 41,* 47-99.

Hadjipetrou, L. P., and Stouthamer, A. H. (1965). Energy production during nitrate respiration by *Aerobacter aerogenes. J. Gen. Microbiol. 38,* 29-34.

Hahn, J. (1974). The North Atlantic Ocean as a source of atmospheric N_2O. *Tellus 26,* 160-168.

Hahn, J. (1975). N_2O measurements in the northeast Atlantic Ocean. *"Meteor" Forschungs ergeb., Reihe A 16,* 1-14.

Haines, J. R., Atlas, R. M., Griffiths, R. P., and Morita, R. Y. (1981). Denitrification and nitrogen fixation in Alaskan continental shelf sediments. *Appl. Environ. Microbiol. 41,* 412-421.

Hamilton, R. D. (1964). Photochemical processes in the inorganic nitrogen cycle of the sea. *Limnol. Oceanogr. 9,* 107-111.

Hart, L. T., Larson, A. D., and McCleskey, C. S. (1965). Denitrification by *Corynebacterium nephridii. J. Bacteriol. 89,* 1104-1108.

Hasan, S. M., and Hall, J. B. (1975). The physiological function of nitrate reduction in *Clostridium perfringens*. *J. Gen. Microbiol. 87,* 120-128.

Hasan, S. M., and Hall, J. B. (1977). Dissimilatory nitrate reduction in *Clostridium tertium*. *Z. Allg. Mikrobiol. 17,* 501-506.

Hattori, A., Wada, E., and Koike, I. (1970). Denitrification in a brackish lake. *In* "Proceedings of the Second Symposium on Nitrogen Fixation and Nitrogen Cycle" (H. Takahashi, ed.), pp. 121-126. Tohoku University, Sendai, Japan.

Hattori, A., Kanamori, S., and Handa, N. (1977). Chemical budgets of biophilic elements in the sea. *In* "Environmental Marine Science" (Y. Horibe, ed.), pp. 25-53. Univ. of Tokyo Press, Tokyo (in Japanese).

Hattori, A., Saijo, Y., Hirano, T., Horikoshi, M., Sakamoto, M., Tatsukawa, R., and Unoki, S. (1979). Coastal ecosystem and biogeochemistry. *In* "Environmental Marine Science" (Y. Horibe, ed.), Vol. 3, pp. 109-205. Univ. of Tokyo Press, Tokyo (in Japanese).

Hauck, R. D., Melsted, S. W., and Yankwich, P. W. (1958). Use of ^{15}N-isotope distribution in nitrogen gas in the study of denitrification. *Soil Sci. 86,* 287-291.

Holland, H. D. (1973). Ocean water, nutrients and atmospheric oxygen. *Proc. Symp. Hydrogeochem. Biogeochem., 1970,* Vol. 1, pp. 68-81.

Iizumi, H., Hattori, A., and McRoy, C. P. (1980). Nitrate and nitrite in interstitial waters of eelgrass beds in relation to the rhizosphere. *J. Exp. Mar. Biol. Ecol. 47,* 191-201.

Ishimoto, M., Umeyama, M., and Chiba, S. (1974). Alteration of fermentation products from butyrate to acetate by nitrate reduction in *Clostridium perfringens*. *Z. Allg. Mikrobiol. 14,* 115-121.

Ivanenkov, N. N., and Rozanov, A. G. (1961). Hydrogen sulfide contamination of the intermediate waters of the Arabian Sea and Bay of Bengal. *Okeanologiya (Moscow) 1,* 443-449.

Kaneko, M., and Ishimoto, M. (1978). A study of nitrate reductase from *Propionibacterium acidi-propionici*. *J. Biochem. (Tokyo) 83,* 191-200.

Kaplan, W., Valiela, I., and Teal, J. M. (1979). Denitrification in a salt marsh ecosystem. *Limnol. Oceanogr. 24,* 726-734.

Knowles, R. (1982). Denitrification. *Microbiol. Rev. 46,* 43-70. *

Kohl, D. H., Vitayathil, F., Whitlow, P., Shearer, G., and Chien, S. H. (1976). Denitrification kinetics in soil systems: The significance of good fits to mathematical forms. *Soil Sci. Soc. Am. Proc. 40,* 249-253.

Koike, I., and Hattori, A. (1975). Energy yield of denitrification: An estimate from growth yield in continuous cultures of *Pseudomonas denitrificans* under nitrate-, nitrite- and nitrous oxide-limited conditions. *J. Gen. Microbiol. 88,* 11-19.

Koike, I., and Hattori, A. (1978a). Denitrification and ammonia formation in anaerobic coastal sediments. *Appl. Environ. Microbiol. 35,* 278-282.

Koike, I., and Hattori, A. (1978b). Simultaneous determinations of nitrification and nitrate reduction in coastal sediments by a ^{15}N dilution technique. *Appl. Environ. Microbiol. 35,* 853-857.

Koike, I., and Hattori, A. (1979). Estimates of denitrification in sediments of the Bering Sea shelf. *Deep-Sea Res. 26,* 409-415.

Koike, I., Wada, E., Tsuji, T., and Hattori, A. (1972). Studies on denitrification in a brakish lake. *Arch. Hydrobiol. 69,* 508-520.

Koike, I., Hattori, A., and Goering, J. J. (1978). Controlled ecosystem pollution experiment: Effect of mercury on enclosed water columns. VI. Denitrification by marine bacteria. *Mar. Sci. Commun. 4,* 1-12.

Martens, C. S., and Berner, R. A. (1977). Interstitial water chemistry of anoxic Long Island Sound sediments. 1. Dissolved gases. *Limnol. Oceanogr. 22,* 10-25.

Matsubara, T. (1971). Studies on denitrification. XIII. Some properties of the N_2O anaerobically grown cell. *J. Biochem. (Tokyo) 69,* 991-1001.

Matsubara, T., and Iwasaki, H. (1971). Enzymatic steps of dissimilatory nitrite reduction in *Alcaligenes faecalis. J. Biochem. (Tokyo) 69,* 859-868.

Miyake, Y., and Wada, E. (1971). The isotope effect on the nitrogen in biochemical oxidation-reduction reactions. *Rec. Oceanogr. Works Jpn. 11,* 1-6.

Miyazaki, T., Wada, E., and Hattori, A. (1973). Capacities of shallow waters of Sagami Bay for oxidation and reduction of inorganic nitrogen. *Deep-Sea Res. 20,* 571-577.

Miyazaki, T., Wada, E., and Hattori, A. (1975). Nitrite production from ammonia and nitrate in the euphotic layer of the western North Pacific Ocean. *Mar. Res. Commun. 1,* 381-394.

Miyazaki, T., Wada, E., and Hattori, A. (1980). Nitrogen-isotope fractionation in the nitrate respiration by the marine bacterium *Serratia marinorubra. Geomicrobiol. J. 2,* 115-126.

Nedwell, D. B. (1975). Inorganic nitrogen metabolism in a eutrophicated tropical mangrove estuary. *Water Res. 9,* 221-231.

Nelson, D. W., and Bremner, J. M. (1970). Gaseous products of nitrite decomposition in soils. *Soil Biol. Biochem. 2,* 203-215.

Nishio, T. (1982). Nitrogen cycling in coastal and estuarine sediments with special reference to nitrate reduction, denitrification and nitrification. Doctoral Dissertation, University of Tokyo, Tokyo.

Nishio, T., Koike, I., and Hattori, A. (1981). N_2/Ar and denitrification in Tama estuary sediments. *Geomicrobiol. J. 2,* 193-209.

Nishio, T., Koike, I., and Hattori, A. (1982). Denitrification, nitrate reduction, and oxygen consumption in coastal and estuarine sediments. *Appl. Environ. Microbiol. 43,* 648-653.

Olson, P. J. (1981). ^{15}N tracer studies of the primary nitrite maximum. *J. Mar. Res. 39,* 203-336.

Oren, A., and Blackburn, T. H. (1979). Estimation of sediment denitrification rates at *in situ* nitrate concentrations. *Appl. Environ. Microbiol. 37,* 174-176.

Packard, T. T., Dugdale, R. C., Goering, J. J., and Barber, R. T. (1978). Nitrate reductase activity in the subsurface waters of the Peru Current. *J. Mar. Res. 36,* 59-76.

Painter, H. A. (1970). A review of literature of inorganic nitrogen metabolism in microorganisms. *Water Res. 4,* 393-450.

Pamatmat, M. M. (1973). Benthic community metabolism on the continental terrace and in the deep sea in the North Pacific. *Int. Rev. Gesamten Hydrobiol. 58,* 345-368.

Payne, W. J. (1973). Reduction of nitrogenous oxides by microorganisms. *Bacteriol. Rev. 37,* 409-452.

Payne, W. J. (1981). "Denitrification." Wiley, New York.

Payne, W. J., and Riley, P. S. (1969). Suppression by nitrate of enzymatic reduction of nitric oxide. *Proc. Soc. Exp. Biol. Med. 132,* 258-260.

Prakash, O., and Sadana, J. C. (1973). Metabolism of nitrate in *Achromobacter fischeri*. *Can. J. Microbiol. 19,* 15-25.

Redfield, A. C., Ketchum, B. H., and Richards, F. A. (1963). The influence of organisms on the composition of sea water. *In* "The Sea" (M. N. Hill, ed.), Vol. 2, pp. 26-77. Wiley (Interscience), New York

Renner, E. D., and Becker, G. L. (1970). Production of nitric oxide and nitrous oxide during denitrification by *Corynebacterium nephridii*. *J. Bacteriol. 101,* 821-826.

Richards, F. A. (1965). Anoxic basins and fjords. *In* "Chemical Oceanography" (J. P. Riley and G. Skirrow, eds.), Vol. 1, pp. 611-645. Academic Press, New York.

Richards, F. A., and Benson, B. B. (1961). Nitrogen/argon and nitrogen isotope ratios in two anaerobic environments, the Cariaco Trench in the Caribbean Sea and Dramsfjord, Norway. *Deep-Sea Res. 7,* 254-264.

Richards, F. A., and Broenkow, W. W. (1971). Chemical changes including nitrate reduction in Darwin Bay, Galapagos Archipelago over a 2-month period, 1969. *Limnol. Oceanogr. 16,* 758-765.

Richards, F. A., Anderson, J. J., and Cline, J. D. (1971). Chemical and physical observations in Golfo Dulce, an anoxic basin on the Pacific coast of Costa Rica. *Limnol. Oceanogr. 16,* 43-50.

Saino, T. (1977). Biological nitrogen fixation in the ocean with emphasis on the nitrogen fixing blue-green alga *Trichodesmium* and its significance in the nitrogen cycling in the low latitude sea areas. Doctoral Dissertation, University of Tokyo, Tokyo.

St. John, R. T., and Hollocher, T. C. (1977). Nitrogen-15 tracer studies on the pathway of denitrification in *Pseudomonas aeruginosa*. *J. Biol. Chem. 252,* 212-218.

Seitzinger, S., Nixon, S., Pilson, M. E. Q., and Burke, S. (1980). Denitrification and N_2O production in near-shore marine sediments. *Geochim. Cosmochim. Acta 44,* 1853-1860.

Seki, S., Hagiwara, M., Kudo, K., and Ishimoto, M. (1979). Studies on nitrate reductase of *Clostridium perfringens.* II. Purification and some properties of ferredoxin. *J. Biochem. (Tokyo) 85,* 833-838.

Söderlund, R., and Svensson, B. H. (1976). The global nitrogen cycle. *Ecol. Bull. 22,* 23-73.

Sørensen, J. (1978a). Capacity for denitrification and reduction of nitrate to ammonia in a coastal marine sediment. *Appl. Environ. Microbiol. 35,* 301-305.

Sørensen, J. (1978b). Denitrification rates in a marine sediment as measured by the acetylene inhibition technique. *Appl. Environ. Microbiol. 36,* 139-143.

Sørensen, J. (1978c). Occurrence of nitric and nitrous oxides in a coastal marine sediment. *Appl. Environ. Microbiol. 36,* 809-813.

Sørensen, J., Jørgensen, B. B., and Revsbech, N. P. (1979). A comparison of oxygen, nitrate, and sulfate respiration in coastal marine sediments. *Mar. Ecol. 5,* 105-115.

Sørensen, J., Tiedje, J. M., and Firestone, R. B. (1980). Inhibition by sulfide of nitric and nitrous oxide reduction by denitrifying *Pseudomonas fluorescens. Appl. Environ. Microbiol. 39,* 105-108.

Stanford, G., Vander Pol, P. A., and Dzienia, S. (1975). Denitrification rates in relation to total and extractable soil carbon. *Soil Sci. Soc. Am. Proc. 39,* 284-289.

Stevenson, F. J., Harrison, R. M., and Leeper, R. A. (1970). Nitrosation of soil organic matter. III. Nature of gases produced by reaction of nitrite with lignins, humic substances, and phenolic constituents under natural and slightly acidic conditions. *Soil Sci. Soc. Am. Proc. 34,* 430-435.

Svedrup, H. U., Johnson, M. W., and Fleming, R. H. (1942). "The Oceans: Their Physics, Chemistry, and General Biology." Prentice-Hall, Englewood Cliffs, New Jersey.

Takahashi, H., Taniguchi, S., and Egami, F. (1963). Inorganic nitrogen compounds distribution and metabolism. *Comp. Biochem. 5,* 91-202.

Thauer, R. K., Jungermann, K., and Decker, K. (1977). Energy conservation in chemotrophic anaerobic bacteria. *Bacteriol. Rev. 41,* 100-180.

Thomas, W. H. (1966). On denitrification in the northeastern tropical Pacific Ocean. *Deep-Sea Res. 13,* 1109-1114.

Tsunogai, S. (1971). Ammonia in the oceanic atmosphere and the cycle of nitrogen compounds through the atmosphere and the hydrosphere. *Geochem. J. 5,* 57-67.

Tsunogai, S. (1972). An estimate of the rate of decomposition of organic matter in the deep water of the Pacific Ocean. *In* "Biological Oceanography of the Northern North Pacific Ocean"

(A. Y. Takenouti, M. Anraku, K. Banse, T. Kawamura, S. Nishizawa, T. R. Parsons, and T. Tsujita, eds.), pp. 517-533. Idemitsu Shoten, Tokyo.

Tsunogai, S., Kusakabe, M., Iizumi, H., Koike, I., and Hattori, A. (1979). Hydrographic features of the deep water of the Bering Sea: The sea of silica. *Deep-Sea Res. 26,* 641-659.

Tuttle, J. H., and Jannasch, H. W. (1972). Occurrence and types of *Thiobacillus*-like bacteria in the sea. *Limnol. Oceanogr. 17,* 532-543.

Tuttle, J. H., and Jannasch, H. W. (1973). Sulfide and thiosulfate oxidizing bacteria in anoxic marine basins. *Mar. Biol. (Berlin) 20,* 64-70.

Vagnai, S., and Klein, D. A. (1974). A study of nitrite-dependent dissimilatory microorganisms isolated from Oregon soils. *Soil Biol. Biochem. 6,* 335-339.

Vanderborght, J.-P., and Billen, G. (1975). Vertical distribution of nitrate concentration in interstitial water of marine sediments with nitrification and denitrification. *Limnol. Oceanogr. 20,* 953-961.

Vanderborght, J.-P., Wollast, R., and Billen, G. (1977). Kinetic models of diagenesis in disturbed sediments. Part 2. Nitrogen diagenesis. *Limnol. Oceanogr. 22,* 794-803.

Van Gent-Ruijters, M. L. W., de Vries, W., and Stouthamer, A. H. (1975). Influence of nitrate on fermentation pattern, molar growth yields and synthesis of cytochrome *b* in *Propionibacterium pentosaceum. J. Gen. Microbiol. 88,* 36-48.

Verhoeven, W. (1956). Some remarks on nitrate metabolism in microorganisms. *In* "A Symposium on Inorganic Nitrogen Metabolism" (W. D. McElroy and B. Glass, eds.), pp. 61-86. Johns Hopkins Press, Baltimore, Maryland.

Wada, E. (1980). Nitrogen isotope fractionation and its significance in biogeochemical processes occurring in marine environments. *In* "Isotope Marine Chemistry" (E. D. Goldberg, Y. Horibe, and K. Saruhashi, eds.), pp. 375-398. Uchida Rokakuho, Tokyo.

Wada, E., and Hattori, A. (1972). Nitrite distribution and nitrate reduction in deep sea waters. *Deep-Sea Res. 19,* 123-132.

Wada, E., and Hattori, A. (1978). Nitrogen isotope effects in the assimilation of inorganic nitrogenous compounds by marine diatoms. *Geomicrobiol. J. 1,* 85-101.

Wada, E., Kadonaga, T., and Matsuo, S. (1975a). ^{15}N abundance in nitrogen of naturally occurring substances and global assessment of denitrification from isotopic viewpoint. *Geochem. J. 9,* 139-148.

Wada, E., Koike, I., and Hattori, A. (1975b). Nitrate metabolism in abyssal waters. *Mar. Biol. (Berlin) 29,* 119-124.

Wellman, R. P., Cook, F. D., and Krouse, H. R. (1968). Nitrogen-15: Microbiological alteration of abundance. *Nature (London) 161,* 269-270.

Wharton, D. C., and Weintraub, S. T. (1980). Identification of nitric oxide and nitrous oxide as products of nitrite reduction by *Pseudomonas* cytochrome oxidase (nitrite reductase). *Biochem. Biophys. Res. Commun. 97,* 236-242.

Whatley, F. R. (1981). Dissimilatory nitrate reduction. *In* "Biology of Inorganic Nitrogen and Sulfur" (H. Bothe and A. Trebst, eds.), pp. 64-77. Springer-Verlag, Berlin and New York.

Whittaker, R. H. (1975). "Communities and Ecosystems," 2nd ed. Macmillan, New York.

Wilson, T. R. S. (1978). Evidence for denitrification in aerobic pelagic sediments. *Nature (London) 274,* 354-356.

Wooster, W. S., Chow, T. J., and Barrett, I. (1965). Nitrite distribution in Peru current waters. *J. Mar. Res. 23,* 210-221.

Wyrtki, K. (1971). "Oceanographic Atlas of the International Indian Ocean Expedition." Nat. Sci. Found., Washington, D.C.

Wyrtki, K. (1977). Advection in the Peru Current as observed by satellite. *JGR, J. Geophys. Res. 82,* 3939-3944.

Yoshinari, T. (1976). Nitrous oxide in the sea. *Mar. Chem. 4,* 189-202.

Yoshinari, T. (1980). N_2O reduction by *Vibrio succinogenes*. *Appl. Environ. Microbiol. 39,* 81-84.

Yoshinari, T., and Knowles, R. (1976). Acetylene inhibition of nitrous oxide reduction by denitrifying bacteria. *Biochem. Biophys. Res. Commun. 69,* 705-710.

Yoshinari, T., Hynes, R., and Knowles, R. (1977). Acetylene inhibition of nitrous oxide reduction and measurement of denitrification and nitrogen fixation in soil. *Soil Biol. Biochem. 9,* 117-183.

Zobell, C. (1946). "Marine Microbiology." Chronica Botanica, Waltham, Massachusetts.

R Zumft, W. G., and Cárdenas, J. (1979). The inorganic biochemistry of nitrogen bioenergetic processes. *Naturwissenschaften 66,* 81-88.

Chapter 7

KINETICS OF INORGANIC NITROGEN UPTAKE BY PHYTOPLANKTON

JOEL C. GOLDMAN
PATRICIA M. GLIBERT
Woods Hole, Oceanographic Institution
Woods Hole, Massachusetts

I. INTRODUCTION

The important roles of NH_4^+ and NO_3^- in the physiological ecology of marine phytoplankton have been well critiqued during the past decade (Morris, 1974; Dugdale, 1976, 1977), culminating in a series of recent reviews on this subject by McCarthy (1980, 1981, 1982) who has updated and analyzed much of the historical information on phytoplankton nitrogenous nutrition. Thus it is not our purpose in this chapter to add another review on the broad subject of inorganic nitrogen uptake, but rather to focus on several specific topics that, in our opinion, are not well understood and are laden with ambiguity. Specifically, we will address the factors that regulate the coupling between inorganic nitrogen uptake and cell growth (Fig. 1). In doing so we will investigate the utility of the Monod equation for describing uptake

ISBN 0-12-160280-X

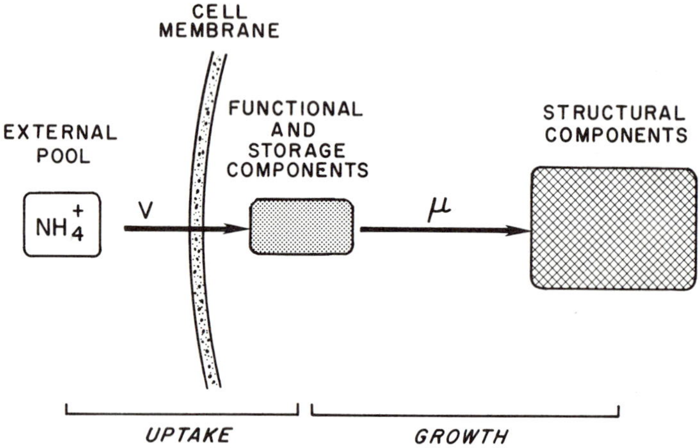

Fig. 1. Schematic diagram of NH_4^+ uptake across phytoplankton cell membrane from external aqueous pool into functional and storage components within cell, which are then used in growth processes leading to the synthesis of cellular structural components.

and growth, and we will show how analytical and experimental limitations restrict our ability to collect the necessary kinetic data for making a meaningful interpretation of the ways in which phytoplankton regulate their physiology to maximize exploitation of available inorganic nitrogen.

II. HISTORICAL PERSPECTIVE

Contemporary ideas on phytoplankton-nitrogen kinetics have their origins, for the most part, in the now classical study by Dugdale (1967) in which he proposed that rates of phytoplankton nutrient uptake could be related to limiting nutrient availability according to a rectangular hyperbolic equation. This concept has had far-reaching implications toward a more complete understanding of how phytoplankton compete for nutrients, which nutrients may be limiting, and how nutrient availability, either through physical transport processes or through biologically regulated regeneration processes, controls the growth and primary production rates of marine phytoplankton populations.

The equation, as originally used by Dugdale (1967) in a mass balance model of phytoplankton nitrogenous-nutrient dynamics, was defined as:

$$V = V_m \left[\frac{N}{K_s + N} \right] \tag{1}$$

in which V is a specific growth rate (time^{-1}) in terms of limiting nitrogen concentration N (mass·volume^{-1}), V_m is the maximum uptake velocity (= maximum growth rate) (time^{-1}) in terms of N, and K_s is the concentration of N for which $V = 0.5\ V_m$ (mass·volume^{-1}). Dugdale (1967) also defined another important term ρ, the absolute nitrogen transport rate, as the product of V and the particulate nitrogen concentration (PN) (mass·volume^{-1}·time^{-1}). Dugdale's model was considered to be a steady state one, and thus it was convenient at that time not to make a distinction between nutrient uptake and growth processes.

Equation (1) originally was developed empirically by Monod (1942) to describe the relationship between bacterial growth rates and limiting concentration of organic substrate in a manner analogous to the way in which the Michaelis-Menten equation is used to describe enzyme-substrate kinetics and the Langmuir isotherm is used to relate physical adsorption processes to available surface area. A fundamental criterion on which the validity of these rectangular hyperbolic equations is based is that substrate concentration or active surface area remains constant over the period during which the rate of the particular process is being measured. For non-steady-state situations, the substrate concentration (or active surface area) is continually changing with time so that the above equations are valid only for describing the instantaneous rate reaction (Powell, 1967). Failure to acknowledge this restriction in applying the above equations has been, in our opinion, a major stumbling block in phytoplankton ecology research.

Nonetheless, the Monod equation has found widespread appeal among phytoplankton ecologists as a simple and quantitative expression for relating nutrient uptake and growth processes to nutrient availability. The literature is now replete with kinetic studies on inorganic nitrogen (and other nutrient) uptake by phytoplankton (see review by McCarthy, 1982). The appeal of Eq. (1) rests primarily in the descriptive power of the two coefficients K_s and V_m for defining quantitatively the affinity a phytoplankton species has for a given nutrient. For example, a combination of low K_s and high V_m conceptually indicates high nutrient affinity, whereas low nutrient affinity would be represented by high K_s and low V_m.

In recent years, however, major ambiguities have arisen in the application of Eq. (1) to studies on inorganic nitrogen kinetics. There are at least two important reasons for this problem. First, and most important, it is well established that only under certain well-defined laboratory conditions that seldom are replicated under natural water conditions are nutrient uptake and growth processes coupled (i.e., in balance and equal) [see reviews by Dugdale (1977) and McCarthy (1982)]. Thus, the original assumption of Dugdale (1967) regarding the equality of specific nitrogen uptake V and specific growth rate μ (as used by Monod) is not generally applicable (Dugdale, 1977). Second, not only are there tremendous intra- and interspecific differences among phytoplankton in their abilities to take up and assimilate different nitrogenous nutrients (e.g., NH_4^+, NO_3^-, NO_2^-, urea, organic N) (Carpenter and Guillard, 1971; Morris, 1974; McCarthy, 1981), but there is great temporal and spatial variation in the availability of these different nitrogen sources in natural waters (see Sharp, this volume, Chapter 1). Thus, it is exceedingly difficult to use K_s and V_m data generated for one particular nitrogen source in an overall assessment of nitrogen dynamics in the marine environment. The limitations in using such kinetic data become evident when consideration is given to the problems inherent in the experimental methodologies that have evolved during recent years for carrying out such experiments; in particular, protocols have tended to be developed more on the basis of analytical considerations than on the best representation of the real world of the phytoplankton (Goldman et al., 1981; Harrison, this volume, Chapter 21).

III. STEADY-STATE KINETICS

A. The Monod Equation

The original Monod equation (Monod, 1942) was defined as:

$$\mu = \hat{\mu}\left[\frac{S}{K_u + S}\right] \tag{2}$$

in which μ is the specific growth rate (time^{-1}), $\hat{\mu}$ is the maximum specific growth rate (time^{-1}), S is the concentration of limiting substrate ($\text{mass} \cdot \text{volume}^{-1}$), and K_u, commonly known as the half-saturation coefficient, is the concentration of S for which $\mu = 0.5\ \hat{\mu}$. The major difference between this equation and Eq. (1) is that they describe dissimilar physiological processes, the latter defining only the rate by which a particular nutrient is incorporated into the cell from the surrounding aqueous medium (uptake), and the former defining the net sum of all physiological processes, including uptake, that lead to cell division and total biomass increase (growth) (Fig. 1). As stated earlier, only under steady-state conditions are the two equations equivalent so that $\mu = V$. It then follows that only under the unique conditions of balanced uptake and growth would $K_s = K_u$ and $V_m = \hat{\mu}$.

B. Steady-State Continuous Cultures

The continuous culture has found widespread application in phytoplankton ecology research as a tool for establishing a steady-state growth rate μ equal to the physical dilution rate D (= medium flow rate$\cdot$culture volume^{-1}) (Herbert et al., 1956; Rhee, 1980).

As long as the condition $D < \hat{\mu}$ is met, a steady state with respect to culture biomass and residual limiting nutrient must occur (Spicer, 1955). Therefore, by simply turning the speed control on a medium feed pump and waiting an appropriate period of time, an experimentalist can establish steady-state growth rates in the range $0 < \mu < \hat{\mu}$, which, in essence, represent a wide spectrum of cell physiological states (Goldman, 1980).

In applying this concept to the study of inorganic nitrogen kinetics, phytoplankton ecologists generally have found that neither NH_4^+ nor NO_3^- limitation in freshwater and marine phytoplankton could be described by Eq. (2) (Williams, 1965; Caperon, 1965; Caperon and Meyer, 1972; Eppley and Renger, 1974; Rhee, 1978; Goldman and McCarthy, 1978). The major difficulty was that it was impossible to generate saturation curves relating residual NH_4^+ or NO_3^- concentration to μ with easily measurable values of K_s [as was found by bacteriologists working with organic substrates (Herbert et al., 1956) and as depicted in Fig. 2A], because residual concentrations of these sources of N commonly were below the detection limit of 0.03 μg·atom liter^{-1} over virtually the entire range of possible steady-state dilution rates (as depicted in Fig. 2B). Goldman and McCarthy (1978), for example, measured steady-state concentrations of residual $NH_4^+ < 0.03$ μg·atom liter^{-1} at all values of $\mu < 0.87\ \hat{\mu}$ in a continuous culture study of NH_4^+ limitation in the marine diatom *Thalassiosira pseudonana* (3H) (Fig. 3), thereby making it impossible to measure the magnitude of K_u.

The impossibility of measuring K_u for NH_4^+ and NO_3^- limitation in marine phytoplankton has negated the major potential application of the Monod equation for identifying the form of limiting nutrient controlling growth rates in natural marine waters, namely, a simple comparison of the measured ambient nutrient concentration with identified K_u values determined from laboratory experiments (McCarthy and Goldman, 1979). In fact, if the kinetic curve shown in Fig. 2B is representative of the affinities natural phytoplankton have for inorganic N, then the absence of measurable NH_4^+ or NO_3^- in

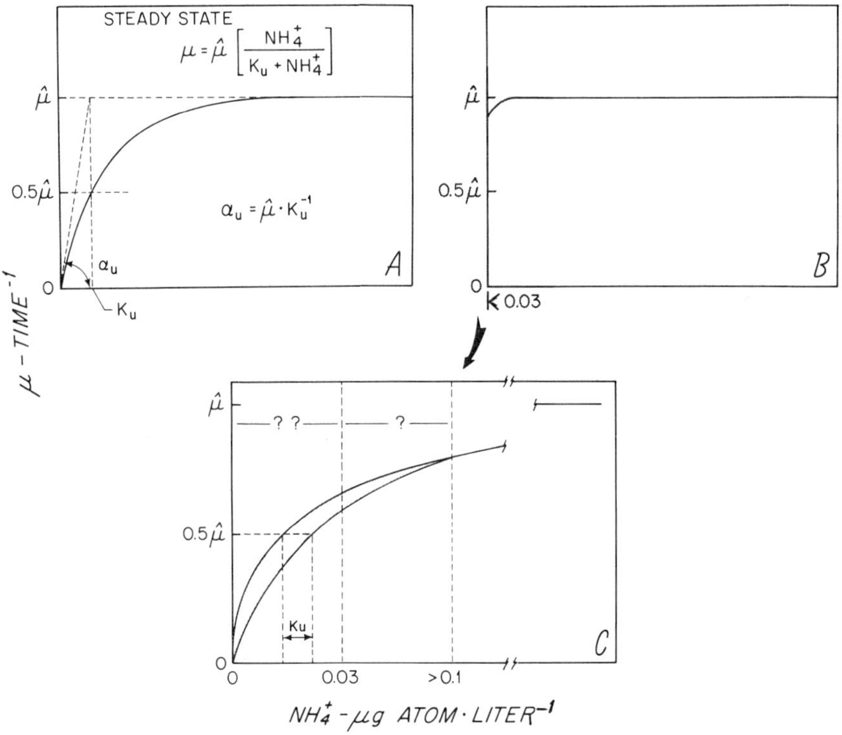

Fig. 2. Graphical depiction of the Monod equation when used to describe the relationship between steady-state growth rate μ and limiting NH_4^+ concentration in continuous cultures of phytoplankton [Eq. (2)]: (A) General shape of steady-state Monod curve; (B) typical shape of steady-state Monod curve found in experimental studies, indicating that K_u is below 0.03 μg·atom liter^{-1} NH_4^+; (C) expanded portion of shaded area from panel B, showing the large uncertainty in K_u because of the analytical constraints in measuring NH_4^+ at concentrations <0.03 μg·atom liter^{-1}.

marine waters provides absolutely no information as to how fast a resident phytoplankton population might be growing (Jackson, 1980). As seen in Fig. 2C, when the scale representing NH_4^+ concentration is expanded to between 0 and 0.1 μg·atom liter^{-1}, a wide range of K_u values between 0 and 0.03 μg·atom liter^{-1} is

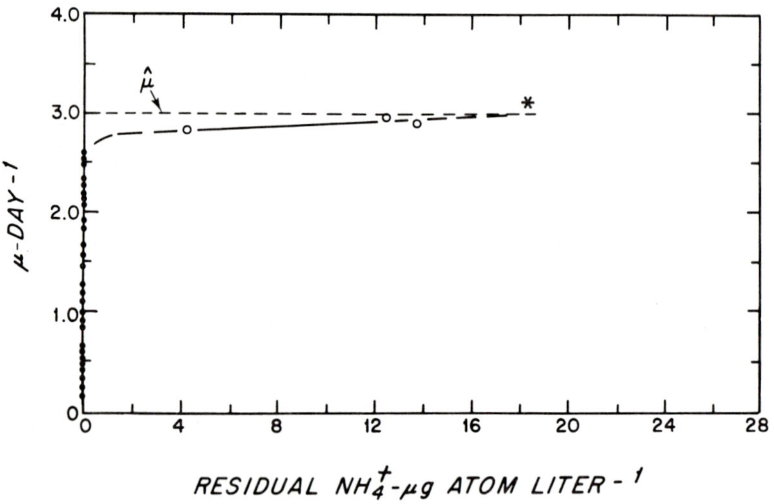

Fig. 3. Relationship between steady-state growth rate μ and residual NH_4^+ concentration in NH_4^+-limited continuous culture study of the marine diatom *Thalassiosira pseudonana* (3H). Key: ●, NH_4^+ concentrations < 0.03 μg·atom·liter^{-1}; o, NH_4^+ concentrations > 0.03 μg·atom liter^{-1}; *, NH_4^+ concentration measured at washout dilution rate of 3.07 day^{-1}; $\hat{\mu}$ = 3.02 day^{-1}. [From Goldman and McCarthy (1978). Reprinted with permission of the American Society of Limnology and Oceanography.]

possible although unobtainable with current methodologies. The limitation in determining an accurate estimate of K_u for NH_4^+ and NO_3^- is therefore not due to an inherent failure of the Monod expression for describing the relationship between growth rate and substrate concentration accurately, but simply to the fact that marine phytoplankton have a tremendous affinity for inorganic N. Analytical limitations in measuring accurately such low levels of inorganic N are thus the major impediment in our ability to characterize the initial part of the kinetic curve seen in Fig. 2C.

C. The Droop Equation

A well-recognized phenomenon that typifies the physiological state of phytoplankton in steady-state continuous cultures is that tremendous variations in cellular chemical composition occur concomitant with changing dilution rate (Goldman, 1980; Rhee, 1980). In particular, the concentration of limiting nutrient consumed by an individual cell, as shown by Droop (1973), generally is a distinct function of μ for a number of limiting nutrients (e.g., vitamin B_{12}, PO_4^{3-}, Fe^{3+}, NO_3^-). This relationship is based on purely empirical grounds and has the form:

$$\mu = \bar{\mu}\left[1 - \frac{k_Q}{Q}\right] \tag{3}$$

in which Q is the cell quota (cellular nutrient mass·total cellular mass^{-1}), k_Q is the minimum cell quota required for growth to proceed, and $\bar{\mu}$ is the unattainable value of μ at infinite Q (Fig. 4A). There is, of course, a maximum Q (Q_m) coincident with the true maximum growth rate $\hat{\mu}$ (Droop, 1973; Goldman and McCarthy, 1978). For those limiting nutrients that make up only a small fraction of total cell mass, such as Fe^{3+}, vitamin B_{12}, and PO_4^{3-}, $Q_M \gg\gg k_Q$ so that $\hat{\mu} \simeq \bar{\mu}$. In contrast, Q_m is only about five times as great as k_Q when nitrogen, which makes up 5-10% of cell mass, is limiting. Then by Eq. (3), $\hat{\mu}$ would be equal to ca. 80% of $\bar{\mu}$ (Goldman and McCarthy, 1978; Goldman and Peavey, 1979).

D. Equivalence of Steady-State Equations

It is not difficult, when viewing Fig. 1, to envision that Eqs. (1)-(3) are equivalent when steady state is attained (i.e., $V = \mu$). This important point has been demonstrated conceptually (Droop, 1973; Burmaster, 1979) and verified experimentally (Goldman, 1977). However, the Droop equation has far more versatility than the Monod expression for describing the manner in which phytoplankton growth rates are controlled by limiting inorganic nitrogen concentration under steady state conditions only because the kinetic

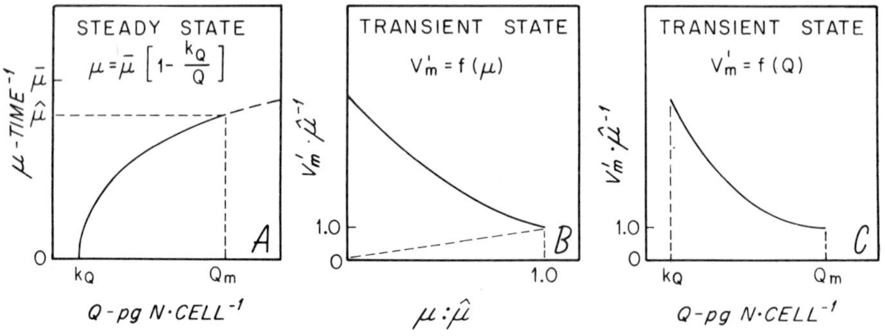

Fig. 4. The effect of phytoplankton cell physiological state on growth and NH_4^+ uptake: (A) Graphical depiction of the Droop equation when used to describe the relationship between steady-state growth rate μ and cell quota Q [Eq. (3)]; (B) conceptual relationship between transient uptake represented by relative maximum uptake rate $V_m':\hat{\mu}$ and steady-state growth rate defined as relative growth rate $\mu:\hat{\mu}$. Dashed line between origin and datum point $V_m':\hat{\mu} = \mu:\hat{\mu} = 1$ represents balanced rates of nutrient uptake and growth, whereas curvilinear increase in $V_m':\hat{\mu}$ with decreasing $\mu:\hat{\mu}$ represents enhanced nutrient uptake relative to growth for a given duration of exposure to saturating nutrient; (C) alternate depiction of curve in panel B in which cell quota Q, derived as a function of growth rate μ according to Eq. (3), is used in place of $\mu:\hat{\mu}$.

coefficients $\bar{\mu}$, k_Q, and Q_m can be determined from a curve-fitting technique based on measurement of the easily quantified parameter Q. The Monod expression, as previously discussed, can not be used conveniently because of analytical limitations in measuring the residual nutrient concentration.

In summary, we reemphasize that uptake and growth processes are equal only under steady-state conditions. Equation (2) differs from Eq. (3) only in that the latter equation is a description of how changes in steady-state growth rate are manifested through the summed physiological and biochemical events that follow the incorporation of limiting nutrient. In contrast, Eq. (1) connects the summed events of uptake and growth with the available external source of limiting nutrient. Obviously, only when $\mu = V$ is the external nutrient concentration in equilibrium with the sum of all storage, functional, and structural components comprising the cellular nutrient mass (Fig. 1).

IV. TRANSIENT UPTAKE KINETICS

A. Cell Physiological State

Although the steady-state continuous culture poorly represents the natural environment of phytoplankton (Jannasch, 1974; see also discussion, Section V), it is possible to define precisely from steady-state data the physiological state of a cell population as a function of growth rate. For example, numerous researchers have shown that the chemical composition of phytoplankton varies tremendously with degree of nutrient limitation (Rhee, 1974; Perry, 1976; Sakshaug and Holm-Hansen, 1977; Goldman et al., 1979).

Goldman et al. (1979) observed that the cellular C:N:P ratio of phytoplankton grown in steady-state continuous cultures under either nitrogen or phosphorus limitation approached the "Redfield" proportions ($C_{106}N_{16}P_1$) only when $\hat{\mu}$ was approached. They used this information in a diagnostic fashion to suggest that because the chemical composition of particulate matter in marine waters often is in the Redfield proportions, natural phytoplankton populations may be growing close to maximal rates. Goldman (1980) carried this concept further by showing that the changes in chemical composition of those phytoplankton species with different maximum growth rates (which could be intrinsic characteristics of species or be regulated by environmental parameters such as light intensity and temperature) occurred in virtually identical fashions when compared on the basis of relative growth rate ($\mu:\hat{\mu}$), rather than on absolute values of $\hat{\mu}$. Thus, chemical composition data can be used to define the physiological state of natural populations, but they do not provide any information on absolute growth rates.

Syrett (1953a,b, 1955, 1956) provided the first clue that cell physiological state played a major role in influencing nitrogen uptake rates by phytoplankton and that tremendous uncoupling between uptake and growth could occur. He found that NH_4^+ and NO_3^- uptake by batch cultures of "nitrogen-starved cells" was much more rapid than by "normal cells" that were nutrient replete. However, it was not until the definitive studies of Eppley and Renger (1974) and McCarthy and Goldman (1979) involving the measurement of maximum NH_4^+ and NO_3^- uptake rates [V_m in Eq. (1)] by cells preconditioned in nitrogen-limited continuous cultures, that a relationship between V_m and μ for NH_4^+ began to emerge. In particular, McCarthy and Goldman (1979) found that the rate of NH_4^+ uptake by the marine diatom *Thalassiosira pseudonana* (3H) during short exposure (5 min) to saturating NH_4^+ concentrations far exceeded the rate necessary for balanced growth at low μ, but that the degree of uncoupling between V_m and μ diminished with increasing μ until uptake and growth were equal at $\hat{\mu}$ (see Fig. 3 in Harrison, this volume, Chapter 21).

With present data we cannot extend to NO_3^- uptake the concept of increasing V_m with decreasing μ observed for NH_4^+ uptake. Indeed, there are some conflicting reports in the literature as to the relationship between V_m and μ for NO_3^-. Eppley and Renger (1974) reported that V_m for NO_3^- and μ were negatively correlated for *T. pseudonana* grown in chemostat on a 12:12 light:dark cycle. In addition, DeManche et al. (1979) and Collos (1980) report initial NO_3^- uptake exceeding growth for nutrient-depleted batch cultures of *Skeletonema costatum* and *Phaeodactylum tricornutum*, respectively. In contrast to the above, Horrigan and McCarthy (1982) have demonstrated, using short-term tracer techniques, no enhancement in V_m above μ for NO_3^- for batch cultures of *T. pseudonana* and *S. costatum* at any nutritional state. Experiments on short-term NO_3^- uptake conducted at rigorously defined steady-state growth rates are necessary before we can relate V_m to μ for NO_3^-. For this reason, our discussion throughout the remainder of the text is restricted only to NH_4^+.

The type of relationship between transient V_m for NH_4^+ and steady-state μ found by McCarthy and Goldman (1979) has been depicted by Goldman and Glibert (1982) as a comparison of either relative maximum uptake rate ($V_m':\hat{\mu}$) to relative growth rate ($\mu:\hat{\mu}$) (Fig. 4B) or relative maximum uptake rate to Q (Fig. 4C). The term V_m' is used in place of V_m to denote that the maximum uptake rate is variable (Dugdale, 1977), and the relative uptake rate (Goldman and Glibert, 1982) along with the relative maximum growth rate (Goldman, 1980) are used to emphasize the significant uncoupling between maximum uptake and growth that takes place when μ decreases. In addition, this format is useful for comparing on a common scale the relationship between uptake and growth by species with different maximum growth rates. Enhanced uptake of NH_4^+ can also be described in relation to variable Q (Fig. 4C) by replacing $\mu:\hat{\mu}$ in Fig. 4B with Q, the latter term being derived as a function of μ according to the Droop equation (Fig. 4A). In either case, the degree of uncoupling between reduced nitrogen uptake and growth (enhanced uptake potential) is seen to increase dramatically as the degree of nitrogen limitation increases.

Goldman and Glibert (1982) more recently found that there are significant species-dependent differences in the relationship between $V_m':\hat{\mu}$ and $\mu:\hat{\mu}$ for NH_4^+ uptake. In particular, some diatoms have a much greater NH_4^+ uptake potential than does *T. pseudonana* (3H) [the test organism of McCarthy and Goldman (1979)] at all values of $\mu:\hat{\mu}$, and, in fact, are capable of sustaining short-term NH_4^+ uptake rates four to eight times greater than required for balanced growth when $\mu:\hat{\mu}$ is in the range 0.80-0.95. As will be discussed, this finding has important bearing on the way in which phytoplankton may obtain their ration of nitrogen in nutrient-poor oceanic waters.

B. Comparative Uptake and Transport Rates

One of the major inconsistencies among researchers working on phytoplankton-nitrogen kinetics is the variable way in which uptake data are reported. Dugdale (1967) clearly distinguished between the specific nitrogen uptake rate V with units of time^{-1} and the absolute nitrogen transport rate ρ with units of mass·volume^{-1}·time^{-1}. In using the Michaelis-Menten equation to relate uptake to nutrient availability, he chose to use V as the uptake parameter [Eq. (1)]. The similarities between Eq. (1) and Eq. (2) are obvious and have the advantage in that uptake and growth rates are compared in the same units of time^{-1}, thus providing a meaningful demonstration of how the two processes are related (Fig. 4B).

Others, in contrast, have chosen to use the absolute transport rate ρ as the uptake parameter in Eq. (1) so that

$$\rho = \rho_m \left[\frac{N}{k_s + N} \right] \tag{4}$$

in which ρ_m is the maximum absolute transport rate (mass·volume^{-1}·time^{-1}), N is as defined in Eq. (1), and k_s is the value of N when $\rho = 0.5\ \rho_m$.

The use of $\rho = V(PN)$ has evolved primarily from the application of ^{15}N methodology to the study of N-uptake dynamics of natural phytoplankton, for which it is currently impossible to distinguish between detrital and phytoplankton nitrogenous biomass (Dugdale and Goering, 1967; also see Harrison, this volume, Chapter 21). The advantage of using $\rho = V(PN)$ as an indicator of uptake is that the measured rate is not influenced by detrital N, whereas use of V leads to an underestimate of uptake proportional to the fraction of detrital N present (Dugdale and Goering, 1967). For laboratory studies involving uptake experiments with single species, the general convention has been to define ρ as the product of V and Q so that ρ has the units mass.cell^{-1}·time^{-1} (Dugdale, 1977).

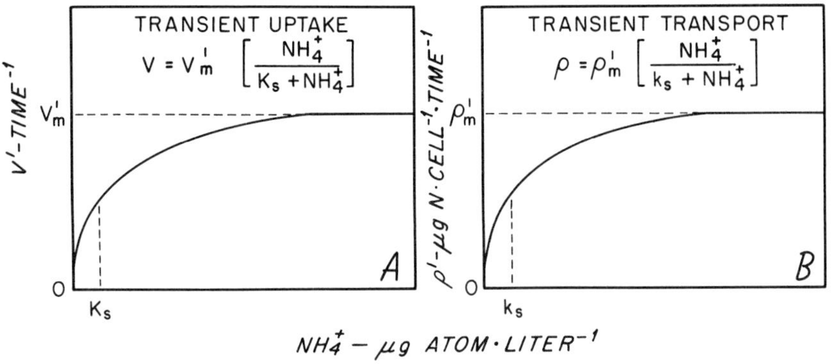

Fig. 5. Graphical depiction of the Monod equation when used to describe the relationship between transient NH_4^+ uptake rate and limiting NH_4^+ concentration: (A) General shape of uptake curve when uptake is defined as specific uptake rate V' [Eq. (1)]; (B) same as panel A, but with transport rate ρ' used in place of V' [Eq. (4)].

Although the similarities between specific uptake and transport kinetics are obvious (Fig. 5), Droop (1978, 1979) has argued that Eq. (4) rather than Eq. (1) is the correct expression for relating uptake to nutrient availability, but only when Q is defined in units of mass of internal nutrient·total cellular mass^{-1}. This apparent confusion stems, in our opinion, primarily from the problem addressed earlier, namely that there has been a gross lack of attention given to the fact that Eqs. (1), (2), and (4) are valid only for describing the instantaneous rate of uptake.

Consider, for example, that each point along the curve in Fig. 4A represents a unique physiological state for a given species and set of environmental conditions within the limits $0 < \mu \leq \hat{\mu}$ and $k_Q \leq Q \leq Q_m$. It follows then that because V'_m varies with $\mu:\hat{\mu}$ (Fig. 4B), a unique saturation curve of instantaneous uptake versus nutrient concentration exists corresponding to each physiological state. Because the instantaneous uptake rate is the defined parameter in Eqs. (1) and (4), it must be assumed that cell nutrient biomass remains unchanged (i.e., Q or any other static indicator of cell nutrient status is constant) during the period over which uptake is measured. Then, contrary to the assertion of Droop (1978), Eq. (1) and (4) must be equivalent since $\rho = VQ$ and $\rho_m = V_m Q$. Inserting these relationships into Eq. (4) leads to Eq. (1) and also to the equality $K_s = k_s$.

C. Temporal Dependence of Uptake

In attempting to develop a quantitative relationship between uptake and growth, Dugdale (1977) considered that the relationship $V_m' = \rho_m Q^{-1}$ could be expanded by defining Q according to the Droop expression. Then

$$V_m' = \frac{\rho_m}{k_Q}\left[1 - \frac{\mu}{\bar{\mu}}\right] \tag{5}$$

Dugdale (1977) also suggested that ρ_m might be constant so that V_m' would be expressed as an inverse linear function of μ according to Eq. (1).

McCarthy (1982), in reviewing the available nitrogen and phosphorus data for ρ_m versus μ, was unable to confirm that ρ_m was constant and suggested that if ρ_m were indeed variable, it would explain the curvilinear increase in V_m' with decreasing μ found in several studies. He proposed a variable maximum rate of uptake per cell for a population in steady state which he defined as ρ_m'. In fact, even if ρ_m' is not constant, it is not necessary to consider this variability in ρ_m' to explain why V_m' for NH_4^+ uptake might increase at a nonlinear rate as μ decreases. For example, there is no *a priori* rationale for assuming that Eq. (5) has general applicability for describing transient nutrient uptake phenomena. Although the Droop equation is adequate for characterizing steady-state growth rate as a function of Q for many nutrients, it has been found to be less than satisfactory for representing similar transient conditions for at least P and Si uptake (Davis et al., 1978; Nyholm, 1978; Burmaster, 1979). The applicability of Eq. (5) is restricted further when dealing with N uptake because transient uptake of reduced N (e.g., NH_4^+, urea) by cells preconditioned with reduced N occurs differently than by those cells preconditioned with NO_3^- or NO_2^- [compare results of Goldman and Glibert (1982) with those of Horrigan and McCarthy (1981, 1982)].

The major drawback of Eq. (5) for describing transient uptake phenomena is that Q is a lumping term for all intracellular nutrient components. When Q is defined in such a manner, there is no way to distinguish between metabolic regulation of transient nutrient uptake, which we define as the process of nutrient transport across the outer cell membrane, and the biochemical processing of these nutrients (metabolic assimilation) that culminates in growth.

Until recently, little consideration was given to the possibility that inorganic nitrogen uptake over time might be nonlinear. In fact, virtually all of the field data and much of the laboratory data on inorganic nitrogen uptake kinetics have been generated from experiments involving single-endpoint measurements over varying incubation periods of either inorganic nitrogen disappearance from the aqueous phase or nitrogen accumulation in the particulate phase; linearity of uptake generally was implied (see discussion by Goldman et al., 1981). However, it is now well recognized that inorganic nitrogen uptake, particularly in the case of NH_4^+, can be nonlinear over time in both laboratory (Conway et al., 1976; Conway and Harrison, 1977; DeManche et al., 1979; McCarthy and Goldman, 1979; Goldman and Glibert, 1982; Horrigan and McCarthy, 1982) and field studies (Glibert and Goldman, 1981; Goldman et al., 1981; Wheeler et al., 1982; Wheeler and McCarthy, 1982).

For the most part, the above laboratory studies have involved pulsed addition of saturating concentrations of NH_4^+ to phytoplankton cultures preconditioned to some well-defined nutritional state, followed by time-course uptake measurement over minutes to hours incubation. Using our earlier time-course results as a model of the temporal as well as growth-rate dependence of uptake (Goldman and Glibert, 1982), we have depicted graphically how transient NH_4^+ uptake ($V_m':\hat{\mu}$) may vary as a function of both $\mu:\hat{\mu}$ and incubation time (Fig. 6). On the plane of $V_m:\hat{\mu}$ versus $\mu:\hat{\mu}$, but at incubation periods just greater than time zero (seconds), enhanced uptake is greatest at 0.25 $\hat{\mu}$ (severe nitrogen limitation) and decreases rapidly as $\mu:\hat{\mu}$ increases until $V_m':\hat{\mu} = \mu:\hat{\mu} = 1.0$ (coupled uptake and growth); this response is identical to the curve shown in Fig. 4B and is typical of that of *T. pseudonana* (3H) (McCarthy and Goldman, 1979). No uptake response is shown for $\mu < 0.25\ \hat{\mu}$ because, as Goldman and Glibert (1982) have found, there is a dramatic decline in the magnitude of $V_m':\hat{\mu}$ below 0.25 $\hat{\mu}$ that is not well understood, but may be the result of problems associated with maintaining the integrity of steady-state continuous cultures at low dilution rates (Pirt, 1975). On the same plane of $V_m':\hat{\mu}$ versus $\mu:\hat{\mu}$, but at long incubation periods (hours), uptake and growth are most likely coupled at all physiological states. In between these temporal limits of seconds to 2 h, there is a surface of enhanced uptake potential that represents maximum decreases in uptake with incubation time at 0.25 $\hat{\mu}$ and coupled uptake and growth ($V_m':\hat{\mu} = 1$) independent of time at $\mu:\hat{\mu} = 1$. The major feature of this model is that tremendous uncoupling between NH_4^+ uptake and growth is possible on both temporal and physiological scales. Thus, depending on the temporal scale chosen to measure NH_4^+ uptake, totally different perspectives emerge as to how uptake and growth are related (Goldman and Glibert, 1982).

D. The Relationship between K_s and K_u

A logical question that comes to mind after considering how dependent enhanced NH_4^+ uptake is on the combination of cell physiological state and duration of exposure to elevated NH_4^+ concentration is that if every point on the uptake surface shown in Fig. 6 represents a unique potential for enhanced uptake (that is, a unique $V_m':\mu$), then how does the magnitude of K_s vary under similar conditions? For example, if we consider only the relationship

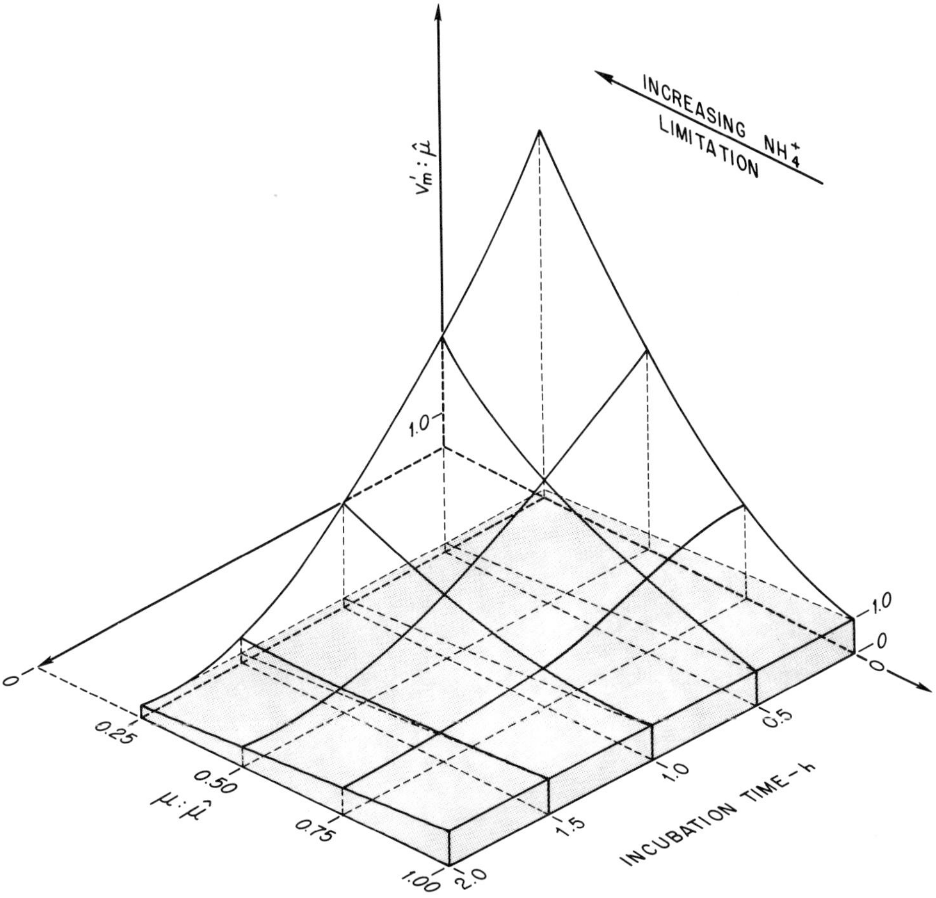

Fig. 6. Conceptual relationship between transient NH_4^+ uptake rate V_m':$\hat{\mu}$ and both physiological state μ:$\hat{\mu}$ and duration of incubation. Resulting surface depicts enhanced uptake rate as a function of μ:$\hat{\mu}$ and incubation period that is greater than coupled uptake and growth (shaded area), as observed for *T. pseudonana* (3H). Curves for other species would be similar in directionality of effects, but not necessarily in magnitude.

between the two variables V_m:$\hat{\mu}$ and μ:$\hat{\mu}$, then the standard saturation curve depicted by Eq. (1), and shown in Fig. 7A, really has the form of a family of saturation curves, each representing a different value of μ:$\hat{\mu}$ in the range 0-1.0 (see McCarthy, 1982). Whether K_s remains constant under these conditions (Fig. 7B) or

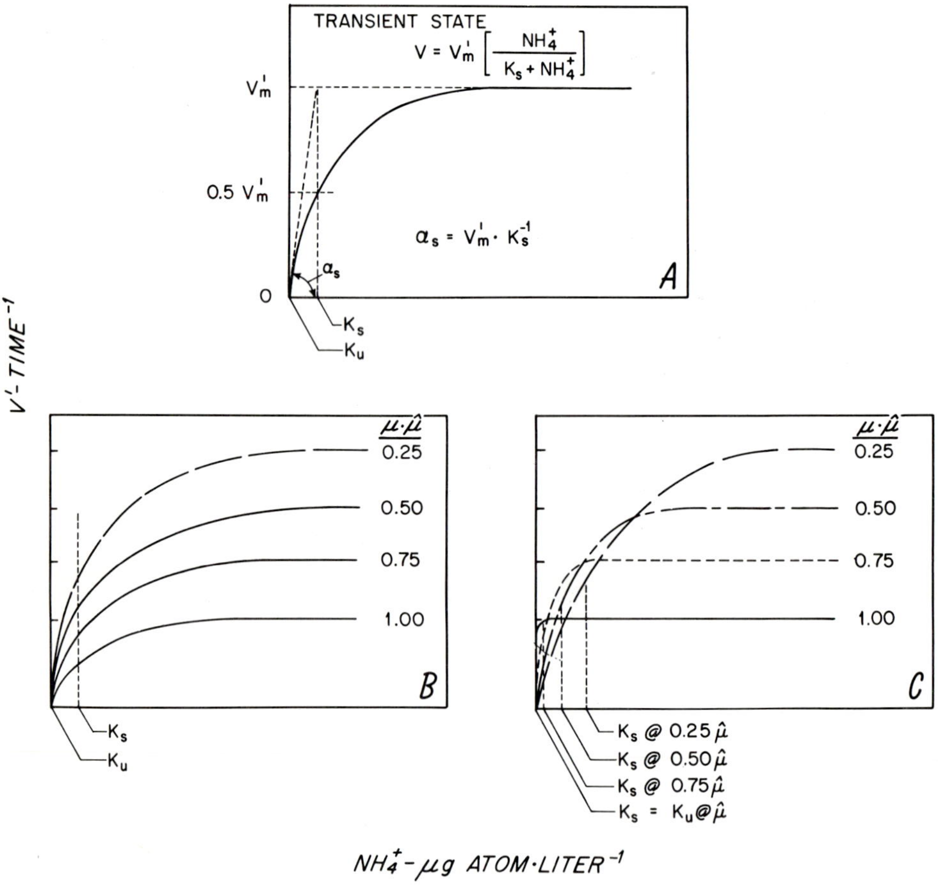

Fig. 7. Effect of cell physiological state ($\mu:\hat{\mu}$) on the NH_4^+ uptake response of phytoplankton when duration of incubation is constant: (A) General shape of Monod saturation curve [Eq. (1)] without consideration for effect of cell physiological state on V'; (B) possible family of Monod saturation curves in which V_m' increases with decreasing $\mu:\hat{\mu}$, but K_s remains constant but greater than K_u; (C) possible family of Monod saturation curves in which both V_m' and K_s increase with decreasing $\mu:\hat{\mu}$. $K_s = K_u$ when $\mu:\hat{\mu} = 1$.

varies proportional to changes in $\mu:\hat{\mu}$ (K_s must equal K_u when $\mu:\hat{\mu} = 1$) (Fig. 7C) is difficult to determine given the paucity of available data on the subject (McCarthy, 1982).

Droop (1973), in attempting to show how K_s and K_u were quantitatively related, algebraically manipulated Eqs. (2)-(4) and the steady-state equation $\rho = \mu Q$ to arrive at the expression

$$K_u = \hat{\mu} k_Q \frac{K_s}{\rho_m} \tag{6}$$

Droop (1973, 1978) and others (Healey, 1980; Rhee, 1980; Turpin et al., 1981) have used this or similar expressions to define the relationship between the steady-state half-saturation coefficient K_u and the half-saturation coefficient K_s that describes a transient phenomenon. In our opinion, however, Eq. (6) is only relevant for describing steady-state kinetics, and not the kinetics of transient NH_4^+ uptake, because it is founded on the premise that the steady-state equation $\rho = \mu Q$, and not the transient analog $\rho = VQ$, is the proper expression for defining the uptake potential that might exist for a given physiological state.

Thus ρ, as used by Droop (1973) in arriving at Eq. (6), is not the same ρ (or V) that describes transient uptake phenomena. If VQ, and not μQ, is used to describe ρ, a relationship between transient K_s and steady-state K_u is indeterminable. To our knowledge, there is no way to derive a quantitative relationship between transient K_s and steady-state K_u with existing empirical equations, other than to acknowledge that the two terms must be equal when $\hat{\mu}$ is attained and there is coupling between growth and uptake.

So far in our discussion of K_s and K_u we have ignored the temporal dependence of uptake. The question that must be addressed is that if V_m' varies as a function of incubation time as well as physiological state, then is there a similar temporal dependence of K_s? As we have stated earlier, Eq. (1) is applicable only for describing the kinetics of instantaneous uptake. Conceptually then, we are faced with the possibility that for a given

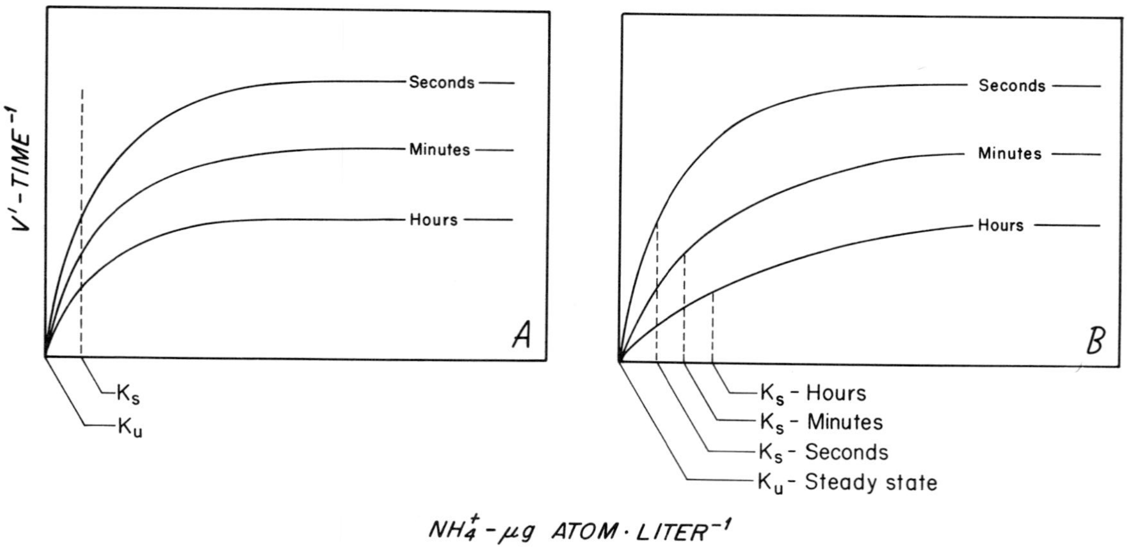

Fig. 8. Possible effects of incubation period on K_s for NH_4^+ uptake when cell physiological state ($\mu:\hat{\mu}$) is constant: (A) Increasing V_m', but constant K_s greater than K_u as incubation period decreases; (B) increasing V', but decreasing K_s approaching K_u as incubation period decreases.

physiological state (the plane of V'_m: versus $\mu:\hat{\mu}$ in Fig. 6), K_s either remains constant as incubation period varies (Fig. 8A) or increases proportional to increasing duration of incubation (Fig. 8B).

In conjunction with our study on transient NH_4^+ kinetics of several species (Goldman and Glibert, 1982), we found (but did not report) that there was a temporal dependency of K_s for at least two marine species, the diatom *Chaetoceros simplex* (BBsm) Fig. 9A) and the chlorophyte *Dunaliella tertiolecta* (Dun) (Fig. 9B). In both cases, increases in K_s (although not statistically significant), along with corresponding decreases in V'_m were noted for increasing incubation periods between 1 and 5 min (Table I). Wheeler et al. (1982) found similar evidence for increasing K_s over incubation periods ranging from 15 to 60 min in short-term NH_4^+-uptake studies with natural populations from the Chesapeake Bay; the differences were statistically real, however, in only one of three experiments (Table I). The two studies discussed above, to our knowledge, represent the only attempts to correlate changes in K_s with duration of incubation; the statistical unreliance of the results points toward the great analytical difficulties inherent in such measurements.

In trying to understand the temporal dependence of K_s, it must be recognized that there are severe analytical limitations in making accurate measurements of K_s (McCarthy, 1982; Harrison, Chapter 21, this volume). Eppley et al. (1977), for example, demonstrated how tremendous distortions in saturation curves, and concomitant overestimates of K_s, could evolve if analytical limitations prevented accurate determination of ambient concentrations of limiting nutrient (see Fig. 2 and accompanying discussion in Harrison, Chapter 21, this volume).

Experimental difficulties in determining K_s fall into two related categories. First, even over very short incubation periods, there may be considerable depletion of substrate, especially at low concentrations (McCarthy, 1982). For example, we found for

TABLE I. Summary of Available Kinetic Data from Laboratory and in Which Incubation Period or Physiological State ($\mu:\hat{\mu}$)

Species or location	$\hat{\mu}$ *(h^{-1})*	$\mu:\hat{\mu}$	*Incubation period (h)*
Chaetoceros simples (BBsm)	0.051	0.16	0.017 0.083
Dunaliella tertiolecta (Dun)	0.038	0.21	0.017 0.083
Thalassiosira pseudonana (13-1)	0.044	0.19 0.40 0.94	0.25-2
Chesapeake Bay			
Station 744-1	?	?	0.017 0.25 1
Station 707 Ø	?	?	0.083 0.50 1
Station 744-2	?	?	0.017 0.25 1

[a] *S.D. when reported.*
[b] *In experiments by Wheeler et al. (1982) saturation was not attained during incubations <5-min duration, and thus kinetic constants were inestimable.*

our laboratory studies discussed above (Table I), that when NH_4^+ uptake was rapid during short-term incubations ($\leq$5 min), 20% of the added NH_4^+ (at concentrations equivalent to the measured K_s value) was consumed during the first minute of incubation, increasing to 30% consumption after 5 min (Fig. 9C). Even when short-term NH_4^+ uptake was relatively sluggish, 10-15% of the added NH_4^+ (at equivalent K_s concentrations) was consumed within the first 5 min (Fig. 9D). These observations thus demonstrate that one of the important assumptions governing the validity of Eq. (1), that of substrate concentration remaining constant, is frequently violated even in short-term experiments.

Field Experiments Involving NH_4^+ Uptake by Marine Phytoplankton Were Variables

K_s (μg·atom liter^{-1})	V'_m (h^{-1})	α_s (liter·h·μg·atom^{-1})	References
1.24 ± 0.298	0.45 ± 0.037[a]	0.36	This study
2.22 ± 0.503	0.17 ± 0.015	0.08	
0.48 ± 0.165	0.10 ± 0.009	0.21	
1.48 ± 0.106	0.07 ± 0.001	0.05	
0.40 ± 0.160	0.58	1.45	Eppley and Renger (1974)
0.66 ± 0.035	0.30	0.45	
0.22 ± 0.080	0.09	0.41	
b	b	∞	
0.27 ± 0.09	0.021 ± 0.001	0.08	Wheeler et al. (1982)
0.40 ± 0.12	0.015 ± 0.001	0.04	
b	b	∞	
0.07 ± 0.03	0.033 ± 0.002	0.47	
0.26 ± 0.07	0.032 ± 0.002	0.12	
b	b	∞	
0.46 ± 0.16	0.017 ± 0.002	0.04	
0.56 ± 0.13	0.019 ± 0.002	0.03	

The second difficulty stems from the fact that the instantaneous rate of uptake is nearly impossible to measure, and in most experiments of this nature, assimilation and growth processes are being measured instead (see Fig. 1). In the studies by Wheeler et al. (1982), kinetics of NH_4^+ uptake by Chesapeake Bay phytoplankton were interpreted in conjunction with data on rates of assimilation of NH_4^+ into proteinaceous material. In experiments of 1-min duration, no saturation in uptake was observed up to 24 μg·atom liter^{-1}; yet saturation was readily apparent in experiments >5 min (see Tables I and II). The nonsaturable initial uptake data could be explained primarily by the rapid labeling of small internal pools

TABLE II. Summary of Reported Values of K_s for NH_4^+ and NO_3^- in ^{15}N Tracer Assay

Location	*Temperature (°C)*	*Incubation duration (h)*
Oligotrophic waters		
Tropical Pacific	N.R.[b]	6-24
Mediterranean	N.R.	24
North Pacific Central gyre	N.R.	24
	N.R.	24
Eutrophic oceanic waters		
Tropical Pacific	N.R.	6-24
Subarctic Pacific	N.R.	6-24
Peru coast	N.R.	24
Coastal waters		
Vineyard Sound, Massachusetts	6.9	0.50
	11.2	0.50
	11.8	0.50
	18.2	0.50
Chesapeake Bay	24	0.25
	24	0.50
	24	1

[a]*Data adapted from stated references. Parentheses indicate number of experiments.*

[b]*N.R., not reported.*

[c]*Data are summarized from data in Table I.*

of NH_4^+ and possibly other soluble metabolites. The saturable kinetics observed in longer incubations appeared to reflect rates of macromolecular synthesis. This view was supported by the observation that the percentage of accumulated ^{15}N label in the soluble fraction often reached a maximum within 15 min, at which time rates of NH_4^+ uptake became limited by rates of assimilation and incorporation into macromolecules rather than transport. Thus, changes in labeling patterns among cellular nitrogen components during the course of uptake experiments violates the assumption that the tracer is homogenously distributed within the particulate fraction, an assumption inherent in the use of tracer techniques to calculate uptake rates (Sheppard, 1962; Wheeler et al., 1982).

Various Marine Waters from Shipboard Experiments Involving Use of

K_s (μg·atom liter^{-1})[a]		
NO_3^-	NH_4^+	References
0.01-0.21(6)	0.10-0.55(3)	MacIsaac and Dugdale (1969)
0.1 - 0.3(4)	<0.1(2)	MacIsaac and Dugdale (1972)
N.R.	0.15(1)	Eppley et al. (1973)
0.00-0.91(11)	0.00-0.13(13)	Eppley et al. (1977)
0.98(1)	1.30(1)	MacIsaac and Dugdale (1969)
4.21(1)	1.30(1)	
N.R.	1.11(4)	MacIsaac and Dugdale (1972)
	0.62	Glibert et al. (1982b)[c]
	0.72	
	0.58	
	0.73	
	0.27-0.46(2)	Wheeler et al. (1982)
	0.71(1)	
	0.26-0.56(3)	

The assumption of homogeneous distribution of label becomes valid during long incubations, but then calculated uptake rates reflect rates of metabolic incorporation of NH_4^+ into macromolecules, and not the maximum potential flux of NH_4^+ across the cell membrane and assimilation into soluble organic nitrogen (Wheeler et al., 1982). It should be noted, however, that in the laboratory studies discussed above (Table I), saturation was obtained within 1 min; these differences may reflect nutrient prehistory, growth rates, and/or species-specific differences. Although considerably more and better data are required to draw sound conclusions as to the relationship between K_s and incubation duration, the above results raise the question as to whether or not much of the K_s data in the literature representing NH_4^+ and NO_3^- uptake by natural populations (Table II) and cultured species (Table III) are overestimates caused, to a large degree, by these types of experimental problems or artifacts.

In summary, aside from the fact that, by definition, $K_s:K_u = 1$ only when $V_m':\hat{\mu} = \mu:\hat{\mu} = 1$, (valid only when K_s is based on measurement of instantaneous uptake rates), there is no *a priori* rationale for deciding how the ratio $K_s:K_u$ for NH_4^+ and NO_3^- uptake might otherwise vary with changing $\mu:\hat{\mu}$ (Fig. 10A). The uncertainty in this relationship is magnified by the limited but conflicting results on the topic. For example, in studies on NH_4^+ and NO_3^- uptake, the ratio $K_s:K_u$ has been found to be invarient with changing $\mu:\hat{\mu}$ (Eppley and Renger, 1974) (Table I and depicted in Fig. 7A), to decrease with increasing $\mu:\hat{\mu}$ (Rhee, 1978) (depicted in Fig. 7B), and even to increase as $\mu:\hat{\mu}$ increases (Zevenbloom and Muir, 1981). The differences in these reported trends may be as much or more due to experimental problems of the type previously discussed, as they are the result of species-specific differences; in none of the three above studies were attempts made to measure the instantaneous uptake rate.

E. The Relationship between K_s and V_m'

An obvious conclusion from our discussion so far is that the transient uptake rate is a function not only of the concentration of limiting N, as defined by Eq. (1), but also of the duration of exposure to the N source and the physiological state of the cell population at the time of exposure. Thus, to characterize V' rather than V_m', the surface of maximum relative uptake response seen in Fig. 6 must be expanded to include the additional variable N so that

$$V':\hat{\mu} = f(N,\ \mu:\hat{\mu},\ T) \tag{7}$$

in which V' is the variable, nonsaturating uptake rate (time^{-1}), $V':\hat{\mu}$ is the relative nonsaturating uptake rate, and T is the in-

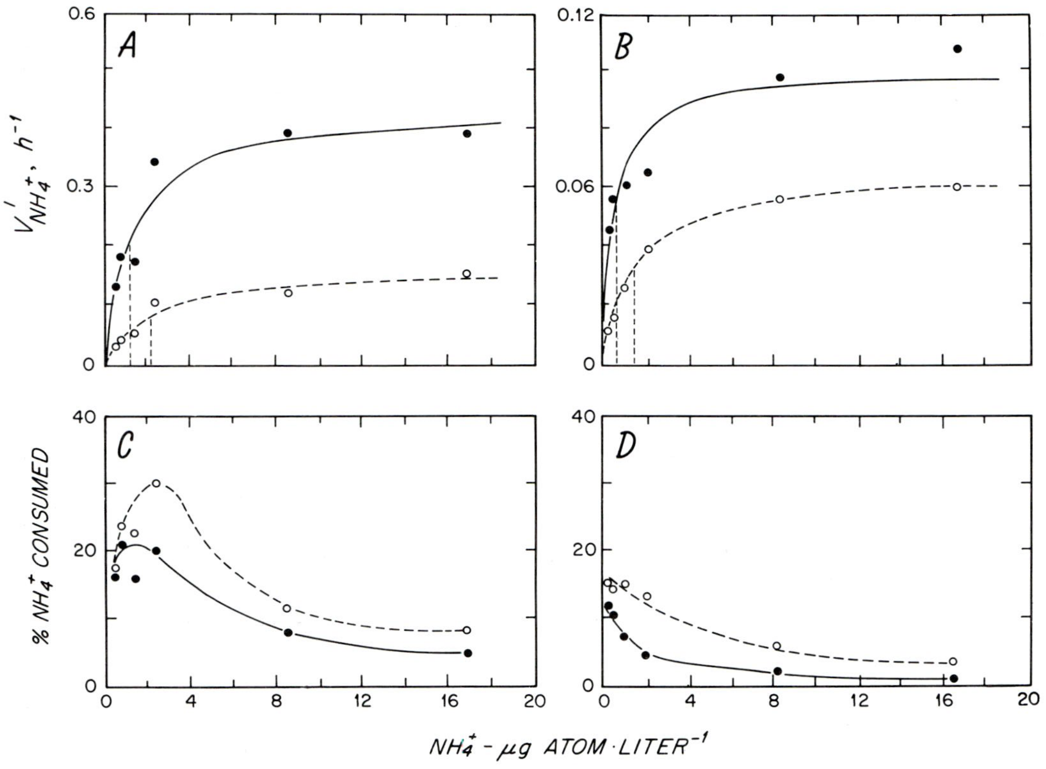

Fig. 9. Effect of incubation period on NH_4^+ uptake kinetics (A,B) and percentage of NH_4^+ consumed (C,D) for varying NH_4^+ concentrations (Key: ●, 1-min incubations, ○, 5-min incubations) for *Chaetoceros simplex* (BBsm) (A,C) and *Dunaliella tertiolecta* (Dun) (B,D). *Chaetoceros simplex* (BBsm) was pregrown in NH_4^+-limited continuous culture at $\mu:\hat{\mu}$ = 0.16 (A,C), and *Dunaliella tertiolecta* (Dun) was pregrown in NH_4^+-limited continuous culture at $\mu:\hat{\mu}$ = 0.21 (B,D). Kinetic data are summarized in Table I.

TABLE III. Summary of Reported Values of K_s and K_u for NH_4^+ and NO_3^- from Laboratory Experiments on Various Cultured Marine Phytoplankton Species

Species	*Preconditioning culture (μ, day⁻¹)*	*Temperature (°C)*	*Incubation type*[a]
Oceanic (4)[b]	Batch (?)	18	T-D
Neritic diatoms (8)[b]			
Neritic or littoral diatoms (5)[b]			
Diatoms			
Asterionella japonica	Batch (?)	18	T-D
Chaetoceros gracilis			
Thalassiosira pseudonana (3)[c]	Batch (?)	20	T-D
Fragilaria pinnata (2)[c]			
Bellerochia sp. (3)[c]			
Skeletonema costatum	Continuous	18	SS
Skeletonema costatum	Continuous (1.2)	18	T-D
Chaetoceros debilis	Continuous (1.2)		
Thalassiosira gravida	Continuous (0.7)		
Thalassiosira pseudonana (3H)	Continuous	18	SS
Thalassiosira pseudonana (3H)	Continuous (1.6)	18	$T\text{-}^{15}N$

[a] *T-D, transient uptake measured by disappearance of nutrient added to culture; SS, steady-state measurements of residual nutrient;* $T\text{-}^{15}N$*, transient uptake measured by accumulation in phytoplankton of* ^{15}N*-labeled nutrient added to culture.*

[b] *Number of species examined.*

[c] *Number of clones of species examined.*

[d] *U, unmeasurable.*

Duration (h)	Half-saturation coefficient (μg·atom liter^{-1}) NO_3^- K_s	K_u	NH_4^+ K_s	K_u	Reference
0.25-2	0.1-0.7		0.1-0.5		Eppley et al. (1969)
	0.4-5.1		0.5-9.3		
	0.1-10.3		0.1-5.7		
0.25-2	0.7,1.3	1-2,1.5			Eppley and Thomas (1969)
	0.3	0.2			
					Carpenter and Guillard (1971)
0.08-0.50		0.4-1.9			
		0.6-1.6			
		0.1-0.9			
3-4				U[d]	Harrison et al. (1976)
6-7			0.5		
6-7			0.5		
6.7			0.5		
					Goldman and McCarthy (1978)
0.08			0.4	<0.03	McCarthy (1981)

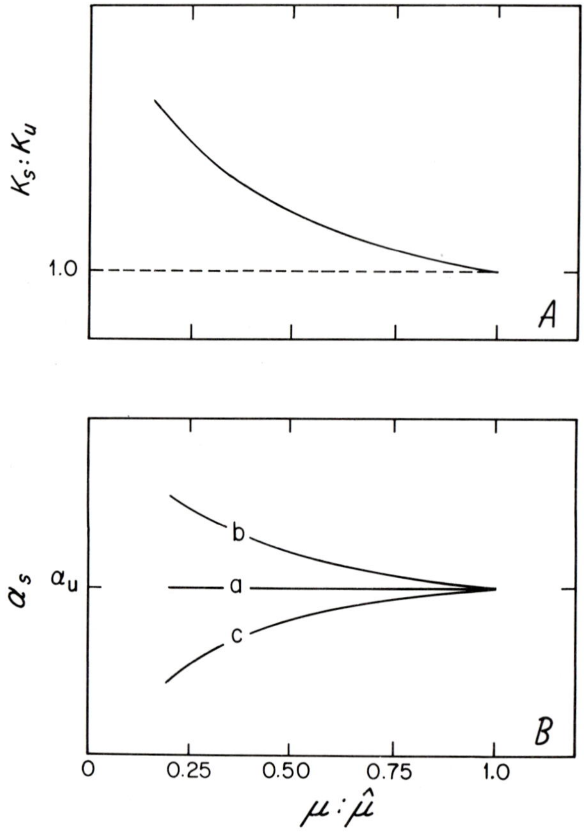

Fig. 10. Possible relationships between steady-state cell physiological state $\mu:\hat{\mu}$ and transient (instantaneous) uptake parameters $K_s:K_u$ (A) and α_s (B).

cubation period (time). Because Eq. (1) is only valid when N remains constant, K_s has little meaning when based on incubations lasting longer than a few seconds and when initial uptake is rapid and nonlinear (Wheeler et al., 1982). Unfortunately, as we have discussed, it is extremely difficult to measure K_s for incubations lasting minutes or less. Aside from analytical limitations, uptake at low concentrations of limiting nutrient, and for short periods, may be diffusion-limited (Gavis, 1976) or possibly controlled by adsorption processes on outer cell surfaces (Goldman and Glibert, 1982), thus further complicating the meaning of measured K_s values.

In principle, the affinity a phytoplankton population has for a particular nutrient is defined as much by the magnitude of V_m' as it is by K_s (Dugdale, 1967). Healey (1980) proposed that the initial slope of the uptake curve defined by Eq. (1) was the best indicator of nutrient affinity. This slope can be quantified (Fig. 7A) as

$$\alpha_s = \frac{V_m'}{K_s} \tag{8}$$

In an analogous manner, the initial slope of the growth curve [Eq. (2)] (Fig. 2A) can be described by

$$\alpha_u = \frac{\hat{\mu}}{K_u} \tag{9}$$

Clearly, Healey's approach provides a far more descriptive picture of nutrient affinity than does either K_s or V_m' alone: the degree of affinity is in direct proportion to the magnitude of α_s for transient uptake (or α_u when steady state is established). Thus, assuming that K_s can be determined accurately, and ignoring for the moment the possible temporal dependency of K_s, a multitude of possible relationships between α_s and $\mu:\hat{\mu}$ can be envisioned (Fig. 10B). For example, curve a in Fig. 10B (constant α_s) would indicate that neither V_m' nor K_s were affected by cell physiological state, or that both parameters increased in the same relative proportion as $\mu:\hat{\mu}$ decreased. In either case, the equality $\alpha_s = \alpha_u$ would be maintained because $\alpha_s = \alpha_u$ must always be true when $\mu:\hat{\mu} = 1$ (Fig. 10A). The other two curves, in contrast, represent situations in which V_m' and K_s vary at different rates relative to changing $\mu:\hat{\mu}$; in one case (curve b), V_m' increases at a steeper rate than K_s, and in the other case (curve c), K_s increases more rapidly relative to V_m'. The assumption is made, purely on empirical grounds (Table I), that neither V_m' nor K_s decrease as $\mu:\hat{\mu}$ decreases.

Summarized in Table I are values of α_s that we calculated from the data in Fig. 9 and from the studies of Eppley and Renger (1974) and Wheeler et al. (1982). It appears, based on the data of Eppley and Renger (1974), that curve b may best describe the relationship between α_s and $\mu:\hat{\mu}$. However, difficulties in measuring K_s may negate the general applicability of this approach, primarily because even small errors in the measurement of K_s [the denominator in Eq. (8)] will lead to larger errors in α_s. The uncertainty as to how K_s varies over incubation period (independent of changes due to nutrient depletion) clouds the meaning of α_s even further (Table I).

V. NEW PARADIGMS OF NITROGENOUS NUTRITION

A common feature of oceanic waters is that, with few exceptions, NH_4^+ and NO_3^- concentrations in the euphotic zone are below the current detection limit of 0.03 μg·atom liter^{-1} (McCarthy and Goldman, 1979). At the same time, we know from laboratory continuous culture studies that many marine phytoplankton have the capacity to grow at rates approaching $\hat{\mu}$, even when available NH_4^+ or NO_3^- is below 0.03 μg·atom liter^{-1} (Fig. 2). The similarities between the surface waters of the open ocean and the laboratory continuous culture are thus apparent (Eppley et al., 1973; Goldman et al., 1979). In both systems, we have no way of knowing how fast the resident phytoplankton population is growing from static measurements of biomass and nutrient concentration. The factor controlling μ in the steady-state continuous culture is the dilution rate and, analogously, in the open ocean it is the rate of herbivore grazing and nutrient regeneration (Goldman et al., 1979). Clearly, for μ to approach $\hat{\mu}$ in oceanic waters, as suggested by Goldman et al. (1979) from their chemical composition hypothesis, requires grazing and nutrient regeneration rates essentially equal to μ. Measurements of nutrient regeneration in oceanic waters covering a range of nutrient concentrations have been made using ^{15}N isotope dilution techniques (see review by Harrison, this volume, Chapter 21), and results commonly demonstrate regeneration rates equal to, and in some cases greater than, nutrient uptake rates (Harrison, 1978; Caperon et al., 1979; Glibert et al., 1982; Glibert, 1982). Identification of the types of microplankton that might be important in such a food chain process and estimation of their feeding rates is a major challenge of modern day biological oceanography (Sieburth et al., 1978).

Perhaps one of the most difficult problems in trying to understand the relationships between phytoplankton and their grazers is in identifying the appropriate temporal and spatial scales that are important to these organisms. For example, the steady-state continuous culture is defined as such because our measurements of steady state are made on temporal scales of hours to days and on spatial scales of tens of milliliters to liters. Yet, from the standpoint of an individual cell that is being mixed inside the culture vessel, the perceived environment is one far removed from steady state.

The common method of medium dispensing in laboratory continuous cultures (peristaltic pumping) places a finite limit on the interval (typically a few seconds) between drops of medium introduced into the culture vessel. Mixing times usually are on the order of tens of seconds (J. C. Goldman, unpublished data). Given the fact that an individual marine phytoplankter is capable of tremendously enhanced NH_4^+ uptake for short periods when exposed to a sudden elevation in NH_4^+ concentration in its surrounding aqueous environment (Fig. 6), a large fraction of tne NH_4^+ available in each individual drop of medium most likely is incorporated by a small percentage of the residing population on times scales much less than those necessary for complete liquid mixing. The steady-state continuous culture when viewed from such a perspective thus becomes very unsteady!

McCarthy and Goldman (1979) first suggested that pulsed and rapid uptake of NH_4^+ that originated from microzones of excreted nutrients on time scales of minutes might be a common mode of phytoplankton nutrition. Such rapid uptake would make larger volume samples seemingly devoid of NH_4^+ when measured with existing methods. Others (Jackson, 1980; Williams and Muir, 1981) have argued that molecular diffusion of NH_4^+ away from point sources such as zooplankton would be so rapid as to prevent micropatches of nutrients from existing long enough to be exploited sufficiently. However, evidence has been accumulating that phytoplankton can effectively exploit microzones of nutrients originating from zooplankton (Lehman and Scavia, 1982), and on time scales of seconds or less (Goldman and Glibert, 1982). Equally important, enhanced uptake capability does not appear to be restricted to phytoplankton growing slowly, as originally observed by McCarthy and Goldman (1979) in their studies on *T. pseudonana* (3H); in some species it may occur even at growth rates approaching, but not at $\hat{\mu}$ (Goldman and Glibert, 1982; Horrigan and McCarthy, 1982).

A common misunderstanding of the arguments of McCarthy and Goldman (1979) and Goldman et al. (1979) is the inference that a nutritional strategy based on exploitation of nutrient micropatches is the only strategy that would allow phytoplankton to grow relatively rapidly in seemingly nutrient-depleted environments (Jackson, 1980; Turpin et al., 1981). With present analytical techniques we simply cannot measure ambient levels of nitrogenous nutrients in oligotrophic waters. These levels may, indeed be detectable in the ng·atom liter^{-1} range, if not in the μg·atom liter^{-1} range. Thus, on the one hand and as seen in Fig. 3, phytoplankton may well be deriving some or all of their nitrogenous nutrition from low ambient levels present in a homogeneous environment. On the other hand, the micropatchy nutrient environment and concomitant pulsed uptake by phytoplankton envisioned by McCarthy and Goldman (1979) may be a common characteristic of nutrient-poor waters. It is those species which can exploit small nutrient patches through temporary elevations in *V'*, which will have an evolutionary advantage in an environment in which such patches may exist (McCarthy and Goldman, 1979; Turpin and Harrison, 1979; Turpin et al., 1981).

In conclusion, we have tried to stress the point that the kinetics of nitrogen uptake is far more complicated than previously envisioned. Traditional kinetic data such as compiled in Tables I-III probably are of limited value in our attempts to understand the important modes of nitrogenous nutrition in the phytoplankton. Newer methods for low-level NH_4^+ analysis and for determining instantaneous rates of nitrogen uptake, both in the laboratory and field, will be required before a more comprehensive understanding of the relationships between uptake and growth emerges. However, without an appreciation of the appropriate temporal scales of phytoplankton response to environmental perturbation, future progress in understanding such relationships will be restricted.

ACKNOWLEDGMENTS

Contribution No. 5126 from the Woods Hole Oceanographic Institution. This work was supported by the National Science Foundation Grant OCE80-24445 (JCG) and a Leopold Schepp Foundation Award (PMG).

REFERENCES

Burmaster, D. E. (1979). The unsteady continuous culture of phosphate-limited *Monochrysis lutheri* Droop: Experimental and theoretical analysis. *J. Exp. Mar. Biol. Ecol. 39*, 1167-1186.

Caperon, J. (1965). The dynamics of nitrate limited growth of *Isochrysis galbana* populations. Ph.D. Thesis, University of California, San Diego.

Caperon, J., and Meyer, J. (1972). Nitrogen-limited growth of marine phytoplankton. I. Changes in population characteristics with steady-state growth rate. *Deep-Sea Res. 19*, 601-618.

Caperon, J., Schell, D., Hirota, J., and Laws, E. (1979). Ammonium excretion rates in Kaneohe Bay, Hawaii, measured by a ^{15}N isotope dilution technique. *Mar. Biol. (Berlin) 54,* 33-40.

Carpenter, E. J., and Guillard, R. R. L. (1971). Intraspecific differences in nitrate half-saturation constants for three species of marine phytoplankton. *Ecology 52,* 183-185.

Collos, Y. (1980). Transient situations in nitrate assimilation by marine diatoms. I. Changes in uptake parameters during nitrogen starvation. *Limnol. Oceanogr. 25,* 1075-1081.

Conway, H. L., and Harrison, P. J. (1977). Marine diatoms grown in chemostats under silicate or ammonium limitation. IV. Transient response of *Chaetoceros debilis, Skeletonema costatum,* and *Thalassiosira gravida* to a single addition of the limiting nutrient. *Mar. Biol. (Berlin) 43,* 33-43.

Conway, H. L., Harrison, P. J., and Davis, C. O. (1976). Marine diatoms grown in chemostats under silicate or ammonium limitation. II. Transient response of *Skeletonema costatum* to a single addition of the limiting nutrient. *Mar. Biol. (Berlin) 35,* 187-199.

Davis, C. O., Breitner, N. F., and Harrison, P. J. (1978). Continuous culture of marine diatoms under silicon limitation. 3. A model of Si-limited diatom growth. *Limnol. Oceanogr. 23,* 41-51.

DeManche, J. M., Curl, H. C., Jr., Lundy, D. W., and Donaghay, P. L. (1979). The rapid response of the marine diatom *Skeletonema costatum* to changes in external and internal nutrient concentration. *Mar. Biol. (Berlin) 53,* 323-333.

Droop, M. R. (1973). Some thoughts on nutrient limitation in algae. *J. Phycol. 9,* 264-272.

Droop, M. R. (1978). Comments on the Davis/Breitner/Harrison model for silicon uptake and utilization by diatoms. *Limnol. Oceanogr. 23,* 383-385.

Droop, M. R. (1979). On the definition of X and of Q in the cell quota model. *J. Exp. Mar. Biol. Ecol. 39,* 203.

Dugdale, R. C. (1967). Nutrient limitation in the sea: Dynamics, identification, and significance. *Limnol. Oceanogr. 12,* 685-695.

Dugdale, R. C. (1976). Nutrient cycles. *In* "The Ecology of the Seas" (D. H. Cushing and J. J. Walsh, eds.), pp. 141-172. Saunders, Philadelphia, Pennsylvania.

Dugdale, R. C. (1977). Modeling. *In* "The Sea: Ideas and Observations on Progress in the Study of the Seas" (E. D. Goldberg, ed.), pp. 789-806. Wiley, New York.

Dugdale, R. C., and Goering, J. J. (1967). Uptake of new and regenerated forms of nitrogen in primary productivity. *Limnol. Oceanogr. 12,* 196-206.

Eppley, R. W., and Renger, E. M. (1974). Nitrogen assimilation of an oceanic diatom in nitrogen-limited continuous culture. *J. Phycol. 10,* 15-23.

Eppley, R. W., and Thomas, W. H. (1969). Comparison of half-saturation constants for growth and nitrate uptake of marine phytoplankton. *J. Phycol. 5,* 375-379.

Eppley, R. W., Roger, J. N., and McCarthy, J. J. (1969). Half-saturation constants for uptake of nitrate and ammonia by marine phytoplankton. *Limnol. Oceanogr. 14,* 912-920.

Eppley, R. W., Renger, E. H., Venrick, E. L., and Mullin, M. M. (1973). A study of plankton dynamics and nutrient cycling in the Central Gyre of the North Pacific Ocean. *Limnol. Oceanogr. 18,* 534-551.

Eppley, R. W., Sharp, J. H., Renger, E. H., Perry, M. J., and Harrison, W. G. (1977). Nitrogen assimilation by phytoplankton and other microorganisms in the surface waters of the central North Pacific Ocean. *Mar. Biol. (Berlin) 39,* 111-120.

Gavis, J. (1976). Munk and Riley revisited: Nutrient diffusion transport and rates of phytoplankton growth. *J. Mar. Res. 34,* 161-179.

Glibert, P. M. (1982). Regional studies of daily, seasonal, and size fraction variability in ammonium remineralization. *Mar. Biol. (Berlin) 70,* 209-222.

Glibert, P. M., and Goldman, J. C. (1981). Rapid ammonium uptake by marine phytoplankton. *Mar. Biol. Lett. 2,* 25-31.

Glibert, P. M., Goldman, J. C., and Carpenter, E. J. (1982). Seasonal variations in the utilization of ammonium and nitrate by phytoplankton in Vineyard Sound, Massachusetts. *Mar. Biol. (Berlin) 70,* 237-249.

Glibert, P. M., Lipschultz, F., McCarthy, J. J., and Altabet, M. A. (1982). Isotope dilution models of uptake and remineralization of ammonium by marine plankton. *Limnol. Oceanogr. 27,* 639-650.

Goldman, J. C. (1977). Steady-state growth of phytoplankton in continuous culture: Comparison of internal and external nutrient equations. *J. Phycol. 13,* 251-258.

Goldman, J. C. (1980). Physiological processes, nutrient availability, and the concept of relative growth rate in marine phytoplankton ecology. *In* "Primary Productivity in the Sea" (P. G. Falkowski, ed.), pp. 179-194. Plenum, New York.

Goldman, J. C., and Glibert, P. M. (1982). Comparative rapid ammonium uptake by four marine phytoplankton species. *Limnol. Oceanogr. 27,* 814-827.

Goldman, J. C., and McCarthy, J. J. (1978). Steady-state growth and ammonium uptake of a fast-growing marine diatom. *Limnol. Oceanogr. 23,* 695-703.

Goldman, J. C., and Peavey, D. G. (1979). Steady-state growth and chemical composition of the marine chlorophyte *Dunaliella tertiolecta* in nitrogen-limited continuous cultures. *Appl. Environ. Microbiol. 38,* 894-901.

Goldman, J. C., McCarthy, J. J., and Peavey, D. G. (1979). Growth rate influence on the chemical composition of phytoplankton in oceanic waters. *Nature (London) 279,* 210-215.

Goldman, J. C., Taylor, C. D., and Glibert, P. M. (1981). Non-time-course uptake of carbon and ammonium by marine phytoplankton. *Mar. Ecol.: Prog. Ser. 6,* 137-148.

Harrison, P. J., Conway, H. L., and Dugdale, R. C. (1976). Marine diatoms grown in chemostats under silicte or ammonium limitation. 1. Cellular chemical composition and steady-state growth kinetics of *Skeletonema costatum. Mar. Biol. (Berlin) 35,* 177-186.

Harrison, W. G. (1978). Experimental measurement of nitrogen remineralization in coastal waters. *Limnol. Oceanogr. 23,* 684-694.

Healey, F. P. (1980). Slope of the Monod equation as an indicator of advantage in nutrient competition. *Microb. Ecol. 5,* 281-286.

Herbert, D., Elsworth, R., and Telling, R. C. (1956). The continuous culture of bacteria; A theoretical and experimental study. *J. Gen. Microbiol. 14,* 601-622.

Horrigan, S. G., and McCarthy, J. J. (1981). Urea uptake by phytoplankton at various stages of nutrient depletion. *J. Plankton Res. 3,* 403-414.

Horrigan, S. G., and McCarthy, J. J. (1982). Phytoplankton uptake of ammonium and urea growth on oxidized forms of nitrogen. *J. Plankton Res. 4,* 379-389.

Jackson, G. A. (1980). Phytoplankton growth and zooplankton grazing in oligotrophic oceans. *Nature (London) 284,* 439-441.

Jannasch, H. W. (1974). Steady state and the chemostat in ecology. *Limnol. Oceanogr. 19,* 716-720.

Lehman, J. T., and Scavia, D. (1982). Microscale patchiness of nutrients in plankton communities: Empirical evidence for presence and utilization. *Science 216,* 729-730.

McCarthy, J. J. (1980). Nitrogen and phytoplankton ecology. *In* "The Physiological Ecology of Phytoplankton" (I. Morris, ed.), pp. 191-233. Blackwell, Oxford.

McCarthy, J. J. (1981). Uptake of major nutrients by estuarine plants. *In* "Estuaries and Nutrients" (B. J. Neilson and L. E. Cronin, eds.), pp. 139-163. Humana Press, Clifton, New Jersey.

McCarthy, J. J. (1982). The kinetics of nutrient utilization. *In* "Physiological Bases of Phytoplankton Ecology" (T. Platt, ed.), Bull. 210, pp. 211-233. *Can. Bull. Fish. Aquat. Sci.,* Ottawa.

McCarthy, J. J., and Goldman, J. C. (1979). Nitrogenous nutrition of marine phytoplankton in nutrient-depleted waters. *Science 203,* 670-672.

MacIsaac, J. J., and Dugdale, R. C. (1969). The kinetics of nitrate and ammonia uptake by natural populations of marine phytoplankton. *Deep-Sea Res. 16,* 45-57.

MacIsaac, J. J., and Dugdale, R. C. (1972). Interactions of light and inorganic nitrogen in controlling nitrogen uptake in the sea. *Deep-Sea Res. 19,* 209-232.

Monod, J. (1942). "Recherches sur la croissance des cultures bactériennes," 2nd ed., Hermann, Paris.

Morris, I. (1974). Nitrogen assimilation and protein synthesis. *In* "Algal Physiology and Biochemistry" (W. D. P. Stewart, ed.), pp. 513-613. Blackwell, Oxford.

Nyholm, N. (1978). Dynamics of phosphate limited algal growth: Simulation of phosphate shocks. *J. Theor. Biol. 70,* 415-425.

Perry, M. J. (1976). Phosphate utilization by an oceanic diatom in phosphorus-limited chemostat culture and in the oligotrophic waters of the Central North Pacific. *Limnol. Oceanogr. 21,* 88-107.

Pirt, S. J. (1975). "Principles of Microbe and Cell Cultivation." Wiley, New York.

Powell, E. O. (1967). Growth rate of microorganisms as a function of substrate concentration. *In* "Microbial Physiology and Continuous Culture" (E. O. Powell, C. G. T. Evans, R. E. Strange, and D. W. Tempest, eds.), pp. 34-55. HM Stationery Office, London.

Rhee, G. Y. (1974). Phosphate uptake under nitrate limitation of *Scendesmus* sp. and its ecological implications. *J. Phycol. 10,* 470-475.

Rhee, G. Y. (1978). Effects of N:P atomic ratios and nitrate limitation on algal growth, cell composition, and nitrate uptake. *Limnol. Oceanogr. 23,* 10-24.

Rhee, G. Y. (1980). Continuous culture in phytoplankton ecology. *Adv. Aquat. Microbiol. 2,* 151-203.

Sakshaug, E., and Holm-Hansen, O. (1977). Chemical composition of *Skeletonema costatum* (Grev.) Cleve and *Pavlova (Monochrysis) lutheri* (Droop) Green as a function of nitrate-, phosphate-, and iron-limited growth. *J. Exp. Mar. Biol. Ecol. 29,* 1-34.

Sheppard, C. W. (1962). "Basic Principles of the Tracer Method." Wiley, New York.

Sieburth, J. M., Smetacek, V., and Lenz, J. (1978). Pelagic ecosystem structure: Heterotrophic compartments of the plankton and their relationship to plankton size fractions. *Limnol. Oceanogr. 23,* 1256-1263.

Spicer, C. C. (1955). The theory of bacterial constant growth apparatus. *Biometrics 11,* 225-230.

Syrett, P. J. (1953a). The assimilation of ammonia by nitrogen-starved cells of *Chlorella vulgaris.* Part I. The correlation of assimilation and respiration. *Ann. Bot. (London)* [N.S.] *27,* 1-19.

Syrett, P. J. (1953b). The assimilation of ammonia by nitrogen-starved cells of *Chlorella vulgaris.* Part II. The assimilation to other compounds. *Ann. Bot. (London)* [N.S.] *27,* 21-36.

Syrett, P. J. (1955). The assimilation of ammonia and nitrate by nitrogen-starved cells of *Chlorella vulgaris.* I. The assimilation of small quantities of nitrogen. *Physiol. Plant. 8,* 924-929.

Syrett, P. J. (1956). The assimilation of ammonia and nitrate by nitrogen-starved cells of *Chlorella vulgaris*. II. The assimilation of large quantities of nitrogen. *Physiol. Plant. 9,* 19-27.

Turpin, D. H., and Harrison, P. J. (1979). Limiting nutrient patchiness and its role in phytoplankton ecology. *J. Exp. Mar. Biol. Ecol. 39,* 151-166.

Turpin, D. H., Parslow, J. S., and Harrison, P. J. (1981). On limiting nutrient patchiness and phytoplankton growth: A conceptual approach. *J. Plankton Res. 3,* 421-431.

Wheeler, P. A., and McCarthy, J. J. (1982). Methylammonium uptake by Chesapeake Bay phytoplankton: Evaluation of the use of the ammonium analogue for field uptake measurements. *Limnol. Oceanogr. 27,* 1129-1140.

Wheeler, P. A., Glibert, P. M., and McCarthy, J. J. (1982). Ammonium uptake and incorporation by Chesapeake Bay phytoplankton: Short-term uptake kinetics. *Limnol. Oceanogr. 27,* 1113-1128.

Williams, F. M. (1965). Population growth and regulation in continuously cultured algae. Ph.D. Thesis, Yale University, New Haven, Connecticut.

Williams, P. J. L., and Muir, L. R. (1981). Diffusion as a constraint on the biological importance of microzones in the sea. *In* "Ecohydrodynamics" (J. C. J. Nihoul, ed.), pp. 209-218. Elsevier, Amsterdam.

Zevenbloom, W., and Muir, L. R. (1981). Ammonium-limited growth and uptake by *Oscillatoria agardhii*. *Arch. Microbiol. 129,* 61-66.

Chapter 8

UPTAKE OF ORGANIC NITROGEN

JOHN H. PAUL
Department of Marine Science
University of South Florida
St. Petersburg, Florida

I. INTRODUCTION

Perhaps the most complex and least understood interactions in the nitrogen cycle in the marine environment involve dissolved organic nitrogen (DON). Due to its inherent complexity, the importance of DON as a nitrogenous nutrient has not yet been fully assessed, even though literature on this topic spans nearly four decades.

The uptake of organic nitrogen in the marine environment is a broad topic encompassing many disciplines and interests, including microbial ecology, enzymology, membrane physiology, and other fields. Necessarily, this chapter has been limited in scope to emphasize microbial processes (some evidence suggests that invertebrates may play a significant role in the cycling of DON, however; see Jorgensen, 1980). Similarly, anaerobic dissimilation of

ISBN 0-12-160280-X

DON will not be treated in this chapter. The emphasis will be on the utilization of DON as a form of nitrogen (nitrogen productivity) rather than a form of carbon or source of energy, although the heterotrophic importance of the carbon in DON is not to be minimized. In this light, the catabolism of organic nitrogen compounds by marine microorganisms (including specific enzymes where known) is briefly examined.

II. AMINO ACIDS

A. Uptake

1. Laboratory Measurements

a. Phytoplankton and other microalgae. For microalgae to compete successfully for amino acids with other microorganisms in the marine environment it is essential that they possess transport systems with kinetic parameters operational in the range of natural concentrations of amino acids. Laboratory measurements with pure cultures have been the source of most kinetic data, although the isolation process and culture conditions may have altered the capacity to take up low (environmentally significant) concentrations of amino acids.

Saks and Kahn (1979) demonstrated that *Cylindrotheca closterium* could compete successfully with an *Aeromonas* sp. for aspartic acid, glutamic acid, glycine, or leucine at 1-10 μM when cultures were maintained in a two-compartment system separated by a 0.45-μm filter. However, kinetic studies employing laboratory cultures of other marine microalgae cast doubt on their capacity to utilize amino acids competitively with bacteria in the environment. The K_s for glutamate in *Navicula pavillardi* was 18 μM (Lewin and Hellebust, 1975), 20 μM for both *Nitzchia anguillaris* (Lewin and Hellebust, 1976), and 30 μM for *N. laevis* (Lewin and Hellebust, 1978).

Nitrogen deprivation long has been known to stimulate amino acid uptake in phytoplankton (Wheeler *et al.*, 1974). North and Stephens (1967) demonstrated that *Platymonas subcordiformis* could rapidly take up glycine at 1 μM and that uptake rates for arginine, glycine, and glutamate were tenfold greater when cells were deprived of nitrogen during growth (North and Stephens, 1971). Kinetic studies indicated that K_s remained constant but V_{max} increased during nitrogen deprivation (North and Stephens, 1971). Under nitrogen deprivation, V_{max} increased for three amino acid transport systems (one each for polybasic, acidic, and neutral amino acids) in *Nitzchia ovalis* (North and Stephens, 1972). The K_s (1.5-2 μM) did not change for the polybasic transport system, while the other transport systems were only detected in nitrogen-poor cultures.

The deprivation of light energy has also been shown to stimulate the uptake of eight amino acids by the benthic diatom *Melosira nummuloides,* resulting in a lower K_s for arginine uptake as compared to illuminated cultures (McLean et al., 1981).

Not surprisingly, the accumulation (and concentration) of amino acids from dilute solutions by microalgae is effected by noninducible, energy-requiring transport systems (Hellebust and Guillard, 1967). Hellebust and Guillard (1967) showed that the uptake of amino acids and the analog α-aminoisobutyric acid (AIB) was inhibited by the energy uncoupler 2,4-dinitrophenol (DNP) in *Melosira nummuloides*. In *Tychodiscus brevis (Gymnodinium breve)*, glycine and methionine were transported via facilitated diffusion systems, while L-valine entered the cell via active transport (Baden and Mende, 1979). Kinetic studies suggested the existence of high- and low-affinity transport systems for similar amino acids in *Cyclotella cryptica* (Liu and Hellebust, 1974).

Although few amino acid transport systems have been found to be inducible in microalgae, glucose induced two amino acid transport systems as well as a hexose transport system in *Chlorella* (Cho et al., 1981). The transport systems include one each for neutral and basic amino acids. Both are also induced by the nonmetabolizable glucose analog 6-deoxyglucose.

The nature of the amino acid recognition molecules and the enzymes involved in the above transport systems have not been well characterized. In mammalian tissues, the best understood transport system is the γ-glutamyl cycle (for review, see Meister, 1980). Kurelec et al. (1977) found four of the five enzyme activities involved in this cycle in extracts of natural phytoplankton populations. These workers postulated that this cycle functions in concentrating amino acids in nutrient-poor waters, and certainly more research is warranted to verify this unique hypothesis.

b. Bacteria. The fact that marine bacteria require Na^+ for growth (MacLeod, 1965) and that this may be due to a sodium dependence for nutrient uptake (Drapeau et al., 1966) has been known for some time. The requirement for sodium, which could not be replaced by other cations, appeared widespread among marine bacteria (Sprott and MacLeod, 1972, 1974; Fein and MacLeon, 1975; Niven and MacLeod, 1980). Another common feature of these transport systems was the sensitivity to inhibitors of cytochrome enzymes and uncouplers such as cyanide and DNP (Drapeau et al., 1966; Sprott and MacLeod, 1974; Geesey and Morita, 1979; Akagi and Taga, 1980). However, arsenate, an inhibitor of energy yielding phosphorylations, and ouabain, an inhibitor of membrane-linked Na^+,K^+-ATPases, had no effect on these systems (Geesey and Morita, 1979; Akagi and Taga, 1980). Sprott and MacLeod (1972, 1974) found similar results with membrane vesicles of a marine pseudomonad, sodium being required for alanine or α-aminoisobutyric acid (AIB) transport, cyanide and DNP strongly inhibiting, and arsenate having no effect. Since an electron donor was required, it was concluded that electron transport was necessary for amino acid transport, but only a short segment of the respiratory complex was involved (Sprott and MacLeod, 1972, 1974).

The work of Lanyi and co-workers (1976a,b; Lanyi and Silverman, 1979) on *Halobacterium halobium* membrane vesicles has put much of this work in perspective and provides a plausible explanation of sodium dependent transport. In general, a sodium gradient is set up across the cell membrane, with $[Na]_{out}/[Na]_{in} >> 1$, via the functioning of a H^+/Na^+ antiport system. The H^+ transmembrane gradient is thought to be generated by respiration (electron transport), perhaps an ATPase, or by light energy on bacteriorhodopsin (Lanyi et al., 1976a,b). The relaxation of the sodium gradient enables the movement of sodium ions and nutrients (in this case amino acids) into cells (or membrane vesicles) via a cotransport or "symport," resulting in the intracellular concentration of nutrients or amino acids.

Tokuda et al. (1982) demonstrated the role of K^+ in the Na-dependent transport of AIB in the marine bacterium *Vibrio alginolyticus*. Intracellular K^+ concentrations were only required to establish the Na^+ electrochemical gradient that drove the transport system.

Although the sodium gradient generated has the capacity to cotransport various amino acids (glutamate, serine, aspartate, and leucine were transported in *Halobium* vesicles), the carriers may be specific for groups or single amino acids, which form the basis for the specificities of the various transport systems (MacDonald et al., 1977). Two sodium-dependent, active transport systems with overlapping specificities were described in a marine pseudomonad by Fein and MacLeod (1975). One, specific for D-amino acids, also transported AIB and glycine (DAG system), while the other was stereospecific for neutral L-amino acids (LIU). AIB was shown to be transported via sodium symport in the former system (Niven and MacLeod, 1980).

Bimodal kinetics for arginine uptake suggested the presence of two energy-dependent transport systems in a marine psychrophile, one operational in the nanomolar range of substrate concentration, the other in the micromolar range (Geesey and Morita, 1979). Chemotaxis, which occurred in the presence of the higher concentrations of arginine, was suggested to be a survival strategy. Although these transport systems were not directly shown to be sodium dependent, uptake was inhibited at reduced salinities. Osmotic shock of these cells led to a release of arginine "binding components" that were partially purified (Geesey and Morita, 1981). Fluorescently labeled antibodies raised against this material were observed to bind to the cell surface, suggesting that these carriers were in the cell membrane (Geesey and Morita, 1981).

Comparison of proline uptake systems between an oligotrophic (i.e., low-nutrient grown; Akagi et al. 1977) and a heterotrophic (i.e., high-nutrient grown) bacterium has shown that the former had a lower K_m for proline and a broad substrate specificity, presumably to utilize low concentrations of a wide range of naturally occurring amino acids (Akagi and Taga, 1980). Both transport systems were inhibited by KCN and unaffected by arsenate, whereas only the oligotrophic system was inhibited by DNP (Akagi and Taga, 1980).

2. *Uptake in the Environment*

The early work of Williams (1970) suggested that bacteria were primarily responsible for amino acid uptake and oxidation in the marine environment, since most of this activity passed through a 1.2-μm filter. However, there may be specific environmental situations (e.g., under low light, in low concentrations of combined nitrogen, and in coastal, hypertrophic situations) where algal amino acid uptake is important. For example, Wheeler et al. (1974) found that marine phytoplankton isolated from the open ocean possessed limited capacity to accumulate ^{14}C-labeled amino acids from dilute solution, whereas isolates from inshore and tidepool habitats possessed a much greater amino acid transport capacity.

North (1975) found that a *Platymonas* species could remove ambient fluorescamine-positive material from water taken from Newport Bay, California. Wheeler et al. (1977) noted that [^{14}C]glycine uptake correlated with bicarbonate fixation and chlorophyll *a* in the 3-25-μm fraction of Newport Bay water. Hollibaugh (1976) found that the rapid liberation of ammonium from natural microbial assemblages incubated with arginine stimulated chlorophyll *a* fluorescence. However, the uptake of $[^{14}C]$glutamate or arginine in these experiments was associated with bacteria as determined by autoradiography. Similarly, Hoppe (1976) found poor labeling of phytoplankton when incubated with ^{3}H-amino acids as determined by autoradiography.

Yet, carbon or tritium label autoradiography may yield limited information about nitrogen metabolism. Stephens and North (1971) observed that nitrogen from $[^{14}C]$arginine was retained by marine phytoplankters, but that the carbon "skeletons" were returned to the medium. This may explain the lack of radiolabeling of phytoplankton observed by Hollibaugh (1976). Paul and Cooksey (1979) described a marine *Chlamydomonas* species that grew well with L-asparagine as a nitrogen source, yet possessed little ability to transport $[^{14}C]$asparagine into the cell. Instead, a cell surface asparaginase deamidated L-asparagine and released ammonium into the medium that supported growth. Schell (1974) employing both ^{15}N- and ^{14}C-labeled amino acids found that the nitrogen of glycine was preferentially incorporated by phytoplankton and the carbon respired. Therefore, a nitrogen isotope is preferable to hydrogen or carbon in assessing phytoplankton amino acid utilization. Unfortunately, autoradiography employing an amino acid isotopically labeled with nitrogen is technically infeasible.

In recent years interest has focused on the rate of removal and regeneration of amino acids in the marine environment (the "kinetic approach"; see Hobbie et al., 1968). Employing the kinetic approach, many investigators have estimated turnover times (T_t) and flux rates (V_n) of dissolved free amino acids (DFAAs) in the marine environment. Turnover times in Alaskan coastal waters for glutamate ranged from 8.8 to 675 h and for glycine from 16 to 1600 h (no correction made for respiration; Schell, 1974). In the Pamlico estuary, turnover times for alanine ranged from 71.3 h in January to 1.36 h in September (Crawford et al., 1974). Flux rates were also measured, the highest measured at 1.36 μg C·liter^{-1}·h^{-1} and the lowest (0.01 μg C·liter^{-1}·h^{-1}) occurring in winter, corrections being made for respiratory losses.

Hollibaugh et al. (1980a) estimated turnover rates and DFAA flux rates in enclosed water columns based on incubations with [^{3}H]leucine. Ambient concentrations (S_n) were estimated in glycine equivalents based on fluorescamine-positive material, and a turnover rate of 2-42% h^{-1} for leucine was determined. Amino acid flux in glycine equivalents was calculated as 0.09-2.42 μ*M* day^{-1}.

Hodson et al. (1981a) found average turnover times for leucine ranging from 124 h on the eastern side to 3070 h on the western side of McMurdo Sound, Antarctica.

As has been previously shown (Williams, 1973), there may be problems with the kinetic approach for the estimation of amino acid turnover by natural microbial populations. Burnison and Morita (1973) showed that even if the biologically available concentration of a particular amino acid is known, the presence of significant quantities of a second amino acid (which is not measured) may competitively inhibit uptake of the first, thereby increasing both turnover time and the sum of the transport constant and the ambient concentration ($K_s + S_n$).

Another problem with the kinetic approach to measuring amino acid uptake is that amino acid uptake rates are often nonlinear, resulting in a departure from saturation kinetics. Hollibaugh (1979) observed increases in amino acid uptake rates in natural water samples when the sample was preincubated with the substrate in question. The increase in uptake rate resembled induction of specific amino acid transport systems or metabolic enzymes since the rate increase showed stereospecificity, was inhibited by inhibitors of protein synthesis, and decayed after a period of time. Although proof of metabolic adaptation or induction of enzymes is strong, selection and growth of specific amino acid metabolizing bacteria cannot be ruled out. Further proof for induction could be obtained by eliciting the same response with nonmetabolizable substrate analogs. Induction of amino acid metabolic capabilities could explain the nonlinear kinetics that are frequently observed in the environment. Also, if prokaryotic utilization of amino acids requires induction resulting in a lag in uptake, this would result in a competitive advantage for the microeukaryotes/phytoplankton, the amino acid transport systems of which are maximally derepressed under nitrogen limitation (North and Stephens, 1971, 1972).

A major problem with determining T_t and V_n is obtaining a reliable analysis of ambient DFAA concentrations. Burnison and Morita (1974) suggested that a significant fraction of the DFAAs detected by cation exchange were not available to heterotrophic plankton, but were somehow released from some "unavailable form" by the sampling and analysis process. Similarly, Dawson and Gocke (1978) found unreasonably high uptake rates of a radiolabeled amino acid mixture devised to reflect ambient concentrations and proportions of DFAAs (i.e., 76.4 $\mu g \cdot liter^{-1} \cdot day^{-1}$ where biomass was only 6.9 μg C $liter^{-1}$) in the Baltic Sea. Their explanation was that the amount of DFAA measured, although accurate, was not "biologically available," and was either chelated or in some macromolecular form (a review of the earlier work concerning the complexation of amino acids appears in Clark et al., 1972). In support of this notion, Lytle and Perdue (1981) found that greater than 96% of the dissolved amino acids in a freshwater system were associated with humic substances, although the biological availability of this material was not measured. The proportion of the dissolved amino acids that are humic bound in marine systems is not known. In any case, if concentrations of DFAAs measured are not biologically available, the amount of labeled substrate added becomes significant, and an overestimation of the uptake rate occurs.

Recently, Iturriaga and Zsolnay (1981) found that natural microbial populations could "transform" (i.e., excrete in an altered form) radiolabeled amino acids as determined by their elution characteristics on Sephadex G-10.

In the open ocean, DFAAs have been found to be concentrated at physical boundaries. Liebezeit et al. (1980) found enrichment of DFAAs and carbohydrates and the highest microbial activity at the upper boundary of the seasonal pycnocline in the Sargasso Sea.

B. Metabolism

1. Algal Growth

Many investigators have examined the potential for a variety of microalgae to utilize amino acids as a carbon or energy source (Lewin and Hellebust, 1975, 1976, 1978; Cooksey and Chansang, 1976) or a nitrogen source (Saks et al., 1976; Berland et al., 1979; Turner, 1979; a review of the earlier literature appears in Antia et al., 1978). The ability to utilize an amino acid as a nitrogen or carbon source does not follow taxonomic division. Many studies indicate that glycine, glutamate, alanine, and glutamine are most widely utilized (Lewin and Hellebust, 1975; Saks et al., 1976). In contrast, Turner (1979) found DL-tryptophan the most widely utilized of a series of amino acids examined. Results obtained with glutamine must be regarded with caution since this amino acid is spontaneously converted to ammonium and pyrrolidone carboxylic acid in the presence of certain anions (Gilbert et al., 1949). In most cases, growth (rate or yield) on amino acids is not as great as with ammonium as sole nitrogen source, and high concentrations are often required to obtain this growth (Berland et al., 1979).

2. Enzymes and Intermediary Metabolism

The majority of the research cited above has dealt with the capacity of algae to utilize amino acids for growth, yet little work has been done on the mechanisms (enzymes and pathways) by which this is accomplished. The remainder of this section will deal with what is known of the biochemistry of amino acid utilization in phytoplankton and other marine microalgae.

Antia and co-workers (1972; Desai et al., 1972) conducted a comparative study of threonine metabolism among seven species of unicellular marine algae. They studied the allosteric enzyme threonine deaminase (L-threonine hydrolase [deaminating] EC 4.2.1.16), the first enzyme in the biosynthetic pathway from threonine to isoleucine (Antia and Kripps, 1978; Lehninger, 1975).

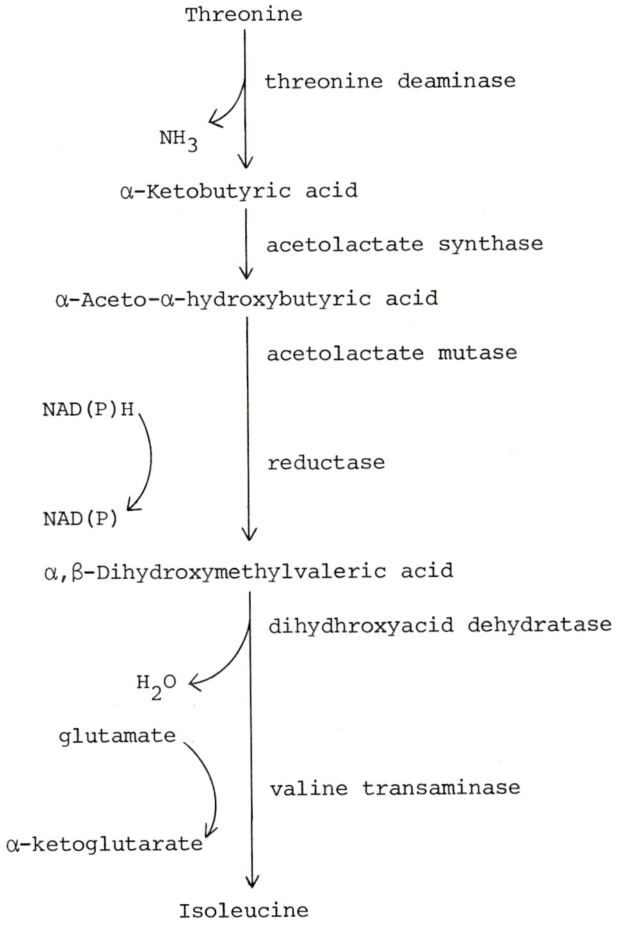

The biosynthetic threonine deaminase is under feedback inhibition by isoleucine. The enzyme was similar to yeast and higher plant enzymes in that it possessed a high K^+ and NH_4^+ requirement (Desai et al., 1972). The enzyme in the two cryptomonads studied had a requirement for both disulfides and sulfhydryls for activity, a property not found in other threonine deaminases (Antia et al., 1972).

Paul and Cooksey (1979) described an L-asparaginase in a marine *Chlamydomonas* species that allowed for growth on L-asparagine comparable to that on inorganic forms of nitrogen (NO_3, NH_4^+). The enzyme possessed a molecular weight of 275,000 and was specific for L-asparagine; D-asparagine hydrolyzed at only 14% of the rate of L-asparagine (Paul, 1982). Both glutamine synthetase and L-asparaginase were repressed by high concentrations of combined nitrogen, while glutamate dehydrogenase activity was stimulated by ammonium and asparagine (Paul and Cooksey, 1981). The relatively high K_m for asparagine (1.1×10^{-4} *M*) casts doubts on its importance in asparagine utilization in the environment (Paul and Cooksey, 1979)

Vose et al. (1971) investigated the capacity of 22 unicellular marine algae to catabolize the aromatic ring of phenylalanine or tyrosine as indicated by $^{14}CO_2$ release from ring-labeled substrates. The amount of radioactivity trapped as CO_2 ranged from 0.05 to 1.5% of the label added after 2-week incubation. Since commercial radiochemicals often contain at least 1% radiolabeled impurities and this proportion increases due to decomposition with storage (J. H. Paul and K. E. Cooksey, unpublished observations), such low amounts of conversion should be regarded with caution. "Aged cultures" were employed so that cells might be in a "catabolic" state, although the viability of these cells might also be questioned. The largest $^{14}CO_2$ producers from ring-labeled phenylalanine and tyrosine, *Navicula incerta* and *Isochrysis galbana*, were shown to grow well on the L-isomers of these two amino acids, but poorly on the D-isomers (Landymore and Antia, 1977). Both organisms metabolized phenylalanine through phenylpyruvate, phenylacetate, mandelate, benzoylformate, and benzoate to *p*-hydroxybenzoate (Landymore et al., 1978). Tyrosine was metabolized through the corresponding *p*-hydroxy derivatives to yield *p*-hydroxybenzoate (Landymore et al., 1978). The aromatic ring was not significantly cleaved to yield CO_2, perhaps explaining the low conversion rates observed earlier (Vose et al., 1971).

C. Ecological Implications of Amino Acid Uptake

A discussion of ecological implications necessitates mentioning sources and sinks of amino acids. In general, it is thought that plant material, whether macrophytes in coastal and estuarine situations or phytoplankton in the open ocean, is the source of amino acids in the marine environment, although exceptions occur in hypertrophic and pollution-stressed situations (Ryther, 1954). Hall et al. (1967) found that the marsh grass *Spartina* had a much greater amino acid content (and a different composition) than suspended solids in a Georgia estuary. Since it was inferred that *Spartina* was a major source of the suspended solids, the difference in amino acid content and composition was thought due to biological breakdown.

Wood and Hayasaka (1981) found a large microbial population on the roots of the eel grass *Zostera marina*. Dominant bacterial isolates showed a chemotactic response to root exudates, which were shown to contain serine, threonine, glycine, alanine, and valine. These amino acids individually were also shown to elicit a chemotactic response by rhizoplane bacterial isolates (Wood and Hayasaka, 1981).

III. UREA

A. Uptake

Laboratory Studies

Since the mechanisms of urea uptake in marine phytoplankton are only partially understood, some work done with the freshwater Chlorophyte *Chlamydomonas reinhardi* will be included in this section.

Williams and Hodson (1977) described two transport systems in *Chlamydomonas reinhardi* operating in different urea concentration ranges. The lower concentration transport system had a K_m of 5.1 μM and was repressed by ammonium. Addition of ammonium, methylamine, or cycloheximide to cells in the process of derepression caused a halt to derepression and a decay of the transport system. However, ammonium did not repress short-term uptake of urea supplied at higher concentrations (3 mM; Hodson et al., 1975). Uncouplers of electron transport or anaerobiosis depressed urea uptake, and no inducer was detected for the transport system (Williams and Hodson, 1977).

Dagestad et al. (1981) described the compartmentalization of urea into two intracellular pools in *C. reinhardi*, one being large and nonmetabolic (presumably located in the chloroplast and resulting from photosynthetically driven active transport) and a small, metabolic pool. Their work suggests that ammonium inhibits transfer of urea from the nonmetabolic pool to the metabolic pool.

The uptake of urea by *Phaeodactylum tricornutum* has been extensively studied by Rees and co-workers (Rees and Syrett, 1979a, b; Rees et al., 1980). As with the *Chlamydomonas* system, ammonium-grown cells could not take up urea, and the capacity to transport urea appeared during nitrogen starvation (Rees et al., 1980). Addition of ammonium during the uptake of urea did not inhibit uptake, however (Rees and Syrett, 1979a). Unlike urea uptake in the freshwater *Chlamydomonas*, there was a sodium requirement (K_{Na} = 71 mM) for urea uptake and derepression of the transport system (Rees et al., 1980). The K_m for urea uptake was 1.0 μM, and uptake rates were slower in the dark (Rees and Syrett, 1979a). Rees and Syrett (1979b) also showed that thiourea competitively inhibited urea uptake in *Phaeodactylum*, suggesting uptake by the same transport system. Except for the K_m (170 μM), characteristics of thiourea uptake were identical to urea, making it a valuable tool for

for the study of urea transport since urea was so rapidly metabolized (Rees and Syrett, 1979b).

B. Metabolism

1. *Enzymes and Pathways of Catabolism*

There are two mechanisms of urea catabolism available to marine phytoplankton: urease (EC 3.5.1.5) and urea amidolyase (UALase, EC 6.3.4.6). Leftley and Syrett (1973) examined 14 species from two algal phyla and found that all Chrysophytes contained urease only, while all Chlorophytes contained UALase only.

The UALase degradation of urea has been shown to proceed by the two-step reaction outlined below (Roon et al., 1972):

$$\text{Urea} + \text{ATP} + HCO_3^- \xleftrightarrow{Mg^{2+},\ K^+,\ \text{biotin}} \text{Allophanate} + \text{ADP} + P_i \quad (1)$$

$$\text{Allophanate} \longleftrightarrow 2\ HCO_3^- + 2\ NH_4^+ \quad (2)$$

The proteins responsible for (1) and (2) are inseparable in yeasts but can be separated by simple protein purification techniques in green algae. Reaction (1) is catalyzed by urea carboxylase (urea: CO_2 ligase), is avidin sensitive, and can also carboxylate formamide, formyl urea, acetaminde, and free D-biotin (Whitney and Cooper, 1973). The enzyme responsible for catalyzing the second step is allophanate hydrolase, which has been recently purified in *C. reinhardi* (Maitz et al., 1982). The enzyme is inhibited by substrates possessing carbonyls or carboxylic acid groups β to a carboxylic acid group, such as oxaloacetic and acetoacetic acids, the nitrogens not necessary for substrate recognition. The UALase enzymes have been shown to be induced by urea and repressed by ammonium (Hodson et al., 1975).

2. *Utilization for growth*

An excellent review of the early literature on urea utilization and ecology appears in McCarthy (1971). Carpenter et al. (1972) described three diatoms from the Sargasso Sea and two inshore species that grew well on urea. Kinetic studies suggested that *Skeletonema costatum* could use ambient urea concentrations in inshore habitats (Carpenter et al., 1972).

Of 26 marine algae examined, 88% showed good growth (often better than that on ammonium) when urea was the sole nitrogen source (Antia et al., 1975). The inability to utilize urea was not related to taxonomy. Antia et al. (1977) have also pointed out that algal urea utilization requires light; no growth can occur in the dark on urea in the absence of another energy source since both mechanisms of urea degradation result in CO_2 formation.

C. Uptake of Urea in the Environment and Ecological Considerations

McCarthy and Eppley (1972) looked for ^{15}N uptake from ammonium, urea, or nitrate in enriched seawater incubations. Urea (and nitrate) uptake was totally suppressed in seawater enriched with ammonium, as has been shown in pure phytoplankton cultures (McCarthy and Eppley, 1972). These findings are not surprising since the uptake and utilization of ammonium is energetically favored over utilization of urea or nitrate.

Remsen et al. (1972) reported urea "decomposition" in estuarine microbiota by measuring $^{14}CO_2$ production from [^{14}C]urea incubations, although uptake may have been underestimated due to the refixing of CO_2 by phytoplankton (see below). Since the majority of urea decomposition occurred in the >20-μm fraction (which represented 15% of the phytoplankton and 39% of the chlorophyll *a*), it was inferred that large-chain diatoms were responsible for urea decomposition. It was also noted that rates of urea decomposition were least in freshwater and increased with increasing salinity (Remsen et al., 1972).

McCarthy (1972) also found that the percentage of total "nitrogen productivity" accounted for by urea averaged 28% and ranged from <1 to >60% for various stations off the coast of Southern California. Again, a high ammonium value was though responsible for inhibition of urea uptake at at least one station.

Webb and Haas (1976) measured both production of $^{14}CO_2$ and incorporation of ^{14}C into particulate fractions resulting from incubation with [^{14}C]urea. The urea uptake/decomposition process was light dependent. The herbicide DCMU primarily inhibited incorporation of [^{14}C]urea into particulate material, while avidin inhibited (by 30%) both incorporation and $^{14}CO_2$ production, suggesting the presence of the UALase pathway of ammonium degradation. Both urea uptake and decomposition was saturated at lower light intensities than photosynthesis, accounting for the C:N enrichment occurring in surface waters.

Mitamura and Saijo (1975) found a parallel response to increasing light intensity between bicarbonate fixation and urea utilization in the coastal waters of Japan, with best rates of both occurring at 12,000 lx. Although Carpenter et al. (1972) reported negligible ^{14}C incorporation from [^{14}C]urea incubations with *Skeletonema costatum,* Mitamura and Saijo (1975) found relatively high incorporation rates (38-84% of the total urea metabolized), although this only represented <1% of the total photosynthetic rate. Mitamura and Saijo (1980) also found that the percent incorporated was greater in the light, while dark uptake of [^{14}C]urea primarily resulted in $^{14}CO_2$ release.

Herbland (1976) found turnover times for urea of less than a day in the mixed layer of the tropical Atlantic Ocean, although Remsen et al. (1974) noted much longer turnover times (58 and 98 days). Herbland (1976), unlike Mitamura and Saijo (1975), found very little incorporation (<5%) of ^{14}C from labeled urea.

Steinmann (1976) studied the distribution of urea- and uric acid-utilizing bacteria in the Baltic Sea by a plate count technique. The greatest numbers of urea-hydrolyzing bacteria occurred in winter (2500 ml^{-1}), when the greatest urea concentrations were present. Turnover times for urea ranged from 3.4 to 60 h. From his studies Steinmann concluded that bacteria were not the only degraders of urea in the Baltic Sea, and that the "plankters" [phytoplankton] may play a role in the uptake of urea.

IV. CONSTITUENTS OF NUCLEIC ACIDS AND RELATED COMPOUNDS

The question of the suitability of nucleic acids or their metabolites to support the growth of marine microorganisms has only been partially answered. Pintner and Provasoli (1968) found that of all the amino acids, amines, purines, and pyrimidines examined as nitrogen sources for the growth of several marine chrysomonads, only "adenylic acid" (AMP) was used by all. The chrysomonad *Pavlova gyrans* could also use guanine as a nitrogen source.

Antia and Landymore (1974) investigated the stability of various purines in sterile seawater-based media and found that uric acid, a common nitrogen excretion product of marine organisms, decomposed rapidly, and that this decomposition was accelerated in the light. Xanthine decomposed slowly in illuminated media, whereas adenine, hypoxanthine, and guanine were all quite stable. Antia et al. (1980a) suggest that the source of hypoxanthine in seawater may be from ciliate excretion.

Hypoxanthine, an intermediate in purine catabolism (see below), supported growth as sole nitrogen source for 69% of 25 marine phytoplankton from 10 taxonomic classes (Antia et al., 1975). However, allantoin, an intermediate in the catabolism of hypoxanthine to urea, only supported good growth of one diatom, *Nitzchia acicularis*, at 0.125 m*M*, when 18 phytoplankton species were examined (Antia et al., 1980b).

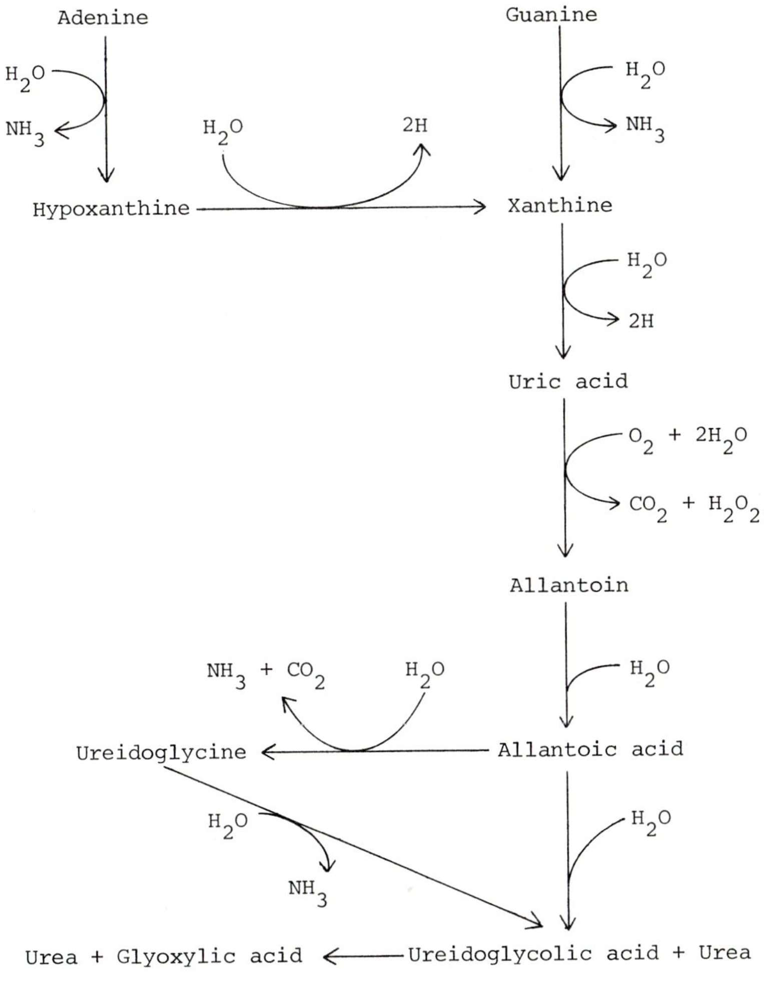

It was concluded that a deficiency in allantoin permeability was responsible for the limited utilization of this compound (Antia et al., 1980b).

Recently, there has been considerable interest in the role of dissolved adenine nucleotides in marine microbial ecology. Azam and Hodson (1977) found relatively high concentrations (0.1-0.6 μg liter^{-1}) of dissolved ATP (DATP) in the coastal waters off southern California and British Columbia. Turnover rates measured with [^{3}H]ATP, ranged from 0.59 to 15.2% h^{-1}, with 80% of the label taken up by the 0.2-0.6-μm fraction. When [U-^{14}C]ATP was employed, over 98% of the label was assimilated with an insignificant amount respired. The capacity to take up DATP may be widespread since 11 of 12 bacterial isolates rapidly assimilated [^{3}H]ATP. Since ambient concentrations of other nutrients are quite small in seawater, Azam and Hodson (1977) proposed that DATP was a significant carbon or ene source in the marine environment. As a sole source of energy, one would expect some respired CO_2 to be produced from uniformly labeled [^{14}C]ATP, unless all energy requirements are to be met from the hydrolysis of the high-energy phosphate bonds. Further work on the metabolic pathways involved in utilizing ATP as a sole carbon or energy source is required to clarify this point.

Hodson et al. (1981a) found a similar range of DATP concentrations in Antarctic coastal waters (89-519 ng liter^{-1}). The average turnover time for DATP was calculated at 16 h, with 84% of the activity in <0.6-μm fraction. Similar DATP concentrations were found in the water off the eastern continental shelf of the United States (Hodson et al., 1981b). Uptake rates, proportional to *in situ* DATP concentrations, were patchy with respect to space and time.

McGrath and Sullivan (1981) measured total adenylates (ATP + AMP + ADP) off the coast of Southern California and estimated adenylate uptake by [^{3}H]AMP uptake ([^{3}H]AMP uptake was shown to be inhibited by unlabeled ATP, ADP, AMP, adenine, and adenosine, suggesting a common transport mechanism). Specific uptake rates ranged from 0.028 to 0.28 n*M* (10^9 cell$^{-1}\cdot$h^{-1}) for December and August, respectively. Over 80% of the DATP uptake occurred in the <1-μm fraction, which represented 20% of the total particulate adenylates. Although total adenlyates comprised a small fraction of DOC (<0.1%), observed uptake rates were comparable to those found for amino acids, the concentration of which exceed that of adenylates by one to two orders of magnitude.

The dephosphorylation of adenine nucleotides by marine bacteria is generally thought to be due to the action of membrane-bound 5'-nucleotidases (Thompson et al., 1969). These enzymes differ from the ATPases of terrestrial organisms in that ATP, ADP, and AMP are all hydrolyzed to adenosine and inorganic phosphate (Thompson et al., 1969). A 5'-nucleotidase purified recently from the moderate halophile *Vibrio costicola* was shown to possess phosphohydrolyzing activity against all purine and pyrimidine 5'-nucleotides except CTP (Bengis-Garber and Kushner, 1981). A role in the transport of adenine was postulated for this enzyme; the enzyme converted adenine 5'-nucleotides to adenosine, which was taken up by a mediated transport system (Bengis-Garber and Kushner, 1982). Whether a similar mechanism exists for non-adenine 5'-nucleotides is not known.

Ammerman and Azam (1981) measured uptake rates and ambient concentrations of cyclic 3', 5'-adenosine monophosphate (cAMP) in seawater collected off Scripps Pier, the Southern California Bight, and in Saanich Inlet, British Columbia, Canada. Even though the concentrations in seawater were minute ($1-35 \times 10^{-12}$ *M*), uptake of ^{32}P- or ^{3}H-labeled cAMP occurred readily and specifically, with no inhibition by a 10^4-fold excess of AMP. As with ATP and AMP, over 80% of the label was taken up in the <0.6-μm fraction. Since [^{32}P] cAMP could be recovered from cell extracts, it was concluded that little catabolism of cAMP was occurring. Kinetic studies indicated the presence of a specific, high-affinity transport system (K_t + S_n < 10 p*M*) and a less specific, lower-affinity system (Ammerman and Azam, 1982). Uptake was thought to occur for "regulatory" reasons, although no culture studies showing a regulatory response elicited from externally supplied cAMP were performed. Recent work by Matsumoto et al. (1982) suggests that mutants of *Saccharomyces cerevisiae* can utilize cAMP as an adenine source.

Hollibaugh et al. (1980b) demonstrated that natural populations of marine bacteria take up [^{3}H]thymidine, and that the majority of the label was incorporated into DNA. This technique was employed to label bacterioplankton for zooplankton feeding experiments, and the suitability of thymidine as a nitrogen source for marine microorganisms was not investigated. A modification of this technique has recently been developed to estimate bacterial production rates (Fuhrman and Azam, 1982).

In spite of the fact that a good deal of the DON in the marine environment is believed to be in a macromolecular form, the importance of this material in the nitrogen nutrition of marine microorganisms has not been determined. It seems likely that DNA might be in this macromolecular fraction, although few measurements to support this notion have been made. Pillai and Ganguly (1972) found 20 μg liter^{-1} in seawater collected from Bombay Harbor, India. Breter et al. (1977) made estimates of DNA based on thymine measurements derived from hydrolyzates of polyanionic material (both particulate and dissolved) in seawater. Total thymine-containing material ranged from 1 to 3 μg liter^{-1}, the majority of this (>60-70%) was particulate material. Measurements of dissolved DNA in sediments are hampered by the binding of DNA to sediment particles (Lorenz et al., 1981).

Maeda and Taga (1974) enumerated DNA-hydrolyzing bacteria in seawater by determining the capacity of individual bacterial isolates from plate counts to hydrolyze DNA. These techniques, albeit selective and perhaps underestimating the DNA-hydrolyzing microbial population, yielded 10^2-10^4 and 10^5-10^7 DNA-hydrolyzing bacteria per 100 ml in oceanic and neritic waters, respectively, in the Western Pacific Ocean. The addition of DNA to the medium of a marine *Vibrio* isolated by these workers stimulated growth, DNA being rapidly hydrolyzed to acid-soluble components. Hypoxanthine concentrations in the medium increased rapidly and reached the highest concentrations of all purines or pyrimidines, whereas cytosine was not found, apparently being utilized for biosynthesis. The metabolism of DNA by DNA-hydrolyzing bacteria may be a source of hypoxanthine in the marine environment, supplementing ciliate excretion as suggested by Antia et al. (1980a).

V. ALKYLAMINES AND OTHER FORMS OF ORGANIC NITROGEN

A. Alkylamines

Alkylamines may be important forms of organic nitrogen in the marine environment from sources such as fish excretion (i.e., trimethylamine and trimethylamine oxide, Lehninger, 1975) and extracellular release by invertebrates (Jorgensen, 1980).

Budd and Spencer (1968) isolated six species of bacteria and one yeast from marine mud that could use alkylamines as sole sources of nitrogen for growth. Studies suggested that an energy-requiring, stepwise demethylation process was involved rather than an amine oxidase in the metabolism of trimethylamine by a *Micrococcus* species.

Wheeler (1979, 1980) has employed methylamine as an ammonium analog for the study of ammonium transport in the kelp *Macrocyctis pyrifera* and the diatom *Cyclotella cryptica*. Methylamine could not be utilized as a nitrogen source by *Cyclotella*. Methylamine inhibited growth on nitrate, nitrite, and arginine in *Cyclotella* but did not affect growth on ammonium or glutamine (Wheeler, 1980). Dimethylamine and ethylamine, accumulated to a lesser degree than methylamine, were believed to be transported by a system distinct from that for methylamine in this organism (Wheeler and Hellebust, 1981). The presence of these transport systems suggests that alkylmines may be utilized by phytoplankton as nitrogen sources, even though this particular species could not grow on methylamine.

B. Other Forms

Glucosamine, widely distributed in the polymers of cell walls and exoskeletons of marine organisms, perhaps may also be an abundant form of nitrogen in the marine environment. Thirty percent of the nitrogen in aged detritus was found to be non-amino nitrogen, and 25-50% of this was found to be glucosamine (Odum et al., 1979). Berland et al. (1976), however, found that many phytoplankton species required a concentration of at least 5 m*M* before utilization as a nitrogen source. Certain species, like *Cyclotella cryptica*, grew best at 2.5 m*M* but were totally inhibited at 5 m*M*. The high concentrations required for growth casts doubt on the importance of glucosamine as a nitrogen source for phytoplankton growth in the marine environment, except perhaps in isolated coastal or sediment environments (Berland et al., 1976).

Anita and Landymore (1975) have argued that since the composition of ~75% of the DON in seawater is unknown and that some of this is believed to be heterocyclic, a portion of this material might be composed of pterins (pterins are substituted pteridines, comprising the bicyclic, aromatic nitrogen-containing moieties of folic acid and the pigment xanthopterin). Half-lives of 2-600 d have been found for various pterins in sterile seawater and seawater-based media (Landymore and Antia, 1978). Evidence that these compounds can be used readily as nitrogen sources by marine microorganisms and that these compounds can be found in appreciable quantities in the marine environment is lacking.

As mentioned earlier, the importance and the nature of the macromolecular forms of organic nitrogen in the marine environment are poorly understood. Dissolved protein may be an important portion of the macromolecular fraction, and certain marine bacteria are known to possess extracellular endopeptidases to degrade proteins into the more utilizable form of DFAAs (see Sipos and Merkel, 1971).

VI. SUMMARY

The wealth of evidence suggests that amino acids are rapidly taken up by the bacterial portion of the microbial population in the marine environment. Simple techniques still do not exist for the determination of the biological availability of individual amino acids nor are there ways to assess the compounds present that might compete with a particular amino acid for uptake. Perhaps measurements of amino acid uptake should be limited to estimations of V_{max}. The induction of amino acid transport and metabolic capabilities of natural bacterial populations (as suggested by Hollibaugh, 1979), as well as any competitive advantage conferred to organisms possessing derepressed (at ambient nutrient concentrations) transport and metabolic capabilities requires further investigation.

Inorganic ion cotransport (particularly sodium) may be found to be a requirement for amino acid uptake as various amino acid transport systems are more closely examined. Virtually nothing is known of the actual membrane proteins involved in the recognition and transport of particular amino acids into cells.

The uptake of urea in the marine environment is an energy-requiring process with a high affinity for urea. Urea transport may require sodium, and is repressed by ammonium. The two enzyme systems involved in urea metabolism both result in CO_2 and ammonia formation, with the CO_2 either being released or refixed by photosynthesis. Urea uptake in the marine environment occurs rapidly and primarily by the action of phytoplankton, although the role of bacteria in urea uptake and degradation is not known.

The purines hypoxanthine, adenine, and guanine can be used readily by many phytoplankton species for growth in culture, while the pyrimidines seem less widely used. Recent research on dissolved adenylates in seawater suggests that these compounds, although occurring in minute quantities, are rapidly taken up by natural microbial populations (primarily <1-μm fraction). Although DNA-hydrolyzing bacteria occur in seawater, the importance of DNA as a nitrogen source is not known.

Alkylamines are perhaps most important in bacterial nitrogen and carbon nutrition, although phytoplankton can also transport these compounds. Glucosamine, although relatively abundant in certain marine environments, is not a good nitrogen source for phytoplankton growth but may be significant in bacterial nutrition.

The fact that conclusions are not easily drawn concerning the uptake of organic nitrogen emphasizes the need for more research in this field. The work cited in this chapter has only scratched the surface of this broad topic and represents a sporadic attack on the problem spanning several decades.

REFERENCES

Akagi, Y., and Taga, N. (1980). Uptake of D-glucose and L-proline by oligotrophic and heterotrophic bacteria. *Can. J. Microbiol. 26*, 454-459.

Akagi, Y., Taga, N., and Simidu, U. (1977). Isolation and distribution of oligotrophic marine bacteria. *Can. J. Microbiol. 23*, 981-987.

Ammerman, J. W., and Azam, F. (1981). Dissolved cyclic adenosine monophosphate (cAMP) in the sea and uptake of cAMP by marine bacteria. *Mar. Ecol.: Prog. Ser. 5*, 85-89.

Ammerman, J. W., and Azam, F. (1982). Uptake of cyclic AMP by natural populations of marine bacteria. *Appl. Environ. Microbiol. 43*, 869-876.

Antia, N. J., and Kripps, R. S. (1978). Threonine deaminase in marine microalgae. *In* "Handbook of Phycological Methods. Physiological and Biochemical Methods" (J. A. Hellebust and J. S. Craigie, eds.), pp. 225-232. Cambridge Univ. Press, London and New York.

Antia, N. J., and Landymore, A. F. (1974). Physiological and ecological significance of the chemical instability of uric acid and related purines in sea water and marine algal culture medium. *J. Fish. Res. Board Can. 31*, 1327-1335.

Antia, N. J., and Landymore, A. F. (1975). The non-biological oxidative degradation of dissolved xanthopterin and 2,4,6-trihydroxypteridine by the pH or salt content of seawater. *Mar. Chem. 3*, 347-363.

Antia, N. J., Kripps, R. S., and Desai, I. D. (1972). L-Threonine deaminase in marine planktonic algae. II. Disulfide and sulfhydryl group requirements of enzyme activity in two Cryptophytes. *J. Phycol. 8*, 283-289.

Antia, N. J., Berland, B. R., Bonin, D. J., and Maestrini, S. Y. (1975). Comparative evaluation of certain organic and inorganic sources of nitrogen for phototrophic growth of marine microalgae. *J. Mar. Biol. Assoc. U. K. 55*, 519-535.

Antia, N. J., Berland, B. R., Bonin, D. J., and Maestrini, S. Y. (1977). Effects of urea concentration in supporting growth of certain marine microplanktonic algae. *Phycologia 16,* 105-111.

Antia, N. J., Berland, B. R., Bonin, D. J., and Maestrini, S. Y. (1978). Utilisation de la matiere organique dissoute entant que substrat par les algues unicellulaires marines. *Subst. Org. Nat. Dissoutes Eau Mer, Colloq. Group. Av. Brochim. Mar., 1976,* pp. 147-178.

Antia, N. J., Berland, B. R., and Bonin, D. J. (1980a). Proposal for an abridged nitrogen turnover cycle in certain marine planktonic systems involving hypoxanthine-guanine excretion by ciliates and their reutilization by phytpolankton. *Mar. Ecol.: Prog. Ser. 2,* 97-103.

Antia, N. J., Berland, B. R., Bonin, D. J., and Maestrini, S. Y. (1980b). Allantoin as a nitrogen source for growth of marine benthic microalgae. *Phycologia 19,* 103-109.

Azam, F., and Hodson, R. E. (1977). Dissolved ATP in the sea and its utilization by marine bacteria. *Nature (London) 267,* 696-698.

Baden, D. G., and Mende, T. J. (1979). Amino acid utilization by *Gymnodinium breve. Phytochemistry 18,* 247-251.

Bengis-Garber, C., and Kushner, D. J. (1981). Purification and properties of a 5'-nucleotidase from the membrane of *Vibrio costicola,* a moderately halophilic bacterium. *J. Bacteriol. 146,* 24-32.

Bengis-Garber, C., and Kushner, D. J. (1981). Role of membrane-bound 5'-nucleotidase in nucleotide uptake by the moderate halopile *Vibrio costicola. J. Bacteriol. 149,* 808-815.

Berland, B. R., Bonin, D. J., Maestrini, S. Y., Lizarraga-Partida, M. L., and Antia, N. J. (1976). The nitrogen concentration requirement of D-glucoseamine for supporting effective growth of marine microalgae. *J. Mar. Biol. Assoc. U.K. 56,* 629-637.

Berland, B. R., Bonin, D. J., Guerin-Ancey, O., and Antia, N. J. (1979). Concentration requirement of glucine as nitrogen source for supporting effective growth of certain marine microplanktonic algae. *Mar. Biol. 55,* 83-82.

Breter, H.-J., Kurelec, B., Muller, W. E. G., and Zahn, R. K. (1977). Thymine content of seawater as a measure of biosynthetic potential. *Mar. Biol. (Berlin) 40,* 1-8.

Budd, J. A., and Spencer, C. P. (1968). The utilization of alkylated amines by marine bacteria. *Mar. Biol. (Berlin) 2,* 92-101.

Burnison, B. K., and Morita, R. Y. (1973). Competitive inhibition for amino acid uptake by the indigenous microflora of upper Klamath Lake. *Appl. Microbiol. 25,* 103-106.

Burnison, B. K., and Morita, R. Y. (1974). Heterotrophic potential for amino acid uptake in a naturally eutrophic lake. *Appl. Microbiol. 27,* 488-495.

Carpenter, E. J., Remsen, C. C., and Watson, S. W. (1972). Utilization of urea by some marine phytoplankters. *Limnol. Oceanogr* *17*, 265-269.

Cho, B.-H., Sauer, N., Komor, E., and Tanner, W. (1981). Glucose induces two amino acid transport systems in *Chlorella*. *Proc. Natl. Acad. Sci. U.S.A.* *78*, 3591-3594.

Clark, M. E., Jackson, G. A., and North, W. J. (1972). Dissolved free amino acids in southern California coastal waters. *Limnol. Oceanogr.* *17*, 749-758.

Cooksey, K. E., and Chansang, H. (1976). Isolation and physiological studies on three isolates of *Amphora* (Bacillariophyceae). *J. Phycol.* *12*, 455-460.

Crawford, C. C., Hobbie, J. E., and Webb, K. L. (1974). The utilization of dissolved free amino acids by estuarine microorganisms. *Ecology* *55*, 551-563.

Dagestad, D., Lien, T., and Knutsen, G. (1981). Degradation and compartmentalization of urea in *Chlamydomonas reinhardii*. *Arch. Microbiol.* *129*, 261-264.

Dawson, R., and Gocke, K. (1978). Heterotrophic activity in comparison to the free amino acid concentrations in Baltic Sea water samples. *Oceanol. Acta* *1*, 45-54.

Desai, I. D., Laub, D., and Antia, N. J. (1972). Comparative characterization of L-threonine dehydratase in seven species of unicellular marine algae. *Phytochemistry* *11*, 277-287.

Drapeau. G. R., Matual, T. I., and MacLeod, R. A. (1966). Nutrition and metabolism of marine bacteria. XV. Relation of Na^+-activated transport to the Na^+ requirement of a marine pseudomonad for growth. *J. Bacteriol.* *92*, 63-71.

Fein, J. E., and MacLeod, R. A. (1975). Characterization of neutral amino acid transport in a marine pseudomonad. *J. Bacteriol.* *124*, 177-1190.

Fuhrman, J. A., and Azam, F. (1982). Thymidine incorporation as a measure of bacterioplankton production in marine surface waters: Evaluation and field results. *Mar. Biol. (Berlin)* *66*, 109-120.

Geesey, G. G., and Morita, R. Y. (1979). Capture of arginine at low concentrations by a marine psychrophilic bacterium. *Appl. Environ. Microbiol.* *38*, 1092-1097.

Geesey, G. G., and Morita, R. Y. (1981). Relationship of cell envelope stability to substrate capture in a marine psychrophilic bacterium. *Appl. Environ. Microbiol.* *42*, 533-540.

Gilbert, J. B., Price, V. E., and Greenstein, J. P. (1949). Effect of anions on non-enzymatic desamidation of glutamine. *J. Biol. Chem.* *180*, 209-218.

Hall, K. H., Weimer, W. C., and Lee, G. F. (1967). Amino acids in an estuarine environment. *Limnol. Oceanogr.* *12*, 162-164.

Hellebust, J. A., and Guillard, R. R. L. (1967). The uptake specificity for organic substrates by the diatom *Melosira nummuloides* *J. Phycol.* *3*, 132-136.

Herbland, A. (1976). *In situ* utilization of urea in the euphotic zone of the tropical Atlantic. *J. Exp. Mar. Biol. Ecol. 21,* 269-277.

Hobbie, J. E., Crawford, C. C., and Webb, K. L. (1968). Amino acid flux in an estuary. *Science 159,* 1463-1464.

Hodson, R. C., Williams, S. K., and Davidson, W. R., Jr. (1975). Metabolic control of urea catabolism in *Chlamydomonas reinhardi* and *Chlorella pyrenoidosa*. *J. Bacteriol. 121,* 1022-1035.

Hodson, R. E., Azam, F., Carlucci, A. F., Fuhrman, J. A., Karl, D. M., and Holm-Hansen, O. (1981a). Microbial uptake of dissolved organic matter in McMurdo Sound, Antarctica. *Mar. Biol. (Berlin) 61,* 89-94.

Hodson, R. E., Maccubbin, A. E., and Pomeroy, L. R. (1981b). Dissolved adenosine triphosphate utilization by free-living and attached bacterioplankton. *Mar. Biol. (Berlin) 64,* 43-51.

Hollibaugh, J. T. (1976). The biological degradation of arginine and glutamic acid in seawater in relation to the growth of phytoplankton. *Mar. Biol. (Berlin) 36,* 303-312.

Hollibaugh, J. T. (1979). Metabolic adaptation in natural bacterial populations supplemented with selected amino acids. *Estuarine Coastal Mar. Sci. 9,* 215-230.

Hollibaugh, J. T., Carruthers, A. B., Fuhrman, J. A., and Azam, F. (1980a). Cycling of organic nitrogen in marine plankton communities studied in enclosed water columns. *Mar. Biol. (Berlin) 59,* 15-21.

Hollibaugh, J. T., Fuhrman, J. A., and Azam, F. (1980b). Radioactively labelling of natural assemblages of bacterioplankton for use in trophic studies. *Limnol. Oceanogr. 25,* 172-181.

Hoppe, H.-G. (1976). Determination and properties of actively metabolizing heterotrophic bacteria in the sea, investigated by means of microautoradiography. *Mar. Biol. (Berlin) 36,* 291-302.

Iturriaga, R., and Zsolnay, A. (1981). Transformation of some dissolved organic compounds by a natural heterotrophic population. *Mar. Biol. (Berlin) 62,* 125-129.

Jorgensen, N. O. G. (1980). Uptake of glycine and release of primary amines by the polychaete *Nereis virens* (Sars) and the mud snail *Hydrobia neglecta* Muus. *J. Exp. Mar. Biol. Ecol. 47,* 281-297.

Kurelec, B., Rijavec, M., Britavic, S., Muller, W. E. G., and Zahn, R. K. (1977). Phytoplankton: Presence of γ-glutamyl cycle enzymes. *Comp. Biochem. Pysiol. B 56B,* 415-419.

Landymore, A. F., and Antia, N. J. (1977). Growth of a marine diatom and a haptophycean alga on phenylalanine and tyrosine serving as sole nitrogen source. *J. Phycol. 13,* 231-238.

Landymore, A. F., and Antia, N. J. (1978). White-light-promoted degradation of leucopterin and related pteridines dissolved in seawater, with evidence for involvement of complexation from major divalent cations of seawater. *Mar. Chem. 6,* 309-325.

Landymore, A. F., Antia, N. J., and Towers, G. H. N. (1978). The catabolism of L-phenylalanine, L-tyrosine, and some related aromatic compounds by two marine species of phytoplankton. *Phycologia 17*, 319-328.

Lanyi, J. K., and Silverman, M. P. (1979). Gating effects in *Halobacterium halobium* membrane transport. *J. Biol. Chem. 254*, 4750-4755.

Lanyi, J. K., Yearwood-Drayton, V., and MacDonald, R. E. (1976a). Light-induced glutamate transport in *Halobacterium halobium* envelope vesicles. I. Kinetics of the light-dependent and the sodium-gradient-dependent uptake. *Biochemistry 15*, 1595-1603.

Lanyi, J. K., Renthal, R., and MacDonald, R. E. (1976b). Light-induced glutamate transport in *Halobacterium halobium* envelope vesicles. II. Evidence that the driving force is a light-dependent sodium gradient. *Biochemistry 15*, 1603-1610.

Leftley, J. W., and Syrett, P. J. (1973). Urease and ATP: Urea amidolyase activity in unicellular algae. *J. Gen. Microbiol. 77*, 109-115.

Lehninger, A. L. (1975). "Biochemistry." Worth Publ., New York.

Lewin, J., and Hellebust, J. A. (1975). Heterotrophic nutrition of the marine pennate diatom *Navicula pavillardi* Hustedt. *Can. J. Microbiol. 21*, 1335-1342.

Lewin, J., and Hellebust, J. A. (1976). Heterotrophic nutrition of the marine pennate diatom *Nitzchia angularis* var. *affinis*. *Mar. Biol. (Berlin) 36*, 313-320.

Lewin, J., and Hellebust, J. A. (1978). Utilization of glutamate and glucose for heterotrophic growth by the marine pennate diatom *Nitzchia laevis*. *Mar. Biol. (Berlin) 47*, 1-7.

Liebezeit, G., Bolter, M., Brown, I. F., and Dawson, R. (1980). Dissolved free amino acids and carbohydrates at pycnocline boundaries in the Sargasso Sea and related microbial activity. *Oceanol. Acta 3*, 357-362.

Liu, M. S., and Hellebust, J. A. (1974). Uptake of amino acids by the marine centric diatom *Cyclotella cryptica*. *Can. J. Microbiol. 20*, 1109-1118.

Lorenz, M. G., Aardema, B. W., and Krumbein, W. E. (1981). Interaction of marine sediment with DNA and DNA availability to nucleases. *Mar. Biol. (Berlin) 64*, 225-230.

Lytle, C. R., and Perdue, E. M. (1981). Free, proteinaceous, and humic-bound amino acids in river water containing high concentrations of aquatic humus. *Environ. Sci. Technol. 15*, 224-228.

McCarthy, J. J. (1971). The role of urea in marine phytoplankton ecology. PhD. Thesis, University of California, San Diego.

McCarthy, J. J. (1972). The uptake of urea by natural populations of marine phytoplankton. *Limnol. Oceanogr. 17*, 738-748.

McCarthy, J. J., and Eppley, R. W. (1972). A comparison of chemical, isotopic, and enzymatic methods for measuring nitrogen assimilation of marine phytoplankton. *Limnol. Oceanogr. 17*, 371-382.

MacDonald, R. E., Greene, R. V., and Lanyi, J. I. (1977). Light-activated acid transportsystems in *Halobacterium halobium* envelope vesicles: Role of the chemical and electrical gradients. *Biochemistry 16*, 3227-3235.

McGrath, S. M., and Sullivan, C. W. (1981). Community metabolism of adenylates by microheterotrophs from the Los Angeles Harbor and Souther California coastal waters. *Mar. Biol. (Berlin) 62*, 217-226.

McLean, R. O., Corrigan, J., and Ebster, J. (1981). Heterotrophic nutrition in *Melosira nummuloides*, a possible role in affecting distribution in the Clyde Estuary. *Br. Pycol. J. 16*, 95-105.

MacLeod, R. A. (1965). The question of the existence of specific marine bacteria. *Bacterol. Rev. 29*, 9-23.

Maeda, M., and Taga, N. (1974). Occurrence and distribution of deoxyribonucleic acid-hydrolyzing bacteria in seawater. *J. Exp. Mar. Biol. Ecol. 14*, 157-169.

Maitz, G. S., Haas, E. M., and Castric, P. A. (1982). Purification and properties of the allophanate hydrolase from *Chlamydomonas reinhardi*. *Biochim. Biophys. Acta*

Matusumoto, K., Uno, I., Toh-e, A., Ishikawa, T., and Oshima, Y. (1982). Cyclic AMP may not be involved in catabolite repression in *Saccharaomyces cerevisiae:* Evidence from mutants capable of utilizing it as an adenine source. *J. Bacteriol. 150*, 277-285.

Meister, A. (1980). Possible relation of the γ-glutamyl cycle to amino acid and peptide transport in microorganisms. *In* "Microorganisms and Nitrogen Sources" (J. W. Payne, ed.), pp. 493-508. Wiley (Interscience), New York.

Mitamura, O., and Saijo, Y. (1975). Decomposition of urea associated with photosynthesis of phytoplankton in coastal waters. *Mar. Biol. (Berlin) 30*, 67-72.

Mitamura, O., and Saijo, Y. (1980). *In situ* measurement of the urea decomposition rate and its turnover rate in the Pacific Ocean. *Mar. Biol. (Berlin) 58*, 147-152.

Niven, D. F., and MacLeod, R. A. (1980). Sodium ion-substrate symport in a marine bacterium. *J. Bacteriol. 142*, 603-607.

North, B. B. (1975). Primary amines in California coastal waters: Utilization by phytoplankton. *Limnol. Oceanogr. 20*, 20-27.

North, B. B., and Stephens, G. C. (1967). Uptake and assimilation of amino acids by *Platymonas*. *Biol. Bull. (Woods Hole, Mass.) 133*, 391-400.

North, B. B., and Stephens, G. C. (1971). Uptake and assimilation of amino acids by *Platymnonas*. II. Increased uptake in nitrogen-deficient cells. *Biol. Bull. (Woods Hole, Mass.) 140*, 242-254.

North, B. B., and Stephens, G. C. (1972). Amino transport in *Nitzchia ovalis* Arnott. *J. Phycol. 8*, 64-68.

Odum, W. E., Kirk, P. W., and Zieman, J. C. (1979). Non-protein nitrogen compounds associated with particles of vascular plant detrtitus. *Oikos 32*, 363-367.

Paul, J. H. (1982). Isolation and characterization of a *Chlamydomonas* L-asparaginase. *Biochem. J. 203,* 109-115.

Paul, J. H., and Cooksey, K. E. (1979). Asparagine and asparaginase activity in a euryhaline *Chlamydomonas* species. *Can. J. Microbiol. 25,* 1443-1451.

Paul, J. H., and Cooksey, K. E. (1981). Regulation of L-asparaginase glutamine synthetase, and glutamate dehydrogenase in response to medium nitrogen concentrations in euryhaline *Chlamydomonas* species. *Plant Physiol. 68,* 1364-1367.

Pillai, T. N. V., and Ganguly, A. K. (1972). Nucleic acid in dissolved constituents of sea water. *J. Mar. Biol. Assoc. India 14,* 384-390.

Pintner, I. J., and Provasoli, L. (1968). Heterotrophy in subdued light of 3 *Chrysochromulina* species. *Bull. Misaki Mar. Biol. Inst., Kyoto Univ. 12,* 25-31.

Rees, T. A. V., and Syrett, P. J. (1979a). The uptake of urea by the diatom *Phaeodactylum. New Phytol. 82,* 169-175.

Rees, T. A. V., and Syrett, P. J. (1979b). Mechanisms for urea uptake by the diatom *Phaeodactylum tricornutum:* The uptake of thiourea. *New Phytol. 83,* 37-48.

Rees, T. A. V., Cresswell, R. C., and Syrett, P. J. (1980). Sodium dependence of urea and nitrate uptake by *Phaeodactylum. Br. Phycol. J. 15,* 199.

Remsen, C. C., Carpenter, E. J., and Schroeder, B. W. (1972). Competition for urea among estuarine microorganisms. *Ecology 53,* 921-926.

Remsen, C. C., Carpenter, E. J., and Schroeder, B. W. (1974). The role of urea in marine microbial ecology. *In* "Effect of the Ocean Environment on Microbial Activities" (R. Colwell and R. Y. Morita, eds.), pp. 289-304. Univ. Park Press, Baltimore, Maryland.

Roon, R. J., Hampshire, J., and Levenberg, B. (1972). Urea amidolyase. The involvement of biotin in urea cleavage. *J. Biol. Chem. 247,* 7539-7545.

Ryther, J. H. (1954). The ecology of phytoplankton blooms in Moriches Bay and Great South Bay, Long Island, New York. *Biol. Bull. 106,* 198-209.

Saks, N. M., and Kahn, E. G. (1979). Substrate competition between a salt marsh diatom and a bacterial population. *J. Phycol. 15,* 17-21.

Saks, N. M., Stone, R. J., and Lee, J. J. (1976). Autotrophic and heterotrophic nutritional budget of salt marsh epiphytic algae. *J. Phycol. 12,* 443-448.

Schell, D. M. (1974). Uptake and regeneration of free amino acids in marine waters of southeast Alaska. *Limnol. Oceanogr. 19,* 260-270.

Sipos, T., and Merkel, J. R. (1971). Marine bacterial proteases. II. Peptide specificity of an endopeptidase produced by *Vibrio* B-30. *Arch. Biochem. Biophys. 145,* 137-142.

Sprott, G. D., and MacLeod, R. A. (1972). Na^+-dependent amino acid transport in isolated membrane vesicles of a marine pseudomonad energized by electron donors. *Biochem. Biophys. Res. Commun.* *47*, 838-845.

Sprott, G. D., and MacLeod, R. A. (1974). Nature of the specificity of alcohol coupling to L-alanine transport into isolated membrane vesicles of a marine pseudomonad. *J. Bacteriol.* *117*, 104-1054.

Steinmann, J. (1976). Untersuchungen uber den bakteriellen Abbau von Harnstoff und Harnsaure in der westlichen Ostee. *Bot. Mar.* *19*, 47-58.

Stephens, G. C., and North, B. B. (1971). Extrusion of carbon accompanying uptake of amino acids by marine phytoplankton. *Limnol. Oceanogr.* *16*, 752-757.

Thompson, J., Green, M. L., and Happold, F. C. (1969). Cation-activated nucleotidase in cell envelopes of a marine bacterium. *J. Bacteriol.* *99*, 834-841.

Tokuda, H., Sugasawa, M., and Unemoto, T. (1982). Roles of Na^+ and K^+ in α-aminoisobutyric acid transport by the marine bacterium *Vibrio alginolyticus*. *J. Biol. Chem.* *257*, 788-794.

Turner, M. F. (1979). Nutrition of some marine microalgae with special reference to vitamin requirements and utilization of nitrogen and carbon sources. *J. Mar. Biol. Assoc. U.K.* *59*, 535-552.

Vose, J. R., Cheng, J. Y., Antia, N. J., and Towers, G. H. N. (1971). The catabolic fission of the aromatic ring of phenylalanine by marine planktonic algae. *Can. J. Bot.* *49*, 259-261.

Webb, K. L., and Haas, L. W. (1976). The significance of urea for phytoplankton nutrition in the York River, Virginia. *In* "Estuarine Processes" (M. Wiley, ed.), Vol. 1, pp. 90-102, Academic Press, New York.

Wheeler, P. A. (1979). Uptake of methylamine (an ammonium analogue) by *Macrocystis pyrifera* (Phaeophyta). *J. Phycol.* *15*, 12-17.

Wheeler, P. A. (1980). Use of methylammonium as an ammonium analogue in nitrogen transport and assimilation studies with *Cyclotella cryptica* (Bacillariophyceae). *J. Phycol.* *16*, 328-334.

Wheeler, P. A., and Hellebust, J. A. (1981). Uptake and concentration of alkylamines by a marine diatom. Effects of H^+ and K^+ and implications for the transport and accumulation of weak bases. *Plant Physiol.* *67*, 367-372.

Wheeler, P. A., North, B. B., and Stephens, G. C. (1974). Amino acid uptake by marine phytoplankters. *Limnol. Oceanogr.* *19*, 249-259.

Wheeler, P., North, B., Littler, M., and Stephens, G. (1977). Uptake of glycine by natural phytoplankton communities. *Limnol. Oceanogr.* *22*, 900-910.

Whitney, P. A., and Cooper, T. (1973). Urea carboxylase from *Saccharomyces cerevisiae*. Evidence for a minimal two-step reaction sequence. *J. Biol. Chem.* *248*, 325-330.

Williams, P. J. LeB. (1970). Heterotrophic utilization of dissolved organic compounds in the sea. I. Size distribution of population and relationship between respiration and incorporation of growth substrates. *J. Mar. Biol. Assoc. U. K. 50,* 859-870.

Williams, P. J. LeB. (1973). The validity of the application of simple kinetic analysis to heterogeneous microbial populations. *Limnol. Oceanogr. 18,* 159-164.

Williams, S. K., and Hodson, R. C. (1977). Transport or urea at low concentrations in *Chlamydomonas reinhardi*. *J. Bacteriol. 130,* 266-273.

Wood, D. C., and Hayasaka, S. S. (1981). Chemotaxis of rhizoplane bacteria to amino acids comprising eelgrass (*Zostera marine* L.) root exudate. *J. Exp. Mar. Biol. Ecol. 50,* 153-161.

Chapter 9

Phytoplankton Nitrogen Metabolism

PATRICIA A. WHEELER
School of Oceanography
Oregon State University
Corvallis, Oregon

I. INTRODUCTION

The uptake of nitrogenous nutrients by oceanic phytoplankton has been the focus of marine research for almost two decades. A major goal has been to determine the extent to which phytoplankton growth is regulated by nitrogen availability. Two approaches taken to answer this question have been: (1) the comparison of nitrogen specific uptake rates to requirements for growth (McCarthy and Goldman, 1979) and (2) the estimation of relative growth rates from elemental composition of particulate material (Goldman *et al.*, 1979).

Although the uptake of nitrogenous nutrients encompasses a series of metabolic processes, only the summation of these processes was considered in early studies (e.g., Caperon, 1967, 1969; Droop, 1968). In subsequent attempts to model the

ISBN 0-12-160280-X

relationship between uptake rates and growth rates, it became apparent that movement of nitrogen through more than one cellular compartment had to be considered. Williams (1967) proposed a two-compartment model and further elaborations were suggested by Grenney *et al.* (1973), Conway (1977), and DeManche *et al.* (1979). These authors recognized the necessity of separating nitrogen uptake into component processes in order to explain and predict both the regulation of net uptake and the relationship of uptake to growth. However, the empirical delineation of the separate processes and their relative importance in contributing to net uptake has not been undertaken. Recent laboratory and field investigations (Goldman *et al.*, 1981; Goldman and Glibert, 1982; Wheeler *et al.*, 1982a; Wheeler and McCarthy, 1982) suggest a common, but previously undetected, occurrence of nonlinear time courses for NH_4^+ uptake. Obvious physiological explanations for nonlinear time courses include the operation of regulatory mechanisms for component metabolic processes responsible for net uptake and changes in the rate-limiting step for uptake over time.

The main goals of the present chapter are to delineate those aspects of phytoplankton nitrogen metabolism that play key roles in regulating uptake and growth rates and to provide a physiological basis for determining the appropriate temporal and spatial scales relevant to investigating phytoplankton nutrient dynamics. The individual reactions involved in phytoplankton nitrogen metabolism will be grouped into four major processes. The relationships between these processes, net uptake, and growth will then be examined by comparison of relative pool sizes for the major components of cellular nitrogen and nitrogen specific rates of the key metabolic processes. Referenced material includes recent work on marine phytoplankton as well as some work on freshwater algae that is expected to apply to phytoplankton in general. Other recent reviews should be consulted for more detailed discussion of issues that will be mentioned only briefly here. Syrett (1981) provides an excellent review of nitrogen metabolism in microalgae, which concentrates on the uptake and assimilation of inorganic and organic nitrogen. Raven (1980) describes the basic cellular energetics for nitrogen uptake in microalgae. Eppley (1981) and McCarthy (1981) have reviewed the general relations between uptake and growth. Directly related reviews in this volume cover the kinetics of inorganic nitrogen uptake (Goldman and Glibert, Chapter 7), the uptake of organic nitrogen (Paul, Chapter 8), and the enzymology of nitrogen assimilation (Falkowski, Chapter 23).

II. MAJOR METABOLIC PATHWAYS OF NITROGEN UTILIZATION

In the present work, four general processes in phytoplankton nitrogen metabolism will be considered (Fig. 1):

1. *Membrane transport*: the movement of a particular form of nitrogen across the plasmalemma into the cell.
2. *Assimilation*: the metabolic conversion of inorganic nitrogen to small organic metabolites.
3. *Incorporation*: the synthesis of macromolecules from small metabolites.
4. *Catabolism*: the breakdown of macromolecules into small metabolites.

Solutes can enter cells by one or more different transport mechanisms that can be distinguished by substrate specificity and uptake kinetics. Nonmediated diffusion is the nonspecific entry of a solute down an electrochemical gradient and is usually characterized by linear uptake kinetics. Mediated diffusion is also down an electrochemical gradient but is substrate-specific and characterized by saturable uptake kinetics. Active transport is both substrate-specific and against an electrochemical gradient and is also characterized by saturable uptake kinetics. Two types of active transport can be distinguished: primary active transport requires a direct input of energy, whereas secondary active transport is driven by an ion gradient established by a

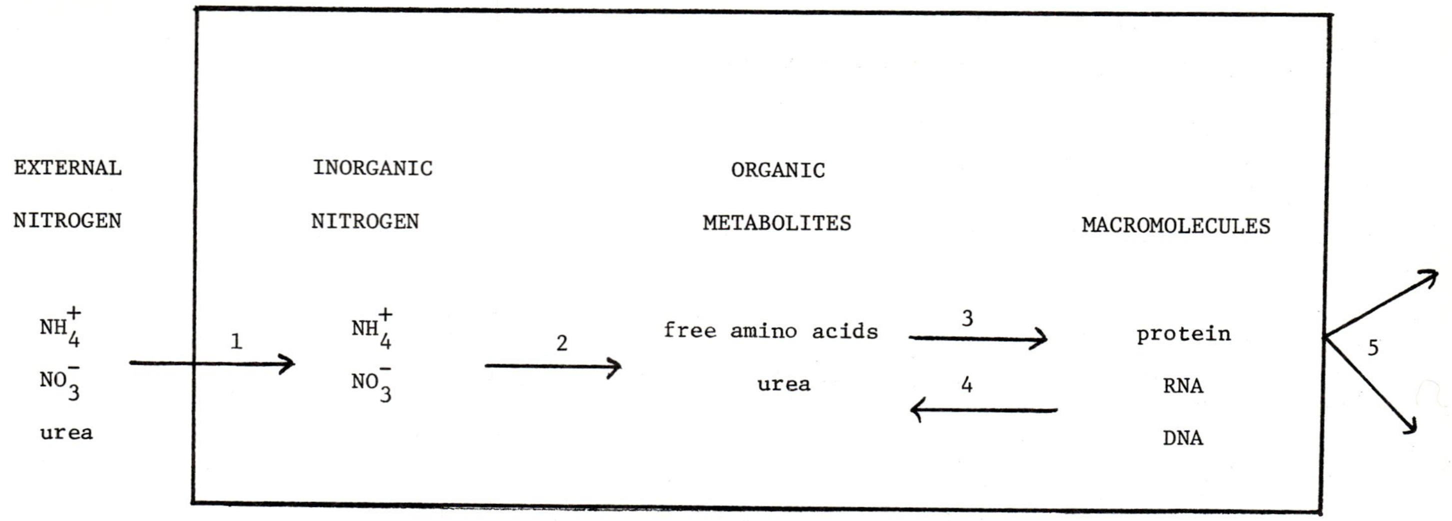

Fig. 1. Schematic diagram of major processes involved in nitrogen uptake. Key: 1, membrane transport; 2, assimilation into organic metabolites; 3, incorporation into macromolecules; 4, catabolism; 5, cell division.

TABLE I. Postulated Transport Mechanisms for Uptake of Nitrogen by Phytoplankton

N Source	*Transport mechanism*	*References*
NH_4^+, NH_3	Diffusion	See Raven, 1980; Stewart, 1980
	Mediated diffusion or secondary active transport	Pelley and Bannister, 1979; Wheeler, 1980a,b; Wheeler and Hellebust, 1981
Urea	Diffusion and secondary active transport	Syrett and Bekheet, 1977; Bekheet and Syrett, 1979; Williams and Hodson, 1977
Amino acids	Secondary active transport	Hellebust, 1978
NO_3^-	Primary active transport	Falkowski, 1975a

primary system. The postulated means of entry for the major nitrogen sources for phytoplankton are listed in Table I.

Significant uptake of nitrogenous nutrients by nonmediated diffusion is feasible only for NH_3 and urea, since these two compounds are very soluble in the lipid phase of membranes (Raven, 1980). Entry (and loss) of NH_3 by diffusion may account for approximately 10% of net uptake (Walker *et al.*, 1979), and theoretically passive accumulation of NH_3 by diffusion and acid trapping could result in $>10^3$-fold internal accumulation in marine diatoms (Wheeler and Hellebust, 1981). Entry of urea by diffusion occurs in *Chlorella* (Syrett and Bekheet, 1977) and *Chlamydomonas* (Kirk and Kirk, 1978), but this uptake is not concentrative.

The small amount of data available suggests that ammonium, urea, and amino acids may enter marine phytoplankton by secondary active transport (cotransport) driven by an Na^+ gradient (see Table I). Transport of NO_3^- and NO_2^-, on the other hand, may be driven directly by a Cl^--stimulated ATPase (Falkowski, 1975a). Considering the negative electrochemical gradient across the cell membrane, the anions are probably the most costly nitrogen sources taken up by phytoplankton.

Assimilation of transported nitrogen into amino acids (the major cellular component of small-molecular-weight organic nitrogen and the precursors for protein) requires complete reduction of the nitrogen atom. Specific enzymes involved in these reactions are discussed in Falkowski (this volume, Chapter 23). Nitrate and nitrite are reduced by nitrate and nitrite reductase, respectively, while urea nitrogen is released by either ATP: urea amidolyase in green algae or urease in most other algae (Syrett and Leftley, 1976). The bulk of NH_3-N appears to be assimilated through the glutamine synthetase/glutamate synthase (GS/GOGAT) pathway, which leads to a net production of glutamate. The major exception to this generalization is the significant activity of glutamate dehydrogenase (GDH) in some green algae when external NH_4^+ is high (Shatilov *et al.*, 1978; Tischner and Lorenzen, 1980). Amino acids are then readily interconverted by several transaminases (see Stewart, 1980; Syrett, 1981).

The two major forms of macromolecular nitrogen are protein and nucleic acids. Under most circumstances the bulk of cellular nitrogen is in the form of protein (see Section III), and if adequate nitrogen is available, phytoplankton essentially function as "protein factories" (Myers, 1980). Early work (Syrett, 1956) suggested that the uptake of nitrogen by nitrogen-starved or nitrogen-limited cells is limited by the rate of protein synthesis. This is substantiated in more recent studies that demonstrate the accumulation of internal pools of NO_3^-, NH_4^+, and free amino acids after the addition of nitrogen to nitrogen-limited cultures (DeManche *et al.*, 1979; DeManche, 1980; Dortch, 1982). Such pools would not accumulate if rates of protein synthesis were equal to or greater than rates of membrane transport and subsequent assimilation to amino acids.

The catabolism and internal turnover of nitrogenous macromolecules during variable nutrient regimes has not been studied extensively in phytoplankton. It is known that many pigments (e.g., chl-*a*, Barrett and Jeffrey, 1971) and enzymes (e.g., nitrate reductase, Johnson, 1979) can turn over rapidly. However, the contribution of catabolism and internal recycling of nitrogen to the maintenance and growth of phytoplankton during conditions of nitrogen deprivation remains to be examined in terms of changes in particular forms or classes of nitrogenous compounds. The two major processes involved in the recycling of cellular nitrogen that have been identified recently are photorespiration (Tolbert, 1980; Cullimore and Sims, 1980; Raven and Beardall, 1981) and the catabolism of specialized nitrogen storage compounds. The latter will be discussed in the following section.

III. INTRACELLULAR NITROGEN POOLS

Phytoplankton cells are composed of approximately 1-3% N as dry weight or a C:N of 6-10, with 1% N and a C:N of 10 representing nitrogen-limited cells. Total cellular nitrogen appears to vary over a fivefold range as a function of growth rate in nitrogen-limited continuous cultures (Goldman and McCarthy, 1978) and is independent of the particular nitrogen source utilized (Goldman and Peavey, 1979). Numerous studies have been published on various aspects of the nitrogenous composition of phytoplankton, but few of these have attempted a complete fractionation of cellular nitrogen. Representative data normalized to concentrations per unit of cell volume or percentage of total cellular nitrogen are presented in Table II.

Cellular pools of inorganic nitrogen tend to be very small: <100 m*M* and <5% of total cellular nitrogen. The major exceptions to this generalization are the large pools of NH_4^+ and NO_3^- in some diatoms during batch culture growth (Conover, 1975) and on addition of NH_4^+ or NO_3^- to nitrogen-limited or nitrogen-starved cultures (DeManche *et al.*, 1979; DeManche, 1980; Dortch, 1982).

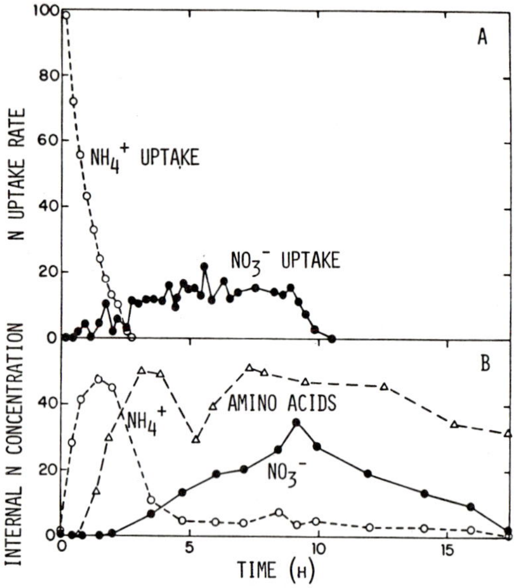

Fig. 2. Formation of transient intracellular nitrogen pools after addition of nitrate and ammonium to nitrogen-starved cultures of Skeletonema costatum. Uptake rates: (A) mmol·liter cell volume^{-1}·h^{-1} and internal pool concentrations (B) mmol·liter cell volume^{-1}. (Fig. 66 from Dortch, 1980.)

These latter increases, however, tend to be transient (Fig. 2). Urea pools are also very small (<5 m*M*), whereas free amino acids account for the largest component of soluble metabolites in all species examined, and levels range from 100-200 m*M* and up to 15-20% of cellular nitrogen. Free amino acid pools also increased significantly in the perturbation studies cited above (see Fig. 2).

It is difficult to estimate the maximal pool sizes for NH_4^+, NO_3^-, and amino acids. The large transient pools reported seem to result from a sudden change in external supply and a temporary imbalance in subsequent metabolism. There may be a physiological limitation on the internal pool size that can be maintained at a given external nutrient concentration and, at elevated internal concentrations, a significant efflux of nitrogenous metabolites

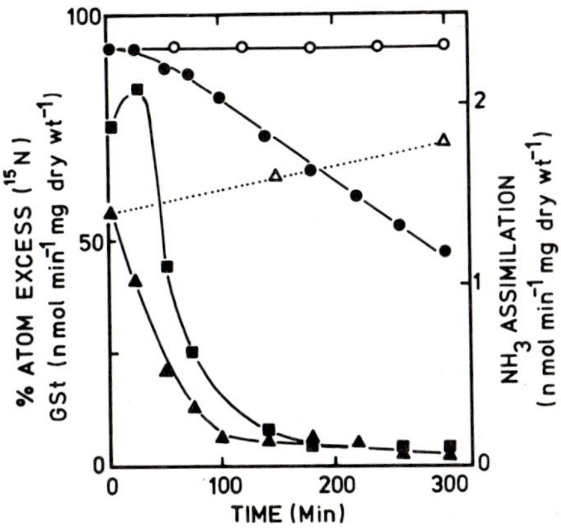

Fig. 3. Efflux of NH_4^+ *from Chlamydomonas.* NH_4^+ *assimilation (■) was prevented by treatment of cultures with MSO, a specific inhibitor of glutamine synthetase activity (▲). Efflux of* NH_4^+ *is demonstrated by dilution of* ^{15}N *in the medium (●) relative to an untreated control (○) in which glutamine synthetase activity remained at normal levels (△). (Fig. 1 from Cullimore and Sims, 1980.)*

may occur. Clear evidence of NH_4^+ efflux is provided in the results of Cullimore and Sims (1980) with *Chlamydomonas* (Fig. 3).

Macromolecular forms of nitrogen account for the largest component of cellular nitrogen. Protein is the dominant form, accounting for 1000-2500 m*M* concentrations or 70-90% of cellular nitrogen. Ribonucleic acid (RNA) is also a significant component (10-15% of cellular nitrogen), whereas deoxyribonucleic acid (DNA) is a very small fraction (1-2% of cellular N). Considering relative pools sizes, the major changes observed in total cellular nitrogen clearly must derive from alterations in the protein and RNA pools (Table II).

Accumulation of surplus nitrogen is inferred from the uncoupling of uptake and growth rates (e.g., Caperon, 1969), but has not been characterized metabolically in any detail. Following the same reasoning presented above, one can presume that any excess or surplus nitrogen is maintained in the cells as macromolecules. Energetically, storage of macromolecules is less costly than maintaining large concentration gradients of small metabolites. Some recently identified types of "storage nitrogen" in specific algal classes are: Cyanophyceae, cyanophycin (Gupta and Carr, 1981) and phycocyanin (Boussiba and Richmond, 1980; Yamanaka and Glazer, 1980); Rhodophyceae, phycoerythrin (Gantt, 1981); Chlorophyceae, guanine (Pettersen, 1975) and arginine (Wheeler, 1977; Wheeler and Stephens, 1977); and Bacillariophyceae, pigment proteins (Perry *et al.*, 1981). These compounds accumulate when surplus nitrogen is available, are specifically degraded on a signal of nitrogen limitation, and (for phycocyanin) can maintain normal growth until depletion of the stored material. Interestingly, many of these storage compounds are involved in photosynthesis, which suggests a close interaction between carbon and nitrogen metabolism.

IV. RATES OF NITROGEN METABOLISM

A. Relations between Nitrogen Availability, Uptake, and Growth

Nitrogen uptake rates are generally described as a hyperbolic function of external nutrient concentration $[S]$ (Dugdale, 1977). Similarly, the relationship between growth rate (μ) and $[S]$, cell nitrogen quota (Q) and $[S]$, and μ and Q are defined within certain limits by rectangular hyperbolae (see Goldman and Glibert, Chapter 7, this volume). The maximum uptake potential (V_{max}) is inversely related to both μ and Q, and this is interpreted as an adaptation of phytoplankton to nitrogen limitation (McCarthy and Goldman, 1979). Although the "enhanced" V_{max} often exceeds μ_{max} by more than an order of magnitude, the duration over which this maximal uptake rate is maintained is variable.

Here, we will examine how the various aspects of nitrogen uptake and metabolism are affected by the nitrogen status of cells.

Comparison of nitrogen-specific rates for the component processes indicates some of the physiological or biochemical limits of "enhanced" uptake responses, provides a means of delineating the rate-limiting step for uptake, and provides a physiological explanation for nonlinear time courses.

B. Nitrogen-Specific Rates of Component Processes

1. Membrane Transport

In view of the very low concentrations of available nitrogen (Chapter 1, this volume) and the common occurrence of saturation kinetics for uptake, it has been presumed that phytoplankton possess efficient membrane transport systems for nitrogenous nutrients (Raven, 1980; Syrett, 1981). Saturation kinetics of uptake are thought to result from the activity of these transport systems (Eppley and Coatsworth, 1968). Unfortunately, metabolic reactions take place rapidly after membrane transport (e.g., see Tischner and Lorenzen, 1979; Cullimore and Sims, 1980; and Wheeler *et al.*, 1982a), making it extremely difficult to measure transport rates per se. Several means of preventing assimilation are available, however, and have provided some information about transport velocities and the regulation of transport activity.

Transport of NO_3^-, in the absence of assimilation, has been studied by assaying the NO_3^- - dependent ATPase (Falkowski, 1975a,b) and by measuring rates of NO_3^- uptake before activation or induction of nitrate reductase (Eppley and Coatsworth, 1968; Tischner and Lorenzen, 1979). Nitrogen-specific transport rates for NO_3^- reach 0.196 h^{-1} for *Chlorella* and 0.310 h^{-1} for diatoms (Table III). Higher rates may occur on nitrogen deprivation, although this has not been determined experimentally.

Urea transport could be measured easily in the absence of assimilation in *Chlamydomonas* because the enzymes necessary for assimilation are not induced until 20-30 min after the addition of urea. In 5-min uptake experiments, Williams and Hodson (1977) demonstrated a high-affinity urea transport system that became

TABLE II. Nitrogenous Cell Composition of Phytoplankton

		Total cellular N		NO_3^-		NH_4^+	
Species (Reference)	*Cell Volume (pl)*	*pg·atom N / cell*	*mg·atom N / liter*	*mg·atom N / liter*	*% cell N*	*mg·atom N / liter*	*% cell N*
Biddulphia aurita (Lui and Roels, 1972)	—	—	—	—	—	—	2–4
Chaetoceros debilis (Dortch, 1980; 1982)	0.199	0.862–2.630	438–714	0–29	<1	0–9	<1
Ditylum brightwellii (Eppley et al., 1967; Eppley and Coatsworth, 1969; Strickland *et al.*, 1969; Eppley and Rogers, 1970)	54	88–121	132	40	23	5–10	<1
Phaeodactylum tricornutum (Rees and Syrett, 1979; Cresswell and Syrett, 1981)	0.148	—	—	0–17	—	—	—
Skeletonema costatum (DeManche, 1980)	—	—	520–1410	3–199	<1–28	1–55	1–6
Skeletonema costatum (Dortch, 1980; 1982)	0.221	0.382–1.315	137–565	0–61	<1	1–47	0–8
Thalassiosira aestivilis (DeManche, 1980)	—	—	1110–1460	95–120	9	—	—
Thalassiosira fluvialitis (Conover, 1975)	1.24	11–30	600–1710	0–415	0-37	0–1773	0–66
Thalassiosira gravida (Dortch, 1980; 1982)	2.664–2.995	14.2–15.6	335–415	73	—	1–17	<1
Dissodinium lunula (Bhovichita and Swift, 1977; Shuter, 1978)	195	16[a]	78[a]	—	<1[a]	1.0	<1[a]
Pyrocystis fusiformis (Bhovichita and Swift, 1977; Shuter, 1978)	8000	171[a]	30[a]	0.2–2.4	1–2[a]	0.1–1.2	<1–6[a]
Pyrocystis noctiluca (Bhovichita and Swift, 1977; Shuter, 1978)	17000	183	14[a]	0.1–0.4	<1[a]	0–0.2	<1–25[a]
Chlamydomonas reinhardii (Cullimore and Sims, 1981a; Dagestad *et al.*, 1981)	0.100	—	—	—	—	—	<1
Chlorella fusca (Bekheet and Syrett, 1979)	0.197	—	—	—	—	—	—
Dunaliella bioculata (Marano *et al.*, 1978)	0.115–0.345	0.110–0.661	958–1915	—	—	—	—
Platymonas striata (Ricketts, 1977)	0.195	0.334–0.773	1713–3759	—	—	—	—
Platymonas subcordiformis (North and Stephens, 1971; Kirst, 1977; Wheeler, 1977)	0.177	0.270–0.977	1525–5520	0–32	<1	12–443	<1–6
Scenedesmus sp. (Rhee, 1973; 1978)	0.054–0.120	0.071–0.214	1315–1783	—	—	—	—

[a]Assuming that the sum of analyzed nitrogen pools is approximately equal to total cellular nitrogen.

Urea		Free amino cells		Protein		RNA		DNA	
mg·atom N / liter	% cell N	mg·atom N / liter	% cell N	mg·atom N / liter	% cell N	mg·atom N / liter	% cell N	mg·atom N / liter	% cell N
—	—	—	4–14	—	67–81	—	—	—	—
—	—	—	—	—	—	—	—	—	—
—	—	—	—	—	—	—	—	7	1.9
1.2	—	—	—	—	—	—	—	—	—
—	—	100	7–10	—	—	—	—	—	—
—	—	1–187	<1–33	176–836	79–92	—	—	—	—
—	—	78–100	5–9	—	—	—	—	—	—
0–15	<1	7–113	1–7	635–1270	24–56	—	—	—	—
—	—	—	—	—	—	—	—	—	
—	—	—	—	77	99[a]	—	—	—	—
—	—	—	—	23–26	77–87[a]	—	—	—	—
—	—	—	—	10–13	71–93[a]	—	—	—	—
—	2–5	—	—	—	—	—	—	—	—
0.1	—	—	—	—	—	—	—	—	—
—	—	—	—	745–1489	78[a]	201–402	21[a]	12–24	1[a]
—	—	—	—	1317–3759	91[a]	360–800	14–21[a]	36–67	1–2[a]
—	—	697–1201	11–23	2842–5914	59–87[a]	—	—	—	—
—	—	50–130	5	222–1192	50–70	611–2625	—	180–130	3–13

TABLE III. Maximum Nitrogen-Specific Rates for Transport and Assimilation of Nitrogenous Compounds and Pool-Specific

Process	*Substrate or enzyme*	*Species*	V_{max} (h^{-1})
Transport	NO_3^- - ATPase	*Ditylum brightwelli*	0.064
		Skeletonema costatum	0.310
		Amphidinium carterae	--
		Dunaliella tertiolecta	0.045
		Eutreptiella gymnastica	0.123
		Chroomonas salina	0.079
		Isochrysis galbana	0.086
	NO_3^-	*Ditylum brightwelli*	0.133
		Chlorella sorokiniana	0.196
	Urea	*Phaeodactylum tricornutum*	0.006
			0.117
		Chlamydomonas reinhardii	0.670
		Chlorella fusca	0.012
	Amino acids:		
	arginine	*Cyclotella cryptica*	0.247
			0.370
	glutamate	*Cyclotella cryptica*	0.053
	(high affinity)		0.151
	glutamate	*Cyclotella cryptica*	0.281
	(low affinity)		0.801
	glutamine	*Cyclotella cryptica*	0.037
			0.198
	proline	*Cyclotella cryptica*	0.023
	(high affinity)		0.105
	proline	*Cyclotella cryptica*	0.105
	(low affinity)		0.408
	alanine	*Cyclotella cryptica*	0.033
			0.180
	arginine	*Nitzschia ovalis*	0.052
			0.690
	glycine	*Platymonas* sp.	0.010
			0.130
	Methylammonium	*Cyclotella cryptica*	0.091
			1.058
		Chlorella pyrenoidosa	0.322
	NH_4^+	*Chaetoceros simplex*	0.675
		Phaeodactylum tricornutum	2.110
		Thalassiosira weissflogii	0.210
		Dunaliella tertiolecta	0.100

Rates for Incorporation of Nitrogen into Major Macromolecular Components

Conditions	*Reference*	*Assumptions made for calculation*
Not given	Falkowski, 1975	
Not given	Falkowski, 1975	
Not given	Falkowski, 1975	
Not given	Falkowski, 1975	
Not given	Falkowski, 1975	
Not given	Falkowski, 1975	
Not given	Falkowski, 1975	
Uptake in dark	Eppley and Coatsworth, 1968	
+N	Tischner and Lorenzen, 1979	
+N	Rees and Syrett, 1979	1.5 *M* cell N
-N	Rees and Syrett, 1979	0.5 *M* cell N
-N	Williams and Hodson, 1977	1.0 *M* cell N
+N	Syrett and Bekheet, 1977	2.0 *M* cell N
+N	Liu and Hellebust, 1974a	[a]
-N	Liu and Hellebust, 1974a	[a]
+N	Liu and Hellebust, 1974a	[a]
-N	Liu and Hellebust, 1974a	[a]
+N	Liu and Hellebust, 1974a	[a]
-N	Liu and Hellebust, 1974a	[a]
+N	Liu and Hellebust, 1974a	[a]
-N	Liu and Hellebust, 1974a	[a]
+N	Liu and Hellebust, 1974a	[a]
-N	Liu and Hellebust, 1974a	[a]
+N	Liu and Hellebust, 1974a	[a]
-N	Liu and Hellebust, 1974a	[a]
+N	Liu and Hellebust, 1974a	[a]
-N	Liu and Hellebust, 1974a	[a]
+N	North and Stephens, 1972	
-N	North and Stephens, 1972	
+N	North and Stephens, 1971	
-N	North and Stephens, 1971	
+N	Wheeler, 1980b	
-N	Wheeler, 1980b	33% decrease in cell N (from Liu and Hellebust, 1974b)
-N	Pelley and Bannister, 1979	0.322 *M* cell N (from Shuter, 1978)
N limited	Goldman and Glibert, 1982	
N limited	Goldman and Glibert, 1982	
N limited	Goldman and Glibert, 1982	
N limited	Goldman and Glibert, 1982	

TABLE III (continued)

Process	Substrate or enzyme	Species	V_{max} (h^{-1})
Assimilation			
	Nitrate reductase	*Ditylum brightwelli*	0.001
			0.003
			0.631
		Thalassiosira pseudonana	0.016
			0.046
			0.092
		Chlamydomonas reinhardii	0.164
		Chlorella pyrenoidosa	0.422
		Chlorella sorokiniana	0.424
		Platymonas striata	0.006
	Nitrite reductase	*Chlamydomonas reinhardii*	0.136
		Chlorella pyrenoidosa	1.830
		Chlorella sorokiniana	1.690
		Platymonas striata	0.016
	Urease	*Skeletonema costatum*	0.180
			0.408
		Thalassiosira pseudonana	0.048
			0.080
	Urea amidolyase	*Ankistrodesmus braunii*	<0.001
	Glutamate dehydrogenase	*Chlamydomonas reinhardii*	0.036
			0.119
	NADP	*Chlorella pyrenoidosa*	0.089
			0.489
			0.044
	NAD	*Chlorella pyrenoidosa*	0.677
			0.608
			0.500
		Chlorella sorokiniana	0.024-0.027
			0.117

Conditions	Reference	Assumptions made for calculation
NO_2^- grown	Eppley and Coatsworth, 1968	
NH_4^+ grown	Eppley and Coatsworth, 1968	
NO_3^- grown	Eppley and Coatsworth, 1968	
increasing	Eppley and Renger, 1974	
nitrogen	Eppley and Renger, 1974	
limitation	Eppley and Renger, 1974	
NO_3^- grown	Cullimore and Sims, 1981b	1.0 *M* cell N
+N	Tischner, 1976	*b*
+N	Tischner and Hütterman, 1980	*b*
+N	Ricketts and Edge, 1977	*b*
NO_3^- grown	Cullimore and Sims, 1981b	1.0 *M* cell N
+N	Tischner, 1976	*b*
+N	Tischner and Lorenzen, 1980	*b*
+N	Ricketts and Edge, 1977	
+N	Horrigan and McCarthy, 1981	5-min rates for release of CO_2 $\simeq$ urease activity
-N	Horrigan and McCarthy, 1981	5-min rates for release of CO_2 $\simeq$ urease activity
+N	Horrigan and McCarthy, 1981	5-min rates for release of CO_2 $\simeq$ urease activity
-N	Horrigan and McCarthy, 1981	5-min rates for release of CO_2 $\simeq$ urease activity
+N	Hipkin and Syrett, 1977	
+N	Cullimore and Sims, 1981a	1.0 *M* cell N
-N	Cullimore and Sims, 1981a	1.0 *M* cell N
-N	Kretovich *et al.*, 1970	*b*
$+NH_4^+$	Kretovich *et al.*, 1970	*b*
$+NO_3^-$	Kretovich *et al.*, 1970	*b*
-N	Kretovich *et al.*, 1970	*b*
$+NH_4^+$	Kretovich *et al.*, 1970	*b*
$+NO_3^-$	Kretovich *et al.*, 1970	*b*
+N, -N	Tischner and Hütterman, 1980	*b*
$NO_3^- \rightarrow NH_4^+$ medium	Tischner and Lorenzen, 1980	1.0 *M* cell N *b* 0.5 *M* cell N

TABLE III (continued)

Process	*Substrate or enzyme*	*Species*	V_{max} (h^{-1})
		Platymonas striata	0.002
			<0.001
	Glutamine synthetase	*Chlamydomonas reinhardii*	0.132
			0.222
		Chlorella sorokiniana	0.076
			0.126
		Platymonas striata	0.002
			0.079
	Glutamate synthase	*Chlamydomonas reinhardii*	0.426
			1.086
		Chlorella sorokiniana	0.360
			0.499
		Platymonas striata	0.017-
			0.029
Incorporation into protein			
		Chlorella pyrenoidosa	0.193
		Dunaliella bioculata	0.095
		Euglena gracilis	0.065
		Platymonas striata	0.285
Incorporation into RNA			
		Dunaliella bioculata	0.106
		Euglena gracilis	0.059
		Platymonas striata	0.051

[a]*A 33% decrease in cell N (liu and Hellebust, 1974b); stimulation of uptake estimated from Table 5 (Liu and Hellebust, 1974a).*

Conditions	Reference	Assumptions made for calculation
+N	Edge and Ricketts, 1978	
-N	Edge and Ricketts, 1978	
+N	Cullimore and Sims, 1981	1 *M* cell N
-N	Cullimore and Sims, 1981	
+N	Tischner and Hütterman, 1980	*b*
-N	Tischner and Hütterman, 1980	
+N	Edge and Ricketts, 1978	
-N	Edge and Ricketts, 1978	
+N	Cullimore and Sims, 1980	1 *M* cell N
-N	Cullimore and Sims, 1980	0.5 *M* cell N
+N	Tischner and Hütterman, 1980	*b*
-N	Tischner and Hütterman, 1980	
+N	Edge and Ricketts, 1978	
+N	Schmidt, 1961	
+N	Marano *et al.*, 1978	
+N	Edmunds, 1965	
+N	Ricketts, 1977	
+N	Marano *et al.*, 1978	
+N	Edmunds, 1965	
+N	Ricketts, 1977	

[b] *Assuming that extracted protein nitrogen is equal to 16% of total cellular nitrogen as was determined for Ankistrodesmus (Hipkin and Syrett, 1977) and Chlorella (Johnson, 1979).*

active only on nitrogen deprivation. Assuming a 1 *M* total nitrogen content, these uptake rates are equivalent to a nitrogen-specific rate of 0.607 h^{-1} (Table III).

Estimates of urea transport by *Phaeodactylum* can be made from the initial rates (5 min) of accumulation of [^{14}C]urea (Rees and Syrett, 1979), which demonstrate a 20-fold increase in V_{max} with nitrogen limitation. However, these values may underestimate the actual capacity of the transport system if significant metabolism occurred. The nonmetabolized analog (thiourea) was used to study urea transport in *Chlorella* (Syrett and Bekheet, 1977). Rates are low (Table III), but were only measured for nitrogen-sufficient cultures.

A considerable amount of information is available on amino acid uptake rates for marine diatoms. Most of the data are from 10-min incubations and may underestimate actual transport rates if significant feedback inhibition occurs. Excretion of $^{14}CO_2$ from labeled amino acids is only significant for alanine (Stephens and North, 1971). Nitrogen limitation results in up to sixfold increases in amino acid uptake for *Cyclotella,* and tenfold increases in *Nitschia* and *Platymonas* (Table III). Maximum nitrogen-specific rates for nitrogen-limited cells range from 0.105 to 0.801 h^{-1}.

Methylammonium, an ammonium analog that is not readily metabolized by phytoplankton, has been used to study the NH_4^+ transport system. A dramatic increase in the V_{max} of the ammonium transport system resulting from nitrogen deprivation has been demonstrated in *Chlorella* (Pelley and Bannister, 1979) and in *Cyclotella* (Wheeler, 1980b). Estimated nitrogen-specific transport rates reach 1.06 h^{-1} (Table III). More recently, 1-min uptake experiments with $^{15}NH_4^+$ (Goldman and Glibert, 1982) indicate V_{max} values of 0.037-2.11 h^{-1} for nitrogen-limited cultures. Significantly, in at least one case there is a substantial decrease in the V_{max} estimated from 5-min incubations (see Fig. 9, Chapter 7, this volume).

2. Nitrogen Assimilation

Nitrate and nitrite reductase activity are inhibited and/or repressed by NH_4^+, induced by NO_3^-, and stimulated by light and by nitrogen deprivation (see Falkowski, Chapter 23). Under optimal conditions nitrogen-specific rates for nitrate and nitrite reductase reach 0.631 and 1.830 h^{-1}, respectively (Table III). In all cases examined nitrite reductase activity exceeds nitrate reductase activity. Hence, the first reduction is the rate-limiting step.

The two pathways for urea assimilation are under different types of regulation. The enzymes of the amidolyase pathway, like those for nitrate reduction, are repressed by NH_4^+ and induced by the appropriate substrate (urea) Hodson *et al.*, 1975; Syrett and Leftley, 1976). Allophanate lyase activity greatly exceeds urea carboxylase activity in both *Chlorella* and *Chlamydomonas*, hence the first step (carboxylase) is rate-limiting (Hodson *et al.*, 1975). Unfortunately, enzyme activities have not been determined under optimal conditions (derepression by nitrogen starvation and induction by urea) and no estimates of maximum nitrogen-specific rates are available. In contrast to the amidolyase pathway, the urease pathway is constitutive and responds instantaneously to urea. Maximal nitrogen-specific rates for urease can be estimated from the data of Horrigan and McCarthy in which 5-min uptake rates are calculated from release of $^{14}CO_2$ from labeled urea. Rates increase about twofold with nitrogen limitation and maximal levels are 0.408 and 0.080 h^{-1} for *Skeletonema* and *Thalassiosira*, respectively (Table III).

Reduced nitrogen is then assimilated into amino acids by one of two pathways: GDH or GS/GOGAT. Among the studied phytoplankton, the GDH pathway (NADP-specific) only appears to play a primary role in NH_4^+ assimilation in *Chlorella*, where it is induced by high concentrations of NH_4^+ (Kretovich *et al.*, 1970; Shatilov *et al.*, 1978). In contrast to other enzymes of nitrogen assimilation, NADP-GDH activity is decreased rather than increased by nitrogen deprivation, and in the presence of high concentrations of NH_4^+, maximal nitrogen-specific rates are 0.489 h^{-1} (Table III). In all other phytoplankton, GS/GOGAT is the major assimilatory pathway for NH_4^+. Activities of both enzymes increase at least several-fold on nitrogen deprivation (Table III). Maximum nitrogen specific rates are 0.168 and 1.086 h^{-1}, respectively for GS and GOGAT (Table III), and in every case examined GS is the rate-limiting enzyme of this pair.

3. Incorporation into Macromolecules

Most information available on rates of macromolecular synthesis is derived from cell cycle studies utilizing synchronized cultures. Although the results of these experiments show effects of the L/D cycle and the cell cycle that cannot be easily distinguished, the data do provide a means of comparison for nitrogen-specific rates. It should be noted however, that these studies are typically conducted with nitrogen-sufficient cultures, and that the effects of nitrogen deprivation on potential rates of incorporation into macromolecules have not been examined. If the potential rates of macromolecular synthesis increase as a function of nitrogen limitation, as has been observed for the other component processes of nitrogen utilization, then the rates presented in Table III should be considered minimum estimates.

Protein synthesis in nitrate-grown, nitrogen-sufficient cultures occurs almost exclusively in the light. For *Euglena* (Edmunds, 1965) and *Chlorella* (Schmidt, 1961) the rate of synthesis is constant through most of the light period; whereas in *Platymonas* (Fig. 4) (Ricketts, 1977) and *Dunaliella* (Marano *et al.*, 1978) the maximum rate of protein synthesis occurs during the first 3-4 h of the light period. Under these conditions, maximum nitrogen-specific rates for protein synthesis range from 0.065 to 0.285 h^{-1} (Table III).

The metabolism of NH_4^+ is much less dependent on light than that of NO_3^- (Syrett, 1981). Rates of uptake of NH_4^+ in the dark are greatly enhanced by nitrogen limitation, and reach nitrogen-specific rates of 0.154 h^{-1} (Fitzgerald, 1968). We have recently

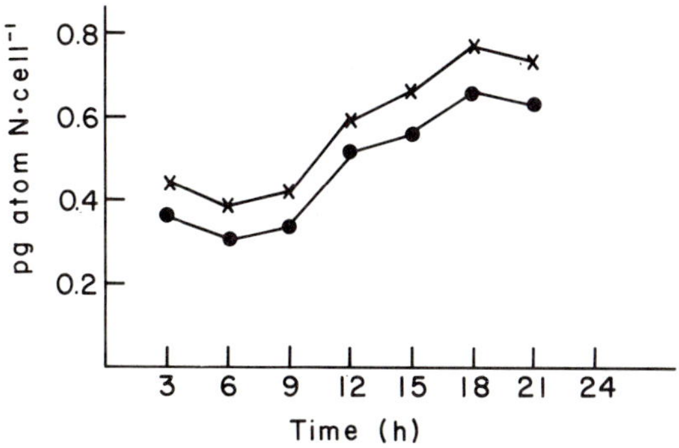

Fig. 4. Increase in total macromolecular nitrogen (×) and protein nitrogen (●) for Platymonas grown on a 14:10 L/D cycle. The light period began at 9.5 h; 60% of total increase in protein occurs during the 9-12-h time interval. (Adapted from Table V of Ricketts, 1977.)

shown that an NH_4^+ pulse supplied to nitrogen-limited cultures of *Amphidinium carterae* and *Hymenomonas carterae* can be completely metabolized and incorporated into protein during the dark portion of the L/D cycle (Wheeler *et al.*, 1982b, 1983). In these experiments, rates of protein synthesis were approximately one order of magnitude faster than mean cellular division rates.

The other form of macromolecular nitrogen that accounts for a significant portion of cellular nitrogen is RNA. The phasing of RNA synthesis appears to vary widely. In *Euglena* it occurs throughout the light period (Edmunds, 1965); in *Platymonas* (Ricketts, 1977) it occurs primarily during 3 h of the dark period; and in *Dunaliella* it occurs during the first 4 h of light. It is noteworthy that maximum RNA synthesis occurs before protein synthesis in *Dunaliella* and after protein synthesis in *Platymonas*. Nitrogen-specific rates of synthesis range from 0.059 to 0.106 h^{-1} (Table III), but have not been examined under conditions of nitrogen deprivation.

These data indicate that macromolecular synthesis is not continuous over the L/D or cell cycle and illustrate the discrete phasing of some metabolic activities. It will be particularly interesting to determine the extent to which nitrogen deprivation increases the potential rate of synthesis of macromolecules, and furthermore how nitrogen deprivation might override normal phasing of nitrogen assimilation and incorporation.

C. Determination of Rate-Limiting Step for Uptake

Nutrient uptake often follows Michaelis-Menten type kinetics (see review by McCarthy (1981) and Goldman and Glibert, this volume, Chapter 7). Although this kinetic model is theoretically justified based on expected kinetic behavior of carrier-mediated membrane transport systems (Dugdale, 1977; Eppley and Coatsworth, 1968), recent laboratory and field studies suggest that kinetics determined from uptake rates measured during typical incubations (1-4 h) do not reflect the kinetic parameters of the membrane transport system for NH_4^+ (Wheeler *et al.*, 1982a; Wheeler and McCarthy, 1982). Since a primary assumption required for analysis of kinetics in terms of the Michaelis-Menten equation is the use of initial uptake rates, it follows that evaluation of nitrogen uptake by phytoplankton as a single process requires the assumption of a single rate-limiting step during the course of the measurements or steady-state conditions in which all rates are equivalent (Dugdale, 1977; Goldman and Glibert, Chapter 7, this volume).

Here we will compare potential (V_{max}) nitrogen-specific rates for membrane transport, assimilation into metabolites, and incorporation into macromolecules. The major regulatory factors for each process will also be reviewed, since these factors control the relevant time scale that must be taken into consideration for each process. It is important to note that the ensuing analysis is based on data derived from a wide variety of experimental organisms, conditions, analytical techniques, and approaches. Furthermore, certain assumptions were required to convert all rates to comparable units. Nonetheless, it is the author's hope that the general conclusion drawn illustrates an intrinsic problem in the application of Michaelis-Menten kinetics to the "net" uptake process.

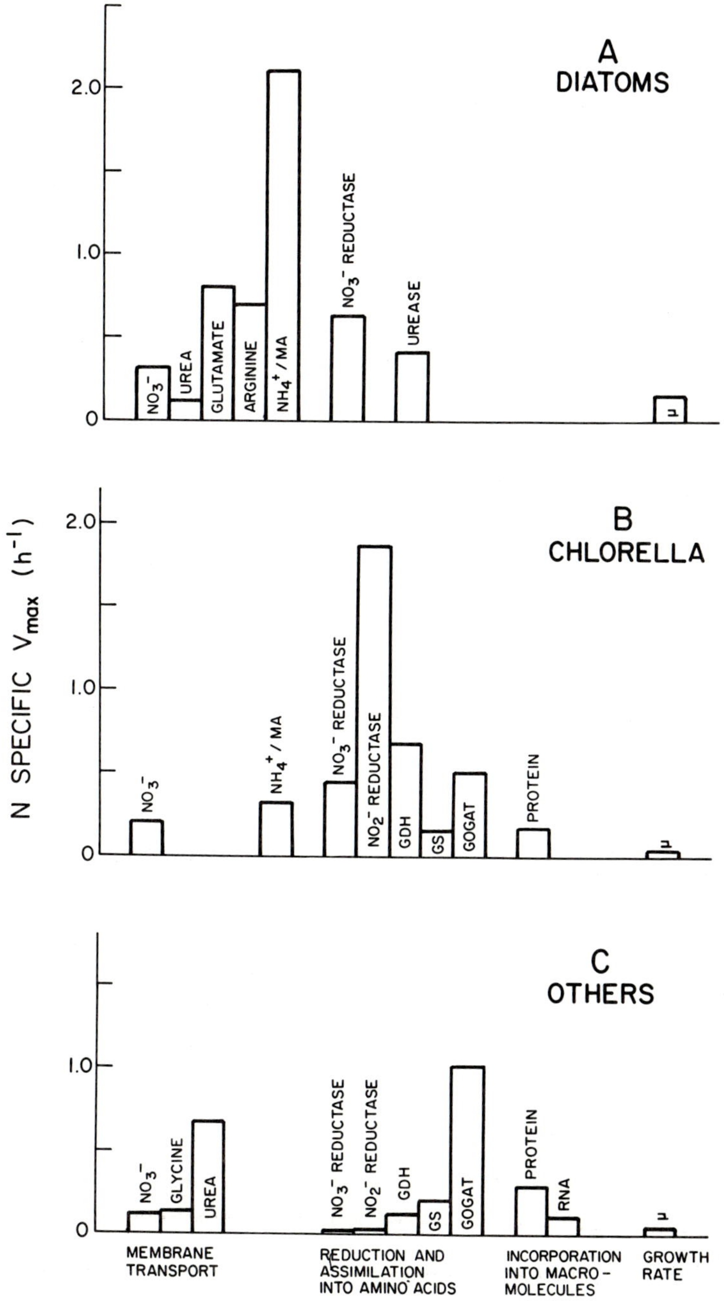

Fig. 5. Comparison of maximum nitrogen-specific rates (from Table III) for transport, assimilation, and incorporation. (A), diatoms; (B) Chlorella; (C) others.

Potential rates of membrane transport (V_{max}) are increased by nitrogen deprivation (Table III) and usually exceed maximum growth rates (Fig. 5). However, transport rates for NH_4^+ can decline within minutes if internal pools accumulate (Wheeler and McCarthy, 1982). Similar studies with bacteria require an incubation period of $\leq$ 30 s to avoid feedback inhibition (Kleiner and Fitzke, 1981). Because uptake rates (and presumably transport rates) can increase by more than an order of magnitude as a result of nitrogen deprivation, but also decline rapidly as a result of feedback inhibition, it is extremely difficult to assess the accuracy of current estimates of transport activity and furthermore to evaluate the actual contribution of the transport process to meeting the nitrogenous requirements of nitrogen-stressed phytoplankton. Nonetheless, it is quite clear that these cells have high-affinity transport systems that are regulated on both a short-term basis by feedback inhibition and on a long-term basis by variations in nitrogen supply.

Rates of NO_3^- and NO_2^- reduction and assimilation of NH_3 into amino acids are all generally stimulated by nitrogen deprivation (Table III). The maximum specific activity of the various enzymes involved vary considerably (Fig. 5). However, for the paired reactions of nitrate/nitrite reductase, urea amidolyase/carboxylase, and GS/GOGAT, the first enzyme is the rate-limiting step (Table III, and references cited therein). In a detailed study of enzyme activities during conditions of varying nitrogen supply for cultures of *Chlorella sorokiniana*, Tischner and Lorenzen (1979) have found that: (1) Highest levels of GS activity are found in cultures grown in or transferred to nitrogen-free medium (Fig. 6A,B,D). (2) On transfer to nitrogen-free or NH_4^+- medium to NO_3^-- medium, assimilation is rate-limited by GS activity (Fig. 6D,E). (3) On transfer from nitrogen-free or NO_3^- medium to NH_4^+ medium, NH_4^+ is assimilated via GDH (Fig. 6C,F). Cullimore and Sims (1981a,b) have also assayed the major suite of nitrogen assimilatory enzymes in *Chlamydomonas reinhardii* and convincingly demonstrate rate limitation of nitrogen assimilation in both NH_4^+-grown and NO_3^--grown cul-

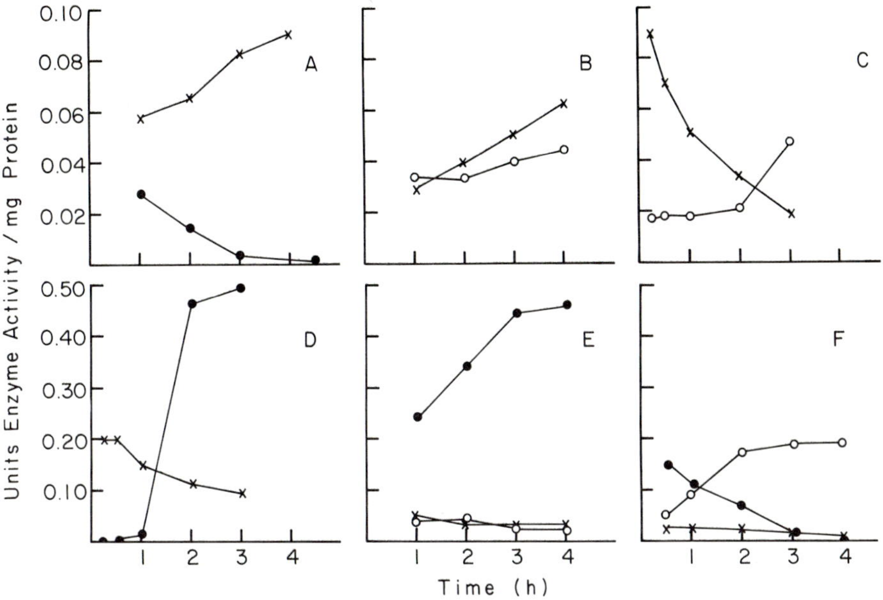

Fig. 6. Comparison of enzyme activities for nitrogen assimilation in Chlorella. At t = o h, the nitrogen supply to the culture was changed as indicated. Key: ×, *glutamine synthetase,* ●, *nitrate reductase, and* o, *glutamate dehydrogenase. (A)* NO_3^-*-grown cells transferred to N-free medium; (B)* NH_4^+*-grown cells transferred to N-free medium; (C) N-starved cells transferred to* NH_4^+ *medium; (D) N-starved cells transferred to* NO_3^- *medium; (E)* NH_4^+*-grown cells transferred to* NO_3^- *medium; (F)* NO_3^-*-grown cells transferred to* NH_4^+ *medium. (Adapted from Tischner and Lorenzen, 1980.)*

tures by GS (Table IV). Moreover, for *Chlamydomonas* the rate of nitrogen uptake measured by accumulation of ^{15}N closely corresponds to the estimates of GS activity (Cullimore and Sims, 1981a,b). It should be noted however, that in some circumstances, e.g., nitrogen deprivation (Fig. 6A) and in the dark (Cullimore and Sims, 1981b), GS activity does exceed nitrate reductase activity.

In summary then, although activities of nitrogen assimilatory enzymes change severalfold within tens of minutes in response to variations in nitrogen supply, GS appears to remain the rate-limiting step for the assimilation of nitrogen into amino acids. Furthermore, the rates of assimilation are generally lower than the estimated potential rates for membrane transport (Fig. 5B,C). Although data on the specific activity of assimilatory enzymes in

TABLE IV. Comparison of the Rates of NO_3^- Assimilation and the Activities of Nitrogen Assimilatory Enzymes in *Chlamydomonas* Grown Phototrophically and Mixotrophically.[a]

	NO_3^- *assimilation or enzyme activity* $(nmol \cdot min^{-1} \cdot mg^{-1}$ *dry wt)*	
	Phototrophic cells	*Mixotrophic cells*
NO_3^- assimilation	4.5	5.0
GS	6.6	6.4
GOGAT (NADH)	15.6	7.2
GOGAT (Fe)	21.7	11.2
GDH (NADPH)	2.0	2.8
NO_3^- reductase	8.2	7.0
NO_2^- reductase	6.8[b]	6.4[b]

[a]*Enzyme levels and rates of assimilation were measured in exponentially dividing cultures (adapted from Cullimore and Sims, 1981b).*

[b]*Enzyme activities determined, except these, were corrected to full substrate saturation.*

diatoms are not presently available, the latter conclusion is supported for this group of phytoplankton by the accumulation of internal NH_4^+ pools (see Section III).

Rate measurements of incorporation of nitrogen into protein by phytoplankton are not abundant. However, the narrow range derived from the cell cycle studies (0.065-0.285 h^{-1}, Table III), is close to the maximum rates reported for GS activity (0.222 h^{-1}, Table III). If potential rates of protein synthesis are increased by nitrogen deprivation, it could be argued that an effective limit on net uptake would be imposed by GS activity (or by GDH activity in NH_4^+-grown *Chlorella*). For diatoms, however, the accumulation of free amino acids suggests that protein synthesis is the rate-limiting step.

Similar conclusions would appear to be valid for the rates of incorporation of nitrogen into nucleic acids. Rates of doubling of protein and RNA are nearly identical in *Dunaliella* and *Euglena*, though RNA synthesis is considerably slower in *Platymonas* (Table III)

V. SUMMARY AND CONCLUSIONS

The potential contribution of a given uptake rate to growth requirements is usually estimated by comparison of nutrient- (element) specific uptake rates to specific growth rates. The available data on nitrogenous composition of phytoplankton cells and nitrogen-specific rates for transport, assimilation, and incorporation support several tentative conclusions concerning the relationships between the component processes of uptake and growth requirements.

For diatoms, maximum membrane transport rates exceed growth rates by more than an order of magnitude (Fig. 5A). The accumulation of internal pools of inorganic nitrogen and of free amino acids indicate that transport $\geq$ assimilation $\geq$ incorporation. Thus, although transport rates can be very high, the contribution of transport per se to net uptake and growth is eventually restricted by the slower rate of assimilation which, in turn, is restricted by the rate of incorporation into macromolecules.

For *Chlorella* and the flagellates, maximum transport rates are not as high as reported for the diatoms but still exceed both growth rates and rates of assimilation via GS (Fig. 5). In contrast to the diatoms however, GS activity appears to limit the rate of incorporation of nitrogen into macromolecules. Thus, in this case we have transport $\geq$ assimilation $\leq$ incorporation. Rates of incorporation into both protein and RNA also can exceed mean daily growth rates (Fig. 5B,C); however, this may result from the fact that synthesis of macromolecules is usually restricted to or concentrated in a specific portion of the cell division cycle for these organisms.

The ultimate goal of predicting growth rates from measured nutrient uptake kinetics has not been attained. The interpretation of kinetic parameters for uptake derived from the application of Michaelis-Menten kinetics to nutrient uptake is ambiguous in light of the occurrence of nonlinear time courses for uptake of nitrogenous compounds in typical incubations (Goldman and Gilbert, this volume, Chapter 7). A very likely explanation for at least part of the observed nonlinearity is a change in the rate-limiting step for uptake during the course of an experimental incubation. If this is the case, a basic assumption required for the application of saturable uptake kinetics to nitrogen uptake is violated. Moreover, a comparison of maximal potential rates for various metabolic steps and transient changes in pool sizes suggests that these shifts in uptake kinetics should be expected particularly under non-steady state conditions.

VI. DIRECTIONS FOR FUTURE RESEARCH

Many gaps are apparent in our knowledge of maximum nitrogen-specific rates for the major processes involved in nitrogen "uptake." As shown in Fig. 5, most of our estimates of membrane transport rates are from experimental work with diatoms, whereas the most complete studies of the assimilatory enzymes and the synthesis of macromolecules are for *Chlorella*, *Chlamydomonas*, and other flagellates. As more complete data become available, the tentative conclusions drawn may require significant modification. However, this review focuses attention on several questions requiring resolution.

1. The validity of a single kinetic description for net uptake needs to be reexamined. Whenever the rate-limiting process changes, the kinetic parameters describing uptake are also likely to change. In particular, we need better estimates of the kinetic parameters for membrane transport. Although kinetic models for uptake were theoretically based on expected kinetics for the transport process, experimental conditions have not been appropriate to yield accurate estimates of the parameters for transport.

2. A coordinated determination of the physiological limits of the rates for each major process involved in nitrogen uptake would

clarify the interpretation of nonlinear time courses for uptake, as well as provide a means of estimating the contribution of each process to meeting cellular nitrogenous requirements for growth. We need to know the kinetics of each process, as well as the range within which these processes are controlled by regulatory mechanisms.

3. The theoretical analysis of uptake and growth kinetics needs to be extended to non-steady-state conditions. Present models require the assumption of *balanced growth*--the simultaneous increase of all cellular components at equal rates (Shuter, 1979; Eppley, 1981). Balanced growth occurs during steady state, unsynchronous growth of cultured phytoplankton, but neither of these conditions is expected to occur in nature. Not only do uptake rates vary widely over short (seconds to minutes) to long (hours) intervals, but instantaneous rates of cell division vary considerably around the mean population growth rate in synchronous or phased popultions (Chisholm *et al.*, 1980; Chisholm, 1981). Eppley (1981) has suggested the use of average values over the 24-h L/D cycle to circumvent these problems in relating uptake and growth rates. Such an approach, however, does not obviate the need for short-term measurements to determine the actual kinetic characteristics of the processes involved in uptake. Non-steady-state models based on the *kinetics* and *regulation* of the key processes involved in uptake should clarify the relationship between uptake and growth by establishing the boundary conditions for uncoupling allowed at each step.

4. The accumulation and utilization of stored or surplus nitrogen by phytoplankton remains to be explored. Preliminary results indicate that specific forms of macromolecular nitrogen serve as nitrogen reserves. Analysis of cellular levels of these forms of nitrogen (e.g., chlorophyll-binding proteins and/or accessory pigments) or, alternatively, the activity of specific proteases could be a useful field assay for the *in situ* determination of the presence or absence of nitrogen limitation.

The kinetics of nitrogen uptake are more complex than previously realized. Current interests in defining relevant spatial and temporal scales in the environment of a marine phytoplankter have pointed up the woeful inadequacies of black-box, steady-state models. Simultaneous advances in plant and cell physiology suggest regulatory mechanisms that form the basis of and set definite limits on phytoplankton responses to variations in nitrogen supply. The extension and application of this knowledge of metabolic regulatory responses should contribute to our progress in understanding the dynamics of the relationship between nitrogenous nutrient availability and phytoplankton productivity.

ACKNOWLEDGMENTS

Support for preparation of this chapter was provided by NSF grants OCE-7826011 (J. J. McCarthy), OCE-8022990 (J. J. McCarthy), OCE-8208873 (P. A. Wheeler), and funds from the School of Oceanography at Oregon State University.

REFERENCES

Barrett, J., and Jeffrey, S. W. (1971). A note on the occurrence of chlorophyllase in marine algae. *J. Exp. Mar. Biol. Ecol.* *7,* 255-262.

Bekheet, I. A., and Syrett, P. J. (1979). The uptake of urea by *Chlorella. New Phytol. 82,* 179-186.

Bhovichitra, M., and Swift, E. (1977). Light and dark uptake of nitrate and ammonium by large oceanic dinoflagellates: *Pyrocystis noctiluca, Pyrocystis fusiformis,* and *Dissodinium lunula. Limnol. Oceanogr. 22,* 73-83.

Boussiba, S., and Richmond, A. E. (1980). C-Phycocyanin as a storage protein in the blue-green alga *Spirulina platensis. Arch. Microbiol. 125,* 143-147.

Caperon, J. (1967). Population growth in microorganisms limited by food supply. *Ecology 48,* 715-722.

Caperon, J. (1969). Time lag in population growth responses of *Isochrysis galbana* to a variable nitrate environment. *Ecology 50,* 188-192.

Chisholm, S. W. (1981). Temporal patterns of cell division in unicellular algae. *In* "Physiological Bases of Phytoplankton Ecology" (T. Platt, ed.), Bull. No. 210, pp. 150-181. Canadian Government Publishing Centre, Hull, Quebec, Canada.

Chisholm, S. W., Morel, F. M. M., and Slocum, W. S. (1980). The phasing and distribution of cell division cycles in marine diatoms. *In* "Primary Productivity in the Sea" (P. G. Falkowski, ed.), pp. 281-299. Plenum, New York.

Conover, S. A. M. (1975). Partitioning of nitrogen and carbon in cultures of the marine diatom *Thalassiorira fluviatilis* supplied with nitrate, ammonium, or urea. *Mar. Biol. (Berlin) 32,* 231-246.

Conway, H. L. (1977). Interactions of inorganic nitrogen in the uptake and assimilation by marine phytoplankton. *Mar. Biol. (Berlin) 39,* 221-232.

Cresswell, R. C., and Syrett, P. J. (1981). Uptake of nitrate by the diatom *Phaeodactylum tricornutum. J. Exp. Bot. 32,* 19-25.

Cullimore, J. V., and Sims, A. P. (1980). An association between photorespiration and protein catabolism: Studies with *Chlamydomonas. Planta 150,* 392-396.

Cullimore, J. V., and Sims, A. P. (1981a). Pathway of ammonia assimilation in illuminated and darkened *Chlamydomonas reinhardii. Phytochemistry 20,* 933-940.

Cullimore, J. V., and Sims, A. P. (1981b). Glutamine synthetase of *Chlamydomonas*: Its role in the control of nitrate assimilation. *Planta 153,* 18-24.

Dagestad, D., Lien, T., and Knutsen, G. (1981). Degradation and compartmentalization of urea in *Chlamydomonas reinhardii. Arch. Microbiol. 129,* 261-264.

DeManche, J. M. (1980). Variations in phytoplankton physiological parameters during transient nitrogen environments. Ph.D. Thesis, Oregon State University, Corvallis.

DeManche, J. M., Curl, H. C., Lundy, D. W., and Donaghay, P. L. (1979). The rapid response of the marine diatom *Skeletonema costatum* to changes in external and internal nutrient concentration. *Mar. Biol. (Berlin) 53,* 323-333.

Dortch, F. Q. (1980). Nitrate and ammonium uptake and assimilation in three marine diatoms. Ph.D. Thesis, University of Washington, Seattle.

Dortch, F. Q. (1982). Effect of growth conditions on accumulation of internal pools of nitrate, ammonium, amino acids, and protein in three marine diatoms. *J. Exp. Mar. Biol. Ecol. 61,* 243-264.

Droop, M. R. (1968). Vitamin B12 and marine ecology. IV. The kinetics of uptake, growth and inhibition in *Monochrysis lutheri. J. Mar. Biol. Assoc. U.K. 48,* 589-733.

Dugdale, R. C. (1977). Modeling. *In* "The Sea: Ideas and Observations on Progress in the Study of the Seas" (E. D. Goldberg, ed.), pp. 789-806. Wiley, New York.

Edge, P. A., and Ricketts, T. R. (1978). Studies on ammonium-assimilating enzymes of *Platymonas striata* Butcher (Prasinophyceae). *Planta 138*, 123-125.

Edmunds, L. N., Jr. (1965). Studies on synchronously dividing cultures of *Euglena gracilis* Klebs (Strain Z). II. Patterns of biosynthesis during the cell cycle. *J. Cell. Comp. Physiol. 66*, 159-182.

Eppley, R. W. (1981). Relations between nutrient assimilation and growth in phytoplankton with a brief review of estimates of growth rate in the ocean. *In* "Physiological Bases of Phytoplankton Ecology" (T. Platt, ed.), Bull. No. 210, pp. 251-263. Canadian Government Publishing Centre, Hull, Quebec, Canada.

Eppley, R. W., and Coatsworth, J. L. (1968). Uptake of nitrate and nitrite by *Ditylum brightwellii*--Kinetics and mechanisms. *J. Phycol. 4*, 151-156.

Eppley, R. W., and Renger, E. H. (1974). Nitrogen assimilation of an oceanic diatom in nitrogen-limited continuous culture. *J. Phycol. 10*, 15-23.

Eppley, R. W., and Rogers, J. N. (1970). Inorganic nitrogen assimilation of *Ditylum brightwellii*, a marine plankton diatom. *J. Phycol. 6*, 344-351.

Eppley, R. W., Holmes, R. W., and Strickland, J. D. H. (1967). Sinking rates of marine phytoplankton measured with a fluorometer. *J. Exp. Mar. Biol. Ecol. 1*, 191-208.

Falkowski, P. G. (1975a). Nitrate uptake in marine phytoplankton (nitrate, chloride) activated adenosine triphosphatase from *Skeletonema costatum* (Bacillariophyceae). *J. Phycol. 11*, 323-326.

Falkowski, P. G. (1975b). Nitrate uptake in marine phytoplankton: Comparisons of half-saturation constants from seven species. *Limnol. Oceanogr. 20*, 412-417.

Fitzgerald, G. P. (1968). Detection of limiting or surplus nitrogen in algae and aquatic weeds. *J. Phycol. 4*, 121-126.

Gantt, E. (1981). Phycobilisomes. *Annu. Rev. Plant Physiol. 32*, 327-347.

Goldman, J. C., and Glibert, P. M. (1982). Comparative rapid ammonium uptake by four marine phytoplankton species. *Limnol. Oceanogr. 27*, 814-827.

Goldman, J. C., and McCarthy, J. J. (1978). Steady-state growth and ammonium uptake of a fast-growing marine diatom. *Limnol. Oceanogr. 23*, 695-730.

Goldman, J. C., and Peavey, D. G. (1979). Steady-state growth and chemical composition of the marine chlorophyte *Dunaliella tertiolecta* in nitrogen-limited continuous culture. *Appl. Environ. Microbiol. 38*, 894-901.

Goldman, J. C., McCarthy, J. J., and Peavey, D. G. (1979). Growth rate influence on the chemical composition of phytoplankton in oceanic waters. *Nature (London) 279*, 210-215.

Goldman, J. C., Taylor, C. D., and Glibert, P. M. (1981). Non-linear time-course uptake of carbon and ammonium by marine phytoplankton. *Mar. Ecol.: Prog. Ser. 6,* 137-148.

Grenney, W. T., Bella, D. A., and Curl, H. C. (1973). A mathematical model of the nutrient dynamics of phytoplankton in a nitrate-limited environment. *Biotechnol. Bioeng. 15,* 331-358.

Gupta, M., and Carr, N. G. (1981). Enzyme activities related to cyanophycin metabolism in heterocysts and vegetative cells of *Anabaena* spp. *J. Gen. Microbiol. 125,* 17-23.

Hellebust, J. A. (1978). Uptake of organic substrates by *Cyclotella cryptica* (Bacillariophyceae): Effects of ions, ionophores, and metabolic and transport inhibitors. *J. Phycol. 14,* 79-83.

Hipkin, C. R., and Syrett, P. J. (1977). Some effects of nitrogen starvation on nitrogen and carbohydrate metabolism in *Ankistrodesmus braunii. Planta 133,* 209-214.

Hodson, R. C., Williams, S. K., II, and Davidson, W. R., Jr. (1975). Metabolic control of urea catabolism in *Chlamydomonas reinhardi* and *Chlorella pyrenoidosa. J. Bacteriol. 121,* 1022-1035.

Horrigan, S. G., and McCarthy, J. J. (1981). Urea uptake by phytoplankton at various stages of nutrient depletion. *J. Plankton Res. 3,* 403-414.

Johnson, C. B. (1979). Activation, synthesis and turnover of nitrate reductase controlled by nitrate and ammonium in *Chlorella vulgaris. Planta 147,* 63-68.

Kirk, D. L., and Kirk, M. M. (1978). Carrier-mediated uptake of arginine and urea by *Chlamydomonas reinhardii. Plant Physiol. 61,* 556-560.

Kirst, G. O. (1977). Ion composition of unicellular marine and fresh-water algae, with special reference to *Platymonas subcordiformis* cultivated in media with different osmotic strengths. *Oecologia 28,* 177-189.

Kleiner, D., and Fitzke, E. (1981). Some properties of a new electrogenic transport system: The ammonium (methylammonium) carrier from *Closteridium pasteurianum. Biochim. Biophys. Acta 641,* 318-147.

Kretovich, W. L., Evstigneeva, Z. G., and Tomova, N. G. (1970). Effects of nitrogen source on glutamate dehydrogenase and alanine dehydrogenase of *Chlorella. Can. J. Bot. 48,* 1179-1183.

Liu, M. S., and Hellebust, J. A. (1974a). Uptake of amino acids by the marine centric diatom *Cyclotella cryptica. Can. J. Microbiol. 20,* 1109-1118.

Liu, M. S., and Hellebust, J. A. (1974b). Utilization of amino acids as nitrogen sources and their effects on nitrate reductase in the marine diatom *Cyclotella cryptica. Can. J. Microbiol. 20,* 1119-1125.

Lui, N. S. T., and Roels, O. A. (1972). Nitrogen metabolism of aquatic organisms. II. The assimilation of nitrate, nitrite, and ammonium by *Biddulphia aurita*. *J. Phycol. 8*, 259-264.
McCarthy, J. J. (1981). The kinetics of nutrient utilization. *In* "Physiological Bases of Phytoplankton Ecology" (T. Platt, ed.), Bull. No. 210, pp. 211-233. Canadian Government Publishing Centre, Hull, Quebec, Canada.
McCarthy, J. J., and Goldman, J. C. (1979). Nitrogenous nutrition of marine phytoplankton in nutrient-depleted waters. *Science 203*, 670-672.
Marano, F., Amancio, S., and Durrand, A. M. (1978). Synchronous growth and synthesis of macromolecules in a naturally wall-less volvocale. *Dunaliella bioculata*. *Protoplasma 95*, 135-144.
Myers, J. E. (1980). On the algae, thoughts about physiology and measurement of efficiency. *In* "Primary Productivity in the Sea" (P. G. Falkowski, ed.), pp. 1-15. Plenum, New York.
North, B. B., and Stephens, G. C. (1971). Uptake and assimilation of amino acids by *Platymonas*. 2. Increased uptake in nitrogen deficient cells. *Biol. Bull. (Woods Hole, Mass.) 140*, 242-254.
North, B. B., and Stephens, G. C. (1972). Amino acid transport in *Nitzschia ovalis* Arnott. *J. Phycol. 8*, 64-68.
Pelley, J. L., and Bannister, T. T. (1979). Methylamine uptake in the green alga *Chlorella pyrenoidosa*. *J. Phycol. 15*, 110-112.
Perry, M. J., Talbot, M. C., and Alberte, R. S. (1981). Photoadaptation in marine phytoplankton: Response of the photosynthetic unit. *Mar. Biol. (Berlin) 62*, 91-101.
Pettersen, R. (1975). Control by ammonium of intercompartmental guanine transport in *Chlorella*. *Z. Pflanzenphysiol. 76*, 213-223.
Raven, J. A. (1980). Nutrient transport in microalgae. *Adv. Microbiol. Physiol. 21*, 48-226.
Raven, J. A., and Beardall, J. (1981). Respiration and photorespiration. *In* "Physiological Bases of Phytoplankton Ecology" (T. Platt, ed.), Bull. No. 210, pp. 55-82. Canadian Government Publishing Centre, Hull, Quebec, Canada.
Rees, T. A. V., and Syrett, P. J. (1979). The uptake of urea by the diatom, *Phaeodactylum*. *New Phytol. 82*, 169-178.
Rhee, G.-Y. (1973). A continuous culture study of phosphate uptake, growth rate, and polyphosphate in *Scenedesmus* sp. *J. Phycol. 9*, 495-506.
Rhee, G.-Y. (1978). Effects of N:P atomic ratios and nitrate limitation on algal growth, cell composition, and nitrate uptake. *Limnol. Oceanogr. 10*, 470-475.
Ricketts, T. R. (1977). Changes in average cell concentrations of various constituents during synchronous division of *Platymonas striata* Butcher (Prasinophyceae). *J. Exp. Bot. 28*, 1278-1288.
Ricketts, T. R., and Edge, P. A. (1977). The effect of nitrogen refeeding on starved cells of *Platymonas striata* Butcher. *Planta 134*, 169-176.

Schmidt, R. R. (1961). Nitrogen and phosphorous metabolism during synchronous growth of *Chlorella pyrenoidosa*. *Exp. Cell Res. 23,* 209-217.

Shatilov, V. R., Sofin, A. V., Kasatkina, T. I., Zabrodina, T. M., Vladimirova, M. G., Semenenka, V. E., and Kretovich, W. L. (1978). Glutamate dehydrogenases of unicellular green algae: Effects of nitrate and ammonium *in vivo*. *Plant Sci. Lett. 11,* 105-114.

Shuter, B. (1978). Size dependence of phosphorous and nitrogen subsistence quotas in unicellular microorganisms. *Limnol. Oceanogr. 23,* 1248-1255.

Shuter, B. (1979). A model of physiological adaptation in unicellular algae. *J. Theor. Biol. 78,* 519-552.

Stephens, G. C., and North, B. B. (1971). Extrusion of carbon accompanying uptake of amino acids by marine phytoplankters. *Limnol. Oceanogr. 16,* 752-757.

Stewart, W. D. P. (1980). Transport and utilization of nitrogen sources by algae. *In* "Microorganisms and Nitrogen Sources" (J. W. Payne, ed.), pp. 577-607. Wiley (Interscience), New York.

Strickland, J. D. H., Holm-Hansen, O., Eppley, R. W., and Linn, R. J. (1969). The use of a deep tank in plankton ecology. I. Studies of the growth and composition of phytoplankton crops at low nutrient levels. *Limnol. Oceanogr. 14,* 23-34.

Syrett, P. J. (1956). The assimilation of ammonia and nitrate by nitrogen-starved cells of *Chlorella vulgaris*. II. The assimilation of large quantities of nitrogen. *Physiol. Plant. 9,* 19-27.

Syrett, P. J. (1981). Nitrogen metabolism of microalgae. *In* "Physiological Bases of Phytoplankton Ecology" (T. Platt, ed.), Bull. No. 210, pp. 182-210. Canadian Government Publishing Centre, Hull, Quebec, Canada.

Syrett, P. J., and Bekheet, I. A. (1977). The uptake of thiourea by *Chlorella*. *New Phytol. 79,* 291-297.

Syrett, P. J., and Leftley, J. W. (1976). Nitrate and urea assimilation by algae. Perspect. Exp. Biol., Proc. Anniv. Meet. Soc. Exp. Biol., 50th, 1974, pp. 221-234.

Tischner, R. (1976). Zur Induktion der Nitrat--und Nitritreductase in Vollsynchronen *Chlorella* Kulturen. *Planta 132,* 285-290.

Tischner, R., and Hütterman, A. (1980). Regulation of glutamine synthetase by light and during synchronous *Chlorella sorokiniana*. *Plant Physiol. 66,* 805-808.

Tischner, R., and Lorenzen, H. (1979). Nitrate uptake and nitrate reduction in synchronous *Chlorella*. *Planta 146,* 287-292.

Tischner, R., and Lorenzen, H. (1980). Changes in the enzyme pattern in synchronous *Chlorella sorokiniana* caused by different nitrogen sources. *Z. Pflanzenphysiol. 100,* 333-341.

Tolbert, N. E. (1980). Photorespiration. *In* "The Biochemistry of Plants" (D. D. Davies, ed.), Vol. 2, pp. 487-523. Academic Press, New York.

Walker, N. A., Smith, F. A., and Beilby, M. J. (1979). Amine uniport at the plasmalemma of charophyte cells. II. Ratio of matter to charge transported and permeability of free base. *J. Membr. Biol. 49,* 283-296.

Wheeler, P. A. (1977). Effect of nitrogen source on *Platymonas* (Chlorophyta) cell composition and amino acid uptake rates. *J. Phycol. 13,* 301-303.

Wheeler, P. A. (1980a). Uptake of methylamine by the marine diatom *Cyclotella cryptica*. *In* "Plant Membrane Transport: Current Conceptual Issues" (R. M. Spanswick, W. J. Lucas, and J. Dainty, eds.), pp. 633-634. Elsevier/North-Holland, Amsterdam.

Wheeler, P. A. (1980b). Use of methylammonium as an ammonium analogue in nitrogen transport and assimilation studies with *Cyclotella cryptica* (Bacillariophyceae). *J. Phycol. 16,* 328-334.

Wheeler, P. A., and Hellebust, J. A. (1981). Uptake and concentration of alkylamines by a marine diatom. Effects of H^+ and K^+ and implications for the transport and accumulation of weak bases. *Plant Physiol. 67,* 367-372.

Wheeler, P. A., and McCarthy, J. J. (1982). Methylammonium uptake by Chesapeake Bay phytoplankton: Evaluation of the use of the ammonium analogue for field uptake measurements. *Limnol. Oceanogr. 27,* 1129-1140.

Wheeler, P. A., and Stephens, G. C. (1977). Metabolic segragation of intracellular free amino acids in *Platymonas* (Chlorophyta). *J. Phycol. 13,* 193-197.

Wheeler, P. A., Glibert, P. M., and McCarthy, J. J. (1982a). Ammonium uptake and incorporation by Chesapeake Bay phytoplankton: Short-term uptake kinetics. *Limnol. Oceanogr. 27,* 1113-1128.

Wheeler, P. A., Olson, R. J., and Chisholm, S. W. (1982b). Dependence of ammonium uptake and assimilation of cell cycle stage. *Eos 63,* 97.

Wheeler, P. A., Olson, R. J., and Chisholm, S. W. (1983). The effects of photocycles and periodic ammonium supply on three marine phytoplankton species. II. NH_4^+ uptake and assimilation. *J. Phycol. 19,* -

Williams, F. M. (1967). A model of cell growth dynamics. *J. Theor. Biol. 15,* 190-207.

Williams, S. K., and Hodson, R. C. (1977). Transport of urea at low concentrations in *Chlamydomonas reinhardi*. *J. Bacteriol. 130,* 266-273.

Yamanaka, G., and Glazer, A. N. (1980). Dynamic aspects of phycobilisome structure. Phycobilisome turnover during nitrogen starvation in *Synechococcus* sp. *Arch. Microbiol. 124,* 39-47.

Chapter 10

NITROGENOUS NUTRITION OF MARINE INVERTEBRATES

MICHAEL R. ROMAN
University of Maryland
Center for Estuarine and Environmental Studies
Cambridge, Maryland

I. INTRODUCTION

The nitrogen requirements of marine invertebrates are poorly understood. Research on the trophic relationships and energy transformations of marine invertebrates usually have focused on calories (e.g., Teal, 1962; Petipa, 1967, 1978) or carbon (e.g., Paffenhöfer, 1976; Parsons et al., 1977b; Parsons and Bawden, 1979). However, when one considers limiting nutrients (Leibig, 1840), nitrogen may better represent one of the specific organic compounds required for invertebrate growth as it is one of the major elements of proteinaceous material. Russell-Hunter (1970) suggested that food sources with C:N ratio (by weight) greater than 17:1 are nitrogen poor and will inhibit heterotrophic growth. As summarized by McCarthy et al. (1975), "Nitrogen provides the simplest perspective from which to view a material balance in

ISBN 0-12-160280-X

plankton ecosystems. Whereas cellular nitrogen is primarily based in structural material, both carbon and phosphorus are largely involved in cellular metabolic activity in addition to their structural roles." However, since most measurements of the biomass and energetic requirements have been expressed in units of dry weight, ash-free dry weight, calories, or carbon, it is presently unclear how the production of marine zooplankton and benthos is affected by the amount of dissolved and particulate organic nitrogen available as food.

II. BIOCHEMICAL COMPOSITION OF FOOD

A. Phytoplankton

The nitrogen content and chemical composition of phytoplankton varies with species (Parsons et al., 1961; Sick, 1976), phase of growth (Fisher and Schwarzenbach, 1978; Scott, 1980), and availability of nutrients (Fogg, 1959; Caperon and Meyer, 1972; Eppley and Renger, 1974; DeManche et al., 1979). The nitrogen content as a percent of dry weight of various phytoplankton species ranged from 4.5 to 9.1% (Parsons et al., 1961) and 4.3% N for a natural population of phytoplankton from St. Margaret's Bay, Nova Scotia (Mayzaud and Martin, 1975). Carbon:nitrogen ratios (by weight) of laboratory phytoplankton can range from 4.4 to 12.0 (Parsons et al., 1961; Gallagher and Mann, 1981). In general, the nitrogen content of phytoplankton increases with nutrient availability and with growth rate (Caperon and Meyer, 1972; Eppley and Renger, 1974; Goldman et al., 1979; Gallagher and Mann, 1981; Dortch, 1982).

The amount of protein in phytoplankton can range from 10 to 70% of dry weight depending on both species composition and culture conditions (Fogg, 1959; Parsons et al., 1961; Morris et al., 1974; Mayzaud and Martin, 1975; Flaak and Epifanio, 1978; Scott, 1980; Dortch, 1982). The ratio of protein:carbohydrate:lipid for various phytoplankton species cultured under the same conditions (ASP media, continuous light) was 4:3:1 (Parsons et al., 1961). In contrast, Scott (1980) found that by controlling growth rates with light intensity, the protein:carbohydrate:lipid content of the chlorophycean alga *Brachiomonas submarina* ranged from 2:5:1 to 0.8:0.5:1. The protein content as a percent of dry weight of *Thalassiosira pseudonana* was positively correlated with relative growth rate, ranging from 12.9% in stationary phase to 69.3% in exponential phase (Flaak and Epifanio, 1978). Dortch (1982) found greater concentration of protein (838 m*M*·liter cell volume^{-1}) in nitrogen-sufficient *Skeletonema costatum* as compared to nitrogen-deficient cells (180-207 m*M*·liter cell volume^{-1}) and nitrogen-starved cells (176 m*M*·liter cell volume^{-1}). Cell protein increased in nitrogen-limited *S. costatum* with addition of nitrate or ammonia.

The composition of phytoplankton amino acids can vary among phytoplankton species as well as the same species under different culture conditions. Similar amino acid compositions were present in *Brachiomonas submarina, Chlorella ellipsoida, Cricosphaera elongata, Cyclotella* sp., *Monochrysis lutheri, Phaeodactylum tricornutum, Prorocentrum triestinum,* and *Skeletonema costatum* (Cowey and Corner, 1966; Okaichi, 1974). However, Chuecas and Riley (1969) found that diatoms contained more serine and 2-aminoisobutyric acid than other classes of marine phytoplankton. Seasonal differences in the quantity of amino acids occur in plankton. For example, Cowey and Corner (1961) found a minimum of 2.0 μg amino acids per 100 mg dry wt in natural phytoplankton collected in Plymouth Sound in August/September and a maximum of 8.0 μg amino acids per 100 mg dry wt in phytoplankton collected in June/July. The amount and composition of amino acids in phytoplankton species is influenced by growth rate and nutrient availability. Picard (1976) demonstrated that the diversity of free amino acids in *Chaetoceros* sp. increased with algal turnover rate, 4, 5, 8, and 13 free amino acids present with turnover rates of

30, 60, 87, and 116% day^{-1}, respectively. The pool size of free amino acids in the diatom *Bellerochea yucatanensis* was greater when the alga was grown on NH_4^+ compared to NO_3^-. When nitrogen-limited cultures were enriched with NH_4^+, the greatest increases in free amino acids occurred in aspartate > glutamine > glutamate > glycine > serine > alanine (Döhler and Roßlenbroich, 1982). Growing the diatom *Skeletonema costatum* on different levels of nitrogen, Dortch (1982) found that free amino acids were 187 m*M*·liter cell volume^{-1} in nitrogen-sufficient cells, 1.1-7.3 m*M*·liter cell volume^{-1} in nitrogen-deficient cells, and 1.3 m*M*·liter cell volume^{-1} in nitrogen-starved cells.

B. Macrophytes

Macrophytes contain large amounts of cellulose and lignin which are relatively nondigestible to many marine invertebrates (Teal, 1962; Boyd, 1968). As a consequence, despite their often high biomass and productivity, macrophytes are not appreciably grazed by invertebrates (Odum and de la Cruz, 1967; Hargrave, 1970; Vicente et al., 1980). As a percent of total organic matter, macrophytes are nitrogen-poor compared to phytoplankton. This lower percent nitrogen in macrophytes is due to the more abundant celluloses and lignins as compared to phytoplankton. Examples of percent nitrogen of dry weight of marine macrophytes are: *Distichlis spicata*, 1.5% (Udell et al., 1969); *Juncus effusus*, 1.1% (Boyd, 1971); *Spartina alterniflora*, 2.2% (Udell et al., 1969); *Thalassia testudinum*, 2.1% (Burkholder et al., 1959); *Ulva lactuca*, 3.3% (Udell et al., 1969); and *Zostera marina*, 2.3% (Udell et al., 1969). In contrast to phytoplankton in which most of the nitrogen is in proteins (84%; Mayzaud and Martin, 1975), over 30% of the nitrogen in macrophytes can be in nonprotein compounds such as amino sugars, phenol proteins, and mucopolysaccarides (Odum et al., 1979). Direct estimates of protein (% of dry weight) for marine angiosperms are 6.8% in *Thalassia testudinum* (Rublee and Roman, 1982) and 10.6% in *Zostera marina* (Harrison and Mann, 1975).

C. Detritus

Detritus includes all forms of organic matter (dissolved and particulate) lost by nonpredatory means from any trophic level (egestion, excretion, secretion), as well as inputs from sources external to the ecosystem (Wetzel et al., 1972). Thus, a detrital food web may be defined as any route by which chemical energy contained within detrital material becomes available to biota. Dissolved organic matter is excreted by plants (Hellebust, 1965; Penhale and Smith, 1977) and animals (Johannes et al., 1969) and leaches from decaying macrophytes (Brylinsky, 1977; Robertson et al., 1982). The pool of dissolved organic matter is usually an order of magnitude greater than particulate matter (Wangerski, 1978). Detritus is the major component of the particulate matter in both coastal waters and in the open ocean (Parsons, 1963; Riley, 1970). Particles of different size, age, and biochemical composition derived from different sources constitute the heterogeneous detrital pool. The dominant types of particulate detritus will vary in different habitats. Suspension-feeding zooplankton and benthos produce large amounts of feces and pseudofeces (Verwey, 1952; Haven and Morales-Alamo, 1966; Frankenberg et al., 1967; Honjo and Roman, 1978; Paffenhöfer and Knowles, 1979), which form a major source of detritus in the sediment, and can be resuspended from the sediment into the water column by tidal currents (Roman and Tenore, 1978). The production of mucus by corals is another form of detritus and can comprise greater than 50% of the particulate matter over reefs (Marshall, 1965; Coles and Strathman, 1973; Bensen and Muscatine, 1974; Gerber and Marshall, 1974; Richman et al., 1975). Decaying macrophytes, both seagrasses and seaweeds, are the major source of detrital material in neritic environments (Mann, 1972). Highly productive salt marshes and submerged seagrasses result in large inputs of detrital material to coastal waters (Teal, 1962; Zieman, 1975).

The nitrogen content and nutritional value of detritus for marine invertebrates can vary because of detrital source, age, and particle size (Tenore et al., 1982). Different types of detritus, because of their origins, have different biochemical compositions. For example, the amino acids found in copepod fecal pellets reflect those present in their food (Cowey and Corner, 1966). The nutritional value of coral mucus detritus can vary in relation to the species of coral. Coles and Strathman (1973) found the C:N ratio (by weight) of mucus produced by *Porites* sp. to be 5.8, *Fungia* sp. 7.5, and *Acropora* sp. to be 8.5. A variety of macrophytes contribute to a heterogeneous detrital pool. Depending on the species and season of production, the nitrogen and protein content of macrophyte detritus will differ. Amino acids in *Distichlis spicata, Juncus roemerianus, Scirpus americanus,* and *Spartina cynosuroides* decreased by over 50% with death of the plants but increased again to levels found in living plants during *in situ* decomposition (de la Cruz and Poe, 1975).

Changes in the nitrogen content and nutritional value of detritus can occur as a result of microbial activity. Detritus is colonized by microfauna, microflora, and meiofauna, all of which interact with each other and the detritus substrate. As microorganisms colonize detritus, they convert structural carbon of the substrate into their body protein, thus adding to the nitrogen content of the detritus. Much of this nitrogen increase may be from mucopolysaccarides excreted by bacteria (Hobbie and Lee, 1980). Assimilation of ambient inorganic nitrogen by bacteria and fungi (Fell et al., 1976), as well as fixation of atmospheric nitrogen by detrital bacteria (Gotto and Taylor, 1976; Capone et al., 1979) are nitrogen inputs to the detrital system. Decreases in the C:N ratio (by weight) with age can occur on organic aggregates

(Riley, 1963), coral mucus (Coles and Strathman, 1973; Richman et al., 1975), abandoned larvacean houses (Alldredge, 1972), *Thalassia testudinum* (Zieman, 1975), *Zostera marina* (Harrison and Mann, 1975), *Juncus roemerianus* and *Rhizophora mangle* (Fell et al., 1976), and *Fucus vesiculosus* (Roman, 1977). Much of the nitrogen increase in decomposing detritus can be from nonprotein nitrogen, not microbial protein (Odum et al., 1979; Rice and Tenore, 1981). Possible non-protein nitrogen compounds include: amino sugars (chitin), phenol protein, N-containing humic acids, complexes of inorganic clays and amino groups, and mucopolysaccharides produced by bacteria. Because of the abundance of these "heteropolycondensates" (Degens, 1968), detrital protein should not be measured indirectly (using the N:protein ratio of 6.25).

D. Animals

The nitrogen content and composition of marine invertebrates is more consistent than that found in phytoplankton, macrophytes, or detritus. Natural populations of zooplankton in Buzzards Bay, Massachusetts, measured monthly for 1 year, had a range of C:N (by weight) of 3.6-4.7 and were 6.0-10.0% nitrogen of dry weight (Roman, 1980). The yearly averages of percent nitrogen of dry weight in zooplankton groups collected in the Sargasso Sea off Bermuda were: copepods, 9.62%; euphausids-mysids, 9.96%; chaetognaths, 7.84%; polychaetes, 8.92%; siphonophores, 2.97%; hydromedusae, 2.89%; and pteropods, 3.75% (Beers, 1966). The biochemical compositions for a variety of marine invertebrates are presented in Table I. Natural populations of the copepod *Calanus finmarchicus* were 9.6-12.3% N and contained amino acids that were similar to those found in the ambient phytoplankton, the total of which ranged from 20 to 35% of dry weight (Cowey and Corner, 1966). The biochemical composition of marine invertebrates can change with the age of the organism. For example, from spat to adult the oyster, *Ostrea edulis* increased from 5 to 30% carbohydrate of dry weight, decreased from 24 to 5% lipid of dry weight,

TABLE I. Biochemical Composition of Representative Invertebrates[a,b]

Phyla Species	*Protein*	*Carbo-hydrate*	*Total lipid*	*Ash*	*Non-protein nitrogen*	*Reference*
Porifera						
Haliclona permollis	34	0.3	6.8	--	2.5	Giese (1966)
Plocamia karykina	17.7	0.9	13.3	--	1.0	Giese (1966)
Rhabdodermella nottingi	6.3	0.2	2.7	--	0.8	Giese (1966)
Coelenterata						
Anthopleura xanthogrammica	51.0 ± 6.0	1.0 ± 0.47	10.1 ± 1.7	--	5.4 ± 0.4	Giese (1966)
Metridium sp.	52.5	1.0	8.0	--	5.1	Giese (1966)
Aglantha digitale	56.5	0.8	3.0	39.2	--	Ikeda (1972)
Chaetognatha						
Sagitta elegans	84.0	0.7	6.7	8.0	--	Ikeda (1972)
Sagitta elegans	54.21	1.51	7.81	6.73	6.40	Mayzaud and Martin (1975)
S. hispida	52.9	3.5	17.0	9.3	5.2	Reeve et al. (1970)
Polychaeta						
Tomopteris septentrionalis	73.8	1.7	8.1	15.1	--	Ikeda (1972)
Phoronopsis viridis	29.4	--	4.9	--	1.9	Giese (1966)
Sabella starki magnificans	31.7	1.5	9.2	--	5.7	Giese (1966)
Glycera rugosa	56.2	1.4	8.3	--	3.3	Giese (1966)
Sipunculoidea						
Phascolosoma agassizii	36.8	0.8	4.1	--	3.9	Towle (1961)
Pteropoda						
Clione limacina	52.7	0.5	17.5	28.2	--	Ikeda (1972)

Bivalvia						
Crassostrea virginica	42.4	25.7	7.98	--	--	Pease (1932)
Teredo pedicelata	13.5	30.0	4.6		--	Greenfield (1953)
Mytilus edulis						
Mytilus californianus						
Foot	66.7	0.6	7.6		3.8	Giese (1966)
Mantle	58.7	8.5	4.5		3.7	Giese (1966)
Viscera	54.2	0.7	5.9		3.8	Giese (1966)
Testis	57.1	2.1	8.6		5.2	Giese (1966)
Ovary	50.7	12.9	7.1		3.8	Giese (1966)
Gastropods						
Acanthina spirata	52.8	1.1	4.8		--	Giese (1966)
Haliotis rufescens						
Foot	46.3 ± 7.1	10.8 ± 3.1	4.1 ± 0.7	--	7.7 ± 1.1	Giese (1966)
Radular muscle	64.1	0.9	6.5		1.4	Giese (1966)
Gut	37.5	0.5	4.0		3.1	Giese (1966)
Digestive gland	50.1	2.3	6.9		3.0	Giese (1966)
Ovary	35.9	3.2 ± 2.4	29.3 ± 2.5		2.0	Giese (1966)
Testis	49.8	0.2	7.5		4.6	Giese (1966)
Copepoda						
Calanus cristatus (V)	50.2	0.6	31.7	14.0		Ikeda (1972)
C. plumchrus (V)	53.5	0.9	38.0	5.4		Ikeda (1972)
C. glacialis	48.6	1.0	40.9	7.4		Ikeda (1972)
Eucalanus bungii bungii	52.5	1.1	25.4	18.3		Ikeda (1972)
C. finmarchicus	44.51	1.69	44.42	3.32	1.93	Mayzaud and Martin (1975)
Euphausiacea						
Thysanoessa raschii	78.6	0.5	7.2	10.9		Ikeda (1972)

Table I (Continued)

Phyla Species	Protein	Carbo-hydrate	Total lipid	Ash	Non-protein nitrogen	Reference
Amphipoda						
Euthemisto libellula	49.4	3.1	21.6	21.1		Ikeda (1972)
Decapoda						
Penaeus japonicus	94.0	--	1.27			Konsu et al. (1959-1959)
Panulirus japonicus	95.3	--	1.46			Konsu et al. (1958-1959)
Cancer pagurus	85.2	0.8	11.3			Vonk (1960)
Cancer antennarius						
Muscle	51.7	0.7	3.8		3.6	Giese (1966)
Digestive gland	27.5	0.4	8.8		3.8	Giese (1966)
Testis	--	--	5.9		--	Giese (1966)
Ovary	59.3	0.5	26.0		1.9	Giese (1966)
Asteroidea						
Pisaster ochraceus						
Body wall	14.1	0.11	3.24		--	Giese (1966)
Body fluid	0.02	0.01	0.02		--	Giese (1966)
Gut	43.7	1.5	15.34		--	Giese (1966)
Gonad	37.8	0.37	37.8		--	Giese (1966)
Caecum	28.3	1.9	34.5		--	Giese (1966)

Echinoidea					
Strongylocentrotus purpuratus					
Body wall	6.23	0.09	2.37	--	Giese (1966)
Body fluid	0.02	0.003	0.03	--	Giese (1966)
Gut	41.72	3.1	18.4	--	Giese (1966)
Gonad	41.03	4.5	19.7	--	Giese (1966)
Lantern	9.33	0.22	1.20	--	Giese (1966)
Tunicata					
Eudistoma ritteri	24.6	1.2	4.5	1.0	Giese (1966)
Amaroucium californicum	24.0	1.3	6.3	1.0	Giese (1966)
Ciona intestinalis					
Body wall	47.2	2.7	11.3	--	Giese (1966)
Gut	40.5	11.0	6.5	--	Giese (1966)
Ovary	53.2	6.2	17.3	--	Giese (1966)
Testis	54.8	7.3	13.3	--	Giese (1966)
Tunic	16.9	2.2	0.9	--	Giese (1966)

[a] *After Conover, 1978; reproduced by permission of Wiley Interscience Publishers.*
[b] *All values in percentage dry weight.*

but only decreased from 64 to 58% in protein content (Holland and Hannant, 1974). The protein content of marine invertebrates may be more conservative than lipid and carbohydrate, which are used by zooplankton and benthos, respectively, as energy reserves. Using different algal diets for the clam *Saxidomus giganteus*, Walne (1973) found that N content as a percent of dry weight remained constant (12-14 μg N·mg dry wt^{-1}) over various diets. In contrast, the carbohydrate content of the clams ranged from 7 to 33 μg glucose·mg dry wt^{-1}. Gallagher and Mann (1981) found similar results for the clam *Tapes japonica*. Using algae of different C:N ratios (by weight) they found that carbohydrate (% of dry weight) in the clam ranged from 4.3 to 9.2% when nitrogen in the clam (% of dry weight) varied from 8.9 to 10.0%.

III. NITROGEN REQUIREMENTS OF MARINE INVERTEBRATES

A. Indirect Evidence--Field Studies

Particulate matter in the natural environment is a heterogeneous mixture of bacteria, phytoplankton, microzooplankton, detritus, and inorganic particles in various proportions, sizes, and biochemical compositions. Because of difficulties in separating these components, it is difficult to assess the nitrogenous nutrition and limitation of specific invertebrates *in situ*. A systems approach was taken by Sutcliffe (1972) who related freshwater input in St. Lawrence River to nitrogen input in St. Margaret's Bay, chlorophyll biomass and the catch of lobster, halibut, and clams in the Gulf of St. Lawrence. This enhancement of food web productivity by nitrogen input has also been demonstrated in controlled mesocosms. Parsons et al. (1977b) compared plankton production in controlled water columns (CEPEX) with and without NO_3^- additions and found greater production of copepods (17%) and ctenophores (85%) in the enclosure that received NO_3^- additions (63 g) as compared to unaltered controls.

Spatial and temporal variations of benthic biomass in neritic waters appear closely linked to suspended living biomass (phytoplankton-N) rather than total particulate nitrogen (Parsons et al., 1977a). Checkley (1980b) reached a similar conclusion for factors controlling egg production of the copepod *Paracalanus parvus* off California. His results indicate that phytoplankton nitrogen accounted for more of the variability in the egg production of *P. parvus* than total particulate nitrogen. The importance of phytoplankton-N for copepod production was also demonstrated by Chervin et al. (1981). They found that copepod communities in the Hudson River plume had higher N:C assimilation ratios and growth efficiencies with increased phytoplankton biomass, suggesting that phytoplankton are more important than detritus as a source of protein.

The inorganic matter and detritus present in seston may not be as nitrogen-rich as phytoplankton (Tenore et al., 1982), however, it should not be assumed that phytoplankton are the major source of nitrogen for all marine invertebrates. After estimating zooplankton biomass over a reef at Bermuda, Johannes et al. (1970) found that there were inadequate zooplankton to support the observed coral growth. While symbiotic zooxanthellae are important in supplying the energy requirements of the corals, the authors hypothesize that corals feed on zooplankton for essential amino acids. D'Elia and Webb (1977) found that the influx of dissolved inorganic nitrogen to the coral *Pocillopora* sp. roughly balanced the efflux of dissolved organic nitrogen. Therefore, for the coral to sustain the observed growth rates, particulate nitrogen must be ingested. Gerber and Marshall (1974) have shown that 70-90% of the food ingested by coral reef zooplankton is nitrogen-rich coral mucus detritus. Over most of the year in Buzzards Bay, Massachusetts, the approximate N assimilated by phytoplankton was lower than the estimated daily ration of the zooplankton community, suggesting that detritus is the dominant particulate nitrogen pool consumed (Roman, 1981).

The nutritional role of bacterial N for detritivores may be important. Fecal pellets produced by the deposit-feeding snail *Hydrobia ulvae* are initially low in nitrogen (Newell, 1965). However, as the fecal pellets age and microbial colonization on the pellets increases, the C:N (by weight) decreases. When the snail ingests the fecal pellets, the nitrogen content of the fecal pellets decreased to the original levels, probably because the microbes were removed by the invertebrates. Levinton and Lopez (1977) suggested that the resource renewal rate, the microbial colonization, and the breakdown of fecal pellets may limit the population size of *Hydrobia* sp. Their assumption was supported by a model that successfully predicted field *Hydrobia* densities from pelletization rates and fecal pellet decomposition rates.

B. Direct Evidence--Laboratory Studies

1. *Bulk Nitrogen*

In several controlled laboratory experiments, the quantity of available food in nitrogen units has been related to the ingestion, assimilation, growth, and reproduction of marine invertebrates. The benthic polychaete *Capitella capitata* was cultured through multiple generations on varied detritus sources derived from vascular plants and seaweed plants (Tenore, 1977a, 1981). The best correlate ($r = 0.80$) of food quality to the saturation density of worms in $0.1 = m^2$ trays after 3 months was the amount of nitrogen supplied rather than the carbon or caloric ration (Fig. 1). Comparing the net incorporation of detritus by *C. capitata,* Tenore (1977b) found maximum incorporation of *Gracilaria* detritus (91 μg dry wt detritus·mg dry wt worm^{-1}·day^{-1}) after 14 days of decomposition, whereas *Zostera* detritus incorporation equaled this level after 120 days of decomposition. Tenore attributed this difference in incorporation to the initial nitrogen content of the *Gracilaria* (>5% N dry wt^{-1}) and *Zostera* (<2% N dry wt^{-1}). *Zostera* detritus nitrogen increased with decomposition so that eventually it's nitrogen content equaled that of

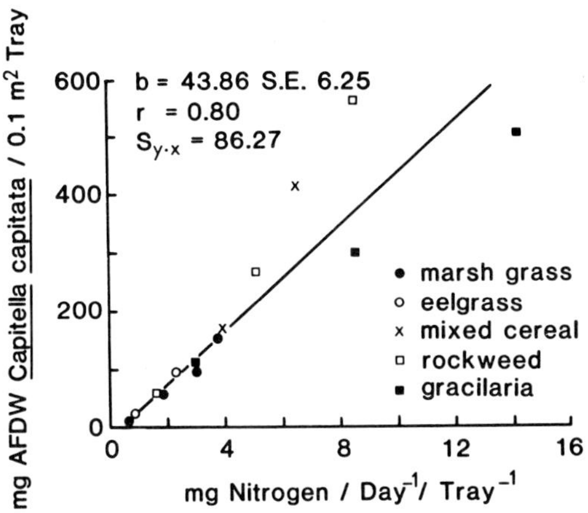

Fig. 1. Biomass of *Capitella capitata* at three food levels of five different types of detritus based on nitrogen. Each value is based on eight replicates, four each at 10° and 20°C. (After Tenore, 1977a; reproduced by permission of the American Society of Limnology and Oceanography.)

Gracillaria detritus. The nitrogen content is not the only regulator of nutritional value of detritus to consumers. The net incorporation of *Spartina* sp. detritus by *C. capitata* did not increase with exogenous organic nitrogen supplements. However, net incorporation by the worm increased exponentially with linear addition of organic N to *Gracilaria* detritus (Tenore et al., 1979). The complex carbohydrates in *Spartina* detritus may make it more resistant to digestion, thus requiring microbial degradation of the structural components to increase caloric "availability" (Tenore, 1981). The nitrogen content of detrital complexes may be low compared to phytoplankton, but is used efficiently by detritivores. The shrimp *Palaemonetes pugio,* when ingesting its own fecal pellets, assimilated 82% of the nitrogen but only 56% of the carbon in the fecal pellets (Johannes and Satomi, 1966). Several phytophagous insects that feed on plants which are low in nitrogen can utilize plant phenols, thereby reducing their requirements for amino nitrogen (Bernays and Woodhead, 1982). Marine invertebrates which feed on nitrogen-poor detritus may have similar ability to utilize phenolic compounds.

Zooplankton growth can also be limited by available nitrogen. Butler et al. (1969) calculated that the gross growth efficiency of *Calanus finmarchicus* in nitrogen units averaged 33.1%, which was higher than previously reported growth efficiencies calculated from units of dry weight or calories (Conover, 1964). The calanoid copepod *Calanus helgolandicus* exhibited asymptotic nitrogen ingestion (Fig. 2) in relation to available particulate nitrogen (Corner et al., 1972). The copepods had a maximum daily ration of 47.5% of their body nitrogen and assimilated 53.8-67.5% of the nitrogen ingested. The complete nitrogen budget was estimated for the harpacticoid copepod *Tigriopus brevicornis* (Harris, 1973). Adult copepods assimilated 75.4% of the nitrogen ingested, excreting 30% of body N daily and losing 5% of body N in molts. The harpacticoids had a gross growth efficiency of 13.0% and a gross efficiency of egg production of 22.1%. Over the entire lifetime of *T. brevicornis,* 72.9% of assimilated nitrogen was used for metabolism; 3.9% for growth; 0.4% in molts, and 26.6% for egg production. A similar nitrogen budget was estimated for the mussel *Geukensia demessa* (Jordon and Valiela, 1982). The mussels assimilated 50% of the particulate nitrogen filtered of which 55% was excreted as NH_4, 4% was used for byssal threads, 20% was used for growth, and 21% went into gamete production. The mussels had a gross growth efficiency of 13.3% and a 10.3% gross efficiency of egg production.

The nitrogen content of phytoplankton can affect zooplankton production. Egg production of the marine copepod *Paracalanus parvus* was significantly related to the available organic nitrogen of four algal species (Checkley, 1980a). Gross efficiency of egg production of *P. parvus* was constant when derived from nitrogen units (37%), but was variable when calculated from carbon units (15-41%). Comparing the growth efficiency of *Artemia salina* on five phytoplankton species, Sick (1976) found that ingestion, growth, and growth efficiency of the brine shrimp were not a simple function of the nitrogen content of algae. The greatest growth efficiency occurred on the algal species with the highest C:N ratios (growth efficiencies 90.4%, 84.3%; C:N of phytoplankton (by weight) 8.8, 7.7). Similarly, Cahoon (1981) found that egg production of the copepod *Acartia tonsa* was not significantly correlated to the nitrogen content of six phytoplankton species offered as food. Cahoon suggested that there are significant food quality effects on the reproduction of *A. tonsa*. Growth of the bivalve *Tapes japonica* was positively correlated with the total quantity of available nitrogen rather than carbon (ANOVA: $P < 0.01$). However, algal diets with C:N (by weight) ratios of 8.4:10.5 resulted in better clam growth than algal cultures with higher or lower C:N ratios (Gallagher and Mann, 1981). This superior growth on algae with intermediate C:N ratios may be the result of a greater proportion of algal carbohydrates. Marine bivalves can rapidly mobilize and preferentially utilize carbohydrates as respiratory substrates (Walne, 1973; Holland and Hannant, 1974).

2. Protein

Proteins are the most abundant organic molecules within invertebrate cells, often comprising 50% or more of the dry weight. Growth rate of the abalone *Haliotus discus* was positively correlated with protein content of the diet (Ogino and Kato, 1964). Juvenile prawn *Macrobrachium rosenbergii* exhibited a similar pattern of increased growth with protein content of food up to a maximum level (40% of dry weight), after which growth decreased with increased protein content of food (Millikin et al., 1980). Sick (1976) found that growth and growth efficiency of the brine shrimp *Artemia salina* was not related to the protein content of the algal diet (15-28% protein·dry wt^{-1}). The oyster *Crassostrea virginica* grew most rapidly when fed diets richer in carbohydrates than proteins (Flaak and Epifanio, 1978). The investigators controlled the biochemical composition of the diatom *Thalassiosira pseudonana*

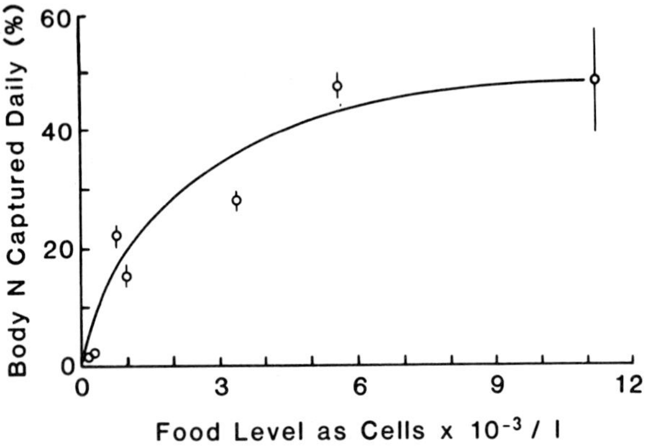

Fig. 2. Daily ration as % body nitrogen ingested by *Calanus helgolandicus* feeding on *Biddulphia sinensis*. Quantities per day estimated from measurements over 4 hours. Vertical lines represent standard error of determination. (After Corner et al., 1972; reproduced by permission of Cambridge Univ. Press.)

by varying photoperiod and light spectra. *Thalassiosira pseudonana* with a generation time of 1.5 days were 13% protein of dry weight (C:N = 7.6) as compared to algae with a generation time of 0.63 days, which were 69% protein of dry weight (C:N = 2.6). The percentage change in dry weight of the oyster ranged from 22 to 61% over the 10-week period, with the algal diets with the highest C:N ratios promoting the greatest oyster tissue growth. The rotifer *Brachionus plicatilus* was fed the chlorophycean alga *Brachiomonas submarina* cultured at different growth rates (Scott, 1980). The protein and lipid content of the algae increased with increased growth rates, whereas the carbohydrate content decreased over higher algal growth rates (Fig. 3). The conversion efficiency of the rotifer (growth ingestion^{-1}) was a parabolic function of the algal growth rates. The maximum conversion efficiency (35%) coincided at the point at which protein, carbohydrate, and lipid content of the algae were in the most equal proportions.

The nutritional effects of the relative compositions of carbohydrate, lipid, and protein have been studied for commercially

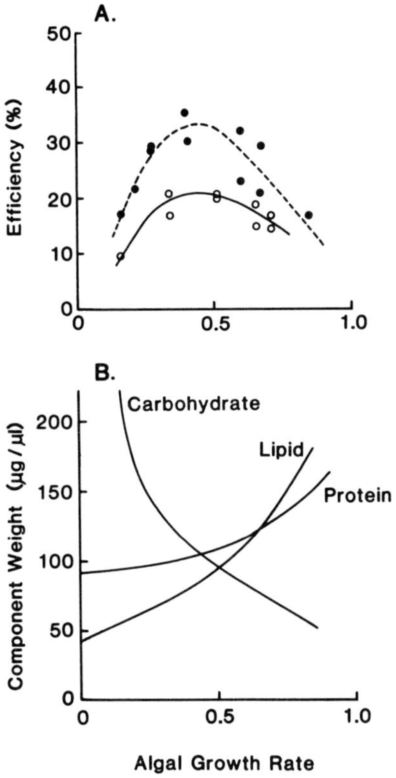

Fig. 3. (A) Rotifer *Brachionus plicatilis* growth/ingestion efficiencies as a function of algal specific growth rate. Key: ●, slow-adapted strain; O, fast-adapted strain. (B) Weights of alga *Brachiomonas submarina* carbohydrate, lipid, and protein as a function of specific growth rate. (After Scott, 1980; reproduced by permission of Cambridge Univ. Press.)

valuable invertebrate species. In aquaculture operations, protein is usually the most expensive ingredient of food ration; therefore, it is desirable to keep protein levels at a minimum and yet satisfy the protein requirements of the invertebrate. Growth of adult lobsters *Homarus americanus* was inversely correlated with the protein:energy ratio (protein cal^{-1}) of the diet when the levels of protein were kept constant (Gallagher et al., 1979). This "protein sparing" effect is the result of carbohydrates and lipids being used

TABLE II. Values for SDA, O:N Ratio, and Protein Efficiency Ratio of Post-Larval Lobsters Fed Different Diets[a,b]

	Protein/carbohydrate ratios of diet			
	1.02	*0.73*	*0.53*	*5.11*
SDA(%)[c]	17.5 (1.0)	17.1 (0.6)	17.6 (0.5)	36.8 (1.5)
O:N ratio[d]	16.2 (0.5)	17.3 (1.4)	23.3 (1.7)	12.9 (0.2)
PER[e]	2.2 (0.1)	2.3 (0.1)	2.7 (0.1)	0.9 (0.1)

[a]*After Capuzzo and Lancaster, 1979; reproduced by permission of Louisiana State University.*

[b]*All values are mean values of stage IV through stage XI Homarus americanus (±1 standard error).*

[c]*Values for percentage increase in respiration rate or specific dynamic action.*

[d]*Atomic ratio of oxygen consumed to* NH_4^+ *excreted.*

[e]*Protein efficiency ratio = wet weight gain (g)/dry weight protein fed (g).*

for maintenance energy, allowing more proteins to be used for growth. Capuzzo and Lancaster (1979) demonstrated that energy produced from protein oxidation in juvenile *H. americanus* is nutritionally "wasteful." Both O:N ratios (O_2 consumed/NH_4 excreted) and protein efficiency ratios (wet weight growth/dry weight protein fed) were inversely correlated with protein:carbohydrate ratios of the diet (Table II). Summarizing recent studies on the relationship of protein catabolism and dietary protein levels and the protein-sparing effect on nonprotein foods in crustacea, Capuzzo (1982) suggests that diets that are 25-30% protein (of dry weight) are sufficient for optimum growth.

The protein content of the diet can affect the biochemical composition of invertebrates as well as their growth rate. The effects of protein, carbohydrate, and lipid levels in diets have been studied for the mysid *Neomysis integer*. The protein content of the mysid decreased from 64 to 14% in 10 days when fed a pure carbohydrate diet (Armitage et al., 1977). The total free amino acid pool did not change significantly over the test period; however, the amino acid composition was altered. No significant change occurred in taurine, aspartate, and arginine, suggesting that these amino acids may be easily synthesized or are preferentially conserved. The amino acids of the protein hydrolysate that exhibited the greatest decrease were aspartic acid, serine, and histidine, whereas concentrations of alanine and valine did not change. When *N. integer* was fed a pure protein diet (albumin), similar in amino composition to itself, both total protein and the amino acid composition of the mysid changed significantly over the 8-day test period (Armitage et al., 1978). The protein amino acids that decreased the most were glutamate and lysine, whereas the amino acids valine and isoleucine changed the least. The free amino acids taurine, aspartate, and arginine were conserved whereas alanine, leucine, tyrosine, and phenylalanine were significantly reduced. Thus, the mysid *N. integer* requires a mixed diet of both carbohydrate and protein to maintain the quality of its protein hydrolysate and free amino acid pool.

3. Amino Acids

a. Protein-amino acids. Marine invertebrates require specific exogenous forms of nitrogen, essential amino acids, for normal growth and reproduction. Most research on the essential amino acids of invertebrates has been conducted on species used in aquaculture. Supplementation of separate crystalline amino acids to isonitrogenous diets of the freshwater shrimp *Macrobrachium rosenbergii* resulted in greater growth and protein conversion efficiency (wet weight gained/protein consumed) for the amino acids: lysine, tryptophan, phenylalanine, leucine, isoleucine, and arginine (Farmanfarmaian and Lauterio, 1979, 1980). Using similar amino acid supplementation techniques, Mason and Castell (1980) found that juvenile lobsters *Homarus americanus* required threonine, histidine, and tryptophan. Another technique to determine essential amino acids uses an isotope precursor to determine amino acid synthesis capabilities. Cowey and Forester (1971) injected [^{14}C]acetate into the prawn *Palaemon serratus* and after 6 days found no labeled arginine, methionine, valine, threonine, isoleucine, leucine, lysine, histidine, phenylalanine, or tryptophan, suggesting that they must be supplied in the diet.

b. Phagostimulants. Besides dietary requirement, amino acids maybe required by many marine invertebrates as phagostimulants. The dissolved amino acids proline and glutathione stimulated polyp expansion in the coral *Cyphastrea ocellina,* while in the coral *Fungia* sp., polyp expansion was elicited by methionine, tyrosine, proline, and glutathione (Mariscal and Lenhoff, 1968). Hamner and Hamner (1977) observed that the planktonic shrimp *Acetes sibogae* followed scent trails of bait. Testing the stimulatory response to 10 amino acids, they found that the shrimp responded to dissolved L-alanine, L-leucine, and L-methionine. Poulet and Marsot (1978) reported that the copepods *Acartia clausii* and *Eurytemora herdmani* preferentially ingested microcapsules, which contained a phytoplankton homogenate, over nonenriched particles. Examining specific compounds that stimulated copepod ingestion, Poulet and Marsot (1980) found greater copepod feeding rate on microcapsules that contained L-leucine as compared to capsules that contained L-methionine, glycolic acid, and nonenriched controls. Among copepods there are species differences with respect to response to dissolved amino acids. Dicarboxylic amino acids elicited the greatest swarming response in *Eurytemora herdmani,* whereas *Acartia hudsonica* responded preferentially to aliphatic amino acids (Poulet and Ouellet, 1982). Reviewing chemical feeding stimulants of numerous decapod crustacean species, Heinen (1980) noted that the most common chemicals that elicited a feeding response were the amino acids L-glutamic acid, glycine, and taurine.

c. Dissolved amino acids. The potential nutritional value of dissolved amino acids for marine invertebrates was first suggested by Pütter (1925). Total dissolved amino acids in seawater usually do not exceed 150 μg liter^{-1} with a range from undetectable to roughly 20 μg $liter^{-1}$ for individual amino acids (Williams, 1975). Stephens and Schinske (1961) studied dissolved amino acid uptake in 35 genera representing 11 phyla by incubating the animals in seawater with added glycine and antibiotics for 16-24 h. Significant reductions in glycine over the incubation period occurred for species of Porifera, Cindaria, Rhynchocela, Ectoprocta, Annelida, Sipunculoidea, Mollusca, Echinodermata, Hemichordata, and Chordata. No reduction in glycine occurred with any Arthropod species. This early attempt to measure dissolved amino acid uptake by animals was criticized (Johannes et al., 1969) because it did not measure the efflux of amino acids, used much greater concentrations of amino acids than occurred in nature, and was indirect by assuming the reduction in amino acid was due to absorption by the animal. Using radioactive tracers, Lewis and Smith (1971) gave direct evidence of the transfer of alanine from zooxanthellae to the tissue of the coral *Porites divaricata*. Increasing the sensitivity of measuring amino acids by using fluorescamine, Stephens (1975) measured changes in free amino acids at natural concentrations and found net influxes ($\mu M \cdot g$ wet wt $worm^{-1} \cdot h^{-1}$) of glycine (0.75), serine (0.85), glutamate (0.82), and aspartate (0.85) in *Capitella capitata* and glycine (0.84) and serine (0.99) in *Nereis diversicolor*. Although there was similar uptake of the four amino acids in *C. capitata*, the acidic amino acids glutamate and aspartate were not significantly absorbed by *N. diversicolor* compared to the neutral amino acids glycine and serine. Siebers and Ehlers (1978) extended this observation to the oligochaete *Enchytraeus albidus*. Using ^{14}C-labeled amino acids, they found that absorption of acidic amino acids was only a few percent of neutral amino acid uptake. Salinity and temperature can also affect amino acid absorption. The maximum uptake (V_{max}) of glycine by the medusae *Aurelia aurita* was a linear function of temperature (r^2 = .92-.98) between 12° and 32°C (Shick, 1975). Uptake of ^{14}C-labeled glycine and α-aminoisobutyric acid was positively correlated with temperature and salinity in both turbellarians and polychaetes (Temple and Westheide, 1980). Autoradiography and protein extraction confirmed dissolved amino acid incorporation in the polychaete *Hesionides arenaria* contributing between 2 and 8% of its daily energy requirement. After 8 weeks at starvation, the number of polyps strobilating in *Aurelia aurita* was reduced 78% compared to animals that were fed brine shrimp, glycine (0.8 μ*M*), or alanine (0.1 μ*M*) (Shick, 1975). Amino acid concentrations as low as 10 n*M* were measured using fluorescent derivatives of amino acids and high-performance liquid chromatography (HPLC) by Manahan et al. (1982) to demonstrate absorption of dissolved amino acids by the mussel *Mytilus edulis*. During a single passage through the mantle cavity of the mussel, 63, 84, and 72% of aspartate, serine, and glycine, respectively, were removed from ambient seawater. Since this reduction in amino acids occurred within seconds, mediated transfer by microbes is unlikely.

4. Nutritional Role of Gut Flora

In recent years, several workers have identified gut flora and activity in marine invertebrates that contribute to their nitrogenous nutrition. Shipworms, like termites, have gut flora that can fix N_2 when the host feeds on woody tissue (Carpenter and Culliney, 1975). Over a 4-month test period, there were no significant differences in the percent nitrogen of dry weight, wood consumption, larval production, and respiration between shipworms *Lyrodus pedicellatus* that were grown on wooden dowls in either phytoplankton-free water or water that contained the flagellate *Isochrysis galbana* (Gallagher et al., 1981). The wood was 0.2% N of dry weight and the worms were 5.8% N of dry weight; therefore, the authors hypothesize that either exogenous or indogenous gut bacteria supply nitrogen to the shipworms. Ammonia excretion by *L. pedicellotus* was not detectable in either treatment, suggesting that the bacterial associations may rapidly utilize ammonia so that nitrogen can be conserved by the shipworms. Macrophytes, like wood, are nitrogen-poor relative to the nutritional requirements of the invertebrates which feed on them. The sea urchin

Strongylocentrotus droebachiensis, which feeds on macrophytes, was found to have a gut flora that can fix N_2 (Guérinot et al., 1977). Significant nitrogenase activity was found in the gut of the urchin compared to the macrophyte diet and control sea urchins with their guts removed. The highest N_2 fixation was found in urchins from *Laminaria digitata* beds, the macrophyte with the lowest percentage of N of dry weight. Nitrogen fixation in these urchins could supply 8-15% of their daily nitrogen requirements. The guts of *S. droebachiensis* contained between 2×10^8 and 6×10^9 bacteria ml^{-1} (Fong and Mann, 1980). When bacteria were present in the sea urchin gut, radioactive carbon fed as either glucose or cellulose appeared in all protein amino acids of the urchin gonads. However, when the gut flora was inhibited by antibiotics, there was no labeled threonine, valine, methionine, isoleucine, leucine, tyrosine, phenylalanine, tryptophan, lysine, histidine, or arginine, suggesting that these essential amino acids are supplied by the gut microflora. The mysid *Mysis stenolepsis* also depends on a gut microflora for its nutrition. Foulds and Mann (1978) fed *M. stenolepsis* ^{14}C-labeled cellulose and hay particles that were either sterile or had epiphytic bacteria. Mysid assimilation efficiencies ranged from 30 to 50% for cellulose and from 20 to 35% for hay, the greatest assimilation occurring on sterile food. Foulds and Mann suggested that the cellulase activity of the mysid is associated with its gut microflora, similar to terrestrial ruminants. Wainwright and Mann (1982) recently confirmed this hypothesis by demonstrating that the mysids ability to assimilate $[^{14}C]$cellulose was lost when they were maintained in seawater containing antibiotics, but the ability to assimilate cellulose was restored when the sterile mysids were fed ground guts of normal mysids. Mysids that were maintained on cellulose-rich diets exhibited over 3 times the cellulose assimilation rates as mysids that were maintained on protein diets. Thus, an acclimation to cellulose may result in a buildup of gut microorganisms containing cellulase enzymes.

IV. SUMMARY

A. Nitrogenous Nutrition of Marine Invertebrates

The growth and reproduction of marine invertebrates are often asymptotic functions of the amount of protein ingested. However, food "quality" may also be important, marine invertebrates requiring an optimum mixture of dietary proteins, carbohydrates, and lipids for maximum growth and reproductive efficiency. Dietary carbohydrates and lipids are used by invertebrates to meet their metabolic demands, thereby "sparing" assimilated protein from catabolism and allowing greater growth and reproduction. Because of this protein sparing, ingestion of nonprotein nitrogen compounds, and the absorption of dissolved nitrogen compounds, it is not possible to designate a minimum weight-specific daily ration of protein for marine invertebrates.

The amino acids most frequently found to be essential for marine invertebrates are: tryptophan, phenylalanine, threonine, isoleucine, leucine, methionine, histidine, arginine, lysine, and valine. Besides being required for anabolism, amino acids are often important chemical cues for marine invertebrates, allowing them to sense food in their environment as well as providing a feedback on the "taste" of food and its nutritional value.

Advances in measurement techniques and the use of radioactive tracers have demonstrated that most marine invertebrates (except Arthropoda) absorb free amino acids in seawater. The extent to which these absorbed amino acids contribute to the nitrogen requirements of the animal are not clear; however, for soft-bodied benthos living in organically rich environments, the nutritional contribution of absorbed amino acids may be important.

Several marine invertebrates have been found to harbor gut flora that can fix N_2 and supply the invertebrate host with essential amino acids. This phenomenon has not been explored widely among marine invertebrates, and the extent to which these associations contribute to the nitrogenous nutrition of the host is not known.

B. Research Needs

In order to advance our understanding of the nitrogenous nutrition of marine invertebrates, the pathways and rates of their protein synthesis must be studied in relation to both abiotic variables and the biochemical composition of ingested food. Understanding these controlling mechanisms and nitrogen turnover times will facilitate efforts to assess the effects of nitrogen limitation by elucidating specific nitrogen storage "pools" in invertebrates. Nitrogen limitation of secondary production in the sea can be studied by relating these biochemical studies to indexes of invertebrate growth and reproduction and environmental variables. Especially important among these environmental parameters is the quantity and quality of available food. Available nitrogen estimates should include not only phytoplankton but also nitrogen in detritus, bacteria, and dissolved nitrogen compounds. Research is needed on the abundance and nutritional value of nonprotein nitrogen compounds. This large group of compounds probably is less labile and has a longer turnover time than protein-nitrogen compounds, but could serve as a storage pool and damp oscillations of total particulate nitrogen in many marine food webs. Analyses of available nitrogen for marine invertebrates should also include contributions from gut flora. Somewhat analogous to the coral-zooxanthellae symbiosis that developed in the nitrogen-pool coral reef environment, marine invertebrates that live in habitats with a large supply of nitrogen-poor detritus may develop a gut flora to supply their host with essential nitrogen compounds. Further examination of this phenomenon will not only help us to understand the nitrogen cycling in marine food webs but also perhaps to identify invertebrate species which, because of reduced exogenous protein requirements, would be ideally suited for aquaculture.

ACKNOWLEDGMENTS

I gratefully acknowledge D. Capone, J. Capuzzo, E. Carpenter, S. Gallagher, R. Mann, and K. Tenore for their helpful criticisms of this review.

REFERENCES

Alldredge, A. L. (1972). Abandoned larvacean houses: A unique food source in the pelagic environment. *Science 177,* 885-887.

Armitage, M. E., Raymont, J. E. G., and Morris, R. J. (1977). The effects of a pure carbohydrate diet on the amino acid composition of *Neomysis integer*. *In* "A Voyage of Discovery" (M. Nagel, ed.), pp. 471-481. Pergamon, Oxford.

Armitage, M. E., Raymont, J. E. G., and Morris, R. J. (1978). The effects of a pure protein diet on the amino acid composition of *Neomysis integer* (Leach). *J. Exp. Mar. Biol. Ecol. 35,* 147-163.

Beers, J. R. (1966). Studies on the chemical composition of the major zooplankton groups in the Sargasso Sea off Bermuda. *Limnol. Oceanogr. 11,* 520-528.

Benson, A. A., and Muscatine, L. (1974). Wax in coral mucus: Energy transfer from corals to reef fishes. *Limnol. Oceanogr. 19,* 810-814.

Bernays, E. A., and Woodhead, S. (1982). Plant phenols utilized as nutrients by a phytophagous insect. *Science 216,* 201-203.

Boyd, C. E. (1968). Fresh-water plants: A potential source of protein. *Econ. Bot. 22,* 359-368.

Boyd, C. E. (1971). The dynamics of dry matter and chemical substances in a *Juncus effusus* population. *Am. Midl. Nat. 86,* 28-45.

Brylinsky, M. (1977). Release of dissolved organic matter by some marine macrophytes. *Mar. Biol. (Berlin) 39,* 213-220.

Burkholder, P. R., Burkholder, L. M., and Rivera, J. A. (1959). Some chemical constituents of turtle grass, *Thalassia testudinum*. *Bull. Torrey Bot. Club 86,* 88-93.

Butler, E. I., Corner, E. D. S., and Marshall, S. M. (1969). On the nutrition and metabolism of zooplankton. VI. Feeding efficiency of *Calanus* in terms of nitrogen and phosphorus. *J. Mar. Biol. Assoc. U.K. 49,* 977-1001.

Cahoon, L. B. (1981). Reproductive response of *Acartia tonsa* to variations in food ration and quality. *Deep-Sea Res. 10,* 1215-1221.

Caperon, J., and Meyer, J. (1972). Nitrogen-limited growth for marine phytoplankton. I. Changes in population characteristics with steady-state growth rate. *Deep-Sea Res. 19,* 601-618.

Capone, D. G., Penhale, P. A., Oremland, R. S., and Taylor, B. F. (1979). Relationship between productivity and $N_2(C_2H_2)$ fixation in a *Thalassia testudinum* community. *Limnol. Oceanogr. 24*, 117-125.

Capuzzo, J. M. (1982). Crustacean bioenergetics: The role of environmental variables and dietary levels of macronutrients on energetic efficiencies. *Proc. Int. Conf. Aquacult. Nutr., 2nd, 1981* (in press).

Capuzzo, J. M., and Lancaster, B. A. (1979). The effects of dietary carbohydrate levels on protein utilization in the American lobster *(Homarus americanus)*. *Proc. Annu. Meet.--World Maric. Soc. 10*, 689-700.

Carpenter, E. J., and Culliney, J. L. (1975). Nitrogen fixation in marine shipworms. *Science 187*, 551-552.

Checkley, D. M. (1980a). The egg production of a marine planktonic copepod in relation to its food supply: Laboratory studies. *Limnol. Oceanogr. 25*, 430-446.

Checkley, D. M. (1980b). Food limitation of egg production by a marine, planktonic copepod in the sea off Southern California. *Limnol. Oceanogr. 25*, 991-998.

Chervin, M. B., Malone, T. C., and Neale, P. J. (1981). Interactions between suspended organic matter and copepod grazing in the plume of the Hudson River. *Estuarine Coastal Shelf Sci. 13*, 169-184.

Chuecas, L., and Riley, J. P. (1969). The component combined amino acids of some marine diatoms. *J. Mar. Biol. Assoc. U.K. 49*, 117-120.

Coles, S. L., and Strathman, R. (1973). Observations on coral mucus "flocs" and their potential trophic significance. *Limnol. Oceanogr. 18*, 673-678.

Conover, R. J. (1964). Food relations and nutrition of zooplankton. *Woods Hole Oceanogr. Inst. Conbrib. No. 1443*, pp. 1-11.

Conover, R. J. (1978). Transformation of organic matter. *In* "Marine Ecology" Vol. 6 (O. Kinne, ed.), pp. 221-500. John Wiley and Sons, New York.

Corner, E. D. S., Head, R. N., and Kilvington, C. C. (1972). On the nutrition and metabolism of zooplankton. VIII. The grazing of *Biddulphia* cells by *Calanus helgolandicus*. *J. Mar. Biol. Assoc. U.K. 52*, 847-862.

Cowey, C. B., and Corner, E. D. S. (1961). The amino acid composition of *Calanus finmarchicus* in relation to that of its food. *Rapp. P. V. Reun., Cons. Int. Explor. Mer 153*, 124-127.

Cowey, C. B., and Corner, E. D. S. (1966). The amino-acid composition of certain unicellular algae, and of the fecal pellets produced by *Calanus finmarchicus* when feeding on them. *In* "Some Contemporary Studies in Marine Science" (H. Barnes, ed.), pp. 225-231. Allen & Unwin, London.

Cowey, C. B., and Forester, J. R. M. (1971). The essential amino acid requirements of the prawn, *Palarmon serratus*. The growth of prawns on diets containing proteins of different amino acid compositions. *Mar. Biol. (Berlin) 10*, 77-81.

Degens, E. T. (1968). Molecular nature of nitrogenous compounds in sea water and recent marine sediments. *In* "Symposium on Organic Matter in Natural Waters" (D. W. Hood, ed.), pp. 77-106. Inst. Mar. Sci., Univ. Alaska, Fairbanks, Alaska.

de la Cruz, A. A., and Poe, W. E. (1975). Amino acids in salt marsh detritus. *Limnol. Oceanogr. 20,* 124-127.

D'Elia, C. F., and Webb, K. L. (1977). The dissolved nitrogen flux of reef corals. *Proc. Int. Coral Reef Symp., 3rd, 1977* Vol. 1, pp. 326-330.

DeManche, J. M., Curl, H. C., Lundy, D. W., and Donaghay, P. C. (1979). The rapid response of the marine diatom *Skeletonema costatum* to changes in external and internal nutrient concentration. *Mar. Biol. (Berlin) 53,* 323-333.

Döhler, G., and Roßlenbroich, H. (1982). Photosynthetic Assimilation of ^{15}N-ammonia and ^{15}N-nitrate in the marine diatoms *Bellerochea yucatanensis* (von Stotch) and *Skeletonema costatum. Z. Naturforsch.* (in press).

Dortch, Q. (1982). Effect of growth conditions on accumulation of internal nitrate, ammonium, amino acids, and protein in three marine diatoms. *J. Exp. Mar. Biol. Ecol. 61,* 243-264.

Eppley, R. W., and Renger, E. H. (1974). Nitrogen assimilation of an oceanic diatom in nitrogen-limited continuous culture. *J. Physiol. 10,* 15-23.

Farmanfarmaian, A., and Lauterio, T. (1979). Amino acid supplementation of feed pellets of the giant shrimp *(Macrobrachium rosenbergii). Proc. Annu. Meet.--World Maric. Soc. 10,* 674-688.

Farmanfarmaian, A., and Lauterio, T. (1980). Amino acid composition of the tail muscle of *Macrobrachium rosenbergii*--Comparison to amino acid patterns of supplemented commercial feed pellets. *Proc. Annu. Meet.--World Maric. Soc. 11,* 454.

Fell, J. W., Hendrix, R. C., Hunter, I. L., Newell, S. Y., and Tallman, A. S. (1976). The role of microorganisms as indicators of changing environmental conditions in mangroves and marsh communities. "Report of the Florida Power and Light Co., Rosenstiel School of Marine Science, University of Miami, Coral Gables, Florida.

Fisher, N. S., and Schwarzenbach, R. P. (1978). Fatty acid dynamics in *Thalassiosira pseudonana* (Bacillariophyceae): Implications for physiological ecology. *J. Physiol. 14*(2), 143-150.

Flaak, A. R., and Epifanio, C. E. (1978). Dietary protein levels and growth of the oyster *Crassostrea virginica. Mar. Biol. (Berlin) 45,* 157-163.

Fogg, G. E. (1959). Nitrogen nutrition and metabolic patterns in algae. *Symp. Soc. Exp. Biol. 13,* 106-125.

Fong, W., and Mann, K. H. (1980). Role of gut flora in the transfer of amino acids through a marine food chain. *Can. J. Fish. Aquat. Sci. 37,* 88-96.

Foulds, J. B., and Mann, K. H. (1978). Cellulose digestion in *Mysis stenolepis* and its ecological implications. *Limnol. Oceanogr. 23,* 760-766.

Frankenberg, D., Coles, S. S. L., and Johannes, R. E. (1967). The potential trophic significance of *Callianassa major* fecal pellets. *Limnol. Oceanogr. 12,* 113-120.

Gallagher, M. L., Bayer, R. C., Leavitt, D. F., and Rittenburg, J. H. (1979). Effects of protein-energy ratios on growth of American lobsters *(Homarus americanus)*. *Proc. Annu. Meet.--World Maric. Soc. 10,* 746-750.

Gallagher, S. M., and Mann, R. (1981). The effect of varying carbon/nitrogen ratios in the phytoplankton *Thalassiosira pseudonana* (3H) on its food value to the bivalve *Tapos japonica*. *Aquaculture 26,* 95-105.

Gallagher, S. M., Turner, R. D., and Berg, C. J. (1981). Physiological aspects of weed consumption, growth, and reproduction in the shipworm *Lyrodus pedicellatus* Quatrefagus (Bivalvia: Teredinae). *J. Exp. Mar. Biol. Ecol. 52,* 63-77.

Gerber, R. P., and Marshall, N. (1974). Ingestion of detritus by the lagoon pelagic community at Eniwetok Atoll. *Limnol. Oceanogr. 19,* 815-824.

Giese, A. C. (1966). Lipid in the economy of marine invertebrates. *Physiol. Rev. 46,* 244-298.

Goldman, J. C., McCarthy, J. J., and Peavey, D. G. (1979). Growth rate influence on the chemical composition of phytoplankton in oceanic waters. *Nature (London) 279,* 210-215.

Gotto, J. W., and Taylor, B. F. (1976). N_2 fixation associated with decaying leaves of the red mangrove *(Rhizophora mangle)*. *Appl. Environ. Microbiol. 31,* 781-783.

Greenfield, L. J. (1953). Observations on the nitrogen and glycogen content of *Teredo (Lyrodus) pedicelata* DE QUATREFAGES at Miami, Florida. *Bull. Mar. Sci. Gulf. Caribb. 2,* 486-496.

Guérinot, M. L., Fong, W., and Patriquin, D. G. (1977). Nitrogen fixation (acetylene reduction) associated with sea urchins *(Strongylacentrotus droebachiensis)* feeding on seaweeds and eelgrass. *J. Fish. Res. Board Can. 34,* 416-420.

Hamner, P., and Hamner, W. M. (1977). Chemosensory tracking of scent trails by the planktonic shrimp *Acetes siboguraustralis*. *Science 195,* 886-888.

Hargrave, B. T. (1970). The utilization of benthic microflora by *Hyalella acieca* (Amphipoda). *J. Anim. Ecol. 39,* 427-437.

Harris, R. P. (1973). Feeding, growth, reproduction and nitrogen utilization by the harpacticoid copepod *Tigriopus brevicornis*. *J. Mar. Biol. Assoc. U.K. 53,* 785-800.

Harrison, P. G., and Mann, K. H. (1975). Detritus formation from eelgrass *(Zostera marina):* The relative effects of fragmentation, leaching and decay. *Limnol. Oceanogr. 29,* 924-934.

Haven, D., and Morales-Alamo, R. (1966). Aspects of biodeposition by oysters and other invertebrate filter feeders. *Limnol. Oceanogr. 11,* 487-498.

Heinen, J. M. (1980). Chemoreception in decapod crustacea and chemical feeding stimulants as potential feed additives. *Proc. Annu. Meet.--World Maric. Soc. 11,* 319-334.

Hellebust, J. A. (1965). Excretion of some organic compounds by marine phytoplankton. *Limnol. Oceanogr. 10,* 192-206.
Hobbie, J. E., and Lee, C. (1980). Microbial production of extracellular material: Importance in benthic ecology. *In* "Marine B enthic Dynamics" (D. R. Tenore and B. C. Coull, eds.), pp. 341-346. Univ. of South Carolina Press, Columbia.
Holland, D. L., and Hannant, P. J. (1974). Biochemical changes during growth of the spat of the oyster *Ostrea edulis* L. *J. Mar. Biol. Assoc. U.K. 54,* 1007-1016.
Honjo, S., and Roman, M. R. (1978). A study of fecal pellets produced by marine copepods. *J. Mar. Res. 36,* 45-57.
Ikeda, T. (1972). Chemical composition and nutrition of zooplankton in the Bering Sea. *In* "Biological Oceanography of the Northern North Pacific Ocean" (A. Y. Takenouti, M. Anraku, K. Banse, T. Kawamura, S. Nishizawa, T. R. Parsons, and T. Tsujita, eds.), pp. 433-442. Idemitsu Shoten, Tokyo.
Johannes, R. E., and Satomi, M. (1966). Composition and nutritive value of fecal pellets of a marine crustacean. *Limnol. Oceanogr. 11,* 191-197.
Johannes, R. E., Coward, S. J., and Webb, K. L. (1969). Are dissolved amino acids an energy source for marine invertebrates? *Comp. Biochem. Physiol. 29,* 283-288.
Johannes, R. E., Coles, S. L., and Kuenzel, N. T. (1970). The role of zooplankton in the nutrition of some scleractinian corals. *Limnol. Oceanogr. 15,* 579-586.
Jordan, T. E., and Valiela, I. (1982). A nitrogen budget of the ribbed mussel, *Geukensia demissa*, and its significance to nitrogen flow in a New England salt marsh. *Limnol. Oceanogr. 27,* 75-90.
Konsu, S., Katori, S., Akiyama, T., and Mori, T. (1958-1959). Amino acid composition of crustacean muscle protein. (Jap.) *Nippon Suisan Gakkaishi 24,* 300-304.
Levinton, J. S., and Lopez, G. R. (1977). A model of renewable resources and limitation of deposit-feeding benthic populations. *Oecologia 31,* 177-190.
Lewis, D. H., and Smith, D. C. (1971). The autotrophic nutrition of symbiotic marine coelenterates with special reference to hermatypic corals. I. Movement of photosynthetic products between the symbionts. *Proc. R. Soc. London, Ser. B 178,* 111-129.
Liebig, J. (1840). "Chemistry in its Application to Agriculture and Physiology." Taylor & Walton, London.
McCarthy, J. J., Taylor, W. R., and Taft, J. L. (1975). The dynamics of nitrogen and phosphorus cycling in the open waters of the Chesapeake Bay. *In* "Marine Chemistry in the Coastal Environment" (T. M. Church, ed.), pp. 664-681. Am. Chem. Soc., Washington, D.C.
Manahan, D. T., Wright, S. H., Stephens, G. C., and Rice, M. A. (1982). Transport of dissolved amino acids by the mussel, *Mytilus edulis*: Demonstration of net uptake from natural seawater. *Science 215,* 1253-1255.

Mann, K. H. (1972). Macrophyte production and detritus food chains in coastal areas. *In* "Detritus and Its Role in Aquatic Ecosystems" (U. Melchiorri-Santolini and J. W. Hopton, eds.), Suppl. 29, pp. 353-382.

Mariscal, R. N., and Lenhoff, H. M. (1968). A chemical control of feeding behavior in *Cyphastrea ocellina* and in other Hawaiian corals. *J. Exp. Biol. 49,* 689-699.

Marshall, N. (1965). Detritus over the reef and its potential contribution to adjacent waters of Eniwetok Atoll. *Ecology 46,* 343-344.

Mason, E. G., and Castell, J. D. (1980). The effects of supplementing purified proteins with limiting essential amino acids on growth and survival of juvenile lobsters *(Homarus americanus). Proc. Annu. Meet.--World Maric. Soc. 11,* 346-354.

Mayzaud, P., and Martin, J. L. (1975). Some aspects of the biochemical and mineral composition of marine plankton. *J. Exp. Mar. Biol. Ecol. 17,* 297-310.

Millikin, M. R., Fortner, A. R., Fair, P. H., and Sick, L. V. (1980). Influence of dietary protein concentration on growth, feed conversion and general metabolism of juvenile prawn *(Macrobrachium rosenbergii). Proc. Annu. Meet.--World Maric. Soc. 11,* 369-382.

Molins, L. R., and Besada Rial, J. R. (1957). Chemical studies on *Mytilus edulis* of the Vigo Estuary. *Bol. Inst. Esp. Oceanogr. 87,* 1-29.

Morris, I., Glover, H. E., and Yentsch, C. S. (1974). Products of photosynthesis by marine phytoplankton: The effect of environmental factors on the relative rates of protein synthesis. *Mar. Biol. (Berlin) 27,* 1-9.

Newell, R. (1965). The role of detritus in the nutrition of two marine deposit feeders, the prosobranch *Hydrobia ulvae* and the bivalve *Macoma balthica. Proc. Zool. Soc. London 144,* 25-45.

Odum, E. P., and de la Cruz, A. A. (1967). Particulate organic detritus in a Georgia salt marsh-estuarine ecosystem. *In* "Estuaries" (G. H. Lauff, ed.), Publ. No. 83, pp. 383-388. Am. Assoc. Adv. Sci., Washington, D.C.

Odum, W. E., Kirk, P. W., and Zieman, J. C. (1979). Non-protein nitrogen compounds associated with particles of vascular plant detritus. *Oikos 32,* 363-367.

Ogino, C., and Kato, N. (1964). Studies on the nutrition of abolone, *Haliotus discus. Bull. Jpn. Soc. Sci. Fish. 30,* 523-526.

Okaichi, T. (1974). Significance of amino acid composition of phytoplankton and suspensoid in marine biological production. *Bull. Jpn. Soc. Sci. Fish. 40,* 471-478.

Paffenhöfer, G. A. (1976). Feeding, growth, and food conversion of the marine planktonic copepod *Calanus helgolandicus. Limnol. Oceanogr. 21,* 39-50.

Paffenhöfer, G. A., and Knowles, S. C. (1979). Ecological implications of fecal pellet size, production and consumption by copepods. *J. Mar. Res. 37,* 35-49.

Parsons, T. R. (1963). Suspended matter in seawater. *Prog. Oceanogr. 1,* 203-239.
Parsons, T. R., and Bawden, C. A. (1979). A controlled ecosystem for the study of the food requirements of amphipod populations. *Estuarine Coastal Mar. Sci. 8,* 547-558.
Parsons, T. R., Stephens, K., and Strickland, J. D. H. (1961). On the chemical composition of eleven species of marine phytoplankters. *J. Fish. Res. Board Can. 18*(6), 1001-1016.
Parsons, T. R., Takahashi, M., and Hargrave, B. (1977a). "Biological Oceanographic Processes," 2nd ed. Pergamon, Oxford.
Parsons, T. R., von Brockel, K., Koeller, P., Takahashi, M., Reeve, M. R., and Holm-Hansen, O. (1977b). The distribution of organic carbon in a marine planktonic food web following nutrient enrichment. *J. Exp. Mar. Biol. Ecol. 26,* 235-247.
Pease, H. D. (1932). The oyster: Modern science comes to the support of an ancient food. *J. Chem. Educ. 9,* 1673-1712.
Penhale, P. A., and Smith, W. O. (1977). Excretion of dissolved organic carbon by eelgrass *(Zostera marina)* and its epiphytes. *Limnol. Oceanogr. 22,* 400-407.
Petipa, T. S. (1967). On the efficiency of utilization of energy in pelagic ecosystems of the Black Sea. *Fish. Res. Board Can.,* Ser. No. 973 (translation).
Petipa, T. S. (1978). Matter accumulation and energy expenditure in planktonic ecosystems at different trophic levels. *Mar. Biol. (Berlin) 49,* 285-293.
Picard, G. A. (1976). Effects of light and dark cycles on the relationship between nitrate uptake and cell growth rates of *Chaetoceros* sp. (STX-105) in continuous culture. Ph.D. Dissertation, City University of New York.
Poulet, S. A., and Marsot, P. (1978). Chemosensory grazing by marine calanoid copepods (Arthropoda: Crustacea). *Science 200,* 1403-1405.
Poulet, S. A., and Marsot, P. (1980). Chemosensory feeding and food-gathering by omnivorous marine copepods. *In* "Evolution and Ecology of Zooplankton Communities" (W. C. Kerfoot, ed.), pp. 198-218. Univ. Press of New England, Hanover, New Hampshire.
Poulet, S. A., and Ouellet, G. (1982). The role of amino acids in the chemosensory swarming and feeding of marine copepods. *J. Plank. Res. 4,* 341-361.
Pütter, A. (1925). Die Ernahrung der Copepoden. *Arch. Hydrobiol. 15,* 70-117.
Reeve, M. R., Raymont, J. E. G., and Raymont, J. K. B. (1970). Seasonal biochemical composition and energy sources of *Sagitta hispida*. *Mar. Biol. (Berlin) 6,* 357-364.
Rice, D. L., and Tenore, K. R. (1981). Dynamics of carbon and nitrogen during the decomposition of detritus derived from estuarine macrophytes. *Estuarine Coastal Shelf Sci. 13,* 681-690.

Richman, S., Loya, Y., and Slobodkin, L. B. (1975). The rate of mucus production by corals and its assimilation by the coral reef copepod *Acartia negligens*. *Limnol. Oceanogr. 20*, 918-923.

Riley, G. A. (1963). Organic aggregates in seawater and the seawater and the dynamics of their formation and utilization. *Limnol. Oceanogr. 8*, 372-381.

Riley, G. A. (1970). Particulate organic matter in seawater. *Adv. Mar. Biol. 8*, 1-118.

Robertson, M. L., Mills, A. L., and Zieman, J. C. (1982). Microbial synthesis of detritus-like particulates from dissolved organic carbon released by tropical seagrasses. *Mar. Ecol.: Prog. Ser. 7*, 279-285.

Roman, M. R. (1977). Feeding of the copepod, *Acartia tonsa*, on the diatom, *Nitzschia closterium* and brown algae *(Fucus vesiculosus)* detritus. *Mar. Biol. (Berlin) 42*, 149-155.

Roman, M. R. (1980). Tidal resuspension in Buzzards Bay, Massachusetts. III. Seasonal cycles of nitrogen and carbon:nitrogen ratios in the seston and zooplankton. *Estuarine Coastal Mar. Sci. 11*, 9-16.

Roman, M. R., and Tenore, K. R. (1978). Tidal resuspension in Buzzards Bay, Massachusetts. I. Seasonal changes in the resuspension of organic carbon and chlorophyll-*a*. *Estuarine Coastal Mar. Sci. 6*, 37-46.

Rublee, P. A., and Roman, M. R. (1982). Decomposition of turtlegrass *(Thalassia testudinum* Konig*)* in flowing seawater tanks and litterbags: Compositional changes and comparison with natural particulate matter. *J. Exp. Mar. Biol. Ecol. 58*, 47-58.

Russell-Hunter, W. D. (1970). "Aquatic Productivity." Macmillan, New York.

Scott, J. M. (1980). Effect of growth rate of the food alga on the growth/ingestion efficiency of a marine herbivore. *J. Mar. Biol. Assoc. U.K. 60*, 681-702.

Shick, J. M. (1975). Uptake and utilization of dissolved glycine by *Aurelia aurita* Scyphistomae: Temperature effects on the uptake process; nutritional role of dissolved amino acids. *Biol. Bull. (Woods Hole, Mass.), 148*, 117-140.

Sick, L. V. (1976). Nutritional effect of five species of marine algae on the growth, development, and survival of the brine shrimp *Artemia salina*. *Mar. Biol. (Berlin) 35*, 69-78.

Siebers, D., and Ehlers, U. (1978). Transintegumentary absorption of acidic amino acids in the oligochaete annelid *Enchytraeus albidus*. *Comp. Biochem. Physiol. A 61A*, 55-60.

Stephens, G. C. (1975). Uptake of naturally occurring primary amines by marine annelids. *Biol. Bull. (Woods Hole, Mass.) 149*, 397-407.

Stephens, G. C., and Schinske, R. A. (1961). Uptake of amino acids by marine invertebrates. *Limnol. Oceanogr. 6*, 175-181.

Sutcliffe, W. H. (1972). Some relations of land drainage, nutrients, particulate material and fish catch in two eastern Canadian bays. *J. Fish. Res. Board Can. 29,* 257-262.
Teal, J. M. (1962). Energy flow in the salt marsh ecosystem of Georgia. *Ecology 43,* 614-624.
Temple, D., and Westheide, W. (1980). Uptake and incorporation of dissolved amino acids by interstitial Turbellaria and Polychaeta and their dependence on temperature and salinity. *Mar. Ecol.: Prog. Ser. 3,* 41-50.
Tenore, K. R. (1977a). Growth of the polychaete, *Capitella capitata,* cultured on various levels of detritus derived from different sources. *Limnol. Oceanogr. 22,* 936-941.
Tenore, K. R. (1977b). Utilization of aged detritus derived from different sources by the polychaete, *Capitalla capitata. Mar. Biol. (Berlin) 44,* 41-55.
Tenore, K. R. (1981). Organic nitrogen and caloric content of detritus. I. Utilization by the deposit-feeding polychaete, *Capitella capitata. Estuarine Coastal Shelf Sci. 12,* 39-47.
Tenore, K. R., Hanson, R. B., Dornseif, B. E., and Wiederhold, C. N. (1979). The effect of organic nitrogen supplement on the utilization of different sources of detritus. *Limnol. Oceanogr. 24,* 350-355.
Tenore, K. R., Cammen, L., Findlay, S. E. G., and Phillips, N. (1982). Perspectives of research on detritus: Do factors controlling the availability of detritus to macroconsumers depend on its source? *J. Mar. Res. 40,* 473-490.
Towle, A. (1961). Physiological changes in *Phascolosoma agassizzi* Kefferstein during the course of an annual reproductive cycle. Ph.D. Dissertation, Stanford University, Stanford, California.
Udell, H. F., Zarudsky, J., Doheny, T. E., and Burkholder, P. R. (1969). Productivity and nutrient values of plants growing in the salt marshes of the Town of Hempstead, Long Island. *Bull. Torrey Bot. Club 96,* 42-51.
Verwey, J. (1952). On the ecology of distribution of cockle and mussel in the Dutch Waddensee, their role in sedimentation, and the source of their food supply, with a short review of the feeding behavior of bivalve mollusks. *Arch. Neerl. Zool. 10,* 172-239.
Vicente, V. P., Arrayo, J. A., and Rivera, J. A. (1980). *Thalassia* as a food source. Importance and potential in the marine and terrestrial environments. *J. Agric. Univ. P.R. 64,* 107-120.
Vonk, H. J. (1960). Digestion and metabolism. *In* "Physiology of Crustacea" (T. H. Waterman, ed.), Vol. 1, pp. 291-316. Academic Press, New York.
Wainwright, P. F., and Mann, K. H. (1982). Effect of antimicrobial substances on the ability of the mysid shrimp *Mysis stenolepis* to digest cellulose. *Mar. Ecol.: Prog. Ser. 7,* 309-313.
Walne, P. R. (1973). Growth rates and nitrogen and carbohydrate contents of juvenile clams, *Saxidomus giganteus,* fed three species of algae. *J. Fish. Res. Board Can. 30,* 1825-1830.

Wangerski, P. J. (1978). Production of dissolved organic matter. *Mar. Ecol. 4,* 115-220.

Wetzel, R. G., Rich, P. H., Miller, M. C., and Allen, H. Z. (1972). Metabolism of dissolved and particulate carbon in a temperate hardwater lake. *In* "Detritus and Its Role in Aquatic Ecosystems" (U. Melchiorri-Santolini and J. W. Hopton, eds.), Suppl. 29, pp. 185-243.

Williams, P. J. Le B. (1975). Biological and chemical aspects of dissolved organic matter in seawater. *In* "Chemical Oceanography" (J. P. Riley and G. Skirrow, eds.), 2nd ed., Vol. 2, pp. 301-363. Academic Press, New York.

Zieman, J. C. (1975). Quantitative and dynamic aspects of the ecology of turtle grass *Thalassia testudinum*. *In* "Estuarine Research" (L. E. Cronin, ed.), Vol. 1, pp. 541-562. Academic Press, New York.

Chapter 11

NITROGEN EXCRETION BY MARINE ZOOPLANKTON

ROBERT R. BIDIGARE
Department of Oceanography
Texas A&M University
College Station, Texas

I. INTRODUCTION

The importance of zooplankton in the marine nitrogen cycle was well documented more than two decades ago in a regional study by Harris (1959) in Long Island Sound. Since then, numerous researchers have confirmed that the release of nitrogenous excretory products by marine zooplankton provides an important source of nutrition for the phytoplankton of coastal and open oceanic waters.

To satisfy their nitrogen requirements, marine phytoplankton can assimilate a variety of inorganic nitrogenous species including ammonium, nitrate, and nitrite, as well as urea, amino acids, and other dissolved organics. McCarthy et al. (1977) reported that the order of nitrogen preference for Chesapeake Bay phytoplankton, as measured by uptake relative to availability, was NH_4^+ > urea > NO_3^- > NO_2^-. In fact, the ability of marine plants to assimilate

ISBN 0-12-160280-X

reduced nitrogenous forms (NH_4^+, urea) preferentially to those in higher oxidation states (NO_2^-, NO_3^-) confers an energetic advantage. For example, the uptake of NO_3^- requires the expenditure of energy for the biosynthesis of nitrate reductase and a reducing potential to convert this nutrient into a usable anabolic metabolite (Eppley et al., 1969). Furthermore, nutrient-limited phytoplankton are capable of rapid ammonium uptake at rates that exceed their total nitrogen requirement (McCarthy and Goldman, 1979).

Ammonium is the primary nitrogenous excretory product of marine zooplankton (Corner and Newell, 1967). Zooplankton, therefore, have the potential to be an important nitrogen source for phytoplankton. Productivity experiments that have measured the uptake of ^{15}N-labeled nitrogen consistently demonstrate that >80% of the nitrogen assimilated by oligotrophic phytoplankton assemblages is provided by epipelagically recycled species of nitrogen (Eppley and Peterson, 1979). Characterization of the flux rates of biogenic ammonium and urea can provide a primary tool for estimating secondary and higher levels of oceanic production (Eppley et al., 1979). However, the accurate determination of zooplankton excretion rates has been difficult because of artifacts associated with the traditional time-course, animal-in-a-jar experiments. The oceanographic application of techniques derived from classical biochemistry such as column chromatography, isotopic dilution, and enzymatic analysis plus improved sampling capabilities, have greatly expanded our understanding of biogenic nutrient fluxes within the marine environment.

This chapter will focus on the biochemical, physiological, and ecological aspects of nitrogen excretion by marine zooplankton. The various forms of nitrogen excreted by zooplankton will be described in relation to food availability and the enzymes that regulate their production. The methods for determining rates of excretion will be discussed with special attention paid to the limitations of each technique. Emphasis will be placed on the effects of size, temperature, and food availability, since these variables have the greatest influence on rates of zooplankton excretion. The significance of microzooplankton and macrozooplankton excretion in maintaining phytoplankton production will be reviewed for several marine environments. This chapter will conclude with a discussion of the current information gaps and problem areas for future research.

II. NITROGENOUS EXCRETORY PRODUCTS AND THEIR BIOCHEMICAL REGULATION

A. *Ammonium*

Marine zooplankton are considered to be primarily ammonotelic, since ammonium makes up more than half of their total nitrogenous excretion (Table I). Since crustaceans predominate in most marine zooplankton collections (Mullin et al., 1975; Longhurst, 1976; Ikeda and Motoda, 1978), many investigators have limited their excretion studies to representative crustacean taxa. Experiments performed with the common North Atlantic copepod *Calanus helgolandicus* have consistently shown that ammonium comprises 60-100% of the total nitrogen excreted by this species (Corner and Newell, 1967; Corner et al., 1972, 1976). Similar results were obtained for copepods sample from the Peru upwelling system (Dagg et al., 1980). The ratio of ammonium to total nitrogen excreted by *C. helgolandicus* appeared to be fairly constant and independent of food quantity (Corner et al., 1976). However, Ikeda et al. (1982a) have recently shown that variations in food quality may alter rates of ammonium excretion.

High percentages of ammonium excretion have been documented for other marine zooplankton as well. Jawed (1969) found that ammonium comprised 82-85% of the total nitrogen excreted by the euphausiid *Euphausia pacifica*. For gelatinous zooplankton, Biggs (1976) and Kremer (1977) reported that ammonium averaged 70% and

TABLE I. Qualitative Aspects of Nitrogen Excretion by Marine Zooplankton[a]

Species	*Ammonium*	*Urea*	*Amino acids*	*Reference*
Calanus helgolandicus	90-91	9-10	0	Corner et al. (1976)
Calanus chilensis	72-80	6-17	11-14	Dagg et al. (1980)
Centropages brachiatus	47	53	0	Dagg et al. (1980)
Eucalanus inermis	57-88	8-28	5-15	Dagg et al. (1980)
Euphausia pacifica	82-85	1	11-12	Jawed (1969)
Mnemiopsis leidyi	54	<5	21	Kremer (1975, 1977)
Neomysis rayii	76-82	0	14-18	Jawed (1969)

[a] *Values are reported as the percent of total nitrogen excreted.*

and 54% of the total Kjeldahl-labile nitrogen excreted by a siphonophore and ctenophore, respectively.

Ammonium is produced as a metabolic endproduct of protein and nucleic acid catabolism. However, only about 5% of the excretory nitrogen is derived from the metabolism of nucleic acid (Pandian, 1975). Protein catabolism occurs as a two-step process in marine crustaceans. Amino acids are first mobilized through the action of proteases, then catabolized in a process known as transdeamination (Fig. 1). In this scheme, amino acids are transaminated with α-ketoglutarate to form glutamate and the corresponding α-keto acid. The newly synthesized glutamate is then deaminated by glutamate dehydrogenase (GDH) to release ammonium and regenerate α-ketoglutarate.

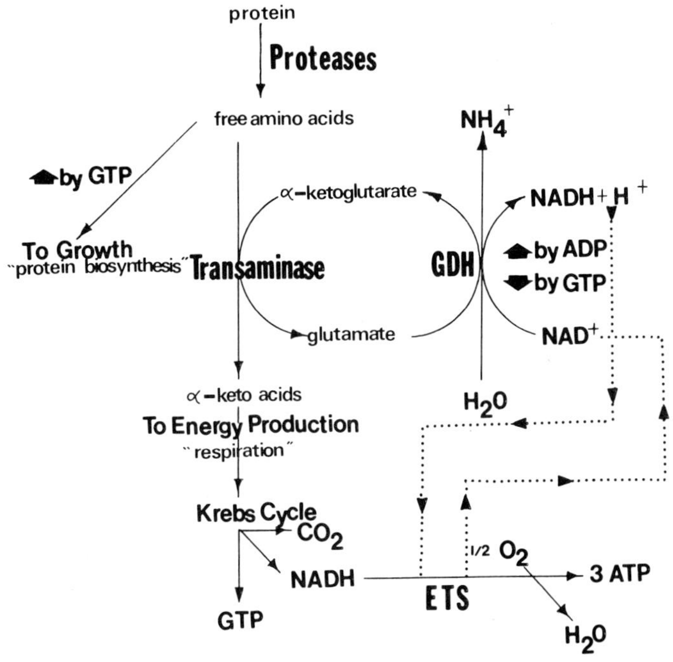

Fig. 1. Possible regulatory role of GDH in ammonium excretion, energy production and growth (Bidigare, 1981). Key: ↑, activation; ↓, inhibition

Significant levels of transaminase and GDH activity have been measured in crustacean zooplankton (Raymont et al., 1968; Bidigare and King, 1981; Bidigare et al., 1982; King, 1984). The kinetic properties of crustacean GDH have been described by Bidigare and King (1981). Crustacean GDH activity is strongly activated by ADP and inhibited by GTP. Such allosteric responses suggest that GDH indirectly may modify rates of growth and energy production in crustacean zooplankton (Fig. 1). However, the enzymes involved in the regulation of ammonium excretion in noncrustacean marine zooplankton (e.g., gelatinous zooplankton, pelagic gastropods) have yet to be studied comparatively.

Once formed, ammonium must be rapidly excreted because of its toxic properties (Pandian, 1975). The primary sites of ammonium excretion are the gills (e.g., euphausiids), maxillary glands, (e.g., copepods) or the body surface (e.g. pteropods, gelatinous zooplankton). Ammonium is freely soluble in seawater and can be lost by simple diffusion across membrane surfaces (Pandian, 1975).

B. Urea

Urea can be considered as a secondary excretory product of marine zooplankton (Table I). Approximately 9% of the total nitrogen excreted by *C. helgolandicus* was in the form of urea (Corner et al., 1976). Significantly lower values were reported for *E. pacifica* (Jawed, 1969) and *Mnemiopsis leidyi* (Kremer, 1975). However, field measurements indicate that urea excretion by marine copepods can be highly variable. Dagg et al. (1980) reported that urea formed 6-53% of the total nitrogen excreted by copepods sampled from the Peru upwelling system.

In mammalian systems, urea is produced via a coupled series of five enzymes collectively known as the urea cycle. In comparison, crustaceans are thought to lack a complete set of urea cycle enzymes (Hartenstein, 1970). However, the excretion of urea requires only the presence of arginase if sufficient quantities of arginine are present in the diet:

$$\text{Arginine} + H_2O \xrightarrow{\text{arginase}} \text{ornithine} + \text{urea}$$

Since crustaceans have been reported to have high levels of arginase activity (Hanlon, 1975), perhaps urea excretion by crustacean zooplankton is regulated by arginase. An examination of the amino acid composition of marine phytoplankton/zooplankton (Cowey and Corner, 1963, 1966; Suyama et al., 1965) reveals that food quality may be responsible for variations in rates of urea excretion. Herbivores would be expected to excrete more urea than carnivores, since arginine levels in marine phytoplankton are approximately twice those found in marine zooplankton.

Little is known about the mechanism of urea release in marine zooplankton. Unlike ammonium, urea is relatively nontoxic. Despite its high solubility in seawater, the diffusion of urea across membranes is slow because of its low oil-water partition coefficient (Pandian, 1975).

C. *Amino Acids*

Like urea, free amino acids can be considered secondary excretory products of marine zooplankton (Table I). Continuous measurements of short-term nitrogen excretion by the copepod *Eucalanus pileatus* have documented that amino acids are released in pulses lasting from 20 to 60 min (Gardner and Paffenhöfer, 1982). In comparison, the release of ammonium over similar time scales was found to be continuous (Gardner and Paffenhöfer, 1982). These results may be useful in interpreting the variability in the ratios of amino acid excretion reported for marine zooplankton. Webb and Johannes (1967) measured the release of dissolved free amino acids by net zooplankton and found the ratio of amino acids to ammonium excreted was approximately 1:4. On the other hand, Corner et al. (1967) were unable to detect any amino acids excreted by *C. helgolandicus*. Free amino acid excretion accounted for 21% and 12% of the total nitrogen excreted by *M. leidyi* (Kremer, 1977) and *E. pacifica* (Jawed, 1969), respectively. Similar values were reported by Dagg et al. (1980) for copepods sampled from the Peru upwelling system. While little is known about the exact excretion mechanism in marine zooplankton, the time-course amino acid release data collected by Gardner and Paffenhöfer (1982) suggest that continuous diffusion is not the major release mechanism in marine copepods.

III. METHODS FOR DETERMINING RATES OF NITROGEN EXCRETION

Nitrogen excretion rates of macrozooplankton (>200 μm) can be measured using three different approaches: water bottle method, kinetic measurement technique, and enzymatic analysis. A higher degree of sensitivity is required for estimating the excretion rates of microzooplankton (<200 μm). These measurements are usually performed using an ^{15}N isotopic dilution technique which is outlined by Harrison in Chapter 21.

A. *Water Bottle Method*

Most macrozooplankton excretion rate measurements are made using some variation of the water bottle method described by Ikeda (1974). Zooplankton are usually presorted and placed in jars containing filtered seawater (0.45 μm), and incubated at a desired temperature for a given length of time. At termination of the experiment, water samples are taken and analyzed for ammonium (Solórzano, 1969; Slawyk and MacIsaac, 1972), urea (DeManche et al., 1973, and primary amines (North, 1975). Differences in the respective nitrogen concentrations between experimental containers and similarly incubated control containers without animals represent the quantities of nitrogen excreted. Historically, excretion rates have been normalized on a protein, nitrogen, carbon, wet weight, dry weight, or per animal basis. The most common choice of normalization is dry weight. Vidal and Whitledge (1982) argued that the best choice of metabolic rate normalization is lipid-free dry weight, since the presence of high lipid levels in some zooplankton (e.g., boreal zooplankton) may lead to erroneous rate comparisons.

The estimation of zooplankton excretion impacts has been difficult because of artifacts associated with the traditional water bottle method. Overestimation of excretion may result from experiments conducted with net-damaged animals (Mullin et al., 1975), while underestimates may be produced by the effects of "crowding" (Christiansen, 1968; Hargrave and Geen, 1968; Smith and Whitledge, 1977).

Starvation is another factor that complicates the evaluation of excretion rates (Christiansen, 1968; Hargrave and Geen, 1968; McCarthy, 1971; Mayzaud, 1976; Ikeda and Skjoldal, 1980). Time-course excretion experiments conducted with zooplankton in filtered seawater show that rates of ammonium production can decrease significantly after 12-h starvation (Mayzaud, 1976).

To reduce these sources of error, Bidigare et al. (1982) conducted excretion experiments with zooplankton collected by nets equipped with solid cod ends to minimize abrasive damage. Zooplankton were used within 15 min of collection in experiments that were terminated after 2 h. Incubations were performed in 4-liter containers containing filtered seawater (Gelman glass fiber filters, Type A) at animal densities recommended by Mullin et al. (1975).

B. *Kinetic Measurement Technique*

A kinetic measurement technique for determining rates of zooplankton ammonium and amino acid excretion has been described by Gardner and Scavia (1981). Their analytical system consists of an incubation flow cell (0.05 mℓ) interfaced with a fluorometric ammonium analyzer. The ammonium analyzer separates ammonium from amino acids by ion-exchange chromatography. Following separation, the ammonium and amino acids are first derivatized with *o*-phthaladehyde then detected fluorometrically. This system allows the determination of individual zooplankton excretion rates over a series of 10-min intervals. This particular technique appears ideal for assessing the variability associated with individual excretion rates. However, unless this system is modified to accommodate larger sample sizes, it would have limited application for estimating the excretion impacts of natural zooplankton assemblages.

C. *Enzymatic Analysis*

In order to avoid the problems associated with making direct rate estimates, Packard (1969) proposed the use of enzymatic assays for determining excretion rates indirectly. In early investigations, nitrogen excretion rates were estimated from zooplankton respiratory electron transport system (ETS) activities (Whitledge and Packard, 1971). However, this enzymatic method will only produce approximate values at best, since zooplankton O:N ratios may vary with size and nutritional history (Mayzaud, 1976; Ikeda, 1977; Ross, 1982).

More recently, Bidigare and King (1981) found a strong positive correlation (r = 0.957) between rates of ammonium excretion and the activity of glutamate dehydrogenase (EC 1.4.1.3; GDH) in the marine mysid *Praunus flexuosus*. Although rates of weight-specific ammonium excretion varied two-fold in *P. flexuosus*, the mean GDH to excretion ratio varied by only 15%. These data suggested the utility of GDH assays for predicting whole-animal ammonium excretion rates.

Bidigare and King (1981) have described a method for determining zooplankton excretion rates by GDH activity measurement. The technique has been slightly modified for concurrent determinations of both GDH and ETS activity in marine zooplankton (Bidigare et al., 1982). The GDH to excretion ratios measured for temperate and subtropical zooplankton have been remarkably consistent. Values averaged 16.8 ± 2.6 (n=5) and 18.7 ± 4.3 (n=11) for *Calanus finmarchicus* and Gulf of Mexico mixed zooplankton, respectively. However, the mean GDH to excretion ratio estimated for Antarctic zooplankton (Bidigare, 1981) was much more variable (18.7 ± 8.5; n=5 categories of Antarctic zooplankton).

While the application of the GDH assay technique to natural zooplankton assemblages has been rather limited (Bidigare et al., 1982; King, 1984), the initial results look encouraging. Bidigare et al. (1982) have measured the ammonium excretion impact of western Gulf of Mexico zooplankton sampled quantitatively with a Multiple Opening and Closing Net and Environmental Sensing System (MOCNESS).

The zooplankton collected at each depth interval were individually homogenized, centrifuged and assayed for specific GDH activity (μmol $NH_4^+ \cdot mg^{-1} \cdot h^{-1}$). The total enzyme activity (μmol NH_4^+ h^{-1}) for each depth stratum was calculated by multiplying the supernatant GDH activity by the total amount of extractable protein in each net haul. Stratified rates of ammonium regeneration (μmol $NH_4^+ \cdot m^{-3} \cdot h^{-1}$) were then calculated by dividing these values by the volumes of water filtered at each depth stratum and a GDH:excretion calibration ratio of 18.7. Representative results are presented in Fig. 2.

As noted earlier, GDH activity has only been documented for crustacean zooplankton and not other groups such as gelatinous zooplankton and pelagic mollusks. Hence, the GDH technique may only be useful for estimating the excretion rates of natural zooplankton assemblages dominated by crustaceans. Despite this, the GDH assay has several advantages over the traditional water bottle method. Since excretion to dry weight ratios can vary 50-100% (Smith and Whitledge, 1977), GDH may offer a two-to-five fold increase in precision over dry weight when used as an index of ammonium excretion. The GDH technique allows rapid data acquisition with a minimal man-hour effort. In addition, King (1984) has shown that the presence of phytoplankton will not interfere with GDH activity measurements.

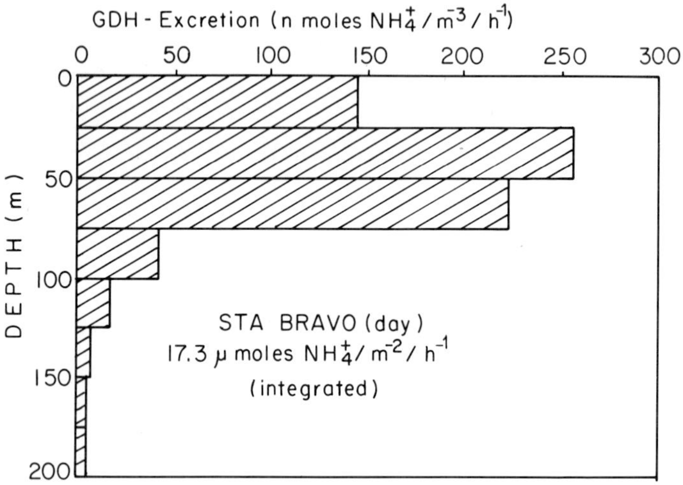

Fig. 2. Ammonium excretion impact of macrozooplankton sampled within a cyclonic circulation feature in the western Gulf of Mexico (Bidigare et al., 1982).

IV. FACTORS AFFECTING RATES OF NITROGEN EXCRETION

Turnover rates of body nitrogen for marine zooplankton are highly variable and dependent on many factors including size, temperature, and food availability. The influence of these variables on rates of nitrogen excretion will be discussed in relation to both field and laboratory investigations. While changes in salinity can alter rates of zooplankton excretion (Raymont et al., 1968; Corkett and McLaren, 1978), the salinity of most oceanic environments is relatively constant and its effects will not be considered here.

A. *Size and Temperature*

The relationship between body size and excretion rate is described by the allometric equation:

$$Y = aW^b$$

where Y = excretion rate, a = species/temperature-dependent constant, W = individual weight, and b = size-dependent constant. For most zooplankton, b is less than 1 and excretion rates increase with decreasing body size (Ikeda, 1974; Biggs, 1977; Ikeda et al., 1982a; Vidal and Whitledge, 1982). However, some zooplankton do not conform to this general rule since their weight-specific excretion rates are independent of body size (Kremer, 1977; Ross, 1982). For closely related species, a is primarily dependent on environmental temperature, and excretion rates increase with increasing temperature.

In assessing the effects of body size and temperature on excretion rates, several precautions must be taken. First, comparisons of excretion rates at different temperatures can only be made between closely related zooplanktonic groups (cf. Vidal and Whitledge, 1982). Second, experimental conditions should be uniform in order to avoid artifacts produced by container effects and food shortage (cf. Ikeda et al., 1982a).

The Q_{10} values for zooplankton ammonium excretion rates are in the range 1.4-4.0 (Table II). When individual species have been studied, lowest values appear to be characteristic of copepods (Ikeda, 1974; Corkett and McLaren, 1978). The excretion rates of some gelatinous zooplankton appear to be very temperature sensitive as evidenced by the Q_{10} value of ~4 reported for *Mnemiopsis leidyi* (Kremer, 1977). The temperature sensitivity of excretion estimated for euphausiids and pooled groups of crustacean zooplankton is intermediate with values in the range 1.8-2.6 (Jawed, 1969; El-Sayed et al., 1978; Ikeda and Hing Fay, 1981; Ross, 1982). If rates of ammonium excretion are regulated by GDH as suggested by Bidigare and King (1981), then one would expect the Q_{10} for GDH activity to be in a similar range. The specific GDH activity measured for zooplankton sampled from the Gulf of Maine (10°C) and Gulf of Mexico (25°C) were 0.9 ± 0.2 (n=5), and 2.1 ± 0.5 (n=11), respectively (R. R. Bidigare, unpublished data). The Q_{10} values calculated from these rates cor-

TABLE II. Q_{10} Values for Rates of Ammonium Excretion by Marine Zooplankton

Sample description	Q_{10}	*Reference*
Pseudocalanus sp.	1.4	Corkett and McLaren (1978)
Euphausiid zooplankton	2.6	El-Sayed et al. (1978)
Crustacean zooplankton	2.6	Ikeda and Hing Fay (1981)
Neomysis rayii	2.3	Jawed (1969)
Euphausia pacifica	2.3	Jawed (1969)
Euphausia pacifica	1.8	Ross (1982)
Coastal zooplankton	2.5	Ikeda et al. (1982a)
Mnemiopsis leidyi	∼4	Kremer (1977)

respond to 1.3-2.4 ($\bar{x}$ = 1.8). Hence, increases in ammonium excretion with increasing temperature may be explained by the temperature-dependence of GDH activity. Amino acid excretion rates appear to be more temperature sensitive than ammonium excretion rates. Jawed (1969) reported Q_{10} values for amino acid excretion by *Neomysis rayii* and *Euphausia pacifica* of 4.3 and 3.0, respectively.

In subtropical/tropical marine environments, the temperature effects on excretion rates would be greatest for those zooplankton that undergo diel migration. On the other hand, temperature-dependent changes in polar zooplankton excretion rates are minimal, since there is little difference between surface- and bottom-water temperatures.

B. *Food Availability*

Rates of ammonium excretion by marine zooplankton are thought to be tightly coupled to rates of primary production (cf. Harrison, 1980). Numerous laboratory studies have demonstrated that starvation produces significant decreases in zooplankton excretion rates (Ikeda, 1977; Mayzaud, 1976; Ikeda and Skjoldal, 1980). However, these effects may be complicated by the artifacts associated with laboratory confinement. The studies that have compared the rates of excretion by zooplankton exposed to different food concentrations are more meaningful.

Corner et al. (1976) found a progressive increase in the rates of ammonium and urea excretion by *C. helgolandicus* fed increasing rations of barnacle nauplii. Likewise, Gardner and Paffenhöfer (1982) measured a significant ($p<0.05$) increase in the ammonium release rate of *Eucalanus pileatus* exposed to high food levels. Similar trends have been reported for *M. mccradyi*, as well. Kremer (1982) found a fourfold increase in the rates of ammonium excretion by this ctenophore when fed and starved specimens were compared.

The ratio of nitrogen excreted (E) to nitrogen ingested (I), E:I, also varies as a function of food availability. Both Corner et al. (1976) and Gardner and Paffenhöfer (1982) have shown that the E:I ratios for marine copepods decrease with increased food rations. Under conditions when food is abundant, a greater percentage of ingested nitrogen can be used for growth and/or reproduction, while the remainder is used to support basal metabolic requirements. Conversely, when food is limiting, a greater percentage of the nitrogen ingested must be used to support basal metabolism. The variations of E:I ratios with respect to food availability observed for zooplankton fits in well with the metabolic scheme presented in Fig. 1.

The use of sophisticated zooplankton sampling equipment, such as the MOCNESS, has improved our understanding of phytoplankton-zooplankton interactions in the marine environment. These devices allow the collection of stratified samples and facilitate fine-scale comparisons of zooplankton biomass with levels of phytoplankton biomass/productivity. Ortner et al. (1980) observed aggregations of zooplankton associated with the Deep Chlorophyll Maximum (DCM) in the northwestern Atlantic Ocean, and concluded that the DCM signals a depth zone of intense trophic activity. Contradictory results have been reported by other investigators working in tropical/subtropical field areas. Longhurst (1976) and Hopkins (1982) found that zooplankton aggregate at depth intervals above the chlorophyll maximum. Concurrent measurements of ^{14}C uptake revealed that the zooplankton biomass often coincides with depths of maximal primary productivity. Such data suggest a strong phytoplankton-zooplankton grazing interaction within these depth strata.

The results of the feeding/excretion experiments outlined above predict that ammonium excretion rates should be elevated within depth intervals of intense trophic activity. Bidigare et al. (1982) have shown that the specific GDH activities of the zooplankton sampled within or above the chlorophyll maximum in the western Gulf of Mexico were elevated two- to fourfold. This probably reflects a biochemical adaptation to the higher food levels available there. Preliminary laboratory studies conducted with euphausiids have demonstrated that starvation produces concurrent decreases of GDH activity and rates of ammonium excretion (Bidigare and Cox, 1982).

V. ECOLOGICAL SIGNIFICANCE OF NITROGEN REGENERATION BY MARINE ZOOPLANKTON

A. *Macrozooplankton*

The macrozooplankton, defined here as >200 μm, supply <50% of the total phytoplankton nitrogen requirement (TPNR) in most oceanic water types (Table III). In upwelling areas, the percentages range from <1 to 36 (Whitledge and Packard, 1971; Smith and Whitledge, 1977; Smith, 1978; Dagg et al., 1980). The major source of nitrogen for the resident phytoplankton of upwelling regions is the vertical flux of nitrate-rich deep water (Dugdale and Goering, 1967). The ammonium excretion rates of macrozooplankton sampled from the Peruvian and North African upwelling systems are in the

TABLE III. Contribution of Macrozooplankton Ammonium Excretion to the Nitrogen Requirements for Primary Production in Various Marine Environments

Water type location	*%*[a]	*NH_4^+ excretion rate*[b]	*Reference*
Upwelling			
NW Africa	12-36	2.25[c]	Smith and Whitledge (1977)
Peru (15°S)	<1	.02-0.35	Whitledge and Packard (1971)
Peru (15°S;bloom)	28	1.74	Smith (1978)
Costa Rica Dome	1-12	.03-0.94	Whitledge and Packard (1971)
Inshore			
Long Island Sound	43-66	4.54-8.42	Harris (1959)
Narragansett Bay	4	3.27	Vargo (1979)
Great Barrier Reef	8-25	0.85-1.61	Ikeda et al. (1982b)
Oceanic/Coastal			
Ross Sea	2	0.11-0.16	Biggs (1982)
Kuroshio and adjacent seas	11-44	0.08-0.48	Ikeda and Motoda (1978)
Bering Sea (inner shelf)	2	0.14-0.18	Dagg et al. (1982)
Bering Sea (mid-shelf)	16	0.85-2.18	Dagg et al. (1982)
Bering Sea (outer shelf)	6-22	0.53-0.99	Dagg et al. (1982)
Offshore waters of Oregon	36	∿ 0.29	Jawed (1973)
Oligotrophic			
Gulf of Mexico	17	0.10-0.26	Bidigare et al. (1982)
North Pacific Central Gyre	16-20[d]	0.04-0.24[d]	Eppley et al. (1973)

[a]*Expressed as a percentage of the total phytoplankton nitrogen requirements.*

[b]*Units, μmol $NH_4^+ \cdot m^{-3}$ seawater$\cdot h^{-1}$.*

[c]*Offshore location only.*

[d]*Corrected by a factor of 2.5 (Harrison, 1980).*

range of 0.02 to 2.25 μmol $NH_4^+ \cdot m^{-3} \cdot h^{-1}$ (Table III). These rates are intermediate between the values reported for inshore and oligotrophic waters.

The highest rates of ammonium regeneration by macrozooplankton are found in inshore waters. For example, Harris (1959) measured rates exceeding 5 μmol $m^{-3}h^{-1}$ in Long Island Sound. The relative contribution of inshore macrozooplankton ammonium excretion to the TPNR lies in the range of 4-66% (Table III). Nutrient regeneration by the benthic communities of some inshore waters is thought to be a more important nitrogen source for phytoplankton production (Vargo, 1979; Ikeda et al., 1982b).

The ammonium excretion impacts of macrozooplankton from oceanic coastal waters (2-44% TPNR) are quite similar to those of upwelling regions (<1-36% TPNR), with ammonium excretion rates in the range of 0.08-2.18 μmol $NH_4^+ \cdot m^{-3} \cdot h^{-1}$ (Table III). The remaining nitrogen deficit is probably supplied by microzooplankton excretion and/or nitrate.

Ammonium regeneration by macrozooplankton of oligotrophic waters supply ∿20% of the TPNR (Table III). The percentage would be slightly higher if rates of urea excretion were included. The remainder is probably supplied by microzooplankton. Support for this comes from a comparison of the total filtration rates for micro- and macrozooplankton. Jackson (1980) calculates that macrozooplankton account for ∿16% of the total grazing in the oligotrophic waters of the North Central Pacific Ocean.

B. *Microzooplankton*

The microzooplankton fraction, defined here as <200 μm, primarily consists of protozoans and the naupliar/postnaupliar stages of copepods. Very little is known about their metabolic processes and nutritional requirements. The ^{15}N isotopic dilution method for determining rates of microzooplankton ammonium excretion has only recently been applied to the study of these rates (Harrison, 1978).

Despite the low biomass of microzooplankton in oceanic waters, their weight-specific excretion rates are severalfold higher than the corresponding rates of macrozooplankton (Harrison, 1980). Recent studies have shown that microzooplankton are capable of supplying >90% of the TPNR in inshore, oceanic coastal and oliogotrophic waters (Harrison, 1978; Caperon et al., 1979; Glibert, 1982). However, in some marine environments, microzooplankton apparently contribute little to the nitrogenous nutrition of marine phytoplankton. Ikeda et al. (1982b) and Paasche and Kristiansen (1982) have reported that microzooplankton provide only 1-4% and 0-28% of the TPNR in the inshore waters of the Great Barrier Reef and the Oslofjord, respectively.

VI. CONCLUSIONS

Ammonium is the primary nitrogenous excretory product of protein metabolism in marine zooplankton. Urea and amino acids are excreted at lower levels and together they represent ∿20% of the total nitrogen released. While increases in food quantity produce elevated excretion rates of these compounds, additional research is needed to assess the effects of food quality on their relative rates of excretion. The biochemical regulation/mechanism of urea and amino acid release by marine zooplankton is poorly understood and should be addressed in future studies.

A standardized method for determining macrozooplankton ammonium excretion is needed (cf. Ikeda et al., 1982a). The method should not be based on the measurement of individual zooplankton excretion rates because of the high degree of variability associated with this manner of data collection (cf. Hassett and Landry, 1982). The GDH method worked well in the Gulf of Mexico (Bidigare et al., 1982) and in the vicinity of Nantucket Shoals (King, 1984) for estimating macrozooplankton excretion rates. With proper calibration, this enzymatic technique may prove useful in other marine environments.

To untangle the complexities associated with phytoplankton-zooplankton interactions, our sampling strategies must be modified. Shipboard sampling should include stratified, time-course measurements of nutrients, phytoplankton biomass/productivity and micro-macrozooplankton excretion rates. The use of the zooplankton laminarinase assay may be useful for defining depths of intense grazing activity (Cox, 1981).

ACKNOWLEDGMENTS

Special thanks are extended to D. C. Biggs and F. D. King with whom I collaborated in oceanographic studies of zooplankton excretion and epipelagic nitrogen cycling in the Gulf of Mexico, the Gulf of Maine, and the Southern Ocean.

REFERENCES

Bidigare, R. R. (1981). Biochemical, physiological and ecological aspects of ammonium regeneration by marine crustaceans. Ph.D. Dissertation, Texas A&M University, College Station.

Bidigare, R. R., and Cox, J. L. (1982). Zooplankton metabolic studies in the Ross Sea. *Antact. J. U. S. 17*, (in press).

Bidigare, R. R., and King, F. D. (1981). The measurement of glutamate dehydrogenase activity in *Praunus flexuosus* and its role in the regulation of ammonium excretion. *Comp. Biochem. Physiol. 70B*, 409-413.

Bidigare, R. R., King, F. D., and Biggs, D. C. (1982). Glutamate dehydrogenase (GDH) and respiratory electron-transport-system (ETS) activities in Gulf of Mexico Zooplankton. *J. Plankton Res. 4*, 895-911.

Biggs, D. C. (1976). Nutritional ecology of *Agalma okeni*. *In* "Coelenterate Ecology and Behavior" (G. O. Mackie, ed.), pp. 201-210. Plenum, New York.

Biggs, D. C. (1977). Respiration and ammonium excretion by open ocean gelatinous zooplankton. *Limnol. Oceanogr. 22*, 108-117.

Biggs, D. C. (1982). Zooplankton excretion and NH_4^+ cycling in near-surface waters of the Southern Ocean. I. Ross Sea, austral summer 1977-1978. *Polar Biol. 1*, 55-67.

Caperon, J., Schell, D., Hirota, J., and Laws, E. (1979). Ammonium excretion rates in Kaneohe Bay, Hawaii, measured by a ^{15}N isotopic dilution technique. *Mar. Biol. 54,* 33-40.

Christiansen, F. E. (1968). Nitrogen excretion by some planktonic copepops of Bras d'Or Lake, Nova Scotia. M.S. Thesis, Dalhousie University, Nova Scotia.

Corkett, C. J., and McLaren, I. A. (1978). The biology of *Pseudocalanus. Adv. Mar. Biol. 15,* 1-231.

Corner, E. D. S., and Newell, B. S. (1967). On the nutrition and metabolism of zooplankton. IV. The forms of nitrogen excreted by *Calanus, J. Mar. Biol. Assoc. U. K. 47,* 113-120.

Corner, E. D. S., Head, R. N., and Kilvington, C. C. (1972). On the nutrition and metabolism of zooplankton. VIII. The grazing of *Biddulphia* cells by *Calanus Helgolandicus. J. Mar. Biol. Assoc. U. K. 52,* 847-861.

Corner, E. D. S., Head, R. N., Kilvington, C. C., and Pennycuick, L. (1976). On the nutrition and metabolism of zooplankton. X. Quantitative aspects of *Calanus Helgolandicus* feeding as a carnivore. *J. Mar. Biol. Assoc. U. K. 56,* 345-358.

Cowey, C. B., and Corner, E. D. S. (1963). On the nutritional and metabolism of zooplankton. II. The relationship between the marine copepod *Calanus helgolandicus* and particulate material in Plymouth seawater, in terms of amino acid composition. *J. Mar. Biol. Assoc. U. K. 43,* 495-511.

Cowey, C. B., and Corner, E. D. S. (1966). The amino acid composition of certain unicellular algae, and of faecal pellets produced by *Calanus finmarchicus* when feeding on them. *In* "Some Contemporary Studies in Marine Science" (H. Barnes, ed.)., pp. 225-231. Allen & Unwin, London.

Cox, J. L. (1981). Laminarinase induction in marine zooplankton and its variability in zooplankton samples. *J. Plankton Res. 3,* 345-356.

Dagg, M., Cowles, T., Whitledge, T., Smith, S., Howe, S., and Judkins, D. (1980). Grazing and excretion by zooplankton in the Peru upwelling system during April 1977. *Deep-Sea Res. 27A,* 43-59.

Dagg, M. J., Vidal, M., Whitledge, T. E., Iverson, R. L., and Goering, J. J. (1982). The feeding, respiration and excretion of zooplankton in the Bering Sea during a spring bloom. *Deep-Sea Res. 29A,* 45-63.

DeManche, J. M., Curl, H., and Coughenower, D. D. (1973). An automated analysis for urea in seawater. *Limnol. Oceanogr. 18,* 686-689.

Dugdale, R. C., and Goering, J. J. (1967). Uptake of new and regenerated form of nitrogen in primary productivity. *Limnol. Oceanogr. 12,* 196-206.

El-Sayed, S. Z., Biggs, D. C., Stockwell, D., Warner, R., and Meyer, M. (1978). Biogeography and metabolism of phytoplankton and zooplankton in the Ross Sea, Antarctica. *Antarct. J. U. S. 13,* 131-133.

Eppley, R. W., and Peterson, B. J. (1979). Particulate organic matter flux and planktonic new production in the deep ocean. *Nature (London) 282,* 677-680.

Eppley, R. W., Coatsworth, J. L., and Solórzano, L. (1969). Studies of nitrate reductase in marine phytoplankton. *Limnol. Oceanogr. 14,* 194-205.

Eppley, R. W., Renger, E. H., Venrick, E. L., and Mullin, M. (1973). A study of plankton dynamics and nutrient cycling in the central gyre of the North Pacific Ocean. *Limnol. Oceanogr. 18,* 534-551.

Eppley, R. W., Renger, E. H., Harrison, W. G., and Cullen, J. J. (1979). Ammonium distribution in southern California coastal waters and its role in the growth of phytoplankton. *Limnol. Oceanogr. 24,* 495-509.

Gardner, W. S., and Paffenhöfer, G. A. (1982). Nitrogen regeneration by the subtropical marine copepod *Eucalanus pileatus*. *J. Plankton Res. 4,* 725-734.

Gardner, W. S., and Scavia, D. (1981). Kinetic examination of nitrogen release by zooplankters. *Limnol. Oceanogr. 26,* 801-810.

Glibert, P. M. (1982). Regional studies of daily, seasonal and size fraction variability in ammonium remineralization. *Mar. Biol. 70,* 209-222.

Hanlon, D. P. (1975). The distribution of arginase and urease in marine invertebrates. *Comp. Biochem. Physiol. B 52B,* 261-264.

Hargarave, B. T., and Geen, G. D. (1968). Phosphorus excretion by zooplankton. *Limnol. Oceanogr. 13,* 332-342.

Harris, E. (1959). The nitrogen cycle in Long Island Sound. *Bull. Bingham Oceanogr. Collect. 17,* 31-65.

Harrison, W. G. (1980). Nutrient regeneration and primary production in the sea. *In* "Primary Productivity in the Sea" (P. Falkowski, ed.), pp. 433-466. Plenum, New York.

Harrison, W. G. (1978). Experimental measurements of nitrogen remineralization in coastal waters. *Limnol. Oceanogr. 23,* 684-694.

Hartenstein, R. (1970). Nitrogen metabolism in non-insect Arthopoda. *In* "Comparative Biochemistry of Nitrogen Metabolism" (J. W. Campbell, ed.), Vol. 1, Chapter 7a, Academic Press, New York.

Hassett, R. P., and Landry, M. R. (1982). Digestive carbohydrase activities in individual marine copepods. *Mar. Biol. Lett. 3,* 211-221.

Hopkins, T. L. (1982). The vertical distribution of zooplankton in the eastern Gulf of Mexico. *Deep-Sea Res. 29A,* 1069-1083.

Ikeda, T. (1974). Nutritional ecology of marine zooplankton. *Mem. Fac. Fish., Hokkaido Univ. 22,* 1-97.

Ikeda, T. (1977). The effects of laboratory conditions on the extrapolation of experimental measurements to the ecology of marine zooplankton. IV. Changes in respiration and excretion rates of boreal zooplankton species maintained under fed and starved conditions. *Mar. Biol. 41,* 241-252.

Ikeda, T., and Hing Fay, E. (1981). Metabolic activity of zooplankton from the Antarctic Ocean. *Aust. J. Mar. Freshwater Res. 32,* 921-930.

Ikeda, T., and Motoda, S. (1978). Estimated zooplankton production and their ammonia excretion in the Kuroshio and adjacent seas. *Fish. Bull. 76,* 357-367.

Ikeda, T., and Skjoldal, H. R. (1980). The effect of laboratory conditions on the extrapolation of experimental measurements to the ecology of marine zooplankton. VI. Changes in physiological activities and biochemical components of *Acetes sibogae australis* and *Acartia australis* after capture. *Mar. Biol. 58,* 285-293.

Ikeda, T., Hing Fay, E., Hutchinson, S. A., and Boto, G. M. (1982a). Ammonia and phosphate excretion by zooplankton from the inshore waters of the Great Barrier Reef, Queensland. I. Relationship between excretion rates and body size. *Aust. J. Mar. Freshwater Res. 33,* 55-70.

Ikeda, T., Carleton, J. H., Mitchell, A. W., and Dixon, P. (1982b). Ammonia and phosphate excretion by zooplankton from the inshore waters of the Great Barrier Reef. II. Their *in situ* contributions to nutrient regeneration. *Aust. J. Mar. Freshwater Res. 33,* 683-698.

Jackson, G. A. (1980). Phytoplankton growth and grazing in oligotrophic oceans. *Nature (London) 284,* 439-441.

Jawed, M. (1969). Body nitrogen and nitrogenous excretion in *Neomysis rayii* Nurdoch and *Euphausia pacifica* Hansen. *Limnol. Oceanogr. 14,* 748-754.

Jawed, M. (1973). Ammonia excretion by zooplankton and its significance to primary productivity during summer. *Mar. Biol. 23,* 115-120.

King, F. D. (1984). The vertical distribution of zooplankton excretion and respiration in relation to chlorophyll in the vicinity of the Nantucket Shoals. *Mar. Biol.* (submitted for publication).

Kremer, P. (1975). Nitrogen regeneration by the ctenophore *Mnemiopsis leidyi. ERDA Symp. Ser.* CONF-740513, 279-290.

Kremer, P. (1977). Respiration and excretion by the ctenophore *Mnemiopsis leidyi. Mar. Biol. 44,* 43-50.

Kremer, P. (1982). Effect of food availability on the metabolism of the ctenophore *Mnemiopsis mccradyi. Mar. Biol. 71,* 149-156.

Longhurst, A. R. (1976). Interactions between zooplankton and phytoplankton profiles in the Eastern Tropical Pacific Ocean. *Deep-Sea Res. 23,* 729-754.

McCarthy, J. J. (1971). The role of urea in marine phytoplankton ecology. Ph. D. Dissertation, University of California, San Diego.

McCarthy, J. J., and Goldman, J. C. (1979). Nitrogenous nutrition of marine phytoplankton in nutrient-depleted waters. *Science 203,* 670-672.

McCarthy, J. J., Taylor, W. R., and Taft, J. L. (1977). Nitrogenous nutrition of the plankton in the Chesapeake Bay. I. Nutrient availability and phytoplankton preferences. *Limnol. Oceanogr. 22,* 996-1011.

Mayzaud, P. (1976). Respiration and nitrogen excretion of zooplankton. IV. The influence of starvation on the metabolism and the biochemical composition of some species. *Mar. Biol. 37,* 47-58.

Mullin, M. M., Perry, J. J., Renger, E. H., and Evans, P. M. (1975). Nutrient regeneration by oceanic zooplankton: A comparison of methods. *Mar. Sci. Commun. 1,* 1-13.

North, B. B. (1975). Primary amines in California coastal waters: Utilization by phytoplankton. *Limnol. Oceanogr. 20,* 20-27.

Ortner, P. B., Wiebe, P. H., and Cox, J. L. (1980). Relationships between oceanic epizooplankton distributions and the seasonal deep chlorophyll maximum in the northwestern Atlantic Ocean. *J. Mar. Res. 38,* 507-531.

Paasche, E., and Kristiansen, S. (1982). Ammonium regeneration by microzooplankton in the Oslofjord. *Mar. Biol. 69,* 55-63.

Packard, T. T. (1969). The estimation of the oxygen utilization rate in seawater from the activity of the respiratory electron transport system in plankton. Ph. D. Dissertation, University of Washington, Seattle.

Pandian, J. J. (1975). Mechanisms of heterotrophy. *In* "Marine Ecology" (O. Kinne, ed.), Vol. 2, pp. 179-188. Wiley, New York.

Raymont, J. E. G., Austin, J., and Linford, E. (1968). Biochemical studies on marine zooplankton. V. The composition of the major biochemical fractions in *Neomysis integr. J. Mar. Biol. Assoc. U. K. 48,* 735-760.

Ross, R. M. (1982). Energetics of *Euphausia pacifica*. I. Effects of body carbon and nitrogen and temperature on measured and predicted production. *Mar. Biol. 68,* 1-13.

Slawyk, A., and MacIsaac, J. J. (1972). Comparison of two automated ammonium methods in a region of coastal upwelling. *Deep-Sea Res. 19,* 521-524.

Smith, S. L. (1978). Nutrient regeneration by zooplankton during a red tide off Peru, with notes on biomass and species composition of zooplankton. *Mar. Biol. 49,* 125-132.

Smith, S. L., and Whitledge, T. E. (1977). The role of zooplankton in the regeneration of nitrogen in a coastal upwelling system of northwest Africa. *Deep-Sea Res. 24,* 49-56.

Solórzano, L. (1969). Determination of ammonia in natural waters by the phenol hypochlorite method. *Limnol. Oceanogr. 14,* 799-801.

Suyama, M., Nakajima, K., and Nonaka, J. (1965). Studies on the protein and non-protein nitrogenous constituents of *Euphasia*. *Bull. J. Soc. Sci. Fish. 31,* 302-306.

Vargo, G. A. (1979). The contribution of ammonia excreted by zooplankton to phytoplankton in Narragansett Bay. *J. Plankton Res. 1,* 75-84.

Vidal, J., and Whitledge, T. E. (1982). Rates of metabolism of planktonic crustaceans as related to body weight and temperature of habitat. *J. Plankton Res. 4,* 77-84.

Webb, K. L., and Johannes, R. E. (1967). Studies of the release of dissolved free amino acids by marine zooplankton. *Limnol. Oceanogr. 12,* 376-382.

Whitledge, T. E., and Packard, T. T. (1971). Nutrient excretion by anchovies and zooplankton in Pacific upwelling regions. *Invest. Pesq. 35,* 243-250.

Chapter 12

Benthic Nitrogen Regeneration

J. VAL KLUMP
Center for Great Lakes Studies
University of Wisconsin-Milwaukee
Milwaukee, Wisconsin

CHRISTOPHER S. MARTENS
Marine Science Program 045A
University of North Carolina
Chapel Hill, North Carolina

Because marine sediments are often major sinks for nitrogen and pelagic systems are generally nitrogen-limited, benthic nitrogen regeneration is an important facet of nitrogen cycling in many marine environments. This chapter attempts to outline the major processes controlling the modes, rates, variability, and extent of nitrogen regeneration in benthic systems. Nitrogen regeneration begun in the water column is continued in the benthos but with some significant differences. Inputs are largely allocthonous, contact with the overlying water is limited to a large degree by diffusive processes, the vertical scales of rapid benthic regeneration are on the order of centimeters rather than tens to hundreds of meters, and the concentrations of particulates are roughly four orders of magnitude greater than those in the overlying water.

ISBN 0-12-160280-X

I. THE DEPOSITION AND REMINERALIZATION OF NITROGENOUS MATERIAL IN THE BENTHIC ENVIRONMENT: PROCESSES AND STOICHIOMETRY

A. Sedimentation and Deposition

Benthic nutrient regeneration and metabolism are controlled by the supply of labile organic matter to the seafloor. Degradable material enters the benthic system as a result of terrestrial, littoral, pelagic, and occasionally *in situ* production (Krumbein et al., 1977; Hartwig, 1978). Organic substances from all but the latter of these sources generally undergo considerable recycling prior to deposition.

Hydrographic conditions, primary production, and sedimentation have been correlated with the deposition, reactivity, and preservation of sedimentary organic matter for pelagic, hemipelagic, and neritic sediments exhibiting accumulation rates varying by four orders of magnitude (from mm 1000 $year^{-1}$ to cm $year^{-1}$) (Hargrave, 1973, 1980; Hartmann et al., 1976; Toth and Lerman, 1977; Berner, 1978; Müller and Suess, 1979; Suess and Müller, 1980; Suess, 1980). The fraction of the organic matter fixed within the photic zone actually reaching the seafloor is most directly related to water depth and varies from roughly 1% in deep sea sediments to 10-50% in coastal sediments to as much as 50-80% in nearshore and estuarine zones (Suess and Müller, 1980; Suess, 1980; Fig. 1). In oceanic systems, the amount of the metabolizable material supplied to the benthos is dominated by the rates of pelagic production, pelagic remineralization, and particle settling. Amino acids, fatty acids, mono- and polysaccharides, and other labile material chemically similar to the organic constituents in phytoplankton (Degens and Mopper, 1975; Tanoue, 1979; Lee and Cronin, 1982) do reach the bottom unreacted via sedimentation processes associated with the rapid fallout of larger particles (>50 μm) and fecal material (Bishop et al., 1977, 1978; Knauer et al., 1979; Honjo and Roman, 1978). This material is rapidly degraded at the sediment-water interface, with less than

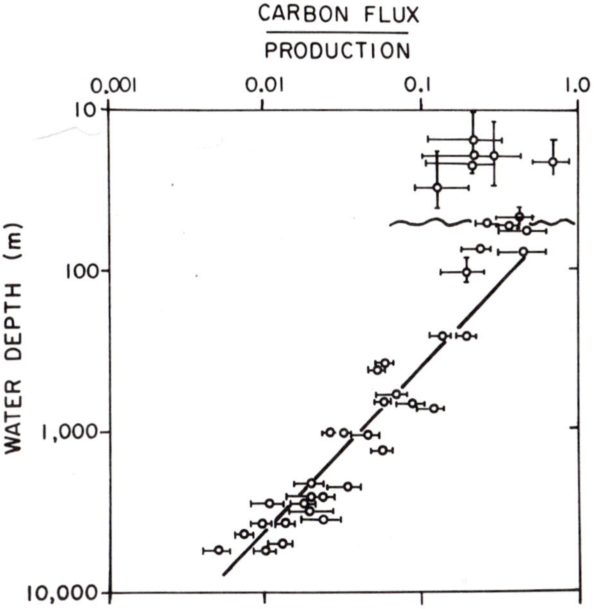

Fig. 1. Organic carbon fluxes with depth in the water column normalized to mean annual primary production rates at the sites of sediment-trap deployments. The linear regression (solid line) is represented by the equation: $C_{flux(z)} = C_{prod}/(0.0238z + 0.212)$. (From Suess, 1980.)

a few percent ultimately being preserved (Müller and Suess, 1979; Cobler and Dymond, 1980). Abyssal benthic nutrient regeneration is of minor importance to nutrient cycling in oceanic systems, largely because the time scales of nutrient dynamics are short relative to the residence times and mixing scales of oceanic water masses (Jørgensen, 1981).

In coastal waters, the time scales of benthic-pelagic interactions are much shorter, the physical regimes much more variable, and the organic matter loading to the benthos greater in both amount and complexity. Largely as a result of the proximity of the bottom, benthic nutrient regeneration plays an important role in the biological dynamics of coastal systems, and simple empirical relationships between the rate of sediment deposition and benthic metabolism, while rare, probably reflect the straightforward reliance of the benthic system on inputs from above (Berner, 1978).

The ratios of C:N:P in sedimentary organic material vary according to the initial chemical composition of the organic matter when formed, the extent of remineralization prior to deposition, and the changes resulting from diagenesis following burial. Increasing C:N ratios in detrital material have been inferred to be a function of the distance offshore in coastal sediments (Martens et al., 1978) and have been observed as a function of depth in the water column at which the material is trapped in the deep sea (Knauer et al., 1979; Lee and Cronin, 1982). The preferential removal of nitrogen (and phosphorus) relative to carbon during the initial stages of organic matter decomposition (Waksman, 1933; Anderson, 1939; Seki et al., 1968; Aller and Yingst, 1980) has been related to the relative bond energies of the principle components of organic matter: carbon, hydrogen, nitrogen, oxygen, and phosphorus. Toth and Lerman (1977) point out that the release of nitrogen and phosphorus via the breaking of C—N, O—P, C—O bonds requires less energy (72.8, 85.1, and 85.5 kcal mol^{-1}, respectively) than the production of CO_2 from the destruction of stronger C—C and C—H bonds (90.7 kcal mol^{-1}).

B. Remineralization of Organic Nitrogen

When living tissues die, nitrogenous polymers such as polyaminosugars, purines, pyrimidines, nucleic acids, and proteins, as well as low-molecular-weight compounds, become available to decomposers. Large polymers cannot generally penetrate cell membranes and the initial breakdown to smaller transportable molecules is frequently catalyzed extracellularly by enzymes (usually hydrolases) excreted by bacteria (Gottschalk, 1979). Amino acids and shorter peptides can be actively taken up by microorganisms and used for growth (Gottschalk, 1979). In the Peruvian upwelling system, amino acids constituted 20-30% of the vertical particulate organic carbon flux and 40-65% of the particulate organic nitrogen flux measured in sediment traps (Lee and Cronin, 1982). While amino acids disappeared more quickly than organic carbon in these waters, the ratio of amino acid flux to particulate organic nitrogen (PON) flux frequently increased with depth, implying that the bulk PON was degraded even more rapidly, perhaps due to the presence of nitrogenous compounds more labile than amino acids, e.g., polyamines, purines, and pyrimidines.

Aerobic bacteria are capable of completely degrading many amino acids to CO_2 and NH_3 (Gottschalk, 1979; Massey et al., 1976). A characteristic of individual anaerobic bacteria, however, is their inability to carry out the complete degradation of organic matter from complex substrates to CO_2 (Zehnder, 1978). In anoxic environments, breakdown frequently occurs in a stepwise fashion, with the metabolite of one organism becoming the growth substrate for another. In the absence of oxygen, amino acids, purine, and pyrimidine bases can serve as substrates for the so-called "acid-forming" bacteria capable of anaerobic fermentation, e.g., *Clostridia* species (Thauer et al., 1977; Brock, 1979). These fermentations can be carried out on single amino acids, e.g.,

$$4\ NH_2CH_2COOH + 2\ H_2O \longrightarrow 4\ NH_3 + 2\ CO_2 + 3\ CH_3COOH \qquad (1)$$

or on mixtures of amino acids in which pairs of amino acids undergo coupled oxidation-reductions, one amino acid (e.g., alanine) being oxidized and the other (e.g., glycine) being reduced (the Strickland reaction):

$$\underset{\text{alanine}}{CH_3CH(NH_2)COOH} + 2\ H_2O \longrightarrow CH_3COOH + CO_2 + NH_3 + 4\ H \qquad (2)$$

$$2\ \underset{\text{glycine}}{NH_2CHCOOH} + 4\ H \longrightarrow 2\ CH_3COOH + 2\ NH_3 \qquad (3)$$

Following anaerobic deamination, decarboxylation can produce volatile fatty acids (VFAs) that can be degraded further by a variety of microorganisms, including sulfate reducers.

Dissolved free amino acids (DFAAs) have been found in the interstitial waters of coastal sediments at concentrations (0.82-5.6 mg $liter^{-1}$), two orders of magnitude greater and with distributions of individual amino acids significantly different from those found in overlying waters (Henrichs and Farrington, 1979). The exact source for these DFAAs is unknown, but presumably the majority result from the metabolism and alteration of detrital proteins and nitrogenous material by benthic micro- and macrofauna.

Hobbie et al. (1980) reported very rapid turnover (every 30-120 min) for labeled proline, alanine, and glutamic acid in the interstitial waters of salt marsh sediments. Respiration of the amino acids, however, was very low, leading to speculation that some of this cycling, particularly that of proline (an osmoregulator), arose from exchange with *Spartina* roots. In studies of ^{14}C-labeled alanine degradation in Limfjorden sediment, Christensen and Blackburn (1980) reported the appearance of the label in the VFA fraction but not in the volatile amine fraction, suggesting deamination preceded decarboxylation. The initial turnover of tracer levels of alanine was rapid but incomplete, with roughly 70% of the alanine removed within 1 h, most of which (70%) was respired to CO_2. An additional ∿20% of the added alanine degraded slowly over several weeks. The remineralization rate calculated for alanine (75 $nmol \cdot cm^{-3} \cdot day^{-1}$) exceeded total NH_4^+ production in the same sediments, leading to the conclusion that the metabolizable pool of alanine was much smaller than the measured pore water concentration. Christensen and Blackburn (1980) interpret the kinetics of amino acid turnover in sediments as being controlled by the existence of two amino acid pools: DFAAs and particle-bound or adsorbed amino acids (Fig. 2). Biologically mediated degradation is assumed to occur via uptake from the DFAA pool only. In the Limfjorden sediments, the particle-bound pool was calculated to be

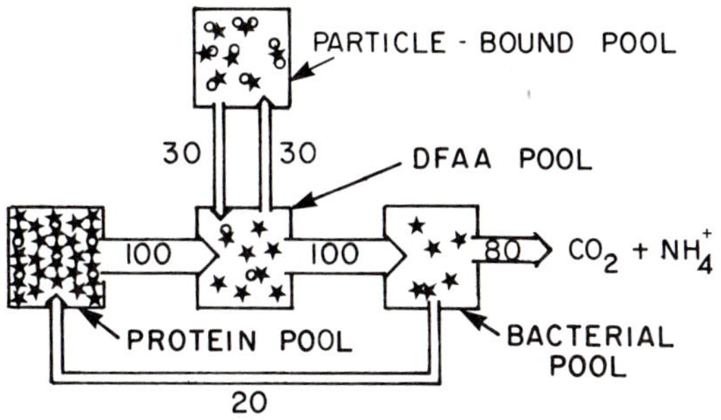

Fig. 2. A model of amino acid turnover in nearshore marine sediments from Christensen and Blackburn (1980). Key: * , amino acid; O , binding site for an amino acid. Protein is degraded to individual amino acids in a dissolved free amino acid (DFAA) pore water pool. The amino acid leaves this pool by biological uptake and adsorption. Some complexing of amino acids may occur in the free pool. The amino acids that are taken up by the bacterial pool are either deaminated and oxidized or incorporated into cellular material which returns to the protein pool. Relative rates for alanine turnover in Aarhus Bay sediments are given. Using tracer experiments and measurements of the DFAA pool, Christensen and Blackburn determined an alanine remineralization rate of 75 $nmol \cdot cm^{-3} \cdot day^{-1}$. This rate exceeded total NH_4^+ production in the same sediments, leading to the conclusion that only a small percentage of the measured dissolved alanine was biologically available.

2500-fold greater than the free pool. Adsorption of amino acids by sediments is apparently both rapid and, in some instances, irreversible (Rosenfeld, 1979b), and corrections must be made for adsorption when calculating rate constants (e.g., Holm et al., 1980). The extent to which particle-bound amino acids are available to microorganisms is unknown, but the availability of even a small fraction of these bound amino acids could have a major impact on the size of the metabolizable pool.

Rosenfeld's (1979b) experiments on Long Island Sound and Florida Bay sediments suggested that organic matter is the predominate site of amino acid binding in clay sediments, but that organic coatings on carbonate grains may inhibit the normally strong adsorption of amino acids on carbonate minerals in carbonate-rich sediments. During diagenesis amino acids are incorporated into humic material via condensation and polymerization processes (Hedges and Parker, 1976; Gagosian and Stuermer, 1977), and as much as 10-50% of sedimentary humic substances by weight may consist of amino acids (Carter and Mitterer, 1978). Sorption of amino acids by sediments, therefore, may be more akin to a chemical reaction than simple equilibrium adsorption.

Studies with a variety of ^{14}C-labeled organic compounds, including amino acids and proteins, show reduced biodegradation of these substances when they become associated with humic acid-type polymers in soils (Martin et al., 1978; Verma et al., 1975). Low C:N ratios in deep-sea sediments have been attributed to the preservation of organic nitrogen by sorption processes, resulting in the protection of nitrogenous material from bacterial attack (Müller, 1977; Waples and Sloan, 1980).

The preferential remineralization of nitrogen during the initial decay of organic matter does not necessarily result in nitrogen depletion relative to carbon in sediments during postdepositional diagenesis. The deposition of nutrient-poor organic material may lead to an immobilization of nitrogen in sediments as microorganisms take up inorganic nitrogen to synthesize biomass at a relatively lower C:N ratio (Fenchel and Blackburn, 1979) and as humification processes sequester organic nitrogen in refractory geopolymers. However, the total nitrogen content of sediments nearly always decreases with depth and, even during the decomposition of low-nitrogen- and phosphorus-containing substrates, net mineralization may occur through the eventual destruction of bacterial biomass and the rerelease of mineral nutrients (Fenchel and Harrison, 1976).

C. Biogeochemical Zonation in Marine Sediments

If the rate of deposition of metabolizable organic matter is sufficiently high, heterotrophic bacteria can no longer completely degrade the labile components of the sediment before they are buried by subsequently deposited material or reworked to depth by macrofauna. Once these organic substrates are sequestered within the sediment, the supply of the principal inorganic electron (hydrogen) acceptors utilized by microbes for the oxidation of organic matter becomes limited by diffusion and exchange across the sediment-water interface. Under these conditions, degradation proceeds via changing modes of oxidation, the sequence of which is apparently determined by competition associated with the relative energy yield of the oxidation-reduction reactions possible for further decomposition (Richards, 1965; Claypool and Kaplan, 1974; Martens, 1978; Froelich et al., 1979) (Table I). Microbes utilizing the most efficient of these available respiration modes appear to dominate, and the result is a spatial and/or temporal succession in the mode of respiration used and in the active microbial community present (Mechalas, 1974). In sediments this manifests itself in a definite biogeochemical zonation as successively favored electron acceptors are sequentially depleted (Table I). Typically, this zonation appears as a vertical stratification in sediments, but examples of horizontal zonation around macrofauna burrow tubes have been observed as a result of the lateral diffusion of electron acceptors from well-irrigated burrow water (Aller, 1977). Microzones may be found throughout sediments and any particular layer should have a mixture of microbial types.

It is generally thought that the most important modes of organic matter degradation in marine sediments are aerobic respiration and simultaneously occurring anaerobic fermentation and sulfate reduction. In coastal sediments, sulfate reducers may oxidize as much organic matter to CO_2 as aerobic bacteria (Jørgensen, 1982). These degradation modes have been represented stoichiometrically by the following idealized reactions (Richards, 1965; Mechalas, 1974; Claypool and Kaplan, 1974):

TABLE I. Biogeochemical Zonation in Sediments

Zone		*Observed sequence of microbially mediated reactions*	ΔG° (kJ mol^{-1} CH_2O)[a]
Aerobic zone	Aerobic respiration:	$(CH_2O)_x(NH_3)_y(H_3PO_4)_z + x\ O_2$	-475
		$\rightarrow x\ CO_2 + x\ H_2O + y\ NH_3 + z\ H_3PO_4$	
	Nitrification:	$NH_4^+ + 2\ O_2 \rightarrow NO_3^- + H_2O + 2\ H^+$	
	Sulfide oxidation:	$HS^- + 2\ O_2 \rightarrow SO_4^{2-} + H^+$	
Nitrate reduction zone	Denitrification:	$5\ CH_2O + 4\ NO_3^- \rightarrow 4\ HCO_3^- + CO_2 + 2N_2 + 3H_2O$	-448
	Manganese oxide reduction:	$CH_2O + 2\ MnO_2 + 3\ CO_2 + H_2O \rightarrow 2\ Mn^{2+} + 4\ HCO_3^-$	-349
	Nitrate reduction:	$2\ CH_2O + NO_3^- + 2\ H^+ \rightarrow 2\ CO_2 + NH_4^+ + H_2O$	-328
	Iron oxide reduction:	$CH_2O + 4\ Fe(OH)_3 + 7\ CO_2 \rightarrow 8\ HCO_3^- + 3\ H_2O + 4\ Fe^{2+}$	-114
Sulfate reduction zone	Sulfate reduction:	$2\ CH_3CHOHCOOH + SO_4^{2-} \rightarrow 2\ CH_3COOH + 2\ HCO_3^- + H_2S$	-77
		$CH_4 + SO_4^{2-} \rightarrow HCO_3^- + HS^- + H_2O$	
		$4\ H_2 + SO_4^{2-} \rightarrow HS^- + OH^- + 3H_2O$	
Methane production zone	Acetate fermentation:	$CH_3COOH \rightarrow CH_4 + CO_2$	-58
	CO_2 reduction:	$CO_2 + 4H_2 \rightarrow CH_4 + 2\ H_2O$	

[a]*Values for standard free energies (at standard conditions and pH 7) taken from Berner (1980) and Stumm and Morgan (1970).*

Aerobic respiration:

$$(CH_2O)_x(NH_3)_y(H_3PO_4)_z + x\ O_2 \longrightarrow x\ CO_2 + x\ H_2O + y\ NH_3 + z\ H_3PO_4 \quad (4)$$

Fermentation:

$$12(CH_2O)_x(NH_3)_y(H_3PO_4)_z \longrightarrow x\ CH_3CH_2COOH + x\ CH_3COOH + 2x\ CH_3CH_2OH + 3x\ CO_2 + x\ H_2 + 12y\ NH_3 + 12z\ H_3PO_4 \quad (5)$$

Sulfate reduction:

$$2(CH_2O)_x(NH_3)_y(H_3PO_4)_z + x\ SO_4^{2-} \longrightarrow 2x\ HCO_3^- + H_2S + 2y\ NH_3 + 2z\ H_3PO_4 \quad (6)$$

The initial composition of the organic material is frequently assumed to follow the Redfield phytoplankton ratio, i.e., an x:y:z of 106:16:1, although with time this ratio may be altered by selective remineralization.

In organic-rich sediments, rapid oxygen depletion resulting from aerobic respiration and the oxidation of reduced chemical species produced anaerobically (e.g., sulfide) may lead to significant degradation occurring via coupled fermentation and sulfate reduction (Jørgensen, 1982). Sulfate reducers may not degrade complex nitrogen-containing compounds but they are apparently capable of completely oxidizing a wide range of fermentation products, primarily the so-called volatile fatty acids (VFAs), carboxylic acids including propionate, butyrate, lactate, and pyruvate (Goldhaber and Kaplan, 1974; Widdel, 1980; Table I). Anaerobic nitrogen remineralization may result primarily from deaminative fermentation, not from sulfate reduction per se, so that a more realistic stoichiometric representation of decomposition would combine Eqs. (5) and (6). Interdependence of these processes is

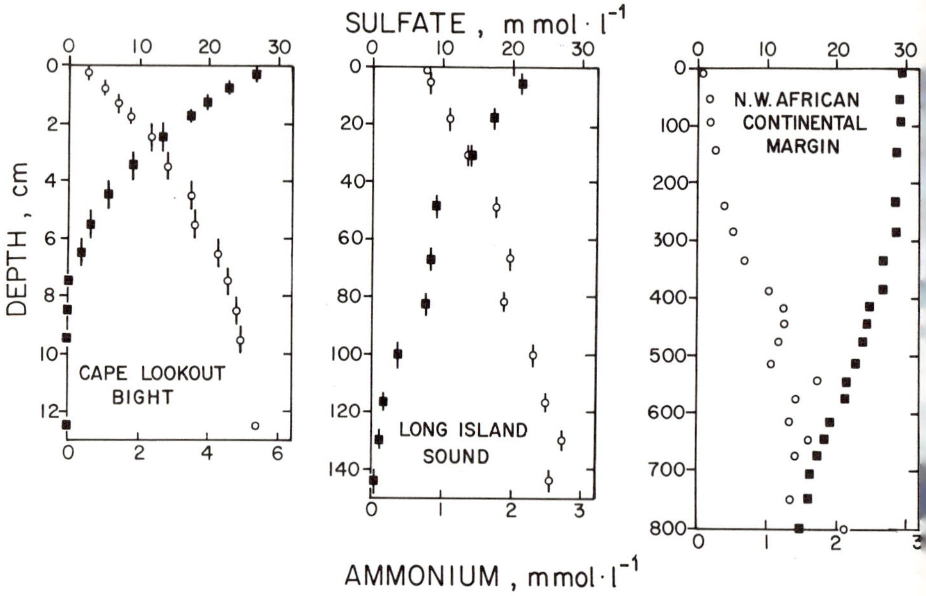

Fig. 3. Dissolved ammonium and sulfate in the pore waters of anoxic marine sediments. Depending on the rate of deposition of degradable organic matter, sulfate depletion and ammonium production may occur at depths from centimeters to meters of the sediment-water interface. Profiles shown: Cape Lookout Bight, North Carolina, ω(sedimentation rate) = 8.4 to 11.6 cm year^{-1}; Long Island Sound, ω = 0.3 cm year^{-1} (Goldhaber et al., 1977); NW African Continental Margin, $\omega \simeq$ 11 cm 1000 year^{-1} (Hartmann et al., 1976).

suggested by the frequent stoichiometric and kinetic correlation between sulfate reduction and ammonium production in anoxic sediments. Observationally, the result is a depletion of sulfate concomitant with an increase in NH_4^+ dissolved in anoxic sedimentary pore water (Fig. 3).

Attempts, therefore, to use benthic oxygen uptake to calculate carbon and nitrogen metabolism in benthic systems by assuming some mole ratio of oxygen consumed to carbon oxidized (frequently called the respiratory quotient), and the C:N ratio of bulk sediment collected at the interface or in a sediment trap face major difficulties. The existence of anaerobic modes of degradation and significantly different C:N ratios (vis-à-vis bulk sediment) in the fraction of organic material actually undergoing degradation frequently renders the stoichiometric coupling of oxygen uptake to nitrogen regeneration undecipherable (Nixon, 1980).

II. THE DYNAMICS OF NITROGEN REGENERATION NEAR THE SEDIMENT-WATER INTERFACE

A. Kinetic Models

There are few direct measurements of the production of remineralized inorganic nitrogen in marine sediments. Difficulties in measuring rates in the deep sea are compounded by the fact that *in situ* benthic metabolism may be an order of magnitude less than measurements made on retrived samples at atmospheric pressure (Smith, 1978; Wirsen and Jannasch, 1976). An approach frequently taken has been to model rates based on pore water distributions of dissolved species. The theoretical basis for this approach has been discussed by Berner (1971, 1974, 1980) through the development of the general one-dimensional diagenetic transport-reaction model.

A simplified version, which ignores compaction and assumes simple linear adsorption [reasonable, e.g., for ammonia in anoxic sediments (Rosenfeld, 1979a; Boatman and Murray, 1982)] and transport by diffusion in pore water only, describes the time-dependent changes in the concentration of a dissolved constituent at a fixed depth via three basic processes: (1) changes due to diffusion into and out of the fixed layer, (2) advective changes due to burial, and (3) changes due to the rapid production or consumption of the constituent as a result of diagenesis and organic matter decomposition (Berner, 1971; Lerman, 1976):

$$\frac{\partial C}{\partial t} = \underbrace{\frac{1}{1+K}\frac{\partial}{Z}\left(D_s\frac{\partial C}{\partial Z}\right)}_{\text{diffusion}} - \underbrace{\omega\frac{\partial C}{\partial Z}}_{\text{burial}} + \underbrace{\frac{\Sigma R}{1+K}}_{\text{reaction}} \tag{7}$$

where C is the concentration of the dissolved species in pore water, D_s the whole-sediment molecular diffusion coefficient for C, z depth below the sediment-water interface, ω the net sediment accumulation rate, K the dimensionless linear adsorption isotherm, and R the sum of all the rates of chemical reaction affecting C. In most fine-grained nearshore environments, the advective term is relatively unimportant for nitrogen diagenesis, hence transport functions and reaction rates are the most critical parameters controlling pore water distributions and benthic fluxes.

Modeling of nitrogen diagenesis from pore water concentration profiles has been used to describe the steady-state vertical distributions of dissolved nitrate in the interstitial water of marine sediments undergoing nitrification and denitrification (Vanderborght and Billen, 1975; Grundmanis and Murray, 1977; Vanderborght et al., 1977b) and to determine rate constants for organic nitrogen decay to ammonium, as well as ammonium distributions in anoxic sediments (Berner, 1974; Goldhaber et al., 1977; Murray et al., 1978; Rosenfeld, 1981; Aller, 1977). Berner (1977) has also shown that if decomposition follows first-order kinetics and is stoichiometric, the C:N ratio of the decomposing organic matter during sulfate reduction may be ascertained from pore water profiles of dissolved NH_4^+ and SO_4^{2-}. The regeneration of nitrogen during sulfate reduction can be expressed as his Eq. (6-59) (Berner, 1980):

$$\frac{G}{N} = \frac{-k_N}{Lk_S}\left[\frac{\omega^2 + k_S D_{sS}}{\omega^2(1-K_N) + k_N D_{sN}}\right]\frac{dC_S}{dC_N} \tag{8}$$

where subscripts S and N refer to SO_4^{2-} and NH_4^+, respectively, and where G/N is the C:N ratio of the decomposing organic matter, L is the ratio of sulfate ions reduced per atom of total carbon oxidized during sulfate reduction, i.e., 0.5 [see Eq. (6)], k is the first-order rate constant for organic matter decay, dC_S/dC_N is the slope of the dissolved NH_4^+ versus SO_4^{2-} plot, and ω, K and D_s are the

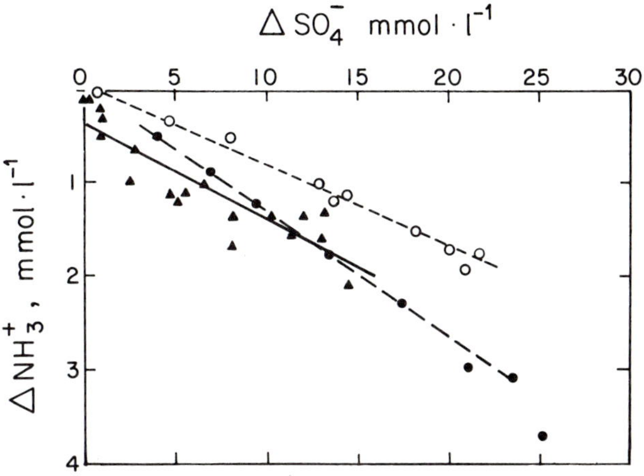

Fig. 4. Linear ΔSO_4^{2-} versus ΔNH_4^+ plots for the profiles given in Fig. 3. Using Eq. (15), the approximate C:N ratio of the decomposing organic matter may be ascertained: Cape Lookout Bight (solid circles) C:N ≃ 106:14; Long Island Sound (open circles) C:N ≃ 106:8.6; NW African Continental Margin (solid triangles) C:N ≃ 106:10.

sedimentation rate, adsorption coefficient ($K_S = 0$), and molecular diffusivities, respectively. If the slope is linear, all parameters on the right side of Eq. (8) are constants (given the model assumptions) and G/N, therefore, must also be a constant. This can be true only if these parameters are independent of depth (Berner, 1980). Linear SO_4^{2-} versus NH_4^+ plots have been observed in a variety of anoxic nearshore and coastal sediments (Sholkovitz, 1973; Hartmann et al., 1976; Suess, 1976; Martens et al., 1978; Klump, 1980; Rosenfeld, 1981; Elderfield et al., 1981b; see Fig. 4), hence, by inference $k_N = k_S$ and the C:N stoichiometry of the decomposing organic matter may be deduced. If the rate constants for ammonium production and sulfate reduction are not equal, then depth dependent terms arise and the situation may no longer be simple. Knowledge of the elemental composition of the metabolizable fraction of sediments should aid our understanding of the nature of the coupling between benthic and pelagic systems. More specific information on the delivery and metabolism of nitrogenous debris in benthic systems will require more detailed characterization of the specific compounds and chemical transformation pathways involved (e.g., Lee and Cronin, 1982).

B. Rates of Nitrogen Regeneration

Overall rates of nitrogen regeneration in sediments are controlled by the rates at which the various remineralization reactions occur within the sediment and at the sediment-water interface. These rates vary both in time, due to the temperature dependence of microbial metabolism, and with depth, both due to the sequential degradation of increasingly refractory material as the more labile compounds are preferentially exhausted and due to the successive use of less efficient anaerobic modes of degradation.

1. Variations with Depth

Bacterial numbers, biomass, and activity in sediments are usually orders of magnitude above that of the overlying water and generally decrease with depth into the sediment (ZoBell and Anderson, 1936; Rittenberg, 1940; Wieser and Zech, 1976; Christensen and Packard, 1977; Christensen and Devol, 1980; Aller and Yingst, 1980; Jørgensen, 1982). Respiratory electron transport system (ETS) activities in the sediments beneath the Northwest Africa upwelling system correlated closely with sedimentary primary amino nitrogen content, and decreased both as a function of the depth in the sediment and of the distance offshore (Christensen and Packard, 1977). Direct measurements of nitrogen remineralization in marine sediments, however, are difficult to make and have been hindered by the lack of generally available nitrogen tracer methodology. In nearshore anoxic sediments, ammonium production rates have been successfully measured in sediments via incubation in closed containers for periods of days to months (Aller and Yingst, 1980; Aller, 1980a; Klump, 1980; Rosenfeld, 1981), and via $^{15}NH_4^+$ dilution techniques on both homogenized and undisturbed sediments (Blackburn, 1979a,b; Blackburn and Henriksen,

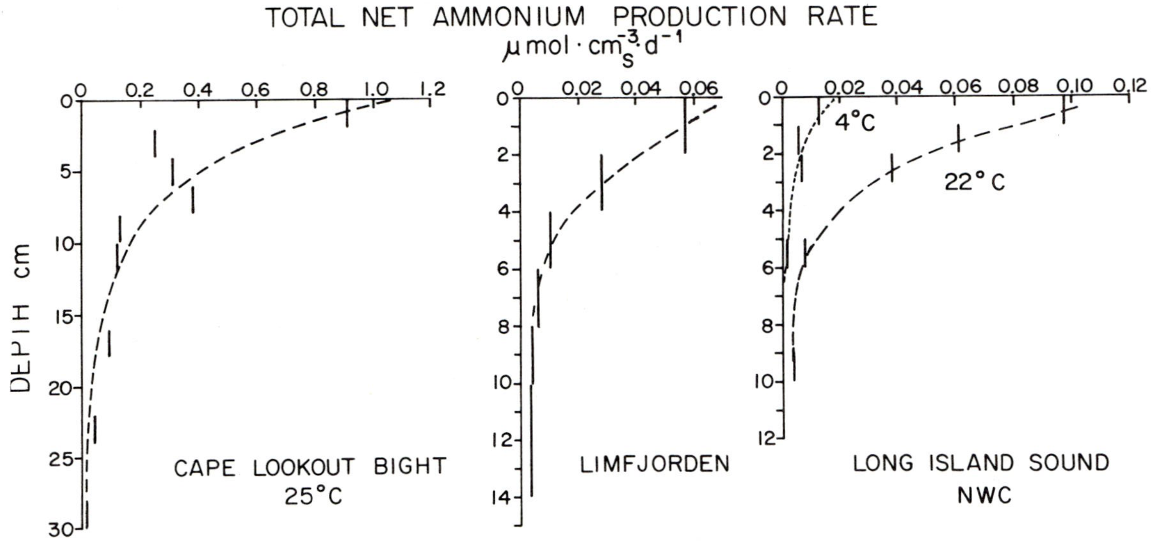

Fig. 5. Ammonium production rates (μmol·cm^{-3} sediment·day^{-1}) in coastal marine sediments as a function of depth for Cape Lookout Bight (Klump, 1980); Limfjorden (Blackburn, 1980); Long Island Sound (Aller and Yingst, 1980).

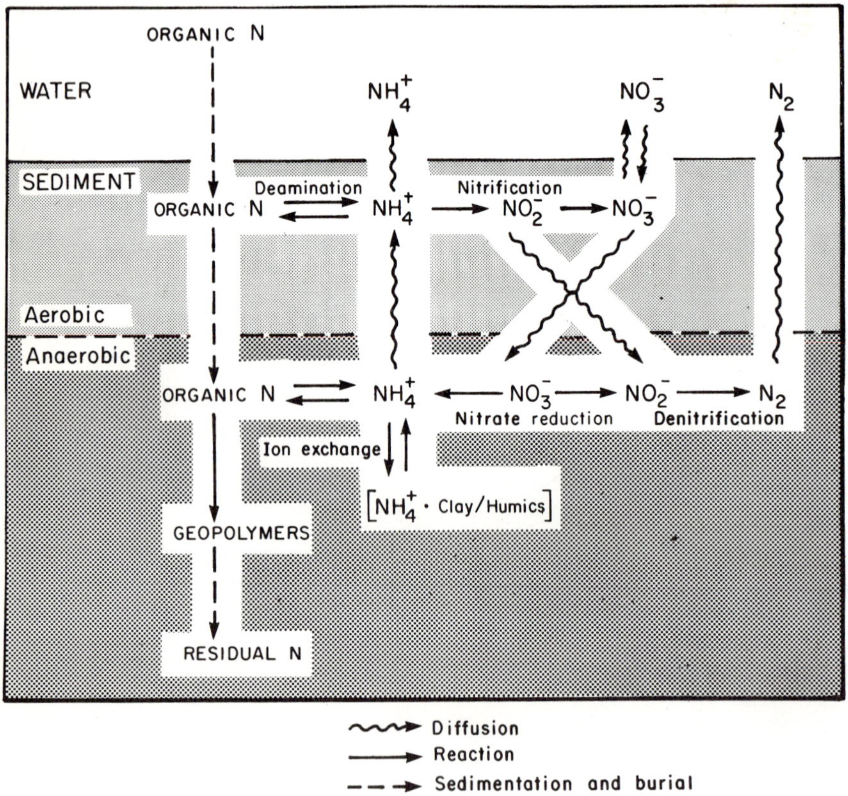

Fig. 6. Nitrogen cycling pathways in marine sediments. (After Fenchel and Blackburn, 1979; Grundmanis and Murray, 1977; Suess et al., 1980.) Key: ∿∿∿> , diffusion; ⟶ , reaction; ---> , sedimentation and burial.

1982). Even in rapidly accumulating or bioturbated sediments ammonium production rates may decrease sharply with depth (Fig. 5), with frequently as much as 75-100% of the total production occurring within the upper 10-15 cm (Blackburn, 1979b; Aller and Yingst, 1980; Aller, 1980a; Klump, 1980; Rosenfeld, 1981).

Measured net ammonium production or consumption must be corrected to account for the rapid adsorption or desorption of ammonium by sediment particles (Rosenfeld, 1981). The exchangeable NH_4^+ capacity of sediments varies according to their relative organic content and clay mineralogy, and one-half to two-thirds or more of the ammonium in organic-rich sediments may be adsorbed on particles (Rosenfeld, 1979a; Boatman and Murray, 1982). Adsorption constants must be determined for all sediments in which ammonium production rates are measured, and estimates of the readily exchangeable ammonium pool made by desorption with monovalent cations (e.g., K^+, Li^+) have, in some instances, been thought to be too high due to the release of hydrolyzable amines during extraction (Boatman and Murray, 1982).

While nitrogen regenerated from remineralized organic matter is probably the most important mode of benthic nitrogen recycling, nitrification, nitrate reduction, and denitrification may play significant roles in determining the form and flux of nitrogen across the sediment-water interface in marine oxic and suboxic sediments (Fig. 6). High concentrations of ammonium suppress the assimilatory reduction of nitrate to ammonium for incorporation into cell material, but do not suppress dissimilatory nitrate-nitrite reduction (Payne, 1973). The production of both N_2 and NH_4^+ from NO_3^- may occur simultaneously in anaerobic sediments, even in the presence of significant quantities of NH_4^+, leading recent investigations to suggest that dissimilatory NO_3^- reduction to NH_4^+ may occur and be of the same magnitude as denitrification in organic-rich sediments (Koike and Hattori, 1978; Sørensen, 1978a). In sandy, organic-poor sediments, however, denitrification may predominate (Koike and Hattori, 1978; Smith, 1982; Hattori, Chapter 6, this volume).

The oxidation of ammonium to nitrate (nitrification) on the other hand, did not appear to be correlated with sedimentary organic content in Danish coastal sediments (Henriksen et al., 1981). The chemoautotrophic nitrifying bacteria are strict aerobes (Painter, 1970), although they may be found in underlying anoxic sediments in an inactive state, and prefer low concentrations of O_2 and high concentrations of NH_4^+. The exact nature of the steps and intermediates produced during nitrification are not entirely known. Hydroxylamine and N_2O, for example, have been detected during the oxidation of ammonium, presumably as an unstable intermediate and a by-product, respectively, of the first phase of nitrification (Rajendran and Venugopalan, 1976; Goreau et al., 1980; Bremner and Blackmer, 1981). However, in simple terms the reaction occurs in essentially two steps: the oxidation of ammonium to nitrite (e.g., by *Nitrosomonas*) and the oxidation of nitrite to nitrate (e.g., by *Nitrobacter*), giving a net reaction of 2 moles of O_2 consumed for each mole of NH_4^+ oxidized (Table I). Generally the reactions go to completion and the intermediates, particularly NO_2^-, accumulate only under conditions more unfavorable to *Nitrobacter* than to *Nitrosomonas*, for example, high pH or low temperature (Alexander, 1965). Since temperature, oxygen penetration, NH_4^+ concentration, H_2S inhibition, and the indigenous bacterial population all appear to control nitrification rates in sediments, rates are best measured using intact, undisturbed cores (Henriksen et al., 1981; Hansen et al., 1981). Actual rates in the range of 13-54 μmol NH_4^+ oxidized$\cdot m^{-2}\cdot h^{-1}$ have been measured in a variety of muddy and sandy Danish coastal sediments in November and July (Henriksen et al., 1981).

For both nearshore and pelagic sediments, Barnes et al. (1975) [California borderland basins], Grundmanis and Murray (1977) [Puget Sound], and Suess et al. (1980) [SW Pacific] proposed a coupling of nitrification and denitrification in surface sediments based on observed pore water profiles for inorganic nitrogen. They suggested that ammonium, either produced *in situ* or diffusing upward into an oxygenated zone, is nitrified through a nitrite intermediate and the nitrate produced is subsequently denitrified in adjacent anoxic layers (Fig. 6).

The role of denitrification in coastal sediments is not well known, but recent studies have indicated that at times, N_2 (and N_2O) release from sediments may constitute a significant amount of the total nitrogen regenerated by the benthos. Rates of denitrification have been measured using a variety of methods and sediments, including the conversion of $^{15}NO_3^-$ to $^{15}N_2$ in sediment slurries (Koike and Hattori, 1979; Oren and Blackburn, 1979) and in undisturbed cores (Nishio et al., 1982), acetylene inhibition and N_2O accumulation (Sørensen, 1978b; Kaspar, 1982), and N_2 production from undisturbed cores incubated in gas tight $He/O_2/CO_2$ purged chambers (Seitzinger et al., 1980) or in bell jars placed on the sediment (Kaplan et al., 1979). Rates on the order of 1-10 μmol $N \cdot m^{-2} \cdot h^{-1}$ have been measured on the Bering Sea shelf (Koike and Hattori, 1979), in Kysing fjord (Oren and Blackburn, 1979), and in Danish coastal sediments (Blackburn and Henriksen, 1982); rates in the range of 10-100 μmol $N \cdot m^{-2} \cdot h^{-1}$ have been measured in Odawa and Tokyo Bays (Nishio et al., 1982), Narragansett Bay (Seitzinger et al., 1980), the Great Sippewissett Marsh (Kaplan et al., 1979), and the North Sea (Billen, 1978); and rates as high as 300-650 μmol $N \cdot m^{-2} \cdot h^{-1}$ have been measured in undisturbed sediment cores taken from the Tama Estuary, Japan (Nishio et al., 1982). In some instances, measurements on chemically or physically altered sediments may reflect potential denitrification rather than *in situ* rates as a result of the stimulation of facultative denitrifiers under incubation conditions. The levels of nitrate, oxygen, sulfide, organic matter, and bacteria in the undisturbed sediment matrix may hold denitrification rates below maximum capacity. Kaspar (1982) found *in situ* denitrification rates (0.8-33 μmol $N \cdot m^{-2} \cdot h^{-1}$) measured on undisturbed cores to be 0.1-2.5% of the measured denitrification potential of slurried sediments on a New Zealand tidal flat. Estimates of the fraction of the total nitrogen remineralized in coastal sediments lost as N_2 and N_2O have been made for the North Sea, 15-20% (Billen, 1978); for Danish coastal waters, 7-26% (Blackburn and Henriksen, 1981); and for Narragansett Bay, 30% (Seitzinger et al., 1980; summer value). Hence, benthic denitrification may represent a pathway for significant losses of combined nitrogen in coastal systems and may lead to low N:P ratios measured for nutrients regenerated within sediments (Nixon, 1981).

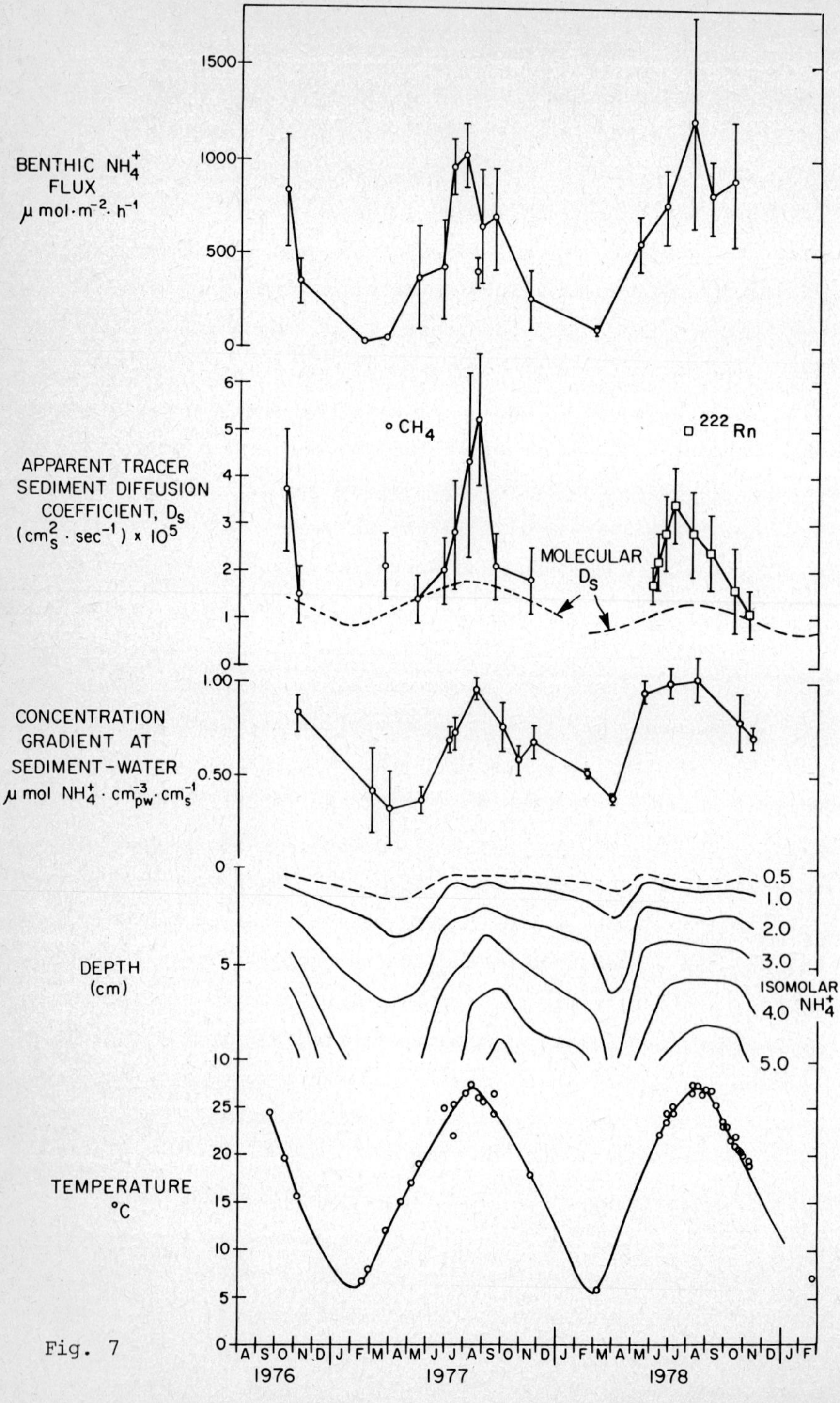

BENTHIC NH4+ FLUX μmol·m-2·h-1
1500
1000
500
0
APPARENT TRACER SEDIMENT DIFFUSION COEFFICIENT, Ds (cm2s · sec-1) x 105
○ CH4
□ 222Rn
MOLECULAR Ds
CONCENTRATION GRADIENT AT SEDIMENT-WATER μmol NH4+ · cm-3pw · cm-1s
1.00
0.50
DEPTH (cm)
0.5
1.0
2.0
3.0
4.0
5.0
ISOMOLAR NH4+
TEMPERATURE °C
A S O N D J F M A M J J A S O N D J F M A M J J A S O N D J F
1976
1977
1978

Fig. 7

2. Variations with Temperature and Season

The paramount physical factor affecting rates at the same site in coastal systems is temperature. Biological rates increase exponentially with temperature, generally showing a two- to four-fold increase for a 10°C temperature rise. In temperate systems exhibiting an ∿20°C temperature range, this imparts a significant seasonality to the microbially mediated regeneration processes. Pore water profiles, concentration gradients, fluxes, and transport processes are all strongly influenced by these seasonal temperature variations (Fig. 7) (Nixon et al., 1976; Aller, 1977; Goldhaber et al., 1977; Martens et al., 1980; Blackburn, 1980; Klump, 1980; Klump and Martens, 1981; Aller and Benninger, 1981). Temperature dependence of rates may be expressed as an Arrhenius function:

$$\text{Rate} = A \exp(-\Delta H/RT) \tag{9}$$

where ΔH, R, and T are the Arrhenius constant, the universal gas constant, and temperature in °K, respectively. Temperature characteristics typical for processes important to benthic regeneration are given in Table II. Because of temperature, seasonal asymmetry in nutrient regeneration may be pronounced (Fig. 7). Seventy percent of the nitrogen and 90% of the phosphorus regenerated by the sediments in Cape Lookout Bight on an annual basis are released during the period of June through October (Klump, 1980; Klump and Martens, 1981). Dissolved reactive phosphorus diffusing from anoxic sediments through an aerobic layer is removed very effectively from solution in these sediments in the winter, whereas dissolved ammonium is not (Klump and Martens, 1981). Complete

Fig. 7. Seasonal variations in temperature, dissolved pore water ammonium, ammonium concentration gradients at the sediment-water interface, measured apparent bulk sediment diffusivities, and ammonium fluxes at the sediment-water interface in Cape Lookout Bight, North Carolina (data from Klump and Martens, 1981). Benthic nitrogen cycling processes are strongly influenced by temperature in coastal systems and may vary by up to 3 orders of magnitude during the course of a year.

TABLE II. Temperature Characteristics for Processes Important in Benthic Nitrogen Regeneration

Process	*Temperature range (°C)*	ΔH *(kcal* mol^{-1}*)*	Q_{10}	*Study site*
NH_4^+ Fluxes across the sediment-water interface	5.5-27.5	22	3.7	Cape Lookout Bight[a]
	5-20	19	3.1	Long Island Sound[b]
	3-24	24	4.1	Narragansett Bay[c]
Bulk sediment NH_4^+ production rates:	5-25	23	3.9	Cape Lookout Bight[a]
		19	2.0	Long Island Sound[d]
Molecular diffusion:	0-30	5	1.4	--[e]

[a] *Klump, 1980; Klump and Martens, 1981.*
[b] *Aller and Benninger, 1981.*
[c] *Nixon et al., 1976.*
[d] *Aller and Yingst, 1980.*
[e] *Li and Gregory, 1974.*

oxygen depletion up to the sediment-water interface in the summer is the primary reason for the greater asymmetry in phosphorus release as reducing conditions apparently remobilize previously oxidized insoluble phosphorus complexes. The fluctuating supply of oxygen and other electron acceptors as a consequence of seasonal variations in oxygen demand, irrigation, and sediment biogenic reworking may result in changes in the relative importance of nitrification, nitrate reduction, and denitrification in sediments both in time and space.

Seasonality is also imparted to benthic systems via fluctuations in the supply of detritus driven by annual cycles in pelagic primary productivity, benthic productivity, and the senescence and die-off of aquatic and terrestrial macrophytes (Davies, 1975; Hargrave, 1980). Even the deep sea, traditionally thought to be one of the least variable environments on earth, has been shown recently to be subject to seasonal fluctuations in particle deposition varying by a factor of 3 or more (Deuser and Ross, 1980; Deuser et al., 1980, 1981). These relatively short-term oscillations in input will not necessarily alter the steady-state appearance of pore water profiles (see Lasaga and Holland, 1976 for discussion), yet they may have a significant effect on the rate and timing of nutrient regeneration by benthos.

Both types of seasonal variations highlight the need for non-steady-state experimental approaches to benthic biogeochemical cycling processes. The widespread variability of benthic nitrogen regeneration and a lack of time-dependent data sets are probably the major reasons why no one has been able to establish a significant correlation between a specific sediment characteristic and NH_4^+ fluxes at a given site. Cyclic and aperiodic variations may affect the magnitude and timing of pelagic primary production and may help to dampen or enhance natural fluctuations in phytoplankton growth in coastal environments.

C. Transport Processes

The major controls on the exchange of dissolved nitrogen between sediments and the overlying water are (1) the rate at which a particular form of dissolved nitrogen is produced or consumed within the sediment, and (2) the mode and rate of transport of dissolved constituents in the benthic system. At steady state, the net depth-integrated rate of production or consumption for a dissolved form of nitrogen should equal its flux across the sediment-water interface. Under non-steady-state conditions, fluxes are controlled by a difficult-to-model coupling of rates and transport processes.

Molecular diffusion is one of the principal mechanisms for solute transport in sediment-water systems and represents the theoretical minimum expected rate of movement. The time T required for a solute to travel a distance z via molecular diffusion is given by

$$T = z^2 D_s^{-1} \tag{10}$$

and the diffusive flux J, resulting from a concentration gradient dC/dz, is given by

$$J = -\phi D_s \ (dC/dz) \tag{11}$$

where D_s is the tortuosity corrected bulk sediment molecular diffusivity and ϕ the porosity (Berner, 1971; Li and Gregory, 1974). However, many studies using mass balance considerations or *in situ* tracer work have observed transport rates and fluxes in excess of those predicted via molecular diffusion (e.g., Hammond and Fuller, 1979; Martens et al., 1980). Nondiffusive processes such as physical stirring by currents and wave action (Webb and Theodor, 1968; Davies, 1975; Hammond et al., 1977; Vanderborght et al., 1977a; Rutgers van der Loeff, 1980a) and by benthic invertebrates (Robbins et al., 1977), irrigation by macrofauna (Goldhaber et al., 1977; Schink and Guinasso, 1977; Grundmanis and Murray, 1977; Aller, 1977; Hammond and Fuller, 1979; McCaffrey et al., 1980;

Henriksen et al., 1980), tidal pumping (Riedl et al., 1972), convective overturn of pore waters due to density changes (Hesslein, 1976; Smetacek et al., 1976; Reeburgh, 1978), and gas ebullition (Martens et al., 1980; Martens and Klump, 1980; Klump and Martens, 1981; Kipphut and Martens, 1982; Fig. 7) have also been shown or hypothesized to modify the mode of sediment-water chemical exchange. Enhancement of *in situ* measured fluxes of ammonium above calculated molecular diffusive fluxes [based on Fick's first law, Eq. (11)] have been observed to be as high as 3 in ebullient sediments (Klump and Martens, 1981) and as high as 20 in irrigated sediments (Callender and Hammond, 1982).

In some instances differences between measured and gradient-supported calculated fluxes may result from rapid remineralization of organic nitrogen and/or rapid removal of dissolved nitrogen at the sediment-water interface (e.g., by benthic algae; Henriksen et al., 1980). If active right at the sediment-water contact, interface processes may not visibly influence the shape of measured vertical pore water concentration gradients and would be detectable only by direct measurement of benthic nutrient fluxes. The actual rate of chemical exchange is ultimately controlled by the rates of production or consumption reactions in sediments; however, rapid irrigation by macrofauna may allow an advective exchange of interstitial and overlying waters that circumvents normal interface processes. Extensive burrowing may also increase the effective area of sediment-water exchange, resulting in greater inputs of oxygen, nitrate, and other dissolved constituents to sediments at depth.

D. Two-Dimensional and Non-Steady-State Modeling

Because of temperature-dependent seasonal variations and non-diffusive solute transport mechanisms, particularly irrigation, one-dimensional steady-state models for nitrogen diagenesis have not always been successful. Recently, Aller (1980b) developed a two-dimensional approach to quantifying solute distributions in

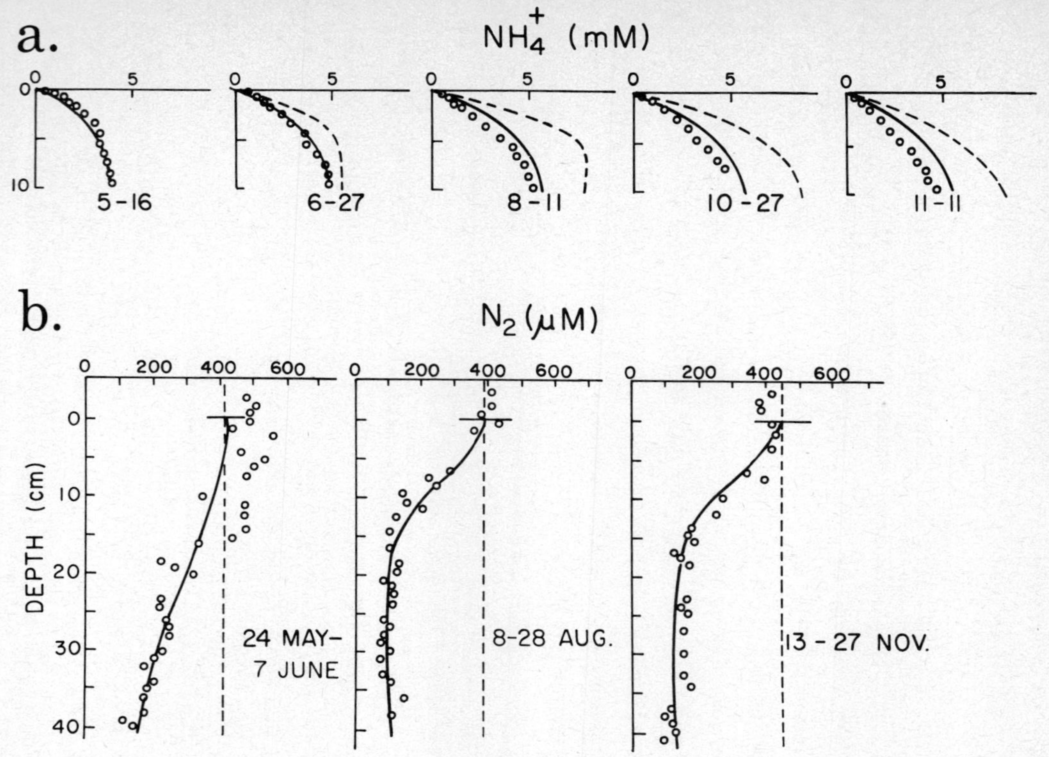

a.
NH_4^+ (mM)
DEPTH (cm)
0
5
10
5-16
6-27
8-11
10-27
11-11
b.
N_2 (μM)
0
200
400
600
DEPTH (cm)
10
20
30
40
24 MAY-
7 JUNE
8-28 AUG.
13 - 27 NOV.

bioturbated sediments through the definition of average microenvironments consisting of single tube-dwelling animals and their surrounding sediments, and represented by close-packed, finite hollow cylinders. He showed that vertical pore water solute distributions, previously modeled using empirical or apparent vertical rates of diffusion and advection, may be explained by changes in the geometry of diffusion and transport induced by macrofauna. The hollow cylinder model incorporates both horizontal and vertical diffusive transport and is constrained by the observed maintenance of near-bottom water solute concentrations within infauna burrows by active irrigation.

Non-steady-state distributions of dissolved inorganic nitrogen have also been recently modeled numerically in organic-rich sediments subject to rapid deposition, compaction, and methane gas ebullition. Klump (1980), using a more general form of Eq. (7),

Fig. 8. (A) Results of a time- and depth-dependent diagenetic model for dissolved pore water ammonium in the sediments of Cape Lookout Bight. Simulation was run under two conditions: (1) using molecular whole-sediment diffusivities only (dashed line); (2) using molecular whole-sediment diffusivities corrected to include the Rn-222 observed transport enhancement for sediment-water exchange (solid line). Open circles represent actual pore water concentrations during this time interval (May-November) (from Klump, 1980). As temperatures increase, ammonium production within the sediment exceeds the flux of ammonium into the overlying water, and ammonium concentrations in pore water increase. As temperatures drop, the reverse occurs and ammonium concentrations decrease, resulting in an annual cycle of expanding and collapsing ammonium profiles coupled to the processes shown in Fig. 7. (B) N_2 distributions in these same sediments during April-November. Dashed line represents dissolved N_2 concentration at equilibrium with the atmosphere. Solid line represents results of a time-dependent diagenetic model which models dissolved gas transport via both diffusion and gas stripping by rising methane bubbles produced below the zone of sulfate reduction (>10 cm). Note N_2 is always below atmospheric equilibrium at depth due to ebullition. Prior to the onset of ebullition in the spring (e.g., May profile), N_2 saturation reaches to 10- to 20-cm depth in the sediment. Once ebullition begins, N_2 is quickly stripped from pore water and is below saturation at all depths (e.g., August profile). (From Kipphut and Martens, 1982.)

combined the observed temperature dependence of nitrogen remineralization rates (Table II) with the observed exponential decrease in those rates with depth (Fig. 5) in an ammonium production rate expression of the form:

$$\mathrm{Rate}_{N(t,z)} = [A \exp(-H/RT)][\exp(-bz)] \quad (12)$$

where

$$T(\text{temperature in } ^{\circ}\mathrm{K}) = x \cos(2/365t) + y \sin(2/365t) + \overline{T} \quad (13)$$

reflecting the sinusoidal annual temperature cycle (Fig. 7). Using measured ammonium adsorption coefficients, sedimentation rates, and depth-dependent porosity changes, observed NH_4^+ distributions were compared with those predicted from both molecular diffusion-controlled transport and ^{222}Rn-tracer observed enhanced transport. During the period of ebullition (May-November), the incorporation of enhanced transport in the zone of gas bubble formation and release significantly improved model predictions of actual concentrations (Fig. 8A). More recently, Kipphut and Martens (1982) successfully modeled the time-dependent N_2 distributions in these same sediments using an expanded transport function that includes both diffusion and gas stripping by rising methane bubbles (Fig. 8B).

III. FLUXES, TURNOVER, AND BUDGETS

The overall impact of the processes discussed above can be directly assessed by constructing benthic nitrogen budgets. Some of the major processes controlling such a budget are illustrated schematically in Fig. 9. Once deposited, organic material suffers one of two fates (assuming no subsequent erosion or migration): It is either remineralized and released back into the overlying water or it is permanently incorporated into the sedimentary

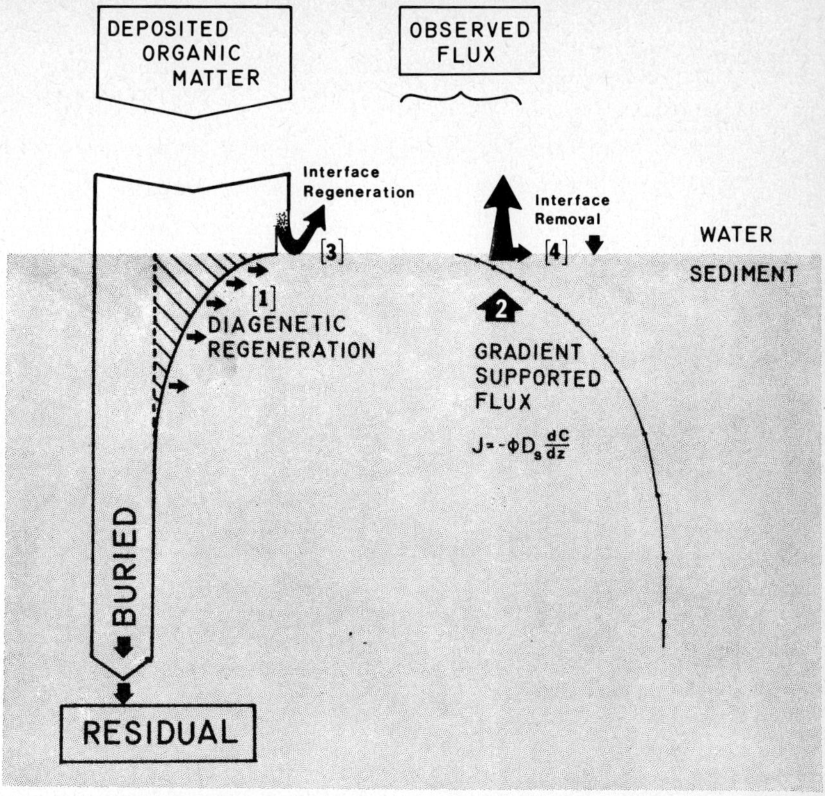

Fig. 9. The flux of remineralized constituents (*C*) across an idealized benthic boundary is the net result of diagenesis (1) and vertical diffusion and transport (2) in the sediment column, and rapid remineralization (3) and removal (4) at the sediment-water interface. Inputs to the benthic system are either regenerated and released back into the overlying water or buried in a residual fraction. (From Klump and Martens, 1981.)

deposit. Budgets for organic matter cycling in sediments are frequently complicated by the difficulty in estimating total input for the construction of chemical mass balances. Rapid diagenesis and regeneration at the sediment-water interface frequently renders surface sediment carbon and nitrogen contents unusable for determining carbon and nitrogen inputs. Sediment-trap collection of inputs may give more accurate estimates of the flux of sedimenting labile material, but they are susceptible to a variety of other problems, e.g., dilution by resuspended material, seasonal or aperiodic fluctuations in input, by-pass by vertically migrating organisms, etc. If long-term burial and release rates are known, cyclic or aperiodic fluctuations in input or *in situ* fixation can be accounted for, and the amount of any component entering the system may be approximated by the sum of the amount lost and the amount retained in the system (Klump, 1980; Martens and Klump, 1983):

$$N_{in} = N_{out} + N_{residual} \tag{14}$$

where N_{in}, N_{out}, and $N_{residual}$ are the rates of nitrogen input, loss, and burial, respectively. The difficulties inherent in measuring the rate of delivery of material to the sediment may be avoided by indirectly estimating inputs through accurate measurements of the output side of this sedimentary chemical mass balance. The net turnover or regeneration can then be cast in terms of Eq. (14):

$$\text{Turnover} = N_{out}/N_{in} = N_{out}/(N_{out}+N_{residual}) \tag{15}$$

where N_{out} is the flux of regenerated nitrogen to the overlying water in both soluble organic and inorganic forms, summed over some period sufficiently long for the system to be considered a steady state (e.g., at least 1 year to account for seasonal variations in rates and fluxes). This, however, is generally a difficult task, particularly where nitrogen fluxes vary seasonally or where denitrification is significant. In at least one environment, Cape Lookout Bight, North Carolina, where sufficient seasonal flux data currently exist to make this estimate, it appears that approximately 40 ± 10% of the total nitrogen input to the sediments is actually released back into the overlying water (Klump, 1980). The remainder is buried, on a time scale of at least decades. Using the same approach, Martens and Klump (1983) estimated that 26 ± 6% of the organic carbon in these sediments is regenerated.

Estimates of burial losses relative to sediment-trap estimated delivery rates of organic material to the benthos range from approximately 30% of the primary flux of carbon in Loch Ewe during the period April through September (Davies, 1975) to 20-25% on the continental slope (Rowe and Gardner, 1979) to 1-10% in deep-sea North Atlantic and equatorial pelagic sediments (Hinga et al., 1979). These estimates follow the general trend observed by many workers (see Müller and Suess, 1979, for review), wherein the relative amount of organic carbon preserved within the sediments correlates with sedimentation rates and pelagic primary productivity.

From the abyssal plain to the coastal shore, rates of nitrogen regeneration by benthic systems rise 1000-fold (Table III). Heightened productivity and increased interaction between benthic and pelagic systems in littoral regions are largely responsible. Coastal zones, particularly estuaries, are very effective traps for aquatic and terrestrially derived material, with as much as 75% of the freshwater-borne particulate organic matter settling out within an estuary (Allen et al., 1976; Wollast and Peters, 1978). A significant fraction of this material is apparently refractory in nature and represents radiometrically old (>1000 year) organic carbon (Benoit et al., 1979). For shallow (<100 m) coastal environments, Nixon (1980) has related benthic remineralization of carbon to the amount of organic matter fixed and imported over an annual cycle for an entire system. The correlation for a variety of systems implies, on the average, roughly 25% of total inputs may be remineralized in the benthos of coastal environments. In Narragansett Bay, for example, 70% of the nitrogen inputs (including fixation, precipitation, runoff, and sewage) are apparently recycled within the system, approximately 50% via benthic regeneration and 20% via pelagic regeneration (predominantly by zooplankton (Nixon, 1980).

While sediments ultimately act as sinks for nitrogen in any depositional environment, benthic nitrogen regeneration is capable of supplying the entire nitrogen requirement of primary producers

Discussion

TABLE III. Fluxes of Dissolved Combined Inorganic Nitrogen Across the Sediment-Water Interface

Location	*Depth (m)*	*Temperature (°C)*	Flux across the sediment-water interface ($\mu mol \cdot m^{-2} \cdot h^{-1}$) NH_4^+	NO_3^-	+	NO_3^-	*Contribution of benthic N regeneration to primary productivity (%)*	*References*
Pelagic and hemi-pelagic sediments								
San Diego trough	1193	4	21.7	(-2.09)		0.78		1
NE Pacific, Patton Escarpment	3815	2	11.8	26.6		0.09		1
NE Atlantic continental slope	2200	3	0.95	(-0.95)		0.06		
	2750	3	0.61	0.80		(-0.01)		2
Coastal sediments								
NW African shelf	278	11	18		36			5
NW African shelf (upwelling)	25		235	160	14		30-40	
North Sea	15	5-20	73	216			60-100	
	35		41	49			30-50	7
	nearshore	--	196	(-41)				
	offshore	--	36	8			10-50	20
Danish coastal waters	10-100	4-14	41	17		--	30-70	10
Sea of Japan	7	2-20	(-7)- 227	0-13		--		21
Buzzards Bay	17	1.5-16	2-125	0-5		(-2)-4	up to 200	8
La Jolla Bight	17	11-17	(-2)- 137	(-2)-27		(-0.2)-4	<5	3
Estuarine zones								
Narragansett Bay	10	3-23	(-4)- 400	(-66)-110		0	25-50	12,13

		summer	58- 750				14,15
		winter	12- 75				
Long Island Sound	3-30	20-22	(-4)- 333			20-25	11
Cape Lookout Bight	9	5-27	20-1200	(-7)-3			16
Great South Bay,							
Long Island	1-3	summer	80-2600			∿30	17
		winter	4- 13				
Potomac River							
estuary		summer	42- 270			∿35	18
White Oak River							
estuary	1-4	25	35- 113				22
San Francisco Bay	2	summer	167				19
Basins and fjords							
Santa Barbara Basin	600	6.3	1.8			0.4	4
Loch Ewe	20-30	6-13	0-80	0	0	70	9
Limfjord	4-12	--	74	40	--	40-80	10

[a]*Key to references: 1. Smith et al., 1979; 2. Smith et al., 1978; 3. Hartwig, 1976; 4. Rittenberg et al., 1955 annual average; 5. Hinga et al., 1979; 6. Rowe et al., 1977; 7. Billen, 1978 annual values; 8. Rowe et al., 1975; 9. Davies, 1975; 10. Blackburn and Henriksen, 1982 annual averages from Great Belt and east and west Kattegat; 11. Aller and Benninger, 1981; 12. Hale, 1975; 13. Nixon et al., 1976; 14. Elderfield et al., 1981a; 15. McCaffrey et al., 1980; 16. Klump and Martens, 1981; 17. Dietz, 1981; 18. Callender and Hammond, 1981; 19. Hammond and Fuller, 1979; 20. Rutgers van der Loeff, 1980b; 21. Propp et al., 1980; 22. J. V. Klump, unpublished data.*

on a short-term basis and may supply from one-fifth to one-half on an annual basis in many coastal and estuarine systems (Table III). The bulk of this nitrogen is released from the benthos as ammonium, which generally constitutes at least half of the total nitrogen regenerated in nearshore sediments, and makes up an increasing fraction of the total as increased organic matter inputs force a larger fraction of the benthic metabolism to be anaerobic. Ammonium uptake by sediments is both less frequently observed and never more than a few $\mu mol \cdot m^{-2} \cdot h^{-1}$, whereas ammonium release rates are on the order of a few hundred $\mu mol \cdot m^{-2} \cdot h^{-1}$ in many coastal sediments at summer temperatures. The direction of nitrate and nitrite exchange across the sediment-water interface is more variable, and is probably governed by the complex set of conditions that control the coupling of nitrification, nitrate reduction, and denitrification in sediments. It is therefore often difficult to estimate nitrogen turnover based on limited flux data.

The total variability observed in benthic nitrogen regeneration, even in similar environments, is considerable. To a degree, this is the result of instantaneous measurements in highly time-dependent systems, but it also reflects the variability and patchiness inherent in nearshore benthic communities.

IV. SUMMARY

Nitrogen is rapidly remineralized from nitrogenous organic material deposited in benthic systems primarily via the metabolic activity of microorganisms. These microbially mediated rates of nitrogen regeneration decrease rapidly with depth in sediments (Fig. 5) and are strongly influenced by temperature, organic matter sedimentation rates, primary production, and water depth. These environmental variables, in turn, affect the pathways for nitrogen transformations in sediments (Table I and Fig. 6), the stoichiometry of nutrient regeneration, and the extent of nitrogen preservation and burial. Benthic nutrient regeneration is of greatest importance in shallow coastal environments where nutrient recycling by the benthos may supply a significant fraction of the nitrogen required for primary production (Table III). Despite the importance of the benthic system in the biogeochemical cycling and the nutrient dynamics of coastal environments, few quantitative, seasonally integrated budgets for nitrogen cycling in sediments currently exist. Accurate *in situ* measurements of detrital inputs, fluxes across the sediment-water interface, and combined nitrogen losses via denitrification and burial are needed for a better understanding of nitrogen cycling as a whole, and of the way in which benthic processes modify, dampen, and enhance pelagic processes.

(Center for Great Lakes Studies Contribution No. 255.)

REFERENCES

Alexander, M. (1965). Nitrification. *Agronomy 10,* 307-343.

Allen, G. P., Savzay, G., and Castaing, S. H. (1976). Transport and deposition of suspended sediment in the Gironde estuary, France. *In* "Estuarine Processes" (M. Wiley, ed.), Vol. 2, pp. 63-81. Academic Press, New York.

Aller, R. C. (1977). The influence of macrobenthos on chemical diagenesis of marine sediments. Ph.D. Dissertation, Yale University, New Haven, Connecticut.

Aller, R. C. (1980a). Relationships of tube-dwelling benthos with sediment and overlying water chemistry. *In* "Marine Benthic Dynamics" (K. R. Tenore and B. C. Coull, eds.), pp. 285-308. Univ. of South Carolina Press, Columbia.

Aller, R. C. (1980b). Quantifying solute distributions in the bioturbated zone of marine sediments by defining an average microenvironment. *Geochim. Cosmochim. Acta 44,* 1955-1965.

Aller, R. C., and Benninger, L. K. (1981). Spatial and temporal patterns of dissolved ammonium, manganese, and silica fluxes from bottom sediments of Long Island Sound, USA. *J. Mar. Res. 39,* 295-314.

Aller, R. C., and Yingst, J. Y. (1980). Relationships between microbial distributions and the anaerobic decomposition of organic matter in surface sediments of Long Island Sound, USA. *Mar. Biol. 56,* 29-42.

Anderson, D. Q. (1939). Distribution of organic matter in marine sediments and its availability to further decomposition. *J. Mar. Res. 2,* 225-235.

Barnes, R. O., Bertine, K. K., and Goldberg, E. D. (1975). N_2:Ar, nitrification and denitrification in southern California borderland basin sediments. *Limnol. Oceanogr. 20,* 962-970.
Benoit, G. J., Turekian, K. K., and Benninger, L. K. (1979). Radiocarbon dating of a core from Long Island Sound. *Estuarine Coastal Mar. Res. 9,* 171-180.
Berner, R. A. (1971). "Principles of Chemical Sedimentology." McGraw-Hill, New York.
Berner, R. A. (1974). Kinetic models for the early diagenesis of nitrogen, sulfur, phosphorus and silicon in anoxic marine sediments. *In* "The Sea" (E. D. Goldberg, ed.), Vol. 5, pp. 427-450. Wiley, New York.
Berner, R. A. (1977). Stoichiometric models for nutrient regeneration in anoxic sediments. *Limnol. Oceanogr. 22,* 781-786.
Berner, R. A. (1978). Sulfate reduction and the rate of deposition of marine sediments. *Earth Planet. Sci. Lett. 37,* 492-498.
Berner, R. A. (1980). "Early Diagenesis." Princeton Univ. Press, Princeton, New Jersey.
Billen, G. (1978). A budget of nitrogen recycling in North Sea sediments off the Belgian coast. *Estuarine Coastal Mar. Sci. 7,* 127-146.
Bishop, J. K. B., Edmond, J. M., Ketten, D. R., Bacon, M. P., and Silkers, W. B. (1977). The chemistry, biology, and vertical flux of particulate matter from the upper 400 m of the equatorial Atlantic Ocean. *Deep-Sea Res. 24,* 511-548.
Bishop, J. K. B., Ketten, D. R., and Edmond, J. M. (1978). The chemistry, biology, and vertical flux of particulate matter from the upper 400 m of the Cape Basin in the SE Atlantic Ocean. *Deep-Sea Res. 25,* 1121-1161.
Blackburn, T. H. (1979a). A method for measuring rates of NH_4^+ turnover in anoxic marine sediments using a $^{15}N-NH_4^+$ dilution technique. *Appl. Environ. Microbiol. 37,* 760-765.
Blackburn, T. H. (1979b). N/C ratios and rates of ammonia turnover in anoxic sediments. *In* "Microbiol Degradation of Pollutants in Marine Environments" (A. W. Bourquin and H. Pritchard, eds.), pp. 148-153. U.S. EPA Office of Research and Development, Washington, D.C.
Blackburn, T. H. (1980). Seasonal variations in the rate of organic-N mineralization in anoxic marine sediments. *In* "Biogéochimie de la matière organique à l'interface eau-sediment marin," pp. 173-183. CNRS, Paris.
Blackburn, T. H., and Henriksen, K. (1982). Nitrogen cycling in different types of sediments from Danish waters. (Manuscript)
Boatman, C. D., and Murray, J. W. (1982). Modeling exchangeable NH_4^+ adsorption in marine sediments: Process and controls of adsorption. *Limnol. Oceanogr. 27,* 99-110.
Bremner, J. M., and Blackmer, A. M. (1981). Terrestrial nitrification as a source of atmospheric nitrous oxide. *In* "Denitrification, Nitrification and Atmospheric Nitrous Oxides" (C. C. Delwiche, ed.), pp. 151-170. Wiley (Interscience), New York.

Brock, T. D. (1979). "Biology of Microorganisms." Prentice-Hall, Englewood Cliffs, New Jersey.

Callender, E., and Hammond, D. E. (1982). Nutrient exchange across the sediment-water interface in the Potomac River estuary. *Estuarine Coastal Mar. Sci. 15,* 395-413.

Carter, P. W., and Mitterer, R. M. (1978). Amino acid composition of organic matter associated with carbonate and noncarbonate sediments. *Geochim. Cosmochim. Acta 42,* 1231-1238.

Christensen, D., and Blackburn, T. H. (1980). Turnover of tracer (^{14}C, ^{3}H labelled) alanine in inshore marine sediments. *Mar. Biol. 58,* 97-103.

Christensen, J. P., and Devol, A. H. (1980). ATP and adenylate energy charge in marine sediments. *Mar. Biol. 56,* 175-182.

Christensen, J. P., and Packard, T. T. (1977). Sediment metabolism from the northwest African upwelling system. *Deep-Sea Res. 24,* 331-343.

Claypool, G., and Kaplan, I. R. (1974). The origin and distribution of methane in marine sediments. *In* "Natural Gases in Marine Sediments" (I. R. Kaplan, ed.), pp. 99-139. Plenum, New York.

Cobler, R., and Dymond, J. (1980). Sediment trap experiment on the Galapagos spreading center, equatorial Pacific. *Science 209,* 801-803.

Davies, J. M. (1975). Energy flow through the benthos in a Scottish Sea lock. *Mar. Biol. 31,* 353-362.

Degens, E. T., and Mopper, K. (1975). Early diagenesis of organic matter in marine soils. *Soil Sci. 119,* 65-72.

Deuser, W. G., and Ross, E. H. (1980). Seasonal change in the flux of organic carbon to the deep Sargasso Sea. *Nature (London) 283,* 364-365.

Deuser, W. G., Ross, E. H., Hemleben, C., and Spindler, M. (1980). Seasonal changes in species composition, numbers, mass, size, and isotopic composition of planktonic formanifera settling into the deep Sargasso Sea. *Palaeogeogr. Palaeoclimatol. Palaeoecol. 33,* 103-127.

Deuser, W. G., Ross, E. H., and Anderson, R. F. (1981). Seasonality in the supply of sediment to the deep Sargasso Sea and implications for rapid transfer of matter to the deep ocean. *Deep-Sea Res. 28A,* 495-505.

Dietz, C. G. (1981). Processes controlling NH_4^+ release from lagoonal sediments: Great South Bay, NY. MS Thesis, SUNY Stony Brook, New York

Elderfield, H., Luedtke, N., McCaffrey, R. J., and Bender, M. (1981a). Benthic flux studies in Narragansett Bay. *Am. J. Sci. 281,* 768-787.

Elderfield, H., McCaffrey, R. J., Luedtke, N., Bender, M., and Truesdale, V. W. (1981b). Chemical diagenesis in Narragansett Bay sediments. *Am. J. Sci. 281,* 1021-1055.

Fenchel, T., and Blackburn, T. H. (1979). "Bacteria and Mineral Cycling." Academic Press, New York.

Fenchel, T., and Harrison, P. (1976). The significance of bacterial grazing and mineral cycling for the decomposition of particulate detritus. *In* "The Role of Terrestrial and Aquatic Organisms in the Decomposition Processes" (J. M. Anderson and A. Macfadyen, eds.), pp. 285-299. Blackwell, Oxford.

Froehlich, P. N., Klinkhammer, G. P., Bender, M. L., Luedtke, N. A., Heath, G. R., Cullen, D., Dauphin, P., Hammond, D., Hartman, B., and Maynard, V. (1979). Early oxidation of organic matter in pelagic sediments of eastern equatorial Atlantic: Suboxic diagenesis. *Geochim. Cosmochim. Acta 43,* 1075-1090.

Gagosian, R. B., and Stuermer, D. H. (1977). The cycling of biogenic compounds and their diagenetically transformed products in seawater. *Mar. Chem. 5,* 605-632.

Goldhaber, M. B., and Kaplan, I. R. (1974). The sulfur cycle. *In* "The Sea" (E. D. Goldberg, ed.), Vol. 5, pp. 569-655. Wiley, New York.

Goldhaber, M. B., Aller, R. C., Cochran, J. K., Rosenfeld, J. K., Martens, C. S., and Berner, R. A. (1977). Sulfate reduction, diffusion and bioturbation in Long Island Sound sediments: Report of the FOAM group. *Am. J. Sci. 277,* 193-237.

Goreau, T. J., Kaplan, W. A., Wofsy, S. C., McElroy, M. B., Valois, F. W., and Watson, S. W. (1980). Production of NO_2^- and N_2O by nitrifying bacteria at reduced concentrations of oxygen. *Appl. Environ. Microbiol. 40,* 526-532.

Gottschalk, G. (1979). "Bacterial Metabolism." Springer-Verglag, Berlin and New York.

Grundmanis, V., and Murray, J. W. (1977). Nitrification and denitrification in marine sediments from Puget Sound. *Limnol. Oceanogr. 22,* 804-813.

Hale, S. S. (1975). The role of benthic communities in the nitrogen and phosphorus cycles of an estuary. *In* "Mineral Cycling in Southeastern Ecosystems" (F. G. Howell, J. B. Gentry, and M. H. Smith, eds.), pp. 291-308. ERDA Symposium Series 1975 (CONF-740513).

Hammond, D. E., and Fuller, C. (1979). The use of radon-222 to estimate benthic exchange and atmospheric exchange rates in San Francisco Bay. *In* "San Francisco Bay: The Urbanized Estuary" (T. J. Conomos, ed.), pp. 213-230. AAAS, San Francisco, California.

Hammond, D. E., Simpson, H. J., and Mathieu, G. (1977). Radon-222 distribution and transport across the sediment-water interface in the Hudson River estuary. *JGR, J. Geophys. Res. 82,* 3913-3920.

Hansen, J. I., Henriksen, K., and Blackburn, T. H. (1981). Seasonal distribution of nitrifying bacteria and rates of nitrification in coastal marine sediments. *Microb. Ecol. 7,* 297-304.

Hargrave, B. T. (1973). Coupling carbon flow through some pelagic and benthic communities. *J. Fish. Res. Board Can. 30,* 1317-1326.
Hargrave, B. T. (1980). Factors affecting the flux of organic matter to sediments in a marine bay. *In* "Marine Benthic Dynamics" (K. R. Tenore and B. C. Coull, eds.), pp. 243-263. Univ. of South Carolina Press, Columbia.
Hartmann, M., Muller, P. J., Suess, E., and van der Weijden, C. H. (1976). Chemistry of late quaternary sediments and their interstitial waters from the NW African continental margin. *"Meteor" Forschungs ergeb., Reihe C 24,* 1-67.
Hartwig, E. O. (1976). The impact of nitrogen and phosphorus release from a siliceous sediment on the overlying water. *In* "Estuarine Processes" (M. Wiley, ed.), Vol. 1, pp. 103-117. Academic Press, New York.
Hartwig, E. O. (1978). Factors affecting respiration and photosynthesis by the benthic community of a subtidal siliceous sediment. *Mar. Biol. 46,* 283-293.
Hedges, J. I., and Parker, P. L. (1976). Land-derived organic matter in surface sediments from the gulf of Mexico. *Geochim. Cosmochim. Acta 40,* 1019-1029.
Henrichs, S. M., and Farrington, J. W. (1979). Amino acids in interstitial waters of marine sediments. *Nature (London) 272,* 319-322.
Henriksen, K., Hansen, J. I., and Blackburn, T. H. (1980). The influence of benthic infauna on exchange rates of inorganic nitrogen between sediment and water. *Ophelia, Suppl. 1,* 249-256.
Henriksen, K., Hansen, J. I., and Blackburn, T. H. (1981). Rates of nitrification, distribution of nitrifying bacteria, and nitrate fluxes in different types of sediment from Danish waters. *Mar. Biol. 61,* 299-304.
Hesslein, R. H. (1976). The fluxes of CH_4, CO_2 and NH_3-N from sediments and their consequent distribution in a small lake. Ph.D. Thesis, Columbia University, New York.
Hinga, K. R., Sieburth, J. McN., and Heath, G. R. (1979). The supply and use of organic material at the deep-sea floor. *J. Mar. Res. 37,* 557-579.
Hobbie, J., Howarth, R., Henrichs, S., and Kilham, P. (1980). Amino acid turnover in sediments of Great Sippewissett Marsh, MA. *Amer. Soc. Limnol. Oceanogr. Annu. Meet., 1980* Abstract No. 9.
Holm, T. R., Anderson, M. A., Stanforth, R. R., and Iverson, D. G. (1980). The influence of adsorption on the rates of microbial degradation of arsenic species in sediments. *Limnol. Oceanogr. 25*(1), 23-30.
Honjo, S., and Roman, M. R. (1978). Marine copepod fecal pellets: Production, preservation and sedimentation. *J. Mar. Res. 36,* 45-57.

Jørgensen, B. B. (1981). Processes at the sediment-water interface. *SCOPE Workshop Interact. Biogeochem. Cycles, 1981.*

Jørgensen, B. B. (1982). Mineralization of organic matter in the sea bed--the role of sulphate reduction. *Nature (London) 296,* 643-645.

Kaplan, W., Valiela, I., and Teal, J. M. (1979). Denitrification in a salt marsh ecosystem. *Limnol. Oceanogr. 24,* 726-734.

Kaspar, H. F. (1982). Denitrification in marine sediment: Measurement of capacity and estimate of *in situ* rate. *Appl. Environ. Microbiol. 43,* 522-527.

Kipphut, G. W., and Martens, C. S. (1982). Biogeochemical cycling in an organic-rich coastal marine basin. 3. Dissolved gas transport in methane-saturated sediments. *Geochim. Cosmochim. Acta 46,* 2049-2060.

Klump, J. V. (1980). Benthic nutrient regeneration and the mechanisms of chemical sediment-water exchange in an organic-rich coastal marine sediment. Ph.D. Thesis, University of North Carolina at Chapel Hill.

Klump, J. V., and Martens, C. S. (1981). Biogeochemical cycling in an organic rich coastal marine basin. II. Nutrient sediment-water exchange processes. *Geochim. Cosmochim. Acta 45,* 101-121.

Knauer, G. A., Martin, J. H., and Bruland, K. W. (1979). Fluxes of particulate carbon, nitrogen and phosphorus in the upper water column of the Northeast Pacific. *Deep-Sea Res. 26A,* 97-108.

Koike, I., and Hattori, A. (1978). Denitrification and ammonia formation in anaerobic coastal sediments. *Appl. Environ. Microbiol. 35,* 278-282.

Koike, I., and Hattori, A. (1979). Estimates of denitrification in sediments of the Bering Sea shelf. *Deep-Sea Res. 26A,* 409-415.

Krumbein, W. E., Cohen, Y., and Shilo, M. (1977). Solar Lake (Sinai). 4. Stromatolitic cyanobacterial mats. *Limnol. Oceanogr. 22,* 635-656.

Lasaga, A. C., and Holland, H. D. (1976). Mathematical aspects of non-steady state diagenesis. *Geochim. Cosmochim. Acta 40,* 257-266.

Lee, C., and Cronin, C. (1982). The vertical flux of particulate organic nitrogen in the sea: Decomposition of amino acids in the Peru upwelling area and the equatorial Atlantic. *J. Mar. Res. 40,* 227-251.

Lerman, A. (1976). Migrational processes and chemical reactions in interstitial waters. *In* "The Sea" (E. D. Goldberg, I. N. McCave, J. J. O'Brien, and J. H. Steele, eds.), Vol. 6, pp. 695-738. Wiley, New York.

Li, Y.-H., and Gregory, S. (1974). Diffusion of ions in seawater and in deep-sea sediments. *Geochim. Cosmochim. Acta 38,* 703-714.

McCaffrey, R. J., Myers, A. C., Davey, E., Morrison, G., Bender, M., Luedtke, N., Cullen, D., Froelich, P., and Klinkhammer, G. (1980). The relation between pore water chemistry and benthic fluxes of nutrients and manganese in Narragansett Bay, Rhode Island. *Limnol. Oceanogr. 25*, 31-44.

Martens, C. S. (1978). Some of the chemical consequences of microbially mediated degradation of organic materials in estuarine sediments. *Biogeochem. Estuarine Sediments,* Proc. UNESCO/SCOR Workshop, 1976, pp. 266-278.

Martens, C. S., and Klump, J. V. (1980). Biogeochemical cycling in Cape Lookout Bight. I. Methane sediment-water exchange processes. *Geochim. Cosmochim. Acta 44*, 471-490.

Martens, C. S., and Klump, J. V. (1983). Biogeochemical cycling in an organic-rich coastal marine basin. 4. An organic carbon budget for sediments dominated by sulfate reduction and methanogenesis. *Geochim. Cosmochim. Acta* (in press).

Martens, C. S., Berner, R. A., and Rosenfeld, J. K. (1978). Interstitial water chemistry of anoxic long Island Sound sediments. 2. Nutrient regeneration and phosphate removal. *Limnol. Oceanogr. 23*, 605-617.

Martens, C. S., Kipphut, G., and Klump, J. V. (1980). Coastal sediment-water chemical exchange traced by *in situ* radon-222 measurements. *Science 208*, 285-288.

Martin, J. P., Parsa, A. A., and Haider, K. (1978). Influence of intimate association with humic polymers on biodegradation of [^{14}C]-labeled organic substrates in soil. *Soil Biol. Biochem. 10*, 483-486.

Massey, L. K., Sokatch, J. R., and Conrad, R. S. (1976). Branched-chain amino acid catabolism in bacteria. *Bacteriol. Rev. 40*, 42-54.

Mechalas, B. J. (1974). Pathways and environmental requirements for biogenic gas production in the ocean. *In* "Natural Gases in Marine Sediments" (I. R. Kaplan, ed.), pp. 11-25. Plenum, New York.

Müller, P. J. (1977). C/N ratios in Pacific deep-sea sediments: Effect of inorganic ammonium and organic nitrogen compounds sorbed by clays. *Geochim. Cosmochim. Acta 41*, 765-776.

Müller, P. J., and Suess, E. (1979). Productivity, sedimentation rate and sedimentary organic matter in the oceans. I. Organic carbon preservation. *Deep-Sea Res. 26*, 1347-1362.

Murray, J. W., Grundmanis, V., and Smethie, W. (1978). Interstitial water chemistry in the sediments of Saanich Inlet. *Geochim. Cosmochim. Acta 42*, 1011-1026.

Nishio, T., Koike, I., and Hattori, A. (1982). Denitrification, nitrate reduction, and oxygen consumption in coastal and estuarine sediments. *Appl. Environ. Microbiol. 43*, 648-653.

Nixon, S. W., Oviatt, C. A., and Hale, S. S. (1976). Nitrogen regeneration and the metabolism of coastal marine bottom communities. *In* "The Role of Terrestrial and Aquatic Organisms in Decomposition Processes" (J. Anderson and A. MacFayden, eds.), pp. 269-283. Blackwell, Oxford.

Nixon, S. W. (1981). Remineralization and nutrient cycling in coastal marine ecosystems. *In* "Nutrient Enrichment in Estuaries" (B. Nelson and L. E. Cronin, eds.), pp. 111-138. Humana Press, Clifton, New Jersey.

Oren, A., and Blackburn, T. H. (1979). Estimation of sediment denitrification rates at *in situ* nitrate concentrations. *Appl. Environ. Microbiol.* *37*, 174-176.

Painter, H. A. (1970). A review of literature on inorganic nitrogen metabolism in microorganisms. *Water Res.* *4*, 393-450.

Payne, W. J. (1973). Reduction of nitrogenous oxides by microorganisms. *Bacteriol. Rev.* *37*, 409-452.

Propp, M. V., Tarasoff, V. G., Cherbadgi, I. I., and Lootzik, N. V. (1980). Benthic-pelagic oxygen and nutrient exchange in a coastal region of the sea of Japan. *In* "Marine Benthic Dynamics" (K. R. Tenore and B. C. Coull, eds.), pp. 265-284. Univ. South Carolina Press, Columbia.

Rajendran, A., and Venugopalan, V. K. (1976). Hydroxylamine formation in laboratory experiments on marine nitrification. *Mar. Chem.* *4*, 93-98.

Reeburgh, W. S. (1978). Convective mixing in sediments. *Biogeochem. Estuarine Sediments, Proc. UNESCO/SCOR Workshop, 1976, pp.* 189-190.

Richards, F. A. (1965). Anoxic basins and fjords. *In* "Chemical Oceanography" (J. P. Riley and G. Skirrow, eds.), Vol. 1, pp. 611-645. Academic Press, New York.

Riedl, R. N., Huang, N., and Machan, R. (1972). The subtidal pump: A mechanism of interstitial water exchange by wave action. *Mar. Biol.* *13*, 210-221.

Rittenberg, S. C. (1940). Bacteriological analysis of some long cores of marine sediments. *J. Mar. Res.* *3*, 191-201.

Rittenberg, S. C., Emery, K. O., and Orr, W. L. (1955). Regeneration of nutrients in sediments of marine basins. *Deep-Sea Res.* *3*, 23-45.

Robbins, J. A., Krezoski, J. R., and Mozley, S. C. (1977). Radioactivity in sediments of the great lakes: Postdepositional redistribution by deposit feeding organisms. *Earth Planet. Sci. Lett.* *36*, 325-333.

Rosenfeld, J. K. (1979a). Ammonium adsorption in nearshore anoxic sediments. *Limnol. Oceanogr.* *24*, 356-364.

Rosenfeld, J. K. (1979b). Amino acid diagenesis and adsorption in nearshore anoxic sediments. *Limnol. Oceanogr.* *24*, 1014-1021.

Rosenfeld, J. K. (1981). Nitrogen diagenesis in Long Island Sound sediments. *Am. J. Sci.* *281*, 436-362.

Rowe, G. T., and Gardner, W. D. (1979). Sedimentation rates in the slope water of the northwest Atlantic Ocean measured directly with sediment traps. *J. Mar. Res.* *37*, 581-600.

Rowe, G. T., Clifford, C. H., Smith, K. L., Jr., and Hamilton, P. L. (1975). Benthic nutrient regeneration and its coupling to primary productivity in coastal waters. *Nature (London)* *255*, 215-217.

Rowe, G. T., Clifford, C. H., Smith, K. L., Jr., and Hamilton, P. L. (1977). Regeneration of nutrients in sediments off Cape Blanc, Spanish Sahara. *Deep-Sea Res. 24,* 57-64.

Rutgers van der Loeff, M. M. (1980a). Time variation in interstitial nutrient concentrations at an exposed subtidal station in the Dutch Wadden Sea. *Neth. J. Sea Res. 14,* 123-143.

Rutgers van der Loeff, M. M. (1980b). Nutrients in the interstitial waters of the southern Bight of the North Sea. *Neth. J. Sea Res. 14,* 144-171.

Schink, D. R., and Guinasso, N. L., Jr. (1977). Effects of bioturbation on sediment-seawater interaction. *Mar. Geol. 23,* 133-154.

Seitzinger, S., Nixon, S., Pilson, M. E. Q., and Burke, S. (1980). Denitrification and N_2O production in nearshore marine sediments. *Geochim. Cosmochim. Acta 44,* 1853-1860.

Seki, H., Skelding, J., and Parsons, T. R. (1968). Observations on the decomposition of a marine sediment. *Limnol. Oceanogr. 13,* 440-447.

Sholkovitz, E. (1973). Interstitial water chemistry of the Santa Barbara Basin sediments. *Geochim. Cosmochim. Acta 37,* 2043-2073.

Smetacek, V., von Bodungen, B., von Brockel, K., and Zeitzschel, B. (1976). The plankton tower. II. Release of nutrients from sediments due to changes in the density of bottom water. *Mar. Biol. 34,* 373-378.

Smith, K. L., Jr. (1978). Benthic community respiration in the NW Atlantic: *In situ* measurements from 40 to 5200 meters. *Mar. Biol. 47,* 337-347.

Smith, K. L., Jr., White, G. A., Laver, M. B., and Haugsness, J. A. (1978). Nutrient exchange and oxygen consumption by deep-sea benthic communities: Preliminary *in situ* measurements. *Limnol. Oceanogr. 23,* 997-1005.

Smith, K. L., Jr., White, G. A., and Laver, M. B. (1979). Oxygen uptake and nutrient exchange of sediments measured *in situ* using a free vehicle grab respirometer. *Deep-Sea Res. 26A,* 337-346.

Smith, M. S. (1982). Dissimilatory reduction of NO_2^- to NH_4^+ and N_2O by a soil *Citrobacter* sp. *Appl. Environ. Microbiol. 43,* 854-860.

Sørensen, J. (1978a). Capacity for denitrification and reduction of nitrate to ammonia in a coastal marine sediment. *Appl. Environ. Microbiol. 35,* 301-305.

Sørensen, J. (1978b). Denitrification rates in a marine sediment as measured by the acetylene inhibition technique. *Appl. Environ. Microbiol. 36,* 139-143.

Stumm, W., and Morgan, J. J. (1970). "Aquatic Chemistry." Wiley, New York.

Suess, E. (1976). Nutrients near the depositional interface. *In* "The Benthic Boundary Layer" (I. N. McCave, ed.), pp. 57-79. Plenum, New York.

Suess, E. (1980). Particulate organic carbon flux in the oceans--Surface productivity and oxygen utilization. *Nature (London) 288*, 260-263.

Suess, E., and Müller, P. J. (1980). Productivity, sedimentation rate and sedimentary organic matter in the oceans. II. Elemental fractionation. *In* "Biogéochimie de la matière organique à l'interface eau-sediment marin, pp. 17-26. CNRS, Paris.

Suess, E., Müller, P. J., Powell, H. S., and Reimers, C. E. (1980). A closer look at nitrification in pelagic sediments. *Geochem. J. 14*, 129-137.

Tanoue, E. (1979). Vertical transportation of organic materials in oceanic environments. Ph.D. Thesis, Nagoya University.

Thauer, R. K., Jungermann, K., and Decker, K. (1977). Energy conservation in chemotrophic anaerobic bacteria. *Bacteriol. Rev. 41*, 100-180.

Toth, D. J., and Lerman, A. (1977). Organic matter reactivity and sedimentation rates in the ocean. *Am. J. Sci. 277*, 465-485.

Vanderborght, J.-P., and Billen, G. (1975). Vertical distribution of nitrate concentration in interstitial water of marine sediments with nitrification and denitrification. *Limnol. Oceanogr. 20*, 953-961.

Vanderborght, J. P., Wollast, R., and Billen, G. (1977a). Kinetic models of diagenesis in disturbed sediments. Part 1. Mass transfer properties and silica diagenesis. *Limnol. Oceanogr. 22*, 787-793.

Vanderborght, J.-P., Wollast, R., and Billen, G. (1977b). Kinetic models of diagenesis in disturbed sediments. Part 2. Nitrogen diagenesis. *Limnol. Oceanogr. 22*, 794-803.

Verma, L., Martin, J. P., and Haider, K. (1975). Decomposition of carbon-14 labeled proteins, peptides and amino acids; free and complexed with humic polymers. *Soil Sci. Soc. Am. Proc. 39*, 279-284.

Waksman, S. A. (1933). On the distribution of organic matter in the sea bottom and the chemical nature and origin of marine humus. *Soil Sci. 36*, 125-147.

Waples, D. W., and Sloan, J. R. (1980). Carbon and nitrogen diagenesis in deep-sea sediments. *Geochim. Cosmochim. Acta 44*, 1463-1470.

Webb, J. E., and Theodor, J. L. (1968). Irrigation of submerged marine sands through wave action. *Nature (London) 220*, 682-683.

Widdel, F. (1980). Anaerober Abbau von Fettsauen und Benzoesauer durch neu isolierte Arten von Sulfat--reduzierenden Bakterien. Ph.D. Thesis, University of Göttingen.

Wieser, W., and Zech, M. (1976). Dehydrogenases as tools in the study of marine sediments. *Mar. Biol. 36*, 113-122.

Wirsen, C. O., and Jannasch, H. W. (1976). Decomposition of solid organic materials in the deep sea. *Environ. Sci. Technol. 10*, 880-886.

Wollast, R., and Peters, J. J. (1978). Biogeochemical properties of an estuarine system: The River Scheldt. *In* "Biogeochemistry of Estuarine Sediments" (E. D. Goldberg, ed.), pp. 279-293. UNESCO, Paris.

Zehnder, A. J. B. (1978). Ecology of methane formation. *In* "Water Pollution Microbiology" (R. Mitchell, ed), Vol. 2, pp. 349-376. Wiley, New York.

ZoBell, C. E., and Anderson, D. Q. (1936). Vertical distribution of bacteria in marine sediments. *Am. Assoc. Pet. Geol. Bull.* *20*, 258-269.

Chapter 13

MAN'S IMPACT ON THE MARINE NITROGEN CYCLE

EDWARD A. LAWS
Department of Oceanography
University of Hawaii
Honolulu, Hawaii

I. OVERVIEW OF POTENTIAL IMPACTS

In considering man's impact on the marine nitrogen cycle, it is necessary to first briefly review those biological and chemical transformations which constitute the marine N cycle, and then to consider how human activities might impact on these processes. Tables I and II summarize what is known about the quantities of N in the oceans and about the fluxes of N to and from the oceans. Although the annual fluxes of fixed N to and from the oceans are not well quantified, it is apparent from Tables I and II that these fluxes are several orders of magnitude smaller than the total marine fixed N pool. Table II indicates that the removal of fixed N in sediments is about an order of magnitude smaller than the input rate of fixed N to the oceans. The discrepancy between fixed N inputs and loss rates in sediments may simply reflect a gradual

ISBN 0-12-160280-X

TABLE I. Major Forms of Nitrogen in the Oceans[a]

Form	*Quantity*[b]
Dissolved N_2	2.0×10^{13}
Fixed N	
Plants	1.7×10^{8}
Animals	1.7×10^{8}
Dead organic matter	4.5×10^{10}
Inorganic N	1.0×10^{11}

[a] *From Delwiche (1981).*

[b] *Units are metric tons.*

TABLE II. Annual Fluxes of Nitrogen to and from the Oceans[a]

	Inputs[b]		*Losses*[b]
N_2 fixation	2.8×10^{7}	in sediments	2.8×10^{6}
Runoff, rainfall	2.8×10^{7}	dentrification	4.2×10^{7}

[a] *From Delwiche (1981).*

[b] *Units are metric tons.*

accumulation of fixed N in the oceans. However, Delwiche (1981) has postulated that denitrification rates may have largely balanced fixed N inputs and losses (Table II), at least until recent years. Although clear evidence of denitrification has been reported in oxygen-minimum regions (Goering, 1968; Goering and Cline, 1970) and anoxic zones (Dugdale et al., 1977), the magnitude of this flux on a global basis is simply not known.

Nitrogen fluxes that have been affected by human activities are summarized in Table III. Some of the nitrogen oxides released to the atmosphere as a result of fossil fuel burning undoubtedly find their way into the ocean (Laska et al., 1980). Release of combined N in sewage discharges amounts to about 10 g N per person per day (Garside et al., 1976). If we assume as an upper bound that the sewage from the entire world population of 4.5×10^9 per-

TABLE III. Nitrogen Fluxes Affected by Human Activities

Flux	*Quantity*[b]
N_2 fixation[a]	
Industrial processes and combustion reactions	5.6×10^7[b]
Natural and legume cultivation	1.5×10^8
Sewage discharge	$<1.6 \times 10^7$

[a]*N_2 fixation rates from Delwiche (1981).*
[b]*Units are metric tons N per year.*

sons is discharged to the oceans, the input rate would be about 1.6×10^7 metric tons N per year, as indicated in Table III. Obviously not all of this nitrogen reaches the oceans, and at least some of this flux has already been accounted for in land runoff (Table II). We conclude from a comparison of the fluxes in Table III with the marine fixed N reservoirs in Table I that the flux of fixed N to the oceans has not been altered by human activity to an extent that the concentration of combined nitrogen in the oceans has significantly changed. On a local scale however, the picture is very different. There is clear evidence of changes in the concentrations of fixed N in coastal systems due to human activities (Ryther and Dunstan, 1971; Garside, 1981), and these changes may well have altered the internal cycling of N within these systems. These internal cycling processes are highly complex, and in most cases it would be impossible to discuss man's impact on these processes in other than a very qualitative manner. However, we now have at least some empirical understanding of how phytoplankton uptake of new and regenerated forms of N (Dugdale and Goering, 1967) is affected by overall levels of production, at least in open ocean systems (Eppley and Peterson, 1979). As I will show below, this empirical correlation can be rationalized to a certain extent by rather simple models of N cycling within pelagic systems. Thus, we have identified an important portion of the marine N cycle where we believe we have accurate data on flux rates and some theoretical understanding of the processes at work. It will therefore be instructive to examine in some detail man's impact on this portion of the marine N cycle.

II. THEORY

Let us consider a trophic level model of a pelagic food chain. Let X_0, X_1, X_2, X_3, $...X_n$ represent the respective concentrations of N as dissolved inorganic N plus urea, phytoplankton, herbivores, primary carnivores, and so forth. Let X_{n+1} represent the concentration of detrital N. We assume that the time rate of change of biomass on trophic level m is controlled by a function f_m describing the rate at which trophic level m consumes trophic level m-1, by a function f_{m+1} describing the rate at which trophic level m+1 consumes trophic level m, and by the efficiency Q_m with which trophic level m converts food into biomass. We assume that of the food consumed by trophic level m, a fraction R_m is respired and a fraction S_m is excreted as detritus. At the top trophic level where there is no grazing, we assume population losses to be described by a function F_n, which represents natural mortality due to old age, disease, parasites, and so forth. The population dynamics of the system can then be described by the following system of equations:

$$\frac{dX_1}{dt} = f_1 - f_2 \tag{1}$$

$$\frac{dX_2}{dt} = Q_2 f_2 - f_3 \tag{2}$$

$$\frac{dX_3}{dt} = Q_3 f_3 - f_4 \tag{3}$$

$$\frac{dX_n}{dt} = Q_n f_n - F_n \tag{4}$$

$$\frac{dX_{n+1}}{dt} = S_2 f_2 + S_3 f_3 + \ldots S_n f_n + F_n - (D+K)X_{n+1} \tag{5}$$

$$\frac{dX_o}{dt} = R_2 f_2 \ R_3 f_3 + \ldots R_n f_n - f_1 + KX_{n+1} + L \tag{6}$$

All symbols used in Eqs. (1)-(6) are defined in Table IV. Note that the definitions of Q_m, R_m, and S_m imply that $Q_m+R_m+S_m = 1$. If we assume a steady state and set all the equations equal to zero, it follows by addition of the equations that $L = DX_{n+1}$. In other words, N inputs are exactly balanced by sinking of detrital N. We have assumed that phytoplankton losses due to sinking are negligible. For our purposes, omission of a phytoplankton sinking term causes no loss in the generality of the model, since, conceptually, such sinking losses could be regarded as the loss of detrital material due to inefficient grazing by herbivores. However, if explicit functional forms for the f_m are assumed and used to calculate equilibrium values for the X_m, then those equilibrium values would be affected by the inclusion or exclusion of a phytoplankton-sinking rate term. However, we are not concerned with the equilibrium values of the X_m in this chapter.

Following the rationale of Dugdale and Goering (1967), the ratio of new production (i.e., production supported by upwelling, land runoff, and rainfall) to total production in this system is given by the ratio L/f_1. Using Eq. (6) to substitute for L, Eq. (5) to substitute for X_{n+1}, and Eqs. (1)-(4) to solve for all the f_m in terms of f_1, we obtain the following expressions for the ration L/f_1.

TABLE IV. Definition of Symbols Used in Equations (1)-(6)

Symbol	*Definition*
X_m $0<m\leq n$	Biomass of nitrogen on trophic level *m* in an *n* trophic level system
f_m	Rate at which trophic level m consumes trophic level *m*-1
Q_m	Fraction of consumed food which trophic level *m* converts to biomass
S_m	Fraction of consumed food which trophic level *m* excretes as detritus
R_m	Fraction of consumed food which trophic level *m* respires
X_{n+1}	Biomass of nitrogen as detritus in a system of *n* trophic levels
F_n	Natural loss rate due to old age, disease, parasites, etc. of trophic level *n* in a system of *n* trophic levels
D	Fractional loss rate of detritus due to sinking
K	Fractional loss rate of detritus due to mineralization
X_o	Concentration of inorganic nitrogen plus urea
L	Loading rate due to upwelling, rainfall, land runoff, etc. of X_o into the system

$$L/f_1=\left[\frac{\alpha}{1+\alpha} \quad 1-R_2-R_3Q_2\right] \qquad n = 3 \qquad (7)$$

$$=\left[\frac{\alpha}{1+\alpha} \quad 1-R_2-R_3Q_2-R_4Q_3Q_2\right] \qquad n = 4 \qquad (8)$$

$$=\left[\frac{\alpha}{1+\alpha} \quad 1-R_2-R_3Q_2-R_4Q_3Q_2-R_5Q_4Q_3Q_2\right] \qquad n = 5 \qquad (9)$$

where $\alpha = D/K$. Thus, the percentage of new production is predicted to be a hyperbolic function of the parameter $D:K$, the ratio of detrital losses by sinking to mineralization. If tissue growth efficiencies are independent of food chain lengths, the model also predicts that the percentage of new production will be negatively correlated with food chain length. The maximum percentage of new production is predicted for short food chains characterized by $D >> K$, in which case all recycling is due to direct excretion of dissolved N by herbivores and primary carnivores. Using values from a typical trophic level model (Caperon, 1975), we have $Q_2 = 0.25$, and $R_2 = R_3 = 0.3$. Using Eq. (7), we calculate the maximum percentage of new production in such a food chain to be 63%. This figure is in rather good agreement with the data summarized by Eppley and Peterson (1979), which show maximum percentages of new production in the range 50-65% for open ocean systems. If we assume all R_m to be 0.3 and all Q_m to be 0.25, Eq. (9) predicts the maximum percentage of new production in a five-trophic-level food chain to be 60%, only slightly less than the value for the three-trophic-level model. Thus, unless there are significant differences in growth efficiencies between short and long food chains, differences in food chain length are expected to have only a small effect on the percentage of recycling. In their 1979 paper, Eppley and Peterson took the fraction of total production accounted for by new production (L/f_1 in the context of our model) to be equal to the ratio of nitrate-N uptake to the sum of nitrate-N plus ammonium-N plus urea-N uptake. When they plotted this new production fraction against total primary production based on $^{14}CO_2$ uptake measurements, they obtained a distinctly hyperbolic curve. This empirical hyperbolic curve relating the percentage of new production to total production is rationalized within the context of our model as reflecting a positive correlation between $D:K$ and total production. Are there any experimental data to suggest that such a correlation exists?

There is no doubt that the composition of the plankton community differs greatly in oligotrophic and eutrophic waters (Ryther, 1969). In oligotrophic regions, as much as 65% of the chlorophyll *a* will plass through a 3-μm filter (Bienfang, 1981), while in upwelling areas phytoplankton tend to be dominated by larger-celled species, chain formers, or colonial species that form gelatinous masses (Ryther, 1969). Such differences in the composition of the phytoplankton community have profound effects on the structure of the heterotrophic community and on the forms of detritus produced by that community. Typical sinking rates of zooplankton fecal pellets are 50-300 m day^{-1} (Turner, 1977), but these sinking rates are markedly affected by diet (Bienfang, 1980). In oligotrophic areas, detrital material collected in particle interceptor traps often consists largely of ill-defined assemblages of small particles, with few of the durable, fast-sinking pellets associated with herbivorous zooplankton excretion in upwelling and coastal areas (Betzer, 1980; Takahashi and Bienfang, 1983). Bienfang and Goodwin (1982) have found the sinking rates of this amorphous detritus to be 0.2-0.5 m day^{-1}, two orders of magnitude lower than the sinking rates of copepod fecal pellets. Furthermore, it seems likely that the encapsulated pellets produced by herbivorous copepods are more durable (i.e., have lower K values) than the amorphous detritus produced by oligotrophic zooplankton. Thus, there is a logical basis for suspecting that the relationship noted by Eppley and Peterson (1979) between the percentage of new production and total production may reflect systematic differences in the $D{:}K$ ratio between oligotrophic and eutrophic areas.

Given this understanding of N cycling in open ocean systems, what sort of impact can be anticipated when N-loading rates are increased as a result of human activities? If the perturbation is sufficiently small, there will be little or no change in the composition of the plankton community, and hence the percentage of new production will remain unchanged. The percentage increase in total production will then be equal to the percentage increase in the N-loading rate. Large and persistent increases in L may of course alter the structure of the plankton community, leading to an increase in the $D:K$ ratio. In that case there will be an increase in the percentage of new production, and the percentage increase in total production will be less than the percentage increase in L. Using Eq. (1) in Eppley and Peterson (1979), one can show that the fractional increase in total production due to a fractional increase β in new production will be $(1+\beta)^{\frac{1}{2}} - 1$. Thus, a 10% increase in new production would result in about a 5% increase in total production. This latter conclusion will evidently apply only in systems where total production is less than 200 $g\ C{\cdot}m^{-2}{\cdot}year^{-1}$. If total production exceeds this figure, then based on Eppley and Peterson's curve, the percentage of new production will be insensitive to total production, and the percentage increase in total production will be equal to the percentage increase in loading rate. We now examine the implications of this theory in several case studies.

III. CASE STUDIES

A. The New York Bight

The New York Bight is an area of approximately $4 \times 10^4\ km^2$, bounded on the north and west by Long Island and the state of New Jersey, respectively, on its seaward edge by the continental shelf break at approximately the 100-fathom isobath, and by arbitrary lines drawn from Cape May and Montauk Point to the shelf break (Fig. 1). The Bight currently receives inputs of both inorganic and organic nitrogen from a variety of anthropogenic sources, but primarily from sewage treatment plants. The total discharge from the New York metropolitan area has been estimated to be about 180 tons N per day, of which about 46% is organic N and 52% is ammonium (Duedall et al., 1977; Atwood et al., 1979). Of the inorganic N,

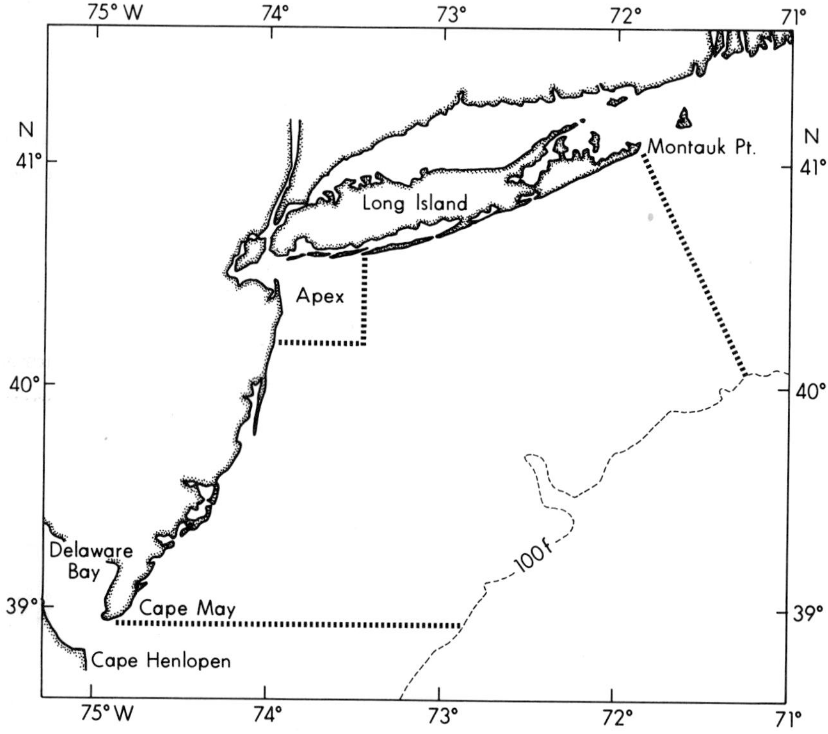

Fig. 1. Approximate boundaries of the New York Bight and Bight Apex.

Duedall et al. (1977) have estimated that the majority of the nitrate is derived from the Hudson and Raritan Rivers, whereas the great majority of the ammonium and nitrite comes from sewer outfalls. If we take the nitrate inputs to be upper bounds to the natural N-loading rate to the Bight, then natural sources amount to no more than 9 tons N per day or 5% of the total N inputs.

Studies by Garside et al. (1976) have shown that the uptake of inorganic N in the estuaries of the Hudson and Raritan rivers is highly seasonal, and is controlled primarily by the availability of light. Inorganic N concentrations in the lower Hudson estuary for example are 10-100 μM, with the surface-water values generally being higher than the concentrations at depth. These concentrations are almost certainly adequate to saturate marine phytoplankton N uptake (Caperon and Meyer, 1972; Eppley et al., 1969). Based on ^{15}N studies, Garside et al. (1976) concluded that N uptake in the estuarine region amounts to about 2.6 and 43 tons N per day in the winter-spring and summer, respectively. Thus, most of the dissolved N is transported to the New York Bight, the fluxes amounting to about 180 tons N per day in the winter-spring and about 140 tons N per day in the summer. Unfortunately, it is not clear what percentage of the dissolved organic nitrogen (DON) discharged to the Bight is actually available for biological uptake, but considering the source it seems likely that much of the DON is in a form (e.g., urea) that can be utilized by phytoplankton. Following Garside (1981), we will assume discharges of biologically available N to the Bight to be 160 tons per day in the winter-spring and 120 tons per day in the summer.

Using ^{15}N techniques, Garside (1981) estimated uptake rates of $NH_4^+ + NO_3^-$ in the apex of the Bight to be 125, 22, and 82 mg $N \cdot m^{-2} \cdot day^{-1}$ in the summer, fall, and winter, respectively. Based on the work of McCarthy (1972) and Eppley et al. (1977), uptake of urea N would probably increase these figures by about 40%. Applying this correction factor to Garside's numbers and assuming a C:N ratio of 5.7 by weight in the phytoplankton (Redfield et al., (1963), we conclude that carbon fixation rates in the apex of the Bight are 1.00, 0.18, and 0.65 g $C \cdot m^{-2} \cdot day^{-1}$ during the summer, fall, and winter, respectively. If we now apply the empirical curve of Eppley and Peterson (1979) to these data, we conclude that the ratio of new to total production during the summer and winter is no more than about 0.5, and in the fall no more than 0.16. These figures are considered upper bounds to the true ratios of new production to total production, because significant recycling of nutrients from sediments may occur in waters <200 m deep (Eppley and Peterson, 1979). If we now assume that Garside's (1981) N uptake numbers are applicable to the entire area of the Bight affected by sewage loading, we can calculate the area affected by sewage loading from the equation

$$\text{Area of Bight affected by sewage loading} = \frac{(\text{Discharge rate of biologically available N})}{\begin{pmatrix}\text{Uptake rate per unit} \\ \text{area of } NO_3^- + NH_4^+ + \text{urea}\end{pmatrix}\begin{pmatrix}\text{Ratio of new to} \\ \text{total production}\end{pmatrix}} \tag{10}$$

Eq. (10) implies that the affected area is at least 2800 km^2 in the winter, 1370 km^2 in the summer, and 32,400 km^2 in the fall. Relevant data are shown in Table V. The former two figures are about 40% greater than numbers calculated by Garside (1981), and the latter figure is about 4.5 times greater, because we assumed that new production accounted for only a fraction of total production and that total N uptake was about 40% higher than $NH_4^+ + NO_3^-$ uptake. During the fall, the affected area would extend to a radial distance of at least 200 km from the apex of the Bight, and would equal over 80% of the area of the Bight (Paine, 1976). In making the calculation for the fall, we assumed the discharge rate of biologically available N to the Bight to be the same as that during the winter-spring, 160 tons per day. However, since the rate of $NH_4^+ + NO_3^-$ uptake in the Bight apex is even lower in the fall than in the winter (Garside, 1981), the actual loading rate and hence the affected area may be even greater in the fall.

How would N cycling in the New York Bight differ if there were no sewage inputs? In that case, dissolved N loading from the Hudson and Raritan rivers would probably be no more than 9 tons N per day. Judging from figures given by Garside et al. (1976), virtually all of this N would be assimilated by phytoplankton in the Hudson and Raritan estuaries during the summer. However, during the winter and spring, as much as 6.4 tons N per day might reach the apex of the Bight. The area of the Bight affected by this discharge is impossible to calculate without a knowledge of the total production rate in the Bight apex. However, assuming production during the winter and spring to be light-limited, we conclude that

TABLE V. Nitrogen Cycling Data for New York Bight

	Winter	*Summer*	*Fall*
N loading rate (tons per day)	160	120	160
N uptake[a] ($mg \cdot m^{-2} \cdot day^{-1}$)	115	175	31
New production / Total production	0.5	0.5	0.16
Area of Bight affected by sewage loading (km^2)[b]	2800	1370	32,400

[a]*Values obtained by multiplying those of Garside (1981) by 1.4 to correct for urea uptake.*

[b]*Calculated using Eq. (10).*

the areal production rate would be the same as that measured by Garside (1981), namely 0.65 g $C \cdot m^{-2} \cdot day^{-1}$. Using this figure, an assumed C:N ratio of 5.7 by weight in the phytoplankton, and the empirical curve in Eppley and Peterson (1979), we calculate that the area of the Bight affected by this discharge would be 112 km^2, or about 4% of the area currently affected.

The conclusion is that sewage N discharges to the New York Bight have not significantly affected the percentage of N recycling in the Bight during the winter, when production is light-limited. However, the rate of carbon fixation and total N uptake has probably been affected over an area of as much as 32,000 km^2 (82% of the Bight area) in the fall. During the summer, N uptake has been stimulated over an area of about 1400 km^2 (3.6% of Bight area), and the percentage of N uptake accounted for by recycling probably has been reduced. The magnitude of this reduction is impossible to estimate without some knowledge of the natural levels of primary production during the summer.

As a final note on conditions in the New York Bight, let us consider the importance of N recycling in determining levels of N loading that might cause anoxic conditions to develop in the water column. The calculation is of more than academic interest, since in early July of 1976 a massive fish kill occurred along roughly a 90-km stretch of New Jersey coastline, evidently the result of the decomposition of a large phytoplankton bloom in the Bight. Although the current conditions and sewage-loading rates which created the bloom and subsequent anoxia were unusual (Paine, 1976), the incident has raised the obvious question as to whether present N-loading rates are pushing the system to dangerous limits.

Atwood et al. (1979) have estimated that the average carbon fixation rate in the Bight during the summer is about 0.5 g $C \cdot m^{-2} \cdot day^{-1}$. They also estimated that a flux of about 0.5 g $C \cdot m^{-2} \cdot day^{-1}$ would be required to produce anoxia in the bottom waters. They indicated that about 50% of the carbon fixed in the surface waters was respired there, a conclusion consistent with Eppley and Peterson's model. At a production rate of 0.5 g $C \cdot m^{-2} \cdot day^{-1}$, the model predicts that 54% of photosynthesis is supported by recycling. Thus, 46% of the fixed carbon would be expected to sink to the bottom, yielding a flux of 0.23 g $C \cdot m^{-2} \cdot day^{-1}$. Atwood et al. (1979) then logically concluded that production would have to double before the flux of carbon to the benthos would approach the 0.5 g $C \cdot m^{-2} \cdot day^{-1}$ figure. This conclusion is correct, because at a total production rate of 1.0 g $C \cdot m^{-2} \cdot day^{-1}$, only slightly more (50% versus 46%) of the fixed carbon would be expected to sink to the bottom. However, assuming a C:N ratio of 5 by weight, Atwood et al. (1979) then concluded that an additional N input of 0.5/5.0 = 0.1 g $N \cdot m^{-2} \cdot day^{-1}$ would be required to increase summer production to 1.0 g $C \cdot m^{-2} \cdot day^{-1}$. This calculation completely ignores the effect of N recycling. Since new production evidently accounts for no more than 50% of total production in the Bight during the summer, the additional N input required to increase production to 1.0 g $C \cdot m^{-2} \cdot day^{-1}$ would be no more than 0.05 g $N \cdot m^{-2} \cdot day^{-1}$. This figure is a little over 20%

of current N-loading rates, assuming (as do Atwood et al., 1979) that the affected area of the Bight is 15,000 km^2. Atwood et al. (1979) then argue that primary production on the shelf would probably become light-limited before the critical loading rate was reached, and that, "It would then be impossible to cause anoxia, as is the case in the apex." However, our own calculations based on Garside's (1981) N uptake numbers indicate that summer production in the apex of the Bight is about 1.0 g $C \cdot m^{-2} \cdot day^{-1}$, a figure which agrees exactly with the summer carbon fixation numbers measured by Garside et al. (1976) in the Hudson River estuary using ^{14}C. If both systems (estuary and apex) are light-limited, then comparable production numbers would be expected. This production rate is of course equal to the critical rate which Atwood et al. (1979) estimated might lead to anoxia in the Bight. Given this realization, we conclude that light limitation may be inadequate to prevent repetitions of the 1976 summer anoxia incident. In fact, anoxia events are common in the Bight, and have occurred in 1968, 1971, 1974, and 1977 as well as 1976 (Sinderman and Swanson, 1979).

B. Ocean Thermal Energy Conversion

Ocean Thermal Energy Conversion (OTEC) plants are currently being operated on an experimental basis but results to date appear promising, and it is possible that large scale OTEC operations may exist in the next 10-30 years. Since these power plants would be operated at deep-water sites, it is logical to use the Eppley and Peterson (1979) model to predict the impact of such operations on marine N cycling.

A typical 1-MW prototype plant would be expected to draw deep water with a NO_3^- concentration of about 36 μM at a rate of 257 m^3 min^{-1} (U. S. Department of Energy, 1979). Prior to release, this deep water will be mixed with a roughly equal volume of surface water, and discharged at a depth of about 40 m. Due to its cooler temperature, this effluent water will gradually sink but will eventually stabilize at depths ranging from 100 to 270 m (U. S. Department of Energy, 1979). Since the euphotic zone in oligotrophic areas may extend to depths as great as 150 m (Bienfang, 1981), the effluent water may actually stabilize within the euphotic zone. To a depth of 150 m, the volume of the water column affected by this discharge is expected to be 4.2 x 10^9 m^3 (U. S. Department of Energy 1979), with the affected parcel of water extending as much as 10 km from the OTEC plant. Since the deep water will be mixed roughly 50:50 with surface water prior to discharge, the effluent water will contain an inoculum of phytoplankton that can be expected to begin rapidly taking up the NO_3^- (McCarthy and Goldman, 1979). What effect will this stimulated uptake have on total production and N cycling?

Natural production rates in deep-water areas average about 50 g $C{\cdot}m^{-2}{\cdot}year^{-1}$ (Eppley et al., 1973; Gilmartin and Revelente, 1974; Gunderson et al., 1976; Bienfang and Gunderson, 1977; Bienfang, 1981). Using the Eppley and Peterson (1979) model, we calculate that new production equals 12.5% of total production. Using this figure and assuming a C:N ratio of 5.7 by weight, we calculate a N input rate of 3 mg $N{\cdot}m^{-2}{\cdot}day^{-1}$. The N input from the OTEC plant to a depth of 150 m would average 6.66 mg $N{\cdot}m^{-2}{\cdot}day^{-1}$.

If a single such plant were operating, the affected area would be only about 30 km^2, and it is unlikely that a significant shift in the plankton community would occur. Thus, one could expect no change in the ratio of new to total production, and the expected percentage increase in total production would be equal to the percentage increase in N inputs. Thus, total production over the affected area would probably average about (1+6.66/3)(50) = 161 g $C{\cdot}m^{-2}{\cdot}year^{-1}$. On the other hand, one might imagine a collection of such plants with a total output of 500-1000 MW. In that case, the affected area might be as large as 15,000-30,000 km^2, and a shift in the plankton community could certainly be anticipated. Applying our earlier logic, we note that the increase in N loading is a factor of 6.66/3 = 2.22, so that the expected fractional increase in total production under these conditions would be $(1 + 2.22)^{\frac{1}{2}} - 1 = 0.79$. Thus, total production could be expected to increase to about 90 g $C \cdot m^{-2} \cdot year^{-1}$, and based on Eppley and Peterson's (1979) model the percentage of new production would become 22%. These calculations illustrate how important the shift in the percentage of new production is in determining the impact on primary production. The assumed change in plankton composition reduces the impact of nutrient loading on primary production by almost a factor of 3 (2.22/0.79 = 2.8).

C. Kaneohe Bay

Kaneohe Bay is the only shallow-water system considered here. Before becoming involved in a discussion of this particular case, a few comments about shallow versus deep systems are in order. First, at a given level of water column production, we can expect that a greater percentage of that production will be supported by recycling in shallow systems than is the case in deep systems. This conclusion follows from the fact that detrital material that sinks to the bottom of shallow systems may be mineralized at the sediment-water interface and rapidly returned to the water column, whereas in deep systems detritus that sinks below the mixed layer is effectively lost. Studies of coastal systems have confirmed that these systems are run primarily by internal nutrient cycling rather than by external inputs (Davies, 1975; Rowe et al., 1975, 1977; Kremer and Nixon, 1978; Nixon et al., 1980), but as much as 50-60% of eutrophic open ocean production may be new production (Eppley and Peterson, 1979). Thus, we expect N recycling to account for a high (>80%) and rather constant percentage of water column N uptake over a wide range of loading rates in coastal systems.

Second, interactions between the benthos and water column may profoundly influence the types of organisms found in a shallow system. Given adequate light, benthic plants appear capable of largely outcompeting phytoplankton for available nutrients, and as a result, the plant community in shallow systems characterized by low nutrient-loading rates tends to be dominated by benthic species. Coral reefs provide an obvious example or such systems (Odum and Odum, 1955). However, as nutrient-loading rates are increased, the biomass of benthic organisms eventually becomes limited by the availability of suitable substrates for attachment, and further increases in biomass occur primarily in the water column (Ling, 1977). As the concentration of plankton in the water column increases, reductions in water clarity may begin to retard the growth of benthic algae. Furthermore, the increase in plankton density will favor the growth of filter-feeding benthic animals, which may ultimately displace the benthic algae as the dominant components of the benthic system (Smith et al., 1981). Thus, one can envision a radical change in the food chains of coastal systems as nutrient-loading rates are increased, with plankton and filter feeders tending to dominate biomass at high loading rates, and benthic algae and their associated predators tending to dominate at low loading rates.

Kaneohe Bay (Fig. 2) is a coastal embayment located on the northeast side of the island of Oahu in the Hawaiian Islands. Prior to 1959, it was the site of the most luxuriant coral reefs in the islands (Edmondson, 1928, 1946), but by 1972 over 75% of the corals in the bay had been killed as a result of sewage discharges and urbanization of the watershed (Maragos, 1972). In 1977 and 1978, sewage discharges were diverted from the bay, giving scientists a chance to study the response of the system to relaxation of the nutrient-loading perturbation.

Studies by Kinsey (1979) have shown that the reefs in the bay were metabolically atypical of healthy coral reefs during the time of sewage enrichment. Table VI summarizes results of Kinsey's studies, and compares characteristics of Kaneohe Bay reefs to typical healthy coral reefs. In three cases production on the Kaneohe

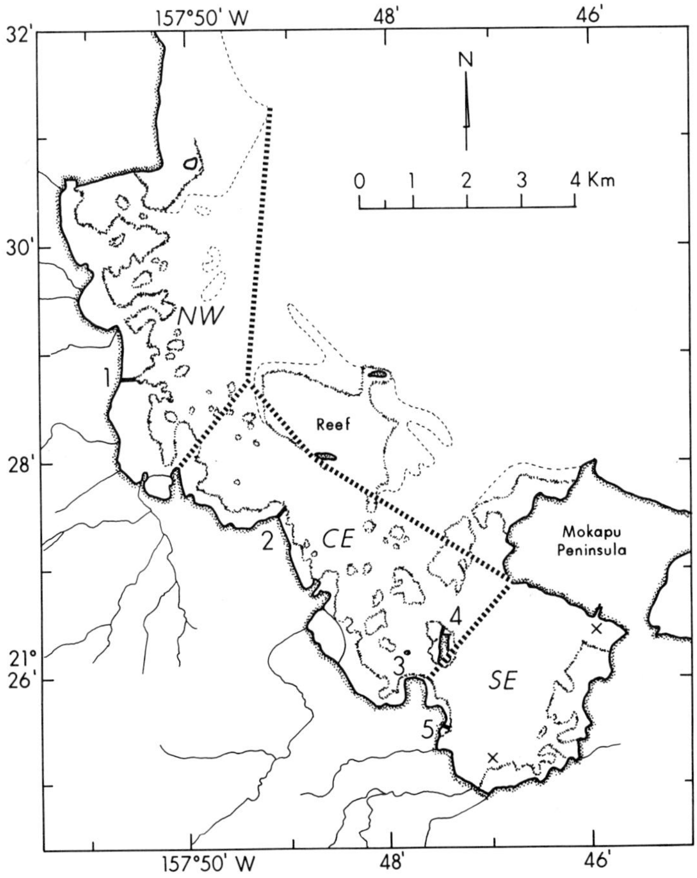

Fig. 2. Kaneohe Bay and approximate division lines of the NW, CE, and SE sectors. X's in SE sector are sewage outfalls. Fringing reef studies were performed at locations 1, 2, and 5; patch reef studies were performed at locations 3 and 4 (refer to Table VI).

reefs was abnormally low, reflecting the effect of high plankton and silt concentrations in the water. With the exception of the NW sector reef, which consisted essentially of an algal mud flat, the production to respiration (P/R) ratio on all reefs was less than 1, reflecting a net input of organic materials (plankton) and a net release of dissolved nutrients. The calcification rate on all reefs except one central sector reef was abnormally low, with

TABLE VI. Community Metabolism and Calcification Rates for Perimeter Zones of Experimental Reefs and Standard Reefs[a]

Reef	Production ($g\ C \cdot m^{-2} \cdot day^{-1}$)	Respiration ($g\ C \cdot m^{-2} \cdot day^{-1}$)	P/R	Calcification ($kg\ CaCo_3 \cdot m^{-2} \cdot year^{-1}$)
1	4.2	2.4	1.8	0.0
2	8.0	9.6	0.8	8.9
3	6.0	13.0	0.5	3.1
4	7.1	10.3	0.7	2.4
5	4.9	5.3	0.9	-6.5
Standard	7.0±0.6	7.0±0.6	1.0±0.1	4.0±0.7

[a]*From Kinsey (1979). See Fig. 2 for experimental reef locations.*

the SE sector reef being literally in the process of dissolving. Kinsey and Davies (1979) found that enrichment of coral reefs with inorganic nutrients for up to 6 months resulted in elevated rates of production and depressed calcification rates. Depressed calcification is apparent in the Kaneohe Bay study, but production was also depressed. Why were both production and calcification depressed in Kaneohe Bay?

The answer is that the inorganic N in the sewage was rapidly taken up from the water within a short distance from the sewer outfalls by phytoplankton. Thus, the signal to the reefs was not inorganic N, but particulate N. This signal stimulated the growth of filter-feeding organisms (sponges, oysters, barnicles, tunicates, gastropods, and annelid worms), many of which contributed to the destruction of the reefs by boring into the reef substratum, particularly in the southeastern portion of the bay (Smith et al., 1981). In the central sector of the bay, water clarity was good enought to allow the development of a large benthic algae population, dominated by *Dictyosphaeria cavernosa*. This species established itsel at the base of coral fronds, growing outward and eventually envelopir the coral heads completely. The reefs in the northwest sector of the bay appear to have been affected primarily by sedimentation from land runoff. The mud flats in the NW sector overlie former coral reefs, and the large biomass of benthic algae on these flats reflects the high inorganic nutrient concentrations in land runoff.

Healthy coral reefs are generally regarded as being net exporters of nitrogen (Johannes and Project Symbiosis Team, 1972; Webb et al., 1975) due to the nitrogen-fixing activities of certain reef algae and the "loose" recycling of N in the system. Maximum reported rates of N fixation of coral reefs are about 2 - 3 $kg \cdot ha^{-1} \cdot day^{-1}$ (Webb et al., 1975; Wiebe et al., 1975). In Kaneohe Bay, the export of inorganic N by the reefs due to the imbalance of production and respiration can be calculated by assuming a C:N ratio of 20 by atoms in the respired biomass (Atkinson, 1981). Using this conversion, the calculated inorganic N export rates for reefs 3, 4 and 5 in Table VI are 4.1, 1.9, and 0.2 $kg \cdot ha^{-1} \cdot day^{-1}$, respectively. Thus, the southeast sector Kaneohe Bay reefs appear to have been converting organic N to inorganic N at rates comparable to the rates of N fixation reported for healthy coral reefs.

In the first year following diversion of the sewage outfalls from Kaneohe Bay, a large scale die-off of benthic organisms occurred in the southeast sector. This die-off, which consisted largely of the aforementioned filter-feeding organisms, amounted to about 400 metric tons dry weight (Smith et al., 1981). This crash in benthic filter feeders reflected a decline by about 50% in plankton concentrations (as measured by ATP or chl-*a*) in the southeast sector (Laws and Redalje, 1981). Assuming the dead biomass contained about 11% N (estimated from data in Vinogradov, 1953), the release of N from this benthic decomposition would have amounted to about 0.14 $kg\ N \cdot ha^{-1} \cdot day^{-1}$ during the first year when averaged over the area of the southeast sector. Following this rapid change in benthic biomass, conditions in Kaneohe Bay have continued to change but at a much more gradual rate. Though metabolic characteristics of the reefs have not been measured recently, visual observations have indicated that the system is slowly returning to a healthy coral reef community.

In summary, sewage discharges to the southeast sector of Kaneohe Bay produced high plankton concentrations in the southeast sector, low water clarity, and a benthic community dominated by filter feeders. In the central portion of the bay, water clarity was good enough to permit algal growth, and the former coral reefs were overgrown with benthic macrophytes as a result of the elevated nutrient levels in the water. Destruction of coral reefs in the northwest sector was less extreme, and appears to have been due mainly to sedimentation. Maragos (1972) estimated that almost 100% of the coral reefs had been killed in the SE sector, 87% in the central sector, and 26% in the NW sector. These changes in the biological community obviously affected the mechanisms of N cycling within the system. The SE sector reefs transformed organic N into inorganic N at rates comparable to the rates of N fixation on healthy reefs.

IV. SUMMARY AND CONCLUSIONS

Human activities have increased the flux of fixed N to the oceans, but the increase has been too small to produce a significant change in average marine fixed N concentrations. On a local scale however, significant alterations in the marine N cycle have occurred. In open ocean systems, the effect of increased N-loading rates can be predicted and at least qualitatively understood with the use of empirical and theoretical models. In this context, it is remarkable that studies purporting to assess man's impact on the oceans have often ignored completely the effect of N recycling in controlling primary production. Economists have known for years that addition of a given amount of money to the national economy produces a much larger change in the gross national product due to a so-called multiplier effect. The multiplier effect accounts for the fact that the added money cycles through many transactions. A similar multiplier effect occurs when nutrients added to the ocean are cycled between producers and consumers. The multiplier effect is especially pronounced in coastal areas, where recycling from the sediments may be quite efficient. Yet, we still find calculations in the literature based on the assumption that if we know the increase in N-loading rate and the C:N ratio in the phytoplankton, we can calculate the change in primary production from the product of the two numbers.

A crucial point about the Eppley and Peterson (1979) model is the fact that the percentage change in new production is greater than the percentage change in total production, when total production is less than about 200 g $C \cdot m^{-2} \cdot year^{-1}$. In other words, the percentage change in primary production will, in general, be smaller than the percentage change in the N-loading rate. However, if our theoretical understanding of the Eppley and Peterson model is correct, this conclusion applies only in cases where the perturbation produces a significant change in the composition of the plankton community. The observed correlation is presumed to reflect changes in the *D*:*K* ratio, resulting from changes in the characteristics of the detritus in the system.

Shallow systems are more complex because of the interactions between the benthos and water column. Our understanding of how changes in N-loading rates affect N cycling in these systems is qualitative at best. Low N-loading rates tend to produce systems in which biomass is dominated by benthic plants and their associated predators, whereas high N-loading rates favor dense plankton concentrations and a benthic community dominated by filter feeders. The pattern of N cycling in these two types of systems differs greatly, as illustrated by the case of Kaneohe Bay.

REFERENCES

Atkinson, M. J. (1981). Phosphate metabolism of coral reefs flats. Ph.D. Dissertation, University of Hawaii, Oceanography Department, Honolulu.

Atwood, D., Brown, D. W., Cabelli, V., Farrington, J., Garside, C., Han, G., Hansen, D. V., Harvey, G., Kamlet, K. S., O'Connor, J., Swanson, L., Seift, D., Thomas, J., Walsh, J., and Whitledge, T. (1979). The New York Bight. *In* "Assimilative Capacity of U. S. Coastal Waters for Pollutants" (E. Goldberg, ed.), pp. 148-178. National Oceanic Atmospheric Administration, Boulder, Colorado.

Betzer, P. R. (1980). Diurnal patterns in particle fluxes and composition in the euphotic zone of the Carribean Sea. *Pap. Geophy. Union Spring Meet., 1980, EDS 61,* 271. Abstract.

Bienfang, P. K. (1980). Herbivore diet affects fecal pellet settling. *Can. J. Fish. Aquat. Sci. 37,* 1352-1357.

Bienfang, P. K. (1981). "Phytoplankton Dynamics in the Oligotrophic Waters off Kahe Point, Oahu, Hawaii." Oceanic Institute, Waimanalo, Hawaii.

Bienfang, P. K., and Goodwin, D. A. (1982). Measurement of biogenic rates in subtropical waters. *Pap. Am. Geophys. Union Oceanogr. Am. Soc. Limnol. Oceanogr. J. Meet. 1982,* p. 000.

Bienfang, P. K., and Gundersen, K. (1977). Light effects on nutrient-limited, oceanic primary production. *Mar. Biol. 43,* 187-191.

Caperon, J. (1975). A trophic level ecosystem model analysis of the plankton community in a shallow-water subtropical estarine embayment. *Estuarine Res. 1,* 693-709.

Caperon, J., and Meyer, J., (1972). Nitrogen-limited growth of marine phytoplankton. *Deep-Sea Res. 19,* 601-632.

Davies, J. M. (1975). Energy flow through the benthos in a Scottish sea loch. *Mar. Biol. 31,* 353-362.

Delwiche, C. C. (1981). The nitrogen cycle and nitrous oxide. *In* "Denitrification, Nitrification and Atmospheric Nitrous Oxide" (C. C. Delwiche, ed.), pp. 1-15. Wiley (Interscience), New York.

Duedall, I. W., O'Connor, H. B., Parker, J. H., Wilson, R. E., and Robbins, A. S. (1977). The abundances, distribution and flux of nutrients and chlorophyll *a* in the New York Bight apex. *Estuarine Coastal Mar. Sci. 5,* 81-105.

Dugdale, R. C., and Goering, J. J. (1967). Uptake of new and regenerated forms of nitrogen in primary productivity. *Limnol. Oceanogr. 12,* 196-206.

Dugdale, R. C., Goering, J. J., Barber, R. T., Smith, R. L., and Packard, T. T. (1977). Denitrification and hydrogen sulfide in the Peru upwelling region during 1976. *Deep-Sea Res. 24,* 601-608.

Edmondson, C. H. (1928). Ecology of a Hawaiian coral reef. *B. P. Bishop Mus. Bull. 45,* 1-64.

Edmondson, C. H. (1946). Reef and shore fauna of Hawaii (2nd ed.). *B. P. Bishop Mus., Spec. Publ. 22,* 1-381.

Eppley, R. W., and Peterson, B. J. (1979). Particulate organic matter flux and planktonic new production in the deep ocean. *Nature (London) 282,* 677-680.

Eppley, R. W., Rogers, J. N., and McCarthy, J. J., (1969). Half-saturation constants for uptake of nitrate and ammonium by marine phytoplankton. *Limnol. Oceanogr. 14,* 912-920.

Eppley, R. W., Renger, E. H., Venrick, E. L., and Mullin, M. M. (1973). A study of plankton dynamics and nutrient cycling in the central gyre of the North Pacific Ocean. *Limnol. Oceanogr. 18,* 534-551.

Eppley, R. W., Sharp, J. H., Renger, E. H., Perry, M. J. and Harrison, W. G. (1977). Nitrogen assimilation by phytoplankton and other microorganisms in the surface waters of the central North Pacific Ocean. *Mar. Biol. 39,* 111-120.

Garside, C. (1981). Nitrate and ammonia uptake in the apex of the New York Bight. *Limnol. Oceanogr. 26,* 731-739.

Garside, C., Malone, T. C., Roels, O. A., and Sharfstein, B. A. (1976). An evaluation of sewage derived nutrients and their influence on the Hudson Estuary and New York Bight. *Estuarine Coastal Mar. Sci. 4,* 281-289.

Gilmartin, M., and Revelente, N. (1974). The island mass effect on the phytoplankton and primary production of the Hawaiian Islands. *J. Exp. Mar. Biol. Ecol. 16,* 181-204.

Goering, J. J. (1968). Denitrification in the oxygen minimum layer of the Eastern Tropical Pacific Ocean. *Deep-Sea Res. 15,* 157-164.

Goering, J. J., and Cline, J. D. (1970). A note on denitrification in seawater. *Limnol. Oceanogr. 15,* 306-309.

Gunderson, K. R., Corbin, J. S., Hanson, C. L., Hanson, M. L., Hanson, R. B., Russell, D. J., Stollar, A., and Yamada, O. (1976). Structure and biological dynamics of the oligotrophic ocean photic zone off the Hawaiian Islands. *Pac. Sci. 30,* 45-68.

Johannes, R. E., and Project Symbiosis Team (1972). The metabolism of some coral reef communities: A team study of nutrient and energy flux at Eniewetok. *Bioscience 22,* 541-543.

Kinsey, D. W. (1979). Carbon turnover and accumulation by coral reefs. Ph.D. Dissertation, University of Hawaii, Oceanography Department, Honolulu.

Kinsey, D. W., and Davies, P. J. (1979). Effects of elevated nitrogen and phosphorus on coral reef growth. *Limnol. Oceanogr. 24,* 935-940.

Kremer, J. N., and Nixon, S. W. (1978). "A Coastal Marine Ecosystems," Ecol. Stud.,Vol. 24. Springer-Verlag, Berlin and New York.

Laska, R., Schaefer, M., Myers, D., Parker, T., Wells, W., Dorset, P., McLean, J., Altschuler, K., Wicker, M., and Scott-Walton, B. (1980). "Acid Rain." Environ. Prot. Agency, U.S. Govt. Printing Office, Washington, D. C.

Laws, E. A., and Redalje, D. G. (1981). Sewage diversion effects on the water column of a subtropical estuary. *Mar. Environ. Res. 6,* 265-279.

Ling, S. W. (1977). Polyculture. *In* "Aquaculture in Southeast Asia: A Historical Overview" (L. Mumaw, ed.), pp. 78-87. Univ. of Washington Press, Seattle.

McCarthy, J. J. (1972). The uptake of urea by natural populations of marine phytoplankton. *Limnol. Oceanogr. 17,* 738-748.

McCarthy, J. J., and Goldman, J. C. (1979). Nitrogenous nutrition of marine phytoplankton in nutrient-depleted waters. *Science 203,* 670-672.

Maragos, J. E. (1972). A study of the ecology of Hawaiian reef corals. Ph.D. Dissertation, University of Hawaii, Oceanography Department, Honolulu.

Nixon, S. W., Kelley, J. R., Furnas, B. N., Oniatt, C. A., and Hale, S. S. (1980). Phosphorus regeneration and the metabolism of coastal marine bottom communities. *In* "Marine Benthic Dynamics" (K. R. Tenore and B. C. Coull, eds.), pp. 219-242. Univ. of South Carolina Press, Columbia.

Odum, H. T., and Odum, E. P. (1955). Trophic structure and productivity of a windward coral reef community on Eniwetok Atoll. *Ecol. Monogr. 25,* 291-320.

Paine, R. (1976). Fifteen thousand square miles of trouble. *NOAA Mag.,* October pp. 4-11.

Redfield, A. C., Ketchum, B. H., and Richards, F. A. (1963). The influence of organisms on the composition of seawater. *In* "The Sea" (M. N. Hill, ed.), Vol. 2, pp. 26-77. (Interscience), Wiley, New York.

Rowe, G. T., Clifford, C. H., Smith, K. L., and Hamilton, P. L. (1975). Benthic nutrient regeneration and its coupling to primary productivity in coastal waters. *Nature (London) 255,* 215-217.

Rowe, G. T., Clifford, C. H., and Smith, K. L. (1977). Nutrient regeneration in sediments off Cap Blanc, Spanish Sahara. *Deep-Sea Res. 24,* 57-63.

Ryther, J. H. (1969). Photosynthesis and fish production in the sea. *Science 166,* 72-76.

Ryther, J. H., and Dunstan, W. M. (1971). Nitrogen, phosphorus, and eutrophication in the coastal marine environment. *Science 171,* 1008-13.

Sinderman, C. J., and Swanson, R. L. (1979). "Oxygen Depletion and Associated Benthic Mortalities in the New York Bight," Prof. Pap. II. National Oceanographic and Atmospheric Administration, Washington, D. C.

Smith, S. V., Kimmerer, W. J., Laws, E. A., Brock, R. E., and Walsh, T. W. (1981). Kaneohe Bay sewage diversion experiment: Perspectives on ecosystem responses to nutritional perturbation. *Pac. Sci. 35,* 279-395.

Takahashi, M., and Bienfang, P. K. (1983). Size structure of phytoplankton biomass and photosynthesis in subtropical Hawaiian waters. *Mar. Biol.* (in press).

Turner, J. T. (1977). Sinking rates of fecal pellets from the marine copepod *Pontella meadii. Mar. Biol. 40,* 249-59.

U. S. Department of Energy (1979). "Ocean Thermal Energy Conversion (OTEC) Program. Preoperational Ocean Test Platform. Environmental Assessment." U. S. Govt. Printing Office, Washington, D. C.

Vinogradov, A. P. (1953). "The Elementary Composition of Marine Organisms," Mem. II. Sears Found. Mar. Res., New Haven, Connecticut.

Webb, K. L., DuPaul, W. D., Wiebe, W., Sottile, W., and Johannes, R. E. (1975). Enewetak (Eniewetok) atoll: Aspects of the nitrogen cycle on a coral reef. *Limnol. Oceanogr. 20,* 198-210.

Wiebe, W. J., Johannes, R. E., and Webb, K. L. (1975). Nitrogen fixation in a coral reef community. *Science 188,* 257-259.

Chapter 14

NITROGEN CYCLING IN NEAR-SURFACE WATERS OF THE OPEN OCEAN

JAMES J. McCARTHY
Museum of Comparative Zoology
Harvard University
Cambridge, Massachusetts

EDWARD J. CARPENTER
Marine Sciences Research Center
State University of New York
Stony Brook, New York

I. INTRODUCTION

In the past two decades we have increased our ability to formulate generalizations regarding the supply of nitrogen to the euphotic zone of oceanic waters. For some of the important water column processes, such as nitrogen fixation, nitrogen remineralization, and denitrification, we have improved substantially the accuracy of our rate estimates. The same can be said for inputs from atmospheric washout. Less progress has been made, however, in quantifying the upward flux of nitrate, the dominant input of new nitrogen into the euphotic zone. Another important nitrogen source for which we lack sufficiently precise information is riverine discharge. This deficiency results from both our incomplete understanding of the fate of this input and insufficient quantitative information regarding the volume of discharge. Failure to improve

ISBN 0-12-160280-X

estimates of these two crucial inputs has hampered efforts to refine our views of the whole-ocean nitrogen cycle. In this chapter, we discuss the impact that recent studies of plankton nutrition and open-ocean processes have had on our overall view of nitrogen cycling in the oceans. Emphasis is placed primarily on the processes that are of importance in the euphotic zone.

II. NITROGEN IN OCEANIC REGIONS

A. As a Limiting Nutrient

Historically, the notion that oceanic phytoplankton are nutrient-limited, and in particular, nitrogen-limited, has arisen from several considerations. Circumstantial evidence came from comparisons between seasonal productivity cycles in coastal waters, particularly from diatom blooms, and nutrient availability (Brandt, 1899; Marshall and Orr, 1928; Lillick, 1937). Phytoplankton biomass and primary productivity were observed to diminish as nutrient concentrations were reduced to low levels in the late spring and early summer. Although grazing by herbivores was known to be important in reducing the standing crop of phytoplankton, nutrient limitation appeared to have been at least partially responsible for reduced rates of productivity (Riley et al., 1949). By analogy, the low biomass and low nutrient concentrations over large areas of the open sea during most of the year led to the presumption that primary productivity in these areas was nutrient-limited (Harvey, 1955).

Nutrient limitation was also implicated by theoretical calculations of the maximum production that could be sustained in the open ocean, assuming typical irradiance and photosynthetic efficiency. It was concluded that light penetrating the euphotic zone could support more productivity than that observed (Ryther, 1959; Vishniac, 1971). If light was not limiting productivity, it was reasonable to assume that nutrients were. Nutrient addition studies in continental shelf waters suggested that the element limiting production was nitrogen (Ryther and Dunstan, 1971).

B. Methodological Problems

Research on nitrogen cycling in the open sea has been hampered by several methodological problems, one being the measurement of ammonium. In most laboratories the limit of detection for ammonium is between 3×10^{-8} and $10^{-7} M$. Higher values were often reported in the older literature, and reasons for this difference may be related to improved procedures for sample handling and analysis (McCarthy, 1980). For one study in the central North Pacific, data from kinetic analyses indicate that the actual concentration of ammonium must have been substantially below the limits of detection (Eppley et al., 1977). The low concentrations of NH_4^+ relative to the sensitivity of measurement techniques and sample contamination problems have combined to make research on NH_4^+ cycling difficult.

Recently, there has been concern that samples held in bottles for the determination of primary productivity may be adversely affected by containment, resulting in an underestimation of production rate (see Eppley, 1980). Some of the criticisms directed at this aspect of ^{14}C methodology for measuring production (Carpenter and Lively, 1980) apply equally to the techniques that use ^{15}N, ^{32}P, and ^{33}P, and other isotopes in studies of plankton nutrition. Nutrients such as ammonium are particularly difficult to study because they are recycled readily by microheterotrophs such as bacteria and protozoans, which are likely to be included with the phytoplankton in bottle studies. Ammonium remineralization has been shown to be substantial in this size fraction (Harrison, 1978; Caperon et al., 1979; Glibert, 1982). Models that include consideration of simultaneous uptake and remineralization indicate that the kinetics of ammonium for natural assemblages can not yet be determined with sufficient precision to yield reliable kinetic parameters (C. Garside, unpublished manuscript).

Another factor that confounds the application of laboratory data to the open ocean is the finding that the specific rate of nutrient uptake (nitrogen taken up per unit phytoplankton nitrogen per unit time) can increase with increasing severity of nutrient deprivation (Conway et al., 1976). Subsequently, this concept was studied on shorter time scales (McCarthy and Goldman, 1979) and with a variety of species (Goldman and Glibert, 1982). The physiological basis and ecological ramifications of this feature of phytoplankton metabolism have been discussed by Wheeler (Chapter 9, this volume), Goldman and Glibert (Chapter 7, this volume), and Harrison (Chapter 21, this volume). There are considerable data that show a predictable relationship between the enhanced rate of ammonium uptake and the nitrogen content of the cell, or cell quota (Q), (McCarthy, 1981). Unfortunately, because of the difficulty of separating phytoplankton nitrogen from total particulates organic nitrogen in field collected material, it is at present impossible to apply this relationship to field situations.

Moreover, as a practical matter it is difficult to say whether the finding of enhanced ammonium uptake in natural populations is evidence for nitrogen limitation on the one hand, or growth on oxidized forms of nitrogen (NO_3^- and NO_2^-) due to insufficient ammonium, on the other. In a study of diatoms grown in batch culture with NO_3^- and NO_2^+, it was noted that an enhanced rate of regenerated nitrogen (ammonium and urea) uptake was clearly evident prior to depletion of ambient nitrate (Horrigan and McCarthy, 1982). The period of time over which this enhanced rate could be maintained was dependent on the extent of cellular nitrogen depletion, although the enhanced rate was observed for short time periods, no matter what the nitrogen content of the cells (Q). Thus, the enhanced uptake of regenerated N [NH_4^+, which is a function of Q when a culture is NH_4^+-limited (McCarthy and Goldman, 1979)], may be independent of Q if NO_3^- or NO_2^- are the only forms of N available for growth and are present in concentrations well in excess of growth-rate-limiting concentrations. Furthermore, the commonly observed ammonium preference phenomenon need not result simply from energetic constraints; rather, it may be a consequence of an opportunistic strategy of these organisms, one that permits the acquisition of ammonium on temporal scales consistent with natural patterns of availability (Horrigan and McCarthy, 1982).

Whereas there seems to be little question that phytoplankton are capable of developing enhanced rates of uptake for ammonium when their growth rate is limited by the availability of nitrogenous nutrient, a similar capability is not evident for nitrate (Horrigan and McCarthy, 1982; Dortch, 1982). It has been suggested that the difference in the nutrient-limited responses for the two nutrients is a function of their temporal scales of supply. Horrigan and McCarthy (1982) have proposed that this may be related to the fact that biologically driven remineralization processes supply ammonium to the euphotic zone on spatial and temporal scales that are small and short, respectively, relative to the physically driven processes that transport nitrate upward in the water column. If the point sources for ammonium are small, and if in a given parcel of water their impact on nutrient availability is transient (Goldman and Glibert, Chapter 7, this volume), then phytoplankton would benefit by being able to assimilate ammonium more rapidly than they would normally synthesize new cellular material. Conversely, the supply of nitrate to the euphotic zone occurs on scales that range from continuous molecular flux along the gradient in the pycnocline to episodic fluxes of large proportions associated with destabilization of the mixed layer as in the case of upwelling or storm-induced mixing. For either process, there would be no obvious advantage associated with a capacity for uptake that is enhanced relative to the average rate at which cells are replicating.

C. The Role of Phosphorus in Nitrogen Fixation

Given the scarcity of all nutrients in the surface waters of oceanic regions, the potential influence of phosphorus availability on the nitrogen cycle deserves close examination. Some geochemists have argued that phosphorus is likely to limit oceanic production because the source of phosphate to the oceans, primarily the riverine transport of weathered terrestrial material, is relatively finite and considerably less proximal to the open ocean than is the reservoir of atmospheric diatomic nitrogen (Broecker, 1974). As far as we know, the primary form of dissolved inorganic phosphorus in the sea, orthophosphate, is universally suitable in meeting the phosphorous requirement of marine phytoplankton. Thus, it is logical to presume that as long as the phytoplankton include species that can fix diatomic nitrogen, the productivity of the open ocean should not be limited by the availability of nitrogenous nutrient. To some degree this same argument applies to freshwater systems, and Schindler (1977) has reviewed the evidence that leads to the general conclusion that phosphorus is the element which most often limits productivity in lakes and rivers. It is important to note, however, that the only phytoplankton that can utilize diatomic nitrogen are nitrogen-fixing cyanobacteria, or blue-green algae. They are known to form nuisance blooms in freshwaters that have been enriched, either naturally or via anthropogenic sources, with phosphorus.

Many factors contribute to the development and maintenance of blue-green algal blooms. Two that may be of importance in this discussion are low salinity and high phosphorous content of the water. Whereas there are numerous genera of nitrogen-fixing blue-green algae that dominate freshwater habitats, only a single genus, *Oscillatoria* or *Trichodesmium*, reaches appreciable densities in the open ocean. It is widely responsible for most of the pelagic nitrogen fixation in the ocean, and although there are occasional reports that high densities and high growth rates have been observed in some regions, the results of most studies have led investigators to conclude that typically this organism replicates at a much slower rate than the average for other oceanic phytoplankton (Carpenter and McCarthy, 1975; McCarthy and Carpenter, 1979). Although numerous isolates of nonphotosynthetic marine bacteria have been shown to be capable of nitrogen fixation, their collective rates in nature are judged to be insignificant relative to those of *Trichodesmium* (see Carpenter, Chapter 3, this volume). For marsh, littoral, and reef habitats, very high rates of fixation have been observed (see Capone, Chapter 4, this volume), but as with riverine input (see below), this material does not readily enter the oceanic realms.

The physiology and ecology of *Trichodesmium* are not well enough understood to know whether the "blooms" that have been reported are analogous to those that develop in freshwaters following phosphorus enrichment. Blooms may simply be the result of reduced loss rates or physical aggregation rather than enhanced growth rates. There is some evidence that *Trichodesmium*, like the nitrogen-fixing freshwater blue-greens, also requires higher phosphorus concentrations than other groups of algae in order for populations to be capable of high rates of growth (McCarthy and Carpenter, 1979). From findings of high levels of alkaline phosphatase activity in this alga (Yentsch et al., 1972; McCarthy and Carpenter, 1979), it is apparent that in addition to orthophosphate, phosphomonoesters can be utilized, but we have no indication of just how significant this source might be. A better understanding of these and other aspects of the physiology of *Trichodesmium* will probably have to await the successful laboratory culture of this organism. Although numerous efforts have failed, recent progress (Ohki and Fujita, 1982) offers promise that this may soon be accomplished.

If *Trichodesmium* production is typically limited by the availability of phosphorus, are there times when this constraint is relaxed? Since the primary input of phosphorus to the sea is via riverine discharge at its perimeter, the only means by which the rate of phosphorus supply to the oceanic regions can be increased is either through more rapid recycling of organic matter in the mixed layer or through increased upward transport of deep ocean reserves. The former would also result in an increased supply of recycled nitrogen, and if total autotrophic production is limited by the availability of phosphorus or nitrogenous nutrients, the net result of more rapid recycling would be to increase autotrophic production without necessarily increasing the concentration of phosphate. If, on the other hand, the supply of phosphorus from deep water were to increase substantially as a result of physical mixing, ambient near-surface phosphate concentrations could rise. Winter mixing increases near-surface phosphate concentrations in the Sargasso Sea (Menzel and Ryther, 1960). Moreover, there is indirect evidence that the euphotic zone of oceanic waters is episodically charged by deep-water inputs of nutrients (McGowan and Hayward, 1978), and the physical processes within eddies propagated by major currents can result in enhanced vertical flux of nutrients from deep reserves (Tranter et al., 1980). However, greater than typical success of *Trichodesmium* has not been observed in conjunction with any of these phenomena. Elevated concentrations of phosphate were reported for the waters overlying the Grenada Banks near the eastern boundary of the eastern Caribbean basin (Leming, 1971), and Carpenter and Price (1977) hypothesized that this source was important in stimulating *Trichodesmium* activity in the Caribbean region. On one occasion, however, we were unable to detect elevation in phosphate concentrations in several near-bottom samples (J. J. McCarthy and E. J. Carpenter, unpublished data). Although the general circulation pattern in this region is known (Stalcup and Metcalf, 1972), its temporal and spatial variability is not, and this may in part explain the different results.

Massive marine phosphate deposits are evidence that, for both extinct and extant seas, the precipitation of apatite minerals represents a significant phosphorus removal process. One single deposit, the Upper Permian Phosphoria formation in the northwestern part of North America, contains six times more phosphorus than is present in the oceans today (Piper and Codispoti, 1975). Doremus (1982) has suggested that constraints of precipitation chemistry would prevent phosphorus concentrations in the ocean from increasing significantly above current levels. Although this certainly may be true for the oceans as a whole, it is the concentration in the euphotic zone that is of concern with regard to nitrogen fixation. If it were to increase to the concentrations typical of deep ocean water (2-3 x 10^{-6} *M*), an increased rate of mineral precipitation would not necessarily follow. Moreover, given the small volume of the euphotic zone compared to that of the entire water column in the oceanic regions, a redistribution of phosphate so that concentration is uniform with depth would reduce the deep ocean phosphate concentration by only a few percent.

Although it is not known whether such an increase in phosphate concentrations in the euphotic zone would stimulate the production of nitrogen-fixing cyanobacteria, it is worth considering whether such stimulation is even possible. If it were to come about through enhanced vertical mixing, the physical process driving this would have to be episodic in order to have the desired effect. Although *Trichodesmium*, like many other cyanobacteria, have gas vesicles (Van Baalen and Brown, 1969) that permit regulation to their position in the water column, these organisms are particularly sensitive to physical disruption. The colony must remain intact in order to fix nitrogen at high rates, and Carpenter and Price (1976) have shown that it only becomes abundant during periods when the seas are calm.

On a much longer time scale, the phosphorus supply to the open sea may vary substantially with the glacial-interglacial cycle. Broecker (1982) argues that with increasing glacial mass and corresponding reduction in sea level, the phosphorus buried in shelf sediments can be released. In the extreme case, that of sea level reduced to the shelf break, phosphorus washed from the exposed shelves by glacia melt effectively will be discharged into the open ocean. Broecker suggests that this addition of nutrient stimulates primary productivity, which in turn results in an increased rate of carbon dioxide removal from the atmosphere. This assumes that primary productivity in the open ocean is limited by the availability of phosphorus. The stimulation of nitrogen fixation proposed herein could occur, however, without a corresponding increase in the rate of total photoautotrophic production. It would, nevertheless, have a profound impact on the marine nitrogen cycle, and could be responsible for a shift from a system predominated by denitrification to one affected more by nitrogen fixation.

III. ACCOUNTING FOR NITROGEN IN THE OCEANIC REALM

A. Riverine Input

In constructing nitrogen budgets, a steady-state condition is often assumed for the marine nitrogen cycle. Terms that cannot be quantified with much precision are estimated by difference, the columns for input and loss often have equal sums, and regional differences are sometimes ignored. The fact that oceanic denitrification occurs primarily in the Eastern Tropical Pacific (Hattori, Chapter 6, this volume) may not matter in an oceanwide budget, but the fact that the combined inorganic nitrogen delivered to the sea by rivers does not reach oceanic regions should not be ignored (Edmond et al., 1981). With improved precision in the estimates for each of the important processes, it becomes apparent that the entire system is usually unbalanced. Piper and Codispoti (1975) argued, for example, that two of the important processes, nitrogen fixation and denitrification, vary in magnitude considerably over intervals of tens of thousands of years. Furthermore, recent anthropogenic modification of some of the important processes in the nitrogen cycle, such as the synthesis of fixed nitrogen fertilizers is now significant in global terms. Our lack of a good quantitative assessment and understanding of the global nitrogen cycle hampers our ability to predict the consequences of these actions.

Although the role of riverine nitrogen in open-ocean cycling of this element is probably small, it is potentially an important term in the marine nitrogen cycle because of its sensitivity to anthropogenic influence. Walsh et al. (1981) have recently suggested that the delivery of anthropogenically fixed nitrogen to the sea via rivers can partially offset the consequence of anthropogenic input of carbon dioxide to the atmosphere. Unfortunately, the riverine discharge values for nitrogen for almost all major rivers are poorly characterized, and there are few long time-series that are of any use. In their effort to obtain evidence to support their contention, Walsh et al. (1981) found it necessary to use data for the Mississippi River that were collected 450 km upstream. It is possible that information at this position has little relevance to the downstream discharge into the sea. Another problem that often arises in interpreting long time-series is that, due to changes in methodology, the quality of the data may be variable throughout the series (F. Knox, personal communication).

Whereas numerous tables of values for marine nitrogen fluxes list a single number for most processes, it is increasingly difficult to justify averaging for some terms such as riverine input. Uncertainties in the transport rates for numerous major rivers require that a global estimate has broad confidence intervals. Values in the recent literature range from 1 to 35 Tg N $year^{-1}$ (Garrels et al., 1975; Delwiche and Likens, 1977; Soderlund and Svensson, 1976). McElroy (1982) estimates the dissolved component to be 24 Tg N $year^{-1}$, but recognizes the need for better data to justify confidence in this value. Richey (1983) has recently reviewed the riverine flux data in the context of a biogeochemical model, and he suggests that because of current uncertainty in regional values, the best global estimate for total nitrogen discharge lies between 14 and 40 Tg N $year^{-1}$. The Amazon, which accounts for 20% of the global riverine flow, carries only 5% of the nitrogen, whereas the rivers of North America and Europe carry proportionately more. Land-use practices, particularly the excessive application of fertilizer in agriculture, have resulted in increased nitrogen discharge rates in these regions. Nevertheless, only a very small but yet unquantified fraction of this material is thought to be transported beyond the estuarine and shelf regions (Edmond et al., 1981).

B. Atmospheric Washout

Estimates of atmospheric washout have similarly broad confidence intervals. Söderlund and Svensson (1976) report ranges of 5-16 Tg N $year^{-1}$ for oxidized forms and 11-25 Tg N $year^{-1}$ for ammonia. High temperatures are required for the oxidation of diatomic nitrogen, and this occurs naturally through the action of lightning and more locally in the vicinity of volcanoes and forest fires. It is known, however, that this is one term that has been dramatically increased by anthropogenic activity. It is estimated that current global washout rates may be as much as five times the natural rate (McElroy, 1976), but because of the relatively short residence times for these oxides, the greatest deposition occurs near the point of origin. The quantity of ammonia that volatilizes with the remineralization of organic nitrogen is also influenced by anthropogenic activity. Due to the basic and well-buffered condition of seawater, most ammonia in the sea is present in the protonated form and is not readily exchanged with the atmosphere. Therefore, concentrations in the atmosphere are highest in the vicinity of land masses, and weighted averages for the oceanic input are difficult to calculate with confidence. Recent data for atmospheric ammonia concentrations over portions of the Southern Ocean that are distant from land are 10-100 times less than those reported for other regions (Ayers and Gras, 1980). With rain data it is possible to estimate the local significance of washout. Carpenter and McCarthy (1975) used the data reported by Menzel and Spaeth (1962) to calculate that, for the western Sargasso Sea, less than 1% of the phytoplankton nitrogen requirement can be met with this source.

C. Denitrification

The global significance of denitrification in sediments of shallow-water habitats is becoming more widely recognized (Hattori, Chapter 6, this volume). It now appears that the contribution of the shallow sediments and the pelagic oceanic habitats are approximately equal (30-50 Tg N year^{-1}, respectively) (Hattori, Chapter 6, this volume). Most of that occurring in the oceanic realm is located in the Eastern Tropical Pacific. McElroy (1982) has reexamined the data for this region, and concludes that the loss of nitrogen through denitrification is about 70 Tg N year^{-1} in the oceanic region alone. Although that occurring in marshes and shelf regions has little impact on the oceanic budget, it may be the fate of much of the combined nitrogen that is discharged by rivers. The Walsh et al. (1981) hypothesis requires multiple recyclings of nitrogen from organic material in order to bury carbon. Capone and Carpenter (1982) have estimated a total nitrogen fixation of 15 Tg N year^{-1} for shallow benthic habitats. This, summed with the riverine discharge value, yields 28-45 Tg N year^{-1}, which at the upper limit is equivalent to the shallow habitat value for denitrification (44 Tg N year^{-1}). For the oceanic region, however, denitrification occurs at a rate that is approximately 10 times that of nitrogen fixation (44 Tg N year^{-1} and 5 Tg N year^{-1}, respectively). The loss due to volatilization of biogenically produced nitrous oxide is small by comparison, and is now estimated to be 1 Tg N year^{-1} (M. B. McElroy, personal communication).

Within the mixed layer, the recycling of organic matter is the major source of combined inorganic nitrogen and, if primary production is nitrogen-limited, the rate of this remineralization should determine the rate of production. Pools of these nutrients have turnover times of hours to days. Consequently, the rate at which nitrogen is introduced via rain and nitrogen fixation is virtually insignificant relative to the rate at which this element is required by the phytoplankton. These sources are of course important on the millenial time scale or the residence time of nitrogen in the sea.

D. Vertical Flux of Nitrate

Practically speaking, the quantity of nitrogen that enters the mixed layer from deep reserves is, at steady state, the quantity that can be exported via the passive sinking of particulate matter and the active transport by migrating animals. The composition of particles collected with sediment traps varies regionally and temporally, and the rates of deposition range greatly. There are no reliable estimates of the loss due to active transport, but it is usually considered to be small relative to the passive loss.

Eppley and Peterson (1979) reviewed the data on the utilization of ^{15}N-labeled substrates and determined that the fractional contribution of nitrate ranged from 6% in oligotrophic central ocean regions to 46% in neritic waters. Intermediate values determined for waters in transition areas between subtropical or subpolar zones and for waters in the regions of equatorial divergence or subpolar zones were 13% and 18%, respectively. The trend of greater dependence on nitrate with increasing production is consistent with ideas put forth earlier by Eppley et al. (1979). It is unfortunate that data of this type are so scarce for oceanic regions.

It is clear from the considerations given above that vertical flux of nitrate by both physical processes is the major source of new nitrogen to the euphotic zone in oceanic regions. Let us consider then how rapidly this new nitrogen enters the euphotic zone in two regions: the Central North Pacific gyre and the Sargasso Sea.

Estimates of these NO_3^- fluxes are difficult to compute with high precision. Basically, the flux (F) via diffusion can be determined from the vertical eddy diffusivity term (K_z) and the gradient of nitrate at the base of the euphotic zone (dN/dz).

$$F = K_z \, dN/dz$$

Vertical eddy diffusivity (K_z) can be calculated using vertical profiles of fission products from thermonuclear bomb tests. For example, Broecker (1966) used several profiles of radioisotopes to calculate mixing across the main thermocline (waters separating the deep cold reservoir of $T < 3^{\circ}C$ and the warm near-surface reservoir of $T > 15^{\circ}C$). The mean eddy diffusion coefficient was estimated at about 1 $cm^2\ s^{-1}$. More recent values from tritium profiles for the North Atlantic main thermocline give a lower K_z of 0.20 ± 0.03 $cm^2 s^{-1}$ for the 200-m depth immediately below the core of 18°C water (Rooth and Ostlund, 1972). In the Pacific at the GEOSECS intercalibration station (28°29'N, 121°38'W), it was found that tritium concentrations were relatively constant in the upper 150 m and decreased dramatically below this depth. At this site, calculations of vertical mixing between 150 m (the approximate base of the euphotic zone) and several hundreds of meters depth gave a relatively low eddy diffusion constant of about 0.15-0.20 $cm^2\ s^{-1}$ (Roether et al., 1970). Similar values for the Central North Pacific (36-44°N and 150-170°W) were noted by White and Bernstein (1981), and ranged from 0.05 to 0.32 $cm^2\ s^{-1}$ for the depth range between 200 and 400 m. There appear to be considerable latitudinal effects in this oceanic region, as the K_z term was larger (> 0.2 $cm^2\ s^{-1}$) south of 40°N as compared with more northerly values. In the Atlantic, latitudinal effects are also observed, with a K_z term at 200 m being 0.42 between 36 and 40°N and 1.3 between 40 and 48°N (Wunch and Minster, 1982). Thus, from these studies, the K_z term for the stratified tropical and subtropical ocean ranges from about 0.1 to 1.0 $cm^2\ s^{-1}$ and is generally in the neighborhood of 0.2-0.4 $cm^2\ s^{-1}$.

Another approach to determining the coefficient of eddy diffusivity is to measure the euphotic zone phytoplankton NO_3^- demand and NO_3^- gradients and solve for K_z. In southern California coastal waters, Eppley et al. (1979) calculated vertical flux from NO_3^-

assimilation rates and vertical concentration gradients and found NO_3^- input to be a major factor in regulating phytoplankton production. They obtained a K_z of 0.8-4.0 $cm^2\ s^{-1}$ in the surface layer and 0.01-0.3 $cm^2\ s^{-1}$ in and below the thermocline. On a global basis, Eppley and Peterson (1979) estimate that new (NO_3^--supported) production is of the order of 3.4 to 4.7 x 10^9 tons of carbon per year or about 18-20% of the total.

For the tropical North Pacific (ca. 5°-20°N, 125°-150°W) Anderson (1978) calculated a vertical flux of 0.18 mg atom $N \cdot m^{-2} \cdot day^{-1}$ somewhat lower than the values of 0.38-1.76 mg atom $N \cdot m^{-2} \cdot day^{-1}$ calculated by King and Devol (1979). Anderson estimated K_z from the empirical formula

$$K_z = 118[(t/z) \times 10^3]^{-1.34} \qquad (2)$$

where t/z is the σ_t gradient at the top of the thermocline. Using this formula, K_z ranged from 0.05 to 0.51 $cm^2\ s^{-1}$ with a mean of 0.14. In this study, the average NO_3^- gradient at the crest of the thermocline was 0.15 mg·atom NO_3^- $N \cdot m^{-4}$.

In determining the importance of upward flux of NO_3^-, it is important that advective processes also be determined. Total NO_3^- flux (J) can be calculated using the advective flux term (w), NO_3^- concentration at the base of the euphotic zone (S) summed with diffusive flux:

$$J = w\,S + K_z(dN/dz) \qquad (3)$$

For example, in the Sargasso Sea near Bermuda, NO_3^- profiles were taken at station "S" at 3400-m depth, and in the summer of 1958 the mean gradient on 7 dates was calculated over the depth range of 50-150 m (Bermuda Biological Station, 1960). In 1961 the mean for nine stations was 0.016 mg·atom NO_3^--N m^{-4} [Woods Hole Oceanographic Institution (WHOI), 1964]. Similarly, Carpenter and McCarthy (1975) observed a mean of 0.013 mg·atom NO_3^--N m^{-4} between

TABLE I. Nitrate Gradients at the Base of the Euphotic Zone Calculated for the Western Sargasso Sea and Central North Pacific Gyre.

Location	*Dates*	*NO_3^- gradient ($mg \cdot atom^{-4}$)*	*Depth range (m)*	*Reference*
Atlantic				
32° 10'N 64° 30'W Bermuda Station "S"	July 2-October 2, 1958 x 7 stations	0.013	50-150	Bermuda Biological Station (1960)
32° 10'N 64° 30'W Bermuda Station "S"	June 7-October 5, 1961 x 9 stations	0.0162	50-150	WHOI Report 64-8 (1964)
Western Sargasso Sea	September 19, 1973 x 9 stations	0.0127	75-150	Carpenter and McCarthy (1975)
20° 13'N 60° 07'W Southern Sargasso Sea	May 23, 1965 1 station	0.0360	150-200	WHOI Report 66-18 (1966)
20° 00'N 60° 00'W	October 26, 1964 1 station	0.0553	125-200	WHOI Report 66-18 (1966)
Pacific				
27° 57'N 154° 42'W	August 29-September 2, 1973	0.0500	125-200	Kiefer et al. (1976)
Near Hawaiian Islands	6 cruises, 1972	0.0350	125-175	Gunderson et al. (1976)

75 and 150 m (Table I). To approximate a NO_3^- flux for the Sargasso Sea, one may use the K_z value of 0.20 $cm^2\ s^{-1}$ (=0.72 $m^2\ h^{-1}$) (Rooth and Ostlund, 1972), which was obtained from tritium profiles in the southwestern Sargasso Sea and the NO_3^- gradient of 0.013 mg•atom NO_3^--N m^{-4}. This gives a mean flux of 0.4 μg•atom NO_3^--N•m^{-2}•h^{-1} or 225 μg•atom NO_3^--N•m^{-2}•day^{-1}.

Similarly, for the Central North Pacific Gyre near Hawaii, NO_3^- gradients can be calculated from the data of Kiefer et al. (1976) and Gunderson et al. (1976). Below 125 m, gradients are 0.05 and 0.35 μg•atom NO_3^--N m^{-4}, respectively. Using the K_z value of 0.20 $cm^2\ s^{-1}$ (Roether et al., 1970) at the Pacific GEOSECS intercalibration station, we obtained 36 and 25 μg•atom NO_3^--N•m^{-2}•h^{-1} or 846 and 600 μg•atom NO_3^--N•m^{-2}•day^{-1}, respectively.

These eddy diffusional flux terms represent estimates made with relatively low K_z values. Pritchard et al. (1971) measured a greater K_z term at the base of the thermocline using vertical profiles of ^{14}C. At 150-m depth in the northeast Pacific Ocean (39°N, between 134° 15'W and 134°59'W) their coefficient of eddy diffusivity was 1.2 $cm^2\ s^{-1}$. Using this, our diffusional flux near Hawaii would be about 5200 and 3600 μg•atom NO_3^--N•m^{-2}•day^{-1}, and if we assume that this is the upper limit K_z value for any oligotrophic region, then vertical flux could be a maximum of 1350 μg•atom NO_3^--N•m^{-2}•day^{-1} in the western Sargasso Sea in summer. Similar results were obtained by Platt et al. (1982) for the Central North Pacific Gyre. They assumed a K_z term of 0.1 $cm^2\ s^{-1}$ or 0.36 $m^2\ h^{-1}$ based on salt and temperature distributions within and below the thermocline (Munk, 1966; Garrett, 1979). The NO_3^- gradient used was 80 μg•atom NO_3^--N m^{-4} for the 200- to 500- m depth strata, and these values yield an estimate of 800 μg•atom NO_3^--N•m^{-2}•day^{-1}. Platt et al. (1982) consider this to be an upper estimate of the NO_3^- flux via eddy diffusion.

Another source of NO_3^- for the euphotic zone is advective flux across the thermocline. To calculate the advective flux, we can take the average concentration of NO_3^- at the midpoint of the NO_3^-

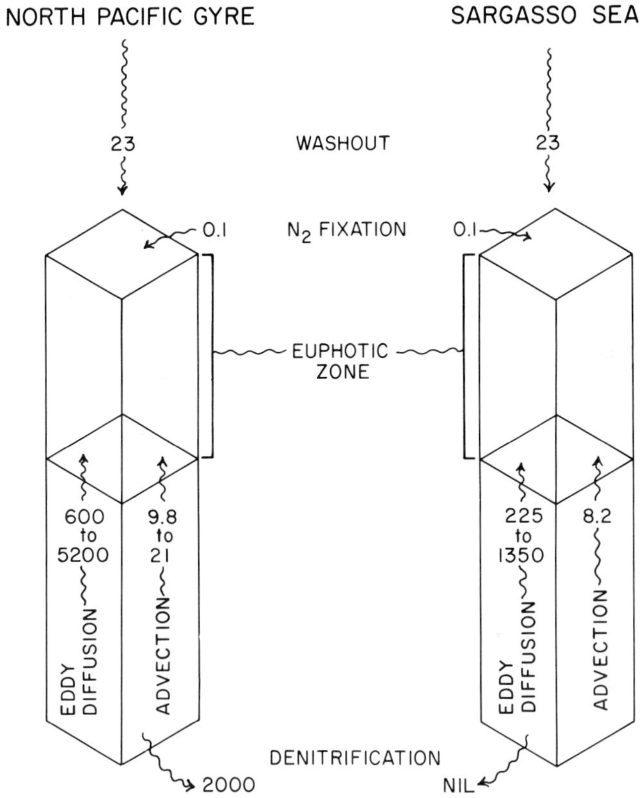

Fig. 1. Nitrogen inputs and losses for the euphotic zones or the Central North Pacific gyre and the Sargasso Sea. Units are as μg•atom•m^{-2}•day^{-1}. *Note that the denitrification estimate (from Hattori, Chapter 6, this volume) is highly dependent on ambient* O_2 *concentration.*

gradient in the Sargasso Sea (0.68 μg•atom NO_3^--N $liter^{-1}$ at 100-m depth), calculated from WHOI (1964) Bermuda Biological Station (1960) and Carpenter and McCarthy (1975). Assuming a *w* value of 1.4×10^{-5} cm s^{-1} (Munk, 1966), we calculate an advective flux of 8.2 μg•atom NO_3^--N•m^{-2}•day^{-1} (Fig. 1). For the Pacific, assuming a NO_3^- concentration at 150 m of 0.82 μg•atom NO_3^--N $liter^{-1}$ (Gunderson et al., 1976) and 1.8 μg•atom NO_3^--N $liter^{-1}$ at 175 m (Kiefer et al., 1976), the advective flux ranges from 9.8 to 21 μg•atom NO_3^--N•m^{-2}•day^{-1}. Obviously, these advective flux terms are low relative to those found in upwelling regions or areas such as the subarctic northeast Pacific Ocean, where there is relatively intense entrainment of deep water into the euphotic zone (Anderson et al., 1969).

In comparison, nitrogen input to the euphotic zone from N_2 fixation via marine *Oscillatoria (Trichodesmium)* on a daily basis (assuming a trichome concentration of $10^3\ m^{-3}$ in the upper 50 m) to both the Pacific and Atlantic is a relatively minor term, about $0.1\ \mu g{\cdot}atom{\cdot}m^{-2}{\cdot}day^{-1}$ (Carpenter, Chapter 3, this volume).

The nitrogen entering as NH_4^+ in rainfall for the Sargasso Sea is calculated to be 23 $\mu g{\cdot}atom\ NH_4^+{-}N{\cdot}m^{-2}{\cdot}day^{-1}$ (Menzel and Spaeth, 1962), and in all probability a similar amount enters the North Pacific Gyre.

These calculations point up the importance of eddy diffusive flux as a source of nitrogen for the euphotic zone (Fig. 1). Unfortunately, the estimations of NO_3^- flux are presently relatively imprecise, being based on indirect measurements of the flux term.

If we sum the calculated nitrogen inputs to the Sargasso Sea euphotic zone, we obtain a daily input of about 600 $\mu g{\cdot}atom\ N\ m^{-2}$ (using the mean of the eddy diffusivity terms), and this value appears high relative to nitrogen requirements of surface waters. For example, Menzel and Ryther (1960) calculate an annual rate of carbon fixation that averages about 140 $mg\ C{\cdot}m^{-2}{\cdot}day^{-1}$ in the Sargasso Sea near Bermuda. This would require (atomic C:N uptake ratio of 6) about 1700 $\mu g{\cdot}atom\ N{\cdot}m^{-2}{\cdot}day^{-1}$, about three times what we estimate enters on a daily basis. However, it is generally accepted that nitrogen cycling is rapid in the euphotic zone, and if we assume an 80% recycling of the newly entered nitrogen, this input of 600 $\mu g{\cdot}atom{\cdot}m^{-2}{\cdot}day^{-1}$ should be able to support a rate of production of about 220 $mg\ C{\cdot}m^{-2}{\cdot}day^{-1}$, about one and one-half times the observed production. This calculation does not include the nitrogen entering during the breakdown of the seasonal thermocline every spring. One possible explanation for this discrepancy is that the rate of primary production in the Sargasso Sea has been underestimated.

IV. SUMMARY

Clearly, we do not have sufficient data to conclude that nitrogen fluxes to and from the oceanic region are either balanced or unbalanced. This uncertainty is due in part to the imprecise nature of our best estimates of some of the key processes. Many of the extrapolated "global" rates are based on small data sets that do not adequately represent, either spatially or temporally, the natural variability in the system. To some degree, the uncertainty is also due to inadequate understanding of exchange processes among various marine habitats. Circulation patterns in coastal waters are typically parallel to the coastline, and even at the mouths of rivers there is little evidence that a significant fraction of the discharged material reaches oceanic regions. Known mechanisms are simply not very effective in transporting nitrogen from the neritic to the open-ocean regions. Meaningful budgets for the marine nitrogen cycle must take this natural partitioning of the oceans into account, although the rates of exchange among compartments can differ considerably on time scales of 1000-10,000 years. At present though, it appears that the substantial imbalance in this cycle exists in the oceanic region.

A consideration of nitrogen inputs from atmospheric washout, N_2 fixation, as well as advective and diffusive processes, illustrates the importance of advection and diffusion as sources of nitrogen to the euphotic zone in stratified oceans. In addition, it is apparent that more effort should be made to quantify all of these terms.

Given the magnitude of certain anthropogenic perturbations to the global nitrogen cycle, it is imperative that we better assess the current state of the system. The oceanic region is an important component in the global cycle of nitrogen as a consequence of the contribution of this area to both global primary production and to denitrification, and because of the magnitude of the deep-ocean reservoir of combined inorganic nitrogen.

REFERENCES

Anderson, G. C., Parsons, T. R., and Stephens, K. (1969). Nitrate distribution in the subarctic Northeast Pacific Ocean. *Deep-Sea Res. 16,* 329-334.

Anderson, J. J. (1978). Deep ocean mining and the ecology of the tropical North Pacific. Dept. of Oceanography. *Univ. Wash. Spec. Rep. 83,* 1-123.

Ayers, G. P., and Gras, J. L. (1980). Ammonia gas concentrations over Southern Oceans. *Nature (London) 284,* 539-540.

Bermuda Biological Station (1960). Physical, chemical and biological observations in the Sargasso Sea off Bermuda, 1957-1960. Part 3. U. S. A. E. C. Rep. Contract AT(30-1)-2078.

Brandt, K. (1899). Uber der Stoffuuechsel im Meere. *Wiss. Meeresunters. Abt. Kiel 4,* 213-230.

Broecker, W. S. (1966). Radioisotopes and the rate of mixing across the main thermocline of the ocean. *J. Geophys. Res. 71,* 5827-5836.

Broecker, W. S. (1974). "Chemical Oceanography". Harcourt, Brace, Jovanovich, New York.

Broecker, W. S. (1982). Glacial to interglacial changes in ocean chemistry. *Prog. Oceanogr. 2,* 151-197.

Caperon, J., Schell, D., Hirota, J., and Laws, E. (1979). Ammonium excretion rates in Kaneohe Bay, Hawaii, measured by a ^{15}N isotope dilution technique. *Mar. Biol. (Berlin) 54,* 33-40.

Capone, D. G., and Carpenter, E. J. (1982). Nitrogen fixation in the marine environment. *Science 217,* 1140-1142.

Carpenter, E. J., and Lively, J. S. (1980). Review of estimates of algal growth using ^{14}C tracer techniques. *In* "Primary Productivity in the Sea" (P. G. Falkowski, ed.), pp. 161-178. Plenum, New York.

Carpenter, E. J., and McCarthy, J. J. (1975). Nitrogen fixation and uptake of combined nitrogenous nutrients by *Oscillatoria (Trichdesmium) thiebautii* in the western Sargasso Sea. *Limnol. Oceanogr. 20,* 389-401.

Carpenter, E. J., and Price, C. C. (1976). Marine *Oscillatoria (Trichodesmium):* Explanation for aerobic nitrogen fixation without heterocysts. *Science 191,* 1278-1280.

Carpenter, E. J., and Price, C. C. (1977). Nitrogen fixation, distribution, and production of *Oscillatoria (Trichodesmium)* spp. in the western Sargasso and Caribbean Seas. *Limnol. Oceanogr. 22,* 60-72.

Conway, H. L., Harrison, P. J., and Davis, C. O. (1976). Marine diatoms grown in chemostats under silicate or ammonium limitation. II. Transient response of *Skeletonema costatum* to a single addition of the limiting nutrient. *Mar. Biol. (Berlin) 35,* 187-199.

Delwiche, C. C., and Likens, G. E. (1977). Biological response to fossil fuel combustion products. *In* "Global Chemical Cycles and their Alteration by Man" (W. Stumm, ed.), pp. 73-88. Dahlem Konferenzen, Berlin.

Doremus, C. (1982). Geochemical control of dinitrogen fixation in the open ocean. *Biol. Oceanogr. 1*(4), 429-436.

Dortch, Q., Clayton, J. R., Jr., Thoreson, S. S., Bressler, S. L., and Ahmed, S. I. (1982). Response of marine phytoplankton to nitrogen deficiency: Decreased nitrate uptake vs. enhanced ammonium uptake. *Mar. Biol. (Berlin) 70,* 13-19.

Edmond, J. M., Boyle, E. A., Grant, B., and Stallard, R. F. (1981). The chemical mass balance in the Amazon plume. I. The nutrients. *Deep-Sea Res. 28A,* 1338-1374.

Eppley, R. W. (1980). Estimating phytoplankton growth rates in the central oligotrophic oceans. *In* "Primary Productivity in the Sea" (P. G. Falkowski, ed.), pp. 231-242. Plenum, New York.

Eppley, R. W., and Peterson, B. J. (1979). Particulate organic matter flux and planktonic new production in the deep ocean. *Nature (London) 282,* 677-680.

Eppley, R. W., Sharp, J. H., Renger, E. H., Perry, M. J., and Harrison, W. G. (1977). Nitrogen assimilation by phytoplankton and other microorganisms in the surface waters of the Central North Pacific Ocean. *Mar. Biol. (Berlin) 39,* 111-120.

Eppley, R. W., Renger, E. H., and Harrison, W. G. (1979). Nitrate and phytoplankton production in southern California coastal waters. *Limnol. Oceanogr. 24*(3), 483-494.

Garrells, R. M., Mackenzie, F. T., and Hunt, C. (1975). "Chemical Cycles and the Global Environment: Assessing Human influences." Wm. Kaufmann, Inc., Los Altos, California.

Garrett, C. (1979). Mixing in the ocean interior. *Dyn. Atmos. Oceans 3,* 239-251.

Glibert, P. M. (1982). Regional studies of daily, seasonal, and size fraction variability in ammonium remineralization. *Mar. Biol. (Berlin) 68,* 209-222.

Goldman, J. C., and Glibert, P. M. (1982). Comparative rapid ammonium uptake by four species of marine phytoplankton. *Limnol. Oceanogr. 27*(5), 814-827.

Gunderson, K. R., Corbin, J. S., Hanson, C. L., Hanson, M. L., Hanson, R. B., Russell, D. J., Stollar, A., and Yamada, O. (1976). Structure and biological dynamics of the oligotrophic ocean photic zone off the Hawaiian Islands. *Pac. Sci. 30,* 45-68.

Harrison, W. G. (1978). Experimental measurement of nitrogen remineralization in coastal waters. *Limnol. Oceanogr. 23,* 684-694.

Harvey, H. W. (1955). "The Chemistry and Fertility of Sea Waters." Cambridge Univ. Press, London and New York.

Horrigan, S. G., and McCarthy, J. J. (1982). Phytoplankton uptake of ammonium and urea during growth on oxidized forms of nitrogen. *J. Plankton Res. 4*(2), 379-435.

Kiefer, D. A., Olson, R. J., and Holm-Hansen, O. (1976). Another look at the nitrite and chlorophyll maxima in the central North Pacific. *Deep-Sea Res. 23,* 1199-1208.

King, F. D., and Devol, A. H. (1979). Estimates of vertical eddy diffusion through the thermocline from phytoplankton nitrate uptake rates in the mixed layer of the Eastern Tropical Pacific. *Limnol. Oceanogr. 24,* 645-651.

Leming, T. D. (1971). Eddies west of the southern Lesser Antilles. *Caribb. Sea Adjacent Reg. Pap., 1968,* pp. 113-120.

Lillick, L. C. 1937. Seasonal studies of the phytoplankton off Woods Hole, Massachusetts. *Biol. Bull. (Woods Hole, Mass.) 73,* 488-503.

McCarthy, J. J. (1980). Nitrogen and phytoplankton ecology. *In* "The Physiological Ecology of Phytoplankton" (I. Morris, ed.), pp. 191-233. Blackwell, Oxford.

McCarthy, J. J. (1981). The kinetics of nutrient utilization. *Can. Bull. Fish. Aquat. Sci. 210,* 211-233.

McCarthy, J. J., and Carpenter, E. J. (1979). *Oscillatoria (Trichodesmium) theibautii* (Cyanophyta) in the central North Atlantic Ocean. *J. Phycol. 15,* 75-82.

McCarthy, J. J., and Goldman, J. C. (1979). Nitrogenous nutrition of marine phytoplankton in nutrient depleted waters. *Science 203,* 670-672.

McElroy, M. B. (1976). Chemical processes in the solar system: A kinetic perspective. *MTP Int. Rev. Sci.* 9(2), 127-211.

McElroy, M. B. (1983). Marine biological controls on atmospheric CO_2 and climate. *Nature (London) 302,* 328-329.

McGowan, J. A., and Hayward, T. L. (1978). Mixing and oceanic productivity. *Deep-Sea Res. 25,* 771-793.

Marshall, S. M., and Orr, A. P. (1928). The photosynthesis of diatom cultures in the sea. *J. Mar. Biol. Assoc. U. K. 15,* 321-360.

Menzel, D. W., and Ryther, J. H. (1960). The annual cycle of primary production in the Sargasso Sea off Bermuda. *Deep-Sea Res. 6,* 351-367.

Menzel, D. W., and Spaeth, J. P. (1962). Occurrence of ammonia in Sargasso Sea waters and in rain water at Bermuda. *Limnol. Oceanogr. 7,* 159-162.

Munk, W. H. (1966). Abyssal recipes. *Deep-Sea Res. 13,* 707-730.

Ohki, K., and Fujita, Y. (1982). Laboratory culture of the pelagic blue-green alga *Trichodesmium thiebautii:* Conditions for unialgal culture. *Mar. Ecol.: Prog. Ser. 7,* 185-190.

Piper, D. Z., and Codispoti, L. A. (1975). Marine phosphorite deposits and the nitrogen cycle. *Science 188,* 15-18.

Platt, T., Lewis, M., and Geider, R. (1982). Thermodynamics of the pelagic ecosystem: Elementary closure conditions for biological production in the open ocean. *In* "Flows of Energy and Materials in Marine Ecosystem: Theory and Practice" (M. J. Fasham, ed.). Plenum, New York.

Pritchard, D. W., Reid, R. O., Okubo, A., and Carter, H. H. (1971). Physical processes of water movement and mixing. *Radioact. Environ.,* pp. 90-136.

Richey, J. E. (1983). Interactions of C, N. P, and S in River Systems: A Biogeochemical Model. *In* "The Interaction of Biogeochemical Cycles Scientific Committee on the Problems of the Environment" (B. Bolin, ed.). Wiley, New York.

Riley, G. A., Stommel, H., and Bumpus, D. F. (1949). Quantitative ecology of the plankton of the western North Atlantic. *Bull. Bingham Oceanogr. Collect. 12,* 1-69.

Roether, W., Munnich, K. O., and Ostlund, H. G. (1970). Tritium profile at the North Pacific (1969) GEOSECS Intercalibration Station. *J. Geophys. Res. 75,* 7672-7675.

Rooth, G., and Ostlund, H. G. (1972). Penetration of tritium into the Atlantic thermocline. *Deep-Sea Res. 19,* 481-492.

Ryther, J. H. (1959). Potential productivity of the sea. *Science 130,* 602-608.

Ryther, J. H., and Dunstan, W. M. (1971). Nitrogen, phosphorus and eutrophication in the marine environment. *Science 171,* 1008-1013.

Schindler, D. W. (1977). Evolution of phosphorus limitation in lakes. *Science 195,* 260-262.

Söderlund, R., and Svensson, B. H. (1976). The global nitrogen cycle. *Ecol. Bull. 22,* 23-73.

Stalcup, M. C., and Metcalf, W. G. (1972). Current measurements in the passages of the lesser Antilles. *J. Geophys. Res. 77,* 1032-1049.

Tranter, D. J., Parker, R. R., and Cresswell, G. R. (1980). Are warmcore eddies unproductive? *Nature (London) 284,* 540-542.

Van Baalen, C., and Brown, R. M., Jr. (1969). The ultrastructure of the marine blue-green alga *Trichodesmium erythraeum,* with special reference to the cell wall, gas vacuoles, and cylindrical bodies. *Arch. Mikrobiol. 69,* 79-91.

Vishniac, W. (1971). Limits of microbial production in the oceans. *In* "Microbes and Biological Productivity" (D. E. Hughes and A. H. Rose, eds.), pp. 355-366. Cambridge Univ. Press, London and New York.

Walsh, J. J., Rowe, G. T., Iverson, R. L., and McRoy, C. P. (1981). Biological export of shelf carbon is a neglected sink of the global CO_2 cycle. *Nature (London) 291,* 196-201.

White, W., and Bernstein, R. (1981). Large-scale vertical eddy diffusion in the main pycnocline of the Central North Pacific. *J. Phys. Oceanogr. 11,* 434-441.

Woods Hole Oceanographic Institution (WHOI)(1964). "Physical, Chemical and Biological Observations in the Sargasso Sea off Bermuda 1960-1963," 64-8, Appendix II. Woods Hole Oceanographic Institution Inst.

Woods Hole Oceanographic Institution (WHOI)(1966). "Biological, Chemical and Radiochemical Studies of Marine Plankton," Rep. 66-18. Woods Hole Oceanographic Institution.

Wunsch, C., and Minster, J. F. (1982). Methods for box models and ocean circulation tracers: Mathematical programming and nonlinear inverse theory. *J. Geophys. Res. 87,* 5647-5662.

Yentsch, C. M., Yentsch, C. S., and Perras, J. P. (1972). Alkaline phosphatase activity in the tropical marine blue-green alga *Oscillatoria erytheraea*. *Limnol. Oceanogr.* *17*, 772-774.

Chapter 15

NITROGEN IN UPWELLING SYSTEMS

LOUIS A. CODISPOTI
Bigelow Laboratory for Ocean Sciences
McKown Point
West Boothbay Harbor, Maine

I. INTRODUCTION

This chapter attempts to provide an introduction to the concepts and literature relevant to a consideration of the nitrogen regimes in zones of enhanced upwelling, a small portion of ocean space in which favorable nutrient and light conditions for phytoplankton growth often coexist. Emphasis on nitrogen is appropriate because this element is most frequently found in "limiting" concentrations* (Ryther and Dunstan, 1971), and because the distribu-

**Some argue that nitrogen limitation may be rare on an individual cell basis (Goldman et al., 1979). Here, we mean limitation of total primary production rates that can occur through changes in species and numbers as well as through changes in the growth rates of individual cells.*

ISBN 0-12-160280-X

tion of nitrogenous chemicals can provide insight into the functioning of these ecosystems (Codispoti and Friederich, 1978; Codispoti and Packard, 1980; Dugdale and Goering, 1967; Whitledge, 1981).

Zones of persistent wind-induced upwelling found on the eastern boundaries of oceans at low and mid-latitudes receive particular attention because of their enhanced productivity. In terms of higher trophic-level production, the differences between these regions and the average ocean are enormous. For example, during 1956 when anchovies were abundant off Peru, their yield to birds and man was one-seventh of the worldwide fish landing, and this catch was taken from an area that comprises only 0.02% of the ocean's surface (Wooster and Reid, 1963). Similarly, Ryther (1969) noted that 50% of the world's fish catch comes from the 0.1% of the ocean's area, where coastal upwelling occurs.

II. BACKGROUND

A. Upwelling as a Physical Process

Although this chapter attempts to focus on the nitrogen regimes in upwelling systems, this subject cannot be understood without some knowledge of the physics of upwelling (Barber and R. L. Smith, 1981; Margalef, 1978).

The major transports of combined nitrogen and most other nutrients into the surface layer are a consequence of vertical mixing and upwelling (Spencer, 1975; Vaccaro, 1965). These processes may co-occur, but they differ in the following important ways:

1. The effect of vertical mixing per se is to produce surface nutrient concentrations intermediate between the values in the initial surface layer and the deeper layers with which it mixes. The effect of upwelling per se is to replace the initial surface layer with a subsurface water parcel that retains the deeper nutrient concentration. Although the following explanation is not totally precise and is, for example, dependent on the spatial scale

selected, it is often useful to think of the ascending and descending motions as occurring within the same space during vertical mixing. Similarly, upwelling is associated with a more organized flow regime in which a net upward motion occurs in the upwelling zone, and the descending motions necessary to preserve continuity occur elsewhere. The organized rising and sinking motions that accompany upwelling produce horizontal counterflows similar to those of an estuary. As a consequence, upwelling regions are "nutrient traps" (see Section III, C). These counterflows may also provide a mechanism for weakly motile organisms to maintain their position in the upwelling ecosystem by means of relatively feeble vertical motions (Barber and R. L. Smith, 1981; Barber and W. O. Smith, Jr., 1981).

2. Vertical mixing is enhanced by strong winds and weak stability and is relatively insensitive to wind direction. Wind-induced upwelling, on the other hand, depends critically on wind direction (Fig. 1) and significant upwelling can occur under modest winds in regions with shallow, mixed-surface layers.

Because of the factors mentioned above, large nutrient additions to the surface layers of the ocean caused by vertical mixing tend to be associated with wintertime conditions in temperate and boreal regions, and are frequently associated with mixed layers that exceed the *compensation depth,* beyond which available light is insufficient to allow phytoplankton growth to exceed metabolic expenditures. After winter, increasing insolation reduces the depth of the mixed surface layer leading to the spring phytoplankton bloom (Sverdrup, 1953); but this bloom may be short-lived compared to the periods of high phytoplankton growth that may be sustained in low-latitude regions of persistent upwelling (e.g., Peru; Barber and R. L. Smith, 1981). Cushing (1978) suggests that fish yields in zones of persistent upwelling may exceed those of "spring bloom" regions. Sverdrup et al. (1942) point out that mid-latitude upwelling is often best developed during the summer months (e.g.,

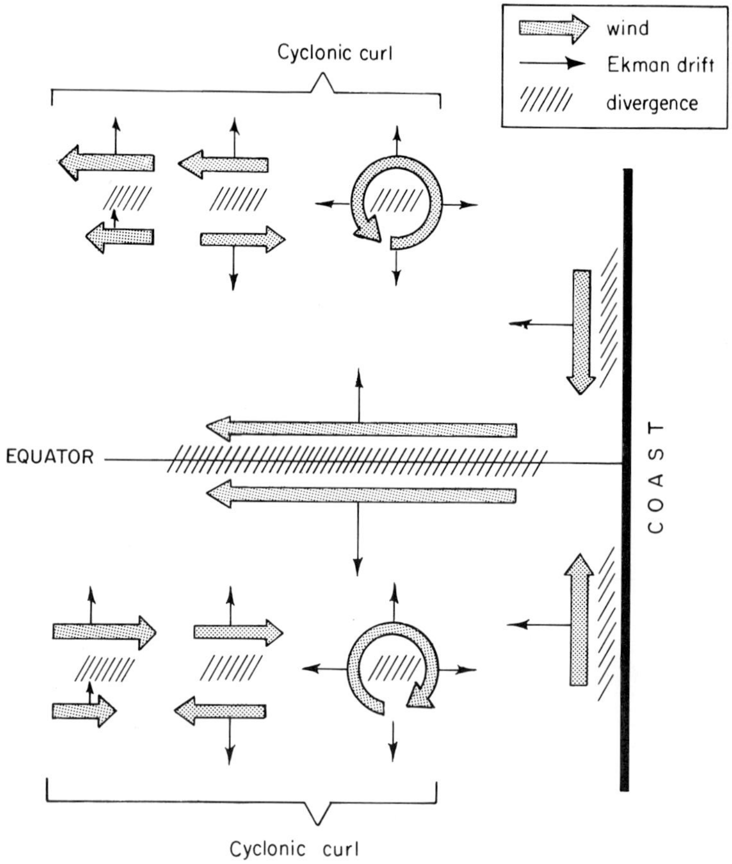

Fig. 1. Schematic diagram showing the different wind patterns that can produce a surface divergence leading to upwelling. Recent studies suggest that the edge of a pack ice field can also induce "coastal" upwelling (Buckley et al., 1979).

off California, Oregon, and Washington) when the stability and light regimes are favorable for phytoplankton growth. Because the stratification of tropical waters tends to be strong, upwelling and the vertical mixing associated with upwelling (Smith, 1968) are probably the most effective mechanisms for enriching the euphotic zones in these regions.

Although wind-driven upwelling systems are emphasized in this chapter, upwelling may occur anywhere in the ocean and can arise from a variety of causes. These include interactions of currents with topographic features (e.g., Blanton et al., 1981; Goering and McRoy, 1981; Hsueh and O'Brien, 1971; Takahashi et al., 1981) and the ascending motions found on the left- (right) hand side of currents in the Northern (Southern) Hemisphere (Yentsch, 1974). The large-scale thermohaline (density-driven) circulation of the ocean contributes to the upwelling of water around Antarctica, although this upwelling is also associated with a divergence in the wind stress (Gordon et al., 1977). Smaller-scale thermohaline circulations can also cause upwelling (Bowden, 1975; Petersen, 1977; Szekielda, 1974), and upwelling can also be caused or modified by long-period waves (Brink et al., 1981; Lee et al., 1981; O'Brien et al., 1981).

Despite the wealth of processes that can cause upwelling, the classic upwelling systems found along the eastern boundaries of oceans at low and mid-latitudes and other important upwelling sites such as the equator and the Arabian coast during the Southwest Moonson, are largely caused by surface divergences produced by the wind (Smith, 1968).

Previous studies of wind-driven upwelling permit the following generalizations:

1. Because the transport of the wind drift layer (the Ekman transport) tends to be to the right of the wind stress in the Northern Hemisphere and to the left in the Southern, upwelling can develop when an equatorward wind blows along the eastern boundary of an ocean or when a westward wind blows along the equator (Fig. 1). In the open ocean, a divergence can be produced when the wind field has a cyclonic curl (e.g., Hidaka, 1972; Hofmann et al., 1981; Smith, 1968). Although upwelling usually develops under these conditions, the effect of the wind can be negated by other forces (e.g., Roden, 1972).

2. Vertical velocities of order 10^{-2}-10^{-1} cm s^{-1} occur during episodes of strong coastal upwelling (Andrews and Hutchings, 1980; Barton et al., 1977), but over the upwelling "season," vertical

velocities averaged over the coastal and equatorial upwelling zones are probably on the order of 10^{-3} cm s^{-1} (Smith, 1968; Wooster, 1981). Vertical velocities induced in the open ocean by the curl of the wind stress are on the order of 10^{-3}-10^{-4} cm s^{-1} (Hofmann et al., 1981; Gordon et al., 1977; Smith and Bottero, 1977).

3. Usually, the waters rise to the surface from depths of less than 250 m (Bennekom, 1978; Smith and Codispoti, 1979; Wyrtki, 1963).

4. Coastal upwelling is usually confined to a narrow band. The extent of these bands is a function of depth, stratification, and the inverse of the Coriolis parameter, but they are often about 25-km wide. If the wind has the proper curl, oceanic upwelling may occur seaward of this strip (Smith, 1968, 1981).

5. The pressure gradients that arise from the divergences and convergences produced by the Ekman drift (Fig. 1) induce currents that flow with the wind (Sverdrup et al., 1942). These currents are normally stronger than those in the upwelling cross-circulation (Fig. 2), but they do not extend to great depths in the equatorial and eastern boundary upwelling regions. Undercurrents flowing against the surface currents are typical features of these regions. The character of these currents may vary markedly from region to region (Hughes and Barton, 1974; Huyer, 1976, 1980; Mittelstaedt et al., 1975; Smith, 1981), but most occur at depths that are shallow enough to allow them to contribute to the waters rising to the surface. This is important because the combined nitrogen content of these waters often differs markedly from that in the water masses flowing with the wind (Fig. 3) and because of the biological and chemical effects of opposing currents (see above).

6. A surface front indicated by the density structure in Fig. 4 occurs frequently at the boundary of a coastal upwelling system (Fig. 2). Regenerated nitrogen in the form of ammonium Fig. 5) may accumulate in the vicinity of these fronts (Hafferty et

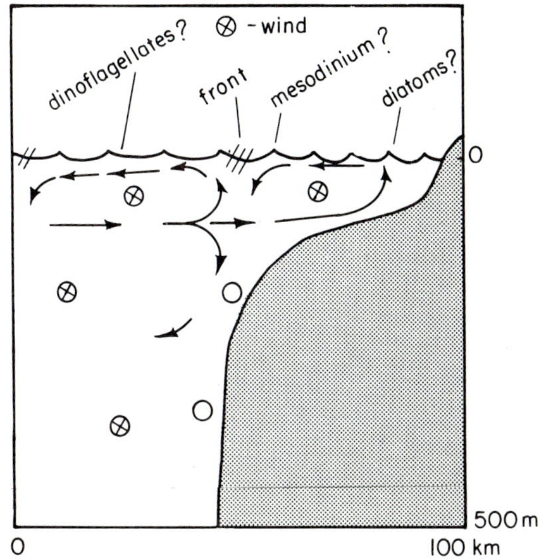

Fig. 2. Highly schematic view of an idealized upwelling system. The "crossed circles" indicate flow with the wind, and the open circles indicate flow against the wind. The arrows are streamlines for the cross circulation, not volume transports. Since upwelling is essentially a three-dimensional process that can be influenced by processes occurring over a range of scales, mass transport balances are unlikely in a particular section. In this conception, the inshore upwelling cell is supposed to indicate strong coastal upwelling, and the offshore cell is taken to be weaker offshore upwelling. The phytoplankton distributions are in line with the speculative and highly generalized scheme suggested in the text.

al., 1979), and they may be sites of downwelling (Andrews and Hutchings, 1980). Sometimes more than one upwelling cell can occur, but this is often not the case within the coastal upwelling zone (Smith, 1981).

7. So far, two-dimensional aspects of upwelling have been emphasized, but a balanced, cross-shelf upwelling cell with onshore flow equal to offshore flow is rare. To a greater or lesser extent, all upwelling systems are three-dimensional and longshore gradients in flow or properties cannot be neglected in any detailed treatment (Smith, 1981). Some upwelling systems are extremely three-dimensional, with upwelling concentrated in plumes (Traganza et al., 1981) whose position may be dictated by topographic features (Shannon et al., 1981).

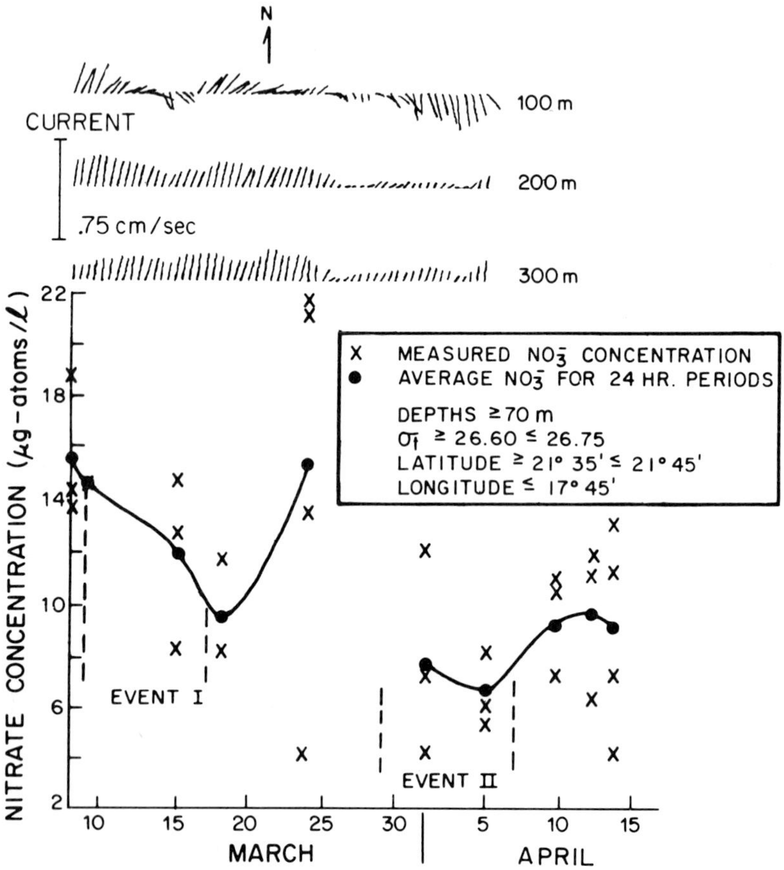

Fig. 3. Data from the 1974 JOINT-I experiment off NW Africa that suggest a relationship between a decrease in the nutrient-rich poleward undercurrent and nitrate concentrations in the upwelling source waters. The data are from the region shown in Figs. 7 and 8, and the current meters were moored over the continental slope portion of this section (from Codispoti and Friederich, 1978).

8. Coastal upwelling is a highly variable process and the variability can occur over a continuum of time scales (Codispoti, 1981; Codispoti et al., 1982a). An example of longer-term (several years) variability is the well-known El Niño phenomenon (Barber et al., 1983; Cushing, 1981; Enfield, 1981; Guillén and

Calienes, 1981; O'Brien et al., 1981) that floods the Peruvian upwelling region with warm waters. Considerable variability occurs over small scales and within the mesoscale time (days to weeks) and space (∿25-250 km) frames. Some of this physical variability can arise from factors such as tides (Codispoti and Friederich, 1978) and trapped coastal waves (Brink et al., 1981), but much of it appears to be forced by variations in the wind field (e.g., Barton et al., 1977; Hopkins, 1974).

There are significant inter- and intraregional differences in the character of the variability found in different coastal upwelling systems (e.g., Codispoti, 1981; Codispoti et al., 1982a; Smith, 1981), and coastal upwelling may be more variable than oceanic upwelling (Smith, 1968). It is important to recognize the types of variability that characterize an upwelling system when considering its nitrogen cycle because ecosystem structure can be importantly influenced by the "details" of the variability in the physical and nutrient fields (Barber and W. O. Smith, Jr., 1981; Margalef, 1978a; Walsh, 1977; Walsh et al., 1980; Weikert, 1977; and below).

B. Distribution of Nitrogen Compounds in Upwelling Systems

Nitrate, nitrite, and ammonium are the nitrogenous compounds that have received the most attention in upwelling systems. Nitrate concentrations[†] at the sea surface may exceed 25 μg·atoms $liter^{-1}$ in some coastal upwelling regions (Table I) and in the Antarctic divergence. In equatorial divergences, maximum values at the surface do not appear to exceed 15 μg·atoms $liter^{-1}$ (Fig. 6; Plank et al., 1973; Walsh, 1976), even in the eastern Pacific where the highest values should occur (see Section III,A). Nitrite and ammonium concentrations are usually below

[†]*Following oceanographic convention, 20 μg·atoms* $liter^{-1}$ *of nitrate really means 20 μg·atoms* $liter^{-1}$ *of nitrate-nitrogen.*

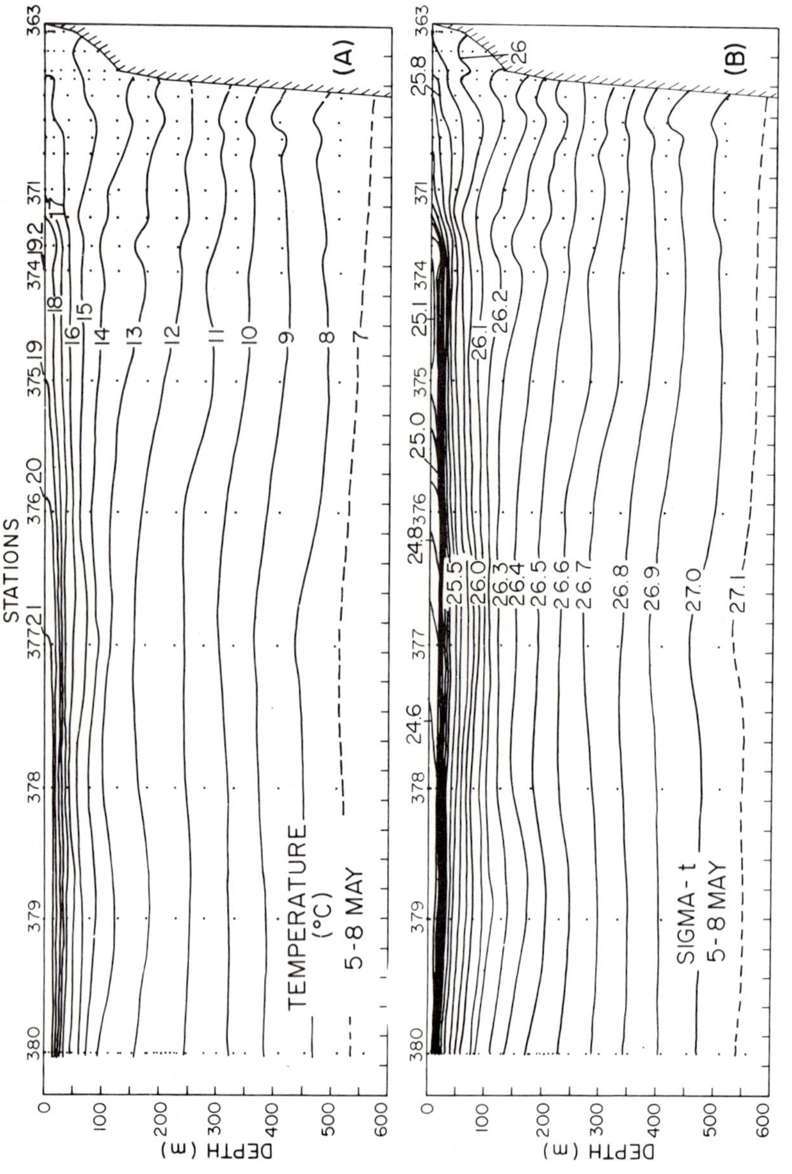
STATIONS
363
371
374
375
376
377
378
379
380
DEPTH (m)
0
100
200
300
400
500
600
(A)
TEMPERATURE
(°C)
5-8 MAY
(B)
SIGMA - t
5-8 MAY

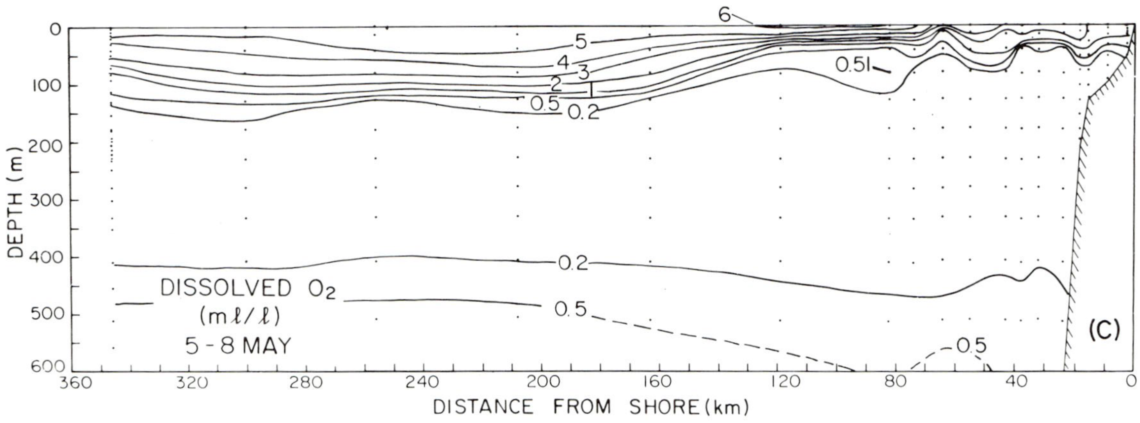

Fig. 4. Temperature, sigma-*t* (a measure of density), and dissolved oxygen concentrations in a section perpendicular to the isobaths at 15°S off Peru. The location of this section is shown in Fig. 9.

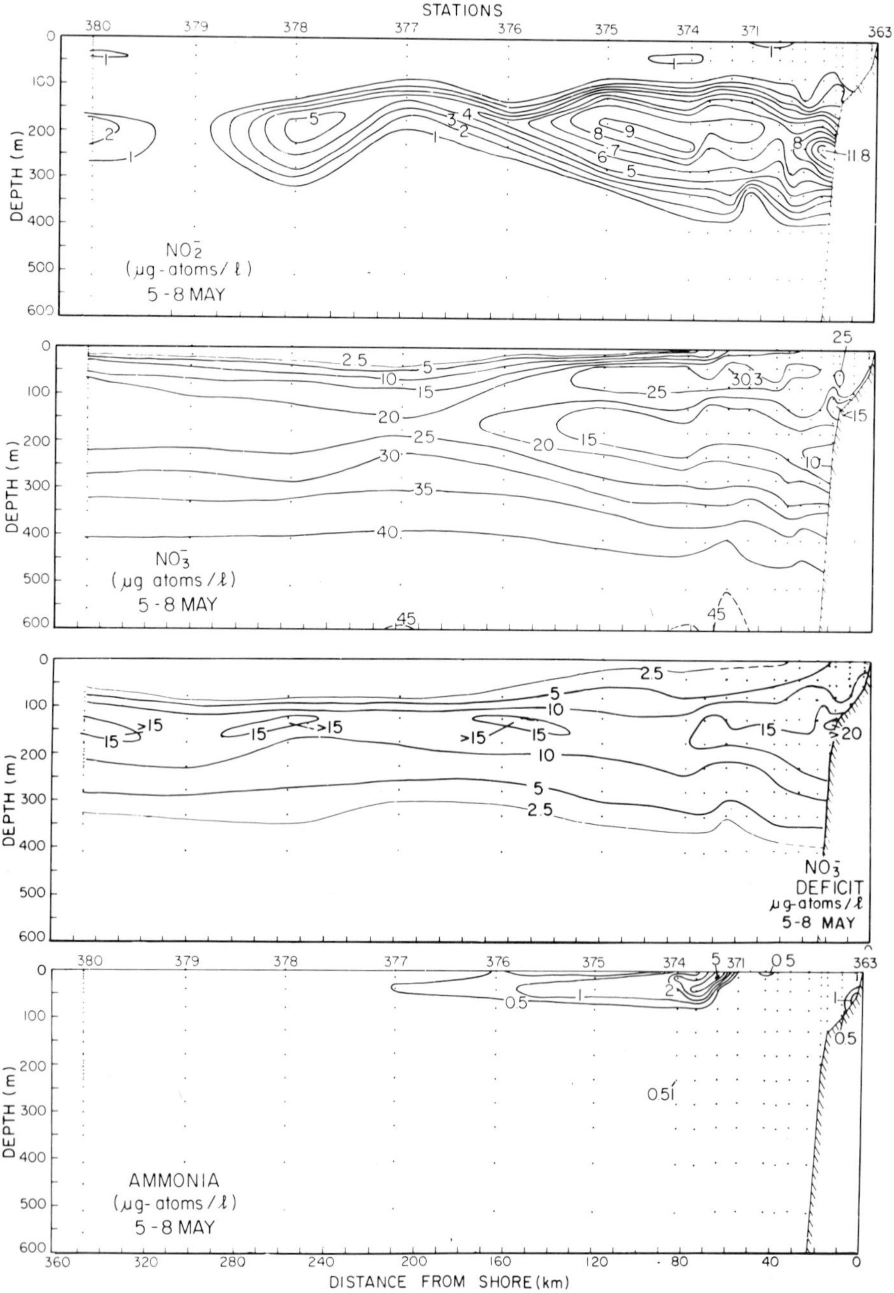

Fig. 5. Nitrate, nitrite, nitrate deficit, and ammonium distributions in the same section shown in Fig. 4. Nitrate deficits are an estimate of the nitrate and nitrite converted to nitrogen gas during denitrification.

TABLE I. High-Nutrient Concentrations Observed at the Surface in Different Upwelling Regions[a]

Region	Reference	NO_3^- (μg·atom liter^{-1})	PO_4^{3-} (μg·atom liter^{-1})	Dissolved Si (μg·atom liter^{-1})	Station
Aleutian Islands ~53°N, 170 W	Hood and Kelley (1976)	30	2.5	70	
Coastal Peru ~15°S	Hafferty et al. (1978)	27	2.2	22	(Sta. 207)
		21[b]	2.2	31	(Sta. 344)
Somali Coast north of 10°12'N	Smith and Codispoti (1979)	20	1.6	16	
NW Africa ~22°N	Friebertshauser et al. (1975)	12	1.0	7	(Sta. 14)
		11	1.0	8	(Sta. 161)
Mediterranean Sea 43°N, 5°E	Minas (1968)	--	0.1	--	

[a]*Although some of the regional differences exemplified in this table could arise from local differences in nutrient regeneration and upwelling patterns, the major differences probably arise from the interocean fractionation process described in the text. For example, the references and discussion herein suggest that upwelling intensities off NW Africa and Somalia were higher than off Peru at the time of these observations, but nutrient concentrations are higher in the latter region.*

[b]*This value may be influenced by denitrification.*

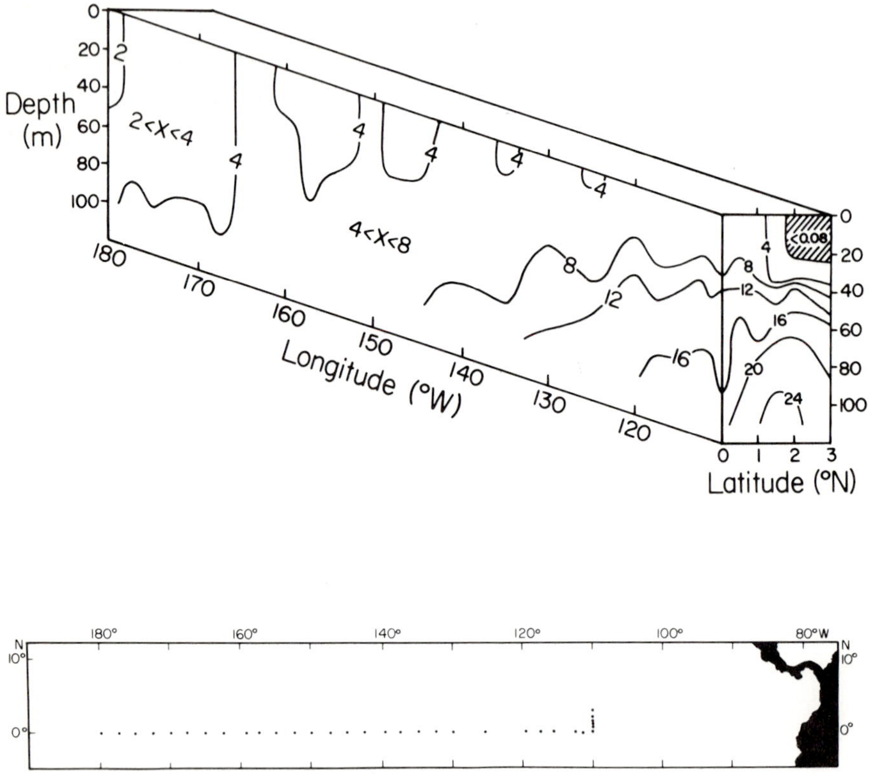

Fig. 6. The upper figure shows nitrate concentrations in an equatorial section taken in the Pacific Ocean. The station locations for this section are shown in the lower figure. Decreasing concentrations toward the west suggest the normal tendency for the highest nutrient concentration to be found in the eastern portion of an ocean. Somewhat higher concentrations would be expected if this section extended further south, but surface values in excess of 15 μg·atoms liter^{-1} are unlikely (e.g., Walsh, 1976). This section was provided by Dr. R. T. Barber.

2 μg·atoms liter^{-1} in well-oxygenated waters, but they may exceed these values under special circumstances. For example, ammonium values sometimes exceeded 2 μg·atoms liter^{-1} in a high-ammonium nearshore zone encountered off NW Africa during the JOINT-I experiment (Figs. 7 and 8). Codispoti and Friederich (1978) and Whitledge (1981) suggest that these high concentrations are a con-

sequence of the turbidity of these nearshore waters: Low light levels in this well-oxygenated nearshore region make it impossible for phytoplankton uptake to match the ammonium regeneration rates. Off Peru, relatively high ammonium concentrations in oxygenated waters are frequently associated with the pycnocline found in the vicinity of the surface front found at the edge of the coastal upwelling system (Figs. 4 and 5). High nitrite concentrations (>9 μg·atoms liter^{-1}) were found in oxygenated waters off Peru during the passage of a trapped coastal wave (Codispoti, 1981). Some idea of the days-to-weeks temporal variability that can occur in nitrate and ammonium distributions off NW Africa and Peru are given in Figs. 7, 8, and 9.

Oxygen-deficient (O_2 < 0.1 ml liter^{-1}) waters are found in the Eastern Tropical Pacific, Arabian Sea, and SW African upwelling regions, and a nitrite maximum with concentrations higher than these in the primary nitrite maximum found near the bottom of the euphotic zone is a normal occurrence in such waters (Richards, 1965). Off Peru, nitrite concentrations in excess of 10 μg·atoms liter^{-1} have been found in the secondary nitrite maximum (Fig. 5; Codispoti and Packard, 1980), but maximum values in the other oxygen deficient zones are usually below 5 μg·atoms liter^{-1} (Calvert and Price, 1971; Codispoti, 1973; Codispoti and Richards, 1976; Deuser et al., 1978).

If nitrate-reducing and denitrifying bacteria consume all of the nitrate and nitrite within an oxygen-deficient water column or within sediments, ammonium concentrations rise in association with sulfate reduction (Richards, 1965). Episodes of sulfate reduction in the water column off Peru have been described, and as a consequence of this and/or the proximity to reducing sediments, high ammonium values (∿5-40 μg·atoms liter^{-1}) have been observed (Doe, 1978; Dugdale et al., 1977; Hafferty et al., 1979; Sorokin, 1978a).

Fewer data have accumulated on urea distributions. Values from Peru (McCarthy, 1970; Remsen, 1971; Whitledge, 1981) lie

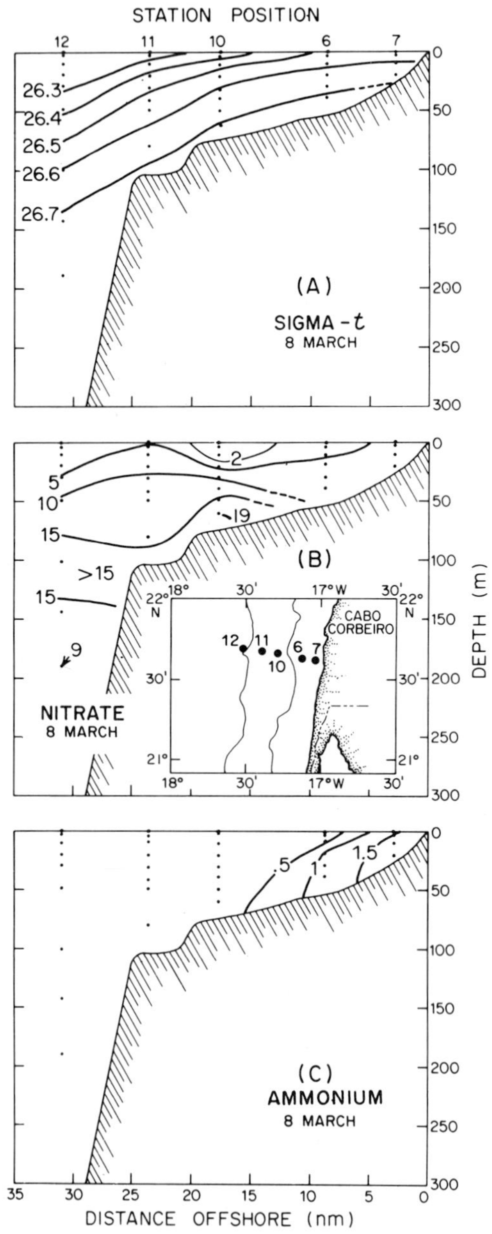

Fig. 7. Distributions of sigma-*t* (A), nitrate (B), and ammonium (C) in a section normal to the coast in the NW African upwelling region at 21° 40'N. Distributions are from just before

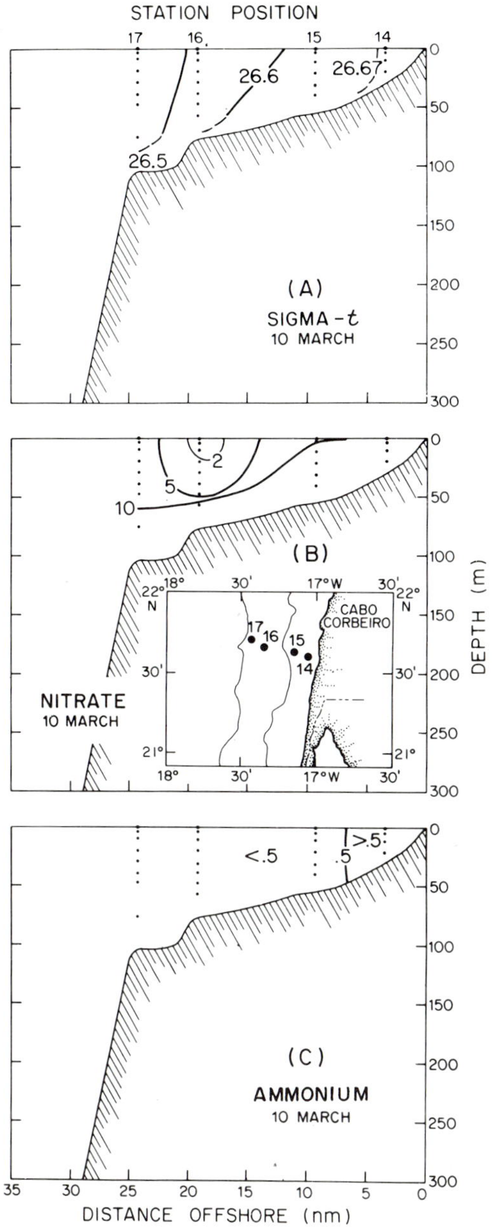

(8 March), and the beginning of a wind-induced upwelling event (10 March). The highest nitrite concentration was a value of 1.25 μg·atoms liter^{-1} found in the deepest sample at station 6.

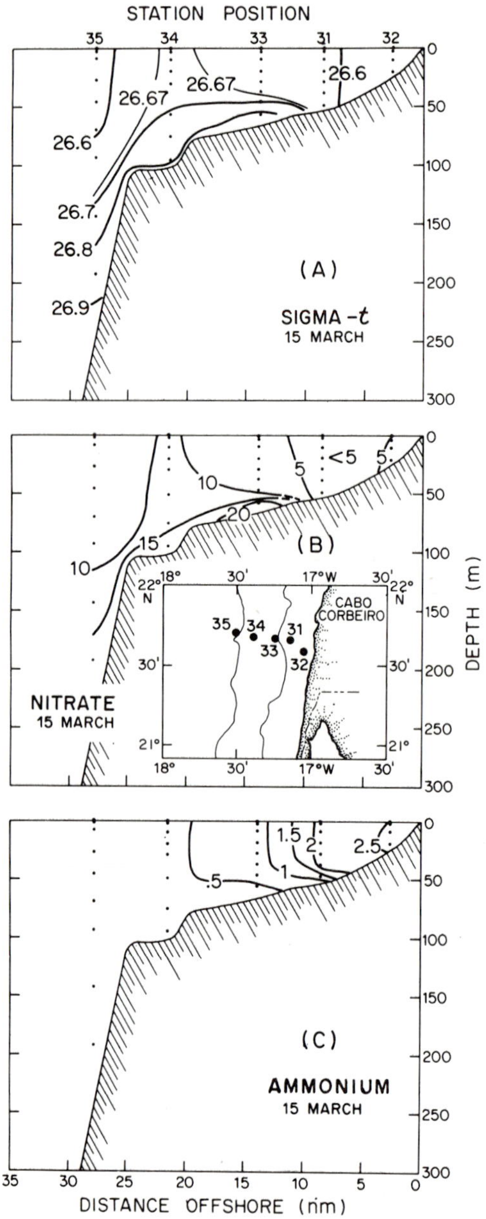

Fig. 8. Distributions of sigma-*t* (A), nitrate (B), and ammonium (C) in a section normal to the coast in the NW African upwelling region at 21° 40'N. Distributions are for the height of an upwelling event (15 March) and during a period when the local

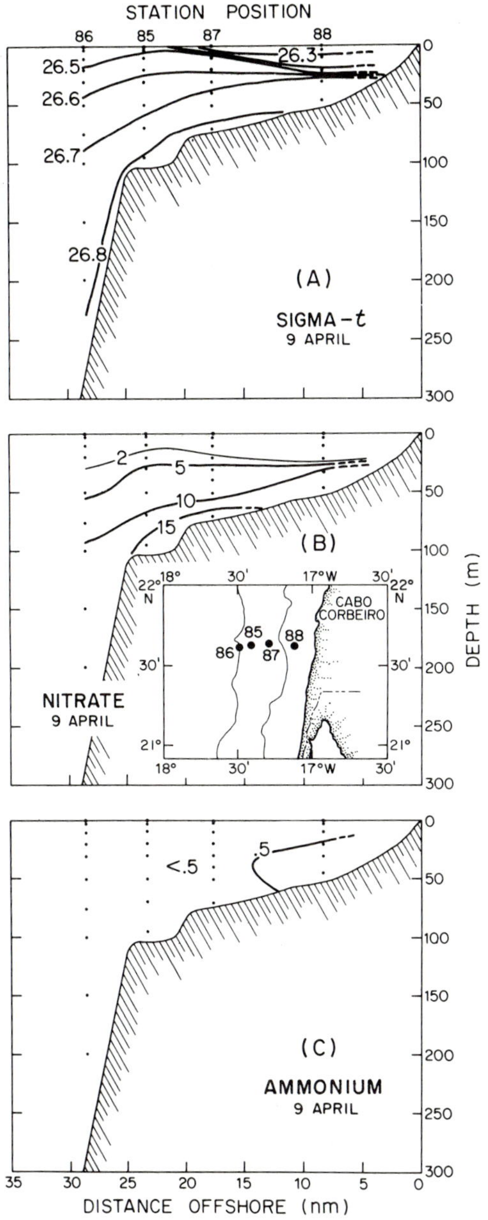

winds suggest that downwelling may have been occurring (9 April). Nitrite never exceeded 0.6 μg·atom liter^{-1} in any of these sections. Detailed discussion of the temporal variability in these sections is given by Codispoti and Friederich (1978).

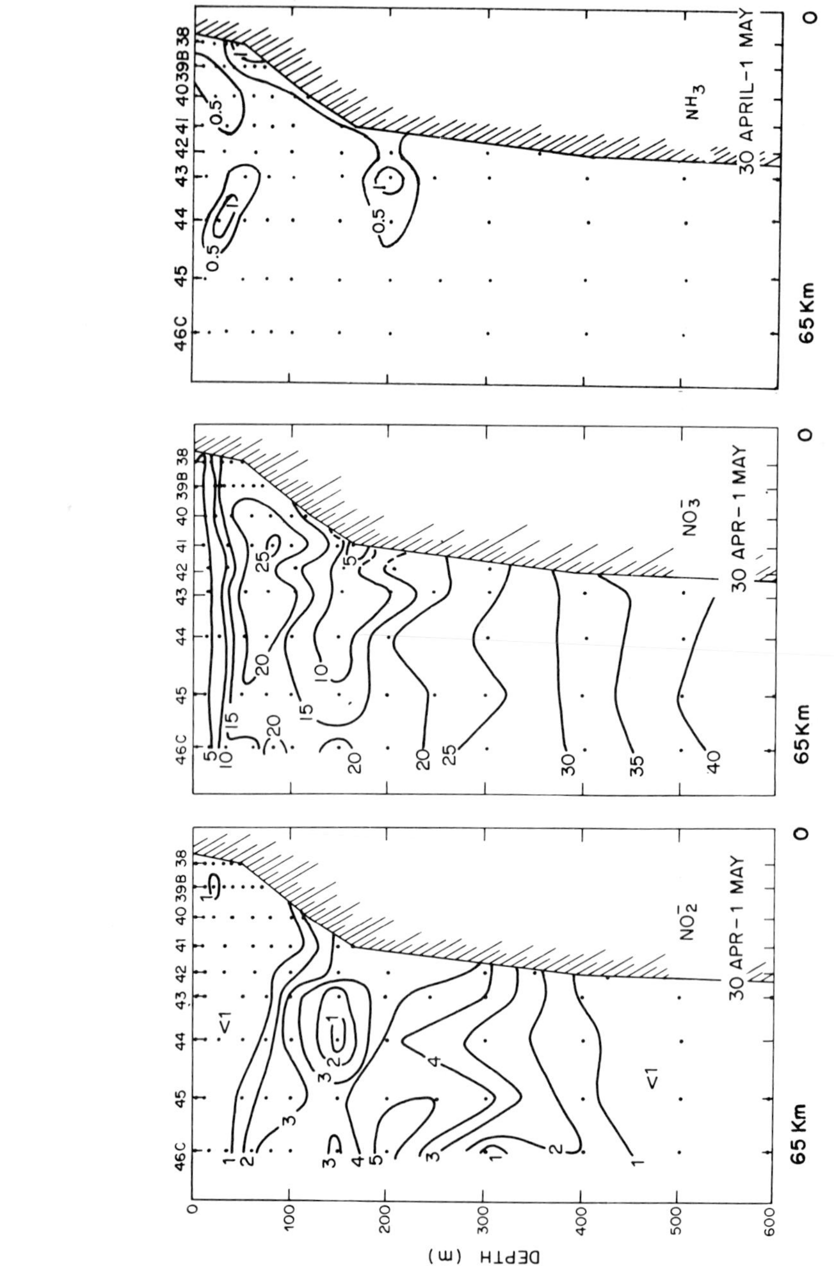

DEPTH (m)
0
100
200
300
400
500
600
46C 45 44 43 42 41 40 39B 38
65 Km
0
NO_2^-
30 APR – 1 MAY
NO_3^-
30 APR – 1 MAY
NH_3
30 APRIL–1 MAY

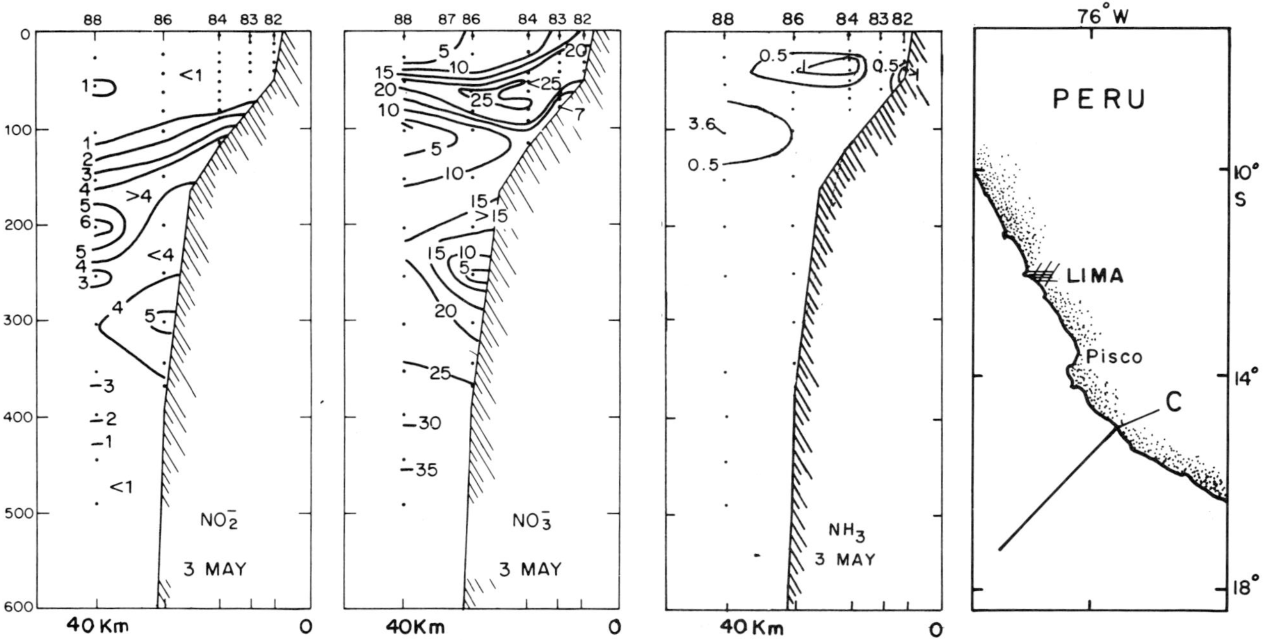

Fig. 9. The possible response of surface nitrate concentrations off Peru to an increase in upwelling favorable winds (from ∿2.5 m s^{-1} during 30 April-1 May to ∿7 m s^{-1} during 3 May). These sections are from the 1976 JOINT-II experiment, and are located on the C-line shown in the lower right corner of this figure. Note the minimum nitrate concentrations (<5 μg·atoms $liter^{-1}$) in the oxygen deficient portion (∿50-500 m) of the water column, and the ammonium plume extending seaward from the continental slope at ∿200 m. These features were not commonly observed during 1977, and point out the unusual nature of 1976. It was an El Niño year and the year of an immense red tide. In addition, hydrogen sulphide was observed in the water column shortly before these sections were taken (Dugdale et al., 1977; Guillén and Calienes, 1981).

between 0-7 μg·atoms liter^{-1}, with most concentrations of urea-N being less than 3 μg·atoms liter^{-1}. Although both ammonium and urea are associated with local regenerative processes, their distributions may differ significantly since ammonium:urea release ratios are species-dependent (McCarthy and Whitledge, 1972; Whitledge, 1981). Creatine is a major excretory product of some fish (Whitledge and Dugdale, 1972; Whitledge and Packard, 1971), but its distribution in seawater is poorly documented. A few observations in the upper ∿100 m off Peru yielded a maximum creatine-N value of 0.22 μg·atom liter^{-1} (Whitledge and Dugdale, 1972).

Systematic differences between methods make it difficult to discuss absolute levels of dissolved organic nitrogen (DON) in the sea (Williams, 1975), but two studies suggest that DON increases arise from the biological processes associated with upwelling. LeBorgne (1978) reported an increase from 5 to 20 μg·atoms liter^{-1} during an ∿1-week period in an upwelled water parcel that was followed after it upwelled south of Cap Timiris (off NW Africa), and Whitledge (1981) presents data that suggest DON values within ∿40 km of the Peru coast (near 15°S) are >10 μg·atoms liter^{-1}, but decrease in the offshore direction to values that are sometimes <5 μg·atoms liter^{-1}.

Particulate organic nitrogen (PON) and particulate protein nitrogen (which normally accounts for most of the PON) distributions in upwelling regions have been discussed by Fraga (1966), Garfield et al. (1979), Packard and Dortch (1975), Slawyk et al. (1978), and Vallespinós and Estrada (1975). These investigations suggest that PON and particulate protein concentrations tend to be positively correlated with biomass indicators (e.g., chlorophyll), and that in the surface waters of upwelling zones, average values of PON and particulate protein may exceed those in the oligotrophic portions of the ocean by more than an order of magnitude. Maximum PON concentrations reported in the above investigations exceed 20 μg·atoms N liter^{-1} (off Baja California and Peru). Garfield et al. (1979) report an average particulate protein value at the surface within ∿350 km of the Peru coast of 4 μg·atoms $liter^{-1}$, decreasing more or less exponentially to concentrations at 500 and 2000 m that are only ∿4 and 1%, respectively, as high as the surface value. Their observations also suggest that a particulate protein maximum was coincident with the secondary nitrite and particle maxima found at 200 m (Fig. 5; Codispoti and Packard, 1980; Kullenberg, 1981; Pak et al., 1980). Within their study region, decreases in particulate protein with distance from the coast could be detected to depths of 370 m, and comparison with data from an oligotrophic region (Siezen and Mague, 1978) suggested that the effects of upwelling on particulate protein levels extends to depths of 1000 m or more.

Nitrous oxide (N_2O) is present in only trace amounts in the ocean and atmosphere, but its effects on atmospheric chemistry and physics are significant (e.g., Crutzen, 1981; Elkins et al., 1978; Lacis et al., 1981). In addition, N_2O accumulates during the ammonium oxidation step of marine nitrification and is consumed during marine denitrification (Cohen, 1978; Cohen and Gordon, 1978; Elkins, 1978; Elkins et al., 1978; Hahn, 1981; Yoshinari, 1980). Consequently, N_2O distributions may provide insight into nitrogen cycling in upwelling systems. Only a small amount of N_2O data have been collected from upwelling systems, but results from Peru (Elkins, 1978; Elkins et al., 1978; Pierotti and Rasmussen, 1980) demonstrate that the combination of high nitrification rates in the waters above the oxygen-deficient zone (Fig. 5) and upwelling leads to surface N_2O values that are many times higher than typical marine values. For example, Elkins (1978) found N_2O-N concentrations as high as ∿0.2 μg·atom $liter^{-1}$ at the sea surface off Peru. With respect to the atmosphere, these concentrations represent supersaturation by more than a factor of 10, in contrast to the N_2O supersaturation factor of ∿1.2 that appears to be characteristic of most of the ocean's surface. These high surface values also rival the highest subsurface values that have been reported (Hahn, 1981). Within the oxygen-deficient waters off

Peru, N_2O concentrations are anomalously low as a consequence of denitrification, and can be less than 0.02 μg·atom N_2O-N liter^{-1} (Elkins, 1978; Elkins et al., 1978).

III. DISCUSSION

A. The Supply of New Nitrogen

Dugdale and Goering (1967, 1970) have made the useful distinction between the new nitrogen being added to the photic zone (primarily by upwelling and vertical mixing) and the nitrogen supplied to plants by recycling within the photic zone. They suggested that the new nitrogen is mostly nitrate** and that recycled nitrogen consists of more reduced compounds such as ammonium and urea (McCarthy, 1972). In most of the ocean, primary production is supported mainly by recycled nutrients (Wangersky, 1977). If recycling is rapid, gross primary production rates may be appreciable, but without a good supply of new nitrogen, large populations of harvestable organisms will not occur. A fundamental distinction between the more productive upwelling systems and the average ocean is that the ratio of new to recycled nitrogen uptake is much higher in the former. The ratios of nitrate-supported primary production to nitrate plus ammonium-supported rates frequently exceeds 0.5 and may be as high as ∿0.8 in upwelling regions, while values of <0.1 have been found in oligotrophic zones (Codispoti et al., 1982a; Dugdale, 1976; Eppley and Peterson, 1979; Eppley et al., 1979). Eppley et al. (1979) have presented data suggesting that

***They point out that nitrogen fixation is also a source of new nitrogen, as are the inputs of combined nitrogen from the atmosphere and runoff. However, these sources are not usually significant in productive upwelling regions (Eppley and Peterson, 1979). While new production is usually supported by nitrate, the basic distinction is between nitrogen supplied by local photic zone regeneration and nitrogen supplied from elsewhere. Near a sewage outfall or near an anoxic zone, significant new production could be supported by ammonium (Eppley et al., 1979).*

total nitrogen productivity is positively correlated with this ratio, so gross (new + recycled) primary production rates may also be higher in upwelling systems.

How different are the supplies of new nitrogen in productive upwelling regions from those in the average ocean? The average supply cannot be calculated from most geochemical models because the surface-layer depth in these models often exceeds the depth of the euphotic zone. Consequently, new nitrogen sources within the surface layer might be neglected. However, multiplying the total marine nitrogen assimilation rate of $\sim 5 \times 10^{15}$ g N year^{-1} (Liu, 1979; Sweeney et al., 1978; Vaccaro, 1965) by a reasonable average ratio of ~25% for new/total production (Eppley and Peterson, 1979), gives an average new nitrogen supply for the ocean's surface (3.6×10^{14} m^2) of $\sim 1 \times 10^{-2}$ μg·atom m^{-2} s^{-1}. Average upwelling velocities in productive coastal upwelling regions are $\sim 2 \times 10^{-3}$ cm s^{-1} (Smith and Bottero, 1977; Wooster and Reid, 1963), and nitrate values in the rising waters are ~10-20 μg·atoms liter^{-1} (Table I) giving a new nitrogen supply of $\sim 2\text{-}4 \times 10^{-1}$ μg·atoms m^{-2} s^{-1} during coastal upwelling.[††]

While these calculations are based on many assumptions, they have some interesting implications if they are even approximately correct. In particular, it is interesting to note that the ratio between the supply of new nitrogen in coastal upwelling zones and the "average ocean" appears to be much less than the 500:1 enhancement in fish yield suggested by Ryther (1969). In agreement with others (Barber and R. L. Smith, 1981; Eppley et al., 1979; Margalef, 1978a; Margalef and Estrada, 1981), we conclude that there are qualitative as well as quantitative differences between upwelling ecosystems and the average ocean.

If the vertical distribution of nitrate were the same everywhere, then the supply of nitrate to the photic zone in a given

[††]*This calculation neglects nutrient inputs due to vertical mixing, but scaling calculations suggest that this term is too small to significantly alter this estimate.*

upwelling system would depend largely on the local dynamics. There is, of course, a good correlation between the dynamics of upwelling and the nitrate supply to the photic zone over limited spatial and temporal scales (e.g., Codispoti and Friederich, 1978), but great differences in the nitrate content of upwelling source waters and in the relationships between nitrate and other nutrients exist between different portions of the world ocean (Table I). There are also significant differences between the water masses that upwell within some regions (Minas et al., 1982). The inter-ocean differences and some intraocean differences (Zentara and Kamykowski, 1981) arise largely from the interaction of the global circulation with nutrient gradients. Since nutrients accumulate at depth, a basin that loses deep waters and receives shallow waters in return will have lower nutrient concentrations than an estuarine basin that receives deep water and gives up shallow water, and this difference will be more pronounced for nutrients with the steepest vertical gradients (e.g., dissolved silicon). The North Atlantic Ocean and the Mediterranean Sea are antiestuaries, and the northern Pacific can be considered to be the head of a global-scale estuary when considering this question. This oceanwide fractionation process has been discussed by a number of authors (Berger, 1970; Broecker, 1974; Codispoti, 1979; Codispoti et al., 1982a; Redfield et al., 1963).

Within the Atlantic and Pacific Oceans at low and mid-latitudes, subsurface nutrient concentrations at the depths associated with upwelling source waters are highest in the eastern boundaries. This situation arises from a piling up of warm, nutrient-poor surface waters on the western boundaries (e.g., Voituriez, 1981; von Arx, 1962) and from enhanced nutrient regeneration in the subsurface waters found in the eastern boundaries (see Section III,C). These factors, when combined with the strength and persistence of upwelling in the classic upwelling systems, help to explain why they have received so much attention. Even if the upwelling physics were identical off Peru and in a low-nutrient environment, such as the Mediterranean Sea or the Colombian-Caribbean Basin (Corredor, 1979), the chemical and biological effects of upwelling should be much greater off Peru due to the intrinsic differences in source water composition.

Variations in the nitrate supply can sometimes be related to subsurface fronts that separate water masses with different nutrient regimes (Fraga, 1974; Fraga and Manríquez, 1975; Friederich and Codispoti, 1981). An example of the latter factor is provided by NW Africa. Some data suggest that the flow of the nutrient-rich undercurrent is weakened when upwelling winds are strong (Codispoti and Friederich, 1978), and this weakening may contribute to reduced nitrate concentrations in the rising waters (Fig. 3).

B. Factors Influencing Nitrogen Uptake Rates in Upwelling Systems

Very high nitrogen uptake rates can be encountered in upwelling systems. For example, during a several-week experiment off Peru, total nitrogen (nitrate + ammonium) productivity averaged 40 mg·atoms N $m^{-2}d^{-1}$, and during the JOINT-I experiment off NW Africa, the average value was 22 mg·atoms N $m^{-2}d^{-1}$ (Codispoti et al., 1982a; Dugdale, 1976). The ratios of nitrate supported (new) to total nitrogen productivity were 75 and 70%, respectively. Some data also suggest a positive correlation between specific growth rates of phytoplankton and nitrate concentrations that extends to values in excess of 20 μg·atoms $liter^{-1}$ (Huntsman and Barber, 1977; Jones, 1978). While these observations suggest that a generous supply of new nitrogen to the photic zone is a necessary condition for maintaining high primary production rates, other factors can also be important. Some of these factors are considered below.

1. Excesses or Deficits of Nonnitrogenous Chemicals

Dugdale (1972) has suggested that the highest nitrogen uptake rates in upwelling systems are associated with diatom populations, and that silicon limitation might therefore be a control on these uptake rates.

The sporadic occurrence of sufficient quantities of major nutrient substances and light in upwelling systems in association with low primary production rates (Strickland et al., 1969) has lead to the suggestion (Harrison et al., 1981) that deficiencies of micronutrients or of agents (e.g., chelators) that can neutralize toxic trace elements such as excess free copper (Barber and Ryther, 1969; Huntsman and Sunda, 1980; Steemann-Nielsen and Wium-Andersen, 1970) may occur. Iron (Johnston, 1963), manganese (Sunda et al., 1981), zinc (Anderson et al., 1978), and various organic compounds (Provosoli, 1963; Swift, 1980) have been suggested as limiting micronutrients under some conditions. An implication of these studies is that upwelling waters must be "conditioned" by mixing with ambient surface water, contact with the bottom, contact with reducing environments, runoff, aeolian transport, or by recent decomposition of organic matter in order to support high primary production rates. Results from Peru and NW Africa suggest that at least one of these conditions is often met (Barber et al., 1971; Smith et al., 1982), but it must be noted that the conditions of these experiments did not exclude the possibility of metal contamination.

2. Circulation and Stratification

If the proper balance exists between factors such as the sinking rate of phytoplankton and the strength of the upwelling circulation, the rising waters can contain significant quantities of living phytoplankton as well as high concentrations of nutrients. However, this is not always the case. During strong upwelling, it is possible that a condition analogous to "wash out" in a chemostat could develop (Barber and W. O. Smith, Jr., 1981; Wroblewski, 1977). This situation may be exacerbated by the colder temperatures associated with strong upwelling, since growth rates of phytoplankton may decrease with decreasing temperature (Jones, 1978). In addition, populations brought rapidly to the surface may be adversely affected by the increased light intensities and require an adjustment period before achieving maximal growth (Yentsch and Lee, 1966).

As noted above, the nutrient-rich undercurrent off NW Africa may be weakened during strong upwelling events leading to nitrate concentration decreases in the rising waters. Strong upwelling events may also be associated with mixed-layer depths that are too great for optimal uptake rates (Huntsman and Barber, 1977) and possibly with enhanced transport of light absorbing organic detritus and minerals into the photic zone.

3. Inadequate Separation between Herbivory and Primary Production

Optimal productivity can only occur if there is sufficient separation between the initial development of a phytoplankton bloom and herbivory. Walsh (1976) has suggested that the relatively low rates of nitrogen uptake in some equatorial upwelling systems may arise because the environment is too steady. In a "predictable" environment, it is possible for grazing pressure to match the growth rate of phytoplankton and prohibit the occurrence of large populations. Conversely, the variability of upwelling at the shelf break off NW Africa has been suggested as the agent responsible for the death of zooplankton (Weikert, 1977). Lack of separation may also result from the life cycle of certain herbivores. In the outer SE Bering Sea* shelf, the migration of juvenile copepods from the depths at which they hatch into the photic zone before the spring bloom may be responsible for the persistence of high surface nitrate concentrations for long periods (Goering and McRoy, 1981; Codispoti et al., 1982b).

4. Differences in the Composition of the Phytoplankton

A number of investigators have considered the conditions that may lead to the dominance of different groups of phytoplankton (Barber and W. O. Smith, Jr., 1981; Blasco et al., 1981; Estrada and Blasco, 1979; Guillard and Kilham, 1977; Huntsman et al.,

**This is primarily a spring bloom region, but some upwelling also occurs (Goering and McRoy, 1981).*

1981; Margalef, 1978b; Margalef et al., 1979; Packard et al., 1978; Smayda, 1980; Smith, 1978). The controlling factors range from the nature of the stratification to biological interactions, and there are undoubtedly many exceptions to the simplified scheme presented below. Nevertheless, it may be useful for illustrating some principles.

i. In general, large forms of phytoplankton are favored in upwelling regions and this can lead to short food webs. This helps to explain why fish yields are greater than would be predicted from the enhanced supply of new nitrogen (Ryther, 1969). Margalef (1978a) suggests that the variability of coastal upwelling systems also leads to short food webs by inhibiting organization of the ecosystem.

ii. In sites of active upwelling, sinking phytoplankton such as diatoms dominate a population if they can avoid washout (Barber and W. O. Smith, Jr., 1981). Diatoms are capable of sustaining high specific nitrogen uptake rates (Codispoti et al., 1982a; Dugdale, 1972).

iii. Under conditions of weak upwelling and enhanced stability, dinoflagellates may dominate (Huntsman et al., 1981). They can migrate vertically at speeds of up to 1 m h^{-1} (Eppley et al., 1967), and they may extend the photosynthetic space of an upwelling system by transporting nutrients that may be abundant at low light levels (e.g., ammonium; Nelson and Conway, 1979) into the euphotic zone. Dinoflagellates are an important food source for the first feeding larvae of the northern and southern anchovy (*Engraulis mordax* and *E. ringens*), which are important components of the Californian and Peruvian upwelling systems (Lasker, 1978; Walsh et al., 1980). Conversely, specific nitrogen uptake rates for dinoflagellates are relatively low (Codispoti et al., 1982a), and once established, dinoflagellates have characteristics that may enable them to outcompete faster growing organisms. Thus, under some conditions it is possible that a dinoflagellate bloom may lower the nitrogen uptake rates in an upwelling system. Massive dinoflagellate blooms have occurred off Peru in 1917 and 1976 (Cushing, 1981).

iv. Recently, the importance of the photosynthetic ciliate *Mesodinium rubrum* to upwelling systems has been investigated (Barber and W. O. Smith, Jr., 1981; Packard et al., 1978; Smith and Barber, 1979). These organisms can migrate at speeds that are an order of magnitude faster than dinoflagellates and they are capable of sustaining high specific carbon (and presumably specific nitrogen) uptake rates. Smith and Barber (1979) suggest that the contribution of these organisms to the productivity of upwelling systems has been seriously underestimated because they can lyse rapidly under the conditions of some standard analyses. Packard et al. (1978) believe that they may be concentrated in and near regions where upwelled waters sink, and Barber and W. O. Smith, Jr. (1981) present a model that suggests their ability to resist washout during periods of vigorous upwelling.

C. Nitrogen Regeneration

The spatial and trophic distributions of nitrogen regeneration depend on both the physics and ecosystem structure of an upwelling system. Among the physical factors that may be important are the nature of the flow regimes, the turbidity of the waters, and the topography. Frequently, plots of nitrate versus temperature and/or salinity show that the source waters in an upwelling region have concentrations that are higher than offshore waters with similar temperatures and salinities (Friederich and Codispoti, 1979; Treguer and LeCorre, 1979). Calvert and Price (1971), in describing such a situation in the SW African upwelling region, suggested that the onshore flow of upwelling source waters was enriched in nutrients by regeneration of the nutrients contained in particles sinking into these waters. They likened this to a nutrient trap resulting from the interaction of sinking particles with an estuarine circulation pattern (Redfield, 1955; Redfield et al.,

1963). Friederich and Codispoti (1981) have demonstrated an enrichment arising from similar causes and occurring along the longshore path of the poleward undercurrent that flows against the surface current and is a source of the upwelling waters off Peru. The existence of such nutrient traps can be an important contributor to the productivity of upwelling regions, since these traps permit a concentration of nutrients in the upwelling waters. In fact, in a simple two-layered upwelling cross-circulation, the relationships described by Redfield et al. (1963) would predict maximum nutrient concentrations just as the waters enter the photic zone. Treguer and LeCorre (1979) have shown that as much as half of the inorganic nitrogen (nitrate + nitrite + ammonia) in some source waters may be due to regeneration that occurs within the coastal upwelling system off Morocco.

As Dugdale (1972) suggests, silicon may sometimes limit the rate of community nitrogen uptake. His data were from Peru where Si:NO_3^- ratios are, in general, higher in source waters for upwelling than they are off NW Africa (Table I), particularly before the occurrence of regeneration within the coastal upwelling system. Paradoxically, off NW Africa during the JOINT-I experiment, inorganic nitrogen (nitrate + nitrite + ammonium) was found in concentrations below 0.5 μg·atom liter^{-1} more frequently than dissolved silicon (Friederich and Codispoti, 1979). This apparent contradiction could arise from differences in the biota's requirement for silicon, but it may also arise from differences in the relative regeneration rates of silicon and nitrogen. Treguer and LeCorre (1979) and Friederich and Codispoti (1979) have shown that silicon regeneration is relatively rapid in the upwelling source waters off NW Africa. Silicon dissolution is fundamentally a chemical process and it increases with increasing temperature and pH and decreasing dissolved silicon concentrations (Hurd, 1972, 1973). All of these variables may have been more favorable for solution in the two regions studied off NW Africa than they were during Dugdale's Peruvian experiment. Another factor that may contribute to more

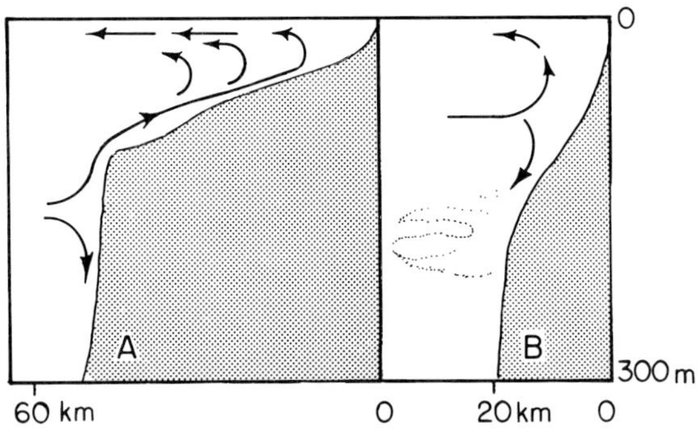

Fig. 10. Bottom topography and generalized cross-shelf flow off NW Africa at 21° 40'N (A), and off Peru at 15°S (B). A particle maximum present at 200 m off Peru is also suggested.

effective silicon regeneration off NW Africa is the difference in shelf morphology and cross-shelf flow. The NW African shelf investigated by Friederich and Codispoti (1979) is wide compared to the peruvian shelf in the region studied by Dugdale (Fig. 10). In addition, the average flow along the bottom off NW Africa is shoreward whereas it is seaward off Peru (Codispoti and Packard, 1980). Since silicon may have a shorter residence time in the euphotic zone than nitrogen and phosphorus (Broecker, 1974), one can envision opaline-rich particles sinking into the offshore flow off Peru and being carried out of the upwelling system, whereas the reverse process would occur in the JOINT-I region off NW Africa.

Studies of nutrient regeneration and/or respiration rates by different populations within upwelling regions (Smith and Whitledge, 1977; Smith, 1978; LeBorgne, 1978; Sorokin, 1978a,b; Setchell and Packard, 1979; Packard, 1979; Sorokin and Kogelschatz, 1979; Suess, 1980; Dagg et al., 1980; von Bröckel, 1981) do not always agree on details, possibly because of systematic errors between methods that, at the current state of knowledge, cannot be fully resolved. Nevertheless, there is agreement on some of the generalities, and these points of agreement may be summarized as follows:

1. Most of the organic matter produced by primary producers in a typical coastal upwelling region is remineralized by the respiratory activities of organisms within the upper 100-200 m and within ∿200 km of the coast. However, the excess of production over respiration in the photic zone of upwelling system is probably higher than average with the greatest excess likely to occur within variable upwelling (Margalef, 1978a).

2. Microplankton (which in practice amount to any organisms that will pass through ∿100-μm net) are important organisms in remineralizing nitrogen, but larger organisms (>100 μm) may sometimes be more important. Smith and Whitledge (1977) and Packard (1979) suggest that zooplankton supplied about half of the ammonium requirement of phytoplankton off NW Africa at 21° 40'N. In contrast, LeBorgne (1978), in drogue studies conducted to the south of this region, found that zooplankton captured in a 200-μm net provided 35-48% of the ammonium produced before the phytoplankton began to increase exponentially, but only 10-16% during the phytoplankton blooms. He suggests that microplankton were responsible for most of the ammonium production during these periods. Whitledge and Packard (1971) present data that suggest that zooplankton might occasionally provide as much as 20% of the total nitrogen requirement of phytoplankton in the Costa Rica Dome (open ocean) upwelling region.

3. Nekton may at times supply important quantities of regenerated nitrogen to the phytoplankton in an upwelling system. Whitledge and Packard (1971) suggest that fish supplied 40% of the ammonium in a study area off Peru and also supplied quantities of creatine and urea that were similar to their ammonium contribution. Whitledge (1981) suggests that nekton recycle more nitrogen in the Peruvian and Baja Californian upwelling systems than zooplankton.

4. Although the studies of Calvert and Price (1971), Treguer and LeCorre (1979), and Friederich and Codispoti (1979) imply that the nitrogen regeneration in and on the sediments underlying some upwelling systems supply important amounts of nutrients to the

local phytoplankton populations, few studies permit separation of the nitrogen regeneration in the waters overlying the sediments from regeneration in and on the sediments. However, Rowe et al. (1977) conducted bell jar experiments in the nearshore portion of the NW African upwelling region that suggest that 30-40% of the total nitrogen requirement of phytoplankton in this zone could be met by sedimentary regeneration. Depending on their oxidation state, sediments can be net suppliers or net removers of inorganic nitrogen from the water column (Bender et al., 1977), so it is difficult to generalize on their relative importance as sources of regenerated nitrogen.

5. Nitrification rate estimates from upwelling systems are rare, but a summary of water column rates from a variety of environments (Ward et al., 1982) suggests a maximum value of $\sim$0.1 μg·atom liter^{-1} day^{-1} for the ammonium oxidation step. Friederich and Codispoti (1981) showed that nitrate concentrations may increase by as much as 8 μg·atoms liter^{-1} on the 26.0 σ_t surface within upwelling source waters in the poleward undercurrent off Peru. This increase occurred over a flow path of $\sim$600 km between 5 and 10°S. A reasonable velocity for the poleward flow at this depth is $\sim$10 cm s^{-1} (Fahrbach et al., 1981; Smith, 1981), and these values (8 μg·atoms liter^{-1} $\div$ 600 km $\times$ 10 cm s^{-1}) permit a rough estimate of $\sim$0.1 μg·atom liter^{-1} day^{-1} for the nitrate production step on this σ_t surface. Sediments in many upwelling systems are oxygen-deficient or anaerobic and are probably net consumers of nitrate and nitrite, but one bell jar experiment in the well-oxygenated JOINT-I region off NW Africa (Rowe et al., 1977) suggested a sedimentary nitrification rate of 0.3 μg·atom liter^{-1} day^{-1} when the sedimentary contribution was averaged over the 25-m water column. A second experiment by these investigators indicated net consumption of nitrate by the sediment, and a small nitrite production rate.

Light may inhibit nitrification within the photic zone (Horrigan et al., 1981; Olson, 1981; Ward et al., 1982), and the water column rates given above are from depths where the available data suggest that high values should occur. Even so, these rates are relatively small compared with maximum nitrogen uptake rates in the euphotic zone, which may exceed 2 μg·atoms $liter^{-1}$ day^{-1} in some cases (Dugdale and Goering, 1970). These facts reinforce the notion that most of the nitrate found in the euphotic zone of upwelling regions is produced elsewhere (Dugdale and Goering, 1967).

D. Denitrification in Upwelling Systems

Denitrification becomes a prominent respiratory process when oxygen concentrations are zero or close to zero (Richards, 1965), and the bacterial conversion of nitrate and nitrite to free nitrogen during this process represents the major marine sink for combined nitrogen (e.g., Emery et al., 1955; Söderlund and Svensson, 1976). Denitrification occurs in poorly ventilated basins (Richards, 1965) and is widespread in sediments (Bender et al., 1977), but a large fraction of the total marine denitrification rate may occur in open-ocean, oxygen-deficient waters that are found in or near low-latitude upwelling regions. The largest of these anomalous water bodies occurs in the Eastern Tropical North Pacific, but denitrification rates in the Eastern Tropical North and South Pacific are similar with a combined rate of $\sim 5 \times 10^{13}$ g N $year^{-1}$ (Cline and Kaplan, 1975; Codispoti and Packard, 1980; Codispoti and Richards, 1976). Denitrification in the oxygen-deficient waters of the Arabian Sea might contribute as much as $\sim 0.5 \times 10^{13}$ g N $year^{-1}$ to the marine denitrification rate (Deuser et al., 1978; Wajih et al., 1982), and some denitrification occurs in relatively small oxygen-deficient zones found in the water column of the SW African and Californian upwelling regions (Calvert and Price, 1971; Liu, 1979).

In addition to contributing significantly to the total marine denitrification rate of $\sim 1 \times 10^{14}$ g N $year^{-1}$ (Liu, 1979), these oxygen-deficient environments may have significant influence on the distribution and cycling of nitrogen compounds within the upwelling systems with which they are associated. For example, significant quantities of combined nitrogen appear to have been removed by denitrification from some of the upwelling waters found off Peru (Fig. 5).

Because the sediments underlying an oxygen-deficient water column will receive little or no oxygen from the overlying waters and reduced amounts of nitrate, their respiratory regime will be dominated by sulfate reduction (Bender et al., 1977), and they may remove less combined nitrogen than the sediments found outside the boundary of the oxygen-deficient zone.

Open-ocean, oxygen-deficient zones in and near upwelling zones are partially a consequence of enhanced subsurface respiration rates driven by the high primary production of the adjoining upwellings and possibly by high export:consumption ratios in the surface layers of these zones (Margalef, 1978a). Other factors may also be important, and one reason why denitrification is not found in the water column off NW Africa may be that the ambient waters in this zone are relatively rich in oxygen and poor in nutrients as a consequence of the interocean fractionation process described above.

Codispoti and Packard (1980) have suggested that the denitrification rate off Peru has increased in recent years as a result of changes in ecosystem structure associated with the decrease in anchoveta.

E. Modeling Coastal Upwelling Ecosystems

Space does not permit a review of the mathematical models that relate to the nitrogen flow in upwelling systems, but such models are incorporated in many of the papers cited above. A good example of an attempt to provide a reasonably comprehensive model of upwelling ecosystems is provided by Walsh (1977) who also discusses the advantages and limitations of such models.

F. Export of Nitrogen from Upwelling Systems

Much of mankind's harvest of nitrogen from coastal upwelling systems is in the form of pelagic schooling fish such as anchoveta, sardines, and pilchards and in the form of the guano deposited by the birds that feed on these fishes. In recent years, there have been serious declines in the stocks of these species and the management failures and natural events that contributed to these declines are summarized by Cushing (1981) and by Glantz and Thompson (1981).

Productive upwelling systems may be relatively inefficient consumers of the organic material produced locally, even though their short food webs make them efficient producers of fish (Margalef and Estrada, 1981). Consequently, it should be expected that they export relatively large amounts of organic nitrogen to the sediments or to the interior of the ocean. Existing estimates of the amount of this export are based on few data, but perhaps some idea of the possible range is given by Walsh's (1981) carbon budgets for Peru. Before the decline of the anchoveta, he estimated that ∿5% of the ^{14}C primary production rate was stored in sediments or advected into the interior of the ocean, and his estimate for the postdecline period is 44%.

IV. CONCLUSION

It should be evident from the above comments that nitrogen productivity and the distribution of nitrogenous compounds in upwelling systems depends on a host of factors with different characteristic spatial and temporal scales, and that the relative importance of these factors may vary from region to region. In attempting to make sense out of these complications, it might be useful to consider the example provided by Peru which, before the severe decline of the anchoveta that began in 1972, was the site of the world's largest fishery (Cushing, 1981; Glantz and Thompson, 1981). It might be fair to say that before the anchoveta decline and during normal (non-El Niño) years, Peru was the most favorable marine site for the production of a harvestable stock of organic nitrogen. Factors that favor Peru include the following:

1. Peru is in a portion of ocean space that benefits from the inter- and intraocean nutrient fractionation processes mentioned above.

2. Upwelling occurs throughout the year off Peru (Walsh, 1977), and since it is located in the tropics, the light regime is usually favorable.

3. Trace chemical conditions may be favorable for phytoplankton growth most of the time (Barber et al., 1971).

4. An effective nutrient trap is produced not only by the cross-shelf circulation, but also by the longshore circulation, a factor that is related to the well-developed and shallow poleward undercurrent off Peru (Friederich and Codispoti, 1981). Most of the rising waters originate in the upper portion of the undercurrent that is usually above the zone of active denitrification.

5. Upwelling velocities off Peru are moderate and the stratification is strong when compared to a region such as NW Africa (Smith, 1981). Consequently, mixed-layer depths off Peru are frequently no greater than the depth of the euphotic zone, and the possibility of phytoplankton "wash-out" and poor light regimes arising from resuspension are reduced.

6. Blooms of dinoflagellates, which are good fodder for the first feeding larvae of the anchoveta (Walsh et al., 1980), and blooms of large and/or chain-forming diatoms, which may lead to short phytoplankton to fish food webs (Barber and Smith, 1981b), are common (de Mendiola, 1981).

7. Significant variability occurs off Peru over a wide range of time scales (Codispoti, 1981), so considerable potential exists for providing a separation between herbivory and primary production and for promoting short food webs (Margalef, 1978a; Walsh, 1976).

While Peru may represent the most favorable "real world" site, it suffers from the following disadvantages which are also instructive:

1. The periodic invasion of low-nutrient waters and/or changes in stratification (Barber et al., 1983) during El Niño years lower primary production and colder-than-average years may also be unfavorable (Guillén and Calienes, 1981), even though upwelling under both situations continues (Enfield, 1981).

2. Some of the upwelling source waters along large portions of the coast have gone beyond the state of optimum nitrogen remineralization. Oxygen concentrations are so low that combined nitrogen is being removed from these waters by denitrification and the extent and rate of denitrification may have increased after the decline of the anchoveta (Codispoti and Packard, 1980).

3. The offshore bottom flow along the shelf and upper slope reduces the effectiveness of local nutrient regeneration and may contribute to the formation of particle (Pak et al., 1980) and protein maxima (Garfield et al., 1979) that lead to increased denitrification rates, and to the migration of biogenic silica to depths that are too deep to supply a significant fraction of the rising waters.

ACKNOWLEDGMENTS

My studies of upwelling began when I was invited to participate in the CUEA program by Drs. R. Barber, R. Dugdale, J. Kelley, T. Packard, R. Smith, and J. Walsh. Since my initial introduction to upwelling systems, I have continued to receive their advice and encouragement. Because the study of upwelling demands a multidisciplinary approach, I have also benefited from the labors and advice of many other colleagues. Although I cannot list all of them here, they have my sincere thanks for contributing to my continuing education on upwelling. I especially wish to thank my colleagues at the Instituto del Mar del Peru for their help during my experiments off Peru and Drs. R. Margalef, J. Wofsy, and C. S. Yentsch for their helpful comments.

Financial support for my research has been provided by the IDOE, polar programs, and chemical oceanography divisions of the National Science Foundation, who have funded my participation in the CUEA, PROBES, and INDEX experiments, respectively. I have also received valuable support from the Office of Naval Research (contract N00014-81-C-0043), which has enabled me to study the processes (e.g., denitrification) occurring beneath the productive surface waters of some upwelling systems. This is contribution number 83023 from the Bigelow Laboratory for Ocean Sciences.

REFERENCES

Anderson, M. A., Morel, F. M. M., and Guillard, R. R. L. (1978). Growth limitation of a coastal diatom by low zinc ion activity. *Nature (London) 276,* 70-71.

Andrews, W. R. H., and Hutchings, L. (1980). Upwelling in the southern Benguela Current South Africa. *Prog. Oceanogr. 9,* 1-81.

Barber, R. T., and Ryther, J. H. (1969). Organic chelators: Factors affecting primary production in the Cromwell current upwelling. *J. Exp. Mar. Biol. Ecol. 3,* 191-199.

Barber, R. T., and Smith, R. L. (1981). Coastal upwelling ecosystems. *In* "Analysis of Marine Ecosystems" (A. R. Longhurst, ed.), pp. 31-68. Academic Press, New York.

Barber, R. T., and Smith, W. O., Jr. (1981). The role of circulation, sinking, and vertical migration in physical sorting of phytoplankton in the upwelling center at 15 S. *In* "Coastal Upwelling" (F. A. Richards, ed.), pp. 366-371. Am. Geophys. Union, Washington, D.C.

Barber, R. T., Dugdale, R. C., MacIsaac, J. J., and Smith, R. L. (1971). Variations in phytoplankton growth associated with the source and conditioning of upwelling water. *Invest. Pesq. 35,* 171-193.

Barber, R. T., Zuta, S., Kogelschatz, J., and Chavez, F. (1983). Temperature and nutrient conditions in the eastern equatorial Pacific. *Trop. Ocean. Atmos. Newsl. 16,* 15-17.

Barton, E. D., Huyer, A., and Smith, R. L. (1977). Temporal variation observed in the hydrographic regime near Cabo Corveiro in the northwest African upwelling region, February to April 1974. *Deep-Sea Res. 24,* 7-23.

Bender, M. L., Fanning, K. A., Froelich, P. N., Heath, G. R., and Maynard, V. (1977). Interstitial nitrate profiles and oxidation of sedimentary organic matter in eastern equatorial Africa. *Science 198,* 605-609.

Bennekom, A. J. (1978). Nutrients on and off the Guyana Shelf related to upwelling and Amazon outflow. *FAO Fish. Rep., 1978,* pp. 233-253.

Berger, W. H. (1970). Biogenous deep-sea sediments: Fractionation by deep-sea circulation. *Geol. Soc. Am. Bull. 81,* 1385-1402.

Blanton, J. O., Atkinson, L. P., Peitrafesa, L. J., and Lee, T. N. (1981). The intrusion of Gulf Stream water across the continental shelf due to topographically induced upwelling. *Deep-Sea Res. 28,* 393-405.

Blasco, D., Estrada, M., and Jones, B. H. (1981). Short-term variability of phytoplankton populations in upwelling regions--The example of northwest Africa. *In* "Coastal Upwelling" (F. A. Richards, ed.), pp. 339-347. Am. Geophys. Union, Washington, D.C.

Bowden, K. F. (1975). Oceanic and estuarine mixing processes. *In* "Chemical Oceanography" (J. P. Riley and G. Skirrow, eds.), 2nd ed., Vol. 1, pp. 1-43. Academic Press, New York.

Brink, K. H., Jones, B. H., VanLeer, J. C., Mooers, C. N. K., Stuart, D. W., Stevenson, M. R., Dugdale, R. C., and Heburn, G. W. (1981). Physical and biological structure and variability in an upwelling center off Peru near 15°S during March 1977. *In* "Coastal Upwelling" (F. A. Richards, ed.), pp. 473-495. Am. Geophys. Union, Washington, D.C.

Broecker, W. S. (1974). "Chemical Oceanography." Harcourt, Brace, Yovanovich, New York.

Buckley, J. R., Gammelsrod, T., Johannessen, J. A., Johannessen, O. M., and Roed, L. P. (1979). Upwelling: Oceanic structure at the edge of the Arctic ice pack in winter. *Science 203,* 165-167.

Calvert, S. E., and Price, N. B. (1971). Upwelling and nutrient regeneration in the Benguela Current, October, 1968. *Deep-Sea Res. 18,* 505-523.

Cline, J. D., and Kaplan, I. R. (1975). Isotopic fractionation of dissolved nitrate during denitrification in the eastern tropical North Pacific Ocean. *Mar. Chem. 3,* 271-299.

Codispoti, L. A. (1973). Denitrification in the eastern tropical North Pacific. Ph.D. Thesis, University of Washington, Seattle.

Codispoti, L. A. (1979). Arctic Ocean processes in relation to the dissolved silicon content of the Atlantic. *Mar. Sci. Commun. 5,* 361-381.

Codispoti, L. A. (1981). Temporal nutrient variability in three different upwelling regions. *In* "Coastal Upwelling" (F. A. Richards, ed.), pp. 209-220. Am. Geophys. Union, Washington, D.C.

Codispoti, L. A., and Friederich, G. E. (1978). Local and meso-scale influences on nutrient variability in the northwest African upwelling region near Cabo Corbeiro. *Deep-Sea Res. 25*, 751-770.
Codispoti, L. A., and Packard, T. T. (1980). Denitrification rates in the eastern tropical South Pacific. *J. Mar. Res. 38*, 453-477.
Codispoti, L. A., and Richards, F. A. (1976). An analysis of the horizontal regime of denitrification in the eastern tropical North Pacific. *Limnol. Oceanogr. 21*, 379-388.
Codispoti, L. A., Dugdale, R. C., and Minas, H. J. (1982a). A comparison of the nutrient regimes off northwest Africa, Peru, and Baja California. *Rapp. P.-V. Reun., Cons. int. Explor. Mer 180*, 177-194.
Codispoti, L. A., Friederich, G. E., Iverson, R. L., and Hood, D. W. (1982b). Temporal changes in the S.E. Bering Sea's inorganic carbon system during spring 1980. *Nature (London) 296*, 242-245.
Cohen, Y. (1978). Consumption of dissolved nitrous oxide in an anoxic basin, Saanich Inlet, British Columbia. *Nature (London) 272*, 235-237.
Cohen, Y., and Gordon, L. I. (1978). Nitrous oxide in the oxygen minimum of the eastern tropical North Pacific. *Deep-Sea Res. 25*, 509-524.
Corredor, J. E. (1979). Phytoplankton response to low level nutrient enrichment through upwelling in the Colombian Caribbean basin. *Deep-Sea Res. 26*, 731-742.
Crutzen, P. J. (1981). Atmospheric chemical processes of the oxides of nitrogen, including nitrous oxide. *In* "Denitrification, Nitrification and Atmospheric Nitrous Oxide" (C. C. Delwiche, ed.), pp. 17-44. Wiley, New York.
Cushing, D. H. (1978). Upper trophic levels in upwelling areas. *In* "Upwelling Ecosystems" (R. Boje and M. Tomczak, eds.), pp. 101-110. Springer-Verlag, Berlin and New York.
Cushing, D. H. (1981). The effect of El Niño upon the Peruvian anchoveta stock. *In* "Coastal Upwelling" (F. A. Richards, ed.), pp. 449-457. Am. Geophys. Union, Washington, D.C.
Dagg, M., Cowles, T., Whitledge, T., Smith, S., Howe, S., and Judkins, D. (1980). Grazing and excretion by zooplankton in the Peru upwelling system during April 1977. *Deep-Sea Res. 27*, 43-59.
de Mendiola, R. B. (1981). Seasonal phytoplankton distribution along the Peruvian coast. *In* "Coastal Upwelling" (F. A. Richards, ed.), pp. 348-356. Am. Geophys. Union, Washington, D.C.
Deuser, W. G., Ross, E. H., and Mlodzinska, Z. J. (1978). Evidence for and rate of denitrification in the Arabian Sea. *Deep-Sea Res. 25*, 431-445.
Doe, L. A. E. (1978). "A Progress and Data Report on a Canada-Peru Study of the Peruvian Anchovy and Its Ecosystem," Proj. ICANE,

Rep. Ser. Bl-R-78-6. Bedford Inst. Oceanogr., Nova Scotia, Canada.
Dugdale, R. C. (1972). Chemical oceanography and primary productivity in upwelling regions. *Geoforum 11,* 47-61.
Dugdale, R. C. (1976). Nutrient cycles. *In* "The Ecology of the Seas" (D. H. Cushing and J. J. Walsh, eds.), pp. 141-172. Saunders, Philadelphia, Pennsylvania.
Dugdale, R. C., and Goering, J. J. (1967). Uptake of new and regenerated forms of nitrogen in primary productivity. *Limnol. Oceanogr. 12,* 196-206.
Dugdale, R. C., and Goering, J. J. (1970). Nutrient limitation and the path of nitrogen in Peru Current production. *In* "Scientific Results of the Southeast Pacific Expedition" (E. Chin, ed.), pp. 5.3-5.8. Texas A&M Univ. Press, College Station.
Dugdale, R. C., Goering, J. J., Barber, R. T., Smith, R. L., and Packard, T. T. (1977). Denitrification and hydrogen sulfide in the Peru upwelling region during 1976. *Deep-Sea Res. 24,* 601-608.
Elkins, J. W. (1978). Aquatic sources and sinks for nitrous oxide. Ph.D. Thesis, Harvard University, Cambridge, Massachusetts.
Elkins, J. W., Wofsy, S. C., McElroy, M. B., Kolb, C. E., and Kaplan, W. A. (1978). Aquatic sources and sinks for nitrous oxide. *Nature (London) 275,* 602-606.
Emery, K. O., Orr, W. L., and Rittenberg, S. C. (1955). Nutrient budgets in the ocean. *In* "Essays in the Natural Sciences in Honor of Captain Alan Hancock," pp. 147-157. University of Southern California, Los Angeles.
Enfield, D. B. (1981). El Niño: Pacific eastern boundary response to interannual forcing. *In* "Resource Management and Environmental Uncertainty: Lessons from Coastal Upwelling Fisheries" (M. H. Glantz and J. D. Thompson, eds.), pp. 213-254. Wiley (Interscience), New York.
Eppley, R. W., and Peterson, B. J. (1979). Particulate organic matter flux and planktonic new production in the deep ocean. *Nature (London) 282,* 677-680.
Eppley, R. W., Holm-Hansen, O., and Strickland, J. D. H. (1967). Some observations on the vertical migration of dinoflagellates. *J. Phycol. 4,* 333-340.
Eppley, R. W., Renger, E. H., and Harrison, W. G. (1979). Nitrate and phytoplankton production in southern California coastal waters. *Limnol. Oceanogr. 24,* 483-494.
Estrada, M., and Blasco, D. (1979). Two phases of the phytoplankton community in the Baja California upwelling. *Limnol. Oceanogr. 24,* 1065-1080.
Fahrbach, E., Brockmann, C., Lostaunau, N., and Urquizo, W. (1981). The northern Peruvian upwelling system during the ESACAN experiment. *In* "Coastal Upwelling" (F. A. Richards, ed.), pp. 134-145. Am. Geophys. Union, Washington, D.C.
Fraga, F. (1966). Distribution of particulate and dissolved nitrogen in the Western Indian Ocean. *Deep-Sea Res. 13,* 413-425.

Fraga, F. (1974). Distribution des masses d'eau dans l'upwelling de Mauritanie. *Tethys 6,* 5-10.

Fraga, F., and Manríquez, M. (1975). Oceanografia quimica de la region de afloramiento del noroeste de Africa. II. Campaña "Atlor II" Marzo 1973. *Res. Exp. Cient. B/O Cornide 4,* 185-217.

Friebertshauser, M. A., Codispoti, L. A., Bishop, D. D., Friederich, G. E., and Westhagen, A. A. (1975). JOINT-I hydrographic station data *R/V Atlantis II* cruise 82. *Coastal Upwelling Ecosyst. Anal. Data Rep. 18,* 1-243.

Friederich, G. E., and Codispoti, L. A. (1979). On some factors influencing dissolved silicon distribution over the northwest African shelf. *J. Mar. Res. 37,* 337-353.

Friederich, G. E., and Codispoti, L. A. (1981). The effects of mixing and regeneration on the nutrient content of upwelling waters off Peru. *In* "Coastal Upwelling" (F. A. Richards, ed.), pp. 221-227. Am. Geophys. Union, Washington, D.C.

Garfield, P. C., Packard, T. T., and Codispoti, L. A. (1979). Particulate protein in the Peru upwelling system. *Deep-Sea Res. 26,* 623-639.

Glantz, M. H., and Thompson, J. D., eds. (1981). "Resource Management and Environmental Uncertainty: Lessons from Coastal Upwelling Fisheries." Wiley (Interscience), New York.

Goering, J. J., and McRoy, C. P. (1981). A synopsis of PROBES. *EOS 62,* 730-731.

Goldman, J. E., McCarthy, J. J., and Peavey, D. G. (1979). Growth rate influence on the chemical composition of phytoplankton in oceanic waters. *Nature (London) 279,* 210-215.

Gordon, A. L., Taylor, H. W., and Georgi, D. T. (1977). Antarctic oceanographic zonation. *In* "Polar Oceans" (M. J. Dunbar, ed.), pp. 45-76. Arct. Inst. North Am., Calgary, Canada.

Guillard, R. R. L., and Kilham, P. (1977). The ecology of planktonic diatoms. *In* "The Biology of Diatoms" (D. Werner, ed.), pp. 372-469. Blackwell, Oxford.

Guillen, O. G., and Calienes, R. Z. (1981). Biological productivity and El Niño. *In* "Resource Management and Environmental Uncertainty: Lessons from Coastal Upwelling Fisheries" (M. H. Glantz and J. D. Thompson, eds.), pp. 255-284. Wiley (Interscience), New York.

Hafferty, A. J., Codispoti, L. A., and Huyer, A. (1978). JOINT-II *R/V Melville* Legs I, II, and IV *R/V Iselin* Leg II bottle data March 1977-May 1977. *Coastal Upwelling Ecosyst. Anal. Data Rep. 45,* 1-779.

Hafferty, A. J., Lowman, D., and Codispoti, L. A. (1979). JOINT-II, *Melville* and *Iselin* bottle data sections, March-May 1977. *Coastal Upwelling Ecosyst. Anal. Tech. Rep. 38,* 1-130.

Hahn, J. (1981). Nitrous oxide in the oceans. *In* "Denitrification, Nitrification and Atmospheric Nitrous Oxide" (C. C. Delwiche, ed.), pp. 191-240. Wiley, New York.

Harrison, W. G., Platt, T., Calienes, R., and Ochoa, N. (1981). Photosynthetic parameters and primary production of phytoplankton populations off the northern coast of Peru. *In* "Coastal Upwelling" (F. A. Richards, ed.), pp. 303-311. Am. Geophys. Union, Washington, D.C.

Hidaka, K. (1972). Physical oceanography of upwelling. *Geoforum 11,* 9-21.

Hofmann, E. E., Busalacchi, A. J., and O'Brien, J. J. (1981). Wind generation of the Costa Rica Dome. *Science 214,* 552-554.

Hood, D. W., and Kelley, J. J. (1976). Evaluation of mean vertical transports in an upwelling system by CO_2 measurements. *Mar. Sci. Commun. 2,* 387-411.

Hopkins, T. S. (1974). The circulation in an upwelling region, the Washington Coast. *Tethys 6,* 375-390.

Horrigan, S. G., Carlucci, A. F., and Williams, P. M. (1981). Light inhibition of nitrification in sea surface films. *J. Mar. Res. 39,* 557-565.

Hsueh, Y., and O'Brien, J. J. (1971). Steady coastal upwelling induced by along-shore current. *J. Phys. Oceanogr. 1,* 180-186.

Hughes, P., and Barton, E. D. (1974). Stratification and water mass structure in the upwelling area off N.W. Africa in April/May 1969. *Deep-Sea Res. 21,* 611-628.

Huntsman, S. A., and Barber, R. T. (1977). Primary production off northwest Africa: The relationship to wind and nutrient conditions. *Deep-Sea Res. 24,* 25-34.

Huntsman, S. A., and Sunda, W. G. (1980). The role of trace metals in regulating phytoplankton growth. *In* "The Physiological Ecology of Phytoplankton" (I. Morris, ed.), pp. 285-328. Blackwell, Oxford.

Huntsman, S. A., Brink, K. H., Barber, R. T., and Blasco, D. (1981). The role of circulation and stability in controlling the relative abundance of dinoflagellates and diatoms over the Peru shelf. *In* "Coastal Upwelling" (F. A. Richards, ed.), pp. 357-365. Am. Geophys. Union, Washington, D.C.

Hurd, D. C. (1972). Factors affecting solution rate of biogenic opal in seawater. *Earth Planet. Sci. Lett. 15,* 411-417.

Hurd, D. C. (1973). Interactions of biogenic opal, sediment and seawater in the Central Equatorial Pacific. *Geochim. Cosmochim. Acta 37,* 2257-2282.

Huyer, A. (1976). A comparison of upwelling events in two locations: Oregon and Northwest Africa. *J. Mar. Res. 34,* 531-546.

Huyer, A. (1980). The offshore structure and subsurface expression of sea level variations off Peru, 1976-1977. *J. Phys. Oceanogr. 10,* 1755-1768.

Johnston, R. (1963). Seawater, the natural medium for phytoplankton. I. General features. *J. Mar. Biol. Assoc. U.K. 43,* 427-456.

Jones, B. H., Jr. (1978). A spatial analysis of the autotrophic response to abiotic forcing in three upwelling ecosystems: Oregon, northwest Africa, and Peru. *Coastal Upwelling Ecosyst. Anal. Tech. Rep. 37,* 1-262.

Kullenberg, G. (1981). A comparison of distributions of suspended matter in the Peru and northwest African upwelling areas. *In* "Coastal Upwelling" (F. A. Richards, ed.), pp. 282-290. Am. Geophys. Union, Washington, D.C.

Lacis, A., Hansen, J., Lee, P., Mitchell, T., and Lebedeff, S. (1981). Greenhouse effect of trace gases, 1970-1980. *Geophys. Res. Lett. 8,* 1035-1038.

Lasker, R. (1978). The relation between oceanographic conditions and larval anchovy food in the California Current: Identification of factors contributing to recruitment failure. *Rapp. P.-V. Reun., Cons. Int. Explor. Mer 173,* 212-230.

LeBorgne, R. P. (1978). Ammonium formation in Cape Timiris (Mauritania) upwelling. *J. Exp. Mar. Biol. Ecol. 31,* 253-265.

Lee, T. N., Atkinson, L. P., and Legeckis, R. (1981). Observations of a Gulf Stream frontal eddy on the Georgia Continental Shelf, April 1977. *Deep-Sea Res. 28,* 347-378.

Liu, K-k. (1979). Geochemistry of inorganic nitrogen compounds in two marine environments: The Santa Barbara Basin and the ocean off Peru. Ph.D. Thesis, University of California, Los Angeles.

McCarthy, J. J. (1970). A urease method for urea in seawater. *Limnol. Oceanogr. 15,* 309-313.

McCarthy, J. J. (1972). The uptake of urea by natural populations of marine phytoplankton. *Limnol. Oceanogr. 17,* 738-748.

McCarthy, J. J., and Whitledge, T. E. (1972). Nitrogen excretion by anochovy (*Engraulis mordax* and *E. ringens*) and jack mackerel (*Trachurus symmetricus*). *Fish. Bull. 70,* 395-401.

Margalef, R. (1978a). What is an upwelling ecosystem? "Upwelling Ecosystems" (R. Boje and M. Tomczak, eds.), pp. 12-14. Springer-Verlag, Berlin and New York.

Margalef, R. (1978b). Life forms of phytoplankton as survival alternatives in an unstable environment. *Oceanol. Acta 1,* 493-509.

Margalef, R., and Estrada, M. (1981). On Upwelling, eutrophic lakes, the primitive biosphere and biological membranes. *In* "Coastal Upwelling" (F. A. Richards, ed.), pp. 522-529. Am. Geophys. Union, Washington, D.C.

Margalef, R., Estrada, M., and Blasco, D. (1979). Functional morphology of organisms involved in red tides, as adapted to decaying turbulence. *In* "Toxic Dinoflagellate Blooms" (D. L. Taylor and H. H. Seliger, eds.), pp. 89-94. Elsevier/North-Holland, New York.

Minas, H. J. (1968). A propos d'une remontée d'eaux "Profondes" dans les parages du Golfe de Marseille (Octobre 1964). Conséquences biologiques. *Cah. Oceanogr. 20,* 648-674.

Minas, H. J., Codispoti, L. A., and Dugdale, R. C. (1982). Nutrients and primary production in the upwelling region off northwest Africa. *Rapp. P.-V. Reun., Cons. Int. Explor. Mer 180,* 141-176.

Mittelstaedt, E., Pillsbury, R. D., and Smith, R. L. (1975). Flow patterns in the northwest African upwelling area. *Dtsch. Hydrogr. Z. 28,* 145-167.

Nelson, D. M., and Conway, H. L. (1979). Effects of the light regime on nutrient assimilation by phytoplankton in the Baja California and northwest Africa upwelling systems. *J. Mar. Res. 37,* 301-318.

O'Brien, J. J., Basalacchi, A., and Kindle, J. (1981). Ocean models of El Niño. *In* "Resource Management and Environmental Uncertainty: Lessons from Coastal Upwelling Fisheries" (M. H. Glantz and J. D. Thompson, eds.), pp. 159-212. Wiley (Interscience), New York.

Olson, R. J. (1981). Differential photoinhibition of marine nitrifying bacteria: A possible mechanism for the formation of the primary nitrite maximum. *J. Mar. Res. 39,* 227-238.

Packard, T. T. (1979). Respiration and respiratory electron transport activity in plankton from the northwest African upwelling area. *J. Mar. Res. 37,* 711-742.

Packard, T. T., and Dortch, Q. (1975). Particulate protein-nitrogen in North Atlantic surface waters. *Mar. Biol. (Berlin) 33,* 347-354.

Packard, T. T., Blasco, D., and Barber, R. T. (1978). *Mesodinium rubrum* in the Baja California upwelling system. *In* "Upwelling Ecosystems" (R. Boje and M. Tomczak, eds.), pp. 73-89. Springer-Verlag, Berlin and New York.

Pak, H., Codispoti, L. A., and Zaneveld, J. R. (1980). On the intermediate particle maxima associated with oxygen-poor water off western South America. *Deep-Sea Res. 27,* 783-797.

Petersen, G. H. (1977). Biological effects of sea ice and icebergs in Greenland. *In* "Polar Oceans" (M. J. Dunbar, ed.), pp. 319-330. Arct. Inst. North Am., Calgary, Canada.

Pierotti, D., and Rasmussen, R. A. (1980). Nitrous oxide measurements in the eastern tropical Pacific Ocean. *Tellus 32,* 56-72.

Plank, W. S., Anderson, J. J., and Pak, H. (1973). Hydrographic, chemical and optical observations from legs 4 and 5, YALOC-71. *Oreg. State Univ., Sch. Oceanogr. Data Rep. 54,* 1-381.

Provasoli, L. (1963). Organic regulation of phytoplankton fertility. *In* "The Sea" (M. N. Hill, ed.), Vol. 2, pp. 165-219. Wiley (Interscience), New York.

Redfield, A. C. (1955). The hydrography of the Gulf of Venezuela. *Pap. Mar. Biol. Oceanogr., Deep-Sea Res. 3,* Suppl., 115-133.

Redfield, A. C., Ketchum, B. H., and Richards, F. A. (1963). The influence of organisms on the composition of seawater. *In* "The Sea" (M. N. Hill, ed.), Vol. 2, pp. 26-77. Wiley (Interscience), New York.

Remsen, C. C. (1971). The distribution of urea in coastal and oceanic waters. *Limnol. Oceanogr. 16,* 732-740.

Richards, F. A. (1965). Anoxic basins and fjords. *In* "Chemical Oceanography" (J. P. Riley and G. Skirrow, eds.), pp. 611-645. Academic Press, New York.

Roden, G. I. (1972). Large-scale upwelling off northwestern Mexico. *J. Phys. Oceanogr. 3,* 184-189.

Rowe, G. T., Clifford, C. H., and Smith, K. L., Jr. (1977). Nutrient regeneration in sediments off Cape Blanc, Spanish Sahara. *Deep-Sea Res. 24,* 57-64.

Ryther, J. H. (1969). Photosynthesis and fish production in the sea. *Science 166,* 71-76.

Ryther, J. H., and Dunstan, W. M. (1971). Nitrogen, phosphorus, and eutrophication in the coastal marine environment. *Science 171,* 1008-1013.

Setchell, F. W., and Packard, T. T. (1979). Phytoplankton respiration in the Peru upwelling. *J. Plankton Res. 1,* 343-354.

Shannon, L. V., Nelson, G., and Juny, M. R. (1981). Hydrological and meteorological aspects of upwelling in the southern Benguela Current. *In* "Coastal Upwelling" (F. A. Richards, ed.), pp. 146-159. Am. Geophys. Union, Washington, D.C.

Siezen, R. J., and Mague, T. H. (1978). Amino acids in suspended particulate matter from oceanic and coastal waters of the Pacific. *Mar. Chem. 6,* 215-231.

Slawyk, G., Collos, Y., Minas, M., and Grall, J. R. (1978). On the relationship between carbon-to-nitrogen composition ratios of the particulate matter and growth rate of marine phytoplankton from the northwest African upwelling region. *J. Exp. Mar. Biol. Ecol. 33,* 119-131.

Smayda, T. J. (1980). Phytoplankton species succession. *In* "The Physiological Ecology of Phytoplankton" (I. Morris, ed.), pp. 493-570. Blackwell, Oxford.

Smith, R. L. (1968). Upwelling. *Oceanogr. Mar. Biol. 6,* 7-46.

Smith, R. L. (1981). A comparison of the structure and variability of the flow field in three coastal upwelling regions: Oregon, Northwest Africa, and Peru. *In* "Coastal Upwelling" (F. A. Richards, ed.), pp. 107-118. Am. Geophys. Union, Washington, D.C.

Smith, R. L., and Bottero, J. S. (1977). On upwelling in the Arabian Sea. *In* "A Voyage of Discovery" (M. Angel, ed.), pp. 291-304. Pergamon, Oxford.

Smith, S. L. (1978). Nutrient regeneration by zooplankton during a red tide off Peru, with notes on biomass and species composition of zooplankton. *Mar. Biol. (Berlin) 49,* 125-132.

Smith, S. L., and Codispoti, L. A. (1979). Southwest monsoon of 1979: Chemical and biological response of Somali coastal waters. *Science 209,* 597-600.

Smith, S. L., and Whitledge, T. E. (1977). The role of zooplankton in the regeneration of nitrogen in a coastal upwelling system off northwest Africa. *Deep-Sea Res. 24,* 49-56.

Smith, W. O., Jr., and Barber, R. T. (1979). A carbon budget for the autotrophic ciliate, *Mesodinium rubrum*. *J. Phycol. 15*, 1-26.

Smith, W. O., Jr., Barber, R. T., and Huntsman, S. A. (1982). The influence of organic ligands on the growth of phytoplankton in the Northwest African upwelling region. *J. Plankton Res. 4*, 651-663.

Söderlund, R., and Svensson, B. H. (1976). The global nitrogen cycle. *In* "Nitrogen, Phosphorus and Sulphur-Global Cycles" (B. H. Svensson and R. Söderlund, eds.), pp. 23-74. Swed. Nat. Sci. Res. Counc., Stockholm.

Sorokin, Y. I. (1978a). Description of primary production and heterotrophic microplankton in the Peruvian upwelling region. *Oceanology 18*, 62-71.

Sorokin, Y. I. (1978b). Decomposition of organic matter and nutrient regeneration. *In* "Marine Ecology" (O. Kenne, ed.), pp. 501-616. Wiley, New York.

Sorokin, Y. I., and Kogelschatz, J. (1979). Analysis of heterotrophic microplankton in an upwelling area. *Hydrobiologia 66*, 195-208.

Spencer, C. P. (1975). The micro-nutrient elements. *In* "Chemical Oceanography" (J. P. Riley and G. Skirrow, eds.), 2nd ed., Vol. 2, pp. 245-300. Academic Press, New York.

Steemann-Nielsen, E., and Wium-Andersen, S. (1970). Copper ions as poison in the sea and in freshwater. *Mar. Biol. (Berlin) 6*, 93-97.

Strickland, J. D. H., Eppley, R. W., and Mendiola, de B. R. (1969). Phytoplankton populations, nutrients and photosynthesis in Peruvian coastal waters. *Bol. Inst. Mar Peru 2*, 1-45.

Suess, E. (1980). Particulate organic carbon flux in the oceans--Surface productivity and oxygen utilization. *Nature (London) 288*, 260-263.

Sunda, W. G., Barber, R. T., and Huntsman, S. A. (1981). Phytoplankton growth in nutrient rich seawater: Importance of copper-manganese cellular interactions. *J. Mar. Res. 39*, 567-586.

Sverdrup, H. U. (1953). On conditions for the vernal blooming of phytoplankton. *J. Cons. Int. Explor. Mer 18*, 287.

Sverdrup, H. U., Johnson, M. W., and Fleming, R. H. (1942). "The Oceans, Their Physics, Chemistry, and Biology." Prentice-Hall, Englewood Cliffs, New Jersey.

Sweeney, R. E., Liu, K-k., and Kaplan, I. R. (1978). Oceanic nitrogen isotopes and their uses in determining the source of sedimentary nitrogen. *In* "Stable Isotopes in the Earth Sciences" (B. W. Robinson, ed.), DSIR Bull., pp. 9-16. Dep. Sci. Ind. Res., Wellington, N.Z.

Swift, D. G. (1980). Vitamins and phytoplankton growth. *In* "The Physiological Ecology of Phytoplankton" (I. Morris, ed.), pp. 329-368. Blackwell, Oxford.

Szekielda, K. H. (1974). The hot spot in the Ross Sea: Upwelling during wintertime. *Tethys 6,* 105-110.

Takahashi, M., Tasuoka, Y., Watanabe, M., Niyazaki, T., and Ichimura, S. (1981). Local upwelling associated with vortex motion off Oshima Island, Japan. *In* "Coastal Upwelling" (F. A. Richards, eds.), pp. 119-124. Am. Geophys. Union, Washington, D.C.

Traganza, E. D., Conrad, J. C., and Breaker, L. C. (1981). Satellite observations of a cyclonic upwelling system and giant plume in the California Current. *In* "Coastal Upwelling" (F. A. Richards, eds.), pp. 228-241. Am. Geophys. Union, Washington, D.C.

Treguer, P., and LeCorre, P. (1979). The ratios of nitrate, phosphate, and silicate during uptake and regeneration phases of the Moroccan upwelling regime. *Deep-Sea Res. 26,* 163-184.

Vaccaro, R. F. (1965). Inorganic nitrogen in sea water. "Chemical Oceanography" (J. P. Riley and G. Skirrow, eds.), pp. 365-408. Academic Press, New York.

Vallespinós, F., and Estrada, M. (1975). Nitrógeno particulado en la región del NW de Africa. Distribución y relación con otros parámetros. *Res. Exp. Cient. B/O Cornide 4,*

Voituriez, B. (1981). Equatorial upwelling in the eastern Atlantic: problems and paradoxes. *In* "Coastal Upwelling" (F. A. Richards, ed.), pp. 95-106. Am. Geophys. Union, Washington, D.C.

von Arx, W. S. (1962). "An Introduction to Physical Oceanography." Addison-Wesley, Reading, Massachusetts.

von Bröckel, K. (1981). A note on short-term production and sedimentation in the upwelling region off Peru. *In* "Coastal Upwelling" (F. A. Richards, ed.), pp. 291-297. Am. Geophys. Union, Washington, D.C.

Wajih, S., Naqui, S., Noronha, R. J., and Reddy, C. V. G. (1982). Denitrification in the Arabian Sea. *Deep-Sea Res. 29,* 459-469.

Walsh, J. J. (1976). Herbivory as a factor in patterns of nutrient utilization in the sea. *Limnol. Oceanogr. 21,* 1-13.

Walsh, J. J. (1977). A biological sketchbook for an eastern boundary current. *In* "The Sea" (E. D. Goldberg, I. N. McCave, J. J. O'Brien, and J. H. Steele, eds.), Vol. 6, pp. 923-968. Wiley (Interscience), New York.

Walsh, J. J. (1981). A carbon budget for overfishing off Peru. *Nature (London) 26,* 300-304.

Walsh, J. J., Whitledge, T. E., Esaias, W. E., Smith, R. L., Huntsman, S. A., Santander, H., and de Mendiola, B. R. (1980). The spawning habitat of the Peruvian anchovy, *Engraulis ringens. Deep-Sea Res. 27,* 1-27.

Wangersky, P. J. (1977). The role of particulate matter in the productivity of surface waters. *Helgol. Wiss. Meeresunters 30,* 546-564.

Ward, B. B., Olson, R. J., and Perry, M. J. (1982). Microbial nitrification rates in the primary nitrite maximum off southern California. *Deep-Sea Res. 29,* 247-255.

Weikert, H. (1977). Copepod carcasses in the upwelling region south of Cape Blanc, N.W. Africa. *Mar. Biol. (Berlin) 42,* 351-355.

Whitledge, T. E. (1981). Nitrogen recycling and biological populations in upwelling ecosystems. *In* "Coastal Upwelling" (F. A. Richards, ed.), pp. 257-273. Am. Geophys. Union, Washington, D.C.

Whitledge, T. E., and Dugdale, R. C. (1972). Creatine in seawater. *Limnol. Oceanogr. 17,* 309-314.

Whitledge, T. E., and Packard, T. T. (1971). Nutrient excretion by anchovies and zooplankton in Pacific upwelling regions. *Invest. Pesq. 35,* 243-250.

Williams, P. J. le B. (1975). Biological and chemical aspects of dissolved organic material in seawater. *In* "Chemical Oceanography" (J. P. Riley and G. Skirrow, eds.), 2nd ed., Vol. 1, pp. 301-363. Academic Press, New York.

Wooster, W. S. (1981). An upwelling mythology. *In* "Coastal Upwelling" (F. A. Richards, ed.), pp. 1-3. Am. Geophys. Union, Washington, D.C.

Wooster, W. S., and Reid, J. L. (1963). Eastern boundary currents. *In* "The Sea (M. N. Hill, ed.), Vol. 2, pp. 253-280. Wiley (Interscience), New York.

Wroblewski, J. S. (1977). A model of phytoplankton plume formation during variable Oregon upwelling. *J. Mar. Res. 35,* 357-394.

Wyrtki, K. (1963). The horizontal and vertical field of motion in the Peru Current. *Bull. Scripps Inst. Oceanogr. 8,* 313-346.

Yentsch, C. S. (1974). The influence of geostrophy on primary production. *Tethys 6,* 111-118.

Yentsch, C. S., and Lee, R. W. (1966). A study of photosynthetic light reactions, a new interpretation of sun and shade phytoplankton. *J. Mar. Res. 24,* 319-337.

Yoshinari, T. (1980). N_2O reduction by Vibrio succinogenes. *Appl. Environ. Microbiol. 39,* 81-84.

Zentara, S. J., and Kamykowski, D. (1981). Geographic variations in the relationship between silicic acid and nitrate in the South Pacific Ocean. *Deep-Sea Res. 28A,* 455-465.

Chapter 16

NITROGEN IN ESTUARINE AND COASTAL MARINE ECOSYSTEMS

SCOTT W. NIXON
MICHAEL E. Q. PILSON
Graduate School of Oceanography
University of Rhode Island
Narragansett, Rhode Island

I. INTRODUCTION

This is an exciting and challenging time to write about nitrogen in estuaries and coastal marine waters. Fifteen years ago the classic symposium volume on "Estuaries" (Lauff, 1967) devoted only one chapter out of seventy to nutrients, and that largely to the role of phosphorus (Ketchum, 1967). The time since has been marked by the appearance of an increasing number and diversity of papers dealing with various aspects of the nitrogen cycle in coastal waters. Much of that work is discussed in detail in other chapters of this book that treat specific nitrogen transformations. Our hope in this chapter is to provide an overview of present knowledge concerning the abundance and distribution of the different forms of nitrogen in a number of estuaries, to relate those patterns of abundance to the major

ISBN 0-12-160280-X

nitrogen cycle processes taking place in the water column and sediments, to assess the relative contributions of nitrogen cycling and nitrogen inputs in various estuaries within the context of annual budgets, and to provide some perspective on the role of nitrogen in estuarine productivity. We recognize that a great deal is lost by focusing on only one element, and that much of the behavior of nitrogen is closely linked with the flow of energy and the cycling of carbon, oxygen, phosphorus, and other materials. But it is ever the problem in ecology to see Nature as Wordsworth did, "all in all", while being constrained by our lesser talents to write about it piece by piece.

II. HISTORICAL PERSPECTIVE

Our understanding of the abundance, distribution, and behavior of nitrogen in estuarine systems began with a very practical concern about the effects of pollution. Some of the earliest systematic measurements of nitrogen in estuarine waters were those obtained by William Joseph Dibdin, chemist to the London Metropolitan Board of Works. Beginning in 1885, he carried out routine monitoring of ammonia levels in the tidal Thames, off the major sewage outfall at Crossness (Wood, 1982). Dibdin's approach to sewage treatment, which included careful measurement of environmental conditions in receiving waters, was probably more advanced than that found elsewhere at the time, and his concern with ammonia and the impact of its oxidation to nitrate on dissolved oxygen levels in the Thames spread through the growing fields of sanitary and environmental engineering (Melosi, 1981). Doubtless this aspect of the nitrogen cycle was soon receiving attention in a number of other urban estuaries where the high concentrations of ammonia (>100 μM) could easily be distilled off and nesslerised according to the analytical techniques of the day. This practical concern with the role of nitrogen in estuarine water quality has continued as one of the two major interests that have stimulated and supported study of the various forms of this element in estuarine ecology. The other has been a continuing inquiry into the role of nitrogen in maintaining and enhancing estuarine productivity.

It is difficult to trace the historical development of current beliefs that so firmly link the productivity of coastal marine waters with the supply and cycling of fixed nitrogen. In discussing their early work on the Romsdalsfjord and other Norwegian coastal waters, Braarud and Klem (1931) noted that, "Since Brandt (1899) published his theory of phosphates and nitrates as minimum substances for production in the sea, the distribution of nutrient substances has been taken into account as an important factor in the effort to understand the uneven development of plankton." It must be appreciated that Brandt's extrapolation of Liebig's (1885) "law of the minimum" from an agricultural context to the sea was an insight that could not have been attained without advances in analytical techniques that made it possible to detect the very low levels of inorganic nutrients dissolved in marine and most estuarine waters. Pomeroy (1974) has emphasized this point with respect to phosphate and the development of the Denigès (1921) method, and it is equally true for nitrogen. While the early methods of Denigès (1911), Harvey (1926, 1928), Cooper (1932), and Atkins (1932) for nitrite and nitrate and of Buch (1923) and Wattenberg (1928) for ammonia have all been improved on considerably (reviewed by Riley, 1975; see also D'Elia, Chapter 20, this volume), they made it possible for the first time to obtain reasonably reliable quantitative measurements of the abundance, distributions, and general seasonal cycles of the different forms of inorganic nitrogen in coastal marine waters (for example, Braarud and Klem, 1931; Cooper, 1933). The analytical difficulties with ammonia were particularly great, however, and it was not until after the publication of Solorzano's direct colorometric technique in 1969 that ammonia measurements became a regular part of estuarine nutrient studies in regions not grossly polluted.

When James Johnstone wrote his influential account on "Conditions of Life in the Sea" in 1908, he included a perceptive and complete chapter on the nitrogen cycle that, almost 75 years later still seems remarkably contemporary. In discussing nutrients and productivity, he concluded on the basis of analyses carried out by Brandt in 1902 that, "It is probable that the abundance of nitrogen compounds in the sea determines the production." However, in spite of this attention to nitrogen, he was careful to consider a larger sample of data, noting that, "It is rather difficult to determine which of the three food-stuffs, nitrogen compounds, phosphoric acid, or silica, is present in the sea in minimum proportion, and so rules the production."

Some years later, Redfield's (1934) observations on the similarity in the nitrogen to phosphorus ratios of marine plankton and oceanic seawater appeared to further confound the analysis of "limiting factors." However, as Redfield noted and as later work made very clear, the nitrogen-to-phosphorus ratio of estuarine and coastal waters is commonly much lower than found in the open sea or in coastal plankton. The early appreciation of this was exaggerated because ammonia, which often accounts for a significant fraction of the dissolved inorganic nitrogen in coastal waters, was usually not included in the early analyses due to the lack of a reliable and practical technique. We now know that the low-nitrogen anomaly remains, however, even when ammonia is included, though a discussion of the reasons for it will be reserved for a later section. Regardless of its origin, the low nitrogen to phosphorus ratio of coastal waters was sufficient evidence for Gordon Riley and his co-workers to emphasize nitrogen cycling in their pioneering ecosystem studies of the productivity of Long Island Sound in the 1950s (Harris and Riley, 1956; Harris, 1959). In spite of Riley's early lead, nitrogen did not come to be a prime focus of estuarine research until after the publication of a report in *Science* by Ryther and Dunstan (1971) which concluded that, "The distribution of inorganic nitrogen and phosphorus and bioassay experiments both show that nitrogen is the critical limiting factor to algal growth and eutrophication in coastal marine waters." Perhaps in response to such an unequivocal statement, the past 10 years have been marked by an intensive effort to study nitrogen dynamics in estuarine ecosystems. The results have been an extensive literature (see, for example, the recent volume on "Estuaries and Nutrients" edited by Neilson and Cronin, 1981), an exciting increase in our understanding of some features of nitrogen cycle processes in estuaries, and a growing appreciation for the complexities and subtleties of estuarine ecosystems that still leave us far from being able to explain or predict their behavior. There have been particular efforts to study nitrogen processes in a variety of special estuarine subsystems such as salt marshes and seagrass beds, and separate chapters in this volume have been devoted to those topics. Our hope here is to provide an overview of the major features of nitrogen dynamics in the open waters of larger estuaries and nearshore coastal waters.

III. THE ABUNDANCE AND DISTRIBUTION OF NITROGEN

With the development of the improved analytical techniques mentioned earlier, a number of reports began to appear throughout the late 1920s and 1930s concerning the concentrations and seasonal cycles in abundance of the various forms of nitrogen in marine waters. For the most part, however, the early emphasis in nutrient chemistry was on the open sea, with less attention to estuarine areas. Because many of the analytical advances were made at the Plymouth Laboratory, much of the pioneering work was carried out in the English Channel (Harvey, 1926, 1928) and, as part of that effort, some nearshore stations were also taken in

Plymouth Sound (Cooper, 1933). While data from the offshore stations were presented more fully, there was also an early interest in comparing nearshore and offshore waters and in the influence of fresh water inputs on the seasonal abundance of nitrogen. While Cooper (1933) found that "land drainage may be of great importance..." for ammonia and nitrate in Plymouth Sound, it was dismissed as a factor of any importance in the larger and (for him) more pressing problem of understanding the nitrogen cycles of the open sea. From the beginning, there has been some controversy over the importance of nitrogen inputs in regulating the concentration of nitrogen and, hence, the productivity of nearshore waters. For example, Johnstone (1908) put it simply: "It is evident that the greater density of plant life near land is directly due to the fact that there is a greater amount of the ultimate food materials, nitrogen compounds and carbon dioxide, there, then far away from land." But after working on the Norwegian fjords, Braarud and Klem (1931) concluded that, "The Atlantic water in this area is richer in nitrates and phosphates than the coastal water, even before the production of plankton begins," and, "The spring maximum of the phytoplankton cannot be due to supplies of nutrient from the land." These differences of opinion about the abundance of nitrogen in nearshore and estuarine waters and the importance of nitrogen inputs from land have continued to trouble our concepts of estuarine productivity and eutrophication for the past 50 years (Nixon, 1981a).

One of the first specific studies of the distribution and seasonal abundance of nitrogen in an estuary appears to have been the work of Nash (1947) who measured nitrate and nitrite during 1936-1937 in the Patuxent River estuary in Chesapeake Bay. While ammonia was not considered in her work, it was included in some of the analyses reported for Chesapeake Bay by Newcombe et al. (1939). As noted earlier, the lack of ammonia data is characteristic of early nitrogen reports (e.g., Lillick, 1937), and it was not until the 1960s that more complete annual inventories of nitrogen in estuarine and coastal waters began to become available [for example, the work of Postma (1966) during 1960-1962 in the Wadden Sea]. At present, the spatial and temporal variations in the concentration of ammonia, nitrite, and nitrate, over at least one annual cycle, have been reasonably well described for perhaps several dozen estuaries, lagoons, and nearshore marine waters around the world. The number of areas for which simultaneous or even similar sampling of dissolved organic nitrogen or particulate organic nitrogen have been reported is very much smaller. Systematic analyses of the dissolved nitrogen gases in estuaries are almost completely lacking, though there have been a number of recent studies of nitrous oxide (McElroy et al., 1978; Deck, 1980; Seitzinger et al., 1983).

A. Seasonal Cycles in the Concentration of Nitrogen in the Water Column

While a winter nitrate maximum is frequently found in temperate estuaries and lagoons as it is in the sea (e.g., Fig. 1), there are enough exceptions that it would probably be more misleading than useful to present "typical" cycles for the various nitrogen species. Moreover, the concentrations of dissolved inorganic nitrogen commonly found in estuaries range from below the limits of detection (perhaps 0.1 μ*M*) to over 100 μ*M*. Some impression of the annual variation in inorganic nitrogen within and among a variety of estuaries can be gained from Fig. 2. In a sample of 12 different systems from around the United States, it is possible to find areas in which inorganic nitrogen is virtually removed from the water at certain times (Narragansett Bay, Rhode Island), areas in which it is always abundant (New York Bay), and areas in which it is always very low (Kaneohe Bay, Hawaii); areas with a strong river influence and a large input of nitrate in late winter, early spring (Potomac estuary; Chesapeake Bay; Apalachicola Bay, Florida), or areas with a large river input but

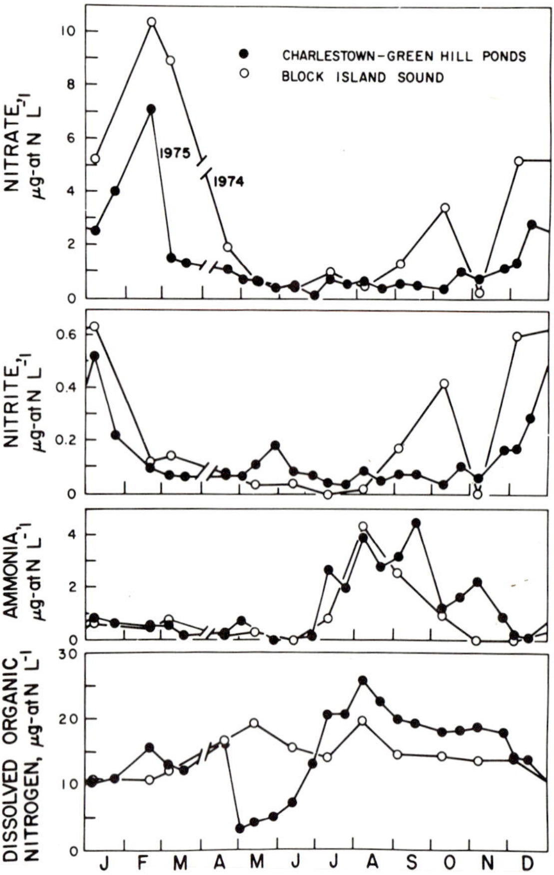

Fig. 1. An annual cycle in the concentrations of nitrate, nitrite, ammonia, and dissolved organic nitrogen averaged over depth in a coastal lagoon and in nearshore seawater off the coast of Rhode Island (USA). Data from Marine Research, Inc., Falmouth Massachusetts, reported by Nixon and Lee (1981).

no marked spring maximum in nitrogen (New York Bay; Mobile Bay, Alabama); areas where nitrate is abundant during summer (Delaware Bay) or where ammonia is most abundant (Pamlico estuary, North Carolina; Barataria Bay, Louisiana); and areas with a large seasonal variation (Narragansett Bay; Patuxent estuary, Maryland) or little variation (South San Francisco Bay; Mobile Bay, Alabama).

At the present time, a lack of data makes it impossible to discover if there are corresponding variations in dissolved and particulate organic nitrogen (DON, PN). It is apparent, however, that within the systems where they have been measured, the amount of nitrogen in these forms undergoes considerable variation during the course of a year, though interpretation of the data as evidence of regular seasonal cycles would be difficult (Fig. 3). In spite of the lack of attention given to PN and DON, it is clear that these forms account for a large fraction of the total nitrogen in estuarine and coastal waters. After comparing data from coastal Cape Cod, Delaware Bay, and the English Channel, Sharp et al. (1982) have suggested that DON alone appears to account for, "a relatively consistent 54 to 62% of the total nitrogen (excluding gases)....". It seems early for such a conclusion, but it is clear that our understanding of the behavior of nitrogen in estuaries has been limited by a lack of information on any aspects of the organic forms, including their composition, abundance, production, consumption, and decomposition.

Before leaving the subject of seasonal cycles, it is important to note that the annual pattern observed in each estuary will doubtless vary somewhat from year to year. At least that has been the case in the relatively few systems where data are available from a sequence of years (for example, Riley, 1955; Whaley et al., 1966; Hobbie, 1974). An analysis of data for dissolved inorganic nitrogen (DIN) in the lower West Passage of Narragansett Bay during the 5 years from 1976-1981 gives a useful perspective on the problem of defining the annual cycle for even one estuary (Fig. 4). Whereas the low values during spring and early summer were remarkably consistent from year to year, the timing and magnitude of the higher values were extremely variable. The mean concentrations regularly show a maximum, occurring in the late fall or early winter, but there is also a high probability of finding very low DIN levels during any particular winter month. Some of this uncertainty arises because Narragansett Bay is characterized by an

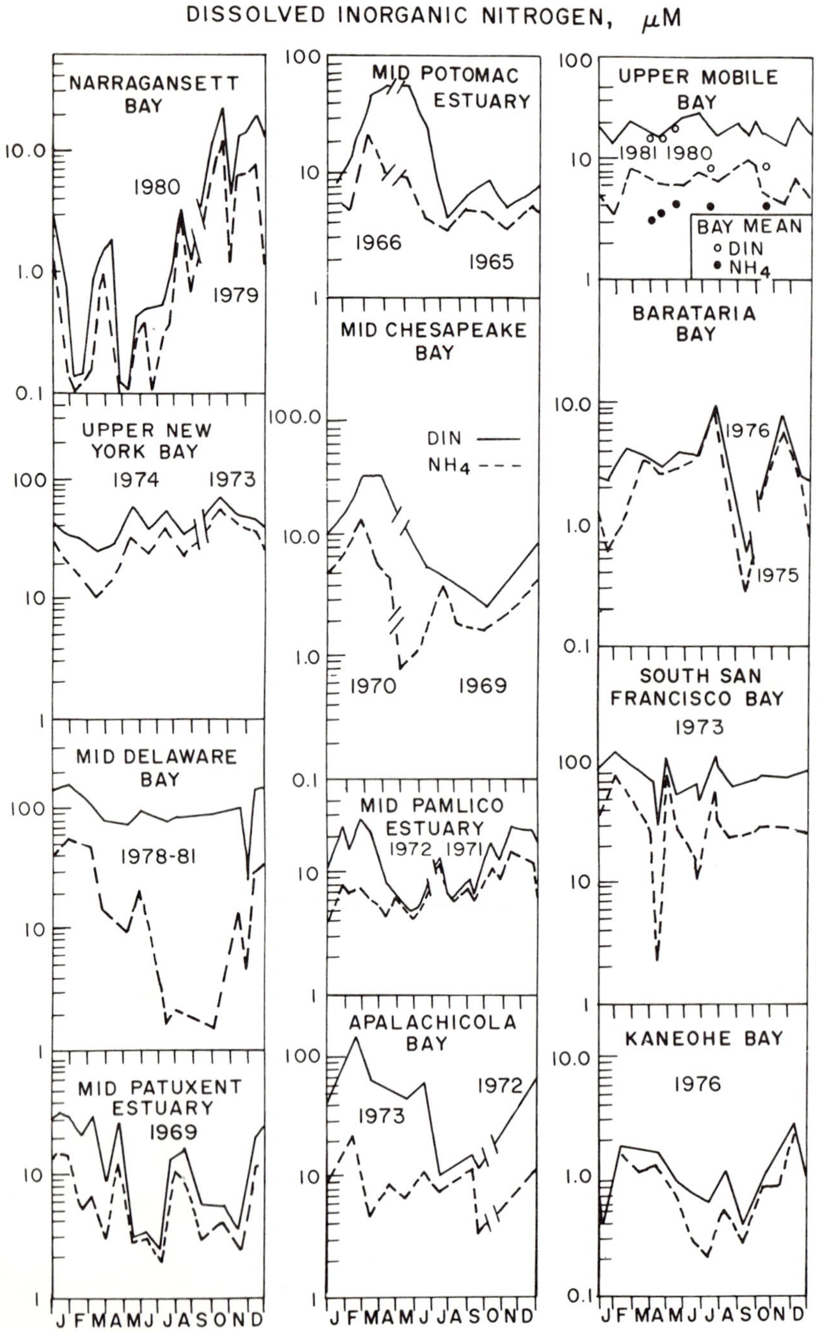
DISSOLVED INORGANIC NITROGEN, μM
NARRAGANSETT BAY
1980
1979
UPPER NEW YORK BAY
1974
1973
MID DELAWARE BAY
1978-81
MID PATUXENT ESTUARY
1969
MID POTOMAC ESTUARY
1966
1965
MID CHESAPEAKE BAY
DIN
NH4
1970
1969
MID PAMLICO ESTUARY
1972
1971
APALACHICOLA BAY
1973
1972
UPPER MOBILE BAY
1981
1980
BAY MEAN
DIN
NH4
BARATARIA BAY
1976
1975
SOUTH SAN FRANCISCO BAY
1973
KANEOHE BAY
1976
J F M A M J J A S O N D

Fig. 2

intense winter phytoplankton bloom, the inception of which varies from month to month in different years, depending on light and other factors (Smayda, 1973; Hitchcock and Smayda, 1977; Nixon et al., 1979). Many features of the variation nevertheless remain unexplained, not only in Narragansett Bay but in other areas where long time-series of data are available. In systems like Chesapeake Bay (Boynton et al., 1982) or Apalachicola Bay (Livingston and Sheridan, 1979), yearly variations in river input may propagate through the nitrogen cycle, while in others the dynamic balance between rapid nitrogen remineralization, intense phytoplankton growth, and animal grazing may make summer nitrogen levels more erratic.

B. Spatial Variation in the Water Column

The nitrogen concentration data shown in the previous section represent conditions at one point or within one region of each estuary at a series of discrete times. At any one time there may also be large spatial variability in any dimension. Because estuaries are often elongated and relatively shallow, the major interest in the spatial variability of nitrogen has been in its distribution along the longitudinal axis, which usually corresponds to the salinity gradient. While geographical displays of concentration isopleths can be useful to illustrate some of the relation-

Fig. 2. An annual cycle in the concentration of dissolved inorganic nitrogen (NO_3^-, NO_2^-, NH_4^+, solid line) and ammonia (broken line) in various estuaries around the United States. Note the logarithmic scale. Mid-Narragansett Bay data from MERL (1980); New York Bay from Malone (1977); mid-Delaware Bay data for various months 1978-1981 from Culberson et al. (1982) and C. H. Culberson, personal communication; Patuxent estuary from Flemer (1970); Potomac estuary from Whaley et al. (1966); Chesapeake Bay from Taylor and Grant (1977); Pamlico estuary from Hobbie (1974); Apalachicola Bay from Estabrook (1973); Upper Mobile Bay from Bault (1972) and South Alabama Regional Planning Commission (1979, and personal communication); Barataria Bay from Barrett et al. (1978); San Francisco Bay from Smith et al. (1979); Kaneohe Bay from Smith et al. (1980).

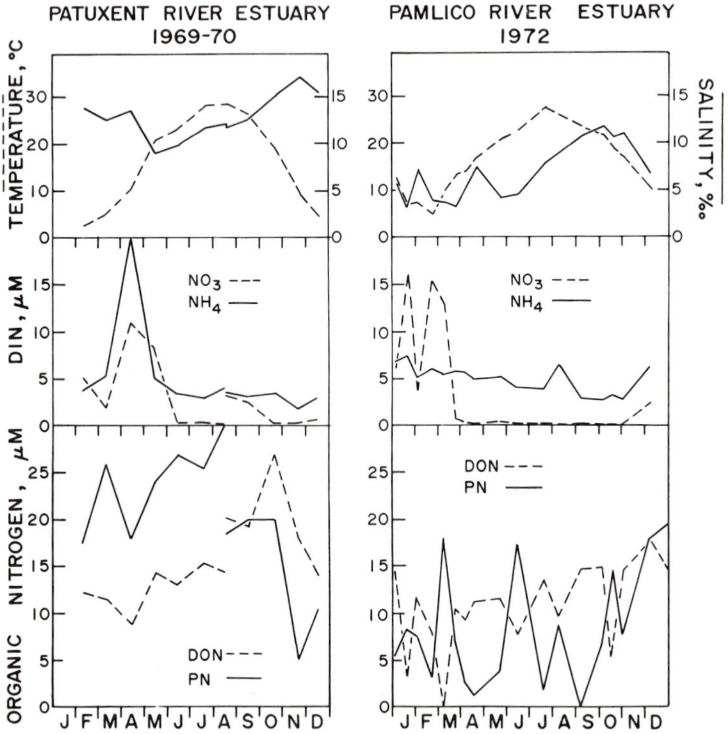

Fig. 3. Changes in the concentration of dissolved inorganic nitrogen (DIN), dissolved organic nitrogen (DON), and particulate organic nitrogen (PN) in the Patuxent River estuary, Maryland (Flemer, 1970) and the Pamlico River estuary, North Carolina (Hobbie, 1974).

ships between nitrogen inputs, local hydrography, and biological activities (Fig. 5), plots of nitrogen concentration as a function of salinity often provide more useful information. In the simplest case, expressing the concentrations of a material against a conservative substance (usually salt) provides unambiguous evidence as to whether the material of interest is being removed within the system, approximately where and to what degree it is being removed (or added), or if it is being carried passively through the estuary. The use of such mixing diagrams has become increasingly popular in studies of estuarine nutrient dynamics

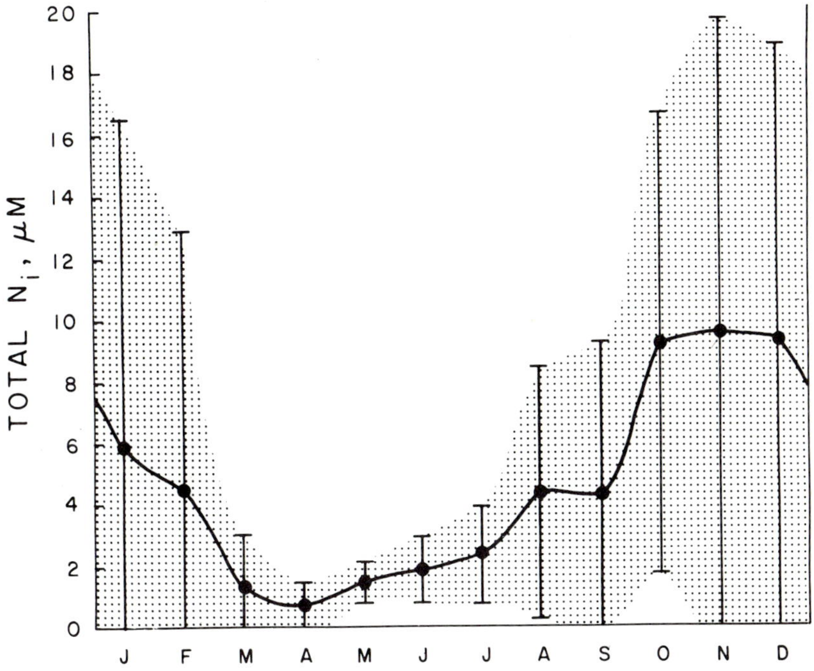

Fig. 4. Dissolved inorganic nitrogen in the lower West Passage of Narragansett Bay, Rhode Island during 1976-1981. The points are average values of the monthly means for each year calculated from weekly measurements; the bars are ±2 S.D. From Pilson (1982).

(Fanning and Pilson, 1973; Liss, 1976; Peterson, 1979; Deck, 1980; Sharp et al., 1982), but their interpretation may be complicated by variations in the fresh- and saltwater "end members" over time scales that are less than the mean residence time of water in the system (Loder and Reichard, 1981). This complication, along with complexities of water circulation, mixing, and biological patchiness, may produce a nitrogen-salinity relationship that is less revealing than might be hoped. Examples with various degrees of clarity are shown in Fig. 6, where each sample plot represents only one point in time. It is evident that, at the time of sampling, nitrate was entering the Squamscott River estuary from both freshwater and the sea, and there was a sink for nitrate in the

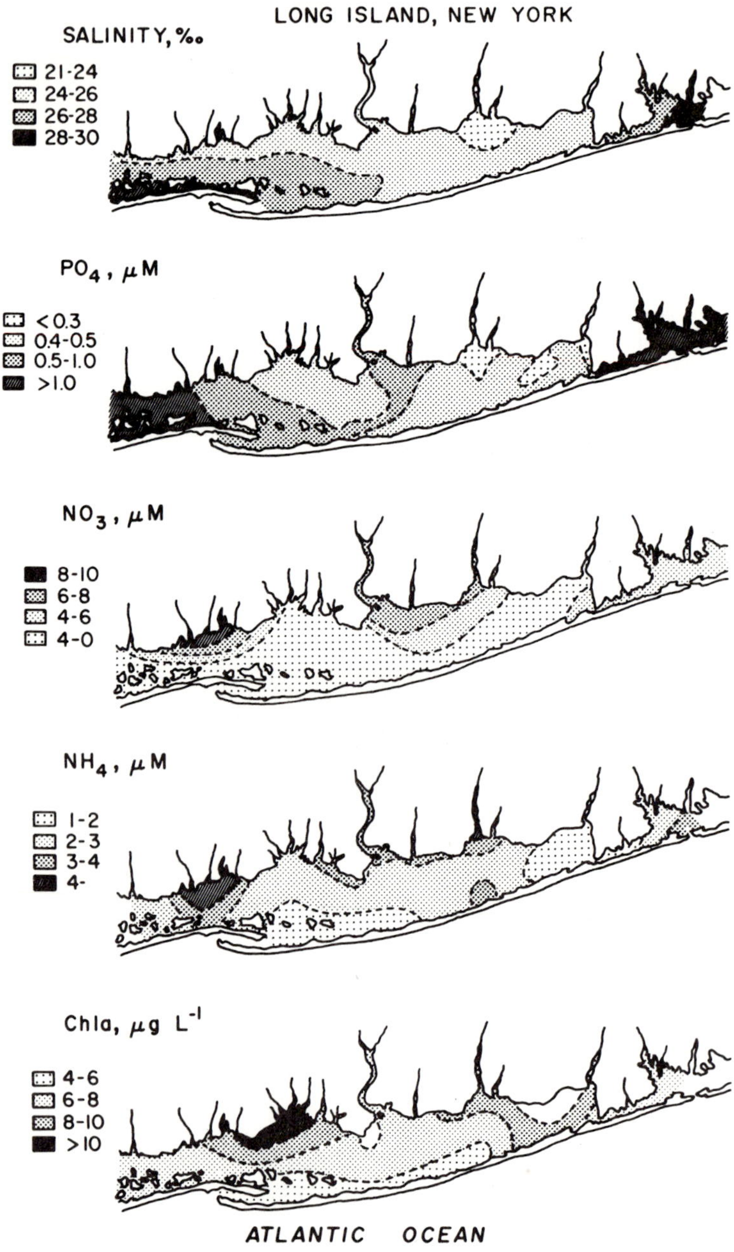

Fig. 5. Display of spatial variability in average summer values of salinity, phosphate, nitrate, ammonia, and pelagic chlorophyll *a* in Great South Bay, Long Island. From Buckner (1973).

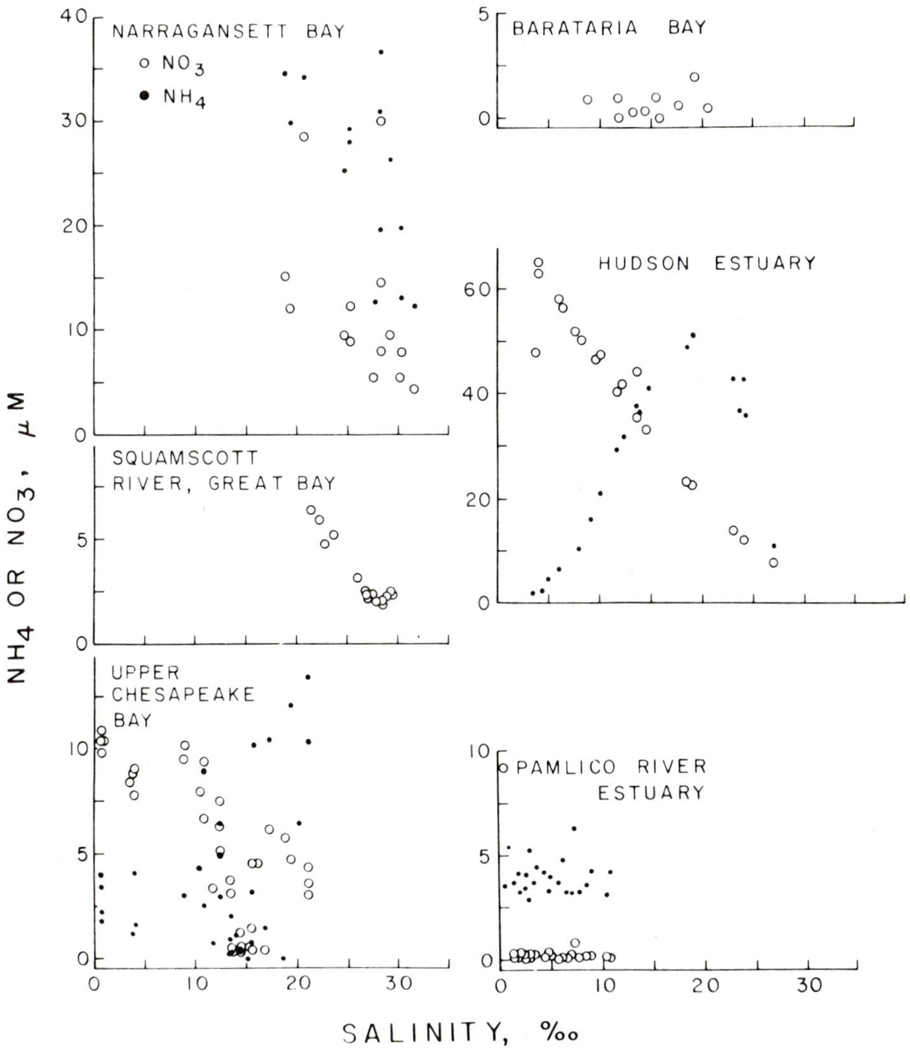

Fig. 6. The concentration of nitrate and ammonia in various estuaries during summer as a function of salinity. See Nixon (1982b) for data sources.

28-29-o/oo zone. In the case of Upper Chesapeake Bay, it appears that nitrate was being added from the Susquehanna River while ammonia was being carried up from the lower bay. The data are much more scattered, however, and other interpretations are possible. Perhaps some of the ammonia was being oxidized to produce nitrate. The situation was clearer in the Hudson estuary, where the residence time of the water is much shorter and there is a large injection of ammonia from New York City sewage that is mixed both upstream and down by the estuarine circulation. On the other hand, in a broad, shallow estuary like Barataria Bay, Louisiana, with no single, strong freshwater source, the mixing diagram contains little information. Neither is it particularly useful in high-salinity Narragansett Bay, where a mixing zone is virtually absent because the freshwater input is small relative to the area of the bay and the seawater exchange. At the other extreme, the behavior of inorganic nitrogen in the Pamlico River, a low-salinity estuary, also seems unexplainable in terms of a simple mixing relationship.

Since the position and salinity of a water mass in the estuary reflect its particular time-history in the system, it is possible in favorable circumstances to use mixing diagrams to show the sequence of processes affecting the distribution and concentration of the various forms of nitrogen. Unfortunately, there are few studies that have followed all of the nitrogen species so that a mass balance could be attempted. One of the most thorough efforts is an ongoing study of Delaware Bay (Sharp et al., 1980; Culberson et al., 1982) from which some recent results have been taken for Fig. 7. In both summer and late fall, it appears that there is a production of nitrate in the very low-salinity waters with, in the colder conditions, an accumulation of nitrite as well. Much of the very rapid ammonia removal at low salinity may be accounted for by nitrification. In his studies of this process in the Scheldt estuary, Billen (1975) found that the activities of the nitrifying bacteria were inhibited by low O_2/E_h conditions below 2 o/oo salinity, but reached a maximum near 2 o/oo and declined markedly at higher salinities. The patterns in the Delaware are not inconsistent with this picture, since oxygen levels in the upper estuary usually remain above 35% of saturation (Sharp et al.,

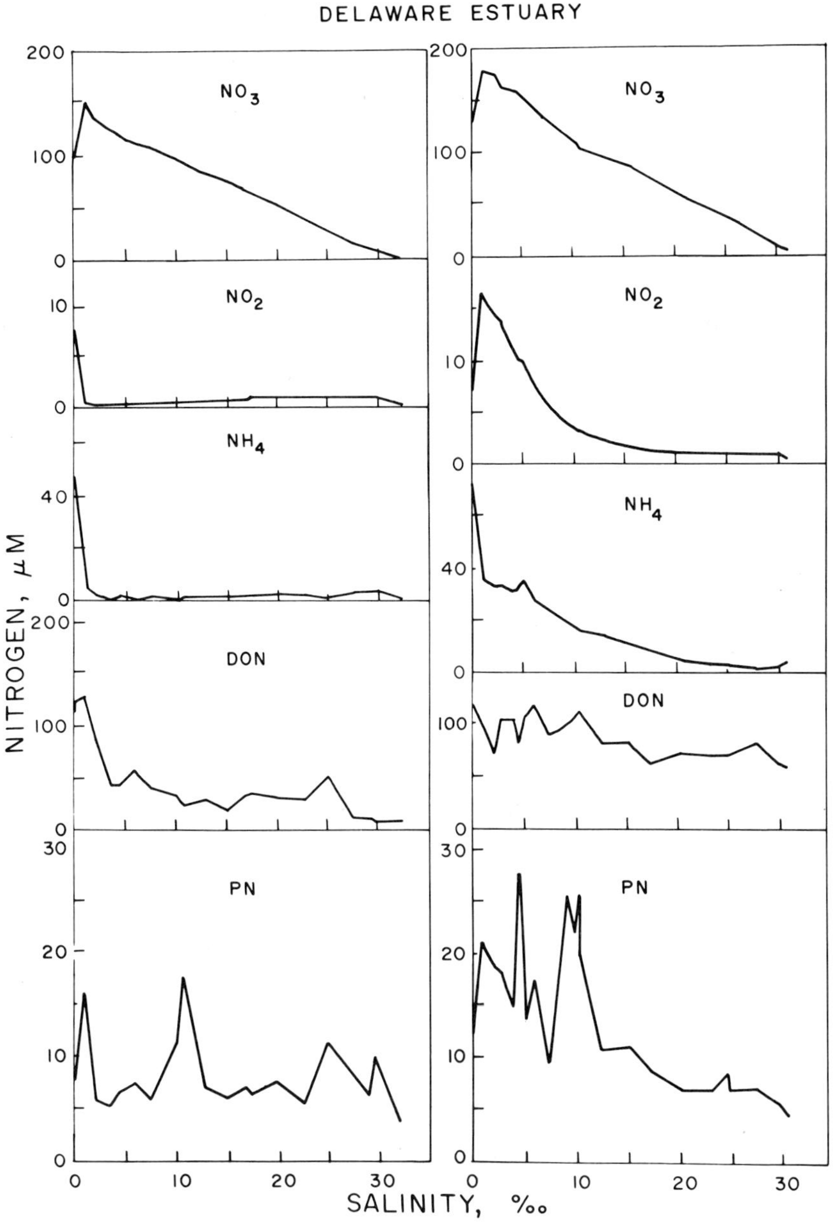

Fig. 7. The concentration of inorganic and organic fixed nitrogen as a function of salinity in Delaware Bay during summer (left panel, 29 June-1 July, 1981) and fall (right panel, 18-20 November, 1980). Data from C. H. Culberson, J. H. Sharp, T. M. Church, and B. W. Lee, personal communication.

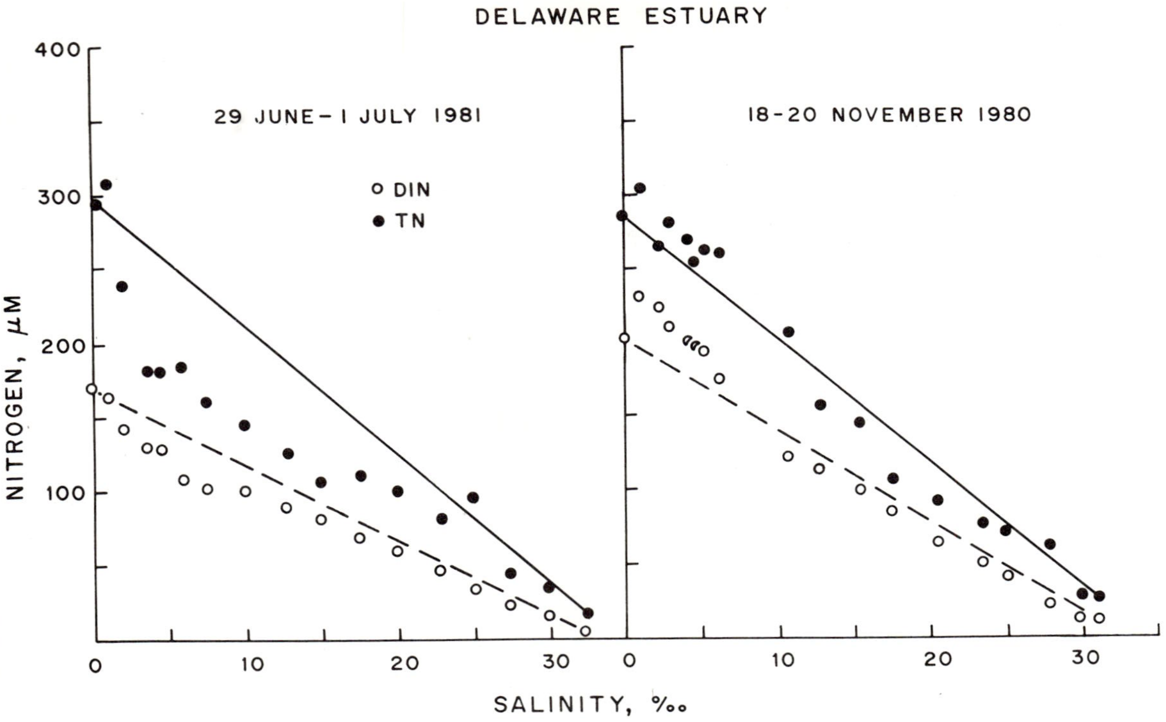

Fig. 8. The concentration of total dissolved inorganic nitrogen (DIN) and total nitrogen (TN, excluding gases) as a function of salinity in the Delaware estuary. Lines are drawn to represent conservative mixing. Data from C. H. Culberson, J. H. Sharp, T. M. Church, and B. W. Lee, personal communication.

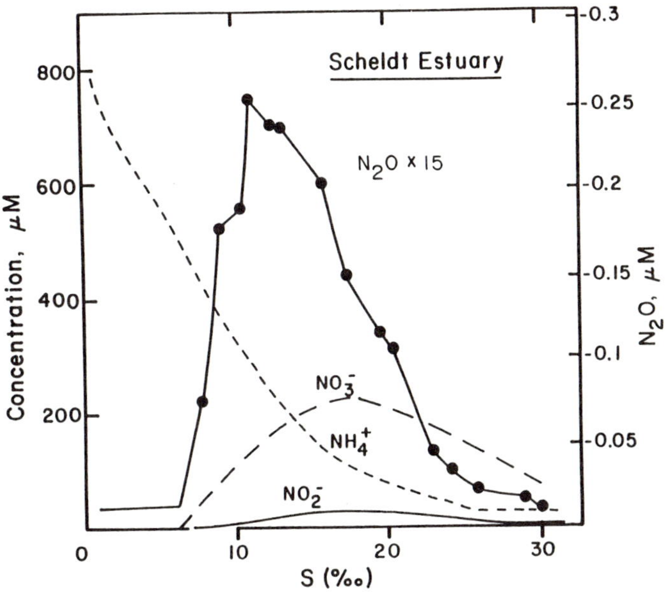

Fig. 9. The distribution of ammonia, nitrite, nitrate, and nitrous oxide as a function of salinity in the Scheldt estuary. From Deck (1980).

1982). During a series of summertime observations, Sharp et al. found that there was, if anything, a slight increase in ammonia and nitrite downbay, and nitrate was mixed conservatively above about 6 o/oo, indicating that any growth of phytoplankton in the lower estuary at that time must have been supported by remineralization of DON or by inputs from the sediment. A common assumption that the large pool of DON in estuaries represents refractory material of little ecological interest (McCarthy et al., 1975) is not supported by the behavior observed in the Delaware, at least during summer.

Since all of the forms of fixed nitrogen were measured in the Delaware study, it is possible to examine the mixing of total inorganic and total nitrogen for comparison with the individual compounds (Fig. 8). The plot of inorganic nitrogen is dominated by the behavior of nitrate, with the summer data suggesting that the estuary was retaining perhaps 20-25% of the input. It is also evident that in the summer the upper estuary was acting as a strong sink for total nitrogen. The late fall samples show a different pattern, with the bay being less of a sink for total nitrogen and possibly the upper bay being a source of inorganic nitrogen.

As a final example of spatial variability, we have included a set of measurements from the Scheldt estuary obtained by Deck (1980) in collaboration with the oceanographic laboratory at Brussels University (Fig. 9). The Scheldt data are unusual because the ammonia, nitrite, and nitrate concentrations are extremely high and because nitrous oxide was measured at the same time. The high concentrations result from a large input of urban sewage from Antwerp. This, combined with a relatively long residence time for the system, makes a number of nitrogen transformations particularly clear. As mentioned earlier, Billen (1975) showed that nitrification in the Scheldt reached a maximum at about 2 o/oo salinity, while the data shown here suggest that at the time of these measurements conditions were such that maximum rates of nitrification were taking place in the region more saline than 2 o/oo. The position of the nitrous oxide curve is consistent with the production of this gas during nitrification, although the N_2O concentration peak occurred upstream from the NO_3^- peak. A similar relationship between nitrous oxide abundance and apparent nitrification has been found in the tidal freshwater portion of the Potomac estuary (McElroy et al., 1978). It is sometimes assumed that the production of nitrous oxide is associated mainly with nitrification in the water column, but direct measurements in Narragansett Bay (Seitzinger et al., 1980) have shown a substantial production from mid-bay sediments and especially high production rates from sediments in eutrophic areas (Seitzinger et al., 1983). The resulting concentrations of nitrous oxide in the water column may be many times the saturation value, so estuaries must be included among the sources of nitrous oxide for the atmosphere.

TABLE I. Approximate Distribution of Nitrogen in the Water Column (9-m deep) and Sediments (to 10 cm) in Mid-Narragansett Bay during Spring[a]

	mmol N m^{-2}	*% of total system N*
Water column		
Dissolved inorganic	10	0.08
Dissolved organic	135	1.16
Plankton	65	0.56
Water column total	210	1.80
Sediments		
Pore water		
DON	8.0	0.07
NH_4^+	7.6	0.06
$NO_2^- + NO_3^-$	0.8	0.00
	16.4	0.14
Extractable		
NH_4^+	30	0.26
$NO_2^- + NO_3^-$	5	0.04
	35	0.30
Detrital organic	11,330	96.9
Living organic		
Bacteria	1.4	0.01
Meiofauna	36	0.31
Macroinfauna	58	0.50
	95.4	0.82
Sediment total	11,477	98.2

[a]*Modified from Garber, 1982.*

C. Nitrogen in Sediments

In spite of the attention given to nitrogen in the water column, most of the fixed nitrogen in estuarine systems is found in the sediments, even if only the top few centimeters of sediment are considered as being *in* the system (Table I). The nitrogen found there is in various forms dissolved in pore water, adsorbed on the sediment, and contained in particulate organic detritus and living organisms of various sizes. Some of these forms are only defined analytically, such as "extractable ammonia" or "total Kjeldahl nitrogen," and their biological and chemical significance remain unclear. The very large pool of detrital organic nitrogen in the sediments must include a great variety of materials of varying nutritional value for the organisms living there.

Some forms of nitrogen in the upper 10-20 cm of the sediments may vary seasonally, (e.g., exchangeable and pore water ammonia; Aller, 1980; Douglas, 1981), and it is clear that there is a significant spatial variability associated with sediment type, the input of organic matter, and other factors (Premuzic, 1980; Douglas, 1981; Figs. 10 and 11). The major interest in sedimentary nitrogen dynamics has focused on the vertical distribution of various chemical species to help in understanding the early diagenesis of organic matter in the sediment (Aller, 1980; Berner, 1980; Fig. 12). While such vertical profiles have provided insight into the chemistry of nitrogen reactions within the sediments, they have been less useful in estuarine systems for calculating the net fluxes of dissolved constituents between the sediments and the overlying water. This failure has been due to the activities of larger animals in bioturbating and irrigating the sediment (Aller, 1980), and to the fact that most of the organic matter coming to the bottom is remineralized rapidly in the upper few millimeters of sediment where it does not contribute to the conventional pore water profile (Kelly, 1982; Kemp et al., 1982).

Compared with our knowledge of the water column, we have information on the distribution and abundance of nitrogen in the sediments of only a very few estuaries. We also know little about nitrogen transformations in sediments and the factors that regulate them. However, as discussed in the following section, the study of nitrogen in estuarine sediments and bottom communities has begun to

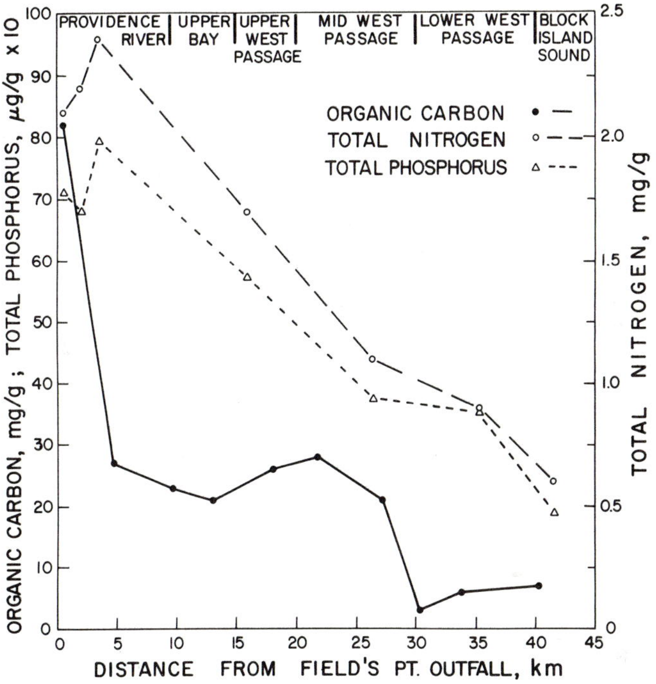

Fig. 10. Concentrations of total nitrogen, phosphorus, and organic carbon in the surface sediments of Narragansett Bay, Rhode Island as a function of distance from a large outfall at Field's Point. Data from Sheith (1974).

advance rapidly during the past 10 years, particularly in terms of the functional coupling between benthic and pelagic phases of coastal marine systems (Zeitschel, 1980; Nixon, 1981b; Kemp et al., 1982).

IV. NITROGEN DYNAMICS

> Thus there is a transmigration of the atoms of the material bodies of organisms through long series of animate and inanimate forms. But while there is an ultimate Nirvana in the transmigrations of souls, the atoms of nitrogen are bound on the wheel

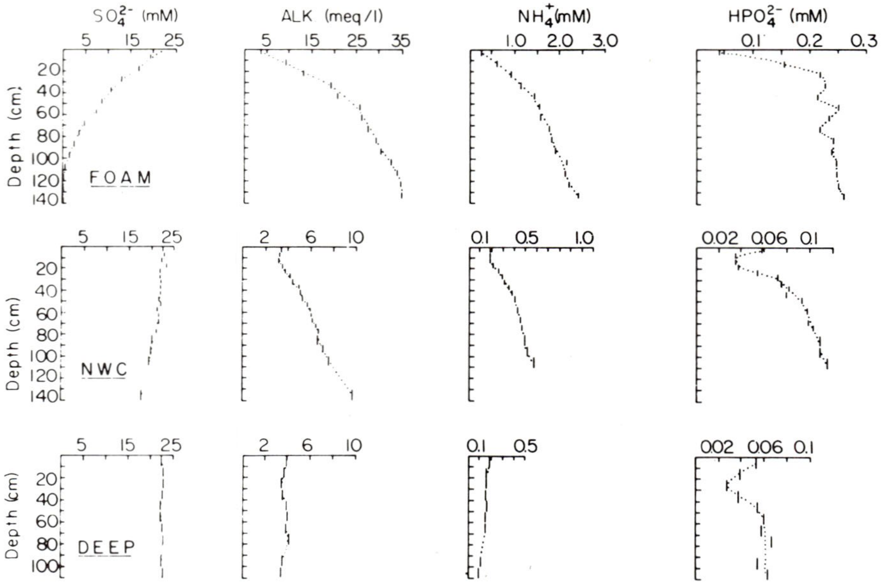

Fig. 11. Concentrations of ammonia and other constituents at various depths in the pore waters of Long Island Sound sediments in shallow (FOAM, 8 m), intermediate (NWC, 15 m), and deep (34 m) water. From Aller (1980).

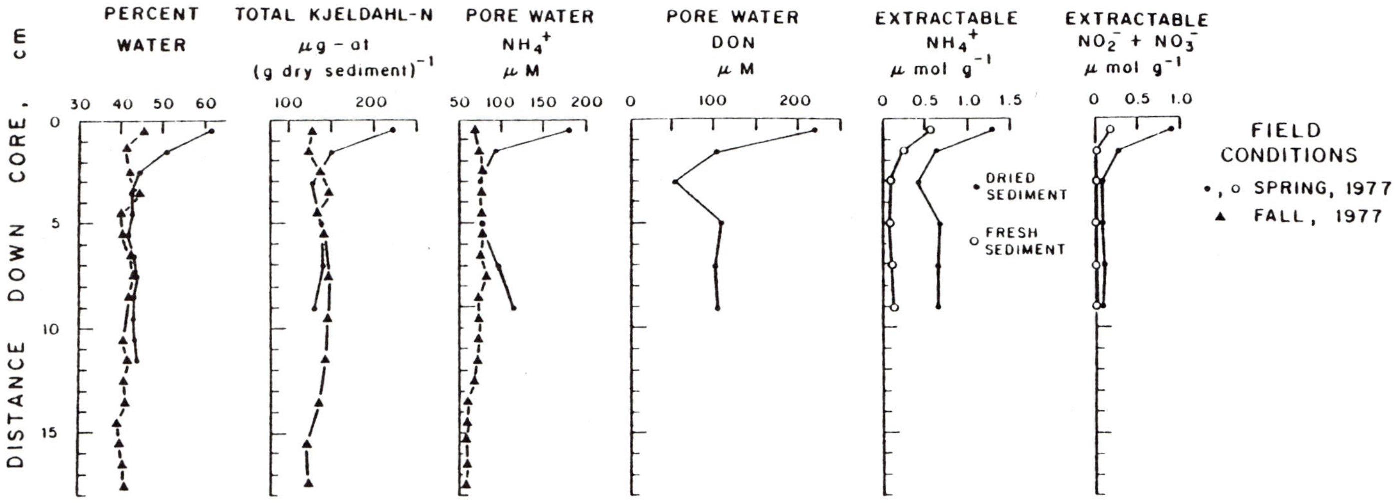

Fig. 12. Distribution of various forms of nitrogen with depth in sediments from the mid-West Passage of Narragansett Bay, Rhode Island. From Garber (1982). Field conditions: ●, o, spring, 1977; ▲, fall, 1977.

> of change, and ceaselessly pass through living and non-living phases. Imperious Caesar dead and turned to clay does not at once fill a crevice in a wall, or stop a bunghole, but may have a place in the architecture of humbler creatures. Perhaps only after many transmutations through living organisms may the nitrogen and carbon of the organism reassume the inorganic phase. But sometime or other, they return to the earth or become dispersed in the sea or air. Dante, in his vision, saw the souls of sinners impelled by a furious wind, and it may be that the material atoms, the substrata of the souls, are carried far and wide in the currents of the atmosphere.
>
> JAMES JOHNSTONE (1908)
> The Circulation of Nitrogen
> *IN* "Conditions of Life in the Sea"

It is often one of the conceits of contemporary ecology that our current perception of the dynamic cycles of materials represent a fundamental break with earlier views of a more static Nature. However, Johnstone's chapters on "Bacteria in the Sea" and "The Circulation of Nitrogen" emphasized the importance of nitrogen cycling in maintaining the productivity of the open ocean and coastal waters. The early analyses of seasonal nutrient cycles by Atkins (1925, 1930), Harvey (1926), Cooper (1933), and others recognized that they were observing only net changes in concentration, and that nitrogen regeneration and uptake were often proceeding at rates they could not measure. It is not a coincidence that studies of decomposition and remineralization in marine systems began along with the first observations showing that the concentrations of nutrients varied widely over time (for example, Waksman et al., 1933; Brand et al., 1937).

A major accomplishment of the past 25 years, however, has been the quantification of cycling rates and the demonstration that very rapid exchanges of nitrogen take place that are not seen as changes in ambient concentrations (see reviews by Pomeroy, 1974; Harrison, 1980; Nixon, 1981b).

A. Processes in the Water Column

Early attempts to measure phytoplankton production in a way more revealing than simply following changes in standing crop relied on detecting changes in dissolved oxygen using the light and dark bottle technique described by Gaarder and Gran (1927). The results gave net consumption and production of organic matter by the plankton community (phytoplankton, smaller zooplankton, bacteria) over a time scale of hours instead of days or weeks. It was not until the introduction of the sensitive ^{14}C-uptake method (Steeman-Nielson, 1952), however, that it became possible to measure directly the very rapid formation of organic matter by the phytoplankton themselves. The emphasis on increasingly shorter measurements of carbon fixation allowed by the ^{14}C technique soon revealed that there was little, if any, correlation between the rate of carbon assimilation and the ambient concentration of dissolved inorganic nitrogen during the incubation (McCarthy et al., 1975; Furnas et al., 1976; Fig. 13).

This apparent paradox was only in part resolved by culture studies showing that it was possible for phytoplankton to grow for some time by using internal storages of nitrogen with a corresponding increase in their carbon-to-nitrogen ratio (Droop, 1973). Recent work, however, has shown that changes in the C:N ratio of nitrogen-limited phytoplankton are more a function of the carbon content per cell than of the nitrogen (McCarthy, 1980). Moreover, the fact that high rates of ^{14}C uptake can be found repeatedly over many weeks while dissolved ammonia and nitrate levels remain very low, suggested that other factors must be operating to sustain the production.

The present view emphasizes the importance of very rapid rates of nitrogen release and uptake in the water column that produce a tight coupling of heterotrophic and autotrophic processes with often virtually no change in free dissolved nitrogen. Quantitative support for this scheme began with the publication of Harris's (1959) measurements of ammonia release by net zooplankton in Long

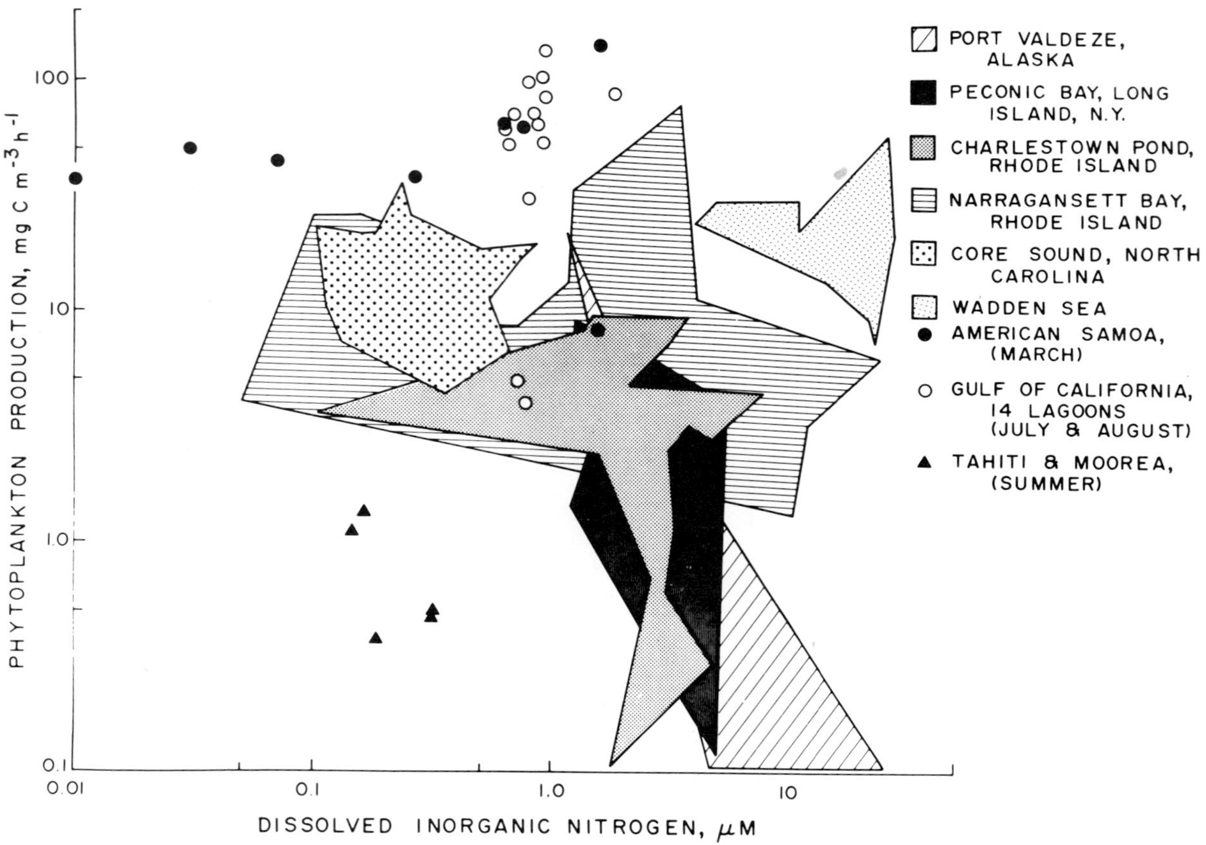
PORT VALDEZE, ALASKA
PECONIC BAY, LONG ISLAND, N.Y.
CHARLESTOWN POND, RHODE ISLAND
NARRAGANSETT BAY, RHODE ISLAND
CORE SOUND, NORTH CAROLINA
WADDEN SEA
AMERICAN SAMOA, (MARCH)
GULF OF CALIFORNIA, 14 LAGOONS (JULY & AUGUST)
TAHITI & MOOREA, (SUMMER)
PHYTOPLANKTON PRODUCTION, mg C m-3 h-1
100
10
1.0
0.1
0.01
0.1
1.0
10
DISSOLVED INORGANIC NITROGEN, μM

Fig. 13

Island Sound. Similar studies by, for example, Martin (1968), Smith (1978), and Vargo (1979) on mesoplankton, by Kremer (1975) on macrozooplankton, and especially by Johannes (1969) on microzooplankton, documented the rapid rates of ammonia excretion compared with the relatively slow remineralization rates found in earlier decomposition experiments (reviewed by Garber, 1982). At the same time, work by many phytoplankton physiologists and ecologists showed that ammonia and urea (both excretion products of zooplankton) were taken up preferentially over nitrate (Fig. 14), and that the uptake of the oxidized forms was inhibited by relatively low ($\sim$1 μM) concentrations of ammonia (Fig. 15). The consequences of this behavior are often evident in the mixing diagrams for various estuaries (e.g., Fig. 7). More recently, experiments using ^{15}N isotope dilution techniques have provided additional evidence that ammonia is produced rapidly in the water column (Harrison, 1978; Caperon et al., 1979; Glibert, 1982). The capacity for phytoplankton to take up this ammonia at very low concentrations and at a very rapid rate has been established only recently through the use of ^{15}N-tracer studies and very precise chemostat experiments (McCarthy and Goldman, 1979; Glibert and Goldman, 1981; Glibert, 1982).

The importance of pelagic nitrogen cycling is not limited to periods when there is virtually no inorganic nitrogen in the water. At the rapid rates of growth often found during summer, the relatively low concentrations of nitrogen seen in many estuaries would be depleted quickly if there were no regeneration. However, it is also evident that the interaction between plankton production and nitrogen must involve more than a rapid cycle. A preoccupation with short-term rate measurements must be balanced by consideration

Fig. 13. Phytoplankton primary production (^{14}C uptake) and the concentrations of dissolved inorganic nitrogen at the time of measurement in a variety of estuaries and coastal lagoons. Polygons describe an annual cycle of measurements. Data sources in Nixon (1983a).

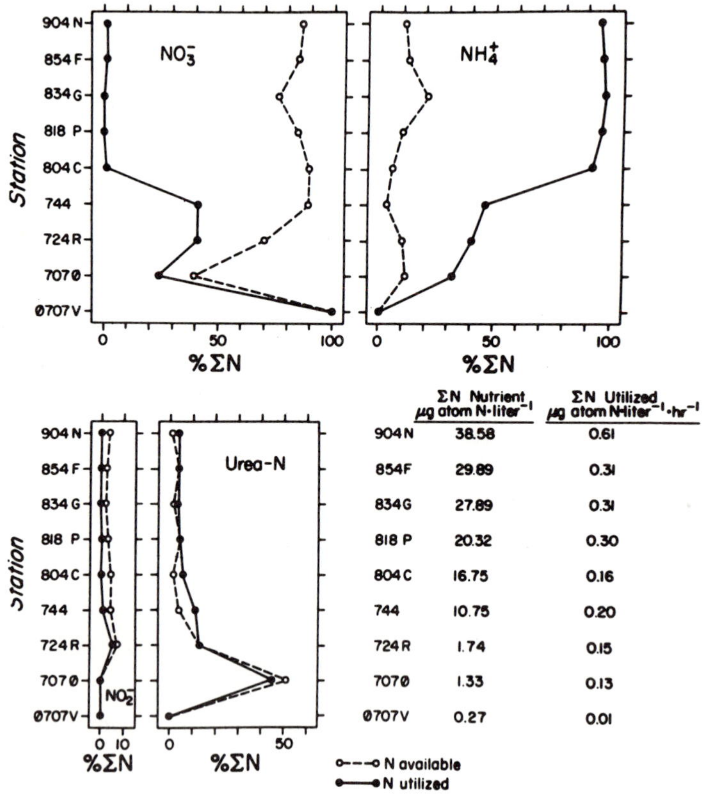

	ΣN Nutrient, µg atom N·liter⁻¹	ΣN Utilized µg atom N·liter⁻¹·hr⁻¹
904 N	38.58	0.61
854 F	29.89	0.31
834 G	27.89	0.31
818 P	20.32	0.30
804 C	16.75	0.16
744	10.75	0.20
724 R	1.74	0.15
7070	1.33	0.13
0707 V	0.27	0.01

Fig. 14. Comparison of the concentration and utilization of different forms of nitrogen in the open waters of Chesapeake Bay during summer. Station 0707V was near the mouth of the bay. From McCarthy et al. (1975).

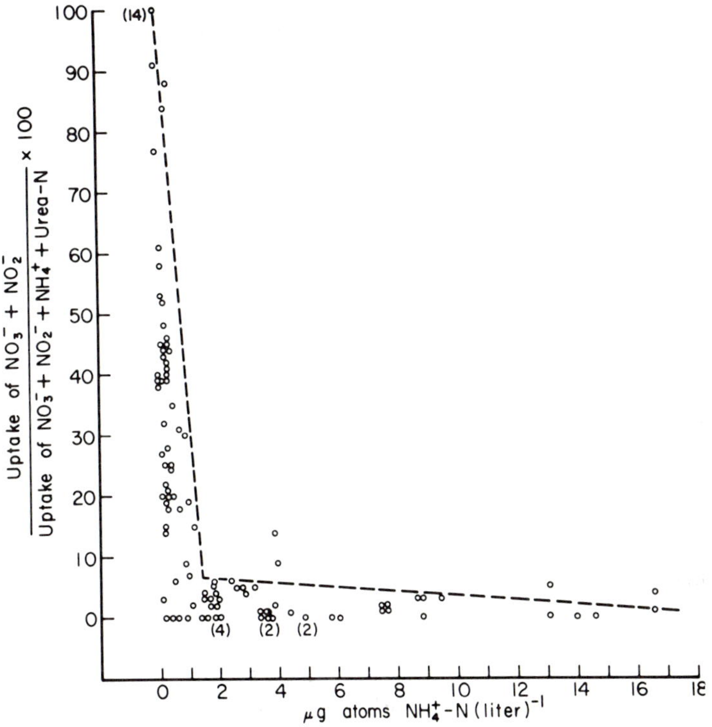

Fig. 15. Effect of the ambient ammonia concentration on the uptake of nitrate and nitrite by phytoplankton in the open waters of Chesapeake Bay over an annual cycle. From McCarthy et al. (1975).

of the changes in standing crop that require longer-term changes in the total mass of nitrogen. We still lack a good mass balance of all the nitrogen forms during a period when the standing crop of plankton is in steady state and then begins to bloom. The relationships between cycling rates, the sizes of various nitrogen pools, and changes in the sizes of those pools remain a challenge.

There are, of course, a number of other processes besides uptake and regeneration that also influence the cycling and storage of nitrogen in the water column (Webb, 1981). The inputs of nitrogen from rivers, runoff, rain, and sewage, and exchanges with sediments and offshore waters are addressed in the following section. Whereas nitrogen fixation by blue-green algae and bacteria is important in some tidal freshwaters and in estuarine subsystems such as salt marshes, we are not aware of any measurements that have shown this process to be generally of any consequence in estuarine or coastal marine waters. Similarly, denitrification in the water column has not been reported as significant in coastal waters except for potential activity at the pycnocline of some strongly stratified estuaries. Density boundaries have also been associated with active nitrification in several estuaries (Indebo et al., 1979; Webb and D'Elia, 1980; Kemp et al., 1982).

As noted in the introduction to this chapter, the process of nitrification has long been of interest in estuaries because of its influence on dissolved oxygen levels. It has also received attention recently because of the production of nitrous oxide (N_2O) during nitrification and the role of this gas in the destruction of stratospheric ozone (Crutzen, 1970; Hahn and Crutzen, 1982) and in the radiative heat budget of the atmosphere (Wang et al., 1976). Whereas nitrification may take place at the sediment surface or in the water column, the latter appears to be much more important in most estuaries. In our earlier discussion of mixing diagrams (Fig. 9), it was noted that ammonia oxidation in the Scheldt appeared to be most active at low salinity, but the complex interactions of factors regulating this nitrogen transformation, including the concentrations of nitrogen, oxygen, and salt, remain to be resolved (Deck, 1980). In their studies of the Potomac River estuary, Elkins et al. (1981) observed that nitrification was inversely correlated with freshwater input. They found a spatial distribution of nitrogen species that appears to provide striking evidence of the importance of ammonia oxidation and the associated production of nitrous oxide (Fig. 16). It is also evident from Fig. 16 that the yield of nitrite and nitrous oxide was low during nitrification, at least under the conditions observed in the Potomac. The low concentrations of nitrite commonly found

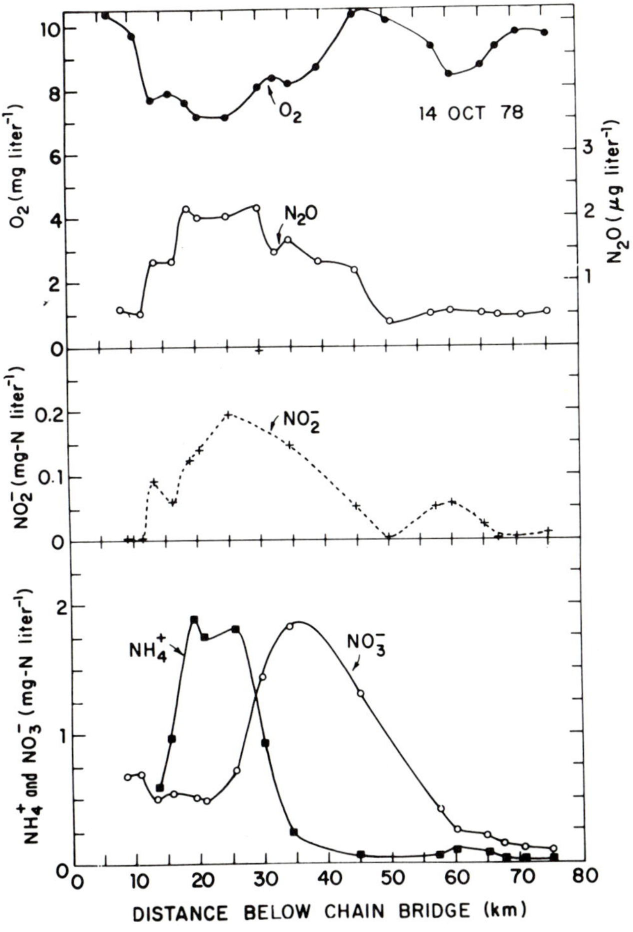

Fig. 16. The distribution of dissolved oxygen and nitrogen along the Potomac estuary on October 14, 1978, at about 18°C. From Elkins et al. (1981).

in estuaries suggest that this behavior is not unique to the Potomac, though the yield of the more reduced forms may increase under low-oxygen conditions.

B. Processes in the Sediments

While all of the nitrogen cycle processes found in the water column may also take place on or within the bottom sediments, the uptake of ammonia and/or other nitrogen forms by plants on the bottom is only of importance in intertidal areas or in waters shallow enough that light is available for photosynthesis. In temperate coastal waters, most of the area of subtidal bottom communities is essentially heterotrophic. As a result, transformations involved in the excretion and remineralization of nitrogen by animals and bacteria dominate nitrogen cycling in most sediments. The presence of low oxygen and anoxic conditions within sediments has also focused attention on denitrification and other aspects of anaerobic metabolism.

Some early efforts were made to study organic decomposition in marine sediments (Waksman and Hotchkiss, 1938; Anderson, 1939), but with the exception of the extensive work in Japanese estuaries by Okuda (1955, 1960) and Kato (1956), little real progress was made in measuring benthic nitrogen cycles until the past 10 years. Much of the Japanese work was far ahead of its time and it is unfortunate that it did not have a wider recognition and influence in marine ecology. Recent ecological studies have focused on direct measurements of the exchanges of nitrogen between sediments and the overlying water (see reviews by Nixon, 1981b; Kemp et al., 1982), whereas those with a geochemical or microbiological interest have tended to assess the nature and rates of transformations within the sediments using sediment sections and ^{15}N tracer or other techniques (Billen, 1978; Sorensen, 1978) or vertical distribution profiles and numerical modeling (Aller, 1980; Berner, 1975; Billen, 1978, 1982).

While it now seems clear that a very large part of benthic nutrient regeneration must take place at the sediment surface, most attention has been given to processes within the body of the sediment. Oxidation, reduction, mineralization, adsorption, desorption, advection, diffusion, and sedimentation are all of interest to those who study sediment diagenesis and these processes appear especially amenable to numerical modeling. Such models have reached a considerable sophistication (e.g., Billen, 1982), but in this chapter we cannot do full justice to them. They are particularly important in bringing the physics of the situation to bear in constraining our understanding of the processes involved and in showing how sensitively the various rates may be affected by changing parameters. The large number of adjustable parameters in current numerical models nevertheless compel us to rely in most situations on measurement if we wish to obtain rates or concentrations as they are, and we have tended to focus on the development of such measurements.

C. Decomposition Rates in the Sediments

The processes of resuspension, active filter feeding, and bioturbation make it difficult, if not impossible, to measure directly the input of organic matter to the bottom of shallow marine areas. The large accumulations of carbon and nitrogen commonly observed in sediments (Fig. 10, Table I) have lead many to suppose that remineralization must be slow. The relatively constant concentrations of particulate organic carbon and nitrogen with depth in the sediment (Fig. 12) support this conclusion, as do calculations based on models of diffusion and the observed gradients of dissolved metabolic products in pore waters (Fig. 11). Experimental measurements of the rate of decomposition and ammonia production at various depths in coastal marine sediments have confirmed that the remineralization rate per unit of organic nitrogen is perhaps 100 times slower within the sediments than in the water column (Table II), and that the rate of ammonia production declines with depth in the near-surface sediment (Fig. 17). The rate of remineralization within the sediment is slowed by the lack of oxygen below the first few millimeters or centimeters and possibly by a decline in the nutritional quality of the organic matter with depth and age in the sediment. In considering this, Marshall (1972) once described the

TABLE II. Approximate Weight-Specific Nitrogen Regeneration Rates for the Water Column and Sediments in Some Coastal Marine Systems

	N Remineralized (% day^{-1})	
Pelagic remineralization		
^{15}N Isotope dilution techniques		
Southern California Bight (Harrison, 1978)[a]	1.5-8.8	
Kaneohe Bay, Hawaii (Caperon et al., 1979)[a]	15-140	
Patricia Bay, British Columbia (Harrison, 1978)[a]	20-30	
Patricia Bay (Hattori et al., 1980)[a]	20-50	
Chesapeake Bay (Glibert, 1983)[b]	17-87	
Decomposition experiments		
Mixed plankton (various authors)[c]	1.7-11.4	
Diatoms (Kamatani, 1969)[c]	2.8-5.1	
Green algae (various authors)[c]	1-5	
Mixed plankton (Garber, 1982)[d]	4.2-25.9	(initial)
	0.6	(final)
Skeletonema costatum (Garber, 1982)[d]	15-24	(initial)
	1.2-2	(final)
Benthic remineralization		
Freshly deposited material (Kelly, 1982)	7.2	
Surface 1 cm of sediment, Narragansett Bay (Kelly, 1982)	0.2-0.7	
Surface 1 cm of sediment, Limfjord, Denmark (Blackburn, 1979)[e]	0.3	
Surface 1 cm of sediment, shallow Long Island Sound (Goldhaber et al., 1977; Rosenfeld, 1977)[e]	0.1	
Surface 1 cm of sediment, Long Island Sound (Aller and Yingst, 1980; Rosenfeld, 1977)[e]	0.2	
Below 1 cm, all sites given above[e]	0.01-0.2	

[a] *Summarized in this form by Garber (1982).*

[b] *Extrapolated from short-term incubations.*

[c] *Summarized by Harrison (1980).*

[d] *Garber showed two distinct phases of remineralization, the first was much more rapid and accounted for about 60% of the total decomposition.*

[e] *Recalculated and expressed in this form by Kelly (1982), see Fig. 17.*

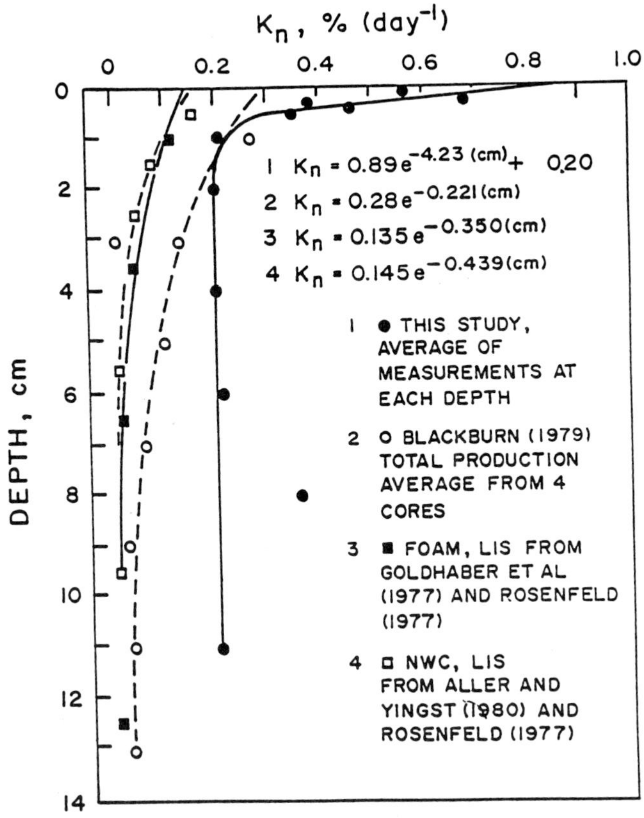

Fig. 17. Remineralization of nitrogen at different depths in the sediment from Narragansett Bay (solid circles), the west coast of Norway (open circles), and two locations in Long Island Sound (squares). From Kelly (1982).

organic matter of shallow water sediments as representing the "ashes of the fire." The question still needs further examination, however. Kelly (1982) exposed slices of silt-clay sediment from various depths to oxygenated water and measured the inorganic nitrogen produced. The results (Fig. 18) suggest that, with the exception of the surface few millimeters containing the most recently deposited material, the near-surface sediments contained organic matter with a rather uniform potential for remineralization. Such a finding is perhaps not surprising, considering the extensive

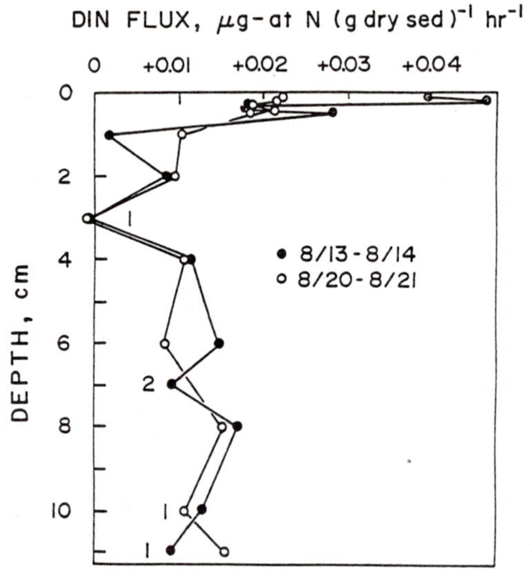

Fig. 18. Production of dissolved inorganic nitrogen by sediment samples taken from different depths in the sediment near mid-Narragansett Bay, Rhode Island and incubated in oxygenated water. From Kelly (1982).

reworking and mixing of these sediments by infauna (Rhodes et al., 1977; Aller, 1980).

Kelly's great care in separating the very surface layers of his sediment cores into approximately 1-mm thick slices, rather than the 1- to 2-cm slices commonly used in sectioning studies, made it possible to show that the very surface material could be remineralized at a rate two to four times greater than material buried only 1-2 cm below the sediment surface. In fact, the consumption of oxygen and release of nitrogen by a surface layer less than 1-cm thick was equal to the net exchange for the entire intact core before it was sectioned. In nature, of course, diffusion and bioadvective flux of deeper pore waters must also contribute to the exchange of oxygen and nutrients at the sediment-water interface. While it is obvious that the results of experiments with core slices cannot simply be summed to give the behavior of intact sediments, they do suggest the importance of processes at the sediment-water interface. The importance of rapid metabolism at this interface is also discussed by Kemp et al. (1982).

The rapid remineralization of surface organic matter was also shown by Kelly's (1982) additional measurements on the decomposition of freshly deposited material collected from a large experimental microcosm tank (Table II) and by experiments carried out by Garber (1982) using intact cores and additions of a ^{15}N-labeled organic matter on the sediment surface. Results of the tracer study suggested that the half-life of organic nitrogen might be 1-2 months in the spring (8°C) and 1-2 weeks in the fall (16°C). Only 5-10% of the labeled organic nitrogen was carried below the surface, although this fraction might be greater in the field where larger infauna would be active. The animals also contribute to benthic decomposition (over the depth range in which they live) by both their own metabolism and by enhancing the transport of particulate and dissolved material (including oxygen) through the sediment.

D. Sediment-Water Column Exchanges

Even though the weight-specific rate of nitrogen remineralization within (or on) the sediments appears much slower than in the pelagic system, the very large amount of material on the bottom makes the exchange of nitrogen between sediments and the overlying water a very important part of estuarine nitrogen cycling. There have now been a sufficient number of measurements of the oxygen consumption of the bottom over an annual cycle in a wide enough variety of estuarine and coastal marine systems to establish that some 20-50% of the organic matter entering these systems each year is respired by the benthos (Fig. 19). In most areas only a small fraction is buried, so that the remaining 50-80% is consumed in the water column or exported. Most of the nitrogen associated with the organic matter reaching the bottom is returned relatively rapidly to the overlying water in the form of ammonia, though the rate of this process is markedly dependent on temperature (Nixon

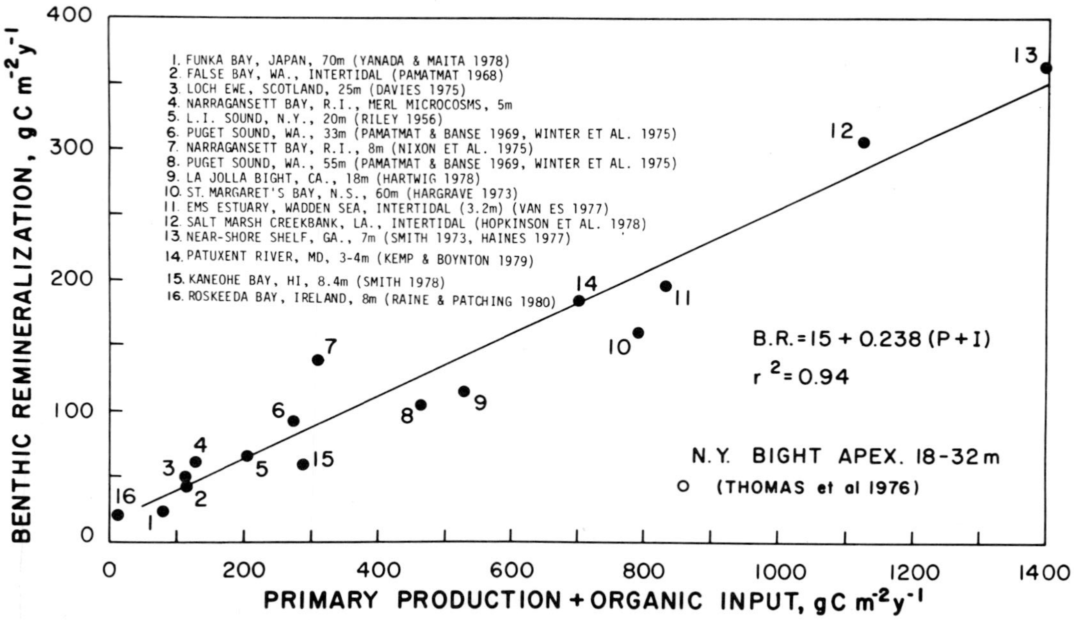

Fig. 19. The relationship between the amount of organic matter fixed and imported in the water column and the amount of organic matter metabolized on the bottom over an annual cycle in a variety of coastal marine systems. Assumptions and data sources in Nixon (1981, 1983).

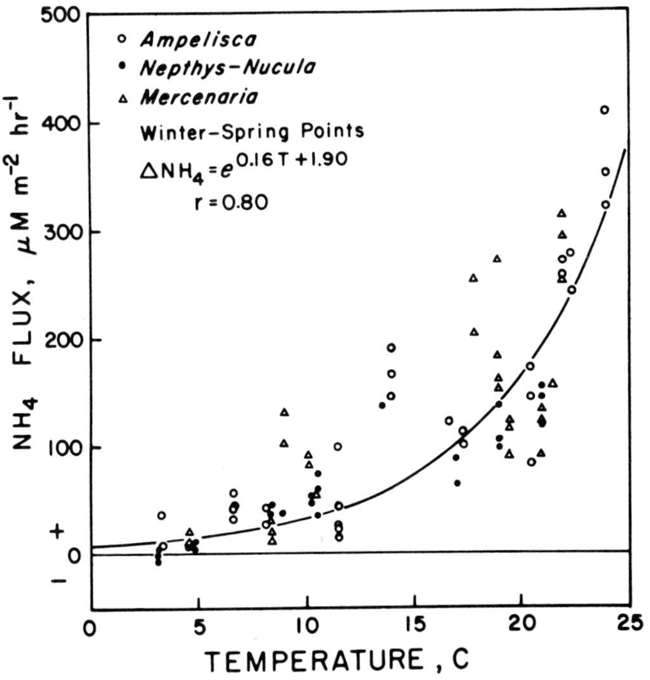

Fig. 20. Flux of ammonia across the sediment-water interface at three locations in Narragansett Bay, Rhode Island as a function of bottom-water temperature. The three stations were dominated by the different macroinfauna shown, and measurements were made over an annual cycle. From Nixon et al. (1976).

et al., 1976; Boynton et al., 1980; Aller, 1980; Fig. 20). The ^{15}N tracer study by Garber (1982) was cited earlier as showing a half-life for freshly added organic matter on the sediment surface of only 1-2 weeks at 16°C. It is also apparent that the flux of ammonia from the sediments to the water responds rapidly to variations in the input of organic matter. For example, cores collected from Narragansett Bay after the winter phytoplankton bloom showed higher ammonia release rates than those taken just before the bloom (Nixon et al., 1980), and in a series of organic input experiments using benthic microcosms, Kelly (1982) found a marked increase in ammonia flux that was proportional to the input.

While measurements of oxygen uptake by estuarine sediments were made in a variety of areas beginning after the mid-1960s (e.g., Carey, 1967; Pamatmat, 1968; Review by Zeitzchel, 1980), the first direct nutrient flux measurements from the bottom over an annual cycle were not reported until some 10 years later (Nixon et al., 1976). In analyzing the data from many measurements at three different stations over an annual cycle in Narragansett Bay, it became evident that whereas the flux of ammonia from the sediments was impressive, it amounted to only about half of what might be expected from the phosphate released and the oxygen consumed. That so much nitrogen appeared to be missing through a whole annual cycle provided compelling evidence that this was a significant feature of benthic nitrogen cycling, rather than a transient response to inputs or differences in the rate of nitrogen decomposition relative to carbon and phosphorus. Measurements in numerous coastal marine systems have since confirmed that the stoichiometry of benthic nutrient regeneration often differs from that in the water column (Nixon, 1981b).

The "missing" nitrogen might be removed from organic matter before it reaches the bottom or it might be lost from the sediment in some form other than ammonia. With respect to the first possibility, it is generally thought that phosphorus is released from dead plankton more rapidly than nitrogen, although data necessary to test this proposition convincingly have not yet appeared (Garber, 1982). There are also very few data on the exchange of dissolved organic nitrogen at the sediment-water interface, but measurements using cores from Narragansett Bay showed an upward flux amounting to only about 10% of the total fixed nitrogen flux (Nixon, 1981b). The missing nitrogen does not appear as nitrite or nitrate. While there may be net fluxes of nitrate from the sediments at some times in some estuaries (Phoel et al., 1981), these fluxes are usually relatively small and erratic (Fig. 21), and in some estuaries there may be a large uptake of nitrate by the sediments (Fig. 22). While a significant portion of the

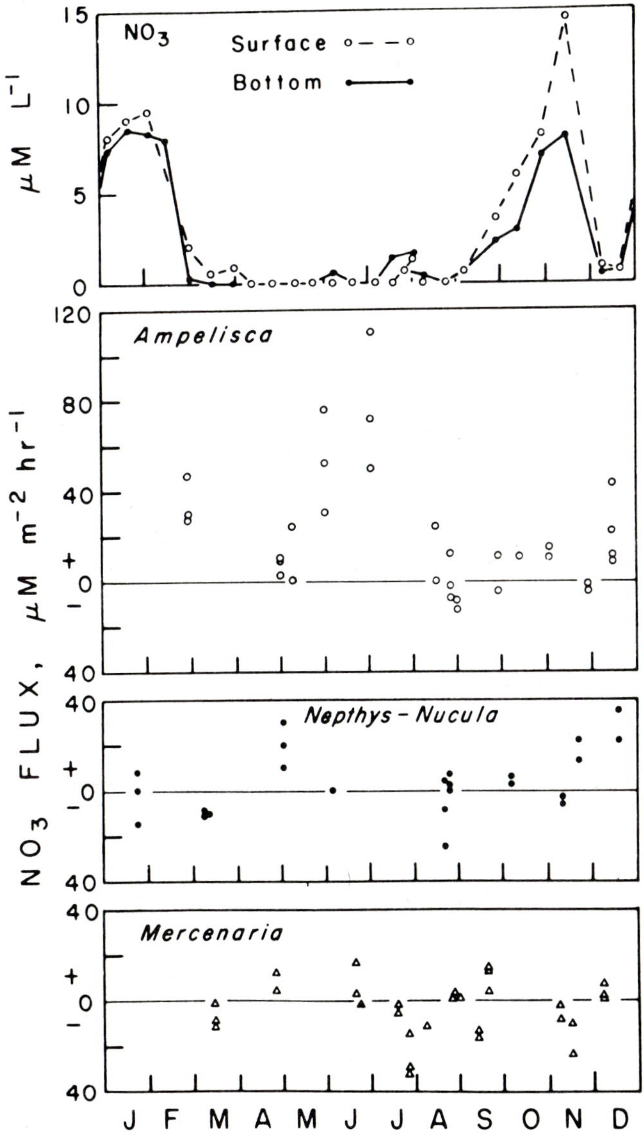

Fig. 21. Exchange of nitrate across the sediment-water interface at three locations in Narragansett Bay, Rhode Island over an annual cycle and the concentration of nitrate in the overlying water. Negative fluxes represent uptake by the sediments. The three locations were dominated by different species of macroinfauna as noted. From Nixon et al. (1976).

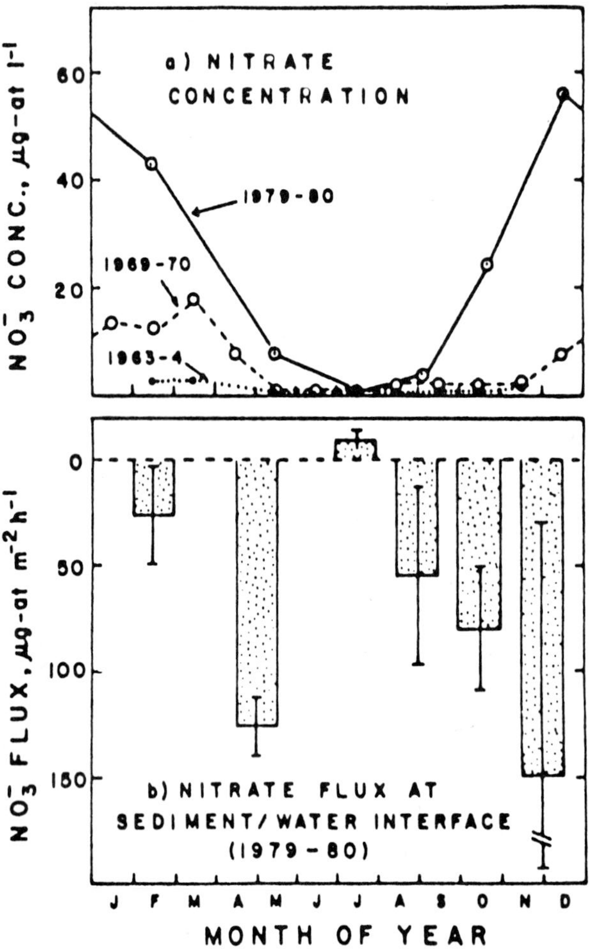

Fig. 22. Exchange of nitrate across the sediment-water interface and the concentration of nitrate in the overlying water during 3 different years in the upper Patuxent River estuary, Maryland. Note the consistently large uptake by the sediment and the high water column concentrations compared with those shown in Fig. 21. From Kemp et al. (1982).

nitrate uptake may be due to bacterial reduction of nitrate to ammonia (Sorensen, 1978), it may also be due to denitrification in the sediments.

The importance of denitrification in the sea was emphasized as early as the turn of the century (Brandt, 1899; Johnstone, 1908)

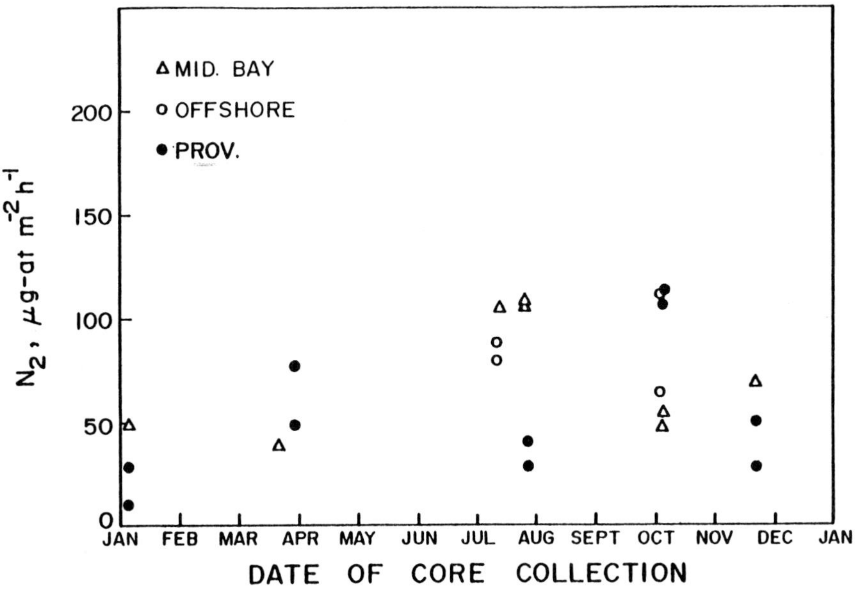

Fig. 23. The flux of nitrogen gas as dissolved N_2 from intact unamended sediment cores collected from mid-Narragansett Bay, Rhode Island, from the highly eutrophic upper bay near the city of Providence, and from the mouth of the bay in Rhode Island Sound. From Seitzinger (1982).

and a number of recent experiments using nitrate enrichment have indicated the great potential for this process in coastal marine sediments (Billen, 1978; Koike and Hattori, 1978; Oren and Blackburn, 1979). It was not until 1980, however, that the first direct measurements of the production of nitrogen gas by a subtidal marine sediment were reported by Seitzinger and co-workers using unamended, intact cores from Narragansett Bay. Those measurements have now been carried out over an annual cycle at three stations in the Bay (Fig. 23), and the results show that approximately 35% of the organic nitrogen remineralized in the sediments passes through the denitrification pathway (Seitzinger, 1982). This value is compatible with the fixed inorganic nitrogen deficit described earlier and, in combination with the small dissolved organic flux, provides a reasonable nitrogen balance for the sediments of the Bay.

Combined with the observation that a large fraction of the organic matter in shallow systems is decomposed on the bottom (Fig. 19), denitrification in the sediments may also help to explain the low N:P ratio of coastal marine waters (Nixon, 1981b). Published estimates of denitrification from a number of other estuarine and freshwater systems suggest the widespread importance of this process in sediment nitrogen cycling (Kemp et al., 1982). It should also be pointed out that the nitrate levels in Narragansett Bay are much lower than in the Patuxent River estuary (Fig. 22), and the high rates of denitrification observed by Seitzinger (1982) must have been supported by nitrification within the sediments rather than by uptake from the overlying water. In fact, nitrification commonly proceeded at a rate sufficient to maintain both denitrification and a nitrate flux of 10-30 $mmol \cdot m^{-2} \cdot h^{-1}$ from the sediments.

There was also a flux of nitrous oxide from the sediments that was correlated with the production of nitrogen gas (though it amounted to less than 10% of the N_2 flux). While nitrous oxide production in the sediments was not a significant term in the total nitrogen budget of Narragansett Bay, it was a major source of this gas in the estuary, and it increased markedly in more eutrophic areas (Seitzinger et al., 1983). Increasing nitrogen inputs to coastal waters (see Fig. 30) may correspondingly increase the importance of estuarine and nearshore sediments in the global nitrous oxide balance.

V. NITROGEN BUDGETS

In his early chapter on the nitrogen cycle, Johnstone (1908) spent some effort trying to quantify the input of nitrogen to the coastal waters of the North Sea so that he could compare it with his estimate of the amount removed with the annual catch of fish. Similar efforts have continued, and while it is easily shown that fisheries do not account for a large flow of nitrogen, it is much harder to develop a detailed and well-constrained budget or mass balance of nitrogen for an estuarine system. The value of such an accounting lies in the perspective it provides for viewing the relative importance of individual nitrogen sources, sinks, and pathways and in the fixed constraint demanded by the conservation of mass. If a balance cannot be obtained, it is immediately evident that there is a flaw in our understanding of the system or in our measurements.

The difficulty of constructing estuarine nitrogen budgets is shown by the fact that very few have been published. As pointed out in the previous sections, many nitrogen cycle processes have only been quantified in the past few years, and most of those rate measurements have been obtained in only a few estuaries. Measurements of the input or loading of nitrogen have become more common, but even they are a difficult and often tedious undertaking. Many of the estimates that have been reported do not include all the forms of nitrogen or do not clearly specify which forms are included (Jaworski, 1981). Inputs via groundwater flow and stormwater runoff are particularly difficult to measure. The loss of nitrogen due to burial in the sediments is probably relatively small in most estuaries, but its measurement is complicated by all of the problems involved in obtaining a reliable sedimentation rate (Santschi, 1980). A much larger flux, the exchange of nitrogen between the estuary and adjacent coastal waters, is seldom estimated and almost never measured because of the sampling problems involved and the common lack of sufficient information on circulation and mixing processes.

In comparing the magnitude and potential importance of nitrogen inputs from freshwater discharge and anthropogenic sources in a number of estuaries, we have compiled data on ammonia, nitrite, and nitrate rather than on "total" nitrogen, since there are many more data for the inorganic forms and they have unquestioned biological availability. The results ranged from 20-30 $\mathrm{mmol \cdot m^{-3} \cdot year^{-1}}$ for

TABLE III. Approximate Annual Input of Dissolved Inorganic Nitrogen (NH_4^+, NO_2^-, NO_3^-) Per Unit Area and Per Unit Volume in Various Estuaries[a,b]

	Land drainage	*Sewage*	*Total*	*Sewage (%)*
Narragansett Bay	560	390	950	41
	60	40	100	
Long Island Sound	130	270	400	67
	10	20	30	
New York Bay	5,700	27,230	31,930	82
	800	3,750	4,550	
Raritan Bay	200	1,260	1,460	86
	50	280	330	
Delaware Bay	650	650	1,300	50
	70	70	140	
Chesapeake Bay	340	170	510	33
	50	30	80	
Patuxent estuary	310	290	600	48
	60	50	110	
Potomac estuary	420	390	810	48
	80	60	140	
Pamlico estuary	860	minor	860	<1
	250	--	250	
Apalachicola Bay	550	10	560	2
	210	3	213	
Mobile Bay	1,206	80	1,280	7
	370	30	400	
Barataria Bay	570	minor	570	<1
	290	--	290	
Northern San Francisco Bay	1,100	910	2,010	45
	160	130	290	
South San Francisco Bay	minor	1,600	1,600	~100
	--	310	310	
Kaneohe Bay	50	180	230	78
	10	30	40	

[a]*From Nixon (in preparation), Estuarine Ecology--A Comparative and Experimental Analysis Using 14 Estuaries and the MERL Microcosms. Data rounded to the nearest 10 units, compiled and calculated for various years from different sources including: Nixon (1981b) for Narragansett Bay; Jay and Bowman (1975), Bowman (1975), and Riley (1959) for Long Island Sound; Garside et al. (1976) and Deck (1980) for New York Bay; H. P. Jeffries (unpublished) and Deck (1980) for Raritan Bay; C. H. Culberson (personal communication) for Delaware Bay; Smullen et al. (1982) for Chesapeake Bay; W. M. Kemp and W. R. Boynton (unpublished) and Flemer (1970) for the Patuxent estuary; Smullen et al. (1983) and Jaworski et al. (1972) for the Potomac estuary; Harrison and Hobbie (1974) for the Pamlico estuary; Boynton (1975) and Estabrook (1973) for Apalachicola Bay; South Alabama Regional Planning Commission (1979) for*

Long Island Sound and Kaneohe Bay, Hawaii up to 4600 mmol$\cdot$m$^{-3}\cdot$year^{-1} for New York Bay, with most estuaries falling between 100 and 300 mmol$\cdot$m$^{-3}\cdot$year^{-1} (Table III). Urban sewage inputs, largely of ammonia, commonly made up about half of the inorganic input, but ranged from 1% or less in the Pamlico River estuary, North Carolina, Apalachicola Bay, Florida, and Barataria Bay, Louisiana, to nearly 100% in South San Francisco Bay.

Unfortunately, there are insufficient data from most of these systems to put the nitrogen-loading rates into the context of a more complete budget. Measurements of annual primary productivity have been reported for all but two of the areas, however (Table IV). With the exception of the Hudson estuary and San Francisco Bay, it is apparent that the productivity of these estuaries cannot be supported by the input of "new" nitrogen alone, but must usually rely to an equal or greater extent on recycled nitrogen. Similar conclusions were reached in earlier analyses of these and other coastal systems (Nixon, 1981a,b; Kemp et al., 1982). The importance of recycling is a remarkable similarity in the dynamics of estuarine and open ocean ecosystems, though in estuaries it seems likely that nitrification is relatively more important in the recycling processes than it is in surface water of the open sea.

More detailed nitrogen budgets that go beyond an analysis of inputs have been reported for a few estuaries, including the Pamlico River estuary (Harrison and Hobbie, 1974), North Sea Francisco Bay (Peterson, 1979), Kaneohe Bay, Hawaii (Smith, 1981) and the Peel-Harvey estuarine system in Australia (Hodgkin et al., 1981). Two of the most complete, however, have been developed for areas with

(Table III continued)
Mobile Bay; Day et al. (1982) for Barataria Bay; Russel et al. (1980), Association of Bay Area Governments (1977) and Peterson (1979) for San Francisco Bay; and Smith (1981) for Kaneohe Bay presewage diversion. Sewage input to Kaneohe Bay postdiversion is about one-third the value shown.

[b]*The top number of each point is in mmol*$\cdot$*m*$^{-2}\cdot$*year*$^{-1}$*; the bottom number is in mmol*$\cdot$*m*$^{-3}\cdot$*year*$^{-1}$*.*

TABLE IV. Estimates of the Amount of Nitrogen[a] Required to Support the Primary Production Measured in Various Estuaries Compared with the Nitrogen Input from Rivers and Sewage (see Table III)

	$g\ C \cdot m^{-2} \cdot year^{-1}$	$mmol\ N \cdot m^{-2} \cdot year^{-1}$ Required	$mmol\ N \cdot m^{-2} \cdot year^{-1}$ Input
Mid-Narragansett Bay (Furnas et al., 1976)	310	3,899	950
Mid-Long Island Sound (Riley, 1956)	205	2,579	400
Lower New York Bay (O'Reilly et al., 1976)	483	6,075	31,930
Lower Delaware Bay (J. Pennock, personal communication)[b]	206	2,590	1,300
Mid-Chesapeake Bay (Boynton et al., 1982)[c]	445	5,597	510
Patuxent estuary (Stross and Stottlemeyer, 1965)	210	2,642	600
Pamlico estuary (Davis et al., 1978; Kuenzler et al., 1979)	200-500	2,516-6,289	860
Apalachicola Bay (Estabrook, 1973)	360	4,528	560
Barataria Bay (Day et al., 1973)[d]	360	4,528	570
San Francisco Bay (Cole, 1982)			
Suisun Bay	95	1,195	2,010
San Pablo Bay	100-130	1,258-1,635	
South Bay	150	1,887	1,600
Kaneohe Bay (Smith, 1981)	165	2,075	230

[a] *Based on the Redfield stoichiometry of 6.625 C:N.*
[b] *Below Leipsic River, 80% of total bay production.*
[c] *Four-year mean (1974-1977).*
[d] *Phytoplankton 165, benthos 195.*

which we are more familiar, Narragansett Bay (Table V) and Charlestown Pond (Table VI), a shallow coastal lagoon on the ocean coast of Rhode Island. Both are relatively high-salinity systems without large freshwater input, though Narragansett Bay receives five small rivers and the lagoon is fed by small streams and

TABLE V. Present State of the Annual Nitrogen Budget for Narragansett Bay, Rhode Island[a]

	PN	*DON*	NH_4^+	$NO_2^- + NO_3^-$	*Total N*
Sources					
Fixation (sediments)[b]	< 0.7	0	0	0	< 0.7
Precipitation[c]	24		30		54
Runoff[d]					60
Rivers[d]	74	258	236	322	890
Sewage[d]	178	485	365	25	1053
Offshore	?	?	?	?	?
				Total input	>2058
Sinks					
Sedimentation[c]	132	0	0	0	132
Denitrification[c]	0	0	0	515	515
Fisheries[e]	<5	0	0	0	<5
Offshore	?	?	?	?	?
				Total output	>652
Recycling					
Microzooplankton excretion	0	?	?	0	?
Mesozooplankton excretion[f]	0	132	242	0	374
Ctenophore excretion[g]	0	14	16	0	30
Menhaden excretion[h]	0		3	0	3
Benthic flux[d]	0	114	886	0	1000
				Total recycled	>1407
Primary production[d]	3900	?	0	0	3900

[a] *Units are mmol* $N \cdot m^{-2} \cdot year^{-1}$.
[b] *From Seitzinger et al. (1978).*
[c] *Calculated by Seitzinger (1982).*
[d] *From Nixon (1981a).*
[e] *Assuming catch is <100 kg* ha^{-1} *(Nixon, 1981b, 1983).*
[f] *Vargo (1976, 1979).*
[g] *Kremer (1975).*
[h] *Durbin (1976).*

groundwater. Narragansett Bay has a mean depth of about 9 m, receives large amounts of urban sewage, supports a plankton-based ecosystem, and exchanges with Rhode Island Sound through two relatively wide, deep passages. The Bay and activities in its watershed are described in detail in Nixon and Kremer (1977), Kremer and Nixon (1978), and Olsen et al. (undated). Charlestown Pond is only about 1-2 m deep, supports large amounts of seagrasses and macroalgae as well as phytoplankton, has no direct

TABLE VI. Present State of the Annual Nitrogen Budget for Charlestown Pond Lagoon, Rhode Island[a]

	PN	*DON*	NH_4^+	$NO_2^-, + NO_3^-$	*Total N*
Sources					
Fixation (sediments, sea grasses)[b]	<30	0	0	0	<30
Precipitation[c]	24		30		54
Runoff[c]					1
Streams[c]	?	?	13		13
Groundwater (including sewage)[c]	0	0	0	290	290
Offshore[d]	0	68	1	104	173
				Total input	561
Sinks					
Sedimentation	?	0	0	0	?
Denitrification				?	?
Fisheries[e]	<2	0	0	0	<2
Offshore[d]	144	110	18	1	273
				Total output	>275
Recycling					
Primary production					
Seagrass[f]	2950				2950
Macroalgae[f]	630				630
Phytoplankton[d]	380				380
				Total production	3960

[a]*Units of mmol* $N \cdot m^{-2} \cdot year^{-1}$.

[b]*Based on data for* Ruppia *during summer in Chesapeake Bay (Lipschultz et al., 1979), summer seagrass biomass data from Thorne-Miller et al. (1983), and a season of 100 days. This is probably a maximum estimate.*

[c]*From Nixon et al. (1982).*

[d]*From Nixon and Lee (1981).*

[e]*From R. Crawford (personal communication).*

[f]*From Thorne-Miller et al. (1983), and Thorne-Miller and Harlin (personal communication).*

sewage discharge (though there is groundwater input of nitrogen from individual on-site sewage disposal systems associated with housing developments), and exchanges with Block Island Sound through a narrow passage in a barrier beach (Lee, 1980; Nixon and Lee, 1981).

While the budgets for these two systems (Tables V and VI) represent a very large effort comprising thousands of measurements from numerous individual studies over many years, they remain far from complete. Some of the entries are also quite uncertain. Many of the individual studies were made for other reasons, with the authors never intending their results to be part of such a budget. A detailed discussion of Tables V and VI is not warranted here, but a few comments may be useful.

For Narragansett Bay, the exchange with offshore is the biggest unknown. The net transport must be the difference between two large fluxes, neither of which is known. The following calculation is made in order to provide some perspective on the general magnitude of these fluxes. If the average depth of the bay is 9 m, the average concentration of total N in the offshore water is 10 μM (not known, since we do not have enough data for DON), and the volume is exchanged 12 times per year (the residence time of water is thought to be about 30 days), the input flux of N would be about 1080 $mmol \cdot m^{-2} \cdot year^{-1}$. The output flux can be similarly estimated, but without adequate spatial representation of the concentrations and the details of the exchange processes, it is not possible to estimate the (expected) net loss of nitrogen by this route.

We do not see why any benthic organism in Narragansett Bay would need to fix nitrogen, and believe that this process does not occur there to any measurable degree.

Losses of DON and NH_3 by sedimentation in Narragansett Bay are not really zero, but round down to zero and are therefore trivial. The estimated loss of particulate nitrogen by sedimentation depends on knowledge of the areally averaged sedimentation rate, which is uncertain by perhaps a factor of 3-5 (see Santschi, 1980), so this loss may be very small or may be of great importance.

In the budget for Charlestown Pond Lagoon (Table IV), all known sources have been associated with some sort of estimate, while the probable major losses by denitrification and sedimentation have not been measured or estimated. This lagoon appears to be a region of net sedimentation (Lee, 1980), so it seems likely that denitrification must be less active here than in Narragansett Bay, or else some major input has been missed.

Despite the gross inadequacies of these budgets, the exercise of bringing them to this stage does seem to demonstrate the major features of nitrogen dynamics in such systems. Some of the results would certainly be different in other systems where large rivers are present, there is intensive agriculture in the watershed, or the water column is strongly stratified. Nevertheless, they are not atypical examples of the present level of accomplishment in attempting mass balances for nitrogen in estuarine systems.

VI. NITROGEN AND THE PRODUCTIVITY OF ESTUARINE AND COASTAL MARINE ECOSYSTEMS

There are at least two features apparent in the modest but growing sample of estuarine and coastal marine primary productivity measurements that have become available during the past 20 years. First, reported values for the production of these areas are higher by a factor of 4 or more than those for most temperate open-ocean waters and, second, the annual production in the great majority of coastal systems falls within the relatively narrow range of 200-400 g $C \cdot m^{-2} \cdot year^{-1}$ (Nixon, 1981a, 1983a,b; Boynton et al., 1982). While the first of these findings would not have surprised Brandt, Johnstone, or many others working at the turn of the last century, it nevertheless is an important confirmation of their impressions, which were, after all, based largely on water color and cell counts. The second observation was not anticipated commonly and may mean that external "forcing," including nutrient input, is less important in regulating estuarine production than once assumed. Taken with the information summarized in the preceding sections, these two features may suggest a conceptual model for the relationship between nitrogen and primary production in coastal waters (Fig. 24).

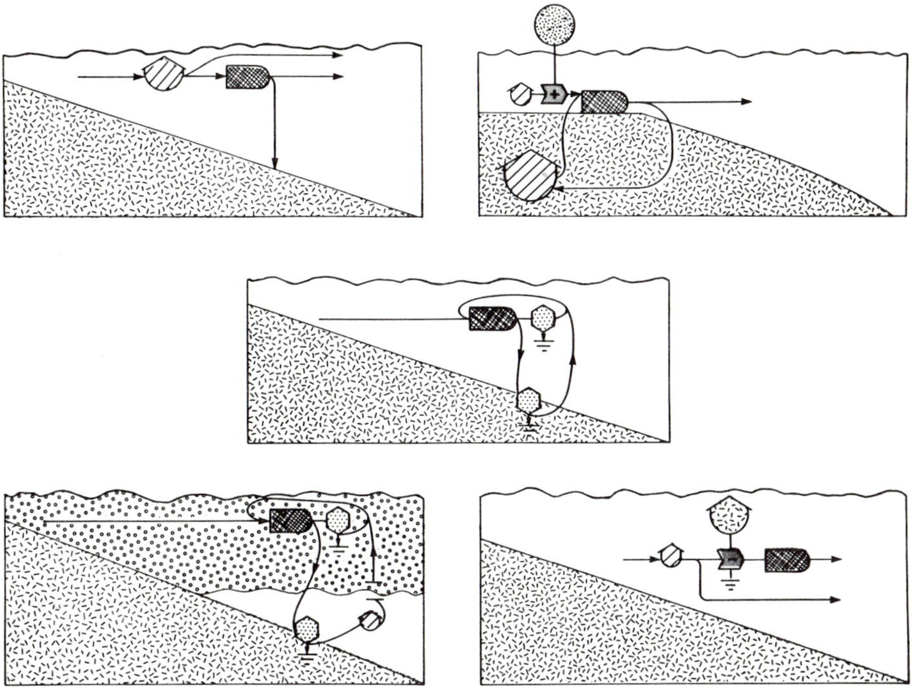

Fig. 24. A conceptual model of the various interactions among nutrients (storage tank symbols), primary producers (bullet-shaped symbols), and heterotrophic nutrient regenerators (hexagons) in estuaries and nearshore waters. Upper right shows highly productive macrophytes using the continuous delivery of nutrients at low concentrations by currents and/or the large pool of nutrients in sediments, while the upper left shows highly productive phytoplankton drawing on the large storage of nutrients supplied by river input, upwelling, or anthropogenic sources. Middle panel represents what may be the most common mode where much of the primary production is supported by recycled nutrients from pelagic and benthic regeneration. Lower right shows systems in which high turbidity keeps primary production low and nutrients relatively abundant. Lower left depicts low production systems in which a substantial part of nutrient regeneration is separated in space and/or time from the primary producers and there are no large nutrient inputs. Examples are given in text. From Nixon (1983).

One of the major differences between virtually all coastal systems and the open sea is their shallow depth. A second is the usually vigorous vertical mixing generated by tidal currents and wind. These characteristics assure that all of the nitrogen regeneration, both pelagic and benthic, will be quickly and completely made available in the euphotic zone. In offshore waters some fraction of the organic matter produced is continually lost from the surface layers to be decomposed in the darkness below. The large amount of benthic and pelagic recycling in most coastal systems has been emphasized in the previous sections, and since the same general types of organisms (bacteria, microzooplankton, microfauna) are responsible for most of this regenerating in all estuaries, the basic similarities in their metabolic rates may be the common feature that constrains primary production in a great variety of environments. The notion that heterotrophic processes may limit autotrophic ones has been expressed before (Odum et al., 1969), but perhaps nowhere as forcefully as by Wangersky (1977) who concluded a discussion of production in the sea by noting that, "The necessity for bacterial control of productivity through nutrient regeneration should be obvious to any oceanographer who has run nutrient analyses in the open ocean." While nutrient concentrations are usually higher in coastal waters, the preceding sections have emphasized the importance of recycled nutrients in these areas as well. However, the relative importance of bacteria, micro- and larger zooplankton, and various classes of benthic animals is still poorly known.

There are, of course, some estuaries where the production exceeds 400 g $C \cdot m^{-2} \cdot year^{-1}$, and some examples deserve comment. Higher rates for phytoplankton appear to be supported by large and more or less continuous inputs of nutrients, as in the lower Hudson River estuary where O'Reilly et al. (1976) found about 800 g $C \cdot m^{-2} \cdot year^{-1}$ in areas impacted by the New York City sewage discharge. Even where nutrient inputs and concentrations are low, similar values may be found if seaweed or seagrass are abundant. By maintaining a fixed position, these plants can take advantage of the energy subsidy of the current to renew constantly the water surrounding their fronds or leaves, thus obtaining a virtually unlimited supply of nutrients (Conover, 1968). Seagrasses may have the additional advantage of accumulating nutrients in the sediments and drawing on this storage with roots. The effectiveness of an Eulerian life-style is evident in St. Margaret's Bay, Nova Scotia, where kelps contribute over 600 g $C \cdot m^{-2} \cdot year^{-1}$ to the entire bay (Mann, 1972a,b), and in Izembeck Lagoon, Alaska, where production by the seagrass *Zostera marina* is over 800 g $C \cdot m^{-2} \cdot year^{-1}$ (Barsdate et al., 1974).

Low levels of production (<200 g $C \cdot m^{-2} \cdot year^{-1}$) appear in systems with a very short growing season (e.g., the Gulf of Bothnia with 50 g $C \cdot m^{-2} \cdot year^{-1}$, Ackefors et al., 1978), in areas where high turbidity limits photosynthesis (e.g., the Dutch Wadden Sea with phytoplankton production of 110 g $C \cdot m^{-2} \cdot year^{-1}$, Cadee and Hegeman, 1974; and the upper Delaware estuary, J. Pennock, personal communication), or where nutrient inputs are not large and a sizeable fraction of the heterotrophic recycling activity is separated in space and/or time from the autotrophic zone. An example of the latter may be phytoplankton production in St. Margaret's Bay, Nova Scotia, a relatively deep arm of the sea with a strong seasonal thermocline and plankton production of 190 g $C \cdot m^{-2} \cdot year^{-1}$ (Platt, 1971).

If we are correct in interpreting the cause of the similarity in estuarine productivity values, most coastal systems are probably best described by the center panel of Fig. 24, which emphasizes recycling, though there are probably also periods such as the spring bloom when nutrient storages and inputs may be more important. What is not yet very clear is the relationship between nutrient inputs and recycling, or the consequences of increasing nutrient inputs so that a system moves from the center toward the upper left panel of Fig. 24. There is apparently a widespread belief that such a shift is occurring in coastal waters, at least along the coastlines of the industrial nations (Neilson and Cronin, 1981; Walsh et al., 1981). In only a few areas, however, such as the Wadden Sea (Postma, 1978) and the Patuxent estuary in

Chesapeake Bay (Fig. 22; Heinle et al., 1983), have quantitative data been available to demonstrate an increase in nutrient concentrations and productivity for the estuary as a whole. For the most part, evidence documenting eutrophication of coastal waters has come from tidal freshwater, from low-salinity, river mouth portions of estuaries, or from areas adjacent to sewage discharges.

While limnologists have been remarkably successful in finding strong correlations between phosphorus inputs or average concentrations and the productivity and algal standing crop in lakes (Vollenweider, 1976; Schindler, 1981), the data to begin similar attempts for nitrogen in estuaries have only recently become available. Since the lack of a relationship between short-term production measurements and inorganic nitrogen concentrations has already been discussed (Fig. 13), we have plotted annual primary production and the standing crop of phytoplankton chlorophyll as a function of dissolved inorganic nitrogen input for a number of estuaries (Figs. 25 and 26). This first attempt includes only a relatively few systems, but the lack of reliable information, especially for nitrogen loading, will make it difficult to increase the sample quickly. With the data at hand, it seems evident that the response of saltwater systems to nitrogen input is much less dramatic than expected from the experience with lakes. This behavior is not due to variation in flushing rates, since attempts to correct for water exchange (an admittedly imperfect calculation) did not fundamentally change the result. A problem with the comparative field data, however, is that in most cases it is not possible to find a complete set of measurements (inputs, production, concentration) for any one year. There may be strong secular trends that make it inappropriate to combine production from one year with nitrogen input from some years earlier or later. We have few descriptions of the year-to-year variability in estuarine nutrient cycles (Fig. 4) or productivity, and it is difficult to know the relative magnitude of differences among systems compared to the annual variation in any one. In a recent analysis of a continuous

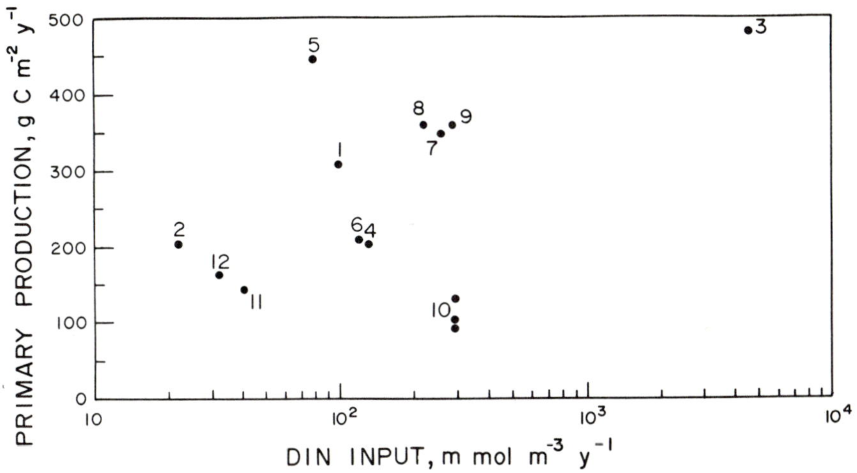

Fig. 25. Annual measurements of primary production as a function of the estimated annual input of dissolved inorganic nitrogen in various estuaries. See Tables III and IV for data sources. (1) Narragansett Bay, Rhode Island; (2) Long Island Sound; (3) Lower New York Bay; (4) Lower Delaware Bay; (5) Chesapeake Bay; (6) Patuxent estuary, Maryland; (7) Pamlico estuary, North Carolina; (8) Apalachicola Bay, Florida; (9) Barataria Bay, Louisiana; (10) North and South San Francisco Bay, California; (11) Kaneohe Bay, Hawaii.

6-year record of chlorophyll *a* and primary production in mid-Chesapeake Bay, Boynton et al. (1982) found a good correlation between summer maxima of each variable and the annual input of nitrogen and phosphorus. This is particularly interesting in Chesapeake Bay, where nutrient inputs per unit volume are very low (Table III), and most of the annual production is supported by recycled nutrients (Table IV).

Given the difficulties in interpreting the field data, it would seem reasonable to turn to the results of a number of experiments in which nutrients alone (apart from the organic matter, sediment, metals, and toxics that are usually associated with them in estuaries) had been added in known amounts to plankton-dominated coastal marine ecosystems. (An extensive series of fertilization experiments on salt marshes is discussed in the

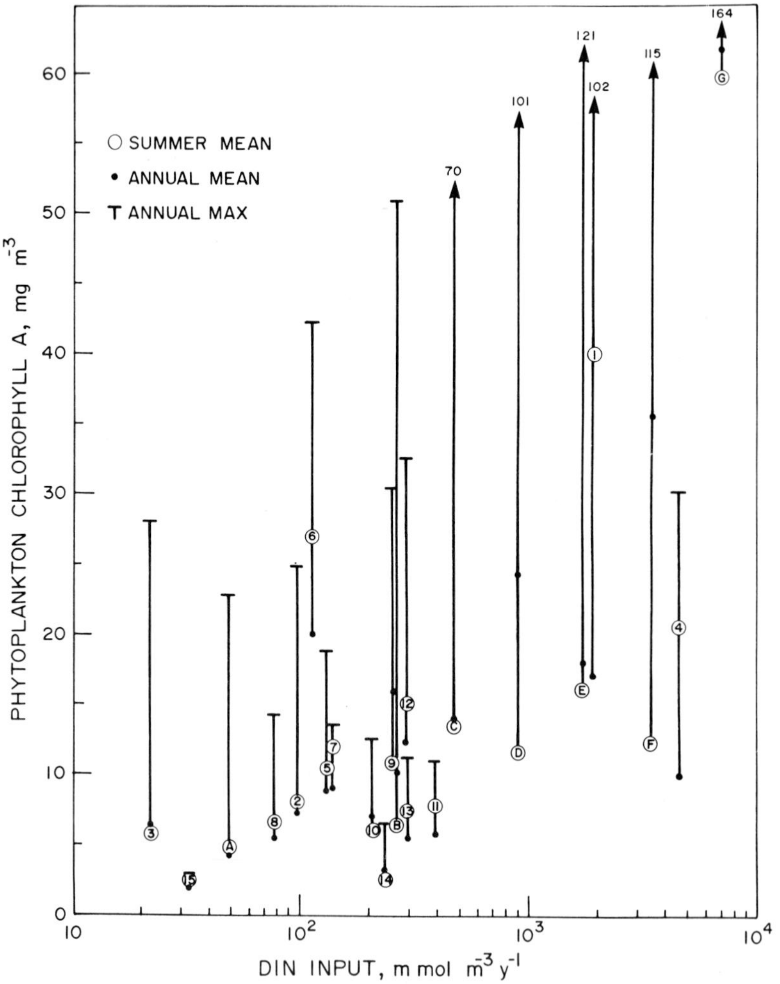

Fig. 26. Concentrations of phytoplankton chlorophyll *a* in the mid-region of various estuaries (1-15) and in the MERL experimental ecosystems (A-G) as a function of the input of dissolved inorganic nitrogen. (1) Providence River estuary, Rhode Island; (2) Narragansett Bay, Rhode Island; (3) Long Island Sound; (4) Lower New York Bay; (5) Delaware Bay; (6) Patuxent River estuary, Maryland; (7) Potomac River estuary, Maryland; (8) Chesapeake Bay; (9) Pamlico River estuary, North Carolina; (10) Apalachicola Bay, Florida; (11) Mobile Bay, Alabama; (12) Barataria Bay, Louisiana; (13) N. San Francisco Bay, California; (14) S. Francisco Bay, California; (15) Kaneohe Bay, Hawaii.

chapter by Valiela, Chapter 17, this volume.) Unfortunately, and in spite of the long-standing interest in production and concern with eutrophication, there appear to have been only four such experiments. The first was carried out in a Norwegian oyster poll or inlet during the summers of 1927-1929 by Gaarder and Sparck (1931). This work was followed by fertilizations of Loch Craighin in Scotland in 1942-1944 (Orr, 1947, and others) and of Loch Sween in 1944-1946 (Gross, 1950, and others). After the sea loch work there appears to have been no further comprehensive experimental study on whole ecosystems for 35 years until an ongoing experiment was begun at the Marine Ecosystems Research Laboratory (MERL) at the University of Rhode Island in the summer of 1981. There has been, of course, a great deal of work on nutrients and phytoplankton physiology, and in a few cases experiments involving nutrient additions have been carried out on enclosed or "captured" water masses containing complex assemblages of plankton (Antia et al., 1963; Strickland, 1967; Takahashi et al., 1982). In most coastal systems, however, the importance of the benthos in nutrient cycling requires that the bottom community be included in meaningful long-term experiments.

The results of the early fertilization work were apparently dramatic with regard to increased standing crops of phytoplankton and macrophytes. Both Gaarder and Sparck (1931) and the Scottish loch groups also became very much aware of the importance of rapid nutrient cycling in the enriched systems. But because these studies were carried out in the field, water exchange was not entirely eliminated or regulated and only one system was available for study, so that it was not possible to have strict controls or a range of treatment levels. The quantitative addition of large amounts of commercial fertilizer to the water was also difficult to achieve and, of course, many of the more sophisticated techniques now used to study nutrient cycles and production were not available. Possibly for these reasons, the results of this work had little impact on marine ecology. This is unfortunate, not only because the results of the sea loch studies appear interesting and important, but because ecosystem level experiments are very much needed to understand the interaction between nutrient inputs, cycles, and production.

The MERL experiment is being carried out in a series of large tanks, each 5-m deep, 2-m in diameter, and containing 13 m^3 of Narragansett Bay water. The tanks are maintained outdoors, and each contains a 30-cm deep layer of natural sediment and associated benthos from the Bay. The tanks are well mixed by a mechanical device and fed with diaphram-pumped bay water to give a mean residence time of 27 days for the water in each tank. A more complete description of the facility and data showing that the mesocosms are good ecological analogs of a coastal marine system are given in Pilson et al. (1980).

The eutrophication experiment consists of daily additions of inorganic N, P, and Si in a ratio of 13:1:1 and at a rate which increases geometrically in six steps from 212 mmol $N \cdot m^{-3} \cdot year^{-1}$ to 6784 mmol $N \cdot m^{-3} \cdot year^{-1}$, an input exceeding that of the lower Hudson estuary (Table III). Three control tanks are also maintained. Some of the results from the first year of the experiment have been plotted along with the field data in Fig. 26. Again, it seems that while the nutrient additions increased the mean and maximum standing crops of chlorophyll, these increases were not in strict proportion to the input. A 16-fold increase produced only about an 8- to 9-fold increase in the annual mean standing crop, while a 32-fold increase in loading resulted in about a 15- to 16-fold increase in standing crop.

There were also increases in the primary production and respiration as a result of nutrient input (Fig. 27), though these responses were much less dramatic than for phytoplankton chlorophyll, and at higher nutrient concentrations there may have been light limitation. At present it is not clear why the increase in standing crop of phytoplankton was so much greater than the increase in apparent production. The tight coupling of autotrophic and

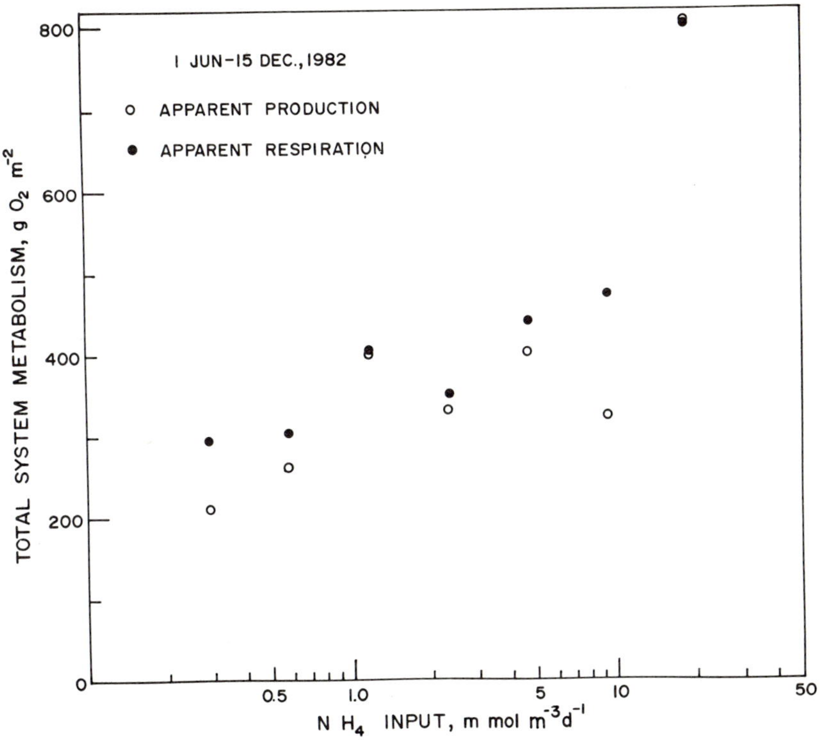

Fig. 27. Integrated total system apparent daytime net production and night respiration during the first 6.5 months of the MERL nutrient addition experiment. Measurements were made using dawn and dusk oxygen determinations and an air-water diffusion correction that may somewhat overestimate production and underestimate respiration.

heterotrophic processes at all the treatment levels is also striking, even when the concentrations of dissolved inorganic nitrogen were very high. It seems difficult to reconcile this behavior with the concept of a cycle in which production is limited by regeneration. Perhaps when the standing crop of organic matter is low, the metabolic rates of the heterotrophic processes are limiting, but with abundant organic matter, the size of the heterotrophic populations rises in proportion to the supply.

Much of the information from the MERL experiment is still being analyzed, and while the preliminary results confirm and clarify many of our impressions from the field data, a number of questions remain. Among them are the long-term effects of nutrient inputs.

Since the addition of nutrients does appear to increase the primary production and standing crop of coastal waters, it is interesting to ask what the effects of anthropogenic nutrient inputs have been. It is a difficult question to answer. There is a certain amount of anecdotal information on algal blooms available for almost every estuary, but long-term quantitative data on nutrient inputs, nutrient concentrations, and primary production are not available. Estuarine sediments can provide some useful evidence, but they can not be read like tree rings. There are, however, detailed historical records of fisheries yields from many coastal areas, and it has been possible to establish a general correlation between total contemporary landings and recent primary productivity measurements in marine systems (Fig. 28). While there is considerable variation in the data, a consistent increase in primary production might reasonably be expected to increase fisheries yields.

In 1734 William Wood published "a true and lively" account of "New England's Prospect" in which he devoted a chapter to fisheries:

> Lobsters be in plenty in most places, very large ones, some being twenty pound in weight. These are taken at a low water amongst the rocks. They are a very good fish, the small ones being the best; their plenty makes them little esteemed and seldom eaten. The Indians get many of them every day for to bait their hooks withal and to eat when they can get no bass. The oysters be great ones in form of a shoehorn; some be a foot long. These breed on certain banks that are bare every spring tide. The fish without the shell is so big that it must admit of a division before you can well get it into your mouth. The periwig is a kind of fish that lieth in the ooze like a head of hair, which being touched conveys itself leaving nothing to be seen but a small round hole. Muscels be in great plenty, left only for the hogs, which if they were in England would be more esteemed of the poorer sort. Clams or

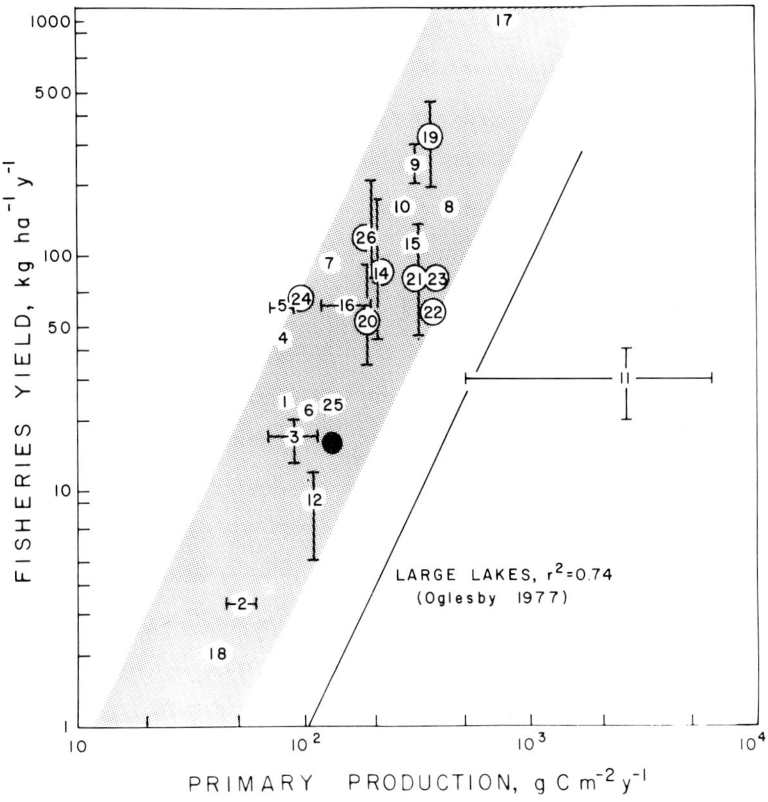

Fig. 28. The relationship between fisheries yield and the primary production of a variety of marine systems compared with the regression line developed by Oglesby (1977) for similar data from large lakes. Range bars have been added where practical. See Nixon (1982) for data sources. (1) Gulf of Finland; (2) Gulf of Bothnia; (3) Adriatic Sea; (4) South Baltic Sea; (5) North Sea; (6) Scotian shelf; (7) Scotian slope; (8) Georges Bank; (9) Peru upwelling; (10) Louisiana nearshore shelf; (11) coral reefs; (12) Black Sea; (13) deleted; (14) Long Island Sound; (15) nearshore Rhode Island; (16) mid-Atlantic Bight; (17) Gulf of Cariaco; (18) Caribbean and Gulf of Mexico; (19) Barataria Bay; (20) Peconic Bay; (21) Charlestown Pond Lagoon; (22) North Carolina Sounds; (23) Apalachicola Bay; (24) Sagami Bay; (25) Seto Inland Sea; (26) Wadden Sea. The heavy point is the world ocean catch, if it is assigned to the total world shelf and slope area.

> clamps is a shellfish not much unlike a cockle; it lieth under the sand, every six or seven of them having a round hole to take air and receive water at. When the tide ebbs and flows, a man running over these clam banks will presently be made all wet by their spouting of water out of those small holes. These fishes be in great plenty in most places of the country, which is a great commodity for the feeding of swine both in winter and summer; for being once used to those places, they will repair to them as duly every ebb as if they were driven to them by keepers. In some places of the country there be clams as big as a penny white loaf, which are great dainties amongst the natives and would be in good esteem amongst the English were it not for better fish.

Impressive accounts such as this may be misleading because they are based largely on impressions of the standing crop of a lightly exploited resource. Nevertheless, the comparison with present conditions is sharp. Among the earliest quantitative accounts of the production of an intensive coastal fishery is the detailed survey carried out for the United States Fish and Fisheries Commission during 1879 by G. B. Goode and associates (1887). The fisheries of the Northeastern states were well developed by that time, with landings consistently in excess of 100 $kg \cdot ha^{-1} \cdot year^{-1}$. The fisheries survey of the many estuaries and coastal lagoons on Long Island (Mather, 1887) was particularly thorough, and provides an excellent example not only of high production but of the variability among systems (Fig. 29). There seems little question that in the past the secondary production of many coastal waters must have been equal to or greater than that found today. The 1879 data are particularly useful because anthropogenic nutrient inputs were not very significant before the turn of the century. According to Melosi (1981), one-third of all urban households had running water and indoor flush toilets by 1880, and by that time "most major cities adopted sewerage systems to accompany or combine with their storm-water systems...." With the construction of sewerage systems, nutrients that had previously been put on fields

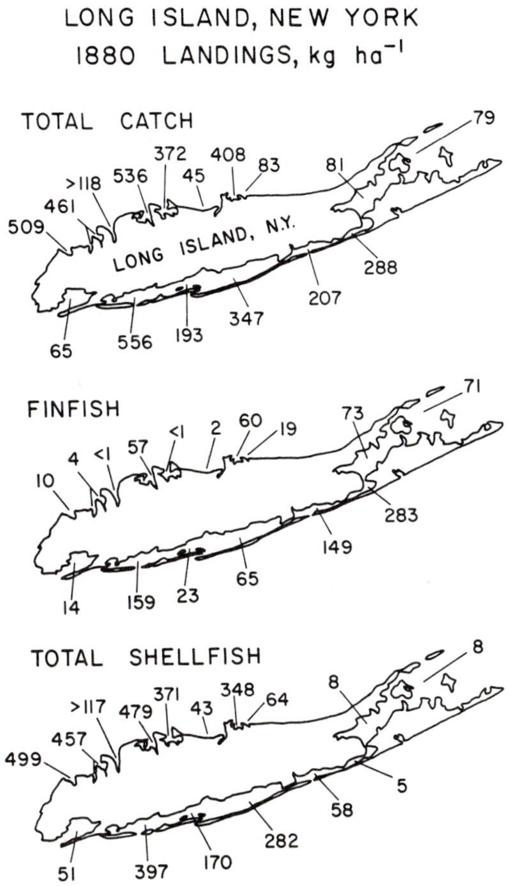

Fig. 29. Landings of finfish and shellfish from the coastal lagoons and embayments around Long Island, in 1879-1880. Total catch reported by Mather (1887). From Nixon (1982).

or into the ground were discharged in large quantity directly into rivers and estuaries. Similarly, the widespread use of commercial fertilizers did not really begin until the late 1800s, but their application then increased at an approximately exponential rate for over 100 years (Fig. 30).

From an historical perspective, large inputs of nitrogen may be a relatively recent feature in many estuaries. Nevertheless,

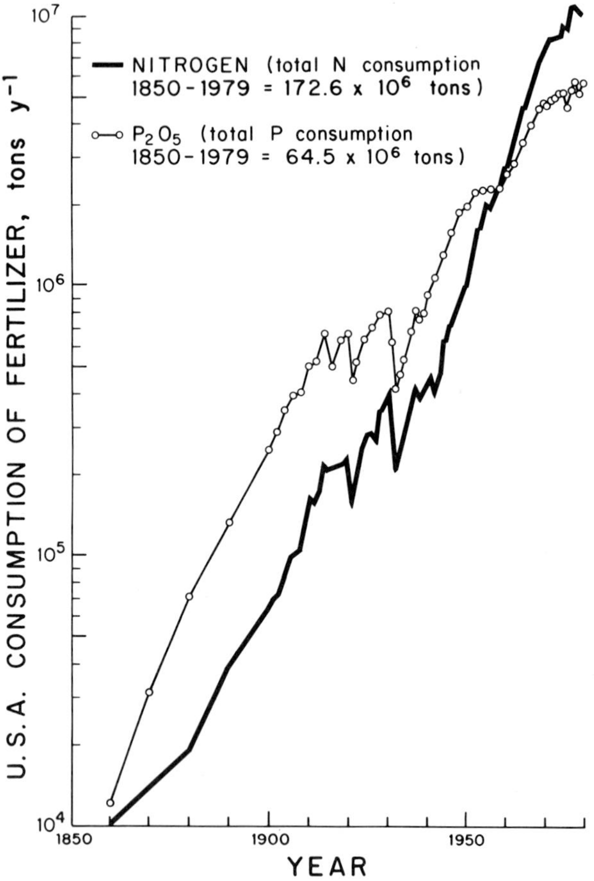

Fig. 30. Historical consumption of nitrogen and phosphorus fertilizer in the United States. Data from the Fertilizer Institute, Washington, D.C.

the record in at least some areas seems long enough that an impact should be evident, yet we are not aware of any evidence to suggest that coastal fisheries have become more productive over the past 100 years. In fact, just the opposite appears to be the case. This situation may, of course, reflect the fact that fisheries respond to changes in markets, technology, and regulations that make inshore harvest less attractive, or it may be that other pollutants have severely reduced the production of fish in coastal waters. In the Scottish sea loch experiments where nutrients alone were added, the growth of plaice and flounder was increased dramatically, and Gross (1947) concluded, "... the vast improvement in growth rate observed may be taken as an indication of a very rapid ultimate conversion of inorganic plant nutrients into fish flesh; or alternatively as an indication that on normal fishing grounds a deficiency in plant nutrients is responsible for both the slow growth and the relatively low yield of fish per acre...."

Secondary production of higher trophic levels is not entirely based on rapidly spinning cycles of the kind often thought to explain much of estuarine primary production. The standing crop of food must also be important, and this feature may relate to the total supply of nitrogen available. The situation is complicated and the relationship between supply and cycling remains unclear. If nitrogen is accumulating in estuarine systems, the mechanisms responsible have not been described adequately, and this must be done before any of the experiments, which are relatively short-term undertakings, can be related to historical and future impacts of human activities on estuaries.

We introduced this chapter by noting that for over 80 years, one of the major reasons for studying nitrogen in estuaries has been the presumed importance of its role in maintaining and enhancing the productivity of coastal waters. Taken "all in all," it is still a good problem on which to work.

VII. SUMMARY

Studies of the role of nitrogen in estuaries began almost 100 years ago with investigations of the effect of sewage-derived ammonia in the tidal Thames and expanded some decades later to include investigations of the factors controlling phytoplankton production. Since then, we have acquired several sets of reasonably complete analyses so that it is possible to evaluate spatial and seasonal variations of typical concentrations of most forms of fixed nitrogen in a few temperate estuaries. There is still no estuary (or for that matter any other marine water body) for which there exists a well-constrained and complete budget of fixed nitrogen.

Many processes involved in the biogeochemical cycling of nitrogen species have been studied extensively, although the importance of some has only recently become evident. These include processes of rapid turnover in the water column, such as uptake, remineralization, and oxidation. In the sediments the processes include burial, remineralization, biological uptake, oxidation, reduction, nitrous oxide production (which also occurs to some extent in the water column), and denitrification. Benthic remineralization and water column production are well correlated and may be closely coupled. Nitrous oxide production appears to be relatively enhanced in highly eutrophic systems, and seems to accompany both nitrification in the water column and denitrification in the sediments. It seems likely that denitrification will be proved very active in estuarine sediments generally, and may be responsible for the typically low N:P ratio in such regions.

The role of anthropogenic inputs of nitrogen in possibly increasing secondary production in estuaries has been speculated on, but the evidence is surprisingly equivocal. There is a good positive correlation between fisheries yields and primary production, both from the field and from limited experimentation, and there is certainly evidence of increased production with increased nutrient input, but there seems, nevertheless, to be a lack of evidence for increasing fisheries yield corresponding to the increased anthropogenic loading of coastal waters with nitrogen. This is still an important subject for investigation.

VIII. ACKNOWLEDGMENTS

We are grateful to a number of people who have provided us with stimulating discussions and data during the preparation of this chapter. At the risk of omitting some, we take the opportunity to thank C. Culberson, J. Sharp, and J. Pennock at the University of Delaware; M. Kemp, W. Boynton, D. Heinle, and J. Stevenson at the University of Maryland; D. Flemer and W. Nehlson at the Environmental Protection Agency Chesapeake Bay Project; W. Schroeder at the Dauplin Island Sea Laboratory; J. Day at Louisiana State University; D. Peterson, J. Cloern, and D. Harman at the United States Geological Survey in San Francisco; S. Smith and W. Kimmerer at the University of Hawaii; H. Postma at Texel; and C. Oviatt, P. Donaghay, S. Seitzinger, J. Kelly, J. Garber, C. Doremus, V. Beronsky, G. Douglas, P. Jeffries, and V. Lee at the University of Rhode Island. The original illustrations were prepared by M. McHugh. We are also grateful to the editors for their patience and to the Office of Sea Grant Programs (NOAA) (Grant NA-81-AA-D-00073 and previous grants), the National Science Foundation (Grants OCE-76-22149, OCE-78-27124 and OCE-78-7820280), and the United States Environmental Protection Agency (Grants CR-810265 and CR-807795) for the support that has enabled us to pursue the study of nitrogen in coastal waters.

REFERENCES

Ackefors, H., Hernroth, L., Lindahl, O., and Wulff, F. (1978). Ecological production studies of the phytoplankton and zooplankton in the Gulf of Bothnia. *Finn. Mar. Res. 244,* 116-126.

Aller, R. C. (1980). Diagenetic processes near the sediment-water interface of Long Island Sound. I. Decomposition and nutrient element geochemistry (S, N, P). *Adv. Geophys. 22,* 238-250.

Aller, R. C., and Yingst, J. Y. (1980). Relationships between microbial distribution and the anaerobic decomposition of organic matter in surface sediments of Long Island Sound, U.S.A. *Mar. Biol. (Berlin) 56,* 29-42.

Anderson, D. Q. (1939). Distribution of organic matter in marine sediments and its availability to further decomposition. *J. Mar. Res. 2,* 225-235.

Antia, N. J., McAllister, C. C., Parsons, J. R., Stevens, K., and Strickland, J. D. H. (1963). Further measurements of primary production using a large volume plastic sphere. *Limnol. Oceanogr. 85,* 166-184.

Atkins, W. R. G. (1925). Seasonal changes in the phosphate content of sea water in relation to the growth of the algal plankton during 1923 and 1924. *J. Mar. Biol. Assoc. U.K.* [N.S.] *13,* 700-720.

Atkins, W. R. G. (1930). Seasonal changes in the nitrite content of sea-water. *J. Mar. Biol. Assoc. U.K.* [N.S.] *16,* 515-518.

Atkins, W. R. G. (1932). Nitrate in sea-water and its estimation by means of diphenylbenzidine. *J. Mar. Biol. Assoc. U.K.* [N.S.] *9, 18,* 167-192.

Association of Bay Area Governments (1977). "Estimated Municipal and Nondiscrete Industrial Wastewater Loads in the San Francisco Bay Region," Tech. Memo. No. 15 (rev.). ABAG, San Francisco, California.

Barrett, B. B., Merrell, J. L., Morrison, T. P., Gillespie, M. C., Ralph, E. J., and Burdon, J. F. (1978). A study of Louisiana's major estuaries and adjacent offshore waters. *La. Dept. Wildl. Fish., Tech. Bull. 27.*

Barsdate, R. J., Nebert, M., and McRoy, C. P. (1974). Lagoon contributions of sediments and water to the Bering Sea. *In* "Oceanography of the Bering Sea" (D. W. Wood and E. J. Kelley, eds.), University of Alaska, Fairbanks.

Bault, E. I. (1972). Hydrology of Alabama estuarine areas-cooperative Gulf of Mexico estuarine inventory. *Ala. Mar. Resour. Bull. 7,* 25.

Berner, R. A. (1975). Diagenetic models of dissolved species in the interstitial waters of compacting sediments. *Am. J. Sci. 275,* 88-96.

Berner, R. A. (1980). "Early Diagenesis," p. 241. Princeton Univ. Press, Princeton, New Jersey.

Billen, G. (1975). Nitrification in the Scheldt estuary (Belgium and The Netherlands). *Estuarine Coastal Mar. Sci. 3,* 79-89.

Billen, G. (1978). A budget of nitrogen recycling in North Sea sediments off the Belgian Coast. *Estuarine Coastal Mar. Sci. 7,* 127-146.

Billen, G. (1982). An idealized model of nitrogen recycling in marine sediments. *Am. J. Sci. 282,* 512-541.

Blackburn, T. H. (1979). Method for measuring rates of NH_4^+ turnover in anoxic marine sediments, using a ^{15}N-NH_4^+ dilution technique. *Appl. Environ. Microbiol. 37,* 760-765.

Bowman, M. J. (1975). Pollution prediction model of Long Island Sound. *Proc. Spec. Conf. Civ. Eng. Oceans Vol. III*, pp. 1084-1103. Amer. Soc. Civil Eng.

Boynton, W. R. (1975). Energy basis of a coastal region: Franklin County and Apalachicola Bay, Florida. Ph.D. Thesis, University of Florida, Gainesville.

Boynton, W. R., Kemp, W. M., and Osborne, C. G. (1980). Nutrient fluxes across the sediment-water interface in the turbid zone of a coastal plain estuary. *In* "Estuarine Perspectives" (V. S. Kennedy, ed.), pp. 93-109. Academic Press, New York.

Boynton, W. R., Kemp, W. M., and Keefe, C. W. (1982). A comparative analysis of nutrients and other factors influencing estuarine phytoplankton production. *In* "Estuarine Comparisons" (V. S. Kennedy, ed.), pp. 69-90. Academic Press, New York.

Braarud, T., and Klem, A. (1931). Hydrographical and chemical investigations in the coastal waters off More and in the Romsdalsfjord. *Hvalradets Skr. 1*, 1-88.

Brandt, K. (1899). Uber den Stoffwechsel im Meere. *Wiss. Meeresunters., Abt. Kiel Bd. 4*, 213-230.

Buch, K. (1923). Methodisches uber die Besgtimmung von stickstoffverbindungen in Wasser. *Merentutkimuslaitoksen Julk. Havsforkningsinst. Skr. 18*, 1-22.

Buckner, S. C. (1973). A nutrient profile of Great South Bay, Long Island, New York, and the effects of its tributary rivers on the nutrient supply. M.S. Thesis, p. 92, Adelphi University, New York.

Cadee, G. C., and Hegeman, J. (1974). Primary production of phytoplankton in the Dutch Wadden Sea. *Neth. J. Sea Res. 8*(2), 240-259.

Caperon, J., Schell, D., Hirota, J., and Laws, L. E. (1979). Ammonium excretion rates in Kaneohe Bay, Hawaii, measured by a ^{15}N isotope dilution technique. *Mar. Biol. (Berlin) 54*, 33-40.

Carey, A. G. (1967). Energetics of the benthos of Long Island Sound. I. Oxygen utilization of sediment. *Bull. Bingham Oceanogr. Collect. 19*, 136-144.

Cole, B. E. (1982). Size-fractionation of phytoplankton production in San Francisco Bay: 1980. *Abstr., Symp. Factors Controlling Biol. Prod. Sacramento-San Joaquin Estuary, Contra Costa County Water District Center, Concord, CA.*

Conover, J. T. (1968). The importance of natural diffusion gradients and transport of substances related to benthic marine plant metabolism. *Bot. Mar. 11*, 1-9.

Cooper, L. H. N. (1932). The reduced strychnine reagent for the determination of nitrate in the sea. *J. Mar. Biol. Assoc. U.K.* [N.S.] *18*, 161-166.

Cooper, L. H. N. (1933). Chemical constituents of biological importance in the English Channel, November, 1930 to January, 1932. Part I. Phosphate, silicate, nitrate, nitrite, ammonia. *J. Mar. Biol. Assoc. U.K.* [N.S.] *18*, 677-728.

Crutzen, P. J. (1970). The influence of nitrogen oxides on the atmospheric ozone content. *Q.J.R. Meteorol. Soc. 96*, 320-325.

Culberson, C. H., Sharp, J. H., Church, T. M., and Lee, B. W. (1982). "Data from the Salsx Cruise, May 1978-July 1980," Oceanogr. Data Rep. No. 2. Delaware Sea Grant College Program, University of Delaware, Newark.

Davis, G. J., Brinson, M. M., and Burke, W. A. (1978). Organic carbon and deoxygenation in the Pamlico River estuary. *Water Resour. Res. Inst. Univ. N.C., Proj. A-088-NC.*

Day, J. W., Smith, W. G., Wagner, P. R., and Stone, W. C. (1973). "Community Structure and Carbon Budget of a Salt Marsh and Shallow Bay Estuarine System in Louisiana," Publ. No. LSU-SG-72-04. Center for Wetlands Resources, Louisiana State University, Baton Rouge.

Day, J. W., Jr., Hopkinson, C. S., and Conner, W. H. (1982). An analysis of environmental factors regulating community metabolism and fisheries production in a Louisiana estuary. *In* "Estuarine Comparisons" (V. S. Kennedy, ed.), pp. 121-136. Academic Press, New York.

Deck, B. (1980). Nutrient element distribution in the Hudson estuary. Ph.D. Thesis, Columbia University, New York.

Denigès, G. (1911). A rapid test for nitrites and nitrates in water by means of a new hydro-strychnine reagent. *Bull. Soc. Chim. Fr. 9*, 544-546.

Denigès, G. (1921). Détermination quantitative de plus faibles quantités de phosphates dans les produites biologiques par la méthod céruléomolybdique. *C. R. Seances Soc. Biol. Ses Fil. 84*, 875-877.

Douglas, G. S. (1981). The distribution of dissolved and adsorbed ammonia in Narragansett Bay sediments. M.S. Thesis, University of Rhode Island, Kingston.

Droop, M. R. (1973). Nutrient limitation in osmotrophic protista. *Am. Zool. 13*, 209-214.

Durbin, A. G. (1976). The role of fish migration in two coastal ecosystems. 1. The Atlantic menhaden, *Brevoortia tyrannus*, in Narragansett Bay, R.I. 2. The anadromous alewife, *Alosa pseudoharengus*, in Rhode Island ponds. Ph.D. Thesis, University of Rhode Island, Kingston.

Elkins, J. W., Wofsy, S. C., McElroy, M. B., and Kaplan, W. A. (1981). Nitrification and production of N_2O in the Potomac: Evidence for variability. *In* "Estuaries and Nutrients" (B. J. Neilson and L. E. Cronin, eds.), pp. 447-464. Humana Press, Clifton, New Jersey.

Estabrook, R. H. (1973). Phytoplankton ecology and hydrography of Apalachicola Bay. M.S. Thesis, Florida State University, College of Arts and Sciences, Tallahassee.

Fanning, K. A., and Pilson, M. E. Q. (1973). The lack of inorganic removal of dissolved silica during river-ocean mixing. *Geochim. Cosmochim. Acta 37*, 2405-2415.

Flemer, D. A. (1970). "Final Report to Office of Water Resources Research on the Effects of Thermal Loading and Water Quality on Estuarine Primary Production." Chesapeake Biological Laboratory, Natural Resources Institute, University of Maryland, College Park.

Furnas, M. J., Hitchcock, G. L., and Smayda, T. J. (1976). Nutrient-phytoplankton relationships in Narragansett Bay during the 1974 summer bloom. *In* "Estuarine Processes" (M. L. Wiley, ed.), Vol. 1, pp. 118-134. Academic Press, New York.

Gaarder, T., and Gran, H. H. (1927). Investigation of the production of plankton in the Oslo Fjord. *Rapp. P.-V. Reun. Cons. Int. Explor. Mer 42,* 1-48.

Gaarder, T., and Sparck, R. (1931). Biochemical and biological investigations of the variations in the productivity of the West Norwegian oyster polls. *Rapp. P.-V. Reun. Cons. Int. Explor. Mer 75,* 47-58.

Garber, J. H. (1982). ^{15}N-tracer and other laboratory studies of nitrogen remineralization in sediments and waters from Narragansett Bay, Rhode Island. Ph.D. Thesis, University of Rhode Island, Kingston.

Garside, C., Malone, T. C., Roels, O. A., and Shartstein, B. A. (1976). On evaluation of sewage-derived nutrients and their influence on the Hudson Estuary and New York Bight. *Estuarine Coastal Mar. Sci. 4,* 281-289.

Glibert, P. M. (1982). Regional studies of daily, seasonal, and size fraction variability in ammonium remineralization. *Mar. Biol. (Berlin) 70,* 209-222.

Glibert, P. M., and Goldman, J. C. (1981). Rapid ammonium uptake by marine phytoplankton. *Mar. Biol. Lett. 2,* 25-31.

Goldhaber, M. B., Aller, R. C., Cochran, J. R., Rosenfeld, J. R., Martens, C. S., and Berner, R. A. (1977). Sulfate reduction, diffusion and bioturbation in Long Island Sound sediments: Report of the FOAM group. *Am. J. Sci. 277,* 193-237.

Goode, G. B., and Associates, eds. (1887). "The Fisheries and Fishery Industries of the United States," p. 786. U.S. Comm. Fish Fish., U.S. Govt. Printing Office, Washington, D.C.

Gross, F. (1947). An experiment in marine fish cultivation. I. Introduction. V. Fish growth in a fertilized sea loch. *Proc. R. Soc. Edinburgh, Sect. B: Biol. 63*(1-2), 56-95.

Gross, F. (1950). Fish cultivation in an arm of a sea-loch. I. Introduction. V. Fish growth in Kyle Scotnish. *Proc. R. Soc. Edinburgh, Sect. B: Biol. 64*(1-4), 109-135.

Hahn, J., and Crutzen, P. J. (1982). The role of fixed nitrogen in atmospheric photochemistry. *Philos. Trans. R. Soc. London, Ser. B 296,* 521-541.

Harris, E. (1959). The nitrogen cycle in Long Island Sound. *Bull. Bingham Oceanogr. Collect. 17,* 31-65.

Harris, E., and Riley, G. A. (1956). Oceanography of Long Island Sound, 1952-1954. VIII. Chemical composition of the plankton. *Bull. Bingham Oceanogr. Collect. 15,* 315-323.

Harrison, W. G. (1978). Experimental measurements of nitrogen remineralization in coastal waters. *Limnol. Oceanogr. 23*(4), 684-694.

Harrison, W. G. (1980). Nutrient regeneration and primary production in the sea. *In* "Primary Productivity in the Sea" (P. G. Falkowski, ed.), pp. 433-460. Plenum, New York.

Harrison, W. G., and Hobbie, J. E. (1974). Nitrogen budget of a North Carolina estuary. *Rep.--Water Resour. Res. Inst. Univ. N.C. 86,* 1-172.

Harvey, H. W. (1926). Nitrate in the sea. *J. Mar. Biol. Assoc. U.K.* [N.S.] *14,* 71-88.

Harvey, H. W. (1928). Concerning methods for estimating phosphates and nitrates in solution in sea water. Section II. Estimation of nitrates and nitrites. *Rapp. P.-V. Reun., Cons. Int. Explor. Mer 53,* 96.

Hattori, A., Koike, I., Ohtsu, M., Goering, J. J., and Boisseau, D. (1980). Uptake and regeneration of nitrogen in controlled aquatic ecosystems and the effects of copper on these processes. *Bull. Mar. Sci. 30,* 431-443.

Heinle, D. R., D'Elia, C. F., Taft, J. L., Wilson, J. S., Cole-Jones, M., Caplins, A. B., and Cronin, L. E. (1983). "Historical Review of Water Quality and Climatic Data from Chesapeake Bay with Emphasis on Effects of Enrichment," Synth. Pap., U.S. E.P.A. Chesapeake Bay Study, Annapolis, Maryland.

Hitchcock, C., and Smayda, T. J. (1977). The importance of light in the initiation of the 1972-1973 winter-spring diatom bloom in Narragansett Bay. *Limnol. Oceanogr. 22,* 126-131.

Hobbie, J. E. (1974). Nutrients and eutrophication in the Pamlico River estuary, North Carolina, 1971-1973. *Water Resour. Res. Inst. Univ. N.C., Contrib. 37.*

Hodgkin, E. P., Birch, P. B., Black, R. E., and Humphries, R. B. (1981). "The Peel-Harvey Estuarine System Study (1976-1980)," Rep. No. 9. Dept. of Conserv. and Environ., Western Australia.

Indebo, G., Pengerud, B., and Dundas, I. (1979). Microbial activities in a permanently stratified estuary. II. Microbial activities at the oxic-anoxic interface. *Mar. Biol. (Berlin) 51,* 305-309.

Jaworski, N. A. (1981). Sources of nutrients and the scale of eutrophication problems in estuaries. *In* "Estuaries and Nutrients" (B. J. Neilson and L. E. Cronin, eds.), pp. 83-110. Humana Press, Clifton, New Jersey.

Jaworski, N. A., Lear, D. W., Jr., and Villa, O., Jr. (1972). Nutrient management in the Potomac estuary. *Spec. Symp.--Am. Soc. Limnol. Oceanogr. 1,* 246-273.

Jay, D. A., and Bowman, M. J. (1975). "The Physical Oceanography and Water Quality of New York Harbor and Western Long Island Sound," Tech. Rep. No. 13, Ref. No. 75-77. Mar. Sci. Res. Cent., State University of New York, Stony Brook.

Jeffries, H. P. (unpublished). Ecological consequences of sewer outfalls in the Raritan estuary: An impact statement comparing six options. University of Rhode Island, Kingston (unpublished report).

Johannes, R. E. (1969). Nutrient regeneration in lakes and oceans. *Adv. Microbiol. Sea 1,* 203-212.

Johnstone, J. (1908). "Conditions of Life in the Sea." Cambridge Univ. Press, London and New York (reprinted by Arno Press, New York, 1977).

Kamitami, A. (1969). Regeneration of inorganic nutrients from diatom decomposition. *J. Oceanogr. Soc. Jpn. 25,* 63-74.

Kato, K. (1956). Chemical investigations on marine humus in bottom sediments. *Mem. Fac. Fish., Hokkaido Univ. 4,* 91-209.

Kelly, J. R. (1982). Benthic-pelagic coupling in Narragansett Bay. Ph.D. Thesis, University of Rhode Island, Kingston.

Kemp, W. M., and Boynton, W. R. (unpublished). Nutrient budgets of the Patuxent Estuary: Sources, sinks and internal cycles. Horn Point Environmental Laboratories, Cambridge, Maryland.

Kemp, W. M., Wetzel, R. L., Boynton, W. R., D'Elia, C. F., and Stevenson, J. C. (1982). Nitrogen cycling and estuarine interfaces: Some current concepts and research directions. *In* "Estuarine Comparisons" (V. Kennedy, ed.), pp. 209-230. Academic Press, New York.

Ketchum, B. H. (1967). Phytoplankton nutrients in estuaries. *In* "Estuaries" (G. H. Lauff, ed.), Publ. No. 83, pp. 329-335. Am. Assoc. Adv. Sci., Washington, D.C.

Koike, I., and Hattori, A. (1978). Simultaneous determinations of nitrification and nitrate reduction in coastal sediments by a ^{15}N dilution technique. *Appl. Environ. Microbiol. 35,* 853-857.

Kremer, J. N., and Nixon, S. W. (1978). "A Coastal Marine Ecosystem, Simulation and Analysis," Ecol. Stud. No. 24. Springer-Verlag, Berlin and New York.

Kremer, P. (1975). The ecology of the ctenophore *Mnemiopsis leidyi* in Narragansett Bay. Ph.D. Thesis, University of Rhode Island, Kingston.

Kuenzler, E. J., Stanley, D. W., and Koenings, J. P. (1979). Nutrient kinetics of phytoplankton in the Pamlico River, North Carolina. *Water Resour. Res. Inst. Univ. N.C., Proj. B-092-NC.*

Lauff, G. H., ed. (1967). "Estuaries," Publ. No. 83. Am. Assoc. Adv. Sci., Washington, D.C.

Lee, V. (1980). "An elusive Compromise: Rhode Island Coastal Ponds and Their People," Mar. Tech. Rep. No. 73. University of Rhode Island, Kingston.

Liebig, J. (1855). "Principles of Agricultural Chemistry with Special Reference to the Late Researches Made in England." Excerpt in "Cycles of Essential Elements" (L. R. Pomeroy, ed.), Dowden, Hutchinson & Ross, Inc., Stroudsburg, Pennsylvania.

Lillick, L. (1937). Seasonal studies of the phytoplankton off Woods Hole, Massachusetts. *Biol. Bull. (Woods Hole, Mass.) 73,* 488-503.

Lipschultz, F., Cunningham, J. J., and Stevenson, J. C. (1979). Nitrogen fixation associated with four species of submerged angiosperms in the central Chesapeake Bay. *Estuarine Coastal Mar. Sci. 9,* 813-818.

Liss, P. S. (1976). Conservative and non-conservative behavior of dissolved constituents during estuarine mixing. *In* "Estuarine Chemistry" (J. D. Burton and P. S. Liss, eds.), pp. 93-130. Academic Press, New York.

Livingston, R. J., and Sheridan, P. F. (1979). Cyclic trophic relationships of fishes in an unpolluted, river-dominated estuary in North Florida. *In* "Ecological Processes in Coastal and Marine Systems" (R. J. Livingston, ed.), pp. 143-161. Plenum, New York.

Loder, T. C., and Reichard, R. P. (1981). The dynamics of conservative mixing in estuaries. *Estuaries 4,* 64-69.

McCarthy, J. J. (1980). Nitrogen. *In* "The Physiological Ecology of Phytoplankton" (I. Morris, ed.), pp. 191-233. Blackwell, Oxford.

McCarthy, J. J., and Goldman, J. C. (1979). Nitrogenous nutrition of marine phytoplankton in nutrient-depleted waters. *Science 203,* 670-672.

McCarthy, J. J., Taylor, W. R., and Taft, J. L. (1975). The dynamics of nitrogen and phosphorus cycling in the open waters of the Chesapeake Bay. *ACS Symp. Ser. 18,* 664-681.

McElroy, M. B., Elkins, J. W., Wofsy, S. C., Kolb, C. E., Duran, A. P., and Kaplan, W. A. (1978). Production and release of N_2O from the Potomac estuary. *Limnol. Oceanogr. 23,* 1168-1182.

Malone, T. C. (1977). Environmental regulation of phytoplankton productivity in the Lower Hudson Estuary. *Estuarine Coastal Mar. Sci. 5,* 157-171.

Mann, K. H. (1972a). Ecological energetics of the seaweed zone in a marine bay on the Atlantic coast of Canada. II. Productivity of the seaweeds. *Mar. Biol. (Berlin) 14,* 199-209.

Mann, K. H. (1972b). Macrophyte production and detritus food chains in coastal waters. *Mem. Ist, Ital. Idrobiol. Dott, Marco de Marchi, Suppl. 29,* 353-383.

Marshall, N. (1972). Interstitial community and sediments of shoal benthic environments. *Mem. Geol. Soc. Am. 133,* 409-415.

Martin, J. H. (1968). Phytoplankton-zooplankton relationships in Narragansett Bay. III. Seasonal changes in zooplankton excretion rates in relation to phytoplankton abundance. *Limnol. Oceanogr. 13,* 63-71.

Mather, F. (1887). New York and its fisheries. *In* "The Fisheries and Fishery Industries of the United States" (G. B. Goode & Associates, eds.), pp. 341-377. U.S. Comm. Fish Fish., U.S. Government Printing Office, Washington, D.C.

Melosi, M. V. (1981). "Garbage in the Cities, Refuse, Reform, and the Environment, 1880-1980." Texas A&M Univ. Press, College Station.

MERL (1980). "Some Aspects of Water Quality in and Pollution Sources to the Providence River." Marine Ecosystems Research Laboratory, University of Rhode Island, Kingston. Report for Region I, EPA, Sept. 1979-Sept. 1980.

Nash, C. B. (1947). Environmental characteristics of a river estuary. *J. Mar. Res. 6,* 147-174.

Neilson, B. J., and Cronin, L. E., eds. (1981). "Estuaries and Nutrients." Humana Press, Clifton, New Jersey.

Newcombe, C. L., Horne, W. A., and Shepherd, B. B. (1939). Studies on the physics and chemistry of estuarine waters in Chesapeake Bay. *J. Mar. Res. 2,* 87-116.

Nixon, S. W. (1981a). Freshwater inputs and estuarine productivity. *In* "Freshwater Inflow to Estuaries" (R. D. Cross and D. L. Williams, eds.). *U.S. Fish Wildl. Serv., Off. Biol. Serv. FWS/OBS FWS/OBS-81/04,* Vol. 1, pp. 31-57.

Nixon, S. W. (1981b). Remineralization and nutrient cycling in coastal marine ecosystems. *In* "Estuaries and Nutrients" (B. J. Neilson and L. E. Cronin, eds.), pp. 111-138. Humana Press, Clifton, New Jersey.

Nixon, S. W. (1982). Nutrient dynamics, primary production and fisheries yields of lagoons. *Oceanol. Acta* Spec. Pub.: 357-371.

Nixon, S. W. (1983). Nutrient dynamics and the productivity of marine coastal waters. *Proc. 1st Arabian Gulf Conf. Environ. Pollut., 1982 Kuwait* (in press).

Nixon, S. W., and Kremer, J. N. (1977). Narragansett Bay--the development of a composite simulation model for a New England estuary. *In* "Ecosystem Modeling in Theory and Practice" (C. Hall and J. Day, eds.), pp. 622-673. Wiley (Interscience), New York.

Nixon, S. W., and Lee, V. (1981). The flux of carbon, nitrogen and phosphorus between coastal lagoons and offshore waters. *In* "Coastal Lagoon Research, Present and Future," pp. 325-348. Div. Mar. Sci., UNESCO, Paris.

Nixon, S. W., Oviatt, C. A., and Hale, S. S. (1976). Nitrogen regeneration and the metabolism of coastal marine bottom communities. *In* "The Role of Terrestrial and Aquatic Organisms in Decomposition Processes" (J. M. Anderson and A. Macfadyen, eds.), pp. 269-283. Blackwell, Oxford.

Nixon, S. W., Oviatt, C. A., Kremer, J. N., and Perez, K. (1979). The use of numerical models and laboratory microcosms in estuarine ecosystem analysis-simulations of a winter phytoplankton bloom. *In* "Marsh-Estuarine Systems Simulation" (R. F. Dame, ed.), pp. 165-188. Univ. of South Carolina Press, Columbia.

Nixon, S. W., Kelley, J. R., Furnas, B. N., Oviatt, C. A., and Hale, S. S. (1980). Phosphorus regeneration and the metabolism of coastal marine bottom communities. *In* "Marine Benthic Dynamics" (K. R. Tenore and B. C. Coull, eds.), pp. 219-242. Univ. of South Carolina Press, Columbia.

Nixon, S. W., Furnas, B. N., Chinman, R., Granger, S., and Heffernan, S. (1982). "Nutrient Inputs to Rhode Island Coastal Lagoons and Salt Ponds," Final Report to R. I. Office of Statewide Planning. University of Rhode Island, Kingston.

Odum, H. T., Nixon, S. W., and DiSalvo, L. H. (1969). Adaptations for photoregenerative cycling. *In* "The Structure and Function of Fresh Water Microbial Communities" (J. Cairns, Jr., ed.), pp. 1-29. Virginia Polytechnic Institute Press, Blacksburg.

Oglesby, R. T. (1977). Relationships of fish yield to lake phytoplankton standing crop, production, and morphoedaphic factors. *J. Fish. Res. Board Can. 34,* 2271-2279.

Okuda, T. (1955). On the soluble nutrients in bay deposits. III. Examinations on the diffusion of soluble nutrients to sea water from mud. *Bull. Tohoku Reg. Fish. Res. Lab. 4,* 215-242.

Okuda, T. (1960). Metabolic circulation of phosphorus and nitrogen in Matsushima Bay (Japan) with special reference to exchange of these elements between sea water and sediments. *Trab. Inst. Oceanogr. Univ. Recife 2,* 7-153.

Olsen, S., Robadue, D. D., Jr., and Lee, V. (undated). "An Interpretive Atlas of Narragansett Bay," Mar. Bull. No. 40. Coastal Resources Center, University of Rhode Island, Kingston.

O'Reilly, J. E., Thomas, J. P., and Evans, C. (1976). Annual primary production (nannoplankton, net plankton, dissolved organic matter) in the Lower New York Bay. *In* "Fourth Symposium on Hudson River Ecology" (W. H. McKeon and G. H. Lauer, eds.), paper No. 19. Hudson River Environ. Soc. Inc., New York.

Oren, A., and Blackburn, J. H. (1979). Estimation of sediment denitrification rates at *in situ* nitrate concentrations. *Appl. Microbiol. 30,* 707-709.

Orr, A. P. (1947). An experiment in marine fish cultivation. II. Some physical and chemical conditions in a fertilized sea-loch (Loch Craiglin, Argyll). *Proc. R. Soc. Edinburgh, Sect. B: Biol. 63,* 3-20.

Pamatmat, M. M. (1968). Ecology and metabolism of a benthic community on an intertidal sandflat. *Int. Rev. Gesamten Hydrobiol. 53*(2), 211-298.

Peterson, D. H. (1979). Sources and sinks of biologically reactive oxygen, carbon, nitrogen, and silica in northern San Francisco Bay. *In* "San Francisco Bay: The Urbanized Estuary" (T. J. Conomos, ed.), pp. 175-193. Allen Press, Lawrence, Kansas.

Phoel, W. C., Webb, K. L., and D'Elia, C. F. (1981). Inorganic nitrogen regeneration and total oxygen consumption by sediments at the mouth of the York River, Virginia. *In* "Estuaries and Nutrients" (B. J. Neilson and L. E. Cronin, eds.), pp. 607-618. Humana Press, Clifton, New Jersey.

Pilson, M. E. Q. (1982). Annual cycles of nutrient concentrations in Narragansett Bay, R.I. *EOS, Trans. Am. Geophys. Union 63,* 347.

Pilson, M. E. Q., Oviatt, C. A., and Nixon, S. W. (1980). Annual nutrient cycles in a marine microcosm. *In* "Microcosms in Ecological Research" (John P. Giesy, ed.), *DOE Symp. Ser.,* CONF 7811-1, 753-778.

Platt, T. (1971). The annual production by phytoplankton in St. Margaret's Bay, Nova Scotia. *J. Cons., Cons. Int. Explor. Mer. 33,* 324-334.

Pomeroy, L. R. (1974). "Cycles of Essential Elements," Benchmark Papers in Ecology, Vol. 1. Dowden, Hutchinson & Ross, Inc., Stroudsburg, Pennsylvania.

Postma, H. (1966). The cycle of nitrogen in the Wadden Sea and adjacent areas. *Neth. J. Sea Res. 3,* 186-221.

Postma, H. (1978). The nutrient contents of North Sea water: Changes in recent years, particularly in the Southern Bight. *Rapp. P.-V. Reun., Cons. Int. Explor. Mer 172,* 350-357.

Premuzic, E. T. (1980). "Organic Carbon and Nitrogen in the Surface Sediments of World Oceans and Seas: Distribution and Bottom Topography." Brookhaven Natl. Lab., U.S. Dept. Energy, Upton, Long Island.

Redfield, A. C. (1934). On the proportions of organic derivatives in sea water and their relation to the composition of plankton. *In* "James Johnstone Memorial Volume," pp. 176-192. Liverpool Univ. Press, Liverpool.

Rhodes, D. C., Aller, R. C., and Goldhaber, M. B. (1977). The influence of colonizing macrobenthos on physical properties and chemical diagenesis of the estuarine seafloor. *In* "Ecology of Marine Benthos" (B. C. Coull, ed.), pp. 113-138. Univ. of South Carolina Press, Columbia.

R Riley, G. A. (1955). Review of the oceanography of Long Island Sound. *Deep-Sea Res. 3,* Suppl., 224-238.

Riley, G. A. (1956). Oceanography of Long Island Sound, 1952-1954. II. Physical oceanography. *Bull. Bingham Oceanogr. Collect. 15,* 15-46.

Riley, G. A. (1959). Oceanography of Long Island Sound, 1954-1955. *Bull. Bingham Oceanogr. Collect. 17,* 9-30.

Riley, J. P. (1975). Analytical chemistry of sea water. *In* "Chemical Oceanography" (J. P. Riley and G. Skirrow, eds.), 2nd ed., Vol. 3, pp. 193-514. Academic Press, New York.

Rosenfeld, J. K. (1977). Nitrogen diagenesis in nearshore anoxic sediments. Ph.D. Thesis, Yale University, New Haven, Connecticut.

Russel, P., Taras, P., Bursztynsky, A., Jackson, L. A., and Leong, E. Y. (1980). Water and waste inputs to San Francisco Bay estuary--A historical perspective. *61st Annu. Meet. Pac. Div./ Am. Assoc. Adv. Sci.*

Ryther, J. H., and Dustan, W. M. (1971). Nitrogen, phosphorus, and eutrophication in the coastal marine environment. *Science 171,* 1008-1013.

Santschi, P. H. (1980). A revised estimate for trace metal fluxes to Narragansett Bay: A comment. *Estuarine Coastal Mar. Sci. 11,* 115-118.

Schindler, D. W. (1981). Studies of eutrophication in lakes and their relevance to the estuarine environment. *In* "Estuaries and Nutrients" (B. J. Neilson and L. E. Cronin, eds.), pp. 71-82. Humana Press, Clifton, New Jersey

Seitzinger, S. P. (1982). The importance of denitrification and nitrous oxide production in the nitrogen dynamics and ecology of Narragansett Bay, Rhode Island. Ph.D. Thesis, University of Rhode Island, Kingston.

Seitzinger, S. P., Burke, S., Garber, J., Nixon, S., and Pilson, M. (1978). Nitrogen fixation and denitrification measurements in Narragansett Bay sediments. *41st Annu. Meet., A.S.L.O.* Abstracts.

Seitzinger, S. P., Nixon, S., Pilson, M. E. Q., and Burke, S. (1980). Denitrification and N_2O production in near-shore marine sediments. *Geochim. Cosmochim. Acta 44,* 1853-1860.

Seitzinger, S. P., Pilson, M. E. Q., and Nixon, S. W. (1983). N_2O production in near-shore marine sediments. *Science* (in press).

Sharp, J. H., Church, T. M., and Culberson, C. H. (1980). "Data from the 1977 Trans X Cruises," Data Rep. No. 1. Coll. Mar. Stud., University of Delaware, Newark.

Sharp, J. H., Frake, A. C., Hillier, G. B., and Underhill, P. A. (1982). Modeling nutrient regeneration in the ocean with an aquarium system. *Mar. Ecol. Prog. Ser. 8,* 15-23.

Sharp, J. H., Culberson, C. H., and Church, T. M. (1982). The chemistry of the Delaware estuary. General considerations. *Limnol. Oceanogr. 27*(6), 1015-1028.

Sheih, M. S. J. (1974). Nutrients in Narragansett Bay sediments. M.S. Thesis, University of Rhode Island, Kingston.

Smayda, T. J. (1973). The growth of *Skeletonema costatum* during a water-spring bloom in Narragansett Bay, Rhode Island. *Norw. J. Bot. 20,* 219-247.

Smith, R. E., Herndon, R. E., and Harman, D. D. (1979). Physical and chemical properties of San Francisco Bay waters, 1969-1976. *U.S. Geol. Surv. Open-File Rep. (U.S.)* 79-511.

Smith, S. L. (1978). The role of zooplankton in the nitrogen dynamics of a shallow estuary. *Estuarine Coastal Mar. Sci. 7,* 555-565.

Smith, S. V. (1981). Responses of Kaneohe Bay, Hawaii, to relaxation of sewage stress. *In* "Estuaries and Nutrients" (B. J. Neilson and L. E. Cronin, eds.), pp. 391-410. Humana Press, Clifton, New Jersey.

Smith, S. V., Kimmerer, W. J., Laws, E. A., Brock, R. E., and Walsh, T. W. (1980). Kaneohe bay sewage diversion experiment: Perspective on ecosystem responses to nutritional perturbation. *Pac. Sci. 35,* 279-395.

Smullen, J. T., Taft, J. L., and Macknis, J. (1982). "Nutrient and Sediment Loads to the Tidal Chesapeake Bay System," pp. 147-162. Synth. Pap. U.S. E.P.A. Chesapeake Bay Project, Annapolis, Maryland.

Solorzano, L. (1969). Determination of ammonia in natural waters by the phenolhypochlorite method. *Limnol. Oceanogr. 14,* 799-801.

Sorensen, J. (1978). Capacity for denitrification and reduction of nitrate to ammonia in a coastal marine sediment. *Appl. Environ. Microbiol. 35,* 301-305.

South Alabama Regional Planning Commission (1979). "Water Quality Management Plan." Mobile and Baldwin Counties, Alabama.

Steeman-Nielson, E. (1952). The use of radioactive carbon for measuring organic production in the sea. *J. Cons., Cons. Int. Explor. Mer 18,* 117-140.

Strickland, J. D. H. (1967). Between beakers and bays. *New Sci. 2,* 276-278.

Stross, R. G., and Stottlemyer, J. R. (1965). Primary production in the Patuxent River. *Chesapeake Sci. 6*(3), 125-140.

Takahashi, M., Koike, I., Iseki, K., Biefang, P. K., and Hattori, A. (1982). Phytoplankton species' response to nutrient changes in experimental closures and coastal waters. *In* "Marine Mesocosms" (G. D. Grice and M. R. Reeve, eds.), pp. 333-340. Springer-Verlag, Berlin and New York.

Taylor, W. R., and Grant, V. (1977). "Phytoplankton Ecology Project, Nutrient and Chlorophyll Data, Aesop Cruises, April 1969 to April 1971," Chesapeake Bay Inst., Spec. Rep. No. 61. Johns Hopkins University, Baltimore, Maryland.

Thorne-Miller, B., Harlin, M. M., Thursby, G. B., Brady-Campbell, M. M., and Dworetzky, B. A. (1983). Variations in the distribution and biomass of submerged macrophytes in five coastal lagoons in Rhode Island, U.S.A. *Bot. Mar. 26,* 231-242.

Vargo, G. A. (1976). The influence of grazing and nutrient excretion by zooplankton on the growth and production of the marine diatom, *Skeletonema costatum* (Greville) Cleve, in Narragansett Bay. Ph.D. Thesis, University of Rhode Island, Kingston.

Vargo, G. A. (1979). The contribution of ammonia excreted by zooplankton to phytoplankton production in Narragansett Bay. *J. Plankton Res. 1,* 78-84.

Vollenweider, R. A. (1976). Advances in defining critical loading levels for phosphorus in lake eutrophication. *Mem. Ist. Ital. Idrobiol. Dott. Marco de Marchi 33,* 53-83.

von Brand, T., Rakestraw, N. W., and Renn, C. E. (1937). The experimental decomposition of nitrogenous organic matter in sea water. *Biol. Bull. (Woods Hole, Mass.) 72,* 165-175.

Waksman, S. A., and Hotchkiss, M. (1938). On the oxidation of organic matter in marine sediments by bacteria. *J. Mar. Res. 1,* 101-118.

Waksman, S. A., Carey, C. L., and Reuszer, W. H. (1933). Marine bacteria and their role in the cycle of life in the sea. I. Decomposition of marine plant and animal residues by bacteria. *Biol. Bull. (Woods Hole, Mass.) 65,* 57-79.

Walsh, J. J., Rowe, G. T., Iverson, R. L., and McRoy, C. P. (1981). Biological export of shelf carbon is a sink of the global CO_2 cycle. *Nature (London) 291,* 196-201.

Wang, W. C., Yung, Y. L., Lacis, A. A., Mo, J., and Hansen, J. E. (1976). Greenhouse effects due to man-made perturbations of trace gases. *Science 194,* 685-690.

Wangersky, P. J. (1977). The role of particulate matter in the productivity of surface waters. *Helgol. Wiss. Meeresunters. 30,* 546-564.

Wattenberg, H. (1928). A simple method for the direct estimation of ammonia in sea-water by the use of Nessler's Reagent. *Rapp. P.-V. Reun., Cons. Int. Explor. Mer 53,* 108-114.

Webb, K. L. (1981). Conceptual models and processes of nutrient cycling in estuaries. *In* "Estuaries and Nutrients" (B. J. Neilson and L. E. Cronin, eds.), pp. 25-46. Humana Press, Clifton, New Jersey.

Webb, K. L., and D'Elia, C. F. (1980). Nutrient and oxygen redistribution during a spring neap tidal cycle in a temperate estuary. *Science 207,* 983-985.

Whaley, R. C., Carpenter, J. H., and Baker, R. L. (1966). "Nutrient Data Summary 1964, 1965, 1966: Upper Chesapeake Bay (Smith Pt. to Turkey Pt.) Potomac, South, Severn, MaGothy, Back, Chester and Miles Rivers and Eastern Bay," Chesapeake Bay Inst. Spec. Rep. No. 12, Ref. No. 66-4. Johns Hopkins University, Baltimore, Maryland.

Wood, L. B. (1982). "The Restoration of the Tidal Thames." Adam Hilger, Ltd., Bristol.

Wood, W. (1634). "New England's Prospect." The Cotes, London (edited by A. T. Vaughan and reprinted by University of Massachusetts Press, Amherst, 1977).

Zeitzchel, B. F. (1980). Sediment-water interactions in nutrient dynamics. *In* "Marine Benthic Dynamics" (K. R. Tenor and B. C. Coull, eds.), pp. 195-212. Univ. of South Carolina Press, Columbia.

Chapter 17

Nitrogen in Salt Marsh Ecosystems

IVAN VALIELA
Boston University Marine Program
Marine Biology Laboratory
Woods Hole, Massachusetts

I. THE IMPORTANCE OF NITROGEN IN SALT MARSHES

A. Impact on Primary Producers

The supply and subsequent availability of nitrogen has fundamental consequences for primary producers in salt marshes. Increased supplies of nitrogen lead to at least five major changes in the plant community: (1) greater standing biomass and productivity of marsh plants (Valiela and Teal, 1974; Sullivan and Daiber, 1974; Gallagher, 1975; Patrick and Delaune, 1976; Tyler, 1967; Broome *et al.*, 1975; Jefferies, 1977; Jefferies and Perkins, 1977; Chalmers, 1979); (2) changes in the morphology of *Spartina alterniflora*, the dominant primary producer of salt marshes of eastern North America, from small, thin plants to considerably taller, robust plants set farther apart (Valiela *et al.*, 1978a);

ISBN 0-12-160280-X

(3) increased percent of plants that bear seed (I. Valiela and J. M. Teal, unpublished data); (4) altered abundances of plant species (Valiela *et al.*, 1975), probably related to differences in the rates of nitrogen uptake of the various species of marsh plants (*Distichlis spicata*, for example, takes over space previously occupied by *S. alterniflora* and *S. patens*); (5) higher nitrogen percent in plant biomass (Vince *et al.*, 1981; Patrick and Delaune, 1976). These are results of a long-term study I have been conducting in Great Sippewissett marsh jointly with J. M. Teal and others in which we chronically have added nutrients to salt marsh plots. We can, however, find natural stands that are very similar to even the most enriched plots. Salt marshes exposed to natural or man-made nitrogen enrichments thus have quite different rates of production, standing stock, species composition, and quality of plant biomass.

Unicellular salt marsh algae also show marked responsed to eutrophication. The species composition changes and production rate of benthic microalgae increases when more nitrogen is available (Van Raalte *et al.*, 1976a,b; Sullivan and Daiber, 1975). The effects of nitrogen fertilization on algae are not as evident as they are for *S. alterniflora* because benthic microalgae are affected by the macrophyte canopy. During the summer, the shading of microalgae by *S. alterniflora* prevents a response of the algae to increased nitrogen after the canopy forms (Van Raalte *et al.*, 1976a).

B. Impact on Animals

Herbivores are limited by the nutritive quality of their food, especially the nitrogen content (Mattson, 1980). In salt marsh plots where the nitrogen content of plants is increased, where is a clear-cut increase in the abundance of the herbivore fauna (Valiela *et al.*, 1982; Vince *et al.*, 1981).

Herbivores in salt marshes consume less than 10% of the annual above-ground primary production (Teal, 1962; Parsons and de la Cruz, 1980), so that salt marsh food webs are principally detrital.* About 70% or so of the annual above-ground production is consumed by decomposers. An amount of particulate matter equivalent to about 20% is exported from the marsh by tides (Valiela *et al.*, 1982) and decomposes elsewhere.

Even though salt marsh detritus is plentiful, its quality as a food for animals is poor. The C:N ratio of salt marsh litter ranges from 20 to 60 (I. Valiela and J. M. Teal, unpublished data). These values are much higher than those of phytoplankton (5-6), proteins (4.5), and bacteria (5.5, Fenchel and Blackburn, 1979), whereas sediments may have a ratio of 10 or more (Parsons and Takahashi, 1973). Animals require a C:N of $\sim$17 as a maintenance ration (Russell-Hunter, 1970), so that salt marsh detritus is a rather poor-quality food.

In part, this low quality comes about because salt marsh grasses are C_4 plants that typically have low nitrogen content (Caswell *et al.*, 1973). After plant senescence, leaching of litter then quickly removes 30% or more of the soluble nitrogen (I. Valiela and J. M. Teal, unpublished data). Eventually, nitrogen does build up in detritus, but a substantial part of this buildup is nitrogen bound in relatively unavailable forms, including chitin and complexes with phenolic compounds (Leatham *et al.*, 1980).

Because detritus usually contains so little easily available nitrogen, detritivores respond markedly to improved food quality, especially increased nitrogen content (Tenore and Rice, 1980). In our fertilization experiments where the nitrogen content of litter increased from 0.9% to 1.5% N, we see a four- to fivefold

In certain localities where overwintering geese congregate, these herbivores may consume about 60% of the production (Smith and Odum, 1981).

increase in abundance of detritivores (I. Valiela and J. M. Teal, unpublished data).

Fish feed on the invertebrate detritivores in salt marshes, and since in our experiments the increased nitrogen supply led to augmented prey biomass, fish ought also to grow more and be more abundant. We do have some evidence for this (Connor, 1980), but by the time the effect of the increased nitrogen goes through three trophic steps, the response to the fertilization is not as clear as is the case with plants, herbivores, and detritivores.

There is a further interaction in addition to increased abundance of prey that also would tend to increase fish growth. The greater spacing between plants in eutrophied stands of *S. alterniflora* allows fish to hunt more freely on the salt marsh surface (Vince *et al.*, 1976). Here again, as in the case of shading of algae by the grasses, we have a potentially important second-order interaction that is affected by nitrogen supply.

C. Impact on Microorganisms

On the marsh surface, bacteria and fungi decompose about 70% of net above-ground production, mainly aerobically, and nitrogen-enriched litter decays faster than control litter (Valiela *et al.*, 1982). Below-ground production is larger than above-ground, and virtually 100% is degraded *in situ* (Valiela *et al.*, 1976), chiefly by fermenters and sulfate reducers (Howarth and Teal, 1980). Denitrification and methanogenesis consume relatively small amounts of organic matter (Kaplan *et al.*, 1979; King and Wiebe, 1978).

The degree of nitrogen enrichment has variable effects on specific microbial processes. The usual notion is that fermenters and sulfate reducers are limited by the supply of organic substrates rather than nitrogen (Howarth and Teal, 1979). Rates of sulfate reduction do not increase in our experimental plots where the below-ground organic matter is enriched in N, but we have evidence from buried litter bags that decay of organic matter is greater in enriched sediments. Since the activity of fermenters and sulfate reducers is enough to process much of the belowground production, this discrepancy requires further study.

Increased nitrogen in interstitial water sharply reduces the activity of nitrogen fixers (Van Raalte *et al.*, 1974; Carpenter *et al.*, 1978) but only slightly increases the rates of denitrification (Kaplan, 1977). Nitrogen oxides (NO, N_2O) inhibit the production of methane in salt marsh sediments, but the mechanism is not understood (Balderston and Payne, 1976).

This litany of effects of nitrogen on plants, animals, and microbes is far from complete, but suffices to show that nitrogen is a key element in many relationships that determine the ecological structure of the salt marsh, including abundance and activity of microorganisms, plants, and animals.

II. INPUTS, OUTPUTS, AND TRANSFORMATION OF NITROGEN IN SALT MARSHES

A. Mechanisms of Nitrogen Exchange

Mechanisms that transport nitrogen in and out of a salt marsh include biological (input through nitrogen fixation, output through denitrification) and physical pathways (gains through precipitation, ground- or surface-water flow, losses into the sediments, and both gains and losses by means of tidal exchanges).

1. *Inputs of Nitrogen through Flow of Freshwater*

Freshwater, either via streams or groundwater flow, can contribute significantly to nitrogen inputs into a salt marsh. River flow may contribute about 3.1 g $N \cdot m^{-2} \cdot year^{-1}$ to southeastern United States salt marshes (Windom *et al.*, 1975). At Great Sippewissett on Cape Cod, groundwater flow may bring about 12.6 g $N \cdot m^{-2} year^{-1}$ (Valiela and Teal, 1979). Flow of groundwater into a salt marsh can be detected by lowered salinities of ebbing tidal waters (Valiela *et al.*, 1978b). In some marshes such as

TABLE I. Annual Inputs into Salt Marshes via Nitrogen Fixation in Various Habitats in Several Locales

Site and Habitat		$g\ N \cdot m^{-2} \cdot year^{-1}$	*Source*
Great Sippewissett Marsh, Massachusetts			
Cyanobacteria in mats			Calculated from data summarized in Carpenter *et al.* (1978), Teal *et al.* (1979), Valiela and Teal (1979)
Not covered by grasses		2.3	
Under tall *Spartina alterniflora*		0.8	
Under short *S. alterniflora*		0.9	
Under *S. patens*		0.3	
Rhizosphere of			Calculated from data summarized in Carpenter *et al.* (1978), Teal *et al.* (1979), Valiela and Teal (1979)
Tall *S. alterniflora*		8.4	
Short *S. alterniflora*		12.1	
		4	
Sediments with no grasses			Calculated from data summarized in Carpenter *et al.* (1978), Teal *et al.* (1979), Valiela and Teal (1979)
Muddy creek bottoms		1.7	
Sandy creek bottom		0.7	
	Total	31.2	
Sapelo Island, Georgia			
Sediment surface		0.4-0.9	Hanson (1977a)
Epiphytes on *S. alterniflora*		0.2-0.5	Hanson (1977a)
Sediment (0-22 cm)		22.6-51.4	Hanson (1977a)
	Total	22.2-52.4	Hanson (1977a)

Bank End, Lancaster, United Kingdom		
Bare mud	0.4	Jones (1974)
Bare mud plus algae	20.1	Jones (1974)
Spartina anglica	17.1	Jones (1974)
Salicornia dolichostachya	2.8	Jones (1974)
Algae on creek banks	31.6	Jones (1974)
Algae in pools	46.2	Jones (1974)
Barataria Bay, Louisiana		
Mud bottom	1.6	Delaune and Patrick (1980)
Tall *S. alterniflora*	15	Delaune and Patrick (1980)
Short *S. alterniflora*	5	Delaune and Patrick (1980)

TABLE II. Qualitative Summary of Studies of Exports and Imports from Salt Marshes to Coastal Waters[a]

	Salinity (o/oo)	NH_4^+	NO_3^-	DON	PN	TN	DOC	POC	TC
Type I									
Crommet Creek, New Hampshire[b]	0-31	I	O						
Flax Pond, New York[c]	26	E	I				O	I	I
Canary Creek, Delaware[d]	10-28	I	I	E	E	E	E	E	E
Sapelo Island, Georgia[e]	20-30	O	O	E	I		E	I	
Stroodorpe Marsh, Netherlands[f]	--				I			I	
Percent of Type I marshes that export:		25	0	100	33	100	67	25	50
Type II									
Great Sippewissett, Massachusetts[g]	25-32	E	E	E	E	E		E	
Gott's Marsh, Maryland[h]	0-9	E	E	E	E			E	
Ware Creek, Virginia[i]	1-7	E	I	E	E	E	E	E	E
Dill Creek, South Carolina[j]	--							E	
Barataria Bay, Louisiana[k]	12-25						E	E	E
Tijuana estuary, California[l]	--	E	E		E			E	
(2 marshes)	--	E	E		E			E	
Type II									
Branford Harbor, Connecticut[m]	16-24	E	I						
Carter Creek, Virginia[i]	2-7	I	I	E	I	E	E	E	E
Percent of Type II and III marshes that export		86	57	100	83	100	100	100	100
Percent of marshes that export		64	36	100	67	100	83	82	80

[a]Marsh types I, II, III are defined in text, but assignation was to an extent arbitrary since it was done mainly from maps or author's descriptions. Key: I, import; E, export; O, no measurable exchange; DON, dissolved organic nitrogen; PN, particulate nitrogen; TN, total nitrogen; DOC, dissolved organic carbon; POC, particulate organic carbon; TC, total carbon.

[b]Daly and Mathieson (1981).

[c]Woodwell et al. (1979).

[d]Lotrich et al. (1979).

[e]Haines et al. (1976).

[f]Wolff et al. (1979).

[g]Valiela et al. (1978b).

[h]Heinle and Flemer (1976).

[i]Axelrad (1974).

[j]Settlemyre and Gardner (1977).

[k]Happ et al. (1977).

[l]Mauriello and Winfield (1978).

[m]B. Welch (unpublished data).

Flax Pond, studied by Woodwell *et al.* (1979), there is no measurable difference in salinity between ebb and flood water. Even in such situations the influence of groundwater should not be discounted because there is often a nearshore depression of salinity relative to deeper coastal waters. The salinity of Flax Pond marsh (about 26°/oo) is, for example, lower than that of the central basin of Long Island Sound (28°/oo; Bowman, 1977). The nitrogen carried by enough freshwater to cause this dilution may be substantial, as it is in Great Sippewissett marsh. Discharge of groundwater into the sea is widespread in many coastal areas (Johannes, 1980; Bokuniewicz, 1980), with major consequences for the coastal environments involved. This is a relatively new topic of research that will surely require study in the future.

2. Inputs through Nitrogen Fixation

Nitrogen fixation rates in the rhizosphere and surface of sediments of various salt marshes (Whitney *et al.*, 1975; Tjepkema and Evans, 1976; Hanson, 1977a,b; Patriquin and McClung, 1978, among others), measured by the acetylene reduction technique, span a similar range. Variability within a marsh is often greater than among marshes (see review by Nixon, 1980), but significant differences emerge when fixation rates are extrapolated to annual values (Table I). In colder climates microbial activity may be restricted during the winter, resulting in lower annual fixation rates. The annual nitrogen fixation rates in salt marshes fall toward the high end of the range of values for a variety of environments (0.007-50 g N m^{-2} $year^{-1}$ (Paul, 1976)). The methods so far used to measure fixation in anaerobic marsh mud are currently being reassessed (Van Berkum and Slager, 1979, Chapter 4, this volume).

3. Exchanges of Nitrogen by Tidal Waters

There are many comments in the literature about the inconsistent behavior of salt marshes with regard to the export of materials, the so-called "outwelling" controversy (Nixon, 1980). This is a specially pertinent issue because export of organic matter and nutrients is a principal and legally accepted argument for the conservation of salt marshes. To an extent, it is not reasonable to expect all salt marshes to do the same thing. Differences in geological setting, rates of sediment delivery, nutrients in water, orientation to current, wind direction, and properties of the surrounding hinterland, all make each salt marsh unique. It is rather the mechanisms involved in the provision and processing of nitrogen that are general; in each case, the local factors, expecially hydrological ones, make for idiosynchratic conditions.

Export and import of a variety of materials have been measured in a number of marshes. The marshes studied include salt and brackish marshes and examples where the vegetation was dominated by different species of plants.

Two studies of exchanges of heavy metals in and out of marshes in Delaware and California have shown net exports of particulates (Pellenbarg and Church, 1979; Lion and Lackie, 1982). There are variable results from marsh to marsh with regard to export of specific nutrients (Table II), but considering the data set as a whole and despite the uniqueness of each marsh, export is the dominant pattern, except for nitrate (bottom row of Table II). High rates of denitrification (see Section II,A,4) and reducing sediments significantly reduce nitrates entering salt marshes. Thus, most of thc nitrogen available for export is in reduced form.

The marshes included in Table II can be roughly classified into three types (Odum *et al.*, 1979): I, those where there is some impairment of free tidal exchange due to long, contorted channels (Crommet Creek, Canary Creek, Stroodorpe marsh), artificial dikes (Sapelo Island marsh), or shallow sills with a deeper lagoon landward (Flax Pond); II, marshes where tidal flow is freer--there may be some constriction in the tidal channels but the channel is not as contorted (Great Sippewissett, Gott's marsh, Ware Creek, Barataria Bay marsh, Tijuana estuary marshes); III, marshes where

TABLE III. Comparison of Several Properties of a Relatively Young (Flax Pond) and an Older Marsh (Great Sippewissett)[a]

	Flax Pond Marsh	*Great Sippewissett Marsh*
Age of marsh (years)	180	2000[b]
Indicators of maturity:		
Average accretion rate (mm $year^{-1}$)		
Expanding part of marsh	1.5-37	14
Established part of marsh	2-6.3	1
(Accretion/net production) 100, in terms of carbon	37%	5.3%
% of area in nonvegetated sediment	47	37
% of area under tall *S. alterniflora*	37	18.8
Average above-ground standing crop of *S. alterniflora* (g m^{-2})	975	350
% of area under *S. patens*	7	18
Number of higher plant species[c]	13	22
Tidal exchanges (E:export; I:import)		
NH_4-N	E > I (>100%)[d]	E > I (74%)
Particulate N	E < I (<100%)[e]	E > I (82%)
Particulate C	E < I (<100%)	E > I (76%)
Total N	-- --	E > I (83%)
Export of N in floating rafts of litter[f]	Significant	Very small

[a]*The age and accretion rates are from Redfield (1972), Redfield and Rubin (1962), Richard (1978), and Flessa et al. (1977). The vegetational data for Great Sippewissett Marsh are from Valiela and Teal (1979); the tidal flux data are from Valiela et al. (1978b). The corresponding results from Flax Pond were kindly provided by R. A. Houghton, C. A. S. Hall, and G. Woodwell. Particulates include only small fragments suspended in the water. In the tidal exchanges, the percentages in parentheses are (tidal import/tidal export) 100.*

[b]*Great Sippewissett was a well-established marsh by 1766 judging from the Atlantic Neptune Atlas, the earliest map available. Earlier dating was not available except by comparison with the age of 4000 years of the nearby Barnstable Marsh (Redfield, 1972). The actual age is less than 4000 and more than 216 years. Some unpublished observations on cores by D. Miller suggest an age of about 2000 years.*

[c]*From Woodwell and Pecan (1973) and C. Cogswell and J. Hartman (personal communication). These totals include species from the lower edge of the S. alterniflora zone up to the marsh edge dominated by Iva frutescens.*

there is seemingly very open flow (Branford Harbor, Carter Creek).

Type I salt marshes (except for dissolved organic nitrogen (DON) and total nitrogen (TN), which all marshes export) tend not to export nutrients and particulates. In contrast, Type II and Type III marshes show a very marked pattern of export, again with the exception of nitrate. The export or import features of salt marshes are therefore related to the geomorphology and hydrology of the site.

As salt marshes age, there are changes in the rate of sedimentation and basin morphology (Redfield, 1972). There are unfortunately only two salt marshes that are studied well enough to enable comparisons (Table III): one (Flax Pond) where the dredging of a channel through a barrier beach in 1803 created a new salt marsh (Flessa *et al.*, 1977); the other (Great Sippewissett marsh) with a longer record of existence as a salt marsh (G. Parris, unpublished manuscript). There are several indicators of relative maturity of marshes. In older marshes, accretion is thought to have caught up largely with the rise in sea level, so that most of the surface of the marsh more or less keeps pace with the increase in sea level (Redfield, 1972). In young salt marshes (Table III), rates of sedimentation and accretion relative to production by plants are high; there is a relatively large area of open water or unvegetated sediment, and the biomass and area

(Table III continued)

[d]The results from Flax Pond are published in a form such that nutrients carried in the ebb and food cannot be separated; the results are reported here only as < or > than 100% of amounts exported.

[e]Assuming that C:N in particles is similar to that of Great Sippewissett; PN was not measured directly in Flax Pond.

[f]Based on measurements by R. A. Houghton on Flax Pond; Great Sippewissett entry based on our observations that in years where ice prevents dead grass thatch from leaving the marsh, the accumulated windrows only amount to about 1% of the above-ground primary production.

covered by the colonizing tall form of *S. alterniflora* are both relatively large. In addition, young marshes show relatively small areas covered by the high marsh dominated by *S. patens*, and the flora of what high marsh there is seems less diverse than that of older marshes.

In terms of the tidal exchanges, both the yound and the older marsh exported ammonium, but for small particulates, the young marsh was an importer while the older marsh was an exporter. This difference must be related to the larger accretion rate of the young marsh, where a large area of expanding creek banks must be supported by inputs of particles. In the older marsh, there is only a small area with high accretion rates, so this marsh is less of a trap for particulates and accretion is due to accumulation of dead, below-ground plant tissues (cf. Section II,A,5).

Wrack in and around salt marshes consists primarily of litter of tall plants of *S. alterniflora*. Since young marshes have proportionately greater areas under this vegetation type, more of the total above-ground production of a young marsh may be converted into floating rafts of wrack. If exports of this material are considered, Flax Pond marsh exports more total organic matter than it imports (G. W. Woodwell and R. A. Houghton, personal communication). In Great Sippewissett marsh, the total amount of wrack is small compared with the above-ground production of the marsh (Table III).

The comparison of Flax Pond and Great Sippewissett should not be extended too far; it unfortunately does not contrast just age extremes, since these two marshes also differ in morphology and hydrology (Table II). In addition to differences in the main tidal channel and in the input of freshwater, there is a relatively deep hole made by dredging for sand and gravel during the 1930s and 1940s (Woodwell and Pecan, 1973; Flessa *et al.*, 1977) and a bed of blue mussels (Nixon, 1980) and oysters in Flax Pond marsh. Both these features can act as sediment traps independent of age or hydrology. There are, however, enough significant contrasts and consistent independent checks and general patterns in Table II and

TABLE IV. Percentage Carbon, Nitrogen and C:N (All as Dry Weight) for Salt Marsh Sediments[a]

Salt marsh	*N(%)*	*C(%)*	*C:N*
Great Sippewissett, Massachusetts[b]	1.4(1.2-2)	27(24-30)	19.3
Providence River, Rhode Island[a]	4.9	5.2	10.6
Block Island Sound, Rhode Island[c]	3.6	4.4	17.1
Narragansett Bay, Rhode Island[c]	2.4	4.1	12.1
Lewes, Delaware[d]	1-1.4	4-24	0.4-24
North Carolina[e]	--	1.6	--
Sapelo Island, Georgia[f]	0.2-0.5	--	1
Sapelo Island, Georgia[g]	0.7	5	7.4
Sapelo Island, Georgia[g]	0.03-0.2	0.2-1.0	6.7-6.9
Sapelo Island, Georgia[g]	0.9	0.6	10.3
Sapelo Island, Georgia[h]	0.2-0.4	--	--
Barataria Bay, Louisiana[i]	0.7(0.5-0.9)	12(8-16)	16.7(12-19.4)

[a]*All values from sediments bearing S. alterniflora, except for %N in Delaware, which is from a Distichlis spicata zone. Ranges in parentheses.*

[b]*Top 15 cm;* [c]*top 5 cm, Nixon and Oviatt, 1973;* [d]*%C from Lord, 1980, %N from Gallagher and Plumley, 1979;* [e]*top 15 cm, Broome et al., 1975;* [f]*Haines et al., 1976;* [g]*Hanson, 1977b;* [h]*Chalmers, 1979;* [i]*DeLaune et al., 1979.*

III that suggest that age of a marsh as well as geomorphology and hydrology influence export and import of materials, including nitrogen. Further comparisons and critical study would be useful.

4. Losses of Nitrogen through Denitrification

Jones (1974) demonstrated that denitrification took place in salt marsh sediment. Measurements of annual rates made in Georgia (12 g $N \cdot m^{-2} \cdot year^{-1}$, Haines *et al.*, 1976) and Massachusetts (14.3 g $N \cdot m^{-2} \cdot year^{-1}$, Kaplan *et al.*, 1979) are quite similar in spite of different methodology. It is very difficult to measure production of N_2 because there is a very large pool of this gas dissolved in the interstitial and tidal water. Development of new methods

TABLE V. Amounts of Nitrogen in Four Major Pools in Great Sippewissett Salt Marsh[a]

	Tidal water	*Animals*[b]	*Plants*[c]	*Sediments*[d]
Total amount	200,000 m^3	17,700 kg	570,000 kg	7.5×10^6 kg
Amount of N (kg)	3	1,700	5,200	110,000 kg

[a]*The average tidal volume is for high tide; at low tide almost all the water leaves the marsh.*

[b]*Animals approximated from Jordan and Valiela (1982).*

[c]*Plant data estimated for peak biomass in later summer.*

[d]*Sediments include only vegetated zones and only to a depth of 15 cm, and contain nonliving roots and rhizomes.*

(Van Raalte and Patriquin, 1979) to measure denitrification would be useful, since the methods used so far were indirect and ignored the possible release of N_2O.

5. *Losses of Nitrogen by Sedimentation*

The percent of nitrogen in salt marsh sediments may reach 5% by weight (Table IV), and, as in most aquatic ecosystems, this sedimentary nitrogen is the largest pool of nitrogen within a salt marsh (Table V). The strongly reduced conditions of salt marsh sediments inhibit the degradation of many organic compounds (Harborne and Van Sumere, 1975), so that peat accumulates. The C:N ratio is not too high compared with other sediments, but the nitrogen may not be easily available to microbes or deposit feeders since it may be immobilized by complexation with lignin, its derivatives, and many other phenolics (Leatham *et al.*, 1980).

As the organic-rich sediments accumulate, some nitrogen is buried beyond the depth where recycling mechanisms can make it available to organisms. About 2.7 g $N \cdot m^{-2} \cdot year^{-1}$ are thus lost from Great Sippewissett marsh (Valiela and Teal, 1979), while 5-20 g $N \cdot m^{-2} \cdot year^{-1}$ may be lost to sediments in other marshes (Nixon, 1980). Nixon further estimated that only 2-4 g $N \cdot m^{-2} \cdot year^{-1}$ was likely to come from particles suspended in the flooding tidal water that became incorporated into the marsh sediment. Most of the accretion must be due to the death of below-ground plant parts. Examination of cores of salt marsh sediments shows that dead roots and rhizomes make up the bulk of the peat accumulated. Based on ^{210}Pb profiles and the continued uniformity of accretion even during periods of lowered rate of rise in sea level, McCaffrey (1977) came to a similar conclusion that peat formation rather than sedimentation is responsible for the major part of accretion.

In young marshes or where *Spartina alterniflora* is rapidly colonizing mud or sand flats, the accumulation of captured suspended particles is much more important. Thus, sediments of salt marsh creek banks are typically much finer and contain less roots and rhizomes and more water than sediments of stable, vegetated marsh. In fact, the difference in water content is great enough that it allowed Redfield (1972) to follow the gradual development of Great Barnstable marsh by tracking the position of creek banks.

B. Budgets and Transformations of Nitrogen

For an ecosystem to have any stable history at all, there ought to be some balance of inputs and outputs of critical limiting factors. We know that, at least in New England, salt marshes change rather little: charts of Cape Cod salt marshes from the 1800s show more or less the same outlines as found today. A 10-year record of standing crops of marsh plants (Valiela *et al.*, 1982) shows that variation from year to year is small, and we therefore expect that in this environment gains and losses of nitrogen ought to be in balance. A nitrogen budget for Great Sippewissett marsh (Valiela and Teal, 1979) does show such a balance (Table VI), in spite of the limitations of the methods used. Every item in the budget was estimated independently and measured as directly as possible.

The nitrogen exchanges in Great Sippewissett marsh are dominated by physical processes (Table VI). Biotic mechanisms

TABLE VI. Nitrogen Budget for Great Sippewissett Marsh (kg N $year^{-1}$)[a]

	Inputs	*Outputs*	*Net exchange*
Precipitation	380		380
Ground water flow	6,120		6,120
N_2 fixation	3,280		3,280
Tidal exchange	26,200	31,600	-5,350
Denitrification		6,940	-6,940
Sedimentation		1,295	-1,295
Other	9	26	-17
	35,990	39,860	-3,870
Percentage of exchanges by biotic mechanisms	9	18	
Percentage of exchanges by physical mechanisms	91	82	

[a]*Adapted from Valiela and Teal (1979). Negative signs indicate net export.*

contribute only 9% of the inputs and 18% of the outputs. Since water transport and sedimentation are so prominent in the nitrogen economy of a marsh, it is no surprise that hydrology and rates of sedimentation are important in export by marshes, as discussed earlier.

The balance between inputs and outputs in the nitrogen budget of Great Sippewissett marsh (Table VI) hides very active transformations within the ecosystem. The total amounts of each nitrogen species entering and leaving the marsh can be obtained from the budget. By back-calculation, application of some knowledge of the nitrogen cycle, and use of data available from Great Sippewissett marsh, the fate of each nitrogen species within the marsh can be traced (Fig. 1).

The nitrogen fixed by microorganisms is soon released as DON and DIN, both of which we believe are mineralized to ammonium in a short time. There is a bit more ammonium contributed by degradation of DON; further amounts of ammonium are contributed by animal excretion and through leaching from live and dead plant tissues.

Salt-marsh grasses take up ammonium preferentially over nitrate (Morris, 1980), and the latter only occurs in low concentrations in reduced sediments. Algae are ignored here since their growth consumes only a small amount of nitrogen compared to grasses, and their turnover time is much faster. About 4000 kg N are taken up by marsh grasses during their annual growth (Fig. 2). The uptake pattern of nitrogen over the growing season largely follows the growth of biomass, even though grasses contain higher amounts of nitrogen in the spring (Vince *et al.*, 1981; Patrick and Delaune, 1976). Most of the nitrogen is taken up below ground (Fig. 2), early in the growing season. Most of the 4000 kg N taken up by plants in the spring is either released through leaching of live and dead tissues (initially about 30% in the fall, probably about 60% for an entire year) (Valiela *et al.*, 1982) or becomes particulate nitrogen (PN) in the form of litter or detritus. The perennial organs of marsh grasses hold only 14-23% of the nitrogen contained in the crop at peak biomass (Fig. 2).

Animals, in the rough calculations of Fig. 1, can consume a very large part of the particulates entering the marsh, and release roughly equivalent amounts of nitrogen as feces and gametes as well as the already mentioned ammonium.

The sum of feces, gametes, litter, and other particles provides almost all the PN that was measured leaving Great Sippewissett marsh. There is a deficit of 430 kg, but this amount cannot be significant given the approximations used to put Fig. 1 together.

All the transformations discussed so far take place with nitrogen in a reduced state. Transformations to the other oxidation states of nitrogen are of course only achieved by activities of microorganisms. The nitrate required to support the measured rate of denitrification (in addition to that leaving the marsh as nitrate) exceeds the nitrate entering the marsh. Nitrification amounting to almost 5000 kg N $year^{-1}$ must therefore take place in Great Sippewissett marsh. This would require a nitrification rate of about 4 mg $N \cdot m^{-2} \cdot day^{-1}$. Even though this is a very rough estimate, the

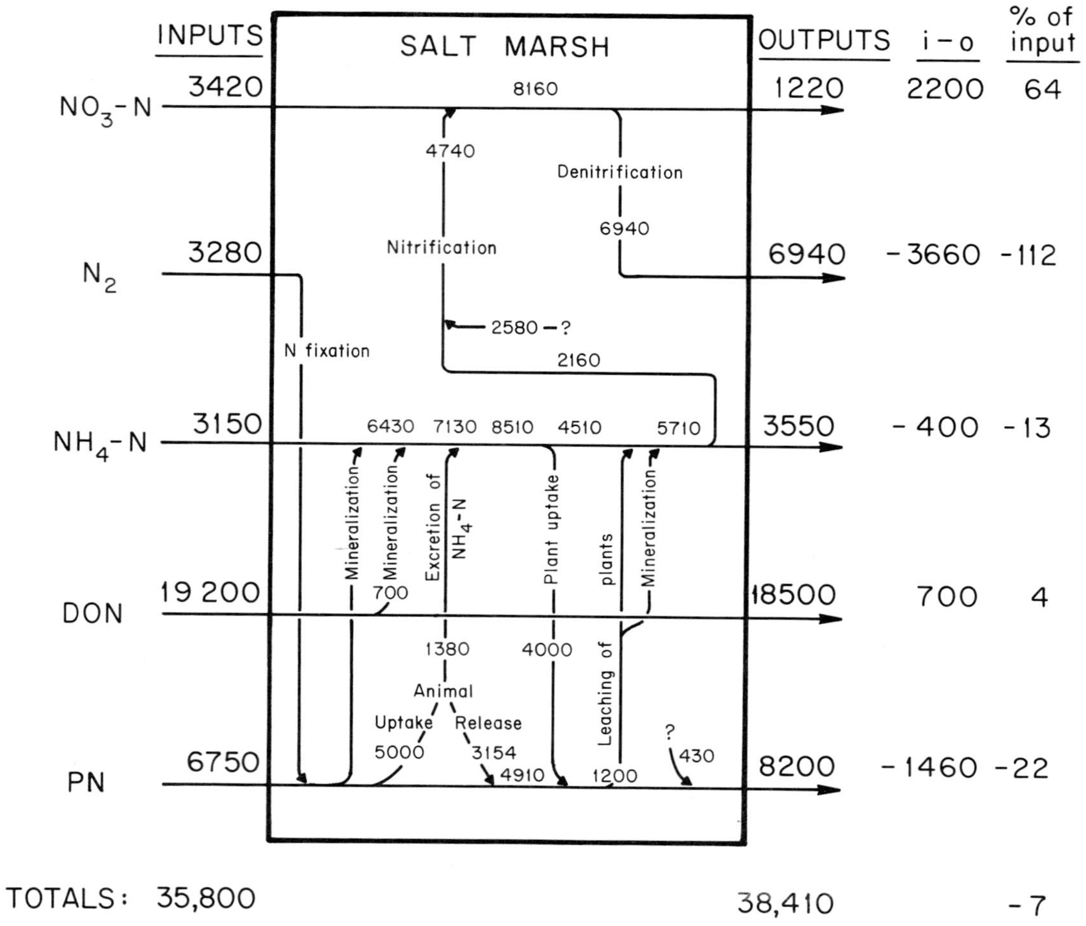

INPUTS
SALT MARSH
OUTPUTS
i - o
% of input
NO3-N
3420
8160
1220
2200
64
4740
Denitrification
6940
Nitrification
N2
3280
6940
-3660
-112
2580 - ?
N fixation
2160
NH4-N
3150
6430
7130
8510
4510
5710
3550
- 400
-13
Mineralization
Mineralization
Excretion of NH4-N
Plant uptake
plants
Mineralization
DON
19 200
700
18500
700
4
1380
4000
Leaching of
Animal
Uptake
Release
5000
3154
?
430
PN
6750
4910
1200
8200
-1460
-22
TOTALS: 35,800
38,410
-7

the calculated value falls within the range of measurements made in similar environments. Henriksen *et al*. (1981) obtained rates of potential nitrification of 3-20 mg $NO_3 \cdot m^{-2} \cdot day^{-1}$ in various shallow water marine sediments. W. Bowdcn (personal communication) estimated a rate of nitrification in a brackish marsh in Massachusetts similar to the value calculated from Fig. 1.

The back-calculation procedure can account for about half of the required ammonium from the sources we have measured. The rest may be due to variation in the year-to-year mineralization of the ample amounts of nitrogen in the sediments (Table V), to variable release of interstitial ammonium, or perhaps due to error in our various measurements. An error of that size could certainly propagate through all the various estimates and back-calculations used to compile Fig. 1.

The values for nitrogen transformations within Great Sippewissett marsh (Fig. 1) are rough estimates: Their intent is to suggest the approximate magnitude of the various processes and to provide a check on closure of the transformations among various nitrogen species. The data on inputs and outputs is more robust, since these are actual measurements. Comparison of the inputs and outputs (Fig. 1) shows net exports of nitrogen gas, ammonium, and particulate nitrogen. Nitrate, on the other hand, is intercepted by the marsh. The salt marsh, therefore, takes up oxidized nitrogen and exports energy-rich, reduced forms. The loss of gaseous nitrogen far exceeds the fixation of N_2 in salt marshes; the presence of anaerobic sediments with a rich supply of organic compounds favors high rates of denitrification. The rates are

Fig. 1. Inputs, outputs, and transformation of nitrogen within Great Sippewissett Marsh. Inputs and outputs from Valiela and Teal (1979) and references therein. Information on animals from Jordan and Valiela (1982) and unpublished data. The transformations were back-calculated from outputs and inputs and relevant information on the biology of nitrogen and decomposition. Values are in kg of nitrogen per year.

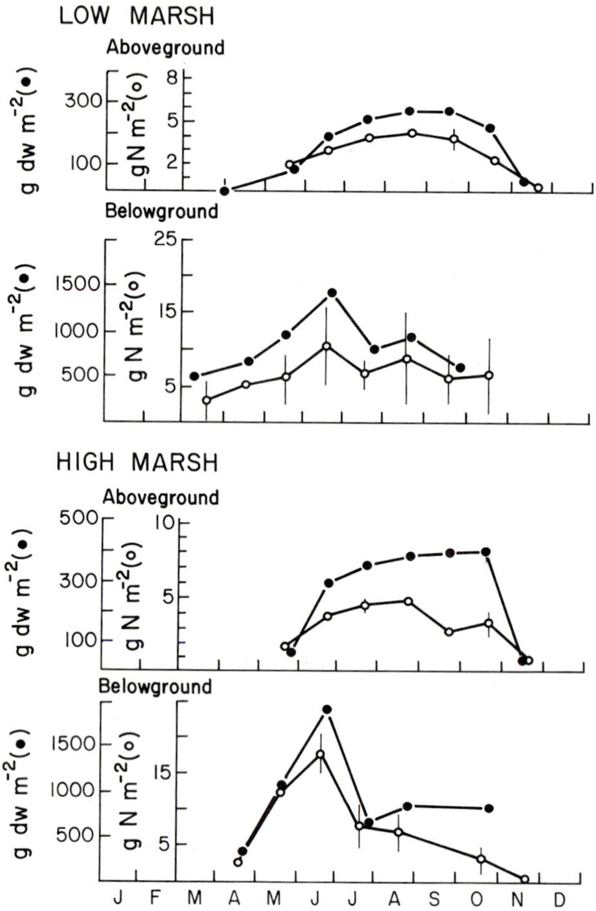

Fig. 2. Biomass and nitrogen in marsh plants (S. alterniflora in low marsh and S. patens in high marsh) above and below ground over a year. The values are pooled from several years, some of which have appeared in Valiela et al. (1975, 1976) where methods are described. Nitrogen data are unpublished and were obtained by analysis by elemental analyzer.

high compared to those of other environments (Kaplan *et al.*, 1979), but not enough is known about how important marsh denitrification is as a source of atmospheric nitrogen. It seems clear, though, that on per unit area, marshes export considerable nitrogen gas.

TABLE VII. Rough Estimates of the Importance of Marsh Exports in Relation to Stock of Dissolved Inorganic and Particulate Nitrogen in the Receiving Waters of Buzzards Bay[a]

	Dissolved inorganic nitrogen (kg)	*Particulate N (kg)*
Annual marsh export	33,100	27,600
Nutrient stock in Buzzards Bay	80,000	400,000
Annual consumption by primary producers	4,100,000	
Annual consumption by zooplankton		2,000,000

[a]*Calculations assume that the nitrogen load of the 13-m water column studied by Roman (1981) is representative of the whole of Buzzards Bay.*

[b]*Data on marsh export from Valiela and Teal (1979).*

[c]*Consumption by algae and zooplankton from Roman (1981).*

To evaluate the importance of exports from salt marshes to the receiving coastal waters, the annual outputs of dissolved inorganic and particulate nitrogen from all the marshes surrounding Buzzards Bay (about 9.2×10^6 m^2) can be calculated. Roughly about 40% of the stock of DIN and 7% of PN in Buzzards Bay are contributed annually by marsh export (Table VII), if we assume that the exports from Great Sippewissett are representative of all these marshes. This may or may not be a good assumption in view of the variable exports of marshes (Table II), but is a best approximation.

The annual consumption of DIN and PN by organisms is about 50 and 5 times greater than the available nutrient stocks (Table VII), so that recycling of these materials in coastal waters is clearly important (Nixon, 1980). In a system where the extant crop is reused so many times a year, the insertion of new available nitrogen by marshes may be significant beyond the mere amount of the addition.

In estuarine ecosystems dominated by river flow, the importance of marshes relative to nitrogen budgets of coastal waters seems much less (Nixon, 1980). It cannot always be claimed that marshes are quantitatively important in contributing to nitrogen supply of coastal waters; it depends, as already discussed, largely on the hydrologic setting.

It is not at all clear what happens to groundwater nitrate in coastal environments with no marshes. We have found (Valiela *et al.*, 1980) that groundwater enters Great Sippewissett marsh principally in the landward edge; the marsh intercepts only a thin slice of groundwater near the groundwater table. This thin layer, however, is substantially richer in nutrients than deeper groundwater. We still need to know how much groundwater flows directly into coastal waters, the vertical distributions of nitrate in the groundwater, and the fate of whatever nitrogen is transported by the groundwater. It could be that denitrification of the usually low concentrations of nitrate takes place in the surface sediments of coastal water bodies as the groundwater flows through the sediment-water interface. If this were the case, little of the nitrogen carried by groundwater would enter coastal waters subtidally. Exports of reduced nitrogen from marshes would therefore be of importance in the nitrogen imports of coastal waters. If, on the other hand, much of the nitrate does cross the sediment-water interface subtidally, the net effect of marshes may reduce the export of nitrogen from terrestrial sources to coastal waters, since denitrification rates in marshes are high and since the particulates exported may be of low food quality for coastal organisms. The actual case may be between these alternatives, but further work is needed.

Consideration of the inputs and outputs of nitrogen into salt marshes points out that ecosystems do not exist as isolated entities. There are significant asymmetrical exchanges of materials among adjoining ecosystems, for example, nitrate from uplands to salt marshes and ammonium and particulate nitrogen from salt marshes to coastal waters. These exchanges can be important enough to alter the structure and function of an ecosystem, for example, the activity and abundance of various guilds of microbes in salt marshes where nitrate from the upland is introduced, or the plankton and benthos of coastal water receiving ammonium and particulate nitrogen from marshes.

ACKNOWLEDGMENTS

This work was supported by grants from the National Science Foundation, the Victoria Foundation, Pew Memorial Trust, and the Sea Grant Program of the Woods Hole Oceanographic Institution. Bruce Peterson provided useful criticism of the manuscript.

REFERENCES

Axelrad, D. M. (1974). Nutrient flux through the salt marsh ecosystem. Ph.D. Thesis, College of William and Mary in Virginia, Williamsburg.

Balderston, W. L., and Payne, W. J. (1976). Inhibition of methanogenesis in salt marsh sediments and whole-cell suspensions of methanogenetic bacteria by nitrogen oxides. *Appl. Environ. Microbiol. 32,* 264-269.

Bokuniewicz, H. (1980). Groundwater seepage into Great South Bay, New York. *Estuarine Coastal Mar. Sci. 10,* 437-444.

Bowman, M. J. (1977). Nutrient distribution and transport in Long Island Sound. *Estuarine Coastal Mar. Sci. 5,* 531-548.

Broome, S. W., Woodhouse, W. W., Jr., and Seneca, D. E. (1975). The relationship of mineral nutrients to growth of *Spartina alterniflora* in North Carolina. I. The effects of N, P, and Fe fertilizers. *Soil Sci. Am. Proc. 39,* 301-307.

Carpenter, E. J., Van Raalte, C. D., and Valiela, I. (1978). Nitrogen fixation by algae in a Massachusetts salt marsh. *Limnol. Oceanogr. 23,* 318-327.

Caswell, H., Reed, F., Stephenson, S. N., and Werner, P. A. (1973). Photosynthetic pathways and selective herbivory: A hypothesis. *Am. Nat. 107,* 465-480.

Chalmers, A. G. (1979). The effects of fertilization on nitrogen distribution in a *Spartina alterniflora* salt marsh. *Estuarine Coastal Mar. Sci. 8,* 327-337.

Connor, M. S. (1980). Snail grazing effects on the composition and metabolism of benthic diatom communities and subsequent effects on fish growth. Ph.D. Dissertation, Woods Hole Oceanographic Institution/Massachusetts Institute of Technology, Woods Hole and Cambridge.

Daly, M. A., and Mathieson, A. C. (1981). Nutrient fluxes within a small north temperate salt marsh. *Mar. Biol. (Berlin) 61,* 337-344.

DeLaune, R. D., and Patrick, W. H., Jr. (1980). Nitrogen and phosphorus cycling in a Gulf Coast salt marsh. *Estuarine Perspect.* [*Proc. Bienn. Int. Estuarine Res. Conf.*], *5th, 1979* pp. 153-162.

DeLaune, R. D., Buresh, R. J., and Patrick, W. H., Jr. (1979). Relationship of soil properties to standing crop biomass of *Spartina alterniflora* in a Louisiana marsh. *Estuarine Coastal Mar. Sci. 8,* 477-487.

Fenchel, T., and Blackburn, T. H. (1979). "Bacteria and Mineral Cycling." Academic Press, New York.

Flessa, K. W., Constantine, K. J., and Cushman, M. K. (1977). Sedimentation rates in a coastal marsh determined from historical records. *Chesapeake Sci. 18,* 172-176.

Gallagher, J. C., and Plumley, F. C. (1979). Underground biomass profiles and productivity in Atlantic coastal marshes. *Am. J. Bot. 66,* 156-161.

Gallagher, J. L. (1975). Effect of an ammonium nitrate pulse on the growth and elemental composition of natural stands of *Spartina alterniflora* and *Juncus roemerianus*. *Am. J. Bot. 62,* 644-648.

Haines, E. B., Chalmers, A., Hanson, R., and Sherr, B. (1976). Nitrogen pools and fluxes on a Georgia salt marsh. *In* "Estuarine Processes" (M. L. Wiley, ed.), Vol. 2, pp. 241-259. Academic Press, New York.

Hanson, R. B. (1977a). Comparison of nitrogen fixation activity in tall and short *Spartina alterniflora* salt marsh soils. *Appl. Environ. Microbiol. 33,* 596-602.

Hanson, R. B. (1977b). Nitrogen fixation (acetylene reduction) in a salt marsh amended with sewage sludge and organic carbon and nitrogen compounds. *Appl. Environ. Microbiol. 33,* 846-852.

Happ, G., Gosselink, J. G., and Day, J. W., Jr. (1977). The seasonal distribution of organic carbon in a Louisiana estuary. *Estuarine Coastal Mar. Sci. 5,* 695-705.

Harborne, J. B., and Van Sumere, C. F. (1975). "The Chemistry and Biochemistry of Plant Proteins." Academic Press, New York.

Heinle, D. R., and Flemer, D. A. (1976). Flows of materials between poorly flooded tidal marshes and an estuary. *Mar. Biol. (Berlin) 35,* 359-373.

Henriksen, K., Hensen, J. I., and Blackburn, T. H. (1981). Rates of nitrification, distribution of nitrifying bacteria, and nitrate fluxes in different types of sediment from Danish waters. *Mar. Biol. (Berlin) 61,* 299-304.

Howarth, R. W., and Teal, J. M. (1979). Sulfate reduction in a New England salt marsh. *Limnol. Oceanogr. 24,* 999-1013.

Howarth, R. W., and Teal, J. M. (1980). Energy flow in a salt marsh ecosystem: The role of reduced inorganic sulfur compounds. *Am. Nat. 116,* 862-872.

Jefferies, R. L. (1977). Growth responses of coastal halophytes to inorganic nitrogen. *J. Ecol.* *65*, 847-865.
Jefferies, R. L., and Perkins, N. (1977). The effects on the vegetation of the additions of inorganic nutrients to salt marsh soils at Stiffkey, Norfolk. *J. Ecol.* *65*, 867-882.
Johannes, R. E. (1980). The ecological significance of the submarine discharge of groundwater. *Mar. Ecol. Prog. Ser.* *3*, 365-373.
Jones, K. (1974). Nitrogen fixation in a salt marsh. *J. Ecol.* *62*, 553-565.
Jordan, T. E., and Valiela, I. (1982). A nitrogen budget of the ribbed mussel, *Geukensia demissa*, and its significance in nitrogen flow in a New England salt marsh. *Limnol. Oceanogr.* *27*, 75-90.
Kaplan, W. (1977). Denitrification in a Massachusetts salt marsh. Ph.D. Thesis, Boston University, Boston, Massachusetts.
Kaplan, W., Valiela, I., and Teal, J. M. (1979). Denitrification of a salt marsh ecosystem. *Limnol. Oceanogr.* *24*, 726-734.
King, G. M., and Wiebe, G. M. (1978). Methane release from soils of a Georgia salt marsh. *Geochim. Cosmochim. Acta* *42*, 343-348.
Leatham, G. F., King, V., and Stahmann, M. A. (1980). *In vitro* protein polymerization by quinones or free radicals generated by plant or fungal oxidative enzymes. *Phytopathology* *70*, 1134-1140.
Lion, L. W., and Leckie, J. O. (1982). Accumulation and transport of Cd, Cu, and Pb in an estuarine salt marsh surface microlayer. *Limnol. Oceanogr.* *27*, 111-125.
Lord, C. J., III (1980). The chemistry and cycling of iron, manganese and sulfur in salt marsh sediments. Ph.D. Dissertation, University of Delaware, Newark.
Lotrich, V. A., Meredith, W. H., Weisberg, S. B., Hurd, L. E., and Daiber, F. C. (1978). Dissolved and particulate nutrient fluxes via tidal exchange between a salt marsh and Lower Delaware Bay. *Jekyll Isl., Georgia, 5th Biennial Int. Estuar. Res. Conf. Abstracts*.
McCaffery, R. J. (1977). A record of the accumulation of sediments and trace metals in a Connecticut, USA, salt marsh. Ph.D. Dissertation, Yale University, New Haven, Connecticut.
Mattson, W. J., Jr. (1980). Herbivory in relation to plant nitrogen content. *Annu. Rev. Ecol. Syst.* *11*, 119-196.
Mauriello, D., and Winfield, T. (1978). Nutrient exchange in the Tijuana estuary. *Coastal Zone 78* *3*, 2221-2238.
Morris, J. T. (1980). The nitrogen uptake kinetics of *Spartina alterniflora* in culture. *Ecology* *61*, 1114-1121.
Nixon, D. W. (1980). Between coastal marshes and coastal waters--A review of twenty years of speculation and research on the role of salt marshes in estuarine productivity and water chemistry. *In* "Estuarine and Wetland Processes" (R. Hamilton and K. B. MacDonald, eds.), pp. 437-425. Plenum, New York.

Nixon, S. W., and Oviatt, C. A. (1973). Ecology of a New England salt marsh. *Ecol. Monogr. 43,* 463-498.

Odum, W. E., Fisher, J. S., and Pickral, J. C. (1979). Factors controlling the flux of particulate organic carbon from estuarine wetlands. *In* "Ecological Processes in Coastal and Marine Systems" (R. J. Livingstone, ed.), pp. 69-80. Plenum, New York.

Parsons, K. A., and de la Cruz, A. (1980). Energy flow and grazing behavior of conocephaline grasshoppers in a *Juncus roemerianus* marsh. *Ecology 61,* 1045-1050.

Parsons, T., and Takahashi, M. (1973). "Biological Oceanographic Processes." Pergamon, Oxford.

Patrick, W. J., Jr., and DeLaune, R. D. (1976). Nitrogen and Phosphorus utilization by *Spartina alterniflora* in a salt marsh in Barataria Bay, Louisiana. *Estuarine Coastal Mar. Sci. 4,* 59-64.

Patriquin, D. G., and McClung, C. R. (1978). Nitrogen accretion and the nature and possible significance of N_2 fixation (acetylene reduction) in a Nova Scotian *Spartina alterniflora* stand. *Mar. Biol. (Berlin) 47,* 227-242.

Paul, E. A. (1976). Nitrogen cycling in terrestrial ecosystems. *In* "Environmental Biogeochemistry" (J. O. Nriagu, ed.), Vol. 1, pp. 225-243. Ann Arbor Sci. Publ., Ann Arbor, Michigan.

Pellenbarg, R. E., and Church, T. M. (1979). The estuarine surface microlayer and trace metal cycling in the salt marsh. *Science 203,* 1010-1012.

Redfield, A. C. (1972). Development of a New England salt marsh. *Ecol. Monogr. 42,* 201-237.

Redfield, A. C., and Rubin, M. (1962). The age of salt marsh peat and its relation to recent changes in sea level at Barnstable, Massachusetts. *Proc. Natl. Acad. Sci. USA 48,* 1728-1735.

Richard, G. A. (1978). Seasonal and environmental variations in sediment accretion in a Long Island salt marsh. *Estuaries 1,* 29-35.

Roman, M. (1981). Tidal resuspension in Buzzards Bay, Massachusetts. III. Seasonal cycles of nitrogen and carbon:nitrogen ratios in the seston and zooplankton. *Estuarine Coastal Mar. Sci. 11,* 9-16.

Russell-Hunter, W. D. (1970). "Aquatic Productivity." Macmillan, New York.

Settlemyre, J. L., and Gardner, L. R. (1977). Suspended sediment flux through a salt marsh drainage basin. *Estuarine Coastal Mar. Sci. 5,* 653-663.

Smith, T. J., III, and Odum, W. E. (1981). The effects of grazing by snow geese on coastal salt marshes. *Ecology 62,* 98-106.

Sullivan, M. J., and Daiber, F. C. (1974). Response in production of cordgrass, *Spartina alterniflora,* to inorganic nitrogen and phosphorus fertilizer. *Chesapeake Sci. 15,* 121-123.

Sullivan, M. J., and Daiber, F. C. (1975). Light, nitrogen and phosphorus limitation of edaphic algae in a Delaware salt marsh. *J. Exp. Mar. Biol. Ecol. 18,* 77-88.

Teal, J. M. (1962). Energy flow in the salt marsh ecosystems of Georgia. *Ecology 43,* 614-624.
Teal, J. M., Valiela, I., and Berlo, I. (1979). Nitrogen fixation by rhizosphere and free-living bacteria in salt marsh sediments. *Limnol. Oceanogr. 24,* 126-132.
Tenore, K. R., and Rice, D. L. (1980). A review of trophic factors affecting secondary production of deposit feeders. *In* "Marine Benthic Dynamics" (K. R. Tenore and B. Coull, eds.), pp. 325-340. Univ. of South Carolina Press, Columbia.
Tjepkema, J. D., and Evans, H. J. (1976). Nitrogen fixation and associated with *Juncus balticus* and other plants of Oregon wetlands. *Soil Biol. Biochem. 8,* 505-509.
Tyler, G. (1967). On the effect of phosphorus and nitrogen, supplied to Baltic shore-meadow vegetation. *Bot. Not. 120,* 433-447.
Valiela, I., and Teal, J. M. (1974). Nutrient limitation in salt marsh vegetation. *In* "Ecology of Halophytes" (R. J. Reimold and W. H. Green, eds.), pp. 547-563. Academic Press, New York.
Valiela, I., and Teal, J. M. (1979). The nitrogen budget of a salt marsh ecosystem. *Nature (London) 280,* 652-656.
Valiela, I., Teal, J. M., and Sass, W. J. (1975). Production and dynamics of salt marsh vegetation and effects of experimental treatment with sewage sludge. *J. Appl. Ecol. 12,* 973-981.
Valiela, I., Teal, J. M., and Persson, N. Y. (1976). Production and dynamics of experimentally enriched salt marsh vegetation: Below-ground biomass. *Limnol. Oceanogr. 21,* 245-252.
Valiela, I., Teal, J. M., and Deuser, W. G. (1978a). The nature of growth forms in the salt marsh grass *Spartina alterniflora. Am. Nat. 112,* 461-470.
Valiela, I., Teal, J. M., Volkmann, S., Shafer, D., and Carpenter, E. J. (1978b). Nutrient and particulate fluxes in a salt marsh ecosystem: Tidal exchanges and inputs by precipitation and groundwater. *Limnol. Oceanogr. 23,* 798-812.
Valiela, I., Teal, J. M., Volkmann, S., Cogswell, C. M., and Harrington, R. A. (1980). On the measurement of tidal exchanges and groundwater flow in salt marshes. *Limnol. Oceanogr. 25,* 187-192.
Valiela, I., Howes, B., Howarth, R., Giblin, A., Foreman, K., Teal, J. M., and Hobbie, J. E. (1982). The regulation of primary production and decomposition in a salt marsh ecosystem. *In* "Wetlands: Ecology and Management" (B. Gopal, R. E. Turner, R. O. Wetzel, and D. F. Whigham, eds.), pp. 151-168. Natl. Inst. Ecol., Jaipur, India.
Van Berkum, P., and Slager, C. (1979). Immediate acetylene reduction by exised grass roots not previously preincubated at low oxygen tension. *Plant Physiol. 64,* 739-743.
Van Raalte, C. D., and Patriquin, D. G. (1979). Use of the "acetylene blockage" technique for assaying denitrification in a salt marsh. *Mar. Biol. (Berlin) 52,* 315-320.

Van Raalte, C. D., Valiela, I., Carpenter, E. J., and Teal, J. M. (1974). Inhibition of nitrogen fixation in salt marshes measured by acetylene reduction. *Estuarine Coastal Mar. Sci.* *2*, 301-305.

Van Raalte, C. D., Valiela, I., and Teal, J. M. (1976a). Production of epibenthic algae: Light and nutrient limitation. *Limnol. Oceanogr.* *21*, 862-872.

Van Raalte, C. D., Valiela, I., and Teal, J. M. (1976b). The effect of fertilization on the species composition of salt marsh diatoms. *Water Res.* *10*, 1-4.

Vince, S. W., Valiela, I., and Teal, J. M. (1981). An experimental study of the structure of herbivorous insect communities in a salt marsh. *Ecology* *62*, 1662-1678.

Whitney, D. E., Woodwell, G. M., and Howarth, R. W. (1975). Nitrogen fixation in Flax Pond, a Long Island salt marsh. *Limnol. Oceanogr.* *4*, 640-643.

Windom, H. L., Dunstan, W. M., and Gardner, W. S. (1975). River input of inorganic phosphorus and nitrogen to the southeastern salt marsh and estuarine environment. *In* "Mineral Cycling in Southeastern Ecosystems" (F. G. Howell, J. B. Gentry, and M. H. Smith, eds.), pp. 309-313.

Wolff, W. J., Van Eeden, M. J., and Lammens, E. (1979). Primary production and import of particulate organic matter on a salt marsh in the Netherlands. *Neth. J. Sea Res.* *13*, 242-255.

Woodwell, G. M., and Pecan, E. V. (1973). Flax Pond: An estuarine marsh. *Brookhaven Natl. Lab.* [*Rep.*] *BNL 50397*.

Woodwell, G. M., Hall, C. A. S., Whitney, D. E., and Houghton, R. A. (1979). The Flax Pond ecosystem study: Exchanges of inorganic nitrogen between an estuarine marsh and Long Island Sound. *Ecology* *60*, 695-702.

Chapter 18

SYMBIOSES

DENNIS L. TAYLOR
Center for Environmental and Estuarine Studies
University of Maryland
Cambridge, Maryland

This universal nitrogen-hunger is a misery that makes strange bedfellows.

F. Keeble, 1910

I. INTRODUCTION

Plant growth and primary production create local nutrient deficits that can only be restored through mechanisms of renewal (importation) and regeneration (conservative cycling). In marine environments, renewal is characterized by physical transport processes (e.g., upwelling ecosystems) and regeneration by its fundamental biological character (e.g., microbial processes, symbioses). For every nutrient there are elementary chemical and physical properties that predetermine both the energetic cost and the com-

ISBN 0-12-160280-X

plexity of these mechanisms. In the case of nitrogen, both cost and complexity are high, and the "nitrogen problem" is related to these effects. It is within this context that symbiotic associations are of particular interest, for they offer uniquely successful solutions to specific problems associated with various steps in the global nitrogen cycle. Most notably, these include mechanisms for the recapture (fixation) of atmospheric nitrogen and the local conservation of organic nitrogen in discreet organismal assemblages. In shallow-water marine ecosystems, both solutions are significant elements sustaining locally high zones of productivity.

Nitrogen distributions in the physical/chemical environment of aquatic ecosystems are believed to be nonhomogeneous (Goldman et al., 1979); current views suggest that organisms experience these variations as both large- and small-scale patches of relatively high nutrient availability, but they are not themselves part of the overall nutrient patchiness observed. Endosymbiotic associations may offer an important exception, since they appear to occupy a unique position relative to overall nutrient availability. Defined as cellular ecosystems (Smith, 1979; Gooday and Doonan, 1980), these associations function to compartmentalize nitrogen and other nutrients away from the physical/chemical environment through various processes effecting acquisition, concentration, and recycling. They are in fact highly discreet nutrient patches supporting closed and semi-closed ecosystems that yield high biomass. Frequently, these endosymbiotic nutrient patches exist in comparatively impoverished physical/chemical environments. This condition is typified by the tropical coral reef and its adjacent waters. While it is true that the water column is impoverished, the system taken as a whole is not. Contrasted with other nonsymbiotically based ecosystems, it is seen that the difference lies in the location of nutrient patches inside rather than outside the biomass.

Whether it is an assertion or an early hypothesis, this view of endosymbiotic associations offers a useful framework for a discussion of their significance within the context of processes effecting nitrogen availability in the marine environment. Symbiosis clearly functions as a mechanism for the regeneration of nitrogen, either through recapture or conservative recycling, the former being characterized by the presence of prokaryotic symbionts, and the latter by the presence of both prokaryotic and eukaryotic symbionts.

II. PARTNERS

"Zusammenleben unleichnamiger Organismen" (DeBary, 1879). Freely translated as, the association of dissimilarly named organisms (Starr, 1975), this definition has proved to be one of the most durable and least well understood in biology. Symbiosis will be used here in its larger sense, as the superclass of all other organismal associations. To distinguish associations of coincidence from associations exhibiting recurrence, it is necessary to insist on the strict application of Koch's postulates. Although symbioses may be either transient or persistent, it is important to recognize the elements of recurrence and significance, and to use these in making a final determination when reporting symbiotic associations. While it may not always be experimentally possible, Koch's postulates provide a useful basis for making a final determination.

The partners of these associations are identified routinely as hosts and symbionts. Decisions in the literature as to which partner is the host and which one is the symbiont appear to be based largely on relative size and topographic position -- symbionts are smaller and lie within hosts. For the bacterial, cyanobacterial, and algal symbioses examined here, symbionts defined in this way are also the primary producers in the association. The following section examines our knowledge of the various species of hosts and symbionts. Whereas the range and variety of hosts are large, it will be seen that the symbionts are relatively few in number.

A. Fixers of Atmospheric Nitrogen

Prokaryotic symbionts capable of fixing atmospheric nitrogen are found among all major physiological groups, including the cyanobacteria, aerobic, facultative and anaerobic heterotrophs, and the photosynthetic bacteria (Beneman and Valentine, 1972). They are all of particular interest since nitrogen fixation is regarded as the rate-limiting step in the global nitrogen cycle (Postgate, 1971) and because these organisms offtimes provide a critical resource of combined nitrogen for their various hosts. In terrestrial environments, symbiotically mediated nitrogen fixation is well-known and has considerable economic and agricultural importance. It is less well studied in marine environments and, with few exceptions poorly understood. When reviewing the literature on symbiotic nitrogen fixation in marine systems, two criteria must be satisfied: First, the existence of a true symbiosis must be established when feasible; and second, nitrogen fixation must be experimentally demonstrated. For many of the reported associations in the literature, nitrogen fixation is merely inferred from the presence of a potential nitrogen-fixing organism, and the association itself may be coincidental and not recurring. This is especially true for symbioses involving invertebrate filter feeders as hosts.

1. Bacteria

The extent and variety of symbiotic interrelationships between bacteria and other bacteria, algae, higher plants, protozoa, invertebrates, and vertebrates is impressive to view but extremely difficult to comprehend. Many of these associations represent important links in aquatic food webs and are thus of concern to ecologists (Pomeroy, 1979). In this regard, the various nitrogen-fixing bacteria are potentially significant contributors to animal nitrogen requirements. They are found most commonly in association with hosts whose diets are noticably deficient with respect to nitrogen. In terrestrial environments, the phenomenon is well known and occurs in both vertebrates and invertebrates. Nitrogen fixation rates are generally low in these associations, but the dietary supplement is significant. Among aquatic hosts, only a few invertebrate associations are reported thus far. These are found to involve unusual nitrogen-fixing species (e.g., Vibrionaceae) and appear to yield relatively high rates of fixation (Carpenter and Culliney, 1975; Guerinot et al., 1977; Guerinot and Patriquin, 1981). It is apparent that microenvironmental conditions play a critical role, influencing activity or nonactivity in the bacterial symbiont. It is in this regard that involvement in symbiosis is uniquely important as a mechanism for providing both a suitable microenvironment and the requisite inputs and outputs that serve to sustain and drive fixation. The following discussion of experimentally verified symbioses is a mere suggestion of the potential.

a. Vibrio. The Vibrionaceae contains gram-negative, polarly flagellated organisms that are usually curved (comma-shaped) rods. They are common, recurring members of the microbial flora of aquatic systems and appear as both free-living and symbiotic *(sensu lato)* forms. *Vibrio cholerae* is a well-known representative found in the water column and various invertebrate and vertebrate hosts (Colwell, 1979). It is an important pathogen in man. Other unidentified species are found in recurring associations with "higher" invertebrates as characteristic "gut-group vibrios" (Colwell and Liston, 1960, 1962; Liston and Colwell, 1963; Beeson and Johnson, 1967; Guerinot et al., 1977; Guerinot and Patriquin, 1981; Hood and Meyers, 1974; Sochard et al., 1979). Similar associations are reported among lower invertebrates, notably sponges (Wilkinson, 1978.) Distinctions between coincidental and recurring relationships in this latter group are made difficult by the filter-feeding habit of the host.

Nitrogen fixation by the Vibrionaceae was unknown prior to reports of nitrogenase activity (C_2H_2 reduction) in symbiotic associations between echinoderms and gut vibrios (Guerinot et al., 1977). In subsequent studies, Guerinot and Patriquin (1981) successfully demonstrated symbiotic nitrogen fixation (^{15}N) (57 ng N_2 fixed·g wet $w^{-1}\cdot h^{-1}$) and determined the recurring nature of the association in several echinoderm species from temperate and tropical habitats. Their data argue strongly for the concept of nitrogen supplementation of a deficient diet in these species. The study is an important model for future work with other associations, since it successfully addresses the question of recurrence in several species covering a broad geographic and climatic range and determines both the rate of production and the degree of incorporation of symbiotically fixed nitrogen by the host. Within the microenvironment of this symbiosis, the host provides a carbon source, near neutral pH, and an anaerobic milieu. The symbiont provides a stable source of dietary nitrogen that ultimately enters the food web via the host.

Similar, gram-negative, curved, rod-shaped bacteria capable of fixing nitrogen (C_2H_2 reduction) under anaerobic conditions are reported from the gut of shipworms (family Teredinidae)(Carpenter and Culliney, 1975). These are also capable of liquefying cellulose and have been identified tentatively as members of the Spirillaceae. They are included here since there is reason to believe that they are also members of the gut-group vibrios discussed above. Nitrogen fixation is reported in associations involving juveniles and adults of four host species, three coastal *(Psiloteredo megotara, Lyrodus pedicellatus, Teredo navalis)* and one collected from the Sargasso Sea *(T. malledus)*. For the three coastal species, nitrogen fixation was inversely related to shipworm dry weight, with juveniles exhibiting the highest rates (81 ng N_2 fixed·mg $dry^{-1}\cdot h^{-1}$). Larvae tested had no associated nitrogen fixation. Adult *T. mal-*

ledus from the Sargasso Sea exhibited high associated nitrogen fixation (1.5 μg N_2 fixed·mg dry $w^{-1}.h^{-1}$). These data are indicative of a nitrogen-fixing capacity that assumes greater importance when dietary deficiencies are manifest in the animal host. This may occur in adult hosts when particulate foods are limiting (e.g. the Sargasso Sea) or in juveniles that are developmentally handicapped for suspension feeding. Under these conditions, nitrogen fixation by bacterial endosymbionts in the gut may be a significant supplementary source for the development and growth of the host.

2. *Cyanobacteria.*

Much of the literature on cyanobacterial symbiosis in marine organisms is observational or incidental in character. As a result, they are poorly known, and until recently there have been no experimental data on nitrogenase (C_2H_2 reduction) activity or nitrogen fixation (^{15}N). Lebour (1930) provides an early illustration of a possibly significant symbiosis occurring in the phytoplankton, the association between *Anabaena* sp. and the diatom *Coscinodiscus concinnus*. More recent work with this symbiosis (D. L. Taylor, unpublished) provides evidence for nitrogenase activity (C_2H_2 reduction). *Coscinodiscus concinnus* is a commonly encountered member of coastal phytoplankton assemblages, and its success may be due in part to a freedom from nitrogen limitation that its symbiosis with a cyanobacterium confers. Symbiosis with phytoplankton is apparently far more common than is generally recognized, despite the fact that it is rare for two photosynthetic organisms to unite symbiotically (Pringsheim, 1958). A stable source of combined nitrogen must be a powerful inducement. Among tropical species, *Richelia intracellularis* is found commonly within cells of *Rhizosolenia* and *Climacodium* and attached to the setae of *Chaetoceros* (Pascher, 1929; Mague et al., 1974, 1977). A further analysis of the recent literature on phytoplankton populations and distributions reveals increasing numbers of reports linking cyanobacteria with various dinoflagellates, cryptomonads, and diatoms (Norris, 1967; F.J.R. Taylor, 1976, 1980). The majority are incidental to the phytoplankton records reported, and exist only because the investigator recognized the presence of a cyanobacterium and its potential role as a symbiont. Since much of contemporary phytoplankton research is geared to other objectives, it is likely that investigators in the field are missing significant numbers of cyanobacterial symbioses. If they are widespread among coastal and open-ocean populations, their existence may seriously impact current thinking on nitrogen availability and utilization. This would be particularly true in the analysis of large, monospecific phytoplankton blooms. The problem clearly deserves closer scrutiny.

Apart from the phytoplankton, cyanobacterial symbioses are rare in other marine protists. This author has recorded an association in the marine ciliate *Sonderia,* but its functional significance is unknown and the extent of cyanobacterial-protistan symbiosis has not been assessed.

Among invertebrates, cyanobacterial symbioses are apparently confined to species of Porifera (Feldmann, 1933, Sara and Liaci, 1964; Sara, 1971; Droop, 1963; Vacelet, 1971; Wilkinson, 1979) and Annelida (Kawaguti, 1971). Nitrogen fixation resulting from these associations has been examined in three tropical sponges using the C_2H_2 reduction technique (Wilkinson and Fay, 1979). Two of these hosts contained cyanobacterial symbionts occupying up to 10% of the ectosomal tissue and contained in specialized amoebocytes, referred to as cyanocytes (Wilkinson, 1978). Acetylene reduction was highest in samples incubated in the light, and was correlated with the presence of symbionts. There are no available data to support an assessment of the role of nitrogen fixation in these associations. However, it is speculated that the additional combined nitrogen resource would be of value in tropical environments that are low in available nitrogen.

B. Assimilators of Combined Nitrogen

A dominant contemporary theme in studies of algal-invertebrate symbiosis is concerned with nitrogen acquisition, conservation, and cycling within the symbiotic unit. These processes are fundamental to both the success and the "raison d'etre" of the majority of associations existing in nutrient-impoverished regimes. As noted previously, they function to concentrate and compartmentalize nutrients within biologically defined "patches." In the associations studied thus far, this process can be initiated by both the algal symbiont (dissolved nitrogen)(Taylor, 1978; D'Elia et al., 1983) and its host (particulate nitrogen). An important factor to consider when assessing degrees of involvement and the relative importance of acquisition, conservation, and cycling to the overall biological impact of the symbiosis is the extent of cellular integration and interdependency. The range of known associations lies along a continuum extending from obligate to facultative interactions. These have been discussed at length elsewhere (Taylor, 1973a,b).

Symbionts include eukaryotic algae and those cyanobacteria that interact with the host primarily through photosynthesis rather than through nitrogen fixation. Our present knowledge of marine symbioses involving cyanobacteria is insufficient to permit a meaningful separation at this time. Accordingly, these have been summarized together in the preceeding section. There is, however, a clear need to evaluate fully both the phylogenetic range and the overall significance of cyanobacterial symbionts and their symbiosis in benthic and planktonic marine ecosystems. In contrast, our knowledge of the phylogenetic range of eukaryotic algal symbionts and their hosts is relatively complete and has actually increased since it was last examined (Taylor, 1974, 1983). Host species may include representatives from nearly all invertebrate phyla, but are most common among the Protozoa and Cnidaria, where an extremely diverse series of relationships are apparent (Taylor, 1973a,b). Symbiont species are restricted to two classes within the Chromophyta (*sensu*, Christensen, 1965), the Dinophyceae and the Bacillariophyceae, and two classes in the Chlorophyta (*sensu*, Christensen, 1965), the Chlorophyceae and the Prasinophyceae. Dinoflagellate symbionts are clearly the most abundant symbionts known, a feature that is possibly due to their successful liasons with various cnidarians (Taylor, 1983). In recent studies, the existence of diatom symbionts has been verified in several host species. These appear to be emerging as an important group engaged in symbiosis with the larger benthic foraminifera (Lee and McEnery, 1979; Lee et al., 1980a,b, c). Species of Chlorophyceae and Prasinophyceae are rare as symbionts, *Chlamydomonas hedleyi* (Lee et al., 1974) being possibly the best known representative of the former, and *Playtmonas convolutae* (Parke and Manton, 1967) being the best known representative of the latter. Despite its comparative rarity, the symbiosis of *P. convolutae* and *Convoluta roscoffensis* is the most fully developed algal-invertebrate association known. It is the essence of an obligate, highly tuned relationship, where internal nutrient resources are conserved and recycled within the cellular ecosystem to sustain a fully autotrophic unit: Keeble's plant-animal (Keeble, 1910). Early studies of this relationship led Keeble and Gamble (1907; also Keeble, 1910) to formulate many of the basic questions regarding symbiotic nitrogen balance. With some modification, they have come down to us in forms that have been subjected to critical review in recent years (Muscatine and Porter, 1977; Muscatine, 1980).

III. PROCESSES

The various processes (fixation, conservation, cycling) effecting nitrogen resources in symbiotic associations are images of the same processes known on a global scale. For fully developed associations, this feature is not surprising, and it serves to reinforce further the concept of cellular ecosystems (Smith, 1979).

A. Atmospheric Nitrogen

The process of biological nitrogen fixation is described elsewhere in this volume, and the reader is referred to those chapters for a discussion (Chapters 3 and 4). Within the symbiotic context, it is also useful to examine how the union of hosts and symbionts supports and sustains this vital process. Inherent to such a discussion is the concept of mutual exchange. From the preceeding discussion of nitrogen-fixing partners, it is apparent that hosts are characterized by nitrogen-poor diets and that nitrogen fixation provides a vital supplementary resource for growth and development (Carpenter and Culliney, 1975; Guérinot and Patriquin, 1981). Gains by symbionts are less clear. Symbionts are likely to receive similar supplements in the form of translocated organic and inorganic carbon. Such exchanges are clearly important, but it is the requirement for anaerobic conditions demanded by the chemistry of nitrogen fixation that is the single most important reciprocal feature of these associations. All of the known examples lead to this conclusion. Gut-group vibrios inhabiting sea urchins and shipworms must be able to find suitable anaerobic microenvironments within their hosts or they would not fix nitrogen at all. Similarly, the nonheterocystis cyanobacteria found in marine sponges must inhabit localized anaerobic zones within the host that serve to sustain their nitrogen-fixing activities. Contrasting but supporting evidence is also found with the heterocystis cyanobacterial symbionts of planktonic diatoms (e.g, *Richelia intracellularis*). These symbionts do not require a suitably anaerobic host milieu because the requirement is met by the heterocysts. This feature explains their survival and success within acellular, photosynthetic hosts.

B. Combined Nitrogen

Symbiosis between marine algae and invertebrates exists on a global scale. It occurs in isolated populations of marine invertebrates in temperate regions (e.g., *Convoluta roscoffensis*) and as the foundation of major ecosystems in the tropics (e.g., coral reefs). The latter has been and continues to be a major ecological event in the earth's history. Within all algal-invertebrate associations, the acquisition, conservation, and recycling of combined nitrogen is of major importance to their survival and sustained high primary production. Much of the literature has focused attention on the nutrient interrelationships existing between cnidarians and dinoflagellates, particularly associations involving hermatypic corals and *Gymnodinium microadriaticum*. This work has been reviewed (Muscatine and Porter, 1977; Muscatine, 1980). The perspective that emerges relates primarily to *open* associations where the host continues to feed, and gives limited insight into the interrelationships that exist in *closed* associations where the host never feeds. Continuing studies of the acoel *Convoluta roscoffensis* and its symbiont *Platymonas convolutae* are important in the latter context. These organisms exist as a complete autotrophic unit, whose mechanisms of external acquisition and internal recycling afford unique opportunities for research (e.g., Smith, 1979; Boyle and Smith, 1975; Gooday and Doonan, 1980).

Acquisition

Regardless of whether an association is classified as open or closed, there is an essential need for an external nitrogen source to provide primary inputs into the symbiosis. Open associations may access both particulate and dissolved nitrogen, the former through host feeding, the latter through processes driven largely by the alga (Muscatine, 1980). Closed associations rely solely upon dissolved sources, with the alga playing a similar role as primary importer (Taylor, 1978). At present, it is difficult to distinguish the relative significance of dissolved and particulate resources. To do so requires accurate measurement of inputs and internal fluxes and the data are lacking. We know considerably more about the uptake, retention, and release of dissolved nitrogen than we do about available particulate sources. We know very little about magnitudes and nutritional suitability; although intuitively, it is apparent that dissolved nitrogen sources provide external nitrogen to symbiotic associations at a significantly lower energetic cost than that associated with the range of available particulates. However, preferential utilization will depend on the relative abundance of dissolved and particulate nitrogen resources. This will vary considerably with the environmental context. It is therefore essential that we know both internal capacities and preferences, as well as environmental availability before reaching any conclusions regarding the primary source of external nitrogen in symbiotic systems.

a. Particulate Nitrogen. The arguments surrounding the importance of particulates as an external source of nitrogen are essentially the same as those that have been raised with respect to the importance of particulates as a source of carbon and calories for symbiotic associations (Muscatine and Porter, 1977; Taylor, 1983). A central issue is the relationship between the particle capture potential of the host and the actual nutritional demand that can be satisfied preferentially from particulate sources. Most cnidarian associations exhibit an impressive particle capture potential (Yonge, 1930) that appears to be underutilized. The few available studies suggest that particles are relatively unimportant to the overall carbon and caloric demand of a symbiosis. It is uncertain whether the same is true for nitrogen. Studies of carbon resource utilization in the coral *Montastrea cavernosa* (Lasker, 1981) lead to the conclusion that the host's machinery for particle capture (polyps and tentacles) is actually utilized to increase surface area for the capture of photons instead of particles. Increased surface area would also facilitate the uptake of dissolved carbon and nitrogen, a feature that argues in favor of dissolved nitrogen resources and against particulates as the most important external resource for this association.

The acoel *Amphiscolops langerhansi* also possesses well-developed abilities for particle capture. Its survival in symbiosis with *Amphidinium klebsii* is dependent on the host's ability to feed (Taylor, 1971). The association is thus an important example of an open symbiosis where particle capture may be relatively more important than the uptake of dissolved materials. Nitrogen utilization is under investigation in this association (D. L. Taylor, unpublished), and the flux of ^{15}N from particulate feeds ingested by the host to the algal symbiont (proteins and chlorophyll *a*) has been measured. Present data suggest a role for particulate nitrogen in this association, but its importance relative to competing dissolved sources is unknown.

b. Dissolved Nitrogen. The uptake and retention of dissolved nitrogen has been examined primarily in open symbioses (notably hermatypic corals), with little emphasis on closed associations (see Muscatine, 1980, for review). There are no conclusive data available on the uptake of dissolved organic nitrogen. Studies thus far have followed the literature on phytoplankton, emphasizing the utilization of inorganic sources (ammonium, nitrate, nitrite). Sources and concentrations of dissolved inorganic nitrogen are variable in the surface waters of reef environments where uptake has been studied most intensively. This variability is likely due to combined hydrographic factors and variations in processes affecting sources of new and regenerated nitrogen.

Analysis of dissolved inorganic nitrogen uptake is a tractable problem that has been examined in several systems. From the variety of reports summarized by Muscatine (1980; see also D'Elia et al., 1983), it is apparent that symbiotic associations are capable of removing significant amounts of dissolved nitrogen from the water column, even when present at extremely low concentrations. There is also clear recognition that these sources alone may be insufficient for the total nutrient demand, and that particulate sources must be an essential part of the total requirement for external nitrogen input to the symbiosis (D'Elia and Webb, 1977; Muscatine, 1980). The total nitrogen demand has not been determined for any of these systems.

Several studies lead to the conclusion that uptake of ammonium, nitrate, and nitrite is mediated by the algal symbiont. Algal assimilation into organic compounds is the important first step whereby dissolved inorganic nitrogen is eventually accessed by the host. The mechanism of the uptake process itself and the intermediate steps leading to the host are poorly known, although there are now available refined data on the kinetics of the process (D'Elia et al., 1983). Research on nitrogen acquisition in algal-invertebrate symbiosis has progressed in the shadow of work on marine phytoplankton (see Dugdale, 1976). In reality, symbiotic algae offer a more accessible experimental system that ought to be exploited in a way that would lead related phytoplankton research rather than follow it. Emphasis on wet chemistry and conformation to classic Michaelis-Menten kinetics does not address the critical questions that are central to an understanding of both the external nutrient requirements of symbiotic associations and their internal conservative processes. The answers require a larger experimental investment that combines the use of ^{15}N and modern approaches to algal/animal nutritional physiology and biochemistry.

IV. CONCLUSIONS

This examination of various symbiotic processes in relation to nitrogen in the marine environment spans a range of associations, biota, and categories of interaction. The unifying theme is, of course, nitrogen. More particularly, it is the "nitrogen-hunger" highlighted by Keeble (1910) several decades ago. Symbiotic interactions effect the global nitrogen cycle as either sources of new or regenerated nitrogen and as discreet organic assemblages capable of concentrating nitrogen in shallow-water benthic environments. For the vast majority of known and potential associations, it is apparent that our knowledge is only cursory. For associations involved in the fixation of atmospheric nitrogen, it is necessary to question the quantitative validity of the C_2H_2 reduction technique that has dominated this area of study. The associations that fix atmospheric N clearly need confirmation by more direct techniques.

Throughout, it is possible to conclude that the analogy of treating a symbiosis as a cellular ecosystem holds up well. Selected symbioses may provide useful models of processes governing nutrient acquisition, exchange, and conservation. This is true for associations that fix atmospheric nitrogen as well as those that concentrate combined nitrogen. The latter are important models of phytoplankton nutrient dynamics and ought to be included in the mainstream of that research. However, it is necessary to examine mechanisms of transport, metabolism, and cycling in detail if the potential is to be realized effectively.

REFERENCES

Beeson, R. J., and Johnson, P. I. (1967). Natural bacterial flora of the bean clam, *Donax gouldi*. *J. Invertebr. Pathol. 9*, 104-110.

Beneman, J. R., and Valentine, R. C. (1972). The pathways of nitrogen fixation. *Adv. Microb. Physiol. 8*, 59-104.

Boyle, J. E., and Smith, D. C. (1975). Biochemical interactions between the symbionts of *Convoluta roscoffensis*. *Proc. R. Soc. London, Ser. B 189*, 121-135.

Carpenter, E. J., and Culliney, J. L. (1975). Nitrogen fixation in marine shipworms. *Science 187*, 551-552.

Christensen, T. (1965). Systematisk Botanik. N. 2. Alger. *Botanika 2*, 1-80.

Colwell, R. R. (1979). Human pathogens in the aquatic environment. *In* "Aquatic Microbial Ecology" (R. R. Colwell and J. Foster, eds.), pp. 337-344. Sea Grant, University of Maryland, College Park.

Colwell, R. R., and Liston, J. (1960). Microbiology of shellfish. Bacteriological study of the natural flora of Pacifica oysters *(Crassostrea gigas)*. *Appl. Microbiol. 8*, 104-109.

Colwell, R. R., and Liston, J. (1962). The natural bacterial flora of certain marine invertebrates. *J. Insect Pathol. 4*, 23-33.

DeBary, A. (1879). "Die Erscheinung der Symbiose." Verlag von Karl J. Trübner.

D'Elia, C. F., and Webb, K. L. (1977). The dissolved nitrogen flux of reef corals. *Proc. Int. Coral Reef Symp. 34d, 1977 Vol. 1,* 325-330.

D'Elia, C. F., Domotor, S. L., and Webb, K. L. (1983). Nutrient uptake kinetics of freshly isolated zooxanthellae. *Mar. Biol.* (in press).

Droop, M. R. (1963). Algae and invertebrates in symbiosis. *Symp. Soc. Gen. Microbiol. 13,* 171-199.

Dugdale, R. D. (1976). Nutrient cycles. *In* "The Ecology of the Seas" (D. H. Cushing and J. J. Walsh, eds.), pp. 141-172. Saunders, Philidelphia, Pennsylvania.

Feldmann, J. (1933). Sur quelques cyanophycees vivant dans le tissu des eponges de Banyuls. *Arch. Zool. Exp. Gen. 75,* 381-404.

Goldman, J. C., McCarthy, J. J., and Peavey, D. G. (1979). Growth rate influence on the chemical composition of phytoplankton in oceanic waters. *Nature (London) 279,* 210-215.

Gooday, G. W., and Doonan, S. A. (1980). The ecology of algal-invertebrate symbioses. *Contemp. Microb. Ecol. (Proc. Int. Symp.) 2nd, 1980,* pp. 377-390.

Guérinot, M. L., and Patriquin, D. G. (1981). The association of N_2-fixing bacteria with sea urchins. *Mar. Biol. (Berlin) 62,* 197-207.

Guérinot, M. L., Fong, W., and Patriquin, D. G. (1977). Nitrogen fixation (acetylene reduction) associated with sea urchins *(Strongylocentrotus droebachiensis)* feeding on seaweeds and eelgrass. *J. Fish. Res. Board Can. 34,* 416-420.

Hood, M. A., and Meyers, S. P. (1974). Microbial aspects of penaeid shrimp digestion. *26th Annu. Meet. Gulf Carribb. Fish.* pp. 81-92.

Kawaguti, S. (1971). *In* "Aspects of the Biology of Symbiosis" (T. C. Cheng, ed.), pp. 265-273.

Keeble, F. W. (1910). "Plant Animals." Cambridge Univ. Press, London and New York.

Keeble, F. W., and Gamble, F. (1907). The origin and nature of the green cells of *Convoluta roscoffensis. Q. J. Microsc. Sci. 51,* 167-219.

Lasker, H. R. (1981). Phenotypic variation in the coral *Montastrea cavernosa* and its effects on colony energetics. *Biol. Bull. (Woods Hole, Mass.) 160,* 292-302.

Lebour, M. V. (1930). "The Planktonic Diatoms of Northern Seas." Ray Society, London.

Lee, J. J., and McEnery, M. E. (1979). Isolation and cultivation of diatom symbionts from larger foraminifera (Protozoa). *Nature (London) 280,* 57-58.

Lee. J. J., Crockett, L. J., Gaen, J., and Stone, R. J. (1974). The taxonomic identity and physiological ecology of *Chlamydomonas hedleyi* sp. nov., algal flagellate symbiont from the foraminifer *Archais angulatus. Br. Phycol. J. 9,* 39-47.

Lee, J. J., McEnery, M. D., Lee, M. J., Reidy, J. J., Garrison, J. R., and Röttger, R. (1980a). Algal symbionts in larger foraminifera. *In* "Endocytobiology, Endosymbiosis and Cell Biology. A Synthesis of Recent Research" (W. Schwemmler and H. E. A. Schenk, eds.), pp. 113-124. de Gruyter, Berlin.

Lee, J. J., Reimer, C. W., and McEnery, M. E. (1980b). The identification of diatoms isolated as endosymbionts from larger foraminifera from the Gulf of Eilat (Red Sea) and the description of two new species, *Fragilaria shiloi* sp. nov. and *Navicula reissi* sp. nov. *Bot. Mar. 23,* 41-48.

Lee, J. J., McEnery, M. E., Röttger, R., and Reimer, C. W. (1980c). The isolation, culture and identification of endosymbiotic diatoms from *Heterostegina depressa* D'Orbigny and *Amphistegina lessonii* D'Orbigny (larger foraminifera) from Hawaii. *Bot. Mar. 23,* 297-302.

Liston, J., and Colwell, R. R. (1963). Host and habitat relationships of marine commensal bacteria. *In* "Symposium on Marine Microbiology" (C. H. Oppenheimer, ed.), pp. 611-624. Thomas, Springfield, Illinois.

Mague, T. H., Weare, N. M., and Holm-Hansen, O. (1974). Nitrogen fixation in the North Pacific Ocean. *Mar. Biol. (Berlin) 24,* 109-119.

Mague, T. H., Mague, F. C., and Holm-Hansen, O. (1977). Physiology and chemical composition of nitrogen-fixing phytoplankton in the Central North Pacific Ocean. *Mar. Biol. (Berlin) 41,* 213-227.

Muscatine, L. (1980). Uptake, retention and release of dissolved inorganic nutrients by marine algal-invertebrate associations. *In* "Cellular Interactions in Symbiosis and Parasitism" (C. B. Cook, P. W. Pappas, and E. D. Rudolph, eds.), pp. 229-244. Ohio State Univ. Press, Columbus.

Muscatine, L., and Porter, J. W. (1977). Reef corals: Mutualistic symbioses adapted to nutrient-poor environments. *BioScience 27,* 454-460.

Norris, R. (1967). Micro-algae in enrichment cultures from Puerto Penasco, Sonora, Mexico. *Bull. South. Calif. Acad. Sci. 66,* 233-250.

Parke, M., and Manton, I. (1967). The specific identity of the algal symbiont in *Convoluta roscoffensis*. *J. Mar. Biol. Assoc. U. K. 47,* 445-464.

Pascher, A. (1929). Studien über Symbiosen. Über einige Endosymbiosen von Blaualgen in Einzellern. *Jahrb. Wiss. Bot. 71,* 386-462.

Pomeroy, L. R. (1979). Microbial roles in aquatic food webs. *In* "Aquatic Microbial Ecology" (R. R. Colwell and J. Foster, eds.), pp. 85-109. Sea Grant, University of Maryland, College Park.

Postgate, J. R. (1971). Relevant aspects of the physiological chemistry of nitrogen fixation. *Symp. Soc. Gen. Microbiol. 21,* 287-307.

Pringsheim, E. G. (1958). Organismen mit blaugrünen Assimilatoren. *In* "Studies in Plant Physiology" (S. Prat, ed.), pp. 421-429. Cesk. Akad. Ved., Praha, Czechoslovakia.

Sará, M. (1971). Ultrastructural aspects of the symbiosis between two species of the genus *Aphanocapsa* (Cyanophyceae) and *Ircinia variabilis* (Demospongiae). *Mar. Biol. (Berlin) 11,* 214-221.

Sará, J., and Liaci, L. (1964). Associzione fra la Cianoficea *Aphanocapsa feldmanni* e alcune Demospongie marine. *Boll. Zool. 31,* 55-65.

Smith, D. C. (1979). From extracellular to intracellular: The establishment of a symbiosis. *Proc. R. Soc. London Ser. B 204,* 115-130.

Sochard, M. R., Wilson, D. F., Austin, B., and Colwell, R. R. (1979). Bacteria associated with the surface and gut of marine copepods. *Appl. Environ. Microbiol. 37,* 750-759.

Starr, M. P. (1975). A generalized scheme for classifying organismic associations. *Symp. Soc. Exp. Biol. 29,* 1-20.

Taylor, D. L. (1971). On the symbiosis between *Amphidinium klebsii* (Dinophyceae) and *Amphiscolops langerhansi* (Turbellaria: Acoela). *J. Mar. Biol. Assoc. U.K. 51,* 301-313.

Taylor, D. L. (1973a). Cellular interactions of algal-invertebrate symbiosis. *Adv. Mar. Biol. 11,* 1-56.

Taylor, D. L. (1973b). Algal symbionts of invertebrates. *Annu. Rev. Microbiol. 27,* 171-187.

Taylor, D. L. (1974). Symbiotic marine algae: Taxonomy and biological fitness. *In* "Symbiosis in the Sea" (W. B. Vernberg and F. J. Vernberg, eds.), pp. 245-262. Univ. of South Carolina Press, Columbia.

Taylor, D. L. (1978). Artificially induced symbiosis between marine flagellates and vertebrate tissues in culture. *J. Protozool. 25,* 77-81.

Taylor, D. L. (1983). Coral/algal symbiosis. *Handb. of Phycol. 3,* Cambridge University Press, Cambridge (in press).

Taylor, F. J. R. (1976). Dinoflagellates of the International Indian Ocean Expedition. *Bibl. Bot. (Stuttgart) 132,* 1-234.

Taylor, F. J. R. (1980). Basic biological features of phytoplankton cells. *In* "The Physiological Ecology of Phytoplankton" (I. Morris, ed.), pp. 3-55. Blackwell, Oxford.

Vacelet, J. (1971). Etude en microscopie electronique de l'association entre une cyanophycee chroococcale et une eponge du genre *Verongia*. *J. Microsc. (Paris) 12,* 363-380.

Wilkinson, C. R. (1978). Microbial associations in sponges. I. Ecology, physiology and microbial populations of coral reef sponges. *Mar. Biol. (Berlin) 49,* 161-167.

Wilkinson, C. R. (1979). Bdellovibrio-like parasite of cyanobacteria symbiotic in marine sponges. *Arch. Microbiol. 123,* 101-103.

Wilkinson, C. R., and Fay, P. (1979). Nitrogen fixation in coral reef sponges with cyanobacteria. *Nature (London) 279,* 527-529.

Yonge, C. M. (1930). Studies on the physiology of corals. I. Feeding mechanisms and food. *Rep. Great Barrier Reef Exped. 1,* 14-57.

Chapter 19

THE NITROGEN RELATIONSHIPS OF MARINE MACROALGAE

M. DENNIS HANISAK
Center for Marine Biotechnology
Harbor Branch Institution
Fort Pierce, Florida

I. INTRODUCTION

Marine macroalgae are an important autotrophic component in many coastal and estuarine ecosystems. These algae are considered to belong to three distinct divisions (Bold and Wynne, 1978): Chlorophycophyta (green algae), Phaeophycophyta (brown algae), and Rhodophycophyta (red algae). While the basis for this systematic delineation is primarily color (i.e., pigmentation), there are also other significant differences in chemical composition among these three groups. Although it is often convenient to group members of these three divisions together under terms such as macroalgae or seaweeds, it is important to remember that there is considerable biochemical, and presumably physiological, diversity among species.

ISBN 0-12-160280-X

The distribution and abundance of macroalgae in coastal and estuarine waters are quite variable from one location to another. Where there is a limited amount of suitable substrate, the macroalgal community is rather sparse, and primary production is due almost exclusively to phytoplankton. In other areas, complex seaweed communities may have high levels of primary production that clearly exceed those of phytoplankton in the same system (e.g., Blinks, 1955; Ryther, 1963; Mann, 1973). On a per unit area basis, the productivity of macroalgae in such systems may exceed that of the phytoplankton by two orders of magnitude. Thus, macroalgae must play important roles in energy transformations and nutrient recycling in these systems.

Despite the important role of macroalgae in coastal and estuarine waters, there has been much less research on their physiology and ecology than that of phytoplankton. This is perhaps due in large part to the relative difficulty of cultivating macroalgae and experimentally manipulating them. Consequently, the average marine biologist or oceanographer is much more familiar with the biology of phytoplankton than that of seaweeds; therefore, it might be useful to consider some basic differences between these two types of algae:

1. Macroalgae are true multicellular organisms whereas phytoplankton are unicellular or colonial. Macroalgae are several orders of magnitude larger than phytoplankton; their thalli are highly differentiated, and, in many cases, they develop highly specialized organs that are somewhat analogous to those of higher plants. Consequently, all cells in a macroalga are not metabolically similar. Often, rates of photosynthesis and nutrient uptake are much higher in meristematic regions of the thallus and/or confined to the outer layer of cells. Internal regions of the thalli are often nonphotosynthetic, and primarily involved in support and/or translocation. Thus, in many ways, higher plants, rather than phytoplankton, may serve as models for physiological investigations of macroalgae.

2. Individual macroalgae have a much longer generation time than phytoplankton. Many macroalgae are perennials, capable of living for several years. They are not quite as dependent on the instantaneous environmental conditions as are phytoplankton; they can store large amounts of nutrient and carbon reserves for use during other times of the year. Thus, while there are changes in the structure of a seaweed community, in general it is much more stable than the phytoplankton community in the same area.

3. Macroalgae are generally attached to substrate. With the exception of some sediment-inhabiting macroalgae that are found primarily in more tropical waters, there is little nutritional interaction between individual plants and their substrate. In general, holdfasts are comparable to the roots of higher plants only in the sense that they anchor the plant; nutrients are derived from the surrounding water as is the case for phytoplankton. Thus, macroalgae and phytoplankton in a given system may compete for nutrient resources; however, their attached mode of existence would indicate that macroalgae have evolved different strategies for acquiring these nutrients than phytoplankton (e.g., Munk and Riley, 1952).

4. The ultimate fate of most macroalgal production is to become part of the detrital food chain, whereas most of the primary production of phytoplankton is consumed by grazing. Mann (1973) estimated that less than 10% of the production of kelps normally is grazed, and the remainder enters the detrital food chain either as particulate or dissolved organic matter. Most seaweeds in temperate areas are probably grazed much less, or even not at all, than kelps; however, in some systems, e.g., tropical reefs, much of the macroalgal production can be consumed by herbivores.

Given the role of macroalgae in many coastal and estuarine systems, it is surprising how little research has focused on their nutrient relationships. Even nitrogen, which is now considered to be the nutrient most frequently limiting to algal growth in the sea, has only recently received much attention. Yet the nitrogen relationships (e.g., uptake, assimilation, storage, and release) of macroalgae should be considered if a complete picture of nitrogen cycling in coastal and estuarine systems is desired. Much of this research on macroalgae still needs to be done. This chapter will provide an overview of what is known about nitrogen and marine macroalgae and suggest topics that require further research.

II. EVIDENCE THAT NITROGEN CAN LIMIT THE GROWTH OF MACROALGAE

It is surprising that as little as 10 years ago it was commonly believed that the growth of macroalgae, unlike that of phytoplankton, was rarely, if ever, limited by the availability of nutrients. In the nearshore environment, where nutrient levels are relatively high compared with offshore waters, there appeared to be sufficient nutrients to sustain high levels of macroalgal production due to the continuous renewal of the surrounding water via water motion, i.e., currents (Munk and Riley, 1952; Ryther, 1963; Mann, 1973). While water motion is no doubt important in providing nutrients to macroalgae (Whitford and Schumacher, 1961; Schumacher and Whitford, 1965; Neushul, 1972; Gerard and Mann, 1979; Parker, 1981; Gerard, 1982b), it has become increasingly clear that the growth of macroalgae is, at times, limited by the availability of nitrogen.

Variations in the nitrogen content of macroalgae have been known for at least 50 years, when Haas and Hill (1933) noted that the "total nitrogen [content of seaweeds] varied with the circumstances of the enviornment." Pronounced seasonal variations in the nitrogen content of macroalgae are well documented for natural populations (e.g., Butler, 1936; Black and Dewar, 1949; Moss, 1950; MacPherson and Young, 1952; Vinogradov, 1953; Iwasaki and Matsudaira, 1954; Larsen and Jensen, 1957; Chapman and Craigie,

1977; Hanisak, 1979b; Gerard and Mann, 1979; Chapman and Lindley, 1980). The most commonly observed seasonal pattern is reduced nitrogen content during the summer when external supplies of nitrogen are generally at their annual minima. Internal nitrogen concentrations during the winter may be two to three times that of the summer. However, low external nutrient concentrations, even if they are correlated with reduced algal growth, do not establish a causal relationship between nitrogen and growth. The utility of measuring external supplies of nitrogen and relating them to macroalgal growth are limited due to the simultaneous variation of many environmental factors, an unknown nitrogen requirement for growth *in situ*, the presence of a continuous supply of nitrogen due to water motion, and the ability of these algae to store large amounts of nitrogen (Hanisak, 1979b).

More convincing evidence for nutrient limitation is the increased growth of macroalgae following the *in situ* addition of nutrients. Considerable increases in the growth of macroalgae are often observed near wastewater outfalls (e.g., Causey et al., 1945; Sawyer, 1956; Subbaramaiah and Parekh, 1966; Tewari, 1972); plants growing in such areas contain higher levels of internal nitrogen than those in unpolluted areas. Stimulation of growth following the addition of nitrogen to natural populations has demonstrated that they are limited by the availability of nitrogen, for at least part of the year (Tseng et al., 1955; Chapman and Craigie, 1977; Harlin and Thorne-Miller, 1981). In addition, the growth of many macroalgae has been stimulated by the enrichment of nitrogen in running seawater systems (e.g., Neish and Shacklock, 1971; Buggeln, 1974; Lapointe et al., 1976; Topinka and Robbins, 1976; DeBoer and Ryther, 1977; Neish et al., 1977; Lapointe and Ryther, 1979; Morgan et al., 1980; Hanisak, 1981; Lapointe and Tenore, 1981). Of course, the stimulation of growth of these macroalgae by the addition of nitrogen in a seawater system does not indicate a nitrogen limitation of *in situ* plants, but all of these studies demonstrate the importance of developing nutrient management

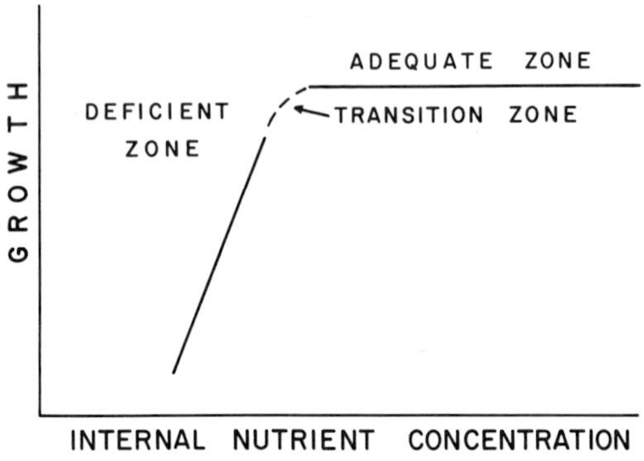

Fig. 1. The theoretical relationship between growth and internal concentration of an essential plant nutrient (modified from Ulrich, 1961). Toxicity may occur at higher concentrations.

strategies in macroalgal mariculture systems and the utility of these systems as experimental tools in macroalgal ecology and physiology.

While the experimental addition of nitrogen to natural populations clearly demonstrates the presence or absence of nitrogen limitation, it is desirable to have a means of making a rapid assessment of the nitrogen status of macroalgal populations or individuals within those populations. Perhaps the best indicator of a macroalga's nitrogen status is its internal nitrogen content (Hanisak, 1979b). This type of tissue analysis is a well-established method of determining the nutrient status of terrestrial plants (Ulrich, 1952; Smith, 1962; Epstein, 1972), and has been used to determine nutrient limitation in aquatic vascular plants (Gerloff and Krombholz, 1966; Gerloff, 1973). Plant tissue analysis allows one to determine a species' critical nutrient concentration, which is the internal nutrient concentration that just limits its maximal growth (Fig. 1). Higher or lower concentrations indicate nutrient reserves or nutrient deficiencies, respectively. Using this method, Hanisak (1979b) measured a critical nitrogen concentration of 1.90% (based on dry weight) for the green alga

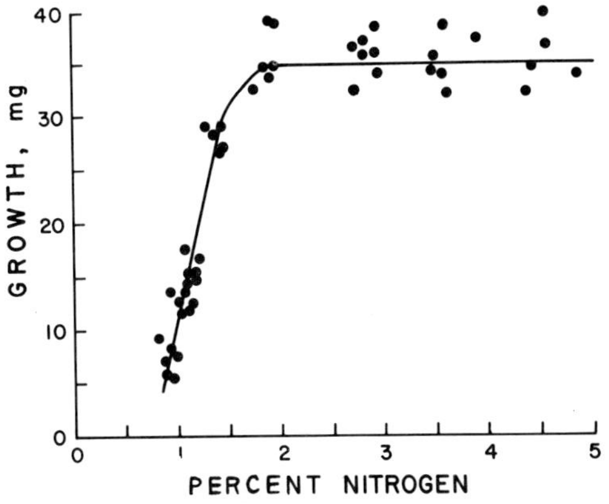

Fig. 2. The relationship of the growth of *Codium fragile* as the increase in dry weight over a 21-day period in culture and its internal nitrogen concentration, as percent of dry weight (from Hanisak, 1979b).

Codium fragile (Fig. 2), and determined that three populations of this species had different degrees of nitrogen limitation during the growing season. Similarly, the red alga *Gracilaria tikvahiae* was found to have a critical nitrogen concentration of ca. 2% (Hanisak, 1981, 1982); this value was used to assess the nitrogen status of this species, both in natural populations and in a mariculture system.

Oceanographers often use carbon:nitrogen ratios (C:N) to assess the physiological state of phytoplankton. C:N values for macroalgae have been reported to vary from 5 to 40 (Niell, 1976; Hanisak, 1979b). Some macroalgal biologists have tried to assess the nitrogen status of macroalgae using these ratios (e.g., Niell, 1976; DeBoer and Ryther, 1977; Jackson, 1977; Hanisak, 1979b; Lapointe and Ryther, 1979). Critical C:N values for macroalgal growth fall between 10 and 15 for the species examined so far; higher C:N values indicate nitrogen limitation while lower values indicate nitrogen storage. Whereas C:N ratios might be useful in interpreting the physiological state of algae (Myers, 1951), it must be remembered that shifts in this ratio can be due to changes in carbon as well as nitrogen metabolism. Thus, a change in C:N ratio of a plant, while of physiological significance, does not directly determine the presence or absence of nitrogen limitation (Hanisak, 1979b); under certain conditions, light intensity can influence C:N ratios as much as nitrogen availability (Lapointe, 1981; Lapointe and Tenore, 1981). Additional difficulties with utilizing C:N ratios for phytoplankton have been reported (Banse, 1974; Donaghy et al., 1978). Instead of C:N ratios, Bird (1983) recently used protein:carbohydrate values to detect nitrogen limitation of *Gracilaria verrucosa*.

Thus, although there is not a uniformly accepted method of rapidly determining the nitrogen status of macroalgae *in situ* short of long-term enrichment studies, there has been a considerable accumulation of evidence within the last few years to indicate that nitrogen is an important limiting nutrient for the growth of macroalgae. For ecological work, the nitrogen status of macroalgae can be assessed readily by using either the internal nitrogen concentration or C:N or protein:carbohydrate ratios. Future assays no doubt will be developed, probably with an enzymatic basis (e.g., see Küppers and Weidner, 1980).

III. THE EFFECT OF NITROGEN ON GROWTH OF MACROALGAE

Like other eukaryotic organisms, macroalgae cannot fix elemental nitrogen (N_2), although some macroalgae may receive a portion of their nitrogen requirement due to their associations with prokaryotic epiphytes that are capable of nitrogen fixation (e.g., Carpenter, 1972; Head and Carpenter, 1975; Capone et al., 1977; Penhale and Capone, 1981). In general, the most important sources of nitrogen for macroalgae are ammonium (NH_4^+) and nitrate (NO_3^-). However, not all macroalgae can grow equally well on these two sources (Table I). Nitrite (NO_2^-) is a third inorganic source of nitrogen that is present normally at much lower levels in seawater

TABLE I. Reported Inorganic Nitrogen Preferences for Selected Macroalgal Species

A. Species that grow equally well on NO_3^- or NH_4^+	
Codium fragile	Hanisak (1979a)
Chondrus crispus	Neish and Fox (1971)
	Neish and Shacklock (1971)
	Prince (1974)
	Laycock et al. (1981)
Fucus spiralis	Topinka and Robbins (1976)
Fucus vesiculosus	Prince (1974)
Gracilaria tikvahiae	Lapointe and Ryther (1978)
Porphyra tenera[a]	Iwasaki (1967)
B. Species that grow better on NO_3^- than on NH_4^+	
Gelidium amansii	Yamada (1961, 1972)
Gonotrichum elegans	Fries (1963)
Nemalion multifidum	Fries (1963)
Palmaria palmata	Morgan and Simpson (1981)
Porphyra tenera[a]	Iwasaki (1967)
C. Species that grow better on NH_4^+ than on NO_3^-	
Chordaria flagelliformis	Probyn (1981)
Gracilaria tikvahiae	DeBoer et al. (1978)
	Lapointe and Ryther (1979)
Neoagardhiella baileyi	DeBoer et al. (1978)
Pterocladia capillacea	Celabrese and Fillicini (1970)

[a]*The preference of P. tenera was shown to depend on the concentration; NO_3^- was preferred at higher concentrations.*

than NH_4^+ and NO_3^-. Hanisak (1979a) found that the green seaweed *Codium fragile* grew as well on NO_2^- as on NH_4^+ or NO_3^-. In contrast, Fries (1963) found that *Goniotrichum elegans* and *Nemalion multifidum* could grow better on NO_2^- than on NO_3^-.

In addition to inorganic sources of nitrogen, many macroalgae apparently can use urea, although often their growth on urea is not as good as on inorganic nitrogen (Table II). Other organic nitrogen compounds can be used as a source of nitrogen for at least some macroalgae, including various amino acids (Fries, 1963; Fries and Pettersson, 1968; Iwasaki, 1967; Berglund, 1969; Puiseux-Dao, 1962; Turner, 1970) and purines and pyrimidines (Iwasaki, 1965). Uncharacterized organic nitrogen, contained in different types of

TABLE II. Reported Preferences of Urea Versus Inorganic Nitrogen Sources for Selected Macroalgal Species

A. Species that grow equally well on urea or inorganic nitrogen	
Codium fragile	Hanisak (1979a)
Enteromorpha linza	Berglund (1969)
Porphyra tenera[a]	Iwasaki (1967)
B. Species that grow better on inorganic nitrogen than on urea	
Asterocytis ramosa	Fries and Pettersson (1968)
Chondrus crispus	Neish and Fox (1971)
Gelidiella acerosa	Probyn (1981)
Gelidium amansii	Rao and Mehta (1973)
Neoagardhiella baileyi	Yamada (1961, 1972)
Porphyra tenera[a]	DeBoer et al. (1978)
Rhodosorus marinus	Iwasaki (1967)
Species that grow better on urea than on inorganic nitrogen	
Pterocladia capillacea	Nasr et al. (1968)
Ulva fasciata	Mohsen et al. (1974)

[a]*The preference of P. tenera was shown to depend on the concentration;* NO_3^- *was preferred over urea at higher concentrations.*

organic wastes, supported the growth of *Gracilaria tikvahiae* and *Neoagardhiella baileyi* (Asare, 1980; Hanisak, 1981).

While no doubt there are real differences among macroalgae in terms of their ability to grow on different nitrogen sources, care must be taken in making generalizations about these abilities for such reasons as:

1. Some of the discrepancies concerning nitrogen utilization studies may be due to toxicity caused by the high levels of added nitrogen normally present in enriched media (Hanisak, 1979a). The recipes of most enriched media contain high levels of NO_3^- (the least likely source of nitrogen to be toxic); the simple substitution of NH_4^+ or organic nitrogen for NO_3^- at these levels could give misleading results about the ability of macroalgae to use *in situ* levels of different nitrogen sources. Iwasaki (1967) noted that the apparent preference of one nitrogen source over another was dependent on the concentration used in the medium.

2. The apparent utilization of different nitrogen sources may depend on other environmental conditions. For example, DeBoer et al. (1978) found that *G. tikvahiae* grew better on NH_4^+ than on NO_3^- under low-light conditions in the laboratory, whereas Lapointe and Ryther (1978) reported that the species grew equally well on these two sources of nitrogen under high-light conditions in outdoor tanks. A second study by Lapointe and Ryther (1979) indicated NH_4^+ is at times a better source of nitrogen for *Gracilaria* than NO_3^- in outdoor cultures and at other times may be toxic.

3. Organic preferences are best determined under axenic conditions when bacterial interactions are eliminated, i.e., bacteria might assimilate urea and release inorganic nitrogen, which is then used by the macroalga for growth. Ecologists may argue that a macroalga and its attached bacteria are an ecological unit and that the exact mechanism for its utilization of nutrients is unimportant; physiologists, of course, would like to identify the mechanism involved.

4. Some older reports of organic nitrogen utilization are suspect due to the method of autoclaving organic nitrogen with other nutrients rather than sterilizing the organic nitrogen separately via filter sterilization and then aseptically adding it to the medium. Autoclaving can cause urea and other organic nitrogen sources to breakdown somewhat, producing NH_4^+ which is then the actual nitrogen source for growth (Fries, 1973).

While it is useful to know what nitrogen sources can be used by a macroalga for optimal growth, it is also important to quantify the relationship of nitrogen and growth rate. Optimal levels for nitrogen have been reported for many macroalgae in batch cultures (e.g., Boalch, 1961; Fries, 1963; Iwasaki, 1967; Mohsen et al., 1974; Steffensen, 1976; Hanisak, 1979b); however, such data have no ecological significance at all because the so-called optimal levels are much greater than levels that are ever found *in situ*.

A more useful approach is the measurement of growth kinetics at concentrations that are more relevant to ambient nitrogen concentrations. Unfortunately, this has not been done often for macroalgae. Since there is a hyperbolic relationship of growth to external nitrogen concentration (Chapman et al., 1978; DeBoer et al., 1978; Hanisak, 1979b; Probyn, 1981), data describing this relationship can be fit readily, as is often done for phytoplankton, to the Michaelis-Menten model (see Goldman and Glibert, Chapter 7, this volume). Several estimates of K_s have been made for macroalgae grown on NO_3^-: 0.2 μ*M* for *Neoagardhiella baileyi* and 0.4 μ*M* for *Gracilaria tikvahiae* (DeBoer et al., 1978); 1.3 μ*M* for *Chordaria flagelliformis* (Probyn, 1981) and 1.4 μ*M* for *Laminaria saccharina* (Chapman et al., 1978). The K_s values for NH_4^+ include 0.2 μ*M* for both *G. tikvahiae* and *N. baileyi* (DeBoer et al., 1978) and 2.8 μ*M* for *C. flagelliformis* (Probyn, 1981). The K_s values for urea include 0.2 μ*M* for *N. baileyi* (DeBoer et al., 1978) and 2.1 μ*M* for *C. flagelliformis* (Probyn, 1981). It is not known how stable K_s may be for a given species, although Chapman et al. (1978) speculated that K_s may be expected to vary in different locations having different ambient nitrogen levels.

IV. THE ACQUISITION OF NITROGEN BY MACROALGAE

Three processes are involved in the acquisition of nitrogen by a macroalga: (1) diffusion across the boundary layer adjacent to the plant surface, (2) uptake across the cell membrane, and (3) assimilation into amino acids either directly (i.e., as NH_4^+) or indirectly (i.e., after the reduction of NO_3^- or NO_2^- to NH_4^+). As with other plants, there have been few measurements of the diffusion process across the unstirred boundary layer, although the importance of water motion to this process is generally recognized (Whitford and Schumacher, 1961, 1964; Schumacher and Whitford, 1965; Neushul, 1972; Gerard and Mann, 1979; Parker, 1981; Gerard, 1982b). Recently, Gerard (1982b) demonstrated that the uptake of NO_3^- by *Macrocystis pyrifera* was saturated at a flow rate equivalent to

2.5 cm s^{-1}, a value that was always exceeded by actual *in situ* measurements of water motion. More research of this type is required on other species of macroalgae and then related to *in situ* conditions.

The main focus of studies on the acquisition of nitrogen by macroalgae has been on the uptake of inorganic nitrogen. It appears that macroalgae have the ability to take up NO_3^- and NH_4^+ simultaneously, although NH_4^+ is usually taken up at a higher rate (Bird, 1976; D'Elia and DeBoer, 1978; Haines and Wheeler, 1978; Hanisak and Harlin, 1978; Topinka, 1978; Morgan and Simpson, 1981; Ryther et al., 1981). In many cases, NO_3^- and/or NO_2^- uptake is inhibited by the presence of NH_4^+ (e.g., D'Elia and DeBoer, 1978; Haines and Wheeler, 1978; Hanisak and Harlin, 1978); the degree of this inhibition is often dependent on the concentration of NH_4^+. Some macroalgae, especially brown algae, can take up NO_3^- without inhibition by the presence of NH_4^+ (Haines and Wheeler, 1978; Harlin and Craigie, 1978; Topinka, 1978). Uptake of NO_2^- has been shown to be inhibited by the presence of NO_3^- when such an interaction has been sought (Hanisak and Harlin, 1978; Harlin and Craigie, 1978).

In many cases where estimates of nitrogen uptake have been made, there have been efforts to invoke saturation uptake kinetics, as has classically been the case for phytoplankton. As with growth kinetics, the uptake of nitrogen is a hyperbolic function of concentration (Fig. 3) and the data are fitted to the Michaelis-Menten equation. The half-saturation constant is often used as a relative estimate of a species' ability to take up a nutrient at low levels, whereas V_{max} estimates the maximal uptake rate that occurs at fairly high concentrations. Such saturation kinetics have been studied for several species of macroalgae and are summarized in Table III. The only obvious trend in these data is that, for a given species, the V_{max} for NH_4^+ is consistently higher than the V_{max} for NO_3^- and NO_2^-. The ranges for these estimates given in the table are due to a number of reasons, including the influence of environmental factors, nitrogen preconditioning, and portion of the

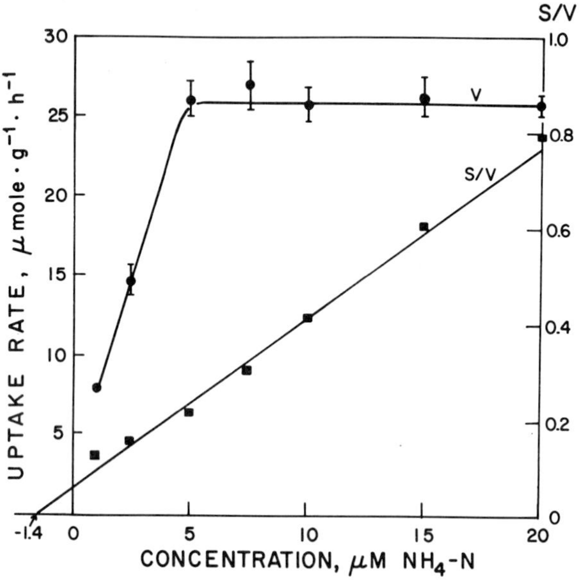

Fig. 3. A representative sample of nitrogen uptake data that follows simple saturation uptake kinetics. These data are for NH_4^+ uptake ($\mu M \cdot g$ dry $wt^{-1} \cdot h^{-1}$) by *Codium fragile* (from Hanisak and Harlin, 1978). Kinetic constants are derived from the linear transformation of the data; in this case, a (S/V) versus (S) plot was used.

thallus used; not shown in this table is the fact that there is frequently a large amount of variation in these estimates, even under the same experimental conditions.

Many environmental factors influence nitrogen uptake rates of macroalgae. Light is an important factor, especially for NO_3^- and NO_2^- uptake; uptake rates are often higher in the light than in the dark (D'Elia and DeBoer, 1978; Hanisak and Harlin, 1978; Harlin, 1978; Harlin and Craigie, 1978; Topinka, 1978). This does not necessarily mean that seaweeds cannot assimilate adequate nitrogen in the dark; for example, the uptake of NH_4^+ by *Codium fragile* in the dark, while reduced much more than that in the light, is still greater than the light uptake of NO_3^- or NO_2^- (Hanisak and Harlin, 1978). The length of the dark period may be important in deter-

TABLE III. Mean Values of V_{max} ($\mu mole \cdot g^{-1} \cdot h^{-1}$) and K_s (μM) for Inorganic Nitrogen Uptake in Selected Macroalgae[a]

Species	NO_3^- uptake		NO_2^- uptake		NH_4^- uptake		Reference
	V_{max}	K_s	V_{max}	K_s	V_{max}	K_s	
Codium fragile	2.8-10.9	1.2-7.6	3.7-9.0	3.1-10.4	13.0-28.0	1.4-2.1	Hanisak and Harlin (1978)
Enteromorpha spp.	129.4	16.6	--	--	--	--	Harlin (1978)
Fucus spiralis	1.4-2.5[b]	5.6-12.8	0.8-2.5[b]	26.3-27.7	1.6-3.2[b]	5.8-9.6	Topinka (1978)
Gracilaria tikvahiae	9.7	2.5	--	--	23.8[c]	1.6	D'Elia and DeBoer (1978)
Hypnea musciformis	28.5	4.9	--	--	*c*	16.6	Haines and Wheeler (1978)
Laminaria longicruris	7.0-9.6	4.1-5.9	--	--	--	--	Harlin and Craigie (1978)
Macrocystis pyrifera	22.4-30.5	8.7-13.1	--	--	*c*	5.3	Haines and Wheeler (1978)
Neoagardhiella baileyi	11.7	2.4	--	--	5.6-30.0[c]	2.3-4.9	D'Elia and DeBoer (1978)

[a]*When ranges are listed, they are for estimates made under different experimental conditions.*

[b]*V_{max} values were originally reported on an areal rather than on a dry weight basis. These values in this table were calculated using the conversion factor given by Topinka (1978).*

[c]*NH_4^+ uptake in these species did not fit simple saturation kinetics. V_{max} values presented here are for the low-affinity system.*

mining the amount of reduction from light uptake; for example, the dark uptake of NH_4^+ by *G. tikvahiae* is initially the same as that in the light, but it decreases with time (Ryther et al., 1981). Nitrogen uptake saturates at fairly low quantum flux, e.g., 7-28 $\mu E \cdot m^{-2} \cdot s^{-1}$ for *C. fragile* (Hanisak and Harlin, 1978). For *Ulva fasciata*, Lapointe and Tenore (1981) found that NO_3^- uptake was inversely proportional to light levels in nitrogen-enriched plants but not in unenriched plants.

Another factor that can affect nitrogen uptake is temperature (Hanisak and Harlin, 1978; Harlin, 1978; Harlin and Craigie, 1978; Topinka, 1978). For example, in *C. fragile*, V_{max} and K_s for NO_3^-, NO_2^-, and NH_4^+ were maximal at intermediate temperatures of 12°-24°C, lower at 30°C, and lowest at 6°C (Hanisak and Harlin, 1978). A similar trend in K_s was observed for *Fucus spiralis* (Topinka, 1978). Other environmental factors that influence nitrogen uptake include water motion (Topinka, 1978; Parker, 1981; Gerard, 1972b), desiccation in intertidal species (Thomas and Turpin, 1980; Thomas, 1982), and other nitrogen sources, i.e., the inhibition of NO_3^- or NO_2^- uptake by NH_4^+, or NO_2^- uptake by NO_3^- as described earlier.

Nitrogen uptake is also influenced by the nitrogen status of a macroalga as well as the part of the thallus used in an experiment. For example, the highest NH_4^+ uptake of *Gracilaria tikvahiae* and *Neoagardhiella baileyi* was found in plants having the highest C:N ratios, i.e., the most nitrogen-starved plants (D'Elia and DeBoer, 1978). Uptake decreased as the C:N ratio decreased down to a C:N of ca. 10; below this value, the C:N ratio did not seem to influence the uptake rate. Similarly, Ryther et al. (1981) observed that NH_4^+ uptake was strongly dependent on the nitrogen content of *G. tikvahiae*. Another interesting observation made by D'Elia and DeBoer (1978) was the presence of a diel pattern in NH_4^+ uptake in nitrogen-replete, but not nitrogen-deplete, plants of *G. tikvahiae*. Although Wheeler (1979), using [^{14}C]methylamine as an NH_4^+ analog, found no relationship between the nitrogen content of *Macrocystis pyrifera* and its uptake rate, these other results indicate that the nutrient preconditioning of macroalgae will influence strongly the measured uptake rates. With this in mind, it is interesting to speculate that nitrogen gradients in macroalgae, such as that observed in *Alaria esculenta* (Buggeln, 1978), could influence uptake rates in different portions of thalli, although obviously other factors also vary throughout macroalgal thalli. For *Fucus spiralis*, the highest NO_3^- and NH_4^+ uptake rates were found to be in apical fronds and whole young plants, and the lowest rates were in slower growing, older fronds and stipes (Topinka, 1978). Wheeler (1979) also demonstrated that NH_4^+ uptake in *M. pyrifera* depends on the portion of the thallus under consideration.

Most of the researchers who have estimated Michaelis-Menten uptake constants have attempted to place ecological significance on them. However, tempting as this is, it probably should be avoided for the present. As shown in the preceding paragraphs, these constants are not constant at all; they are influenced by environmental conditions and the nutrient and physiological state of the tissue being examined. In addition, there are vast differences in the way researchers have performed the experiments needed to estimate these constants, making comparisons between species studied by different investigators difficult. There has been much emphasis on K_s, but, as D'Elia and DeBoer (1978) point out, V_{max} may also have ecological significance, especially if nitrogen is made available in short pulses (e.g., animal excretion). Finally, the applicability of simple saturation kinetics to all uptake systems is questionable; indeed, many of the observed data do not fit the Michaelis-Menten equation. For example, D'Elia and DeBoer (1978) determined that there was a dual-phasic system involved in NH_4^+ uptake by *G. tikvahiae* and *N. baileyi*. Below 10 μ*M*, a high-affinity (low K_s) system was operative, while at higher concentrations, there was evidence for either a low-affinity (high K_s) system or a strong diffusive element. Similarly, MacFarlane and Smith (1982) have recently presented evidence for a dual-phasic NH_4^+ uptake system in *Ulva rigida*; at low concentrations uptake is via an amine cation uptake "porter" that is rate-limited by diffusion across the unstirred boundary layer and the second phase is due to diffusion of NH_4^+ into the tissue. Most of the other studies on nitrogen uptake data have not observed such a system, perhaps because a broad enough range of concentrations was not always used. However, Haines and Wheeler (1978) observed a phenomenon similar to that of D'Elia and DeBoer (1978), and Topinka's data (1978) for *F. spiralis* also seem to depart from simple saturation kinetics. Such multiphasic patterns of uptake are common in higher plants (Epstein, 1972; Nissen, 1974) and should be looked for more often in macroalgae.

There is little known about the actual mechanisms of nitrogen uptake in macroalgae; it is generally assumed to be via active transport systems similar to that of other plants. Such a system appears to operate for NH_4^+ uptake in at least some macroalgae (Wheeler, 1979; MacFarlane and Smith, 1982), but more research on this process is required (see DeBoer, 1981, for a discussion of active transport and other possible uptake mechanisms). Similarly, the mechanisms of assimilation have received little attention, although they are generally assumed to be similar to those of phytoplankton (e.g., Falkowski, 1978; Chapter 23, this volume)

V. NITROGEN POOLS AND STORAGE IN MACROALGAE

There has been little characterization of nitrogen pools in macroalgae. What research has been done suggests that there are large amounts of variation in both the quality and quantity of these pools. It is clear that macroalgae can build up substantial nitrogen reserves and then use them for growth during periods of low external supplies (e.g., Chapman and Craigie, 1977; Jackson, 1977; Gerard and Mann, 1979; Hanisak, 1979b; Chapman and Lindley, 1980; Gerard, 1982a; Rosenberg and Ramus, 1982b). Under favorable external nitrogen conditions, considerably more nitrogen is taken up by macroalgae than can be used at that time for growth. This is dramatically demonstrated by *G. tikvahiae*, which can take up enough nitrogen in 6 h to provide for up to 2 weeks of non-nitrogen-limited growth (Ryther et al., 1981).

The nature of this nitrogen storage varies among the macroalgae. For example, there can be a large NO_3^- pool in *Laminaria longicruris* (Chapman and Craigie, 1977), with a maximal NO_3^- content of 150 μmol·g wet wt^{-1}. This is a concentration factor of 28,000 over ambient external levels and amounts to 2.1% of the dry weight. This NO_3^- reserve decreased as external supplies declined, but it was adequate to sustain growth for 2 months after external levels were depleted. Large NO_3^- pools appear typical for *Laminaria* (Black and Dewar, 1949; Larsen and Jensen, 1957; Chapman et al., 1978; Gerard and Mann, 1979; Chapman and Lindley, 1980); presumably this inorganic pool is located in the vacuole, as in the case for other macroalgae having a large NO_3^- pool (Jacques and Osterhout, 1938).

Other macroalgae that have been examined have much lower inorganic nitrogen pools, typically 5-10% of the total nitrogen. In such species, it appears that amino acids and proteins can function as major nitrogen storage pools. For example, half of the organic nitrogen in *Chondrus crispus* can be the dipeptide citrullinylarginine, an amount that is ten times that of any free amino acid (Laycock and Craigie, 1977). The seasonal variation of this compound was such that it appeared to function as a nitrogen reserve for *Chondrus*. Amino acids are also important nitrogen reserves for *G. tikvahiae* (Bird et al., 1982); some of this reserve may be in the form of citrullinylarginine, which has been found in this species as well as in *Chondrus* (Laycock and Craigie, 1977). Interestingly, the amino acid composition of macroalgae can be influenced by the form of nitrogen that is available externally (Nasr et al., 1968). A considerable amount of nitrogen can be found in algal pigments. Most prominent in this regard are phycobiliproteins (i.e., phycoerythrins and phycocyanins), which may function as a nitrogen reserve in *G. tikvahiae* (Lapointe, 1981; Bird et al., 1982).

Juvenile sporophytes of *Macrocystis* are believed to have a limited storage capacity for nitrogen (Wheeler and North, 1980), based on the observation that growth rates increased linearly with increases in tissue nitrogen from 1-3% (dry weight). Saturation of growth as a function of this nitrogen did not occur, even though internal nitrogen levels were saturated as a function of external nitrogen. However, for older sporophytes there do appear to be some nitrogen reserves (Gerard, 1982a), presumably mainly proteins and amino acids that can be translocated throughout the thallus (Schmitz and Srivastava, 1979; Gerard, 1982a).

Although much work is required on nitrogen composition and pools in macroalgae, it may be helpful to conceptualize three "major nitrogen pools" (Fig. 4). For convenience, these may be designated as structural, physiological, and storage pools. For any given species, these pools can be further divided into numerous smaller pools. The size of these pools depends on the growth rate of the organism (which is controlled by the interaction of numerous environmental factors) and the internal nitrogen concentration (which reflects previous external nitrogen availability). The structural pool is rather conservative, i.e., it does not increase significantly when additional nitrogen is available; it consists of molecules that are absolutely required for life to exist, e.g., nucleic acids, structural proteins of membranes. For *G. tikvahiae*, this value is ca. 0.8% of the dry weight (M. D. Hanisak, unpublished). The physiological pool consists of nitrogen that is important to the physiological or metabolic processes of the macroalga, e.g., enzymes, photosynthetic pigments. Given the constraints of the enviornment, a macroalga will grow at a certain rate that will require a certain physiological pool size (i.e., the critical internal nitrogen concentration). Any nitrogen taken up in addition to what is required to sustain that growth will enter the storage pool. When external nitrogen levels decrease so that the alga can no longer assimilate as much nitrogen as is required for growth, it must draw on stored nitrogen to meet its

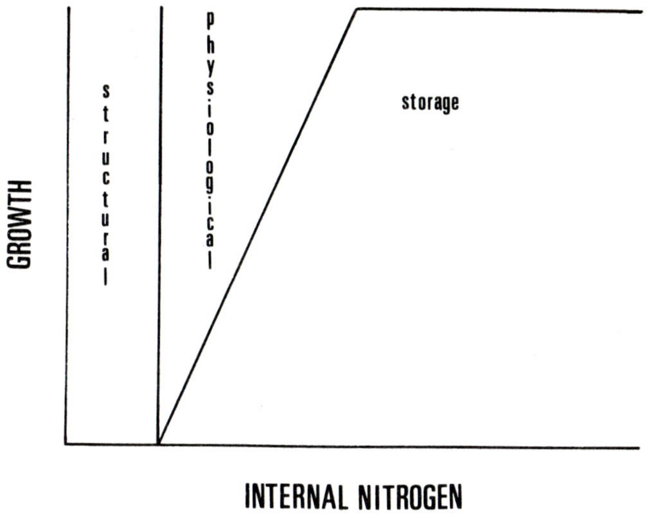

Fig. 4. Major nitrogen pools and their relationship to internal nitrogen concentration (e.g., percent of dry weight) and growth rate. The size of the "structural pool" is conservative and independent of growth rate; the size of the "physiological pool" increases as both internal nitrogen and growth rate increase; the "storage pool" contains nitrogen that is assimilated in excess of structural and physiological requirements.

structural and physiological requirements. Similarly, shifts in the growth rate of the alga caused by changes in other environmental factors can alter the balance between the physiological and storage pools. Perhaps the most interesting physiological work on nitrogen and macroalgae in the future will be the interactions among the external environment, growth processes, and various nitrogen pools.

VI. EFFECTS OF NITROGEN ON THE METABOLISM AND CHEMICAL COMPOSITION OF MACROALGAE

Because nitrogen is an important component of proteins, nucleic acids, and other cellular components, it is to be expected that the amount of nitrogen available to macroalgae will have substantial impact on their metabolism. For example, nitrogen availability can influence photosynthetic capacity due to effects on carboxylating enzymes and photosynthetic pigments (e.g., Calabrese and Felicini, 1970; Neish et al., 1977; DeBoer and Ryther, 1977; Chapman et al., 1978; Lapointe and Ryther, 1979; Küppers and Weidner, 1980; Morgan et al., 1980; Lapointe, 1981; Lapointe and Tenore, 1981; Ramus, 1981; Rosenberg and Ramus, 1982a). Chapman et al. (1978) found that photosynthesis of *Laminaria saccharina* increased with increasing levels of NO_3^-. Nitrate uptake was found to compete with photosynthesis in *Ulva fasciata* (Lapointe and Tenore, 1981); and high levels of NH_4^+ can reduce photosynthesis in some macroalgae, presumably due to toxic effects (Waite and Mitchell, 1972). Bird et al. (1980) found that nitrogen-limited thalli of *Gracilaria verucosa* will fix more carbon in the dark than when they are nitrogen-enriched. This fixed carbon was mostly in the form of amino acids, primarily citrulline. Bird et al. (1980) speculated that *G. verucosa* can use this nonphotosynthetic carbon fixation pathway for the rapid assimilation of nitrogen into amino acids and proteins when new nitrogen sources become available during periods of nitrogen limitation. Dark respiration was also enhanced by nitrogen limitation. Similarly, nitrogen limitation stimulates dark respiration in *Fucus spiralis* (Topinka and Robbins, 1976). Further studies of how nitrogen metabolism interacts with other aspects of macroalgal metabolism are needed.

As would be expected, amino acids, proteins, and nucleic acids can decrease in abundance when nitrogen becomes limiting (e.g., Nasr et al., 1968; Mohsen et al., 1974; Bird et al., 1982), although nucleic acids are much more likely to be conserved than amino acids and protein (Bird et al., 1982). As protein levels fall due to nitrogen limitation, it is common to observe in algae a concomittant increase in carbohydrates and/or lipids (Fogg, 1959; Syrett, 1962); such a response occurs in all macroalgae that have been examined. Perhaps this has been most dramatically shown for macroalgal phycocolloids; for example, there is a strong inverse relationship between carrageenan and protein levels or nitrogen availability

(Neish and Shacklock, 1971; Shacklock et al., 1973; Mathieson and Tveter, 1975; Neish et al., 1977). This relationship in macroalgae has often been called the "Neish effect;" similar inverse relationships between carbohydrates and nitrogen availability have often been reported (e.g., Tewari, 1972; Dawes et al., 1974; Chapman and Craigie, 1978; DeBoer, 1979; Chapman and Lindley, 1980; Gerard, 1982a; Rosenberg and Ramus, 1982b). As discussed earlier, this relationship is often measured by C:N or protein:carbohydrate ratios.

There are many other ways that nitrogen can influence macroalgal metabolism, as evident from numerous effects of nitrogen or reproduction and morphology (see DeBoer, 1981, for a discussion of these effects). The interaction of nitrogen with the uptake or metabolism of other nutrients remains to be explored; the report by Rice and Lapointe (1981) that nitrogen uptake in *Ulva fasciata* can regulate the uptake of micronutrients suggests that this will be an interesting, albeit complex, area of future research.

VII. CONCLUDING REMARKS

Although there has been an explosive interest in the nitrogen relationships of marine macroalgae during the last few years, much more work is needed on all aspects of this subject. Only a handful of species have received much attention, and there is no single pattern for these; obviously there is much physiological diversity in macroalgae, and many more species from different types of habitats should be studied. There must be greater standardization of methods to facilitate comparisons between species, particularly with regard to uptake kinetics; a greater appreciation of the effects of preconditioning on kinetic constants is necessary. Despite the previously described limitations of applying these "constants" in an ecological context, there will no doubt be further attempts to do so. If so, experiments should be performed *in situ* or at least under conditions that closely approximate those *in situ*. In this regard, the unpublished study of nitrogen uptake by Cowper (1981) using ^{15}N to demonstrate NH_4^+ uptake from the sediments by *Caulerpa cupressoides* is innovative. Also, macroalgal biologists should examine the environment more carefully than they have in the past; a single monthly sampling does not yield a very complete picture of nutrient dynamics in seawater. There are significant diel and tidal variations in nitrogen availability, as well as the more familiar seasonal ones. Diel patterns in metabolism of macroalgae, as well as in nutrient availability, should be considered. Similarly, the role of water motion in providing nutrients to macroalgae requires field studies, such as Gerard (1982b) has recently performed.

Little is known about the fate of nitrogen contained in macroalgae. Presumably, as mentioned earlier, in most systems, only a small amount of their production is consumed by grazing, and the bulk goes into the detrital food chain. With some exceptions, this latter food chain is not well defined. It is reasonable to assume that the nitrogen content of macroalgae has considerable influence on consumers that rely on macroalgae, either directly or indirectly, for their nutrition. There is also the possibility that macroalgae can release nitrogen in the form of dissolved organic nitrogen (e.g., DeBoer et al., 1978), although this too has not been studied to much extent.

It is clear that the productivity of macroalgae is an important component in some marine ecosystems and they can take up, assimilate, and store large amounts of nitrogen. Thus, macroalgae are important in the cycling and transformation of nitrogen in these systems. If so, it is interesting to speculate on the following question: To what extent are macroalgae competing with phytoplankton for nitrogen resources of their environment? Although such competition was first suggested by Marshall and Orr (1949), this question has received little attention. Although there are many factors involved in algal competition other than nitrogen, macroalgae appear to have some advantages over phytoplankton in their competition for nitrogen (Hanisak and Harlin, 1978). They include: simultaneous uptake of several forms of nitrogen at high uptake rates, large internal storage pools of nitrogen, and an increased supply of nitrogen due to currents. Since nitrogen is considered to be the most likely nutrient to limit algal growth in coastal areas, competition for nitrogen between macroalgae and phytoplankton may be important in these areas. Future research will elucidate the diverse strategies that exist for marine algae to partition the nitrogen resources of the marine environment.

REFERENCES

Asare, S. O. (1980). Animal waste as a nitrogen source for *Gracilaria tikvahiae* and *Neoagardhiella baileyi* in culture. *Aquaculture 21*, 87-91.

Banse, K. (1974). On the interpretation of data for the carbon-to-nitrogen ratio of phytoplankton. *Limnol. Oceanogr. 19*, 695-699.

Berglund, H. (1969). On the cultivation of multicellular marine green algae in axenic culture. *Sv. Bot. Tidskr. 63*, 251-264.

Bird, K. T. (1976). Simultaneous assimilation of ammonium and nitrate by *Gelidium nudifrons* (Gelidiales: Rhodophyta). *J. Phycol. 12*, 238-241.

Bird, K. T. (1983). Seasonal variation in protein:carbohydrate ratios in a subtropical estuarine alga, *Gracilaria verrucosa*, and the determination of nitrogen limitation status using these ratios. Unpublished manuscript.

Bird, K. T., Dawes, C. J., and Romeo, J. T. (1980). Patterns of non-photosynthetic carbon fixation in dark held, respiring thalli of *Gracilaria verrucosa*. *Z. Pflanzenphysiol. 98*, 359-364.

Bird, K. T., Habig, C., and DeBusk, T. (1982). Nitrogen allocation and storage patterns in *Gracilaria tikvahiae* (Rhodophyta). *J. Phycol. 18*, 344-348.

Black, W. A. P., and DeWar, E. T. (1949). Correlation of some of the physical and chemical properties of the sea with the chemical composition of the algae. *J. Mar. Biol. Assoc. U. K. 28*, 673-699.

Blinks, L. R. (1955). Photosynthesis and productivity of littoral marine algae. *J. Mar. Res. 14*, 363-373.

Boalch, G. T. (1961). Studies on *Ectocarpus* in culture. II. Growth and nutrition of a bacteria-free culture. *J. Mar. Biol. Assoc. U. K. 41*, 287-304.

Bold, H. C., and Wynne, M. J. (1978). "Introduction to the Algae: Structure and Reproduction." Prentice-Hall, Englewood Cliffs, New Jersey.

Buggeln, R. G. (1974). Physiological investigations on *Alaria esculenta* (L). Grev. (Laminariales). I. Elongation of the blade. *J. Phycol. 10,* 283-288.
Buggeln, R. G. (1978). Physiological investigations on *Alaria esculenta* (Laminariales, Phaeophyceae). IV. Inorganic and organic nitrogen in the blade. *J. Phycol. 14,* 156-160.
Butler, M. R. (1936). Seasonal variations in *Chondrus crispus. Biochem. J. 30,* 1338-1344.
Calabrese, G., and Felicini, G. P. (1970). Research on the red algae pigments. I. Pigments of *Pterocladia capillacea* cultured with some nitrogenous materials. *G. Bot. Ital. 104,* 81-89.
Capone, D. G., Taylor, D. L., and Taylor, B. F. (1977). Nitrogen fixation (acetylene reduction) associated with macroalgae in a coral-reef community in the Bahamas. *Mar. Biol. (Berlin) 40,* 29-32.
Carpenter, E. J. (1972). Nitrogen fixation by a blue-green epiphyte on pelagic *Sargassum. Science 178,* 1207-1209.
Causey, N. B., Prytherck, J. P., McCaskill, J., Humm, H. J., and Wolf, F. A. (1945). Influence of environmental factors upon the growth of *Gracilaria confervoides. Bull. Duke Univ. Mar. Stn. 3,* 19-24.
Chapman, A. R. O., and Craigie, J. S. (1977). Seasonal growth in *Laminaria longicruris*: Relations with dissolved inorganic nutrients and internal reserves of nitrogen. *Mar. Biol. (Berlin) 40,* 197-205.
Chapman, A. R. O., and Craigie, J. S. (1978). Seasonal growth in *Laminaria longicruris*: Relations with reserve carbohydrate storage and production. *Mar. Biol. (Berlin) 46,* 209-213.
Chapman, A. R. O., and Lindley, J. E. (1980). Seasonal growth of *Laminaria solidungula* in the Canadian High Arctic in relation to irradiance and dissolved nutrient concentrations. *Mar. Biol. (Berlin) 57,* 1-5.
Chapman, A. R. O., Markham, J. W., and Lüning, K. (1978). Effects of nitrate concentration on the growth and physiology of *Laminaria saccharina* (Phaeophyta) in culture. *J. Phycol. 14,* 195-198.
Cowper, S. W. (1981). Uptake of NH_3 by the rhizoids of *Caulerpa cupressoides* and translocation: Use of sediment nutrient sources. *J. Phycol. 17*(s), 4 (abstr. only).
Dawes, C. J., Lawrence, J. M., Cheney, D. P., and Mathieson, A. C. (1974). Ecological studies of Floridian *Eucheuma* (Rhodophyta, Gigartinales). III. Seasonal variation of carrageenan, total carbohydrate, protein, and lipid. *Bull. Mar. Sci. 24,* 286-299.
DeBoer, J. A. (1979). Effects of nitrogen enrichment of growth rate and phycocolloid content in *Gracilaria foliifera* and *Neoagardhiella baileyi* (Florideophyceae). *Proc. Int. Seaweed Symp., 9th, 1977,* pp. 263-271.
DeBoer, J. A. (1981). Nutrients. *In* "The Biology of Seaweeds" (C. S. Lobban and M. J. Wynne, eds.), pp. 356-392. Univ. of California Press, Berkeley.

DeBoer, J. A., and Ryther, J. H. (1977). Potential yields from a waste-recycling algal mariculture system. *In* "The Marine Plant Biomass of the Pacific Northwest Coast" (R. Krauss, ed.), pp. 231-249. Oregon State Univ. Press, Corvallis.

DeBoer, J. A., Guigli, H. J., Israel, T. L., and D'Elia, C. F. (1978). Nutritional studies of two red algae. I. Growth rate as a function of nitrogen source and concentration. *J. Phycol. 14,* 261-266.

D'Elia, C., and DeBoer, J. (1978). Nutritional studies of two red algae. II. Kinetics of ammonium and nitrate uptake. *J. Phycol. 14,* 266-272.

Donaghay, P. L., DeManche, J. M., and Small, L. F. (1978). On predicting phytoplankton growth rates from carbon:nitrogen ratios. *Limnol. Oceanogr. 23,* 359-362.

Epstein, E. (1972). "Mineral Nutrition of Plants: Principles and Perspectives." Wiley, New York.

Falkowski, P. G. (1978). The regulation of nitrate assimilation in lower plants. *In* "Nitrogen in the Environment" (D. R. Nielsen and J. G. MacDonald, eds.), Vol. 2, pp. 143-155. Academic Press, New York.

Fogg, G. E. (1959). Nitrogen nutrition and metabolic patterns in algae. *Symp. Soc. Exp. Biol. 13,* 106-125.

Fries, L. (1963). On the cultivation of axenic red algae. *Physiol. Plant. 16,* 695-708.

Fries, L. (1973). Requirements for organic substances in seaweeds. *Bot. Mar. 16,* 19-31.

Fries, L., and Pettersson, H. (1968). On the physiology of the red alga *Asterocytis ramosa* in axenic culture. *Br. Phycol. Bull. 3,* 417-422.

Gerard, V. A. (1982a). Growth and utilization of internal nitrogen reserves by the giant kelp *Macrocystis pyrifera* in a low-nitrogen environment. *Mar. Biol. (Berlin) 66,* 27-35.

Gerard, V. A. (1982b). *In situ* water motion and nutrient uptake by the giant kelp *Macrocystis pyrifera. Mar. Biol. (Berlin) 69,* 51-54.

Gerard, V. A., and Mann, K. H. (1979). Growth and production of *Laminaria longicruris* (Phaeophyta) populations exposed to different intensities of water movement. *J. Phycol. 15,* 33-41.

Gerloff, G. C. (1973). "Plant Analysis for Nutrient Assay of Natural Waters." U.S. Environ. Prot. Agency, Environ. Health Eff. Res. Ser., Washington, D.C.

Gerloff, G. C., and Krombholz, P. H. (1966). Tissue analysis as a measurement of nutrient availability for the growth of angiosperm aquatic plants. *Limnol. Oceanogr. 11,* 529-537.

Haas, P., and Hill, T. G. (1933). Observations on the metabolism of certain seaweeds. *Ann. Bot. (London)* [N.S.] *47,* 55-67.

Haines, K. C., and Wheeler, P. A. (1978). Ammonium and nitrate uptake by the marine macrophytes *Hypnea musciformis* (Rhodophyta) and *Macrocystis pyrifera* (Phaeophyta). *J. Phycol. 14,* 319-324.

Hanisak, M. D. (1979a). Growth patterns of *Codium fragile* ssp. *tomentosoides* in response to temperature, irradiance, salinity, and nitrogen source. *Mar. Biol. (Berlin) 50*, 319-332.
Hanisak, M. D. (1979b). Nitrogen limitation of *Codium fragile* ssp. *tomentosoides* as determined by tissue analysis. *Mar. Biol. (Berlin) 50*, 333-337.
Hanisak, M. D. (1981). Recycling the residues from anaerobic digesters as a nutrient source for seaweed growth. *Bot. Mar. 24*, 57-61.
Hanisak, M. D. (1982). The nitrogen status of *Gracilaria tikvahiae* in natural and mariculture systems as determined by tissue analysis. *Int. Phycol. Congr. 1st, 1982.* Unpublished paper.
Hanisak, M. D., and Harlin, M. M. (1978). Uptake of inorganic nitrogen by *Codium fragile* subsp. *tomentosoides* (Chlorophyta). *J. Phycol. 14*, 450-454.
Harlin, M. M. (1978). Nitrate uptake by *Enteromorpha* spp. (Chlorophyceae): Applications to aquaculture systems. *Aquaculture 15*, 373-376.
Harlin, M. M., and Craigie, J. S. (1978). Nitrate uptake by *Laminaria longicruris* (Phaeophyceae). *J. Phycol. 14*, 464-467.
Harlin, M. M., and Thorne-Miller, B. (1981). Nutrient enrichment of seagrass beds in a Rhode Island coastal lagoon. *Mar. Biol. (Berlin) 65*, 221-229.
Head, W. D., and Carpenter, E. J. (1975). Nitrogen fixation associated with the marine macroalga *Codium fragile*. *Limnol. Oceanogr. 20*, 815-823.
Iwasaki, H. (1965). Nutritional studies of the edible seaweed *Porphyra tenera*. I. The influence of different B_{12} analogues, plant hormones, purines, and pyrimidines on the growth of *Conchocelis*. *Plant Cell Physiol. 6*, 325-336.
Iwasaki, H. (1967). Nutritional studies of the edible seaweed *Porphyra tenera*. II. Nutrition of *Conchocelis*. *J. Phycol. 3*, 30-34.
Iwasaki, H., and Matsudaira, C. (1954). Studies of cultural grounds of a laver *Porphyra tenera* Kjellman in Matsukawa-Ura Inlet. I. Environmental characteristics affecting upon nitrogen and phosphorus contents of laver. *Bull. Jpn. Soc. Sci. Fish. 20*, 112-119.
Jackson, G. A. (1977). Nutrients and production of giant kelp, *Macrocystis pyrifera*, off southern California. *Limnol. Oceanogr. 22*, 979-995.
Jacques, A. G., and Osterhout, W. J. V. (1938). The accumulation of electrolytes. XI. Accumulation of nitrate by *Valonia and Halicystis*. *J. Gen. Physiol. 21*, 767-773.
Küppers, U., and Weidner, M. (1980). Seasonal variation of enzyme activities in *Laminaria hyperborea*. *Planta 148*, 222-230.
Lapointe, B. E. (1981). The effects of light and nitrogen on growth, pigment content, and biochemical composition of *Gracilaria foliifera* v. *angustissima* (Gigartinales, Rhodophyta). *J. Phycol. 17*, 90-95.

Lapointe, B. E., and Ryther, J. H. (1978). Some aspects of the growth and yield of *Gracilaria tikvahiae* in culture. *Aquaculture 15,* 185-193.

Lapointe, B. E., and Ryther, J. H. (1979). The effects of nitrogen and seawater flow rate on the growth and biochemical composition of *Gracilaria foliifera* v. *angustissima* in mass outdoor cultures. *Bot. Mar. 22,* 529-537.

Lapointe, B. E., and Tenore, K. R. (1981). Experimental outdoor studies with *Ulva fasciata* Delile. I. Interaction of light and nitrogen on nutrient uptake, growth, and biochemical composition. *J. Exp. Mar. Biol. Ecol. 53,* 135-152.

Lapointe, B. E., Williams, L. D., Goldman, J. C., and Ryther, J. H. (1976). The mass outdoor culture of macroscopic marine algae. *Aquaculture 8,* 9-21.

Larsen, B., and Jensen, A. (1957). The determination of nitrate and nitrite in algal material, and the seasonal variations in the nitrate content of some Norwegian seaweeds. *Rep. Norw. Inst. Seaweed Res. 15,* 1-22.

Laycock, M. V., and Craigie, J. S. (1977). The occurrence and seasonal variation of gigartine and L-citrullinyl-L-arginine in *Chondrus crispus* Stackh. *Can. J. Biochem. 55,* 27-30.

Laycock, M. V., Morgan, K. C., and Craigie, J. S. (1981). Physiological factors affecting the accumulation of L-citrullinyl-L-arginine in *Chondrus crispus*. *Can. J. Bot. 59,* 522-527.

MacFarlane, J. J., and Smith, F. A. (1982). Uptake of methylamine by *Ulva rigida*: Transport of cations and diffusion of free base. *J. Exp. Bot. 33,* 195-207.

MacPherson, M. G., and Young, E. G. (1952). Seasonal variation in the chemical composition of the Fucaceae in the maritime provinces. *Can. J. Bot. 30,* 67-77.

Mann, K. H. (1973). Seaweeds: Their productivity and strategy for growth. *Science 182,* 975-981.

Marshall, S. M., and Orr, A. P. (1949). Further experiments in the fertilization of a sea loch (Loch Craiglin). The effect of different plant nutrients on the phytoplankton. *J. Mar. Biol. Assoc. U. K. 27,* 360-379.

Mathieson, A. C., and Tveter, E. (1975). Carrageenan ecology of *Chondrus crispus* Stackhouse. *Aquat. Bot. 1,* 25-43.

Mohsen, A. F., Khaleafa, A. F., Hashem, M. A., and Metwalli, A. (1974). Effect of different nitrogen sources on growth, reproduction, amino acid, fat and sugar contents in *Ulva fasciata* Delile. *Bot. Mar. 17,* 218-222.

Morgan, K. C., and Simpson, F. J. (1981). Cultivation of *Palmaria (Rhodymenia) palmata*: Effect of high concentrations of nitrate and ammonium on growth and nitrogen uptake. *Aquat. Bot. 11,* 167-171.

Morgan, K. C., Shacklock, P. F., and Simpson, F. J. (1980). Some aspects of the cultivation of *Palmaria palmata* in greenhouse tanks. *Bot. Mar. 23,* 765-770.

Moss, B. (1950). Studies in the genus *Fucus*. II. The anatomical structure and chemical composition of receptacles of *Fucus vesiculosus* from three contrasting habitats. *Ann. Bot. (London)* [N.S.] *14,* 395-410.

Munk, W. H., and Riley, G. A. (1952). Absorption of nutrients by aquatic plants. *J. Mar. Res. 11,* 215-240.

Myers, J. (1951). Physiology of the algae. *Annu. Rev. Microbiol. 5,* 157-180.

Nasr, A. H., Bekheet, I. A., and Ibrahim, R. K. (1968). The effect of different nitrogen and carbon sources on amino acid synthesis in *Ulva, Dictyota,* and *Pterocladia*. *Hydrobiologia 31,* 7-16.

Neish, A. C., and Fox, C. H. (1971). "Greenhouse Experiments on the Vegetative Propagation of *Chondrus crispus* (Irish Moss)," Atl. Reg. Tech. Rep. Ser. No. 12. Nat. Res. Counc., Halifax, Canada.

Neish, A. C., and Shacklock, P. F. (1971). "Greenhouse Experiments on the Propagation of Strain T4 of Irish Moss," Atl. Reg. Tech. Rep. Ser. No. 14. Nat. Res. Counc., Halifax, Canada.

Neish, A. C., Shacklock, P. F., Fox, C. H., and Simpson, F. J. (1977). The cultivation of *Chondrus crispus*. Factors affecting growth under greenhouse conditions. *Can. J. Bot. 55,* 2263-2271.

Neushul, M. (1972). Functional interpretation of benthic marine algal morphology. *In* "Contributions to the Systematics of Benthic Marine Algae of the North Pacific" (I. A. Abbott and M. Kurogi, eds.), pp. 47-74. Jpn. Soc. Phycol., Kobe.

Niell, F. X. (1976). C:N ratio in some marine macrophytes and its possible ecological significance. *Bot. Mar. 19,* 347-350.

Nissen, P. (1974). Uptake mechanisms: Inorganic and organic. *Annu. Rev. Plant Physiol. 25,* 53-79.

Parker, H. S. (1981). Influence of relative water motion on the growth, ammonium uptake and carbon and nitrogen composition of *Ulva lactuca* (Chlorophyta). *Mar. Biol. (Berlin) 63,* 309-318.

Penhale, P. A., and Capone, D. G. (1981). Primary productivity and nitrogen fixation in two macroalgae-cyanobacteria associations. *Bull. Mar. Sci. 31,* 164-169.

Prince, J. S. (1974). Nutrient assimilation and growth of some seaweeds in mixtures of sea water and secondary sewage treatment effluents. *Aquaculture 4,* 69-79.

Probyn, T. A. (1981). Aspects of the light and nitrogenous nutrient requirement for growth of *Chordaria flagelliformis* (O. F. Mull.) C. Ag. *Proc. Int. Seaweed Symp., 10th, 1980,* pp. 339-344.

Puiseux-Dao, S. (1962). Récherches biologiques et physiologiques sur quelques Dasycladacées, en particulier le *Batophora Oerstedi* J. Ag. et l'*Acetabularia mediterranea* Lam. *Rev. Gen. Bot. 819-820,* 409-503.

Ramus, J. (1981). The capture and transduction of light energy. *In* "The Biology of Seaweeds" (C. S. Lobban and M. J. Wynne, eds.), pp. 458-492. Univ. of California Press, Berkeley.

Rao, P. S., and Mehta, V. B. (1973). Physiological ecology of *Gelidiella acerosa* (Forsskal) Feldmann et Hamel. *J. Phycol. 9,* 333-335.

Rice, D. L., and Lapointe, B. E. (1981). Experimental outdoor studies with *Ulva fasciata* Delile. II. Trace metal chemistry. *J. Exp. Mar. Biol. Ecol. 54,* 1-11.

Rosenberg, G., and Ramus, J. (1982a). Ecological growth strategies in the seaweeds *Gracilaria foliifera* (Rhodophyceae) and *Ulva* sp. (Chlorophyceae): Photosynthesis and antenna composition. *Mar. Ecol.: Prog. Ser. 8,* 233-241.

Rosenberg, G., and Ramus, J. (1982b). Ecological growth strategies in the seaweeds *Gracilaria foliifera* (Rhodophyceae) and *Ulva* sp. (Chlorophyceae): Soluble nitrogen and reserve carbohydrates. *Mar. Biol. (Berlin) 66,* 251-259.

Ryther, J. H. (1963). Geographic variations in productivity. "The Sea" (M. N. Hill, ed.), Vol. 2, pp. 347-380. Wiley, New York.

Ryther, J. H., Corwin, N., DeBusk, T. A., and Williams, L. D. (1981). Nitrogen uptake and storage by the red alga *Gracilaria tikvahiae* (McLachlan, 1979). *Aquaculture 26,* 107-115.

Sawyer, C. N. (1956). The sea lettuce problem in Boston Harbor. *J. Water Pollut. Control Fed. 37,* 1122-1133.

Schmitz, K., and Srivastava, L. M. (1979). Long-distance transport in *Macrocystis integrifolia*. I. Translocation of ^{14}C-labeled assimilates. *Plant Physiol. 63,* 995-1002.

Schumacher, G. J., and Whitford, L. A. (1965). Respiration and P^{32} uptake in various species of freshwater algae as affected by a current. *J. Phycol. 1,* 78-80.

Shacklock, P. F., Robson, D., Forsyth, I., and Neish, A. C. (1973). "Further Experiments on the Vegetative Propagation of *Chondrus crispus* T4," Atl. Reg. Tech. Rep. Ser. No. 18. Nat. Res. Counc., Halifax, Canada.

Smith, P. F. (1962). Mineral analysis of plant tissue. *Annu. Rev. Plant Physiol. 13,* 81-108.

Steffensen, D. A. (1976). The effect of nutrient enrichment and temperature on the growth in culture of *Ulva lactuca* L. *Aquat. Bot. 2,* 337-351.

Subbaramaiah, K., and Parekh, R. G. (1966). Observations on a crop of *Ulva fasciata* Delile growing in polluted sea water. *Sci. Cult. 32,* 370.

Syrett, P. J. (1962). Nitrogen assimilation. *In* "Physiology and Biochemistry of Algae" (R. A. Lewin, ed.), pp. 171-189. Academic Press, New York.

Tewari, A. (1972). The effect of sewage pollution on *Enteromorpha prolifera* var. *tubulosa* growing under natural habitat. *Bot. Mar. 15,* 167.

Thomas, T. E. (1982). Nitrogen uptake and assimilation in intertidal *Gracilaria verrucosa*. *Int. Phycol. Congr. 1st, 1982.* Unpublished paper.

Thomas, T. E., and Turpin, D. H. (1980). Desiccation enhanced nutrient uptake rates in the intertidal alga *Fucus distichus*. *Bot. Mar. 23*, 479-481.

Topinka, J. A. (1978). Nitrogen uptake by *Fucus spiralis* (Phaeophyceae). *J. Phycol. 14*, 241-247.

Topinka, J. A., and Robbins, J. V. (1976). Effects of nitrate and ammonium enrichment on growth and nitrogen physiology in *Fucus spiralis*. *Limnol. Oceanogr. 21*, 659-664.

Tseng, C. K., Sun, K. Y., and Wu, C. Y. (1955). Studies on fertilizer application in the cultivation of haidai *(Laminaria japonica* Aresch.*)*. *Acta Bot. Sin. 4*, 375-392.

Turner, M. F. (1970). A note on the nutrition of *Rhodella*. *Br. Phycol. J. 5*, 15-18.

Ulrich, A. (1952). Physiological basis for assessing the nutritional requirements of plants. *Annu. Rev. Plant. Physiol. 3*, 207-228.

Ulrich, A. (1961). Plant analysis in sugarbeet nutrition. *In* "Plant Analysis and Fertilizer Problems" (W. Reuther, ed.), pp. 190-211. Pub. 8, AIBS, Washington.

Vinogradov, A. P. (1953). "The Elementary Chemical Composition of Marine Organisms." Sears Found. Mar. Res., Yale University, New Haven, Connecticut.

Waite, T., and Mitchell, R. (1972). The effect of nutrient fertilization on the benthic alga *Ulva lactuca*. *Bot. Mar. 15*, 151-156.

Wheeler, P. A. (1979). Uptake of methylamine (an ammonium analogue) by *Macrocystis pyrifera* (Phaeophyta). *J. Phycol. 15*, 12-17.

Wheeler, P. A., and North, W. J. (1980). Effect of nitrogen supply on nitrogen content and growth rate of juvenile *Macrocystis pyrifera* (Phaeophyta) sporophytes. *J. Phycol. 16*, 577-582.

Whitford, L. A., and Schumacher, G. J. (1961). Effect of current on mineral uptake and respiration by a fresh-water alga. *Limnol. Oceanogr. 6*, 423-425.

Whitford, L. A., and Schumacher, G. J. (1964). Effect of a current on respiration and mineral uptake in *Spirogyra* and *Oedogonium*. *Ecology 45*, 168-170.

Yamada, N. (1961). Studies on the manure for seaweed. I. On the change of nitrogenous component of *Gelidium amansii* Lmx. cultured with different nitrogen sources. *Bull. Japn. Soc. Sci. Fish. 27*, 953-957.

Yamada, N. (1972). Manuring for *Gelidium*. *Proc. Int. Seaweed Symp., 7th, 1971*, pp. 385-390.

Chapter 20

NITROGEN DETERMINATION IN SEAWATER

CHRISTOPHER F. D'ELIA
Chesapeake Biological Laboratory
University of Maryland
Center for Environmental and Estuarine Studies
Solomons, Maryland

I. INTRODUCTION

Promising new methods have been developed in the past decade that have enhanced our abilities to analyze quantitatively different forms of nitrogen in seawater. Nonetheless, the most significant changes in the practice of nitrogen determination result not as much from widespread adoption of new methods as from widespread acceptance and improvement of automated methods adapted from the traditional manual ones. Accordingly, this chapter is not limited in scope to new methods alone, but includes information especially germane to the automation of traditional methods. Space precludes a thorough discussion here of the historical evolution of the traditional methods; interested readers can find such information elsewhere (e.g., Harvey, 1945; Barnes, 1959; Strickland and Parsons, 1972; Grasshoff, 1976; Riley, 1965, 1975). However, I have cited liberally from the significant body of "gray" litera-

ISBN 0-12-160280-X

ture (e.g., widely distributed institution publications), since it contains much information regarding present analytical practices and automation of manual methods. I devote particular attention in this chapter to the determination of those forms of nitrogen of greatest importance in the nitrogen cycle in the water column.

II. UNITS, NOMENCLATURE, AND PRECISION

Leading scientific journals such as *Analytical Chemistry* (cf. Vol. 54(1), 1982, p. 158) advocate the use of the International System of Units (SI) but recognize that practical use dictates that some "acceptable exceptions" to SI units exist. Nutrient concentrations can be acceptably reported in a variety of ways (Table I), of which molar or gram-atomic units are preferable because they allow the ready comparison of the ratio, by atoms, of one nutrient element to another. The use of weight per volume units (e.g., μg per liter) can be confusing and is discouraged because the weight basis, the molecule or ion (e.g., nitrate), or the atom of interest (e.g., nitrate-N), is not always clearly specified.

Nitrogen compounds are often classified by their size or chemical composition. For purposes of analytical chemistry of seawater and by convention, the substances passing through filters with pore sizes of approximately 0.2-1.2 μm are referred to as "dissolved," whereas those retained on these filters are referred to as "particulate." It is well known, however, that substances pass through a filter that are not truly dissolved in the chemical sense (Strickland and Parsons, 1972; Sharp, 1973).

Many analysts refer to nutrient forms that have passed through a filter and have been determined by wet chemistry and photometry as "soluble, reactive." This appellation is applied less often to nitrogen compounds than phosphorus compounds, but is unfortunate terminology in either case because it is both cumbersome and redundant. Compounds are dissolved by the criterion that they pass

TABLE I. Acceptable Units for Expressing Nutrient Concentrations, Their Meanings, and Their Abbreviations

Unit	*Abbreviation*	*Meaning*
Microgram-atoms per liter	μg•atom $liter^{-1}$	10^{-6} g•atoms per liter
Milligram-atoms per cubic meter	mg•atom m^{-3}	10^{-3} g•atoms per cubic meter (equivalent to μg•atom $liter^{-1}$)
Micromolar[a]	μ*M* (*not* μ*M* $liter^{-1}$)	10^{-6} moles per liter
Micromoles per liter	μmol $liter^{-1}$ (*not* μ*M* $liter^{-1}$)	10^{-6} moles per liter[b] (equivalent to μ*M*)
Millimoles per cubic meter	mmol m^{-3} (*not* m*M* m^{-3})	10^{-3} moles per cubic meter (equivalent to μ*M*)
Micrograms per liter	μg $liter^{-1}$	10^{-6} grams per liter (sometimes referred to as parts per billion)
Milligrams per liter	mg $liter^{-1}$	10^{-3} grams per liter (sometimes referred to as parts per million)

[a]*All molar-related expressions properly refer to dissolved not particulate nitrogen; gram-atomic units can refer to either.*

[b]*Present preferred by most standard-setting organizations, e.g., American Society for Testing and Materials (1982).*

through a filter; it is no more precise to refer to them as soluble, for chemically insoluble substances of appropriate size can also pass through. Reactive seems redundant because implicit in any wet chemical determination is a certain lack of specificity such that only those compounds that react with the reagents are measured. Thus, in this chapter I use the terms "nitrate," "nitrite," and "ammonia" without further qualification.

Terms relating to statistical properties of a given determination are not used consistently in the literature, and current practices in environmental analytical chemistry vary so much that results are often not deemed useful for interlaboratory comparison (American Chemical Society Committee on Environmental Improvement

[ACS/CIE], 1980). Accordingly, it is difficult to summarize and compare the precision, accuracy, and limit of detection reported by different authors for their methods.

"Precision" properly refers to the reproducibility of measurements within a set. Most measures of precision are based on, but are not equivalent to, standard deviation, which is typically calculated in one of two ways:

$$S = \sqrt{\Sigma(X_i - \overline{X}) / (N-1)} \qquad (1)$$

where S is the standard deviation, $\overline{X}_i$ is the ith replicate determination, $\overline{X}$ is the mean of replicates, and N is the number of replicates, or:

$$S = \sqrt{\Sigma d^2 / 2N} \qquad (2)$$

where S and N are as above and d is the difference between duplicates (Youden, 1959). I prefer Eq. (2) because it can be calculated on a running basis for day-to-day evaluation of routine samples: S determined by the first equation tends to be applied under ideal conditions on few samples, and thus may give an unrealistically optimistic estimate.

In analytical chemistry, S often increases, although not proportionally, as $\overline{X}$ increases, thus "relative standard deviation" or "coefficient of variation" (i.e., the standard deviation expressed as a fraction of the mean $S/\overline{X}$ is useful in evaluating precision over a wide range of concentrations. These terms are used with reasonable consistency in the literature, although they are not always reported.

"Limit of detection" is quite often expressed ambiguously, and great care should be used when comparing detection limits reported by different analysts. Limit of detection is typically calculated as a function of the standard deviation of the blank (S_b); 2 S_b or

TABLE II. Reported Standard Deviations, or My Estimates Thereof, for Selected Procedures for Nitrogen Determination

Method	*Type*[a]	*Standard deviation* *(μmol N liter^{-1})*	*Reference*
Ammonia	M,P	0.07	Solórzano (1969)
	M,P	0.05	Strickland and Parsons (1972)
	M,P	0.04	Liddicoat et al. (1975)
	M,E	0.01	Garside et al. (1978)
	M,F	0.06	Gardner (1978a)
	A,P	0.02	Le Corre and Treguer (1976)
	A,P	0.06	Helder and De Vries (1979)
	A,P	0.05	Grasshoff and Johannsen (1972)
	A,P	0.07	Reusch Berg and Abdullah (1977)
	A,P	0.05	Head (1970)
	A,P	0.02	Folkard (1978)
	A,P	0.076	Slawyk and MacIsaac (1972)
	A,P	0.01	Tréguer and Le Corre (1975)
Nitrite	M,P	0.0115	Strickland and Parsons (1972)
	A,P	0.04	Folkard (1978)
	A,P	0.01	Tréguer and Le Corre (1975)
	M,C	0.0005	Garside (1983)
Nitrate	M,P	0.025	Strickland and Parsons (1972)
	M,C	0.002	Garside (1983)
	A,P	0.014	Folkard (1978)
	A,P	0.05	Tréguer and Le Corre (1975)
Total Kjeldahl nitrogen	M,P	0.6	Strickland and Parsons (1972)
	S,P	8.6	Adamski (1976)

(Continued)

(Table II continued)

Method	*Type*[a]	*Standard deviation* ($\mu mol\ N\ liter^{-1}$)	*Reference*
Total dissolved nitrogen			
1. Photo-oxidation	M,P	0.125	Strickland and Parsons (1972)
	A,P	0.05	Lowry and Mancy (1978)
2. Alkaline persulfate	M,P	1.2	D'Elia et al. (1977)
	M,P	0.4	Solórzano and Sharp (1980)

[a] *M, manual; S, semiautomated; A, automated; P, photometric, E, electrometric; F, fluorometric.*

[b] *Standard deviations were chosen from the lowest concentration range reported by the author.*

3 S_b is commonly reported, the latter being recommended by ASC/CIE.

Table II presents standard deviations at low analyte concentrations reported for the selected methods I will discuss in this chapter. A rough notion of the precision of different methods can be obtained from this table, but one must keep in mind the lack of conformity by which such data were gathered and reduced by different authors. Thus, the methods with the smallest standard deviations are not necessarily the best or most precise.

III. DISSOLVED INORGANIC NITROGEN

Nitrate, nitrite, and ammonia are the most important "combined" or "bound" forms of dissolved inorganic nitrogen (DIN) found in the sea, and for purposes of the present discussion, DIN concentration is considered to be the sum of their concentrations. Determinations of these forms have been performed on seawater for over 50 years (Riley, 1965). However, of these, only nitrite has been determined with ease and precision for most of that time period. Nitrate has been the next most reliably determined analyte of the three, but even now the most widely used analytical procedures are not completely satisfactory. Convenient and reasonably precise procedures for ammonia determination in seawater have been developed in the past 20 years, largely as a consequence of the full recognition by biological oceanographers of the importance of ammonia in the sea; nonetheless, the new procedures have some inadequacies.

A. Ammonia

Ammonia (NH_3) exists at seawater pH primarily in the cationic ammonium (NH_4^+) form. Since analytical procedures for ammonium determination in seawater typically buffer pH at levels where ammonium deprotonates to ammonia, I follow the convention of Riley (1975) and use the term "ammonia" here to represent the total of both.

1. Distillation Methods

Early determination of ammonia relied on a distillation procedure to separate it from a seawater sample where alkaline earth metals pose an interference problem (Barnes, 1959). The distillation process is tedious and difficult to perform aboard ship, and there is a serious risk of liberating combined nitrogen as ammonia from organic nitrogen compounds unless the pH is carefully controlled. Data obtained using this method and reported in the early literature must be viewed with caution.

2. Photometric Methods

Most laboratories now employ photometric methods for the determination of ammonia (e.g., Atlas et al., 1971; Dal Pont et al.,

1974; Folkard, 1978; Loder and Gilbert, 1977; Glibert and Loder, 1977; Ryle et al., 1981; Tréguer and Le Corre, 1975; Zimmermann et al., 1977). Riley (1975) discussed four genre of photometric methods for ammonia determination: those based on (1) indophenol blue formation; (2) rubazoic acid formation; (3) oxidation of ammonia to nitrite; and (4) hypobromite oxidation. Riley concluded that of these, the indophenol blue method was the most satisfactory. I discuss this method in some detail below; for further details about the other three, consult Riley (1975).

The indophenol blue method has virtually supplanted other photometric methods. The color-forming reaction was discovered by Berthelot in 1859, but was difficult to adapt to seawater until it was found how to avoid precipitating magnesium and calcium compounds at high pH. Present methods commonly accomplish this by adding a complexing agent before the pH is made alkaline (e.g., Solórzano, 1969; Koroleff, 1970b). Despite the interest in this procedure and its wide use, the Berthelot reaction is not satisfactorily understood, no doubt because its mechanism is so complex (Patton and Crouch, 1977; Krom, 1980). Virtually everyone using a procedure based on it encounters occasional problems with reproducibility and with high and variable blanks. The ubiquity of ammonia as a contaminant often accounts for the latter problem; thus, careful attention must be paid to technique and to producing ammonia-free water for reagents Modifications of the Solórzano (1969) technique have been developed to reduce its blank and improve its reproducibility (Liddicoat et al., 1975; Gravitz and Gleye, 1975; Patton and Crouch, 1977; Dal Pont et al., 1974; Koroleff, 1976a). These modifications involve a variety of factors from choice of reagents used to holding samples under specified conditions of irradiance during color development. Other problems such as contamination and poor reproducibility have been encountered in ammonia determinations that relate to sampling and storage procedures (Degobbis, 1973; Eaton and Grant, 1979).

Color formation in indophenol blue procedures is subject to various interfering compounds, in particular sulfide (Grasshoff and Johannsen, 1972; Hampson, 1977; Helder and De Vries, 1979; Ngo et al., 1982), nitrite (Hampson, 1977; Helder and De Vries, 1979), and mercury (Harwood and Huyser, 1970). Internal standards should be analyzed routinely to detect such interferences.

To overcome some of the disadvantages (odor and toxicity) of the phenolic reagent used in the indophenol blue procedure, Verdouw et al. (1978) substituted sodium salicylate for phenol and found it to be but slightly less sensitive. Krom (1980), who recently studied the kinetics and reactions of the salicylate-modified Berthelot reaction, emphasized the importance of optimizing pH for each combination of reagents used, and recommended, as have others (e.g., Grasshoff and Johannsen, 1972), that the sodium hypochlorite reagent be replaced by a dilute solution of sodium dichloroisocyanurate (dichloro-*s*-triacine 2,4,6-trione), which is more stable--a characteristic particularly important in minimizing drift in automated chemistries. The salicylate modification, yet to become widely used for seawater ammonia determinations, has potential as a substitute procedure.

Segmented flow analysis (SFA) is a type of continuous flow analysis (Snyder et al., 1976; Salpeter and LaPerch, 1981) in which bubbles are introduced into the sample stream to minimize cross-contamination and to improve mixing of reagents and sample; it has gained wide acceptance in the last 10 years, and although once considered novel (Armstrong et al., 1967), it is now not uncommon to see an SFA-based "autoanalyzer" in use at sea. Recently developed ammonia determinations employing SFA offer precision similar to that of the manual Solorzano (1969) method, with most, but not all such procedures being based on the phenol-hypochlorite reaction (that of Le Corre and Tréguer, 1976, is one of the few exceptions). However, there are some specialized problems encountered particularly in automated analysis that affect precision and accuracy.

The Berthelot reaction does not proceed rapidly enough to achieve adequate color formation in continuous flow procedures, hence the reaction mixture is usually heated to accelerate the reaction (e.g.,

Head, 1970). The temperatures employed vary widely (up to 80°C, Slawyk and MacIsaac, 1972); at the higher temperatures serious interference from amino acids may occur (Grasshoff and Johannsen, 1972; Reusch Berg and Abdullah, 1977).

Also of importance are salinity effects that may be substantial for some methods and instruments. Grasshoff and Johannsen (1972) encountered an unavoidable interference from salinity, and Harwood and Huyser (1970) found that pH must be buffered adequately to ensure constant sensitivity of the phenol-hypochlorite at different sample pH. Heider and De Vries's (1979) procedure requires that the pH of the reaction mixture be adjusted in accordance with sample salinity. Refractive index changes at different salinities are also substantial and must be corrected for (Froelich and Pilson, 1978; Loder and Gilbert, 1977); they result from the flow cells in the colorimeters of autoanalyzers typically having curved ends to improve flow characteristics. Refractive index problems must also be reconciled in the automated colorimetric determination of other nutrients in seawater.

Establishing the value of the blank in SFA-based ammonia determinations can be difficult for several reasons. First, reaction products (precipitates) of wetting agents can affect blanks (Loder and Gilbert, 1977)--Ryle et al. (1981) add 1% EDTA into the flow stream to prevent formation of precipitates. Second, colored substances in the seawater sample also cause error (Loder and Glibert, 1977). Crowther and Evans (1980) corrected for this problem in freshwater samples using a synchronized reference stream in a dual-flow cell colorimeter. Finally, wash water used to establish the baseline is easily contaminated. To prevent this Folkard (1978) put an ion-exchange column in the wash-water feed line, and others (e.g., Whitledge et al., 1981) have used acid scrubbers in the air bubble intake.

Other automated ammonia photometric procedures may see use in the future. Automated batch analysis (Hale, 1979) seems promising and is adaptable to a variety of chemistries. Flow injection analysis (FIA), a type of continuous flow analysis in which samples are injected into a flowing, unsegmented reagent stream and offers faster response time than SFA (Ranger, 1981), has been developed (Krug et al., 1979), but ammonia levels in seawater are typically too low for precision of present FIA techniques. FIA precision will undoubtedly improve in the near future.

3. Electrometric Methods

Electrometric methods offer simplicity and portability that seem appealing for work at sea. Alkali is added to a sample of water for electrometric ammonia determination to convert ammonium ion to ammonia that, in turn, diffuses through a hydrophobic, gas-permeable membrane to shift the pH in an internal filling solution containing ammonium chloride. Although ammonia electrode performance is improving (Driscoll et al., 1980), early work (Thomas and Booth, 1973) suggested that this direct measurement technique was not particularly sensitive and that direct measurements were subject to an appreciable salt effect (Merks, 1975). Gilbert and Clay (1973), who added a standard amount of ammonium solution to all samples to improve response time, found no salt effect. Garside et al. (1978) used a standard addition technique for a gas-sensing electrode containing a modified filling solution and reported achieving a detection limit of 0.2 μmol NH_3-N $liter^{-1}$ (a precision of 0.1 μmol NH_3-N $liter^{-1}$) and a rapid enough response time to process 15 samples an hour manually. Unfortunately, a fairly large sample size was required (150 mℓ) and those authors felt that even a reduced sample volume, although possible, would exceed by many times that required for use in an autoanalyzer.

Future improvements in electrode precision seem probable. New types of electrodes are now being developed that may ultimately lead to better precision; for example, one type incorporates immobilized nitrifying bacteria (Hikuma et al., 1980).

4. *Fluorometric Methods*

Gardner (1978a) developed a novel but somewhat cumbersome fluorometric ammonia determination in which he chromatographically separated ammonium ion from amino acids for reaction with a reagent commonly used for amino acids, *o*-phthalaldehyde, that also produces a fluorescent product when reacted with ammonia. His method is precise and requires only a small amount of sample.

B. Nitrite

1. *Photometric Methods*

The earliest method for nitrite determination was based on the classical photometric Griess-Ilosvay diazotization procedure. This involves first nitrosation and diazonium ion formation when nitrite diazotizes an aromatic primary amine; it next involves the coupling of that ion with another aromatic amine to form a pink azo dye. A variety of different aniline and naphthylene derivatives can be used, with the amount of diazo pigment being formed dependent on a number of factors that Fox (1979) summarizes. The sensitivity, reproducibility, and simplicity of the diazotization procedure have made it the procedure of choice since nitrite determinations of seawater were first made (cf. Riley, 1965).

Most laboratories currently employ a procedure based on the method of Shinn (1941), as modified for seawater analysis by Bendschneider and Robinson (1952), in which sulphanilamide and *N*-(1-naphthyl)-ethylenediamine are the diazotizing and coupling reagents, respectively. The lower limit of detection can be extended by carrying out the reaction with a large volume of sample and concentrating the dye for photometry by ion exchange (Wada and Hattori, 1971) or solvent extraction (Zeller, 1955; Macchi and Cescon, 1970). The nitrite procedure has been adapted easily to continuous flow analysis (Technicon Industrial Systems, 1973).

2. *Electrometric Methods*

An electrode sensitive to nitrogen dioxide in aqueous samples can be used for nitrite determination (Orion Research, 1978); nitrite is converted to nitrogen dioxide by treatment with acid. However, the detection limit (∿4 μmol NO_2^--N liter^{-1}) is inadequate for most oceanographic work.

3. *Chemiluminescence Methods*

Detection of nitric oxide in seawater by chemiluminescence is extremely sensitive (Zafirou and McFarland, 1980); it is possible to extend this analytical sensitivity to other analytes by using a variety of chemical agents to convert them to nitric oxide (Cox, 1980). Garside (1982) has developed an analytical technique sensitive at the nanomolar level in which nitrite in seawater is reduced to nitric oxide for determination using a commercial nitrogen oxide analyzer. Sample sizes required are reasonably small (∿10 mℓ) and determinations can be made aboard ship.

4. *Gas Chromatographic Methods*

The procedure of Funazo et al. (1980), or a modification of it, should also be adaptable to nitrite determination of seawater. In this procedure, nitrite is quantitatively reacted with *p*-bromoaniline in a reaction forming *p*-bromochlorobenzene that is extracted into toluene and determined by gas chromatography with electron capture. Detection limits for this procedure approximate those of present photometric methods.

C. Nitrate

1. *Photometric Methods*

The nitrate ion has an intense absorption band in the far ultraviolet, but interferences from dissolved organic compounds and the bromide ion preclude its direct photometric determination by ultraviolet photometry (Riley, 1975). Thus, procedures for determination of nitrate in seawater have involved reacting nitrate to produce a compound absorbing at wavelengths not subject to such interference.

Armstrong (1963) mixed equal volumes of seawater sample and concentrated sulphuric acid to produce an absorbance band at approximately 230 nm, which he attributed to the formation of nitrosyl chloride. The strong acid used in this procedure has discouraged its use. Similarly unpopular are procedures such as the "brucine" method (cf. United States Environmental Protection Agency, 1979) in which nitrate in strongly acidic medium oxidizes organic compounds to produce colored products for photometric determination. The brucine method suffers from the additional disadvantage of yielding a nonlinear standard curve.

Mullin and Riley (1955) developed a procedure in which nitrate is reduced to nitrite by hydrazine, with nitrite determination by a Greiss reaction-based procedure. Because of its various drawbacks (cf. Riley, 1975), it is little used for seawater analysis today, although satisfactory automated procedures based on improvements of it are still in use for freshwater analysis (e.g., Downes, 1978) and may prove applicable to seawater. Presently, nitrate determination typically relies on the reduction of nitrate to nitrite by cadmium (Morris and Riley, 1963; Grasshoff, 1964); nitrite is determined by diazotization and colorimetry (Bendschneider and Robinson, 1952). A number of different forms of cadmium have been used: copper-coupled cadmium filings packed in reduction columns (Wood et al., 1967; Strickland and Parsons, 1972), copper-treated cadmium wire (Stainton, 1974), electrolytically precipitated cadmium (Nydahl, 1976), granulated cadmium (Nydahl, 1976), spongy cadmium (Davison and Woof, 1978), and cadmium powder (Davison and Woof, 1978).

Despite the considerable interest in developing a reliable technique in which virtually 100% of the nitrate is reduced to nitrite and no further, there have been relatively few systematic studies of the factors affecting reduction. Nydahl (1976) evaluated the effects on column performance of flow rate (expressed as bed volumes/min), pH, oxygen, temperature, chloride concentration, type of reductor cadmium, and type of buffer. He showed that two requirements dictate the choice of buffer: proper pH for reduction of nitrate to nitrite and effective complexing of cadmium hydroxide produced by the reduction of nitrate and oxygen. With increasing pH (to pH 9.5), a maximum yield of nitrate is obtained over a broader range of flow rates at the sacrifice of reaction speed and complexing ability. Buffer choice and concentration affect cadmium-complexing ability, with the complexing effect of chloride in seawater helping in this regard at the expense of retarding the rate of reduction. The complex effect of chloride is probably responsible for the popularity of copperized cadmium for seawater nitrate analysis; it gives the fastest reduction rates and can thus be used at faster flow rates. Nydahl (1976) notes however, that reduction and overreduction of nitrate are difficult to control with this form of cadmium, except in a narrow range of column conditions and flow rates. He also explains why precautions recommended by Wood et al. (1967) to prevent cadmium from being exposed to air during its preparation are superfluous.

Cadmium reduction of nitrate is subject to several interfering compounds, including phosphate (Olson, 1980; Davison and Woof, 1979), silicate (Davison and Woof, 1979; Solorzano and Sharp, 1980), and sulfide (Davison and Woof, 1979; Satsmadjis, 1978). Phosphate at a concentration of 2.5 μM may inhibit reduction by 10% and at a concentration of 25 μM, by 40% (Olson, 1980). Interference is gradual, reversible, and more pronounced for old columns. Phosphorus-based surfactants used to improve flow characteristics in continuous flow analysis systems may degrade with age and cause interference. Thus, column efficiency should be checked frequently with internal standards, and samples such as interstitial pore waters that contain high levels of interfering compounds may have to be analyzed by another technique. Ion chromatography, a form of liquid chromatography in which ionic constituents are separated by ion exchange followed by electrical conductivity detection (Small et al., 1975), may prove particularly useful in this regard, providing nitrate concentrations exceed about 20 μmol NO_3^--N liter^{-1} (Dr. A. Fitchett, Dionex Corporation, Sunnyvale, California, personal communication).

Automated analytical procedures have been particularly welcomed for three reasons by those performing nitrate determinations by the cadmium reduction procedure. First, all samples pass through the same reductor column, precluding column-to-column standardization problems. Second, flow rates through the column are more precisely controlled than with gravity feed columns; this increases reproducibility of reduction and thus precision. Third, automation of nitrate analysis represents a considerable relief to the analyst by reducing the amount of time spent in preparing and standardizing columns, in sample handling, and in performing between-sample column rinses. Automated procedures of particular interest here have been published by Brewer and Riley (1965), Strickland and Parsons (1972), Henriksen (1965), Armstrong et al. (1967), and Stainton (1974).

2. *Other Methods*

The most promising new approach to nitrate determination in seawater involves its reduction to nitric oxide, which is in turn detected using chemiluminescence (Garside, 1982; see above, Section III,B,3). Garside reports a precision of up to 2 nmol NO_2^--N liter^{-1}. There are two electrometric methods: In one, nitrate is reduced to ammonia for determination by ammonia probe (Mertens et al., 1975); in the other, nitrate is converted to nitrite by cadmium reduction and the nitrite to nitrogen dioxide by acidification for detection by probe (Orion Research, 1978). The lower detection limit of such techniques does not yet appear sensitive enough for many seawater samples.

A new procedure, enzyme amplification laser fluorometry, has been proposed as a promising possibility for extremely sensitive nitrate and other determinations on small samples (Imasaka and Zare, 1979). This procedure involves the sensitive fluorometric detection of coenzymes such as NADPH that are produced during enzymatic reactions involving the analyte of interest. According to Imasaka and Zare (1979), microanalysis may be possible because of the ability to focus laser light on very small samples, and those authors proposed, but did not test, the use of their technique to quantify inorganic substances.

IV. DISSOLVED ORGANIC NITROGEN

Dissolved organic nitrogen (DON) is determined by difference between total dissolved nitrogen (i.e., nitrate + nitrite + ammonia + organic nitrogen) and dissolved inorganic nitrogen (DIN)(i.e., nitrate + nitrite + ammonia), or by difference between Kjeldahl nitrogen (ammonia + dissolved organic nitrogen) and ammonia. I will limit discussion here primarily to those forms of nitrogen in organic compounds of particular interest in the nitrogen cycle and trophic interactions of the water column.

A. Wet Oxidation Procedures

1. Kjeldahl Oxidation

Most of the earlier procedures for DON determination lacked adequate sensitivity and involved the tedious Kjeldahl wet oxidation procedure (Kjeldahl, 1883), which lends itself neither to shipboard use or to automation. This procedure consists of an initial evaporation followed by an oxidation with concentrated sulfuric acid. The ammonium produced by the digestion process was determined in early work by titration (Barnes, 1959), whereas colorimetric procedures have been used more recently (Strickland and Parsons, 1972; Webb et al., 1975; Webb, 1978, Nicholls, 1975). A number of semiautomated procedures are in use in which samples are oxidized by a manual Kjeldahl procedure, with subsequent ammonia determination on the digests being performed by autonalysis using photometric (Faithfull, 1971; Scheiner, 1976; Jirka et al., 1976; Conetta et al., 1976; Adamski, 1976) or electrometric procedures (Stevens, 1976).

2. *Photooxidation*

The photochemical oxidation procedure (Armstrong et al., 1966), for seawater samples at least, has generally superceded the Kjeldahl oxidation procedure. A small quantity of hydrogen peroxide is added to a sample contained in a quartz reaction vessel and high-wattage mercury lamps are used to produce ultraviolet light to photooxidize organic nitrogen, nitrite, and ammonia to nitrate; nitrate is then determined as described previously. The procedure is considerably less tedious than the Kjeldahl procedure, it can be performed at sea, and unlike other procedures for DON oxidation, is relatively easy to automate (Afghan et al., 1971; Lowry and Mancy, 1978). However, it does have some shortcomings. Workers testing it in freshwaters have found that the photochemical reaction is pH sensitive and may not oxidize completely compounds such as ammonia and urea (Afghan et al., 1971; Henriksen, 1970; Lowry and Mancy, 1978). Lowry and Mancy (1978) found that ultraviolet digestion gave good results decomposing C-N but not N-N bonds, yet felt that most compounds implicated in biological processes would be recovered satisfactorily. Obviously, for samples containing a large amount of nitrate plus nitrite, such as those from the deep ocean, the precision of DON determination by use of photooxidation will be less than that of a modern Kjeldahl procedure.

3. *Persulfate Oxidation*

Koroleff (1970a, 1976b) developed an alternative wet oxidation procedure for total nitrogen determinations that is becoming used more widely. He found that under alkaline conditions at 100°C and in the presence of excess potassium persulfate, combined nitrogen in a seawater sample is oxidized to nitrate. Nitrate is then determined by standard photometric procedures described above. D'Elia et al. (1977) and Smart et al. (1981) have shown that organic nitrogen determinations by the persulfate and Kjeldahl techniques yield comparable results and precision for both sea- and freshwater samples; these authors also discuss the advantages and disadvantages of persulfate oxidation relative to Kjeldahl oxidation and photooxidation. Nydahl (1978) and Solórzano and Sharp (1980) have suggested some improvements to Koroleff's original procedure that alter reaction pH, lower blanks, and provide for the requisite excess of persulfate. Nydahl (1978) notes that errors may result when using persulfate oxidation on turbid samples; he also provides an in-depth study of reaction kinetics and percentage recovery at varying oxidation temperatures. Valderrama (1981) reports the simultaneous determination of total N and total P using alkaline persulfate oxidation. Goulden and Anthony (1978) have studied kinetics of the oxidation of organic material using persulfate, and have thus provided a basis for still further refinement of the procedure, such that simultaneous determination of C, N, and P may ultimately be possible on the same sample. As in the case of photochemical oxidation, determination of DON by the persulfate technique will have poor precision in the presence of large quantities of nitrate or nitrite.

B. Dry Combustion Procedures

Dry combustion procedures generally have been disappointing or impractical for determining DON. Gordon and Sutcliff (1973) reported a dry combustion procedure in which a seawater sample is freeze-dried and the salt residues subsequently ignited in a CHN analyzer. The obvious disadvantage of this is the necessity of access to a freeze drier and the time involved in sample preparation. Other procedures have been developed in which small volumes of sample are injected directly into a combusion tube for evaporation and combustion (Van Hall et al., 1963; Fabbro et al., 1971; Hernandez, 1981), but have not found wide use by oceanographers because expensive and specialized equipment is required and because sea salts cause problems in the combustion chamber.

C. Urea and Amino Acids

Low-molecular-weight N-containing compounds such as urea and amino acids may be important as potential sources of N for phytoplankton (Wheeler et al., 1974, 1977; North, 1975; Paul, Chapter 8, this volume) and bacteria (Williams, 1970). Thus, there has been continuing interest in the quantitative analysis of these compounds.

1. Urea

Both direct and enzymatic techniques have been developed for photometric urea analysis. Newell et al. (1967) developed a manual procedure based on a method commonly used for blood urea determination: Urea in seawater is reacted with diacetylmonoxime, the product of which is then reacted with semicarbazide to form the chromaphore semicarbazone. Emmet (1969) developed a manual urea determination procedure based on a modification of the phenol-hypochlorite ammonia method. McCarthy (1970) added the enzyme urease to seawater samples, and measured the ammonia liberated by means of the Solórzano (1969) method. Providing ammonia levels in the samples are low, the latter seems the most sensitive manual procedure of the three.

The manual urea determination of Newell et al. (1967) has been adapted for use in continuous flow analysis by DeManche et al. (1973). T. E. Whitledge (personal communication) has also developed an automated urea determination based on Newell et al.'s (1967) procedure and incorporates modifications of Rahmatullah and Boyd (1980). Ferric chloride is used as a catalyst to improve reproducibility and enhance the color development of semicarbazone. Although it should be possible to develop a urea determination employing bound urease, I am not aware of its having been done for seawater samples.

2. *Amino Acids*

Riley (1975) wrote a comprehensive review of the qualitative and quantitative analysis of organic compounds in seawater. He stressed that most analytical methods up to the time of his review were insufficiently sensitive for the determination of organic compounds at the very low concentrations at which they are encountered in the sea; consequently, organics such as amino acids had to be determined by using a bioassay procedure (e.g., Crawford et al., 1974) or by a separation procedure. Separation procedures employed included solvent extraction and sorption onto synthetic resins and have generally been superceded by HPLC techniques (see below). Of these, the easiest and most quantitative seemed to be to remove the amino acids from the seawater sample with a copper-loaded chelating resin, to elute them from the column with a small volume of ammonia, and to detect them with ninhydrin (e.g., Webb and Wood, 1967). Coughenower and Curl (1975) developed an automated technique for total dissolved free amino acids using a photometric, ninhydrin-based procedure.

The advent of fluorometric procedures has greatly improved our ability to detect minute quantities of primary amino acids ($R\text{-}NH_2$). Two reagents, *o*-phthaldialdehyde (Roth, 1971) and fluorescamine (Udenfriend et al., 1976) have been used for amino acid analysis at enhanced sensitivities (Udenfriend et al., 1972; Georgiadis and Coffey, 1973; Benson and Hare, 1975; Gardner, 1978b; Lindroth and Mopper, 1979) to determine total primary amines in seawater and estuarine water directly and rapidly without separation procedures (North, 1975; Jorgensen, 1979), and to assay for proteins in biological samples (Bohlen et al., 1973) and in seawater samples (Packard and Dortch, 1975). A sensitive procedure employing fluorescamine has also been developed for the determination of secondary amines (R_1-NH-R_2) (Nakamura and Tamura, 1980), but to my knowledge it has not been modified for use with seawater samples.

High-pressure (performance) liquid chromatography (HPLC) has gained wide acceptance in the last decade and has also played a role with the fluorogenic reagents in advancing the state-of-the-art of amino acid analysis in seawater. HPLC methods seem likely to supplant most older methods completely. HPLC helps improve both the resolution and the speed of processing of chromatographic sample and the instrumentation for it is sufficiently compact and portable to be used aboard ship. Lindroth and Mopper (1979) and Mopper and Lindroth (1982) have developed and refined a precolumn fluorescence derivatization technique for seawater in which a sample for quantitative amino acid analysis is prereacted with phthaldialdehyde; the fluorescent amino acid derivatives are then rapidly separated and quantified using HPLC with a fluorometric detector. Garrasi et al. (1979) and Mopper and Lindroth (1982) report evidence that preservation and storage of samples produce artifacts, and recommend analysis of freshly collected samples, if possible.

V. PARTICULATE NITROGEN

Particulate nitrogen determination can be done satisfactorily, although somewhat tediously, on most seawater samples. Samples are typically obtained by filtration onto glass fiber filters; sensitivity is not generally a problem since enough water is filtered to collect an adequate amount of nitrogen (∿1 μg; Gordon and Sutcliffe, 1974) for detection. Wet or dry combustion methods can be used. Wet combustion techniques generally employ Kjeldahl analysis (see Section IV,A,1). A number of commercially available instruments have been used in the past for dry combustion techniques, e.g., the Coleman model 29 uses manometry to detect nitrogen gas produced by Dumas combustion. Recently, the Perkin-Elmer model 240 (Gordon and

Sutcliffe, 1974), the Hewlett Packard model 185B (Sharp, 1974) and the Carlo Erba model 1106 that detect nitrogen gas produced from combustion by thermal conductivity have gained popularity, primarily because particulate carbon values can be obtained on the same sample. Also available are other commercial instruments using somewhat different dry combustion procedures (cf. Fabbro et al., 1971; Hernandez, 1981).

V. CONCLUSIONS

The past decade has seen significant advances in the analytical determination of nitrogen forms in seawater. Some sensitive new analytical procedures have been introduced, but most of the changes in the practice of analytical chemistry of nitrogen in seawater have involved widespread adoption of automated versions of traditional manual procedures.

Automated procedures now offer the analyst the ability to collect considerably more data in a shorter period of time. In providing more data, they also require more sophisticated procedures for data management and great care in assuring that the methods are properly used. Fortunately, low-cost yet powerful and portable small computers are becoming available to facilitate analysis and data management and to reduce operator error. For example, Whitledge et al. (1981) have interfaced a desktop computer to a continuous flow analysis system to allow computer control of sample analysis, precision checking, data storage, and data and geographical coordinate plotting. Such sophisticated systems will undoubtedly see wider use in the future.

HPLC is being used to provide better and more rapid separations for dissolved amino acids, in particular, and other chromatographic methods, such as gas chromatography-mass spectrometry, will certainly play a role in the analytical chemistry of specific trace-level,

N-containing organic compounds that are beyond the scope of the present discussion.

Fluorometric and chemiluminescence techniques are being developed that improve detection limits up to three orders of magnitude for some analytes. Increased analytical sensitivity will be necessary for research in newly emerging subjects of interest such as microscale processes, in which it will be necessary to detect minute quantities of analytes in small volumes.

ACKNOWLEDGMENTS

This chapter was prepared with support from the University of Maryland Sea Grant Program. I thank J. H. Sharp, T. E. Whitledge, C. W. Keefe, B. J. Rothschild, and J. G. Sanders for their comments; C. Garside for providing me with an unpublished manuscript; and C. Gilmour, S. Domotor, and T. Johnson for logistical assistance. This chapter is contribution No. 1416, Center for Environmental and Estuarine Studies of the University of Maryland.

REFERENCES

Adamski, J. M. (1976). Simplified Kjeldahl nitrogen determination for seawater by a semiautomated persulfate digestion method. *Anal. Chem. 48*, 1194-1197.

Afghan, B. K., Goulden, P. D., and Ryan, J. F. (1971). "Use of Ultraviolet Irradiation in the Determination of Nutrients in Water with Special Reference to Nitrogen," Tech. Bull. No. 40. Inland Waters Branch, Department of Energy, Mines and Resources, Ottawa, Canada.

American Chemical Society Committee on Environmental Improvement (ACS/CIE)(1980). Guidelines for data acquisition and data quality evaluation in environmental Chemistry. *Anal. Chem. 52*, 2242-2249.

American Society for Testing and Materials (ASTM)(1982). "Standard for Metric Practice." ASTM Designation E380-82. Philadelphia, Pennsylvania.

Armstrong, F. A. J. (1963). Determination of nitrate in water by ultraviolet spectrophotometry. *Anal. Chem. 35*, 1292-1294.

Armstrong, F. A. J., Williams, P. M., and Strickland, J. D. H. (1966). Photooxidation of organic matter in sea water by ultra-violet radiation, analytical and other applications. *Nature (London) 211*, 481.

Armstrong, F. A. J., Stearns, C. R., and Strickland, J. D. H. (1967). The measurement of upwelling and subsequent biological processes by means of the Technicon AutoAnalyzer and associated equipment. *Deep-Sea Res. 14*, 381-389.

Atlas, E. L., Hager, S. W., Gordon, L. I., and Park, P. K. (1971). "A Practical Manual for Use of the Technicon AutoAnalyzer in Seawater Nutrient Analysis," Tech. Rep. No. 215, Oregon State University, Corvallis.

Barnes, H. (1959). "Apparatus and Methods of Oceanography. Part One: Chemical." Allen & Unwin, London.

Bendschneider, K., and Robinson, R. J. (1952). A new spectrophotometric method for the determination of nitrite in sea water. *J. Mar. Res. 11*, 87-96.

Benson, J. R., and Hare, P. E. (1975). *o*-Phthalaldehyde: Fluorogenic detection of primary amines in the picomole range. Comparison with fluorescamine and ninhydrin. *Proc. Natl. Acad. Sci. U.S.A. 72*, 619-622.

Berthelot, M. P. E. (1859). *Rép. Chim. Appl. 1*, 282-284.

Bohlen, P., Stein, S., Dairman, W., and Udenfriend, S. (1973). Fluorometric assay of proteins in the nanogram range. *Arch. Biochem. Biophys. 155*, 213-220.

Brewer, P. G., and Riley, J. P. (1965). The automatic determination of nitrate in seawater. *Deep-Sea Res. 12*, 765.

Conetta, A., Buccafuri, A., and Jansen, J. (1976). Wet digestion of water samples for total Kjeldahl N and total P. *Am. Lab. 8*, 103-106.

Coughenower, D. D., and Curl, H. C. (1975). An automated technique for total dissolved free amino acids in seawater. *Limnol. Oceanogr. 20*, 128-131.

Cox, R. D. (1980). Determination of nitrate and nitrite at the parts per billion level by chemiluminescence. *Anal. Chem. 52*, 332-335.

Crawford, C. C., Hobbie, J. E., and Webb, K. L. (1974). The utilization of dissolved free amino acids by estuarine microorganisms. *Ecology 55*, 551-563.

Crowther, J., and Evans, J. (1980). Blanking system for the spectrophotometric determination of ammonia in surface waters. *Analyst 105*, 849-854.

Dal Pont, G., Hogan, M., and Newell, B. (1974). "Laboratory Techniques in Marine Chemistry. II. Determination of Ammonia in Sea Water and the Preservation of Samples for Nitrate Analysis," Rep. No. 55. Marine Laboratory, Cronulla, N. S. W., Div. Fish. Oceanogr., C. S. I. R. O., Australia.

Davison, W., and Woof, C. (1978). Comparison of different forms of cadmium as reducing agents for the batch determination of nitrate. *Analyst 103*, 403-406.

Davison, W., and Woof, C. (1979). Interference studies in the batch determination of nitrate in freshwater by reduction with cadmium to nitrite. *Analyst 104,* 385-390.

Degobbis, D. (1973). On the storage of seawater samples for ammonia determination. *Limnol. Oceanogr. 18,* 146-150.

D'Elia, C. F., Steudler, P. A., and Corwin, N. (1977). Determination of total nitrogen in aqueous samples using persulfate digestion. *Limnol. Oceanogr. 22,* 760-764.

DeManche, J. M., Curl, H., Jr., and Coughenower, D. D. (1973). An automated analysis for urea in seawater. *Limnol. Oceanogr. 18,* 686-689.

Downes, M. T. (1978). An improved hydrazine reduction method for the automated determination of low nitrate levels in freshwater. *Water Res. 12,* 673-675.

Driscoll, J. N., Atwood, E. S., Fowler, J. E., and Selig, W. (1980). Electrodes measure dissolved ammonia and carbon dioxide. *Ind. Res. Dev. 22,* 91-93.

Eaton, A. D., and Grant, V. (1979). Sorption of ammonium by glass frits and filters: Implications for analyses of brackish and freshwater. *Limnol. Oceanogr. 24,* 397-399.

Emmet, R. T. (1969). Spectrophotometric determinations of urea and ammonia in natural waters with hypochlorite and phenol. *Anal. Chem. 41,* 64-65.

Fabbro, L. A., Filachek, L. A., Iannacone, R. L., Joyce, R. J., Takahashi, Y., and Riddle, M. E. (1971). Extension of the microcoulometric determination of total bound nitrogen in hydrocarbons and water. *Anal. Chem. 43,* 1671-1678.

Faithfull, N. T. (1971). Automated simultaneous determination of nitrogen, phosphorus, potassium and calcium on the same herbage digest solution. *Lab. Pract. 20,* 41-44.

Folkard, A. R. (1978). "Automatic analysis of sea water nutrients." *Fish. Res., Tech. Rep. (U. K., Dir. Fish. Res.) 46,* 1-23.

Fox, J. B. (1979). Kinetics and mechanism of the Griess reaction. *Anal. Chem. 51,* 1493-1502.

Froelich, P. N., and Pilson, M. E. Q. (1978). Systematic absorbance errors with Technicon AutoAnalyzer II colorimeters. *Water Res. 12,* 599-603.

Funazo, K., Tanaka, M., and Shono, T. (1980). Determination of nitrite at parts-per-billion levels by derivatization and electron capture gas chromatography. *Anal. Chem. 52,* 1222-1224.

Gardner, W. S. (1978a). Microfluorometric method to measure ammonium in natural waters. *Limnol. Oceanogr. 23,* 1069-1072.

Gardner, W. S. (1978b). Sensitive fluorometric procedure to determine individual amino acids in marine waters. *Mar. Chem. 6,* 15-26.

Garrasi, C., Degens, E. T., and Mopper, K. (1979). The free amino acid composition of seawater obtained without desalting and preconcentration. *Mar. Chem. 8,* 71-85.

Garside, C. (1982). A chemiluminescent technique for the determination of nanomolar concentrations of nitrate and nitrite in seawater. *Mar. Chem. 11,* 159-167.

Garside, C., Hull, G., and Murray, S. (1978). Determination of submicromolar concentration of ammonia in natural waters by a standard addition method using a gas-sensing electrode. *Limnol. Oceanogr. 23,* 1073-1075.

Georgiadis, A., and Coffey, J. (1973). Single column analysis of amino acids in protein hydrolysates utilizing the fluorescamine reaction. *Anal. Biochem. 56,* 121-125.

Gilbert, T. R., and Clay, A. M. (1973). Determination of ammonia in aquaria and in sea water using the ammonia electrode. *Anal. Chem. 45,* 1757-1759.

Glibert, P. M., and Loder, T. C. (1977). "Automated Analysis of Nutrient Seawater: Manual of Techniques," Tech. Rep. No. WHOI-77-47. Woods Hole Oceanogr. Inst., Woods Hole, Massachusetts.

Gordon, D. C., and Sutcliffe, W. H., Jr. (1973). A new dry combustion method for the simultaneous determination of total organic carbon and nitrogen in seawater. *Mar. Chem. 1,* 231-244.

Gordon, D. C., and Sutcliffe, W. H., Jr. (1974). Filtration of seawater using silver filters for particulate nitrogen and carbon analysis. *Limnol. Oceanogr. 19,* 989-993.

Goulden, P. D., and Anthony, D. H. J. (1978). Kinetics of uncatalyzed peroxydisulfate oxidation of organic material in fresh water. *Anal. Chem. 50,* 953-958.

Grasshoff, K. (1964). Zur Bestimmung von Nitrat in Meer- und Trinkwasser. *Kiel. Meeresforsch. 20,* 5-11.

Grasshoff, K., Ed. (1976). "Methods of Seawater Analysis." Verlag Chemie. Weinheim.

Grasshoff, K., and Johannsen, H. (1972). A new sensitive and direct method for the automatic determination of ammonia in sea water. *J. Cons., Cons. Int. Explor. Mer 3,* 516-521.

Gravitz, N., and Gleye, L. (1975). A photochemical side reaction that interferes with the phenolhypochlorite assay for ammonia. *Limnol. Oceanogr. 20,* 1015-1017.

Hale, D. R. (1979). Performance evaluation of an automated batch analyzer. *Am. Lab. 11,* 117-130.

Hampson, B. L. (1977). The analysis of ammonia in polluted sea water. *Water Res. 11,* 305-308.

Harvey, H. W. (1945). "Recent Advances in the Chemistry and Biology of Sea-Water," 2nd ed. Macmillan, New York.

Harwood, J. E., and Huyser, D. J. (1970). Automated analysis of ammonia in water. *Water Res. 4,* 695-704.

Head, P. C. (1971). An automated phenolhypochlorite method for the determination of ammonia in sea water. *Deep-Sea Res. 18,* 531-532.

Helder, W., and De Vries, R. T. P. (1979). An automatic phenolhypochlorite method for the determination of ammonia in sea- and brackish waters. *Neth. J. Sea Res. 13,* 154-160.

Henriksen, A. (1965). An automated method for determining nitrate and nitrite in fresh and saline waters. *Analyst 90,* 83-88.
Henriksen, A. (1970). Determination of total nitrogen, phosphorus and iron in fresh water by photooxidation with ultraviolet radiation. *Analyst 95,* 601-608.
Hernandez, H. A. (1981). Total bound nitrogen determination by pyrochemiluminescence. *Am. Lab. 13,* 72-76.
Hikuma, M., Kubo, T., Yasuda, T., Karube, T., and Suzuki, I. (1980). Ammonia electrode with immobilized nitrifying bacteria. *Anal. Chem. 52,* 1020-1024.
Imasaka, T., and Zare, R. N. (1979). Enzyme amplification laser fluorimetry. *Anal. Chem. 51,* 2082-2085.
Jirka, A. M., Carter, M. J., May, D., and Fuller, F. D. (1976). Ultramicro semiautomated method for simultaneous determination of total phosphorus and total Kjeldahl nitrogen in wastewaters. *Environ. Sci. Technol. 10,* 1038-1044.
Jorgensen, N. O. G. (1979). Annual variation of dissolved free primary amines in estuarine water and sediment. *Oecologia 40,* 207-217.
Kjeldahl, J. (1883). A new method for the determination of nitrogen in organic matter. *Z. Anal. Chem. 22,* 366.
Koroleff, F. (1970a). Revised version of "Determination of Total Nitrogen in Natural Waters by Means of Persulfate Oxidation," Int. Counc. Explor. Sea (ICES), Paper C. M. 1969/C:8. ICES, Charlottenlund, Denmark.
Koroleff, F. (1970b). Revised version of "Direct Determination of Ammonia in Natural Waters as Indophenol Blue," Int. Counc. Explor. Sea, Paper C. M. 1969/C:9. ICES, Charlottenlund, Denmark.
Koroleff, F. (1976a). Determination of ammonia. *In* "Methods of Seawater Analysis" (K. Grasshoff, ed.), pp. 126-133. Verlag Chemie, Weinheim.
Krom, M. D. (1980). Spectrophotometric determination of ammonia: A study of a modified Berthelot reaction using salicylate and dichloroisocyanurate. *Water Res. 105,* 305-316.
Krug, F. J., Ruzicka, J., and Hansen, E. H. (1979). Determination of ammonia in low concentrations with Nessler's reagent by flow injection analysis. *Analyst 104,* 47-54.
Le Corre, P., and Tréguer, P. (1976). Dosage de l'ammonium dans l'eau de mer: Comparaison entre deux methodes d'analyse automatique. *J. Cons., Cons. Int. Explor. Mer 38,* 147-153.
Liddicoat, M. I., Tibbitts, S., and Butler, E. I. (1975). The determination of ammonia in seawater. *Limnol. Oceanogr. 20,* 131-132.
Lindroth, P., and Mopper, K. (1979). High performance liquid chromatographic determination of subpicomole amounts of amino acids by precolumn fluorescence derivatization with *o*-phthaldialdehyde. *Anal. Chem. 51,* 1667-1674.
Loder, T. C., and Gilbert, P. M. (1977). "Blank and Salinity Corrections for Automated Nutrient Analysis of Estuarine and Sea Waters," Contrib. No. UNH-SG-JR-101. Univ. of New Hampshire Marine Advisory Program, Durham.

Lowry, J. H., and Mancy, K. H. (1978). A rapid automated system for the analysis of dissolved total organic nitrogen in aqueous solutions. *Water Res. 12,* 471-475.

McCarthy, J. J. (1970). A urease method for urea in seawater. *Limnol. Oceanogr. 15,* 309-313.

Macchi, G. R., and Cescon, B. S. (1970). Spectrophotometric submicro determination of nitrites after double extraction of the azo compound produced. *Anal. Chem. 42,* 1809-1810.

Merks, A. G. A. (1975). Determination of ammonia in sea-water with an ion selective electrode. *Neth. J. Sea Res. 9,* 371-375.

Mertens, J., Van Den Winkle, P., and Massert, D. L. (1975). Determination of nitrate in water with an ammonia probe. *Anal. Chem. 47,* 522-526.

Mopper, K., and Lindroth, P. (1982). Diel and depth variations in dissolved free amino acids and ammonium in the Baltic Sea determined by shipboard HPLC analysis. *Limnol. Oceanogr. 27,* 336-347.

Morris, A. W., and Riley, J. P. (1963). The determination of nitrate in sea water. *Anal. Chim. Acta 29,* 272.

Mullin, J. B., and Riley, J. P. (1955). The spectrometric determination of nitrate in natural waters, with particular reference to sea-water. *Anal. Chim. Acta 12,* 464-480.

Nakamura, H., and Tamura, Z. (1980). Fluorometric determination of secondary amines based on reaction with fluorescamine. *Anal. Chem. 52,* 2087-2092.

Newell, B. S., Morgan, B., and Cundy, J. (1967). The determination of urea in seawater. *J. Mar. Res. 25,* 201-202.

Ngo, T. T., Phan, A. P. H., Yam, C. F., and Lenhoff, H. M. (1982). Interference in determination of ammonia with the hypochlorite-alkaline phenol method of Berthelot. *Anal. Chem. 54,* 4649.

Nicholls, K. (1975). A single digestion procedure for rapid manual determinations of Kjeldahl nitrogen and total phosphorus in natural waters. *Anal. Chim. Acta 76,* 208-212.

North, B. B. (1975). Primary amines in California coastal waters: Utilization by phytoplankton. *Limnol. Oceanogr. 20,* 20-27.

Nydahl, F. (1976). On the optimum conditions for the reduction of nitrate to nitrite by cadmium. *Talanta 23,* 349-357.

Nydahl, F. (1978). On the peroxodisulphate oxidation of total nitrogen in waters to nitrate. *Water Res. 12,* 1123-1130.

Olson, R. J. (1980). Phosphate interference in the cadmium reduction analysis of nitrate. *Limnol. Oceanogr. 25,* 758-760.

Orion Research (1978). "Analytical Methods Guide," 9th ed. Orion Research, Inc., Cambridge, Massachusetts.

Packard, T. T., and Dortch, Q. (1975). Particulate protein-nitrogen in North Atlantic surface waters. *Mar. Biol. 33,* 347-354.

Patton, C. J., and Crouch, S. R. (1977). Spectrophotometric and kinetics investigation of the Berthelot reaction for the determination of ammonia. *Anal. Chem. 49,* 464-469.

Rahmatullah, M., and Boyd, T. R. C. (1980). Improvements in the determination of urea using diacetyl monoxime; methods without deproteinisation. *Clin. Chim. Acta 107,* 39.

Ranger, C. B. (1981). Flow injection analysis: Principles, techniques applications, and design. *Anal. Chem. 53,* 20A-30A.

Reusch Berg, B., and Abdullah, M. I. (1977). An automatic method for the determination of ammonia in sea water. *Water Res. 11,* 637-638.

Riley, J. P. (1965). Historical introduction. *In* "Chemical Oceanography" (J. P. Riley and G. Skirrow, eds.), Vol. 1, pp. 1-41. Academic Press, New York.

Riley, J. P. (1975). Analytical chemistry of sea water. *In* "Chemical Oceanography" (J. P. Riley and G. Skirrow, eds.), 2nd ed., Vol. 3, pp. 193-514. Academic Press, New York.

Roth, M. (1971). Fluorescence reaction for amino acids. *Anal. Chem. 43,* 880-882.

Ryle, V. D., Mueller, H. R., and Gentien, P. (1981). "Automated Analysis of Nutrients in Tropical Sea Water," Tech. Bull. Oceanogr. Ser. No. 3. Aust. Inst. Mar. Sci., Townsville, Queensland.

Salpeter, J., and LaPerch, F. (1981). Automated segmented continuous-flow colorimetric analysis. *Am. Lab. 13,* 78-85.

Satsmadjis, J. (1978). The simultaneous determination of nutrients by autoanalyzer in Greek coastal waters. *Thalassographica 2,* 173-190.

Scheiner, D. (1976). Determination of ammonia and Kjeldahl nitrogen by indophenol method. *Water Res. 10,* 31-36.

Sharp, J. H. (1973). Size classes of organic carbon in seawater. *Limnol. Oceanogr. 18,* 441-477.

Sharp, J. H. (1974). Improved analysis for "particulate" organic carbon and nitrogen from seawater. *Limnol. Oceanogr. 19,* 984-989.

Shinn, M. B. (1941). A colorimetric method for the determination of nitrite. *Ind. Eng. Chem., Anal. Ed. 13,* 33-35.

Slawyk, G., and MacIsaac, J. J. (1972). Comparison of two automated ammonium methods in a region of coastal upwelling. *Deep-Sea Res. 19,* 521-524.

Small, H., Stevens, T. S., and Bauman, W. C. (1975). Novel ion exchange chromatographic method using conductimetric detection. *Anal. Chem. 47,* 1801-1809.

Smart, M. M., Reid, F. A., and Jones, J. R. (1981). A comparison of a persulfate digestion and the Kjeldahl procedure for determination of total nitrogen in freshwater samples. *Water Res. 15,* 919-921.

Snyder, L., Levine, J., Stoy, R., and Conetta, A. (1976). Automated chemical analysis: update on continuous-flow approach. *Anal. Chem. 48,* 942A-956A.

Solórzano, L. (1969). Determination of ammonia in natural waters by the phenolhypochlorite method. *Limnol. Oceanogr. 14,* 799-801.

Solórzano, L., and Sharp, J. H. (1980). Determination of total dissolved nitrogen in natural waters. *Limnol. Oceanogr. 25,* 751-754.

R Stainton, M. P. (1974). Simple, efficient reduction column for use in the automated determination of nitrate in water. *Anal. Chem. 46,* 1616.

Stevens, R. J. (1976). Semi-automated ammonia probe determination of Kjeldahl nitrogen in freshwaters. *Water Res. 10,* 171-175.

Strickland, J. D. H., and Parsons, T. R. (1972). "A Practical Manual of Sea-Water Analysis," 2nd ed. Tech. Bull. 167, Fish. Res. Board Can., Ottawa.

Technicon Industrial Systems (1973). "Nitrite in Water and Seawater," Ind. Method No. 161-71W. Technicon Corp., Tarrytown, New York.

Thomas, R. F., and Booth, R. L. (1973). Selective electrode measurement of ammonia in water and wastes. *Environ. Sci. Technol. 7,* 523-526.

Tréguer, P., and Le Corre, P. (1975). "Manuel d'analyse des seis nutritifs dans l'eau de mer." Laboratoire d'Oceanologie Chimique, Universite de Bretagne Occidentale, Brest, France.

Udenfriend, S., Stein, S., Bohien, P., and Dairman, W. (1972). Fluorescamine: A reagent for assay of amino acids, peptides, proteins and primary amines in the picomole range. *Science 178,* 871-872.

U. S. Enviornmental Protection Agency (1979). "Methods for Chemical Analysis of Water and Wastes," EPA-600/4-79-020. Off. Res. Dev., Cincinnati, Ohio.

Valderrama, J. (1981). The simultaneous analysis of total nitrogen and total phosphorus in natural waters. *Mar. Chem. 10,* 109-122.

Van Hall, C. E., Safranko, J., and Stenger, V. A. (1963). Rapid combustion method for the determination of organic substrances in aqueous solutions. *Anal. Chem. 35,* 315-319.

Verdouw, H., van Echteid, C. J. A., and Dekkers, E. M. J. (1978). Ammonia determination based on indophenol formation with sodium salicylate. *Water Res. 12,* 399-402.

Wada, E., and Hattori, A. (1971). Spectrophotometric determination of traces of nitrite by concentration of azo dye on an anion-exchange resin. Application to sea waters. *Anal. Chim. Acta 56,* 233-240.

Webb, K. L. (1978). Nitrogen determination. *In* "Coral Reefs: Research Methods" (D. R. Stoddart and R. E. Johannes, eds.), Monogr. Oceanogr. Methodol. Vol. 5, pp. 413-419. SCOR/UNESCO, Paris.

Webb, K. L., and Wood, L. (1967). Improved techniques for analysis of free amino acids in seawater. *Automa. Analy. Chem., Technicon Symp., 1966* Vol. 1, pp. 440-444.

Webb, K. L., DuPaul, W. D., Wiebe, W., Sottile, W., and Johannes, R. E. (1975). Enewetak (Eniwetok) Atoll: Aspects of the nitrogen cycle on a coral reef. *Limnol. Oceanogr. 20,* 198-210.

Wheeler, P. A., North, B. B., and Stephens, G. C. (1974). Amino acid uptake by marine phytoplankters. *Limnol. Oceanogr. 19,* 249-259.

Wheeler, P. A., North, B. B., Littler, M., and Stephens, G. C. (1977). Uptake of glycine by natural phytoplankton communities. *Limnol. Oceanogr. 22,* 900-910.

Whitledge, T. E., Malloy, S. C., Patton, C. J., and Wirick, C. D. (1981). "Automated Nutrient Analysis in Seawater," Environ.

Control Technol. Earth Sci. TIC-4500. Department of Energy and Environment, Brookhaven Natl. Lab., Upton, New York (available from NTIS, Springfield, Virginia).

Williams, P. J. (1970). Heterotrophic utilization of dissolved organic compounds in the sea. I. Size distribution and relationship between respiration and incorporation of growth substances. *J. Mar. Biol. Assoc. U. K. 50,* 859-870.

Wood, E. D., Armstrong, F. A. J., and Richards, F. A. (1967). Determination of nitrate in sea water by cadmium-copper reduction to nitrite. *J. Mar. Biol. Assoc. U. K. 47,* 23-31.

Youden, W. J. (1959). Accuracy and precision: Evolution and evaluation of analytical data. *In* "Treatise on Analytical Chemistry," (I. M. Kolthoff and P. J. Elving, eds.), Part 1, Vol. 1, pp. 47-66. Wiley (Interscience), New York.

Zafirou, O. C., and McFarland, M. (1980). Determination of trace levels of nitrite oxide in aqueous solution. *Anal. Chem. 52,* 1662-1667.

Zeller, H. D. (1955). Modification of the I-naphtyhlamine-sulfanilic acid method for the determination of nitrites in low concentration. *Analyst 80,* 632-633.

Zimmermann, C., Price, M., and Montgomery, J. (1977). "Operation, Methods, and Quality Control of the Technicon AutoAnalyzer II Systems for Nutrient Determinations in Seawater," Tech. Rep. No. 11, Harbor Branch Foundation, Inc., Fort Pierce, Florida.

Chapter 21

NITROGEN IN THE MARINE ENVIRONMENT
IV.2 USE OF ISOTOPES

WILLIAM G. HARRISON
Marine Ecology Laboratory
Bedford Institute of Oceanography
Dartmouth, Nova Scotia, Canada

I. INTRODUCTION

The importance of nitrogen in the biological production cycle of the marine environment has been the subject of 50 years of research (Harvey, 1969). However, our present knowledge about the utilization and recycling of nitrogen compounds is based largely on recent studies using isotope tracer techniques (Dugdale, 1976; McCarthy, 1980). Of most significance has been the introduction and use of the stable nitrogen isotope ^{15}N (Dugdale and Goering, 1967).

Earliest ^{15}N studies described the N_2-fixation capacity of the marine cyanbacterium *Trichodesmium* (Dugdale et al., 1961) and later focused on the assimilation of dissolved inorganic nitrogen compounds (principally, NO_3^- and NH_4^+) by marine phytoplankton (Goering et al., 1964, 1966; Dugdale and Goering, 1967). During

ISBN 0-12-160280-X

this period, Dugdale (1967) developed the theoretical basis for subsequent studies of the role of nitrogen in marine primary productivity and introduced the concept of "new" and "regenerated" production based on early $^{15}NO_3^-$ and $^{15}NH_4^+$ assimilation work (Dugdale and Goering, 1967). For the next decade, nitrogen tracer studies were directed toward understanding the environmental regulation of nitrogen assimilation by phytoplankton. The utility of the Michaelis-Menten formulation for describing concentration-dependent uptake kinetics was tested extensively in field studies (MacIsaac and Dugdale, 1969, 1972) and also applied to light-dependent uptake relationships (MacIsaac and Dugdale, 1972). Subsequent studies emphasized the nutritional role of dissolved organic nitrogen compounds, e.g., urea (McCarthy, 1972) and amino acids (Schell, 1974), and investigated nutrient interactions (MacIsaac and Dugdale, 1972; McCarthy, 1972; McCarthy et al., 1977; Conway, 1977). From these studies, the concept of nutrient preference was established (McCarthy et al., 1977). Other nitrogen transformations have also been studied using ^{15}N tracers, for example, denitrification (Goering, 1968; Koike and Hattori, 1978a,b, 1979), nitrification (Miyazaki et al., 1973, 1975; Olson, 1980), dissimilatory nitrate reduction (Koike and Hattori, 1978a,b), and ammonification (Harrison, 1978; Blackburn, 1979; Caperon et al., 1979; Glibert et al., 1982; Paasche and Kristiansen, 1982b).

More recent research has focused on the nitrogen uptake characteristics of nutrient-starved phytoplankton under laboratory (McCarthy and Goldman, 1979) and field conditions (Glibert and Goldman, 1981; Goldman et al., 1981). Results from these studies have prompted a reevaluation of widely accepted theories about the steady-state nature of nutrient supply and primary production (Dugdale, 1967), and have suggested the need for a serious reassessment of the ^{15}N tracer methods used for the past 20 years.

This chapter will therefore deal with the methodology of nitrogen tracer experiments, emphasizing the assumptions used and some of the limitations inherent in the techniques. Because of the preponderance of research using ^{15}N as a tracer, other isotopes used either directly (^{13}N) or indirectly (^{3}H, ^{14}C, ^{35}S, etc.) in nitrogen studies are discussed only briefly. Emphasis is also placed on field measurements of nitrogen assimilation and remineralization by plankton. For other components of the marine ecosystem, for example, macrophytes (Owen et al., 1979) and the benthos (Blackburn, 1979; Koike and Hattori, 1978a,b, 1979), many of the same principles (assumptions and limitations) for use of isotopic tracers apply. The chapter concludes with an assessment of the sources and magnitudes of errors in past and currently used nitrogen tracer methods and with some speculation on the interpretation of historical data.

II. TRACER METHODOLOGY

Early attempts to measure nitrogen transformations in seawater relied on relatively imprecise chemical methods and, hence, were generally restricted to studies in artificial laboratory systems (Vaccaro, 1965). The later use of isotopic tracer techniques with superior analytical precision made possible reliable measurements of the fluxes (transformations) of nitrogen under natural field conditions. Furthermore, by the selective labeling of compounds, it was possible to study various transformational pathways.

A. Experimental Design

Most nitrogen tracer studies in the marine environment have dealt with the utilization of dissolved nitrogen compounds by planktonic microorganisms. The general experimental design has consisted of: (1) confinement of a seawater sample, usually in a bottle; (2) inoculation with the labeled nitrogen form (substrate); (3) incubation; (4) collection of the particulate organic material

Table I. Nitrogen Transformations in the Marine Environment Studied by Use of Isotopic Tracers

Pathway	*Transformation*[a]	*Isotopes used*
Nitrogen assimilation		
Combined inorganic	$(NO_3^-, NO_2^-, NH_4^+) \rightarrow$ PON	^{15}N, ^{13}N, ^{36}Cl (^{14}C, ^{35}S)[b]
Organic	(urea, amino-N) $\rightarrow$ PON	^{15}N, ^{14}C, ^{3}H
N_2 fixation	$N_2 \rightarrow$ PON	^{15}N
Nitrogen remineralization		
Organic breakdown/release	PON $\rightarrow$ DON	^{15}N, ^{14}C, ^{125}I
Ammonification	(PON, DON) $\rightarrow NH_4^+$	^{15}N
Nitrification	$NH_4^+ \rightarrow NO_2^- \rightarrow NO_3^-$	^{15}N
Denitrification	$NO_3^- \rightarrow NO_2^- \rightarrow NO \rightarrow N_2O \rightarrow N_2$	^{15}N, ^{13}N
Dissimilatory NO_3^- reduction	$NO_3^- \rightarrow NO_2^- \rightarrow NH_4^+$	^{15}N

[a] *PON, particulate organic nitrogen; DON, dissolved organic nitrogen.*

[b] *Used to estimate protein synthesis.*

TABLE II. Characteristics and Detection of Isotopes[a]

Isotope	Mass no.	Natural abundance (%)	Principal radiation (MeV)	Half-life	Analyzer[b]	Labeled compounds[c]
Hydrogen	3	--	β- (0.186)	12.33 years	S	Organic-N (amino acids, etc.)
Carbon	14	--	β- (0.156)	5730 years	S	Organic-N (amino acids, methlamine[d], etc.)
Nitrogen	13[e]	--	β+ (1.19)	9.99 months	G	N.A.[f]
	15	0.37	--	--	M,E	N_2, N_2O, NH_3, NO_2^-, NH_4^+, urea, amino acids, etc.
Sulfur	35	--	β- (0.167)	87.2 days	S	Principally salts (H_2SO_4, $NaSO_4$, etc.)
Chlorine	36	--	β- (0.709), β+ (0.12)	3.07×10^5 years	S	HCL[g]
Iodine	125	--	γ (0.035)	59.7 days	G	Organic N (nonspecific binding to proteins, etc.)

[a]*From Table I.*

[b]*S = Scintillation spectrometry, M = mass spectrometry, E = emission spectrometry, G = gamma counter.*

[c]*Commercially available forms listed by New England Nuclear, Boston, Massachusetts (radioisotopes) and ICN Pharmaceuticals Inc., Cleveland, Ohio (stable isotopes).*

[d]*Analog for* NH_4^+.

[e]*Nitrogen-13 labeled* NO_3^-, N_2, *and* NH_3 *have been produced (Gersberg et al., 1976).*

[f]*N.A. = Not commercially available.*

[g]*Analog for nitrate when in form of* $^{36}ClO_3^-$ *(not commercially available).*

(plankton), normally by filtration; and (5) measurement of isotopic tracer incorporation. For seagoing studies, incubation conditions (light, temperature) have most often been chosen to simulate, as closely as possible, those of the natural environment. To accomplish this, samples have been incubated in bottles attenuated to correspond to the light depths from which they are taken and kept at constant temperature with flowing, near-surface seawater (MacIsaac and Dugdale, 1972). Results from "simulated" *in situ* incubations have compared reasonably well with true *in situ* incubations (Slawyk et al., 1976), although extensive comparisons have not been made. Most features of the above experimental protocol are also applicable to other nitrogen transformations. These are discussed in more detail below (Section III).

Recently, a variation of this protocol has been employed in studies of nitrogen remineralization. Here, isotopic composition of the labeled substrate is followed over time. Decreases in the relative abundance of the tracer or "isotopic dilution" is used to estimate the production rates of unlabeled substrate (see Section III,C).

B. Isotopes

Isotopes used in nitrogen studies are listed in Tables I and II. Most of the major pathways of nitrogen cycling have been studied using tracers, but oceanographic work has dealt principally with combined, inorganic nitrogen assimilation. The stable isotope ^{15}N remains the most widely used tracer. However, a number of radioisotope applications have been described (McCarthy, 1980).

Radioactive ^{13}N has been used recently in studies of denitrification (Gersberg et al., 1976) and nitrate uptake by natural phytoplankton populations (Gersberg et al., 1978; Axler and Goldman, 1981). Because of its short half-life (10 min), however, its application, particularly in field studies, is severely limited.

McCarthy (1980) described both culture and field studies of ^{14}C-labeled organic nitrogen (urea, amino acids) utilization by marine phytoplankton. The use of these labeled compounds has increased, and studies have focused on the size-partitioning of urea utilization and decomposition in field populations (e.g., Herbland, 1976; Savidge and Hutley, 1977) and on enhanced urea uptake characteristics of nutrient-depleted phytoplankton cultures (Horrigan and McCarthy, 1981). Another line of research has investigated the utility of [^{14}C]methylamine as an effective analog of ammonium in studies of nitrogen assimilation by phytoplankton (Vincent, 1979; Wheeler, 1980). Ditullio and Laws (1983) have suggested using the incorporation of $H^{14}CO_3^-$ into phytoplankton protein (Morris et al., 1974) as an indicator of nitrogen assimilation, and Hollibaugh and Azam (1980) have used both ^{14}C- and ^{125}I-labeled protein to study the mechanisms and kinetics of bacterial protolysis.

Bates (1981) took a different approach by adapting a method used in freshwater microbial work (Monheimer, 1974) to measure phytoplankton protein synthesis based on the incorporation of $^{35}SO_4^{2-}$. Despite some methodological problems, comparisons of $^{35}SO_4^{2-}$ and $^{15}NO_3^-$ + $^{15}NH_4^+$ uptake rates in field populations have given similar estimates of protein synthesis (Bates, 1982). The combined measurements of $H^{14}CO_3^-$ and $^{35}SO_4^{2-}$ incorporation to determine phytoplankton nitrogen assimilation rates are currently under investigation (E. Laws, personal communication).

Tromballa (1970) and Tromballa and Broda (1971) showed that $^{36}ClO_3^-$ reduction by cultured algae is competitively inhibited by NO_3^- and NO_2^- and recommended its use as an analog for estimating NO_3^- and NO_2^- assimilation. However, procedures for labeling and purifying ClO_3^- are tedious (Tromballa, 1970). There have been no published investigations using $^{36}ClO_3^-$ with marine microorganisms.

C. Analytical Procedures

Preparation of samples for isotopic analysis generally follows three steps: (1) sample isolation (extraction or purification), (2) sample conversion (e.g., in the case of ^{15}N analysis, conversion to N_2 gas), and (3) isotope detection (e.g., mass spectrographic or radioisotopic analysis) (Table II).

1. Nitrogen-15

Bremner (1965) summarized methods commonly used for the extraction and interconversion of a variety of forms of nitrogen for ^{15}N analysis by mass spectrometry. Fiedler and Proksch (1975) have reviewed procedures for emission spectrometry.

In oceanographic studies, particulate sample material is usually collected by filtration (Section II,A); two procedures for sample conversion have been employed. In earlier studies (Dugdale et al., 1961), Kjeldahl digestion was used to convert particulate organic nitrogen to NH_4^+, which was subsequently steam-distilled and converted to N_2 by reaction with $NaOBr^-$ (Rittenburg method). The second, a dry combustion method (Dumas procedure), has been most commonly used. In this procedure, particulate nitrogen is converted directly to N_2 by heating (>500°C) in the presence of a copper (Cu + CuO) catalyst. This procedure has been automated for conversion and direct sample delivery into the isotope analyzer (Barsdate and Dugdale, 1965; Desaty et al., 1969). Refinements in this conversion scheme were later described (Wada et al., 1977), including the capability for simultaneous determination of isotope ratios and total nitrogen content (Pavlov et al., 1974). Recently, it has been suggested that incomplete sample combustion can occur in the automated procedure, resulting in analytical errors (cited in McCarthy, 1980). Batch sample preparation has been employed also (e.g., Grunseich et al., 1982), and is most commonly used in emission spectrometric analysis (Fiedler and Proksch, 1975; Kristiansen and Paasche, 1982).

Nitrogen-15 analysis of compounds in liquid phase (NH_4^+, NO_2^-, NO_3^-, dissolved organic N) can be accomplished after: (1) conversion to NH_4^+ (by chemical reduction or digestion) and subsequent conversion to N_2 by the Rittenburg method (Bremner, 1965), or after (2) selective extraction using ion-exchange procedures (Miyazaki et al., 1973). Ammonium dissolved in seawater has been recovered by direct distillation and converted to N_2 by either the Rittenburg method (Harrison, 1978; Caperon et al., 1979) or Dumas

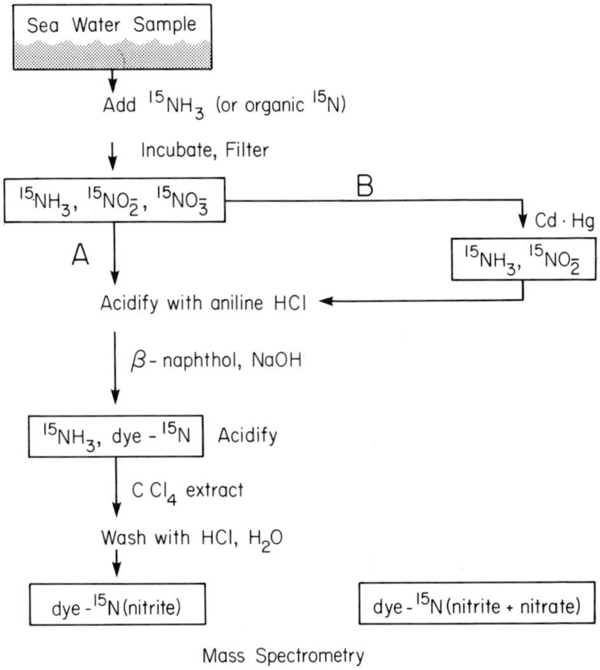

Fig. 1. Flow diagram for the extraction of NO_3^--N and NO_2^--N from aqueous solutions for mass spectrometric analysis. Procedure A isolates only NO_2^--N; B isolates both NO_3^--N and NO_2^--N (Schell, 1978).

combustion of the NH_4^+ trapped in capillary tubes (Blackburn, 1979) or on glass fiber filters (Glibert et al., 1982). Nitrite has been selectively extracted by cation exchange (Dowex 1×8) after conversion to an azo dye (Miyazaki et al., 1973). The azo dye is subsequently isolated and converted (through NH_4^+) to N_2 for isotope analysis. This procedure has been improved to reduce the problem of isotope dilution from reagent nitrogen and to include steps for the recovery of nitrate (Fig. 1). Chan and Campbell (1978) have described a procedure for extracting dissolved organic N from lake water for ^{15}N analysis using cation-exchange chromatography. However, the applicability of this method to seawater has not been tested.

Procedures for extracting gas-phase samples (N_2, N_2O, etc.) for ^{15}N analysis are direct and relatively simple. Goering and Dugdale (1966) described a gas extraction-bulb procedure for recovering labeled nitrogen gas from lake water samples. Later, Goering and Pamatmat (1971) modified this method and recovered gases dissolved in seawater by purging with CO_2. Oxides of nitrogen, carrier gas (CO_2), and contaminants can be eliminated by utilizing an in-line liquid N_2 trap prior to delivery of the sample into the isotope analyzer.

Once the nitrogen compound is purified and converted to N_2, isotopic content is determined, most commonly, by mass analysis. The theory and use of mass spectrometry for measuring ^{15}N have been thoroughly reviewed (San Pietro, 1957; Bremner, 1965; Fiedler and Proksch, 1975) and are discussed here only briefly. The abundance of ^{15}N is calculated using the expressions

$$\%\ ^{15}N = 100/2\ R + 1 \text{ when } \%\ ^{15}N < 50\% \tag{1}$$

or

$$\%\ ^{15}N = 100/\ R' + 1 \text{ when } \%\ ^{15}N > 50\% \tag{2}$$

where $R = I_{28}/I_{29}$ and $R' = I_{29}/2 \cdot I_{30}$ and I_{28}, I_{29}, I_{30} are the ion currents corresponding to the nitrogen masses 28 (^{14}N, ^{14}N), 29 (^{15}N, ^{14}N), and 30 (^{15}N, ^{15}N). Since relative mass abundances are measured, sample recovery need not be quantitative. Fiedler and Proksch (1975) have suggested that as a general rule, 30 μg N is the minimum sample size required for most mass analyzers. This is comparable to requirements noted in oceanographic amplications (e.g., MacIsaac and Dugdale, 1969), however, the analysis of smaller samples is possible (J. McCarthy, personal communication). Although instrumentation is costly and maintenance requirements are high, analytical precision is generally superior to alternative ^{15}N analyzers (e.g., emission spectrometry). Analytical precision for unenriched (^{15}N natural abundance = 0.37%) as well as ^{15}N-enriched samples is generally characterized by a coefficient of variation (cv) less than 0.5% on replicate determinations (e.g. Goering, 1968). However, an ^{15}N enrichment of 0.01 atom % is usually considered analytically significant (Dugdale and Goering, 1967; McCarthy et al., 1977). Particulate ^{15}N enrichments in nitrogen assimilation studies, for example, commonly exceed this value by one or two orders of magnitude.

The most widely used alternative to mass spectrometry in ^{15}N analysis is optical emission spectrometry. Fiedler and Proksch (1975) have reviewed the principles of use, sample preparation, and performance comparisons with mass spectrometry. Emission spectrometers are less costly, generally require less maintenance, and have the capability of analyzing substantially smaller samples than mass spectrometers. Fiedler and Proksch (1975), for example, suggest that samples ranging in size from 0.2 to 10 μg N can be analyzed. In oceanographic applications, a lower limit of 2 μg N per sample is more representative (Kristiansen and Paasche, 1982). Analysis of smaller samples, however, can be made but special precautions against contamination and lengthier sample preparation procedures are required (e.g., Goleb and Middelboe, 1968). Precision of emission spectrometry is not as good as that of mass spectrometric analysis, particularly at low enrichments, although accuracy appears to be comparable (Keeney and Tedesco, 1973). In our laboratory, samples with low ^{15}N enrichment (0.1 atom % or less) generally have a cv of about 5% for replicate determinations. This decreases to 1-2% when enrichments exceed 1 atom % ^{15}N (see also Keeney and Tedesco, 1973).

Benedetti et al. (1976) have described an alternative to mass and emission spectrometries for analysis of ^{15}N. This procedure employs a gas chromatograph equipped with a katharometer to measure ^{15}N differentially, by comparing sample nitrogen with carrier gas having the natural ^{15}N abundance. Precision of this method is relatively poor (cv = 10% or greater with 0.1 atom % ^{15}N enrichments), and therefore enrichment levels required for reliable estimates are considerably higher than those required for mass spectrometry (Wada et al., 1977). Small sample-size requirements, however, may be an advantage with this system if ^{15}N enrichments are relatively high.

2. Radioisotopes

Preparation and analysis of samples labeled with radioisotopic tracers will vary considerably, depending on the characteristics of the radioisotope and the experimental material labeled. Procedural details are not discussed here since a number of comprehensive treatments of the theory, application, and analysis of radioisotopes have been published (e.g., Kobayashi and Maudsley, 1974). Radioisotopes used in nitrogen studies have generally followed experimental protocols similar to ^{15}N studies (Section II,A); sample collection is similar. In the case of radioisotopes, however, sample preparation is generally less complicated, since instruments with capabilities of measuring radioisotopes in solid, liquid, or gas phase are common. Radioisotope analysis involves an *absolute* measure of tracer abundance, in contrast to a relative abundance measurement in stable isotope analysis. For that reason, quantitative recovery in processing of radioactive samples is usually required.

D. Theory and Assumptions

Before isotope tracer experiments can be interpreted, a number of basic assumptions about the behavior of the tracer relative to that of the biological system under study are necessary. Atkins (1969), for example, has identified four characteristics of the behavior of an "ideal" tracer. They are:

1. The tracer undergoes the same metabolic changes as the unlabeled substance, i.e., "isotope effects" are negligible (or quantifiable).

2. The tracer does not disturb the *steady state* existing in the system as a whole, i.e., there are no significant perturbations to compartments or their transformations.

3. The tracer is initially not in equilibrium with the system studied and its changes over time are quantifiable and reflect transfers (or transformations) of the traced substance.

4. There is no exchange of the isotope between the labeled substance and other substances in the system.

For isotopes used in nitrogen studies (Section II,B), assumption (1) is not seriously violated nor is (4), except in the case of some tritiated compounds (Azam and Holm-Hansen, 1973). For assumption (2), however, true *tracer* additions of ^{15}N-labeled compounds often have not been achievable in oceanographic studies (Dugdale and Goering, 1967). As a consequence, changes in substrate concentration and fluxes may be significant. For radioisotopes (e.g., ^{13}N, ^{14}C), tracer-level additions are generally possible. The validity of assumption (3) may also not hold for many nitrogen tracer studies. This may be a particular problem when long exposure times (experimental incubations) are used and tracer transformation rates, relative concentration of tracer, and absolute concentration of labeled substrate can change substantially. Violation of assumption (2) also becomes significant under these circumstances. These problems are discussed in more detail in Section IV.

III. APPLICATIONS

Most nitrogen transformations in the marine environment have been studied using stable or radioactive tracers (Table I). However, the majority of studies have concentrated on the assimilative fluxes of nitrogen compounds by planktonic microorganisms and more recently, on nitrogen recycling.

Table III. Uptake of Combined Inorganic and Organic Forms of Nitrogen Using ^{15}N as a Tracer

Region	^{15}N *form*	*Tracer concentration*[a]	*Sample depth*[b]	*Prescreen*[c]
Oceanic	NO_3^-, NH_4^+	1-2 μM	surface	?[g]
	NO_3^-, NH_4^+	2 μM	100, 50	
	Amino-N	10 μg liter^{-1}	30% I_o	233 μm[h]
	NO_3^-, NH_4^+	5-20% A	15-20 m	?
	NO_3^-, NH_4^+	∿10% A	100, 50	?
	NO_3^-, NH_4^+	K 0.1-2.0 μM	25% I_o Ze	?
	NO_3^-, NO_2^-	10, 4.5	Ze	?
	NH_4^-	10 μM, K		
	NO_3^-, NH_4^+	∿10% A	Ze	?
	NO_3^-, NH_4^+, Urea	0.1 μM, K	Ze	183 μm
	NO_3^-, NO_2^-	0.067 μM	Ze	330 μm[h]
	NO_3^-, NH_4^+ Urea	1 μM	Ze	?
	NO_3^-, NH_4^+	5-10% A	Ze	200 μm
	NO_3^-, NH_4^+	10, 1 μM	10% I_o	?
	NO_3^-, NH_4^+	0.1 μM	Ze	(-)
Coastal	NO_3^-, NH_4^+, Urea	0.1-1 μM	Ze	183 μm
	NO_3^-, NH_4^+, Amino-N	5-10 μM	?	?

in the Marine Environment: Representative Experimental Conditions

Reaction vessel Type	Reaction vessel volume (liters)	Incubation[e] (h) IS	Incubation[e] (h) SIS	Analyzer[f]	Reference
P	4		4	M	Goering et al., 1964
P	1		3-6	M	Goering et al., 1966
P	12		24	M	Dugdale and Goering, 1967
P	4		4-24	M	MacIsaac and Dugdale, 1969, 1972
P	?[g]		24	M	Goering et al., 1970
P	1	10	10	M	Hattori and Wada, 1972
?	?		24	M	McRoy et al., 1972
P	4		24	M	Eppley et al., 1973, 1977a
P	0.02		3	M	Carpenter and McCarthy, 1975
P	4		24	M	Mague et al., 1977
P	10		24	M	Slawyk, 1979
P	4-9		12-24	M	Olson, 1980
P	1		24	E	Harrison et al., 1982
P	4		24	M	McCarthy, 1972
P	4-100		2-6	M	Schell, 1974

Table III (Continued)

Region	^{15}N form	Tracer concentration[a]	Sample depth[b]	Prescreen[c]
	NO_3^-, NH_4^+	0.5, 0.1 μ*M*	Ze	?
		5-10 μ*M*		
	NO_3^-, NO_2^-, NH_4^+, urea	10-20% A	75, 50% I_o	35 μm
	NO_3^-, NH_4^+	20 μ*M*	Ze	?
	NO_3^-, NH_4^+	0.1 μ*M*	Ze	183 μm
	NO_3^-, NH_4^+	30-100% A	0-20 m	?
	NO_3^-, NH_4^+	15 μ*M*	Ze	253 μm
	NO_3^-, NH_4^+	?	Ze	?
	NO_3^-, NH_4^+	10 μ*M*	100, 4% I_o	?
	NO_3^-, NH_4^+	0.1 μ*M*	50% I_o	(-)
	NO_3^-, NH_4^+	10 μ*M*	Ze	212 μm
	NH_4^+	0.06, 12.6 μ*M*	1 m	202, 35 μm
	NO_3^-, NH_4^+	0.4, 4 μ*M*	0-2 m	300 μm

[a] *A = Ambient concentration, K = kinetics experiments (multiple concentration additions).*

[b] *I_o = Surface light intensity, Ze = samples usually taken from 5 to 6 depths between surface and 1% light levels.*

[c] *(-) = no prescreening.*

Reaction vessel Type	Reaction vessel volume (liters)	Incubation[e] (h) IS	Incubation[e] (h) SIS	Analyzer[f]	Reference
P	4		24	M	MacIsaac et al., 1974
P	1		2-4	M	McCarthy et al., 1977
P	2		6	M	Nelson and Conway, 1979
P	1		24	M	Eppley et al., 1979
PE	4		24	M	MacIsaac et al., 1979
P	2-4		6-8	M	Conway and Whitledge, 1979
P	1		6	M	Stanley and Hobbie, 1981
P	2		2-8	M	Garside, 1981
P	1		4	E	LaRoche, 1981
P	4		6	M	MacIsaac, 1978
P	1		0.08-2	M	Glibert and Goldman, 1981
P	1		3-5	E	Paasche and Kristiansen, 1982a

[d]P = Pyrex glass, PE = polyethylene.
[e]IS = In situ, SIS = "simulated" in situ.
[f]M = mass spectrometer, E = emission spectrometer.
[g]? = information not given in source reference.
[h]Screens used to concentrate Oscillatoria colonies.

A. Nitrogen Assimilation

1. *Nitrogen-15*

The assimilation of combined inorganic forms of nitrogen (NO_3^-, NO_2^-, NH_4^+) by phytoplankton using ^{15}N-labeled compounds has been studied most extensively in the marine environment. Most published studies have followed the experimental protocol described in Section II,A. However, details have varied considerably (Table III). For example, isotope additions have usually been made at either *tracer* levels (submicromolar) or *substrate-saturated* levels (several micromoles). Dugdale and Goering (1967) suggested that reliable estimates of *in situ* uptake rates required isotope additions to be 10% or less of the ambient substrate concentration. However, use of this protocol requires knowledge of the ambient substrate levels before uptake experiments are done (McCarthy et al., 1977). Fixed, small additions (0.1 μM) have been used as an alternative when substrate concentrations were not immediately known (e.g., Eppley et al., 1973, 1979a). In other experiments, large isotope additions have been used (e.g., MacIsaac and Dugdale, 1969; Hattori and Wada, 1972). These "maximum" or "potential" uptake rate measurements have been the method of choice in many recent studies (e.g., Conway and Whitledge, 1979; Garside, 1981).

The choice of incubation periods has also varied. Most studies have used long incubations (24 h), primarily to avoid diurnal effects (e.g., McCarthy, 1972), although shorter incubations have been used, particularly in coastal waters (Table III). With few exceptions, most experiments have employed "simulated" *in situ* insubation systems (see Section II,A).

After sample collection and processing for ^{15}N analysis, uptake rates are calculated using the formulation given by Dugdale and Goering (1967):

$$\nu = [d(a_N)/dt]/(a_S - a_N) \tag{3}$$

where ν is the specific uptake rate in units of time^{-1}, a_N is the atom % ^{15}N in the particulate nitrogen (N), and a_S is the atom % ^{15}N in the substrate. For incubations of duraction t, this equation can be rewritten as:

$$\nu = \rho/N = a_t/R \cdot t \qquad (4)$$

where a_t is the atom % excess (atom % ^{15}N minus 0.37, the natural abundance of ^{15}N) of the particulate nitrogen at time t, R is the atom % enrichment of the substrate $[(100 \cdot S_L)/(S_L+S_u)]$ where S_L is the amount of labeled substrate added and S_u is the amount unlabeled substrate present, and ρ is the absolute uptake or transport rate in units of mass$\cdot$(unit volume)$^{-1}\cdot$time^{-1}. Equation (4) assumes constancy in the atom % enrichment of the substrate (R) during the incubation period. Also, since the particulate nitrogen in the sample (N) is often determined at the beginning of experiments, any changes during incubation will affect the estimate of ρ since $\rho = \nu \cdot N$.

This potential error, however, has been circumvented by determining the nitrogen content of samples after incubation (Pavlov et al., 1974; Kristiansen and Paasche, 1982; Grunseich et al., 1982). Dugdale and Goering (1967) discussed other systematic errors, for example, the underestimation of specific uptake rate (ν) resulting from detrital N contamination. Grunseich et al. (1982) have recently made a detailed analysis of these computational errors.

The early single substrate-level experiments were later supplemented with studies of substrate concentration-dependent uptake rates. MacIsaac and Dugdale (1969) began nutrient kinetics studies using the hyberbolic Michaelis-Menten relationship as a model:

$$\nu = (V_{max} \cdot S)/(k_t+S) \qquad (5)$$

where ν is the specific uptake rate, S the substrate concentration, V_{max} the maximum uptake rate at substrate saturation, and K_t the "half-saturation constant" or substrate concentration where $\nu = V_{max}/2$. In this formulation, as in the case of Eq. (4), ν and V_{max} are underestimated if detritus is present; K_t is unaffected.

MacIsaac and Dugdale (1972) extended their kinetics studies to include light-dependent uptake relationships using this same model. In these studies, S represented light intensity, and the half-saturation constant (K_{Lt}) represented the light level (in absolute or relative units) where the uptake rate was half that at light saturation.

More recent studies have employed *tracer* level and *substrate-saturated* additions to investigate the rapid uptake response of nitrogen-starved phytoplankton in culture and in the field (e.g. McCarthy and Goldman, 1979; Glibert and Goldman, 1981). This work has been extended to include a more general assessment of time-varying ^{15}N uptake rates under a variety of laboratory and field conditions (Goldman et al., 1981).

2. Radioisotopes

Studies of the assimilation of nitrogenous compounds using radioisotopes have followed the same experimental protocol as used for ^{15}N, with some exceptions. Typically, smaller incubation chambers and shorter incubation times have been used. Also because of the availability of high specific activity substrates, tracer-level substrate additions are the rule. Uptake can be calculated using the general expression:

$$\rho = \frac{(dpm_t - dpm_0) \cdot f \cdot S}{dpm_a \cdot t} \tag{6}$$

where ρ is the uptake rate in units of mass$\cdot$(unit volume)$^{-1}\cdot$time^{-1}; dpm_0, dpm_t = radioactivity of sample at times 0 and t, dpm_0 may also be "killed" control, dpm_a is the radioactivity of substrate added; f the isotope discrimination factor; S the substrate concentration in units of mass (unit volume)$^{-1}$; and t the incubation time. For radioisotopes with a short half-life (e.g., ^{13}N), the inclusion of an isotope decay factor is also required (Gersberg et al., 1976).

B. Nitrogen (N_2) Fixation

Nitrogen fixation in marine and freshwater environments has been measured most often using the sensitive acetylene reduction technique (Stewart, 1971; Mague, 1977). However, unequivocal evidence for N_2 fixation and calibration of the C_2H_2 method is provided by directly measuring the reduction of N_2 and its incorporation into cellular organic matter; this is done using $^{15}N_2$ as a tracer (Mague, 1978). In ocean waters, $^{15}N_2$ fixation by planktonic cyanobacteria (primarily *Oscillatoria*) has been studied extensively (Dugdale et al., 1961, 1964; Mague et al., 1974, 1977; Carpenter and Price, 1977). Experimental protocol has basically followed that first described by Neess et al. (1962) and later by Mague (1978). Briefly, $^{14}N_2$ is removed from samples by purging ($Ar/O_2/CO_2$) and replaced with highly enriched $^{15}N_2$. Samples are subsequently incubated and then "killed" (usually with H_2SO_4 or TCA). Following this, both gas and particulate samples are analyzed for ^{15}N enrichment. Fixation rates can then be calculated using a derivation of Eq. (4).

C. Nitrogen Remineralization

In situ nutrient recycling has been recognized to be an important process for maintaining phytoplankton growth in low-nutrient waters since the earliest studies of nitrogen in the marine environment (Harrison, 1980). However, not until recently have isotope tracers been used to quantify regenerative fluxes.

1. *Ammonification*

Studies of the rates and biological sources of NH_4^+ regeneration in oceanic waters are new. The most straightforward measurements have been based on mass balance considerations using an equation of the form:

$$r = (S_t - S_0)/t + \rho \qquad (7)$$

where is the regeneration rate, S_0, S_t the substrate concentra-

TABLE IV. Nitrogen Remineralization in the Marine Environment:

Region	^{15}N form	Tracer concentration	Sample depth	Prescreen
Oceanic	NH_4^+	0.05 μM, 10% A	100, 60% I_0	130, 10 μm
Coastal	NH_4^+	1 μM	Ze^c	183 μm
	NH_4^+	10 μM	surf.	333, 35 μm
	NH_4^+	20 μM	0-100 m	?
	NH_4^+	1 μM	50% I_0	(-)

[a]*Symbols as in Table III.*
[b]*PC = polycarbonate.*
[c]*"pooled" samples from 1 to 100%* I_0.

tion at times 0 and *t*, and ρ the substrate uptake rate. This method has been applied in coastal waters (Harrison et al., 1977) and more recently in lakes (Axler et al., 1981). In both studies, ρ was computed from $^{15}NH_4^+$ uptake [Eq. (4)]. The applicability of this method, however, is limited and depends on reliable measurements of ρ and substrate concentrations. It would be unfeasible to use this method in oligotrophic waters, for example, where substrate concentrations (e.g., NH_4^+) are difficult to measure (Eppley al., 1977a).

A more generally applicable and direct method, proposed by Alexander (1970), is the $^{15}NH_4^+$ isotope dilution technique. Ammonium regeneration rates are determined from the change in specific activity (% ^{15}N) of the NH_4^+ pool with time (see Section II,A). These changes result from "dilution" of the isotope as a consequence of biological production of $^{14}NH_4^+$. This method has recently been used in the marine environment to measure planktonic (Harrison, 1978; Caperon et al., 1979; LaRoche, 1981; Glibert et al., 1982; Paasche and Kristiansen, 1982b) and sedi-

Representative Experimental Conditions Using ^{15}N as a Tracer[a]

Reaction vessel		Incubation (h)			
Type	volume (liters)	IS	SIS	Analyzer	Reference
PC[b]	2.5		4	M	Glibert et al., 1982
P	1		24	M	Harrison, 1978
P	4	7-10		M	Caperon et al., 1979
PE	4		24	M	Olson, 1980
P	1		4	E	LaRoche, 1981

ment NH_4^+ regeneration rates (Blackburn, 1979). In most experiments (Table IV), relatively large concentrations (1-20 μ*M*) of labeled substrate have been added, and incubations have been long (24 h). More recent studies, however, have used shorter incubations (LaRoche, 1981) and tracer level additions of labeled substrate (Glibert et al., 1982).

Alexander (1970) and Harrison (1978) calculated regeneration rates using a modification of Hevesy's isotope dilution principle:

$$r = \frac{S_L \cdot R_L (R_0 - R_t)}{R_0 \cdot R_t} \tag{8}$$

where r is the regeneration rate, S_L the amount of $^{15}NH_4^+$ added, R_L the atom % excess ^{15}N of $^{15}NH_4^+$ added, R_0 and R_t the atom % excess ^{15}N in the total NH_4^+ pool at times 0 and t, respectively. This equation is based on the assumptions that only $^{14}NH_4^+$ is regenerated during the experiment and that the dilution rate (dR/dt) is constant. This model has limited usefulness, giving reliable estimates only when $^{15}NH_4^+$ additions are large relative to ambient $^{14}NH_4^+$ and when the change in R is small (Caperon et al., 1979). A more generally applicable, linear differential equation model was introduced by Blackburn (1979) and Caperon et al. (1979) that has the capability of estimating both regeneration and uptake rates from the change in concentration and specific activity of the substrate. This is accomplished by the simultaneous solution of the following two equations:

$$S_t = S_0 + (r-\rho) \tag{9}$$

and

$$\ln(R_t) = \ln(R_0) - r/(r-\rho)\cdot\ln(S_t/S_0) \tag{10}$$

where S is the substrate concentration, r the regeneration rate, ρ the uptake rate, and R the atom % excess ^{15}N in the NH_4^+ pool. It is assumed in this model that uptake and regeneration rates are constant with time and that only $^{14}NH_4^+$ is regenerated. Under some circumstances, for example, in oligotrophic ocean waters, application of this model poses some problems, particularly if substrate concentrations do not change measurably or if substrate concentrations are at or below the analytical limit of detection. Glibert et al. (1982) have addressed these special cases and proposed some modification to the above equations. One of their proposed modifications involves correcting conventional NH_4^+ uptake calculations [Eq. (4)] when changes in substrate specific activity (atom % enrichment) are known. This is accomplished by substituting the atom % enrichment (R) of Eq. (4) with the "average" atom % enrichment over the incubation time:

$$\bar{R} = (R_0/k_t)(1-e^{-kt}) \tag{11}$$

where R_0 is the atom % enrichment at time 0 and k is the first-order rate constant describing the time-dependent change in (R).

One promising development emerging from NH_4^+ regeneration experiments has been application of the isotope dilution principle (Johansen and Middelboe, 1976) to measure more precisely ambient substrate (NH_4^+) concentrations, particularly in oceanic waters where levels are below conventional water chemistry detection limits (Glibert et al., 1982).

2. *Organic Decomposition Intermediates*

Few tracer studies have dealt directly with intermediates produced during the mineralization of nitrogenous organic matter. Schell (1974) made limited measurements of ^{15}N-labeled dissolved organic matter production in his studies of amino acid metabolism by microplankton communities. More recently, Chan and Champbell (1978) have studied the extracellular production of ^{15}N-labeled dissolved organic nitrogen by lake phytoplankton. Comparable studies, however, have not been performed in the marine environment. In recent radioisotopic studies, [^{3}H]leucine metabolism was used to estimate turnover of dissolved free amino acids in coastal marine waters (Hollibaugh et al., 1980). Also, the production of TCA-soluble ^{14}C and ^{125}I compounds from previously labeled proteins was used to study microbial protein degradation (Hollibaugh and Azam, 1980).

D. Nitrification/Denitrification/Dissimilatory NO_3^- Reduction

Few attempts have been made to quantify nitrification rates in seawater using isotope methods. Miyazaki et al. (1973, 1975) measured the production of $^{15}NO_2^-$ in coastal and oceanic waters enriched with high levels (10 μ*M*) of $^{15}NH_4^+$. More recently, Olson (1980) made similar measurements in addition to measurements of $^{15}NO_3^-$ production. Koike and Hattori (1978b) measured nitrification rates in marine sediments using an isotope dilution method. Because of elevated substrate additions in most cases, nitrification rates based on ^{15}N tracer studies performed to date should probably be considered "potential" rates.

Direct measurements of denitrification and dissimilatory NO_3^- reduction have been equally sparce. Goering (1968) employed a ^{15}N tracer method described for lake denitrification studies (Goering and Dugdale, 1966) to investigate $^{15}N_2$ production in the O_2-minimum layer of the Eastern Tropical Pacific and in coastal sediments (Goering and Pamatmat, 1971). More recent marine sediment studies have included estimates of NO_2^- and NH_4^+ formation (Koike and Hattori, 1978a,b) and extended denitrification measurements (Koike and Hattori, 1979). In these studies also, relatively large substrate additions were used and incubation times were long, particularly in the earlier work. Gersberg et al. (1976) have recently used the radioisotope $^{13}NO_3^-$ to study denitrification in soils. This method avoids high substrate additions and long incubations, but the utility of this short half-life isotope for ocean-going denitrification studies is doubtful.

IV. VIOLATIONS OF TRACER ASSUMPTIONS AND SIGNIFICANCE

McCarthy (1980) identified two general constraints in the application of isotopic tracer methods (bottle experiments) to studies of nitrogen utilization by phytoplankton: (1) the disruption of natural *loss* processes, i.e., sinking and grazing (particularly in prescreened samples) fluxes, and (2) the disruption of *supply* processes, i.e., nutrients from animal excretions, etc. Furthermore, the seriousness of these problems are likely to increase with confinement (incubation) time. The impact of these "bottle effects" and other methodological problems (e.g., large tracer additions) on the assumptions used in tracer studies may be significant.

A. Nitrogen-15

1. *Substrate Enrichment*

Dugdale and Goering (1967) recognized early the difficulties in adding tracer concentrations of ^{15}N-labeled substrates when, particularly in oceanic waters, ambient levels were vanishingly low. To address this problem, one of the objectives of nutrient kinetics experiments (MacIsaac and Dugdale, 1969) was to correct for enrichment errors by the use of empirically derived nutrient concentration uptake relationships. MacIsaac and Dugdale (1969) estimated uptake rates at ambient concentration using the Michaelis-Menten Eq. (5), where S, k_t, and V_{max} were derived from kinetics experiments. Eppley et al. (1973) employed a similar correction for NH_4^+ uptake rates in the Central North Pacific Gyre. However, subsequent studies demonstrated inadequacies in this model, particularly where determinations of ambient substrate concentrations were unreliable (Eppley et al., 1977a; Fisher et al., 1981). Eppley et al. (1977a) clearly showed this for NH_4^+ and urea uptake experiments in the oligotrophic Pacific where substrate levels were below the level of analytical detection (Fig. 2). As a consequence, Eppley suggested two alternative methods for estimating ambient uptake rates. The first, called an "adjusted" uptake rate, was based on the presumption that increases in uptake rate from isotope addition were proportional to the increases in substrate concentration:

$$\rho = \frac{a_t \cdot N \cdot S_u}{100 \cdot S_L \cdot t} \tag{12}$$

using the notation given for Eq. (4). The second, called a "conservative" rate, assumed that substrate concentration was zero:

$$\rho = (a_t \cdot N)/(100 \cdot t) \tag{13}$$

The more general implication of unreliable estimates of ambient substrate concentrations for determining uptake rates is obvious [Eq. (4)] and, moreover, this difficulty is considered one of the most serious drawbacks to our present interpretation of ^{15}N tracer data (McCarthy, 1982).

Quantifying *enrichment* effects in nutrient-depleted waters can be further complicated. McCarthy and Goldman (1979), for example, showed that phytoplankton in culture could rapidly take up $^{15}NH_4^+$,

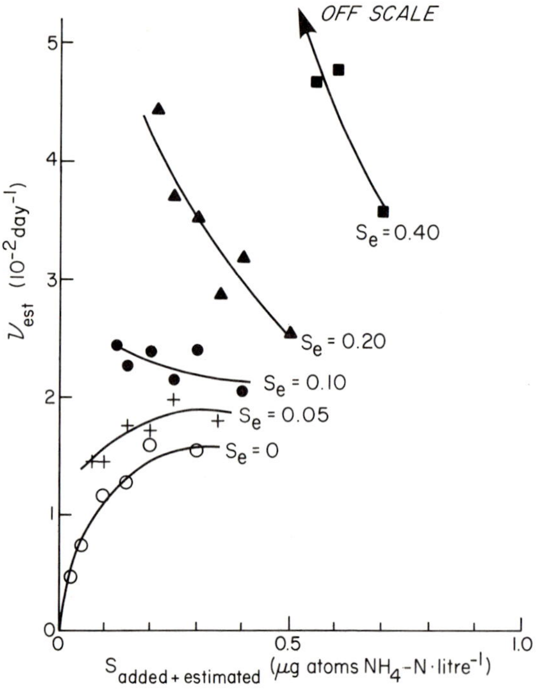

Fig. 2. Effects of varying the estimated ambient substrate concentration (S_e) on calculated uptake rates. Data are from $^{15}NH_4^+$ uptake experiments in near-surface waters of the Central North Pacific Ocean (Eppley et al., 1977a).

at a rate as much as 30 times their steady-state growth requirement when nitrogen-starved (Fig. 3). Similar transient uptake responses have been observed in field populations where as much as 50% of the substrate taken up occurred in the first few minutes of incubation (Glibert and Goldman, 1981). Enhanced urea uptake has also been demonstrated for nitrogen-starved phytoplankton cultures (Horrigan and McCarthy, 1981). The differentiation of artificially enhanced uptake rates (resulting from elevated substrate additions) from rapid uptake (resulting from nutrient depletion) would be difficult to make under these circumstances. Clearly, conventional nutrient kinetics analysis [Eq. (5)] would not be applicable here (McCarthy, 1982).

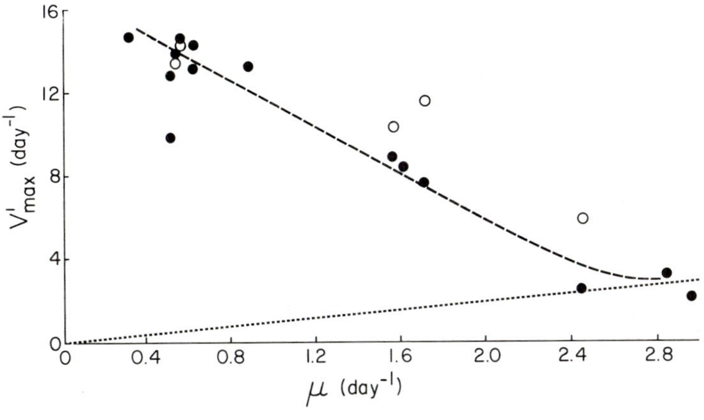

Fig. 3. Maximum rate of NH_4^+ uptake per unit plant nitrogen (V'_{max}) at specific growth rates (μ) for Thalassiosira pseudonana (clone 3H) grown in a NH_4^+-limited chemostat. The circles represent rates determined immediately and 20 min after the nutrient supply ceased. The triangles represent rates determined 120 min after the nutrient supply ceased. Rates of NH_4^+ uptake adequate to meet the requirements of growth are depicted by the dotted line (McCarthy and Goldman, 1979).

Other nitrogen transformations measured by ^{15}N are also subject to enrichment errors (e.g., nitrification, denitrification, ammonification) as a result of high, labeled substrate additions. However, these have not been quantitatively addressed. Measurements of nitrogen-fixation rates, on the other hand, are probably not affected, since substrate concentrations (N_2) are generally far in excess of biological requirements.

2. *Substrate Depletion*

Substrate depletion has been considered a potential problem for ^{15}N tracer experiments in low-nutrient waters, particularly when incubations are long (e.g., MacIsaac and Dugdale, 1969). However, a quantitative assessment of associated errors have only recently been undertaken. Fisher et al. (1981) have recently noted that "anamolous" NH_4^+ uptake kinetics are often observed in eutrophic coastal waters and result primarily from substrate depletion during incubation. Goldman et al. (1981) have investigated the consequence of substrate depletion on the time-course of NH_4^+ uptake by phytoplankton cultures and field populations. Their results have shown pronounced nonlinearity in NH_4^+ uptake rates over time that was attributable directly to NH_4^+ depletion in the incubation bottles. Moreover, they have suggested that this may be a common problem for experimental work in low-nutrient waters where long incubation times are used. In fact, their data suggest that under these circumstances, estimates of nitrogen turnover from long incubations (24 h) could be in error (underestimated) by an order of magnitude when compared with estimates from short incubations (2 h or less), where uptake rates are typically linear. Nitrogen depletion in incubation bottles probably also contributes to the frequently observed high C:N assimilation ratios of near-surface plankton populations (G. Jackson, personal communication). There have been a number of recent studies published, however, that demonstrate substantial nitrogen (NH_4^+) recycling in incubation bottles (e.g., Harrison, 1978; Cameron et al., 1979; Glibert et al., 1982), and therefore suggest that depletion may not be as serious as Goldman et al. (1981) speculate, at least for NH_4^+.

3. *Substrate Recycling*

Nonlinear incorporation of ^{15}N and high C:N assimilation ratios can also result from nitrogen recycling in bottles (G. Jackson, personal communication). This is more evident when one considers relative pool sizes. In California coastal waters, for example, particulate nitrogen is often 10 times the dissolved inorganic N pool. By comparison, particulate carbon rarely exceeds 5% of the dissolved HCO_4^- pool (Eppley et al., 1977b). With rapid nitrogen turnover in bottles, isotopic equilibration might easily occur within the time of typical incubations; such would rarely be the case when carbon (Marra et al., 1981). Harrison (1978), Caperon et al. (1979), and more recently Glibert et al. (1982) have, in fact, documented rapid NH_4^+-recycling rates in coastal and oceanic waters. Furthermore, Glibert and co-workers were first to determine the magnitude of the isotope dilution errors in uptake estimates (see Section III,C). This was done by comparing uptake rates using Eq. (4), where substrate enrichment (*R*) is assumed constant with estimates where the measured change in enrichment [$\bar{R}$, Eq. (11)] is used. Their results suggest that conventional estimates [Eq. (4)] are low by a factor of 2 or more in coastal and oceanic waters. However, since Glibert's experimental protocol was different from that of most previous ^{15}N uptake studies (Tables III, IV), it is difficult to evaluate the degree of underestimation in earlier work.

I have attempted to assess these errors in a more general context by employing a closed, two-compartment, steady-state model (Riggs, 1970) to simulate the exchange of ^{15}N between the dissolved (NH_4^+) and particulate (PON) pools under conditions where ammonium recycling rates are high and comparable to uptake rates, i.e., nutrient depletion does not occur. Because this is a closed system model, the tracer ^{15}N is allowed to recycle, a condition not permitted in models employed to date but realistic, particularly when compartment turnover times are short relative to incubation time. The model output can be used to evaluate the magnitude of recycling errors for general or specific cases, e.g., where compartment sizes and experimental conditions are chosen to approximate those for marine environments where NH_4^+ uptake rates have been previously measured (Table V).

Using the notation of Glibert et al. (1982), comparisons of the uncorrected uptake rates ρ [calculated from the model output using Eq. (4)] and the "true" uptake rates from the model input *P* [approximately equivalent to Glibert's "corrected rho" from Eq. (11)] are summarized in Fig. 4. In this figure, the ratio P/ρ is shown as a function of substrate turnover time (S/ρ or S/r) for two incubations periods, 4 and 24 h. It is clear from this analysis that uptake rates calculated using conventional assumptions are least reliable (i.e., $P/\rho \gg 1$) when substrate turnover times are short and incubations long; in other words, under conditions when the "non-equilibrium" assumption does not hold (see

TABLE V. Parameter Estimates Used in Steady-State, Two-Compartment Closed Model[a]

Ocean region	*Ammonium-N*[b]		*Particulate-N*		*N-Flux*[c]	Reference
	$(NH_4^+)_u$	$(NH_4^+)_L$	ΣPON	Phyto-N	$(mg \cdot atom \cdot m^{-3} \cdot h^{-1})$	
Oceanic	0.05	0.10	0.20	0.08	0.003	Eppley et al., 1973, 1977a
	0.05	0.10	0.20	0.08	(0.0005)[d]	
Coastal	0.35	0.10	2.00	0.80	0.033	Eppley et al., 1977a, 1979
Estuarine	4.50	0.50	10.00	10.00	0.417	McCarthy et al., 1977

[a]*Compartment values (in units of* $mg \cdot atom \cdot m^{-3}$*) are considered representative conditions based on references cited.*

[b]$(\ )_u$ *= labeled substrate present,* $(\ )_L$ *= labeled substrate added.*

[c]*Equivalent to 1 doubling of phyto-N* day^{-1}.

[d]*Equivalent to 0.2 doublings* day^{-1}.

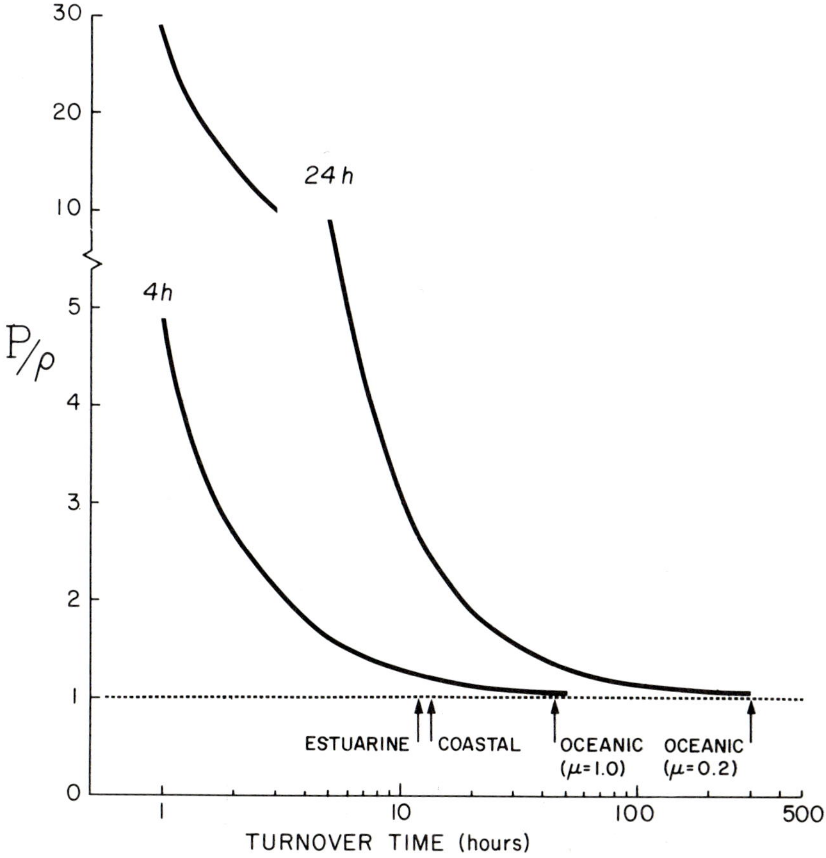

Fig. 4. Effects of incubation time (4 h, 24 h) and substrate turnover time (S/ρ) on the relationship between true (P) and calculated (ρ) NH_4^+ uptake rates, generated from a steady-state, two-compartment, closed model. Numbers in parentheses are phytoplankton growth rates in doublings of cell nitrogen per day. Oceanic, coastal, and estuarine data from Table V.

Section II, D). For the specific examples chosen (Table V, Fig. 4), turnover times were computed based on the assumption that fluxes (*P* and *r*) in each environment were equivalent to one doubling of phytoplankton N per day (rates being constant over 24 h). Interestingly, the model predicts least errors under oceanic conditions

(P/ρ = 1.5 for t = 24 h and μ = 1.0). G. Jackson (personal communication) has come to similar conclusions in his analysis of the methodological problems of ^{15}N measurements. Furthermore, if oceanic growth rates are as low as Sharp et al. (1980) suggest, the error would be negligible, even for 24 h incubations (P/ρ < 1.1). Glibert et al. (1982) have shown from actual measurements in oceanic waters that the errors are, in fact, much greater (P/ρ = 2-3) than this model would predict. However, experimental measurements include errors other than those associated with isotopic dilution and recycling, i.e., ^{15}N "losses" from the particulates not recovered in the NH_4^+ pool. Mass balance calculated from Glibert's et al.'s data (1982, their Table 1) show that about 40% of the ^{15}N initially present in the NH_4^+ and particulate nitrogen fractions was not recovered at the end of their experiments. This suggests measurable losses of label to some unrecovered pool (e.g., dissolved organic nitrogen), which may result from either excretion or possibly from cell leakage during filtration of the particulate matter. Glibert et al. (1982) have also noted discrepancies between NH_4^+ uptake rates calculated from incorporation of ^{15}N into the particulates [Eqs. (4) and (11)] and rates calculated from changes in the substrate concentration and specific activity [Eqs. (9) and (10)]: Rates based on particulate enrichment were consistently lower. Unrecovered ^{15}N may largely account for these differences. Few ^{15}N studies, however, have dealt quantitatively with these tracer losses (see Section III, C).

For the conditions outlined in Table V, the model predicts that uptake rates calculated from 24 h incubations in coastal and estuarine waters could be underestimated as much as a factor of 2-3 from isotopic equilibrium considerations alone. If, however, true turnover times in coastal or in oceanic water are much shorter than used in these examples, i.e., due to rapid uptake phenomena (McCarthy and Goldman, 1979; Glibert and Goldman, 1981) or to much higher actual growth rates (e.g., Sheldon and Sutcliffe, 1978), then the errors will increase significantly (Fig. 4).

The effects of these recycling errors on uptake estimates for other nitrogen substrates (e.g. urea) may also be significant although comparable measurements have not been made. However, because of the relatively slow recycling of oxidized forms, NO_3^-, NO_2^- (Olson, 1980), errors in uptake measurements from isotopic dilution or equilibration would probably be minimal. However, substrate-depletion problems may be relatively more significant for these slowly recycling oxidized nitrogen forms than for reduced forms (NH_4^+), which exhibit high recycling rates.

B. Radioisotopes

The capability of making true tracer-level additions of labeled substrate is clearly an advantage in radioisotope studies of nitrogen transformations. However, a number of limitations in the use of radioisotopes are evident. For example, ^{14}C-labeled urea has been used to follow short-term urea-N incorporation rates in cultured phytoplankton (Horrigan and McCarthy, 1981). In this analysis, uptake rates were based on the nearly quantitative (>98%) decomposition and excretion of the ^{14}C (as $^{14}CO_2$) during assimilation of the nitrogen. In field investigations, however, ^{14}C retention by the particulate matter can be significant and highly variable (e.g., Mitamura and Saijo, 1975, 1980), and an assessment of urea-N uptake based on $^{14}CO_2$ production could be misleading. In a like manner, the utility of other laboratory-tested radioactive tracers under field conditions, such as the nitrogen analogs ([^{14}C]methylamine, $^{36}ClO_3^-$), would require verification. Despite superior analytical technology for radioisotope detection and, as a consequence, fewer violations of the assumptions used in tracer studies, reliable results are still contingent on the ability to measure accurately ambient substrate levels and on the severity of problems associated with population confinement in bottles (McCarthy, 1980).

V. CONCLUSIONS

Isotopic tracers have been used extensively in studies of nitrogen cycling in the marine environment since their introduction in the early 1960s. The stable isotope ^{15}N has been used in most cases. Because of the analytical requirements for relatively large samples and because of relatively low instrument precision (compared to radioisotope detection), a number of the assumptions inherent in the use of ^{15}N as an isotopic tracer have been violated in experimental work done to date. These problems appear most significant in marine waters characterized by low substrate concentration and high turnover rates. Historically, ^{15}N studies have involved additions of labeled substrate at concentrations that are high relative to ambient substrate levels. Corrections have been proposed for these "enrichment" errors using empirically derived, concentration uptake relationships. However, these methods are often unsatisfactory because of undetectable substrate concentrations, nutrient depletion in bottles, or transient, rapid, uptake phenomena. More recent studies have, in fact, suggested that substrate depletion and recycling during incubations may be a more serious problem than substrate enrichment.

Many of the problems associated with ^{15}N tracer experiments appear to be manageable and can be reduced by modifying existing experimental protocols, e.g., by using smaller isotope additions, using shorter incubation times and time-course measurements, and by monitoring isotope changes, when possible, in dissolved and particulate components of the system under investigation. The implementation of such modifications, however, will require considerably more effort than was required in the past. Despite possible improvements in experimental design, there still remains a large area of uncertainty regarding the interpretation of tracer experiments, where bottle containment effects may be substantial and are still unquantifiable (Eppley, 1980; Peterson, 1980; McCarthy, 1980). The large-enclosure concept has been suggested as a possible solution (e.g., Grice and Reeve, 1981) and has been employed in some nitrogen tracer studies (e.g., Harrison and Davies, 1977; Hattori et al., 1980), but has significant practical constraints for general use.

Based on our present knowledge of quantifiable errors, nitrogen fluxes determined by long incubations may be underestimated substantially in environments where substrate concentrations are low and biological activity high. Ammonium, for example, is subject to rapid recycling in bottles (Harrison, 1978; Caperon et al., 1979; Glibert et al., 1982), and may reach isotopic equilibrium within the time span of typical incubations (<24 h).

The impact of these findings in the global context, for example, relative to current estimates of "new" and "regenerated" production (Dugdale and Goering, 1967; Eppley and Peterson, 1979) may be significant. If regenerated production were to be doubled as implied in recent studies (Glibert et al., 1982), revised estimates of total nitrogen productivity could increase by a factor of 1.3 in coastal waters and by almost 2.0 in oligotrophic oceanic waters (where regenerated production may be 90% or more of the total, Eppley and Peterson, 1979). On the other hand, proportionally less new production would be required to maintain observed primary productivity levels than previously believed. Finally, it is rather frustrating to note that such revised new and regenerated production estimates only complicate present attempts to balance high export (sinking) fluxes and low new production in oligotrophic ocean waters (Eppley, 1980).

ACKNOWLEDGMENTS

I wish to thank Drs. G. Jackson, E. Paasche, E. Laws, T. Hollibaugh, and P. Glibert for access to unpublished materials. I benefited especially from discussions with Drs. G. Jackson, W. Li, and W. Silvert on conceptual and methodological problems of tracer studies. Drs. J. Cullen, W. Li, and P. Glibert provided helpful comments in the preparation of this manuscript.

REFERENCES

Alexander, V. (1970). Relationships between turnover rates in the biological nitrogen cycle and algal productivity. *Proc. Ind. Waste Conf. 25,* 1-7.

Atkins, G. L. (1969). "Multicompartment Models for Biological Systems." Methuen, London.

Axler, R. P., and Goldman, C. R. (1981). Isotope tracer methods for investigations of nitrogen deficiency in Castle Lake, California. *Water Res. 15,* 627-632.

Axler, R. P., Redfield, G. W., and Goldman, C. R. (1981). The importance of regenerated nitrogen to phytoplankton productivity in a subalpine lake. *Ecology 62,* 345-354.

Azam, F., and Holm-Hansen, O. (1973). Use of tritiated substrates in the study of heterotrophy in seawater. *Mar. Biol. 23,* 191-196.

Barsdate, R. J., and Dugdale, R. C. (1965). Rapid conversion of organic nitrogen to N_2 for mass spectrometry: An automated Dumas procedure. *Anal. Biochem. 13,* 1-5.

Bates, S. S. (1981). Determination of the physiological state of marine phytoplankton by use of radiosulfate incorporation. *J. Exp. Mar. Biol. Ecol. 51,* 219-239.

Bates, S. S. (1982). Physiological state and rate of protein synthesis, phytoplankton off the coast of Peru as measured by sulfur-35 incorporation. *Bol. Inst. Mar Peru-Callao* (in press).

Benedetti, M. S., Perczel, S., Strolin, P., Quaglia, U., Larue, D., and Assandri, A. (1976). A new method for the determination of ^{15}N and ^{13}C for excretion balance studies of foreign compounds. *Proc. Int. Conf. Stable Isot., 2nd., 1975,* pp. 511-517.

Blackburn, T. H. (1979). Method for measuring rates of NH_4^+ turnover in anoxic marine sediments, using a ^{15}N-NH_4^+ dilution technique. *Appl. Environ. Microbiol. 37,* 760-765.

Bremner, J. M. (1965). Isotope-ratio analysis of nitrogen in nitrogen-15 tracer investigations. *Agronomy 9,* 1257-1286.

Caperon, J., Schell, D., Hirota, J., and Laws, E. (1979). Ammonium excretion rates in Kaneoke Bay, Hawaii, measured by a ^{15}N isotope dilution technique. *Mar. Biol. 54,* 33-40.

Carpenter, E. J., and McCarthy, J. J. (1975). Nitrogen fixation and uptake of combined nitrogenous nutrients by *Oscillatoria (Trichodesmium) thiebautii* in the western Sargasso Sea. *Limnol. Oceanogr. 20,* 389-401.

Carpenter, E. J., and Price, C. C. (1977). Nitrogen fixation, distribution and production of *Oscillatoria (Tridrodesmium)* spp. in the western Sargasso and Caribbean Seas. *Limnol. Oceanogr. 22,* 60-72.

Chan, Y. K., and Campbell, N. E. R. (1978). Phytoplankton uptake and excretion of assimilated nitrate in a small Canadian Shield Lake. *Appl. Environ. Microbiol. 35,* 1052-1060.

Conway, H. L. (1977). Interactions of inorganic nitrogen in the uptake and assimilation by marine phytoplankton. *Mar. Biol. 39,* 221-232.

Conway, H. L., and Whitledge, T. E. (1979). Distribution, fluxes and biological utilization of inorganic nitrogen during a spring bloom in the New York Bight. *J. Mar. Res. 37,* 657-668.

Desaty, D., McGrath, R., and Vining, L. C. (1969). Mass spectrometric measurement of ^{15}N enrichment in nitrogen obtained by Dumas combustion. *Anal. Biochem. 29,* 22-30.

Ditullio, G. R., and Laws, E. A. (1983). Estimates of phytoplankton N uptake based on $^{14}CO_2$ incorporation into protein. *Limnol. Oceanogr. 28,* 177-185.

Dugdale, R. C. (1967). Nutrient limitation in the sea: Dynamics, identification and significance. *Limnol. Oceanogr. 12,* 685-695.

Dugdale, R. C. (1976). Nutrient cycles. *In* "The Ecology of the Seas" (D. H. Cushing and J. J. Walsh, eds.), pp. 141-172. Saunders, Philadelphia, Pennsylvania.

Dugdale, R. C., and Goering, J. J. (1967). Uptake of new and regenerated forms of nitrogen in primary productivity. *Limnol. Oceanogr. 12,* 196-206.

Dugdale, R. C., Menzel, D. W., and Ryther, J. H. (1961). Nitrogen fixation in the Sargasso Sea. *Deep-Sea Res. 7,* 298-300.

Dugdale, R. C., Goering, J. J., and Ryther, J. H. (1964). High Nigrogen fixation rates in the Sargasso Sea and the Arabian Sea. *Limnol. Oceanogr. 9,* 507-510.

Eppley, R. W. (1980). Estimating phytoplankton growth rates in the central oligotrophic oceans. *In* "Primary Productivity in the Sea" (P. G. Falkowski, ed.), pp. 231-242. Plenum, New York.

Eppley, R. W., and Peterson, B. J. (1979). Particulate organic matter flux and planktonic new production in the deep ocean. *Nature (London) 282,* 677-680.

Eppley, R. W., Renger, E. H., Venrich, E. L., and Mullin, M. M. (1973). A study of plankton dynamics and nutrient cycling in the central gyre of the north Pacific Ocean. *Limnol. Oceanogr. 18,* 534-551.

Eppley, R. W., Sharp, J. H., Renger, E. H., Perry, M. J., and Harrison, W. G. (1977a). Nitrogen assimilation by phytoplankton and other microorganisms in the surface waters of the Central North Pacific Ocean. *Mar. Biol. 39,* 111-120.

Eppley, R. W., Harrison, W. G., Chisholm, S. W., and Stewart, E. (1977b). Particulate organic matter in surface waters off Southern California and its relationship to phytoplankton. *J. Mar. Res. 35,* 671-696.

Eppley, R. W., Renger, E. H., and Harrison, W. G. (1979). Nitrate and phytoplankton production in southern California coastal waters. *Limnol. Oceanogr. 24,* 483-494.

Fiedler, R., and Proksch, G. (1975). The determination of nitrogen-15 by emission and mass spectrometry in biochemical analysis: A review. *Anal. Chim. Acta 78,* 1-62.

Fisher, T. R., Carlson, P. R., and Barber, R. T. (1981). Some problems in the interpretations of ammonium uptake kinetics. *Mar. Biol. Lett. 2,* 33-44.

Garside, C. (1981). Nitrate and ammonium uptake in the apex of the New York Bight. *Limnol. Oceanogr. 26,* 731-739.

Gersberg, R. M., Krohn, K., Peek, N., and Goldman, C. R. (1976). Denitrification studies using ^{13}N-labelled nitrate. *Science 192,* 1229-1231.

Gersberg, R. M., Axler, R. P., Krohn, K., Peek, N., and Goldman, C. R. (1978). Nitrate uptake by phytoplankton: Measurements utilizing the radiotracer ^{13}N. *Verh.-Int. Ver. Theor. Angew. Limnol. 20,* 388-392.

Glibert, P. M., and Goldman, J. C. (1981). Rapid ammonium uptake by marine phytoplankton. *Mar. Biol. Lett. 2,* 25-31.

Glibert, P. M., Lipschultz, F., McCarthy, J. J., and Altabet, M. A. (1982). Isotope dilution models of uptake and remineralization of ammonium by marine plankton. *Limnol. Oceanogr. 27,* 639-650.

Goering, J. J. (1968). Denitrification in the oxygen minimum layer of the eastern tropical Pacific Ocean. *Deep-Sea Res. 15,* 157-164.

Goering, J. J., and Dugdale, V. A. (1966). Estimates of the rates of denitrification in a subarctic lake. *Limnol. Oceanogr. 11,* 113-117.

Goering, J. J., and Pamatmat, M. M. (1971). Denitrification in sediments of the sea off Peru. *Invest. Pesq. 35,* 233-242.

Goering, J. J., Dugdale, R. C., and Menzel, D. W. (1964). Cyclic diurnal variations in the uptake of ammonia and nitrate by photosynthetic organisms in the Sargasso Sea. *Limnol. Oceanogr. 9,* 448-451.

Goering, J. J., Dugdale, R. C., and Menzel, D. W. (1966). Estimates of *in situ* rates of nitrogen uptake by *Trichodesmium* in the tropical Atlantic Ocean. *Limnol. Oceanogr. 11,* 614-620.

Goering, J. J., Wallen, D. D., and Nauman, R. M. (1970). Nitrogen uptake by phytoplankton in the discontinuity layer of the Eastern subtropical Pacific Ocean. *Limnol. Oceanogr. 15,* 789-796.

Goldman, J. C., Taylor, C. D., and Glibert, P. M. (1981). Nonlinear time-course uptake of carbon and ammonium by marine phytoplankton. *Mar. Ecol.: Prog. Ser. 6,* 137-148.

Goleb, J. A., and Middelboe, V. (1968). Optical nitrogen-15 analysis of small nitrogen samples with a mixture of helium and xenon to sustain the discharge in an electrodeless tube. *Anal. Chim. Acta 43,* 229-234.

Grice, G. D., and Reeve, M. R. (1981). "Marine Mesocosms." Springer-Verlag, Berlin and New York.

Grunseich, G. S., Dugdale, R. C., Breitner, N. F., and MacIsaac, J. J. (1982). Sample conversion, mass spectrometry, and calculations for ^{15}N analysis of phytoplankton nutrient uptake. Coastal Upwelling Ecosystems Analysis (CUEA). Tech. Rep. No. 44. International Decade of Ocean Exploration.

Harrison, W. G. (1978). Experimental measurements of nitrogen remineralization in coastal waters. *Limnol. Oceanogr. 23,* 684-694.

Harrison, W. G. (1980). Nutrient regeneration and primary production in the sea. *In* "Primary Productivity in the Sea" (P. G. Falkowski, ed.), pp. 433-460. Plenum, New York.

Harrison, W. G., and Davies, J. M. (1977). Nitrogen cycling in a marine planktonic food chain: Nitrogen fluxes through the principal components and the effects of adding copper. *Mar. Biol. 43,* 299-306.

Harrison, W. G., Eppley, R. W., and Renger, E. H. (1977). Phytoplankton nitrogen metabolism, nitrogen budgets and observations on copper toxicity: Controlled ecosystem pollution experiment. *Bull. Mar. Sci. 27,* 44-57.

Harrison, W. G., Platt, T., and Irwin, B. (1982). Primary production and nutrient assimilation by natural phytoplankton populations of the Eastern Canadian Arctic. *Can. J. Fish. Aquat. Sci. 39,* 335-345.

Harvey, H. W. (1969). "The Chemistry and Fertility of Sea Waters." Cambridge Univ. Press, London and New York.

Hattori, A., and Wada, E. (1972). Assimilation of inorganic nitrogen in the euphotic layer of the north Pacific Ocean. *In* "Biological Oceanography of the Northern North Pacific Ocean" (A. Y. Takenouti, ed.), pp. 279-287. Idemitsu Shoten, Tokyo.

Hattori, A., Koike, I., Ohtou, M., Goering, J. J., and Boisseau, D. (1980). Uptake and regeneration of nitrogen in controlled aquatic ecosystems and the effects of copper on these processes. *Bull. Mar. Sci. 30,* 431-443.

Herbland, A. (1976). *In situ* utilization of urea in the euphotic zone of the Tropical Atlantic. *J. Exp. Mar. Biol. Ecol. 21,* 269-277.

Hollibaugh, J. T., and Azam, F. (1980). How do marine bacteria degrade proteins? *Abstr., Am. Soc. Limnol. Oceanogr., Winter Meet.* No. 3.

Hollibaugh, J. T., Carruthers, A. B., Fuhrman, J. A., and Azam, F. (1980). Cycling of organic nitrogen in marine plankton communities studied in enclosed water columns. *Mar. Biol. 59,* 15-21.

Horrigan, S. G., and McCarthy, J. J. (1981). Urea uptake by phytoplankton at various stages of nutrient depletion. *J. Plankton Res. 3,* 403-414.

Johansen, H. S., and Middelboe, V. (1976). Trace analysis for total nitrogen in a water sample by isotopic dilution. *Int. J. Appl. Radiat. Isot. 27,* 591-592.

Keeney, D. R., and Tedesco, M. J. (1973). Sample preparation for and nitrogen isotope analysis by the NOI-4 emission spectrometer. *Anal. Chim. Acta 65,* 19-34.

Kobayashi, Y., and Maudsley, D. V. (1974). "Biological Applications of Liquid Scintillation Counting." Academic Press, New York.

Koike, I., and Hattori, A. (1978a). Denitrification and ammonia formation in anaerobic coastal sediments. *Appl. Environ. Microbiol. 35,* 278-282.

Koike, I., and Hattori, A. (1978b). Simultaneous determinations of nitrification and nitrate reduction in coastal sediments by a ^{15}N dilution technique. *Appl. Environ. Microbiol. 35,* 853-857.

Koike, I., and Hattori, A. (1979). Estimates of denitrification in sediments of the Bering Sea shelf. *Deep-Sea Res. 26,* 409-415.

Kristiansen, S., and Paasche, E. (1982). Preparation of ^{15}N-labelled phytoplankton samples for optical emission spectrometry. *Limnol. Oceanogr. 27,* 373-375.

LaRoche, J. (1981). Ammonium regeneration: Its contribution to phytoplankton nitrogen requirements in an eutrophic environment. M.Sc. Thesis, Dalhousie Univ., Halifax, Nova Scotia.

McCarthy, J. J. (1972). The uptake of urea by natural populations of marine phytoplankton. *Limnol. Oceanogr. 17,* 738-748.

McCarthy, J. J. (1980). Nitrogen. *In* "The Physiological Ecology of Phytoplankton" (I. Morris, ed.), pp. 191-233. Blackwell, Oxford.

McCarthy, J. J. (1982). The kinetics of nutrient utilization. *In* "Physiological Bases of Phytoplankton Ecology" (T. Platt, ed.), Bull. No. 210, pp. 211-233. Canadian Government Publishing Centre, Hull, Quebec, Canada.

McCarthy, J. J., and Goldman, J. C. (1979). Nitrogenous nutrition of marine phytoplankton in nutrient-depleted waters. *Science 203,* 670-672.

McCarthy, J. J., Taylor, W. R., and Taft, J. L. (1977). Nitrogenous nutrition of the plankton in the Chesapeake Bay. 1. Nutrient availability and phytoplankton preferences. *Limnol. Oceanogr. 22,* 996-1011.

MacIsaac, J. J. (1978). Diel cycles of inorganic nitrogen uptake in a natural phytoplankton population dominated by *Gonyaulax polyedra*. *Limnol. Oceanogr. 23,* 1-9.

MacIsaac, J. J., and Dugdale, R. C. (1969). The kinetics of nitrate and ammonia uptake by natural populations of marine phytoplankton. *Deep-Sea Res. 16,* 45-57.

MacIsaac, J. J., and Dugdale, R. C. (1972). Interactions of light and inorganic nitrogen in controlling of nitrogen uptake in the sea. *Deep-Sea Res. 19,* 209-232.

MacIsaac, J. J., Dugdale, R. C., and Slawyk, G. (1974). Nitrogen uptake in the northwest Africa upwelling area: Results from the Cineca-Charcot II cruise. *Tethys 6,* 69-76.

MacIsaac, J. J., Dugdale, R. C., Huntsman, S. W., and Conway, H. L. (1979). The effect of sewage on uptake of inorganic nitrogen and carbon by natural populations of marine phytoplankton. *J. Mar. Res. 37,* 51-66.
McRoy, C. P., Goering, J. J., and Shiels, W. S. (1972). Studies of primary productivity in the eastern Bering Sea. *In* "Biological Oceanography of the Northern North Pacific Ocean" (A. Y. Takenouti, ed.), pp. 199-216. Idemitsu Shoten, Tokyo.
Mague, T. H. (1977). Ecological aspects of dinitrogen fixation by bluegreen algae. *In* "Dinitrogen Fixation" (R. Hardy, ed.), Vol. II, pp. 85-140. Wiley, New York.
Mague, T. H. (1978). Nitrogen fixation. *In* "Handbook of Phycological Methods" (J. A. Hellebust and J. S. Craigie, eds.), pp. 363-378. Cambridge Univ. Press, London and New York.
Mague, T. H., Weare, N. M., and Holm Hansen, O. (1974). Nitrogen fixation in the north Pacific Ocean. *Mar. Biol. 24,* 109-119.
Mague, T. H., Mague, F. C., and Holm-Hansen, O. (1977). Physiology and chemical composition of nitrogen-fixing phytoplankton in the central north Pacific Ocean. *Mar. Biol. 41,* 213-227.
Marra, J., Landrian, G., Jr., and Ducklow, H. W. (1981). Tracer kinetics and plankton rate processes in oligotrophic oceans. *Mar. Biol. Lett. 2,* 215-223.
Mitamura, O., and Saijo, Y. (1975). Decomposition of urea associated with photosynthesis of phytoplankton in coastal waters. *Mar. Biol. 30,* 67-72.
Mitamura, O., and Saijo, Y. (1980). *In situ* measurement of the urea decomposition rate and its turnover rate in the Pacific Ocean. *Mar. Biol. 58,* 147-152.
Miyazaki, T., Wada, E., and Hattori, A. (1973). Capacities of shallow waters of Sagami Bay for oxidation and reduction of inorganic nitrogen. *Deep-Sea Res. 20,* 571-577.
Miyazaki, T., Wada, E., and Hattori, A. (1975). Nitrite production from ammonia and nitrate in the euphotic layer of the western north Pacific. *Mar. Sci. Scommun. 1,* 381-394.
Monheimer, R. H. (1974). Sulfate uptake as a measure of planktonic microbial production in freshwater ecosystems. *Can. J. Microbiol. 20,* 825-831.
Morris, I., Glover, H. E., and Yentsch, C. S. (1974). Products of photosynthesis by marine phytoplankton: The effect of environmental factors on the relative rates of protein synthesis. *Mar. Biol. 27,* 1-9.
Neess, J. C., Dugdale, R. C., Dugdale, V. A., and Goering, J. J. (1962). Nitrogen metabolism in lakes. I. Measurement of nitrogen fixation with ^{15}N. *Limnol. Oceanogr. 7,* 163-169.
Nelson, D. M., and Conway, H. L. (1979). Effects of the light regime on nutrient assimilation by phytoplankton in the Baja California and northwest Africa upwelling systems. *J. Mar. Res. 37,* 301-318.
Olson, R. J. (1980). Studies of biological nitrogen cycle processes in the upper waters of the ocean, with special reference to the primary nitrite maximum. Ph.D. Thesis, University of California, San Diego.

Owen, N. J. P., Christofi, N., and Stewart, W. D. P. (1979). Primary production and nitrogen cycling in an estuarine environment. *In* "Cycling Phenomena in Marine Plants and Animals" (E. Naylor and R. Hartnoll, eds.), pp. 249-258. Pergamon, Oxford.

Paasche, E., and Kristiansen, S. (1982a). Nitrogen nutrition of the phytoplankton in the Oglofjord. *Estuarine, Coastal Shelf Sci. 14,* 237-249.

Paasche, E., and Kristiansen, S. (1982b). Ammonium regeneration by microzooplankton in the Oslofjord. *Mar. Biol. 69,* 55-63.

Pavlov, S. P., Friederich, G. E., and MacIsaac, J. J. (1974). Quantitative determination of total organic nitrogen and isotope enrichment in marine phytoplankton. *Anal. Biochem. 61,* 16-24.

Peterson, B. J. (1980). Aquatic primary productivity and the ^{14}C-CO_2 method: A history of the productivity problem. *Annu. Rev. Ecol. Syst. 11,* 359-385.

Riggs, D. S. (1970). "The Mathematical Approach to Physiological Problems." M.I.T. Press, Cambridge, Massachusetts.

San Pietro, A. (1957). The measurement of stable isotopes. *In* "Methods of Enzymology" (S. P. Colowick and N. O. Kaplan, eds.), Vol. 4, pp. 473-488. Academic Press, New York.

Savidge, G., and Hutley, H. T. (1977). Rates of remineralization and assimilation of urea by fractionated plankton populations in coastal waters. *J. Exp. Mar. Biol. Ecol. 28,* 1-16.

Schell, D. M. (1974). Uptake and regeneration of free amino acids in marine waters of southeast Alaska. *Limnol. Oceanogr. 19,* 260-270.

Schell, D. M. (1978). Chemical and isotopic methods in nitrification studies. *In* "Microbiology--1978" (D. Schlessinger, ed.), pp. 292-295. Am. Soc. Microbiol., Washington, D.C.

Sharp, J. H., Perry, M. J., Renger, E. H., and Eppley, R. W. (1980). Phytoplankton rate processes in the oligotrophic waters of the central north Pacific Ocean. *J. Plankton Res. 2,* 335-353.

Sheldon, R. W., and Sutcliffe, W. H., Jr. (1978). Generation times of 3h for Sargasso Sea microplankton determined by ATP analysis. *Limnol. Oceanogr. 23,* 1051-1055.

Slawyk, G. (1979). ^{13}C and ^{15}N uptake by phytoplankton in the Antarctic upwelling area: Results from the Antripod I cruise in the Indean Ocean sector. *Aust. J. Mar. Freshwater Res. 30,* 431-448.

Slawyk, G., MacIsaac, J. J., and Dugdale, R. C. (1976). Inorganic nitrogen uptake by marine phytoplankton under *in situ* and simulated *in situ* incubation conditions: Results from the northwest African upwelling region. *Limnol. Oceanogr. 21,* 149-152.

Stanley, D. W., and Hobbie, J. E. (1981). Nitrogen recycling in a North Carolina coastal river. *Limnol. Oceanogr. 26,* 30-42.

Stewart, W. D. P. (1971). Nitrogen fixation in the sea. *In* "Fertility of the Sea" (J. D. Costlow, ed.), Vol. 2, pp. 537-564. Gordon & Breach, London.

Tromballa, H. W. (1970). Preparation and determination of ^{36}Cl-labelled chloride, chlorate and perchlorate. *Radiochem. Radioanal. Lett. 4,* 285-292.

Tromballa, H. W., and Broda, E. (1971). Das verhalten von *Chlorella fusca* gegenuber perchlorat und chlorat. *Arch. Mikrobiol. 78,* 214-223.

Vaccaro, R. F. (1965). Inorganic nitrogen in sea water. *In* "Chemical Oceanography" (J. P. Riley and G. Skirrow, eds.), pp. 365-408. Academic Press, New York.

Vincent, W. F. (1979). Uptake of (^{14}C) methyl-ammonium by plankton communities: A comparative assay for ammonium transport systems in natural waters. *Can. J. Microbiol. 25,* 1401-1407.

Wada, E., Tsuji, T., Saino, T., and Hattori, A. (1977). A simple procedure for mass spectrometric microanalysis of ^{15}N in particulate organic matter with special reference to ^{15}N-tracer experiments. *Anal. Biochem. 80,* 312-318.

Wheeler, P. A. (1980). Use of methylammonium as an analogue in nitrogen transport and assimilation studies with *Cyclotella cryptica* (Bacillariophyceae). *J. Phycol. 16,* 328-334.

Chapter 22

ASSAYS OF MICROBIAL NITROGEN TRANSFORMATIONS

BARRIE F. TAYLOR
Division of Marine and Atmospheric Chemistry
Rosenstiel School of Marine and Atmospheric Sciences
University of Miami
Miami, Florida

I. INTRODUCTION

This chapter concerns itself with the measurement of nitrogen fixation, nitrification, denitrification, and dissimilatory nitrate reduction. The methods used to assay these bacterially mediated transformations are not peculiar to marine science, since most were developed with terrigenous organisms or samples. These techniques require the exposure of the microbes to a labeled substrate, substrate analog, or inhibitor. A desirable but often unattainable objective in investigations of marine systems is the measurement of *in situ* rates of the transformations. However, the simultaneous and uniform exposure of bacteria in a natural habitat to an optimal concentration of an added compound is almost impossible without disturbing the microcosm and its interactions. Assays with planktonic microbes may be relatively free from this problem, but if the ma-

ISBN 0-12-160280-X

terial has to be concentrated by filtration or centrifugation, a physical stress or damage may result. Portions of macroalgae, macrophyte leaves, coral rubble, rocks, etc., are assayed easily for their associated activities without significantly disturbing their attached microflora, but sediments and the root systems of plants, such as mangroves and seagrasses, present technical problems. *In situ* assays for the latter systems that minimally disturb the environment, such as chambers or the use of intact cores, usually present diffusional barriers to substrates, inhibitors, and products. Procedures that minimize diffusional problems in which portions of sediment and/or root systems are incubated in a container (usually with agitation) or perfused incur a different set of imponderables. Low initial rates of activity followed by higher rates are often characteristic of these procedures and it is difficult, if not impossible, to determine which rate, if either, reflects the environmental situation. The low initial rates may be due to a shock or shocks received by the microbes during collection and manipulation of the natural samples, whereas the subsequently higher rates may result from growth, derepression of enzymes, and/or the accessibility of substrates normally remote in the original microcosm. Furthermore, compounds in exudates from excised underground tissues may have complicated and conflicting effects on microbial activities in a sample. The evaluation of microbial transformations in the environment, in general and not just in relation to nitrogenous compounds, continues to challenge the microbial ecologist.

The methods developed for measuring the principal bacterial transformations of inorganic nitrogen compounds have utilized isotopically labeled substrates, substrate analogs, inhibitors, and procedures for enumerating the populations of specific bacteria. This chapter discusses and evaluates these assays, with an emphasis on examples involving their use in marine regions or with marine samples.

II. ISOTOPIC ASSAYS

A. ^{15}N-Labeled Substrates

The abundance of ^{15}N, relative to ^{14}N, in natural N compounds is about 0.365 atom %; mass discrimination effects only slightly alter this value in nature (Bergersen, 1980). The commercial availability of nitrogen compounds containing up to 99 atom % ^{15}N consequently has allowed the assay of the main steps in the nitrogen cycle, and examples from marine studies with ^{15}N-labeled compounds include: nitrate dissimilation (Koike and Hattori, 1978a,b; Sørensen, 1978a), nitrification (Koike and Hattori, 1978b; Olson, 1981; Ward et al., 1982), denitrification (Goering and Pamatmat, 1971; Iizumi et al., 1980; Koike and Hattori, 1978a; Nishio et al., 1982), and N_2 fixation (Dugdale et al., 1961; Potts et al., 1978; Stewart, 1971).

Detailed instructions for working with ^{15}N, with an emphasis on N_2 fixation, are available in several excellent reviews (Bergersen, 1980; Burris, 1972, 1974; Burris and Wilson, 1957; Fiedler and Proksch, 1975).

In essence, the methodology involves the determination of the ^{15}N content of the substrate (i.e., the added tracer and the endogenous substrate), and a determination of the total amount of product and its ^{15}N content. The ratio of the atom % excess of ^{15}N, i.e., that in excess of its natural abundance of 0.365 atom %, in the product relative to the substrate, multiplied by the total nitrogen content of the product gives the amount of nitrogen transformed. The analysis of nitrogen compounds for their ^{15}N content requires either a mass spectrometer or an optical emission spectroscope. Thorough accounts of mass spectrometry have been given by Bergersen (1980), Fiedler and Proksch (1975), and San Pietro (1957). The optical emission spectroscope has only come into use in the past decade or so. Its operation depends on the fact that N_2 molecules with different masses, when excited by high-frequency (microwave) radiation, emit light with slightly different wavelength maxima ($^{14}N_2$, 297.7 nm; $^{14}N^{15}N$, 298.3 nm; $^{15}N_2$, 298.9 nm), and the emission intensities are proportional to the concentrations of the different molecules. The accuracy and precision with optical emission spectroscopy is not as good as for mass spectrometry, but much less technical expertise is required in its operation. More information on optical emission spectroscopy is available in several technical articles (Fiedler and Proksch, 1975; Keeney and Tedesco, 1973; Meyer et al., 1974). The samples for analysis, by both mass spectrometry and optical emission spectroscopy, must be gaseous and a variety of methods have been devised for converting nitrogen compounds into N_2.

1. N_2 Fixation

$^{15}N_2$ for assays of N_2 fixation is commercially available but it is cheaper and easier to generate $^{15}N_2$ from $^{15}NH_4^+$ salts by their oxidation with alkaline hypobromite (Rittenberg, 1946):

$$2NH_3 + 3NaOBr \rightarrow N_2 + 3NaBr + 3H_2O$$

The hypobromite oxidation method is ideal for field work because the $^{15}N_2$ can be prepared in syringes, serum bottles, or vacutainers (Burris, 1974; Potts et al., 1978); $^{15}N_2$ can also be obtained by the oxidation of $^{15}NH_3$, liberated under alkaline conditions from $^{15}NH_4^+$ salts in the Dumas reaction with CuO at about 600°C (Burris and Wilson, 1957). It is possible that $^{15}N_2$ may contain traces of the gaseous oxides of nitrogen that can be removed with alkaline permanganate and NH_3, which is easily eliminated by storing the $^{15}N_2$ over dilute acid (Bergersen, 1980; Burris, 1972). Assays of N_2 fixation, as for all assays with ^{15}N-labeled compounds, are usually carried out in small serum bottles sealed with rubber stoppers, or even in Saran plastic bags (Burris, 1974), to minimize the $^{15}N_2$ used. The gas phase is usually Ar, or Ar with subambient O_2 levels, and N_2 is added to give a partial pressure (0.3 atm) that is subsaturating for nitrogenase; this is again to lower the cost but controls should routinely be performed with a PN_2 of 0.8 atm (Burris, 1974). The gas phase is sampled directly for ^{15}N analysis but the fixed nitrogen must be converted to N_2 by Kjeldhal digestion, followed by distillation and acid trapping of the NH_3 before its conversion into N_2 by reaction with alkaline hypobromite in an apparatus originally devised by Rittenberg (1946; Bergersen, 1980; Burris, 1972). Simpler methods for the oxidation of combined nitrogen compounds into N_2 entail the use of an automated Dumas apparatus (Barsdate and Dugdale, 1965; Wada et al., 1977).

2. *Nitrification*

In nitrification studies the samples are incubated aerobically with $^{15}NH_4^+$, and the residual substrate at the end of an assay must be removed before analyzing the NO_3^- + NO_2^- fraction for ^{15}N. This removal is often accomplished by steam distillation, and any inefficiency in the process is serious because residual traces of $^{15}NH_4^+$ will inflate the ^{15}N content of the NO_3^- + NO_2^- fraction when it is converted to N_2 by the Dumas method, or via the Rittenberg procedure after reduction to NH_3 with Devarda's alloy. Schell (1978) devised a novel procedure to obviate this problem by isolating the NO_3^- + NO_2^- fraction, after its reduction to NO_2^- with a Cu-Cd column (Strickland and Parsons, 1972), via diazotization and coupling reactions to form a dye that was extracted in $CC\ell_4$. The dye, after evaporation of the $CC\ell_4$, was converted to N_2 by the Dumas procedure. Another way to overcome the problem is to convert the NO_3^- + NO_2^-, again after reduction of the NO_3^- to NO_2^- with Cu-Cd column, to N_2 in the reaction with sulfamic acid (Koike and Hattori, 1978b):

$$NO_2^- + NH_2SO_3H \rightarrow N_2 + H_2SO_4 + OH^-$$

3. *Dissimilatory Nitrate Reduction*

Incubations for denitrification are carried out anaerobically and usually with minimal gas-phase volumes, and the head space is sampled for the direct analysis of the $^{15}N_2$ produced. The reductive dissimilation of NO_3^- to NH_4^+ also occurs anaerobically and the product can be recovered by steam distillation (Bremner and Keeney, 1965) and converted to N_2 by the Dumas or Rittenberg procedures. The concentrations of NH_4^+, NO_3^-, and NO_2^- as products in assays are determined by routine chemical procedures (Solorzano, 1969; Strickland and Parsons, 1972).

B. Assays Using ^{13}N

The radioisotope ^{13}N with a half-life of 9.96 min has been used to a limited extent in studies of nitrogen metabolism. It is generated by the bombardment of various targets with protons in a cyclotron; $^{13}NO_3^-$ is made with H_2O as the target (Tiedje et al., 1979), and with ^{13}C as the target both $^{13}N_2$ and $^{13}NH_4^+$ have been prepared (Wolk et al., 1974; Thomas et al., 1977). The high specific activity of the products has proved beneficial in studies of the initial phases of N_2 fixation and NH_4^+ assimilation in cyanobacteria (Meeks et al., 1978), and in measuring the initial rates of denitrification and dissimilatory nitrate reduction in soils, sewage sludge, and rumen fluid (Gersberg et al., 1976; Kaspar and Tiedje, 1981; Kaspar et al., 1981; Smith et al., 1978).

III. SUBSTRATE ANALOGS AND INHIBITORS

A. C_2H_2 Reduction Assay

After C_2H_2 was established as an inhibitor and substrate of nitrogenase (Dilworth, 1966; Schöllhorn and Burris, 1966), an assay for N_2 fixation, based on the reduction of C_2H_2 to C_2H_4, was rapidly developed for laboratory and field studies (Hardy et al., 1968; Stewart et al., 1967). The C_2H_2 reduction method is the most widely and frequently used assay of a bacterial activity, and many excellent reviews have been written concerning its technical aspects and its suitability as a measure of N_2 fixation (Burris, 1972, 1974; Knowles, 1980; Postgate, 1972; Turner and Gibson, 1980; van Berkum and Bohlool, 1980). The method has been used extensively to examine N_2 fixation in a variety of marine ecosystems as illustrated by the following list which is representative rather than comprehensive: seagrasses (Capone and Taylor, 1980; Patriquin and Knowles, 1972), macroalgae (Carpenter, 1972; Head and Carpenter, 1975), mangroves (Gotto and Taylor, 1976; Zuberer and Silver, 1978), salt marshes (Smith, 1980; Van Raalte et al., 1974), coral reefs (Capone et al., 1977; Wiebe et al., 1975), lagoons (Potts and Whitton, 1977), planktonic algae (Carpenter and Price, 1977; Mague et al., 1974), and marine sediments (Haines et al., 1981; see Capone, Chapter 4, this volume). This description of a rather wellworn subject is restricted to a brief account of the methodology and an examination of the limitations or uncertainties still associated with the assay.

The attraction of the C_2H_2 reduction technique resides in its speed, technical simplicity, sensitivity, and the cheapness and portability of gas chromatographs relative to mass spectrometers and cyclotrons. Typically, the material to be assayed is exposed to 10-20% C_2H_2 in an aerobic, microaerobic, or anoxic gas (N_2, He, or Ar), the latter being introduced by purging or, more economically, by repeated evacuation and filling with the gas. A procedure that avoids a gas space and that is more realistic for many aquatic situations prescribes the introduction of C_2H_2 to the extent that it will almost completely dissolve when agitated to initiate the assay (Flett et al., 1976). Acetylene is commercially available or can be prepared by adding water to CaC_2 (Postgate, 1972), a procedure especially useful in the field. Acetylene may contain impurities (e.g., PH_3, NH_3, H_2S, acetone) but these can be removed by passing it through concentrated H_2SO_4 and water (Burris, 1974; Turner and Gibson, 1980). The assay vessels are closed with, or have ports closed with, rubber stoppers or seals through which the C_2H_2 is injected and the gas phase sampled. Assays can be terminated with solutions of NaOH or metabolic inhibitors, but preferably not acidic solutions if seawater is present because of CO_2 evolution. Gas samples are analyzed by gas chromatography, typically using metal columns (2-3 m in length by 3-mm diameter) packed with Porapak R or N and operated isothermally (at temperatures of between 40 and 80°C), with high-purity N_2 as the carrier gas at a flow rate of 20-30 mℓ min^{-1}. Acetylene and C_2H_4 are quantified by flame ionization detection, and each assay occupies only 2-4 min and may be accomplished in less than 1 min if ammoniacal $AgNO_3$ is used to remove the C_2H_2 prior to chromatography (David et al., 1980).

Continuing uncertainties about the C_2H_2 reduction assay principally relate to the conversion of C_2H_2 reduction rates into N_2 fixation rates, the influence of C_2H_2 on N_2 fixation and on other microbial activities in a sample or microcosm and the cause and significance of lag phases often encountered in the assays. To convert C_2H_2 reduction rates into N_2 fixation rates, the ratio between the two activities must be determined. With electron transfer to only C_2H_2 or N_2, a molar ratio of 3:1 should apply but, with pure cultures and natural samples, a higher ratio is usually obtained probably because some H_2 evolution accompanies N_2 fixation but not C_2H_2 reduction (Andersen and Shanmugan, 1977; Bothe et al., 1977; Schubert and Evans, 1976). In carefully controlled comparative studies with cyanobacteria, the mean molar ratios fell between 4:1 and 5:1 (Peterson and Burris, 1976; Potts et al., 1978). The higher values sometimes reported may be due to variations in the efficiency of different N_2-fixing microbes to recapture H_2 and, in flooded soils or aquatic sediments, the saturation of nitrogenase by C_2H_2 but not by N_2 (Knowles, 1980; Rice and Paul, 1971).

Acetylene interferes with several bacterial transformations in the nitrogen and carbon cycles. Nitrogen (N_2) fixation is blocked by C_2H_2, which restricts the growth of N_2-fixing bacteria in N-depleted situations (Brouzes and Knowles, 1973) and, more significantly, may fully derepress nitrogenase and result in misleadingly high rates of C_2H_2 reduction (David and Fay, 1977). Acetylene also disturbs microbial interactions in the nitrogen cycle by inhibiting denitrification (Section IIIB), nitrification (Hynes and Knowles, 1978), and NH_4^+ oxidation by *Clostridium* (Brouzes and Knowles, 1971). Methanogenesis is sensitive to low levels of C_2H_2 (Oremland and Taylor, 1975; Raimbault, 1975), and because the oxidation of CH_4 and some alkanes is blocked by C_2H_2 (DeBont and Mulder, 1974, 1976), N_2 fixation supported by these substrates cannot be assessed with the C_2H_2 reduction assay.

The microbial metabolism of C_2H_2, other than via nitrogenase, and C_2H_4 may give misleading results in the C_2H_2 reduction assay. Many species of bacteria and fungi synthesize C_2H_4 either in the presence or absence of O_2 (Primrose, 1979), and it is routine practice with the C_2H_2 reduction assay to include controls lacking C_2H_2 to compensate for this possibility. Ethylene is aerobically oxidized but this activity is blocked by C_2H_2 (DeBont, 1976; Flett et al., 1975), which means that the endogenous generation of C_2H_4 in a system can lead to artificially high rates of C_2H_4 production in the C_2H_2 reduction assay; this situation has been observed in soil and $^{14}C_2H_2$ is required to obtain correct data (Witty, 1979). Acetylene is subject to microbial attack, either aerobically or anaerobically, and this could conceivably influence the results in both the C_2H_2 reduction and C_2H_2 blockage assays, but only after rather extended incubations (Culbertson et al., 1981; DeBont and Peck, 1980; Kanner and Bartha, 1979; Watanabe and deGuzman, 1980). It is worth noting the C_2H_4 is soluble in water and corrections for this factor are necessary, especially when the aqueous phase is large relative to the gas phase (Flett et al., 1976).

Initial low periods of activity that preface higher rates of C_2H_2 reduction continue to plague investigators in assays of natural materials. The dilemma of the lag phase has been discussed *in minutiae* by many authors (Patriquin, 1978; Patriquin and Denike, 1978; Smith, 1980; van Berkum and Bohlool, 1980). Diffusional problems, especially with *in situ* assays using enclosures, and O_2 inhibition of microaerophilic and anaerobic diazotrophs, particularly for sediment and root samples, can account for the lag phase. If information on the environmental rates of activity is desired, then the higher rates of C_2H_2 reduction obtained after a lag or preincubation period must be viewed with suspicion because of the possibility of mechanisms that deplete the sample of combined nitrogen and so derepress nitrogenase (Capone and Carpenter, 1982; Dicker and Smith, 1980a; van Berkum, 1980). Clearly, each individual situation requires careful investigation if the lag phase is to be understood.

B. C_2H_2 Blockage of Denitrification

The accumulation of N_2O during denitrification in the presence of C_2H_2 was first reported by Federova et al. (1973). Acetylene was subsequently found to block the reduction of N_2O to N_2 by pure cultures of denitrifying bacteria (Balderston et al., 1976; Yoshinari and Knowles, 1976), and to inhibit noncompetitively N_2O reductase, with a K_i value of about 28 μM (Kristijansson and Hollocher, 1980). Acetylene inhibits N_2O reduction not only in "classical" denitrifiers but in the strictly anaerobic *Vibrio succinogenes*, which dissimilates NO_3^- to NH_4^+ and also reduces N_2O to N_2 (Yoshinari, 1980). The interference of C_2H_2 with N_2O metabolism has allowed the development of a relatively simple and sensitive assay for denitrification. Although denitrification assays that measure the liberation of N_2 (Kaplan et al., 1977; Payne, 1973; Seitzinger et al., 1980) or the consumption of N_2O (Garcia, 1974,

1975; Sherr and Payne, 1978) are more natural than the C_2H_2 blockage assay (because no inhibitor is added), they suffer from some virtually unavoidable technical disadvantages. The measurement of N_2 evolution rates is complicated by the occlusion of N_2, even when the gas phase is replaced by an inert gas (He, Ar), and the accurate determination of small decreases in the N_2O added for an assay is difficult. Also, as suggested by Knowles (1981), the environmental aqueous concentrations of N_2O may often not saturate N_2O reductase. A K_m value for N_2O of about 5 μM has been reported for the N_2O reductase from *Paracoccus denitrificans* (Krisjansson and Hollocher, 1980), and N_2O levels of <2 μM were detected in the pore waters of a coastal marine sediment in contrast to NO_3^- concentrations of up to 200 μM (Sørensen, 1978b).

In the C_2H_2 blockage procedure, C_2H_2 is added either to the gas or liquid phase (as a solution) and the rate of N_2O accumulation determined under anaerobic conditions. The recent interest in N_2O because of its effect on stratospheric ozone levels (Crutzen, 1981) has stimulated the development of sensitive techniques for quantifying N_2O in atmospheric and aqueous samples. Knowles (1981) in a recent review that emphasizes agronomic aspects presents a useful tabulation of the gas chromatographic methods that have been employed to separate and assay N_2O. Helium ionization detection is highly sensitive for N_2O but, although it has been used in marine studies (Yoshinari, 1976), it is much less commonly used than thermal conductivity (TCD) or electron capture detection (ECD) because of its extreme sensitivity to contaminants. The development of high-temperature electron capture detectors (Wentworth and Freeman, 1973) that take advantage of the high electron absorption coefficient of N_2O at about 300°C, has enabled the direct determination of N_2O in air samples (Rasmussen et al., 1976). Tritium and ^3H-scandium detectors (Cohen, 1977) have been used but ^{63}Ni is most commonly employed. Acetylene reacts with ^{63}Ni, impairing the ECD function, and so must be vented off prior to the detector (Chan and Knowles, 1979). In general, gas samples are analyzed by chromatography on Porapak Q in a stainless-steel column, which allows the optimal resolution of N_2O and CO_2; the addition of another column, either in series or parallel, containing a molecular sieve (e.g., type 5A) also permits the resolution of N_2, O_2, Ar, and H_2 (Kaspar and Tiedje, 1980; Knowles, 1981). The chromatographic separations are usually carried out isothermally, at temperatures from about 25 to 100°C, with a carrier gas flow rate in the region of 30 mℓ min^{-1}. Helium is the carrier gas of choice for TCD, whereas 95% Ar + 5% CH_4 is most often selected for ECD. The addition of 100 ppm of O_2 to the carrier gas for ECD improves the response to CO_2 to a level similar to that for TCD, thereby allowing the simultaneous determination of N_2O and CO_2 (Simmonds, 1978). The C_2H_2 blockage assay permits the simultaneous assay of N_2 fixation (C_2H_2 reduction) and one advantage of TCD over ECD, if its lower sensitivity can be tolerated, is that N_2O and C_2H_4 can be measured in a single injection if a column splitter is used to divert part of the sample for resolution in a separate column with FID detection (Knowles, 1980; Nelson and Knowles, 1978).

When only the gas phase is sampled with a gas-tight syringe or sampling loop (Kaspar and Tiedje, 1980), the dissolved N_2O must be calculated using solubility tables (Markham and Kobe, 1941). Alternatively, corrections for N_2O in the liquid phase can be assessed by injecting a known amount of N_2O into the assay vessels after the microbial activity has been terminated with a metabolic inhibitor, and determining the subsequent increase in N_2O concentration in the gas phase after its equilibration between the phases. If aqueous samples must be analyzed, then the N_2O can be purged with an inert gas as described by Swinnerton and Linnenbom (1967) for the extraction of low-molecular-weight hydrocarbons from seawater, and trapped in a column of molecular sieve type 13X at 0°C (Cohen, 1977; Yoshinari 1976) or in a tube immersed in liquid N_2 (Sørensen, 1978b). The extracted gas is dried (e.g., with Drierite and Ascarite) either before or after trapping and is released from the trap by elevating its temperature and swept by a stream of inert gas into the gas chromatograph. As observed by Knowles (1981), N_2O may be recovered from aqueous samples by simpler methods such as equilibrium extraction with He in the sample container (Chan and Campbell, 1974) or a multiple equilibration extraction method (Chan and Knowles, 1979; McAuliffe, 1971).

Initial comparisons of the C_2H_2 blockage technique versus methods using $^{15}NO_3^-$ or $^{13}NO_3^-$ indicated its validity in short-term assays (minutes to hours) of denitrification in marine sediments (Sørensen, 1978a) and soils (Smith et al., 1978; Tiedje, 1978). However, in long-term incubations (days) C_2H_2, although not metabolized, sometimes became ineffective for preventing N_2O utilization (Van Raalte and Patriquin, 1979; Yeomans and Beauchamp, 1978). This phenomenon is not yet fully understood, but several environmental factors contribute to its development, namely the presence of S^{2-}, low NO_3^- concentrations, and high bacterial populations (Kaspar et al., 1981; Sørensen et al., 1980; Tam and Knowles, 1979). The effect of S^{2-}, somewhat unexpected since S^{2-} inhibits the reduction of N_2O to N_2 (Sørensen et al., 1980; Tam and Knowles, 1979), is especially pertinent to marine situations because seawater contains about 30 m*M* SO_4^{2-}, and sulfate-reducing bacteria are the dominant anaerobes in marine sediments (Jørgensen, 1977). Acetylene is usually employed at a concentration of 10-20% in the gas phase of denitrification assays, but studies with digested sewage sludge showed that as NO_3^- levels decreased progressively, higher levels of C_2H_2 were necessary to observe N_2O accumulation, and at 1 μM NO_3^- no N_2O accumulated, even with a 80% C_2H_2 atmosphere; $^{13}NO_3^-$ assays, however, demonstrated that denitrification occurred at 1 μM NO_3^- (Kaspar et al., 1981). Recent work with marine sediments has also revealed an apparent failure of the C_2H_2 blockage assay at low, but natural, nitrate concentrations (Kaspar, 1982).

Further problems with the C_2H_2 blockage assay involve the production of N_2O during the dissimilative reduction of NO_3^- to NH_4^+, which is probably quantitatively insignificant but deserves further examination, and the inhibition of dissimilative NO_3^- reduction by

C_2H_2 (Kaspar and Tiedje, 1981). Five percent C_2H_2 halved the rate of dissimilative reduction of NO_3^- by rumen microflora, and this suggests that C_2H_2 may divert NO_3^- to the denitrification pathway and thereby give misleadingly high rates for denitrification. Another complication with the use of C_2H_2 is that it inhibits nitrification (Hynes and Knowles, 1978) and so may eliminate the natural supply of substrates for denitrification (Knowles, 1978, 1979).

C. Inhibitors of Nitrification

Two inhibitors have been used to develop assays for nitrification, i.e., chlorate and 2-chloro-6-(trichloromethyl)pyridine (N-Serve, nitrapyrin). N-Serve blocks the oxidation of NH_4^+ to NO_2^- at the initial oxygen fixation step that yields NH_2OH (Campbell and Aleem, 1965), whereas $C\ell O_3^-$ interferes with the conversion of NO_2^- to NO_3^- (Lees and Simpson, 1957). Lees and Simpson (1957) found that the growth of autotrophic NO_2^--oxidizing bacteria was prevented by low concentrations of $C\ell O_3^-$ (about 10 μM) but that levels up 10 m*M* were required to block NO_2^- oxidation. Belser and Mays (1980) recently examined the effect of $C\ell O_3^-$ on NH_4^+ and NO_2^- oxidation by pure cultures, including a marine *Nitrosomonas* strain, and slurries of soil and intertidal sediments. The effect of 10 m*M* chlorate was slight on the growth and activity of autotrophic NH_4^+-oxidizing bacteria, but it did not completely block NO_2^- metabolism by the slurries if the rate of NO_2^- oxidation ($C\ell O_3^-$ absent) was higher than that for NH_4^+. Chlorate is not a specific inhibitor for NO_2^- oxidation since it is also reduced to $C\ell O_2$ (which is toxic) by bacteria that possess dissimilatory NO_3^- reductase (Pichinoty, 1973). In spite of these drawbacks, the $C\ell O_3^-$ assay deserves further investigation, especially with natural samples, because the NO_2^- that accumulates is measured so sensitively and easily (Strickland and Parson, 1972).

Nitrapyrin inhibits chemolithotrophic but not heterotrophic nitrification (Goring, 1962; Shattuck and Alexander, 1963), and is used in agriculture to retard the loss of fertilizer nitrogen via nitrification and its subsequent leaching and denitrification (Gasser, 1970; Huber et al., 1977). Nitrapyrin is an effective inhibitor, for both pure cultures and natural samples, at concentrations of about 2-20 $\mu g/m\ell^{-1}$ (Belser and Schmidt, 1981). Samples are incubated with and without N-Serve and colorimetric methods are used to determine decreases in NH_4^+ utilization rates, preferably in conjunction with changes in NO_3^- + NO_2^- production rates, due to the inhibitors. Increased sensitivity in the assay with much shorter incubation times (hours rather than days) has been attained by measuring the fall in dark rates of $^{14}CO_2$ fixation caused by the addition of nitrapyrin (Billen, 1976, Indrebø et al., 1979; Somville, 1978). The $^{14}CO_2$ nitrapyrin method assumes that a constant carbon fixed-to-nitrogen oxidized (C:N) ratio is applicable. This is a questionable assumption (Hall, 1982; see Kaplan, Chapter 5, this volume) because it was shown some time ago that the C:N ratio varies with the environmental conditions and physiological state of the organism (Carlucci and McNally, 1969). The determination of nitrification rates, except in assays of sediment slurries shaken and enriched with NH_4^+ (Henriksen et al., 1981), usually requires several days incubation to obtain detectable changes. Long incubations with nitrapyrin present a problem because it is slowly hydrolyzed to 6-chloropicolinate, which does not inhibit NH_4^+ oxidation (Bremner et al., 1978; Redemann et al., 1964). Henriksen (1980), in assays of sediment cores lasting several days, had to replace the pore water every 1-2 days with fresh seawater containing nitrapyrin. Nitrapyrin hydrolysis apparently can be detected by smell (Henriksen, 1980) but more quantitative procedures exist. Nitrapyrin can be extracted into acetone:hexane (1:1, by volume) and then assayed by gas chromatography with FID (Briggs, 1975), or aqueous samples can be directly quantified for nitrapyrin and 6-chloropicolinate by high-performance liquid chromatography (HPLC) (Salvas and Taylor, 1980). At the low levels used in nitrification assays, nitrapyrin does not inhibit ammonification, denitrification, or NH_4^+ assimilation by bacteria (Henriksen, 1980) but it does block sulfate reduction (Somville, 1978), methanogenesis (Salvas and Taylor, 1980), and, more importantly, methane oxidation and CO_2 fixation by methanotrophs (Topp and Knowles, 1982). The presence of methanotrophs in samples assayed by the $^{14}CO_2$- nitrapyrin method may thus cause an overestimation of autotrophic nitrifying activity (Topp and Knowles, 1982). Whether or not nitrapyin inhibits nitrification by methanotrophs remains to be determined. The inability of nitrapyrin to prevent heterotrophic nitrification is not generally relevant to marine situations since the process is only considered significant in unusual environments (Focht and Verstraete, 1977).

IV. ENUMERATION OF BACTERIAL POPULATIONS

Plate counts and most probable number (MPN) determinations on selective media have often been carried out in studies of the marine environment. Problems with these methods include the difficulty of providing the correct culture conditions and the lack of information obtained on the physiological condition of the organisms *in situ*, i. e., whether dormant or the degree of activity. Procedures that determine a particular activity in the MPN tubes showing growth, e.g., C_2H_2 reduction or N_2O production in the C_2H_2 blockage assay (Patriquin and McClung, 1978), are preferable to assuming that a particular genus carries out the activity under study (Dicker and Smith, 1980b). The MPN method has often been used in nitrification studies for want of a better method, but its use is not recommended because it requires long incubation periods (up to 3 months) and certainly underestimates populations of nitrifiers (Alexander, 1965; Belser, 1979; Belser and Mays, 1982; Indrebø et al., 1979).

The immunofluorescent assay (IFA) technique is simple in principle but in practice requires technical skill and experience. Pure cultures of the bacterium under study are injected into a small mammal (usually a rabbit) to induce antibody production to the organism. Antibodies in the serum obtained from the rabbit are rendered fluorescent by conjugation with a fluorescent dye, usually fluorescein isothiocyanate. Bacteria that react with the fluorescent antibodies are fluorescent when viewed with the appropriate illumination and filters in a microscope. The technique has been employed extensively in some areas of microbial ecology (Bohlool and Schmidt, 1980; Schmidt, 1978), and procedures have been described for the enumeration of specific bacterial populations in natural habitats (Schmidt, 1974). The method has received increasing attention over the last few years. Webb and Wiebe (1975) used an IFA technique with *Nitrobacter agilis* to determine the populations of NO_2^--oxidizing bacteria in a study at Enewetak Atoll. There was good agreement between nitrification rates assayed by the nitrapyrin method and total activities derived from the IFA population determinations and published activities. More recently, Payne et al. (1981) used the immunofluorescent technique to enumerate NO_2^--oxidizing bacteria in salt-marsh sediments. The FA technique is usually specific for a genus or even species or strain, rather than for a group of organisms that catalyze a particular transformation such as NH_4^+ or NO_2^- oxidation. However, for the NH_4^+-oxidizing autotrophs at least, the specificity of the immunofluorescent method may not be a problem in marine studies. Only two genera of NH_4^+ oxidizers have been isolated from marine samples and *Nitrosomonas* predominates over *Nitrosococcus* (Ward et al., 1982). Antibodies were prepared against both *Nitrosomonas marinus* and *Nitrosococcus oceanus* and their use indicated realistic numbers relative to activity measurements employing $^{15}NH_4^+$ (Ward and Perry, 1980; Ward et al., 1982).

V. CONCLUSIONS

The C_2H_2 reduction and C_2H_2 blockage assays are attractive because of their sensitivity and relative technical simplicity, especially for field work. Their principal disadvantage is their lack of specificity for the transformation under study. Inhibitors that may be more selective, such as gaseous alkynes and alkenes (e.g., methyl acetylene, cyclopropene, allene) (Hardy, 1979), should be investigated as useful alternative indicators of nitrogenase activity. Although various nitriles, 2-propyn-1-ol, and acetylene dicarboxylic acid do not inhibit N_2O reductase (Payne, 1981), it is worth examining other small triple-bonded molecules, e.g., cyanato and thiocyanato compounds, and N-oxides as inhibitors of the enzymes: An inhibitor maybe found that is more effective than C_2H_2 at low NO_3^- concentrations and a nongaseous inhibitor would be especially useful in analyses using ECD.

Substrates labeled with ^{15}N permit the evaluation of all the transformations in the nitrogen cycle, but the assays often require long incubations (days) and the amount of isotope added often significantly raises the natural level of the substrate. The latter problem can be overcome to a degree by measuring rates at several substrate concentrations and extrapolating to obtain a rate at the *in situ* level (Oren and Blackburn, 1979). However, the use of ^{15}N is reaching its limits of resolution for environmental work (Schell, 1978) and a move toward studies with ^{13}N is desirable. This radioisotope provides insights into initial rates of transformation without affecting natural concentrations of nitrogenous compounds, and would be especially helpful in studies of nitrification and understanding the flow of NO_3^- and NO_2^- into the dissimilative and denitrifying pathways. The use of ^{13}N necessitates the collaboration of personnel from a variety of scientific disciplines (as achieved at Michigan State University) and preferably a coastal or island location of a cyclotron. However, studies with laboratory-based model ecosystems are possible. As a final comment on methods in microbial ecology, it is evident that the immunofluorescent technique will find increasing use in determining the interplay between plants and bacteria, with respect to carbon and nitrogen, in the root zones of marsh plants and seagrasses (Patriquin et al., 1981; McClung and Patriquin, 1980; Payne et al., 1981). The procedure might attain greater breadth if antibodies to enzymes (e.g., NH_4^+ oxygenase) rather than organism were employed.

REFERENCES

Alexander, M. (1965). Most probable number method for microbial populations. *In* "Methods in Soil Analysis" (C. A. Black, ed.), pp. 1467-1472. Am. Soc. Agron., Madison, Wisconsin.

Andersen, K., and Shanmugan, K. T. (1977). Energetics of biological nitrogen fixation: Determination of the ratio of formation of H_2 to NH_4^+ catalyzed by nitrogenase of *Klebsiella pneumoniae in vivo*. *J. Gen. Microbiol. 103*, 107-122.

Balderston, W. L., Sherr, B., and Payne, W. J. (1976). Blockage by acetylene of nitrous oxide reduction in *Pseudomonas perfectomarinus*. *Appl. Environ. Microbiol. 31*, 504-508.

Barsdate, R. J., and Dugdale, R. C. (1965). Rapid conversion of organic nitrogen to N_2 for mass spectrometry: An automated Dumas procedure. *Anal. Biochem. 13*, 1-5.

Belser, L. W. (1979). Population ecology of nitrifying bacteria. *Annu. Rev. Microbiol. 33*, 309-333.

Belser, L. W., and Mays, E. L. (1980). Specific inhibition of nitrite oxidation by chlorate and its use in assessing nitrification in soils and sediments. *Appl. Environ. Microbiol. 39*, 505-510.

Belser, L. W., and Mays, E. L. (1982). Use of nitrifier activity measurements to estimate the efficiency of viable nitrifier counts in soils and sediments. *Appl. Environ. Microbiol. 43*, 945-948.

Belser, L. W., and Schmidt, E. L. (1981). Inhibitory effect of nitrapyrin on three genera of ammonia-oxidizing nitrifiers. *Appl. Environ. Microbiol. 41*, 819-821.

Bergersen, F. J. (1980). Measurement of nitrogen fixation by direct means. *In* "Methods for Evaluating Biological Nitrogen Fixation" (F. J. Bergersen, ed.), pp. 65-110. Wiley, New York.

Billen, G. (1976). Evaluations of nitrifying activity in sediments by dark ^{14}C-bicarbonate incorporation. *Water Res. 10*, 51-57.

Bohlool, B. B., and Schmidt, E. L. (1980). The immuno fluorescence approach in microbial ecology. *Adv. Microb. Ecol. 4*, 203-241.

Bothe, H., Tennigkeit, J., Eisbrenner, G., and Yates, M. G. (1977). The hydrogenase-nitrogenase relationship in the blue-green alga *Anabaena cylindrica*. *Planta 133*, 237-242.

Bremner, J. W., and Keeney, D. R. (1965). Steam distillation methods for determination of ammonium, nitrate and nitrite. *Anal. Chim. Acta 32*, 485-495.

Bremner, J. W., Blackmer, A. M., and Bundy, L. G. (1978). Problems in the use of nitrapyrin (N-Serve) to inhibit nitrification in soils. *Soil Biol. Biochem. 10,* 441-442.

Briggs, G. G. (1975). The behavior of the nitrification inhibitor N-Serve in broadcast and incorporated applications to soil. *J. Sci. Food Agric. 26,* 1083-1092.

Brouzes, R., and Knowles, R. (1971). Inhibition of growth of *Clostridium pasteurianium* by acetylene: Implication for nitrogen-fixation assay. *Can. J. Microbiol. 17,* 1483-1489.

Brouzes, R., and Knowles, R. (1973). Kinetics of nitrogen fixation in a glucose-amended, anaerobically incubated soil. *Soil Biol. Biochem. 5,* 223-229.

Burris, R. H. (1972). Nitrogen fixation-assay methods and techniques. *In* "Methods in Enzymology" (A. San Pietro, ed.), Vol. 24, Part B, pp. 415-431. Academic Press, New York.

Burris, R. H. (1974). Methodology. *In* "The Biology of Nitrogen Fixation" (A. Quispel, ed.), pp. 9-33. North-Holland Publ., Amsterdam.

Burris, R. H., and Wilson, P. W. (1957). Methods for measurement of nitrogen fixation. *In* "Methods in Enzymology" (S. P. Colowick and N. O. Kaplan, eds.), Vol. 4, pp. 355-366. Academic Press, New York.

Campbell, N. E. R., and Aleem, M. I. H. (1965). The effect of 2-chloro-6-(trichloromethyl)pyridine and the chemoautotrophic metabolism of nitrifying bacteria. *Antonie van Leeuwenhoek 31,* 124-136.

Capone, D. G., and Carpenter, E. J. (1982). Perfusion method for assaying microbial activities in sediments: Applicability to studies of N_2 fixation by C_2H_2 reduction. *Appl. Environ. Microbiol. 43,* 1400-1405.

Capone, D. G., and Taylor, B. F. (1980). Microbial nitrogen cycling in a seagrass community. *In* "Estuarine Perspectives" (V. S. Kennedy, ed.), pp. 153-161. Academic Press, New York.

Capone, D. G., Taylor, D. L., and Taylor, B. F. (1977). $N_2(C_2H_2)$ fixation associated with macroalgae in a coral reef community in the Bahamas. *Mar. Biol. 41,* 29-32.

Carlucci, A. F., and McNally, P. M. (1969). Nitrification by marine bacteria in low concentrations of substrate and oxygen. *Limnol. Oceanogr. 14,* 736-739.

Carpenter, E. J. (1972). Nitrogen fixation by a blue-green epiphyte on pelagic *Sargassum*. *Science 178,* 1207-1209.

Carpenter, E. J., and Price, C. C. (1977). Nitrogen fixation, distribution, and production of *Oscillatoria (Trichodesmium)* spp. in the western Sargasso and Caribbean Seas. *Limnol. Oceanogr. 22,* 60-72.

Chan, Y.-K., and Campbell, N. E. R. (1974). A rapid gas-extraction technique for the quantitative study of denitrification in aquatic systems by N-isotope ratio analysis. *Can. J. Microbiol. 20,* 275-281.

Chan, Y.-K., and Knowles, R. (1979). Measurement of denitrification in two freshwater sediments by an *in situ* acetylene inhibition method. *Appl. Environ. Microbiol. 37,* 1067-1072.

Cohen, Y. (1977). Shipboard measurement of dissolved nitrous oxide in seawater by electron capture gas chromatography. *Anal. Chem. 49,* 1238-1240.

Crutzen, P. J. (1981). Atmospheric chemical processes of the oxides of nitrogen, including nitrous oxide. *In* "Denitrification, Nitrification and Atmospheric Nitrous Oxide" (C. C. Delwiche, ed.), pp. 17-44. Wiley, New York.

Culbertson, C. W., Zehnder, A. J. B., and Oremland, R. S. (1981). Anaerobic oxidation of acetylene by estuarine sediments and enrichment cultures. *Appl. Environ. Microbiol. 41,* 396-403.

David, K. A. V., and Fay, P. (1977). Effects of longterm treatment with acetylene on nitrogen-fixing microorganisms. *Appl. Environ. Microbiol. 34,* 640-646.

David, K. A. V., Apte, S. K., Banerji, A., and Thomas, J. (1980). Acetylene reduction assay for nitrogenase activity: Gas chromatographic determination of ethylene per sample in less than one minute. *Appl. Environ. Microbiol. 39,* 1079-1080.

DeBont, J. A. M. (1976). Bacterial degradation of ethylene and the acetylene reduction test. *Can. J. Microbiol. 22,* 1060-1062.

DeBont, J. A. M., and Mulder, E. G. (1974). Nitrogen fixation and co-oxidation of ethylene by a methane-utilizing bacterium. *J. Gen. Microbiol. 83,* 113-121.

DeBont, J. A. M., and Mulder, E. G. (1976). Invalidity of the acetylene reduction assay in alkane-utilizing, nitrogen-fixing bacteria. *Appl. Environ. Microbiol. 31,* 640-647.

DeBont, J. A. M., and Peck, M. (1980). Metabolism of acetylene by *Rhodococcus* A1. *Arch. Microbiol. 127,* 99-104.

Dicker, H. J., and Smith, D. W. (1980a). Physiological ecology of acetylene reduction (nitrogen fixation) in a Delaware salt marsh. *Microb. Ecol. 6,* 161-171.

Dicker, H. J., and Smith, D. W. (1980b). Enumeration and relative importance of acetylene-reducing (nitrogen-fixing) bacteria in a Delaware salt marsh. *Appl. Environ. Microbiol. 39,* 1019-1025.

Dilworth, M. J. (1966). Acetylene reduction by nitrogen-fixing preparations from *Clostridium pasteurianum*. *Biochim. Biophys. Acta 127,* 285-294.

Dugdale, R. C., Menzel, D. W., and Ryther, J. H. (1961). Nitrogen fixation in the Sargasso Sea. *Deep-Sea Res. 7,* 293-300.

Federova, R. I., Milekhina, E. I., and L'yukhina, I. (1973). Evaluation of the method of "gas metabolism" for detecting extraterrestrial life. Identification of nitrogen-fixing microorganisms. *Izv. Akad. Nauk. SSSR, Ser. Biol. 6,* 797-806.

Fiedler, R., and Proksch, G. (1975). The determination of nitrogen-15 by emission and mass spectrometry in biochemical analysis: A review. *Anal. Chim. Acta 78,* 1-62.

Flett, R. J., Rudd, J. W. M., and Hamilton, R. D. (1975). Acetylene reduction assays for nitrogen fixation in freshwater: A note of caution. *Appl. Environ. Microbiol. 29,* 580-583.

Flett, R. J., Hamilton, R. D., and Campbell, N. E. R. (1976). Aquatic acetylene-reduction technique: Solutions to general problems. *Can. J. Microbiol. 22,* 43-51.

Focht, D. C., and Verstraete, W. (1977). Biochemical ecology of nitrification and denitrification. *Adv. Microb. Ecol. 1,* 135-214.

Garcia, J.-L. (1974). Reduction de l'oxyde nitreux dans le sols de rizieres du Senegal: Mesure de l'activite denitrifiante. *Soil Biol. Biochem. 6,* 79-84.

Garcia, J.-L. (1975). Evaluation de la denitrification dans les rizieres par methode de reduction de N_2O. *Soil Biol. Biochem. 7,* 251-256.

Gasser, J. K. R. (1970). Nitrification inhibitors--Their occurrence, production, and effects of their use on crop yields and composition. *Soils Fert. 33,* 547-554.

Gersberg, R., Krohn, K., Peek, N., and Goldman, E. R. (1976). Denitrification studies with ^{13}N-labeled nitrate. *Science 192,* 1229-1231.

Goering, J. J., and Pamatmat, M. M. (1971). Denitrification in sediments of the sea off Peru. *Invest. Pesq. 35,* 233-242.

Goring, C. A. I. (1962). Control of nitrification by 2-chloro-6-(trichloromethyl) pyridine. *Soil Sci. 93,* 211-218.

Gotto, J. W., and Taylor, B. F. (1976). N_2-fixation associated with decaying leaves of the red mangrove *(Rhizophora mangle). Appl. Environ. Microbiol. 31,* 781-783.

Haines, J. R., Atlas, R. M., Griffiths, R. P., and Morita, R. Y. (1981). Denitrification and nitrogen fixation in Alaskan Continental Shelf sediments. *Appl. Environ. Microbiol. 41,* 412-421.

Hall, G. H. (1982). Apparent and measured rates of nitrification in the hypolimnion of a mesotrophic lake. *Appl. Environ. Microbiol. 43,* 542-547.

Hardy, R. W. F. (1979). Reducible substrates of nitrogenase. *In* "A Treatise on Dinitrogen Fixation" (R. W. F. Hardy, ed.), pp. 515-568. Wiley, New York.

Hardy, R. W. F., Holsten, R. D., Jackson, E. K., and Burns, R. C. (1968). The acetylene-ethylene assay for N_2 fixation: Laboratory and field evaluation. *Plant Physiol. 43,* 1185-1207.

Head, W. D., and Carpenter, E. J. (1975). Nitrogen fixation associated with the marine macroalga *Codium fragile. Limnol. Oceanogr. 20,* 815-823.

Henriksen, K. (1980). Measurement of *in situ* rates of nitrification in sediment. *Microb. Ecol. 6,* 329-337.

Henriksen, K., Hansen, J. I., and Blackburn, T. H. (1981). Rates of nitrification, distribution of nitrifying bacteria, and nitrate fluxes in different types of sediment from Danish waters. *Mar. Biol. 61,* 299-304.

Huber, D. M., Watten, H. L., Nelson, D. W., and Tsai, C. Y. (1977). Nitrification inhibitors--New tools for food production. *Bio-Science 27,* 523-529.

Hynes, R. K., and Knowles, R. (1978). Inhibition by acetylene of ammonia oxidation in *Nitrosomonas europaea. FEMS Microbiol. Lett. 4,* 319-321.

Iizumi, H., Hattori, A., and McRoy, C. P. (1980). Nitrate and nitrite in interstitial waters of Eelgrass bads in relation to the rhizosphere. *J. Exp. Mar. Biol. Ecol. 47,* 191-201.

Indrebø, G., Pengerud, B., and Dundas, I. (1979). Microbial activities in a permanently stratified estuary. II. Microbial activities at the oxic-anoxic interface. *Mar. Biol. 51,* 305-309.

Jørgensen, B. B. (1977). The sulfur cycle of a coastal marine sediment (Limfjorden, Denmark). *Limnol. Oceanogr. 22,* 814-832.

Kanner, D., and Bartha, R. (1979). Growth of *Nocardia rhodochrous* on acetylene gas. *J. Bacteriol. 139,* 225-230.

Kaplan, W. A., Teal, J. M., and Valiela, I. (1977). Denitrification in salt marsh sediments: Evidence for seasonal temperature selection among populations of denitrifiers. *Microb. Ecol. 3,* 193-204.

Kasper, H. F. (1982). Denitrification in marine sediment: Measurement of capacity and estimate of *in situ* rate. *Appl. Environ. Microbiol. 43,* 522-527.

Kaspar, H. F., and Tiedje, J. M. (1980). Response of the electron-capture detector to hydrogen, oxygen, nitrogen, carbon dioxide, nitric oxide and nitrous oxide. *J. Chromatogr. 193,* 142-147.

Kaspar, H. F., and Tiedje, J. M. (1981). Dissimilatory reduction of nitrate and nitrite in the bovine rumen: Nitrous oxide production and effect of acetylene. *Appl. Environ. Microbiol. 41,* 705-709.

Kaspar, H. F., Tiedje, J. M., and Firestone, R. B. (1981). Denitrification and dissimilatory nitrate reduction to ammonium in digested sludge. *Can. J. Microbiol. 27,* 878-885.

Keeney, D. R., and Tedesco, M. J. (1973). Sample preparation for and nitrogen isotope analysis by the NOI-4 emission spectrocope. *Anal. Chim. Acta 65,* 19-34.

Knowles, R. (1978). Common intermediates of nitrification and denitrification, and the metabolism of nitrous oxide. *In* "Microbiology--1978" (D. Schlessinger, ed.), pp. 367-371. *Am. Soc. Microbiol.,* Washington, D. C.

Knowles, R. (1979). Denitrification, acetylene reduction, and methane metabolism in lake sediments exposed to acetylene. *Appl. Environ. Microbiol. 38,* 486-493.

Knowles, R. (1980). Nitrogen fixation in natural plant communities and soils. *In* "Methods for Evaluating Biological Nitrogen Fixation" (F. J. Bergersen, ed.), pp. 557-582. Wiley, New York.

Knowles, R. (1981). Denitrification. *Soil Biochem. 5,* 323-369.

Koike, I., and Hattori, A. (1978a). Denitrification and ammonia formation in anaerobic coastal sediments. *Appl. Environ. Microbiol. 35,* 278-282.

Koike, I., and Hattori, A. (1978b). Simultaneous determinations of nitrification and nitrate reduction in coastal sediments by a ^{15}N dilution technique. *Appl. Environ. Microbiol. 35,* 853-857.

Kristjansson, J. K., and Hollocher, T. C. (1980). First practical assay for soluble nitrous oxide reductase of denitrifying bacteria and a partial kinetic characterization. *J. Biol. Chem. 255,* 704-707.

Lees, H., and Simpson, J. R. (1957). The biochemistry of the nitrifying organisms. 5. Nitrite oxidation by *Nitrobacter*. *Biochem. J. 65,* 297-305.

McAuliffe, C. (1971). GC determination of solute by multiple phase equilibration. *Chem. Technol. 1,* 46-51.

McClung, C. R., and Patriquin, D. G. (1980). Isolation of a nitrogen-fixing *Campylobacter* species from the roots of *Spartina alterniflora*. *Can. J. Microbiol. 26,* 881-886.

Mague, T. H., Weare, N. M., and Holm-Hansen, O. (1974). Nitrogen fixation in the North Pacific Ocean. *Mar. Biol. 4,* 109-119.

Markham, A. E., and Kobe, K. A. (1941). The solubility of carbon dioxide and nitrous oxide in aqueous salt solutions. *J. Am. Chem. Soc. 63,* 449-454.

Meeks, J. C., Wolk, C. P., Lockau, W., Schilling, N., Shafer, P. W., and Chien, W.-S. (1978). Pathways of assimilation of [^{13}N] N_2 and $^{13}NH_4{}^+$ by cyanobacteria with and without heterocysts. J. Bacteriol. 134, 125-130.

Meyer, G. W., McCaslin, B. D., and Gast, R. G. (1974). Sample preparation and 15-N analysis using a Statron NOI-5 optical analyzer. *Soil Sci. 117,* 378-385.

Nelson, L. M., and Knowles, R. (1978). Effect of oxygen and nitrate on nitrogen fixation and denitrification by *Azospirillum brasilense* grown in continuous culture. *Can. J. Microbiol. 24,* 1395-1403.

Nishio, T., Koike, I., and Hattori, A. (1982). Denitrification, nitrate reduction, and oxygen consumption in coastal and estuarine sediments. *Appl. Environ. Microbiol. 43,* 648-653.

Olson, R. J. (1981). ^{15}N tracer studies of the primary nitrite maximum. *J. Mar. Res. 39,* 203-226.

Oremland, R. S., and Taylor, B. F. (1975). Inhibition of methanogenesis in marine sediments by acetylene and ethylene: Validity of the acetylene reduction assay for anaerobic microcosms. *Appl. Microbiol. 30,* 707-709.

Oren, A., and Blackburn, T. H. (1979). Estimation of sediment denitrification rates at *in situ* nitrate concentrations. *Appl. Environ. Microbiol. 37,* 174-176.

Patriquin, D. G. (1978). Factors affecting nitrogenase activity (acetylene reducing activity) associated with excised roots of the emergent halophyte *Spartina alterniflora* Loisel. *Aquat. Bot. 4,* 193-210.

Patriquin, D. G., and Denike, D. (1978). *In situ* acetylene reduction assays of nitrogenase activity associated with the emergent halophyte *Spartina alterniflora* Loisel: Methodological problems. *Aquat. Bot. 4,* 211-226.

Patriquin, D. G. and Knowles, R. (1972). Nitrogen fixation in the rhizosphere of marine angiosperms. *Mar. Biol. 16,* 49-58.

Patriquin, D. G., and McClung, C. R. (1978). Nitrogen accretion, and the nature and possible significance of N_2 fixation (acetylene reduction) in a Nova Scotian *Spartina alterniflora* stand. *Mar. Biol. 47,* 227-242.

Patriquin, D. G., Boule, C. D., Livingstone, D. C., and McClung, C. R. (1981). Physiology of the diazotrophic rhizocoenosis in salt marsh cord grass, *Spartina alterniflora* Loisel. *In* "Associative N_2-Fixation" (P. B. Vose and A. P. Ruschel, eds.), Vol. 2, pp. 11-26. CRC Press, Boca Raton, Florida.

Payne, W. J. (1973). The use of gas chromatography for studies of denitrification in ecosystems. *Bull. Ecol. Res. Commun. 17,* 263-268.

Payne, W. J. (1981). The status of nitric oxide and nitrous oxide as intermediates in denitrification. *In* "Denitrification, Nitrification and Atmospheric Nitrous Oxide" (C. C. Delwiche, ed.), pp. 85-103. Wiley, New York.

Payne, W. J., Sherr, B. F., and Chalmers, A. (1981). Nitrification-denitrification associated with plant roots. *In* "Associative N_2-Fixation" (P. B. Vose and A. P. Ruschel, eds.), Vol. 1, pp. 37-48. CRC Press, Boca Raton, Florida.

Peterson, R. B., and Burris, R. H. (1976). Conversion of acetylene reduction rates to nitrogen fixation rates in natural populations of blue-green algae. *Anal. Biochem. 73,* 404-410.

Pichinoty, F. (1973). La reduction bacterienne des composes oxygenes mineraux de l'azote. *Bull. Inst. Pasteur (Paris) 71,* 317-395.

Postgate, J. R. (1972). The acetylene reduction test for nitrogen fixation. *In* "Methods in Microbiology" (J. R. Norris and D. W. Ribbons, eds.), Vol. 6B, pp. 343-356. Academic Press, New York.

Potts, M., and Whitton, B. A. (1977). Nitrogen fixation by blue-green algal communities in the intertidal zone of the lagoon of Aldabra Atoll. *Oecologia 27,* 275-283.

Potts, M., Krumbein, W. E., and Metzger, J. (1978). Nitrogen fixation rates in anaerobic sediments determined by acetylene reduction, a new ^{15}N field assay, and simultaneous total N ^{15}N determination. *In* "Environmental Biogeochemistry and Geomicrobiology" (W. E. Krumbein, ed.), Vol. 3, pp. 753-769. Ann Arbor Sci. Publ., Ann Arbor, Michigan.

Primrose, S. B. (1979). Ethylene and agriculture: The role of the microbe. *J. Appl. Bacteriol. 46,* 1-25.

Raimbault, M. (1975). Etude de l'influence inhibitrice de l'acetylene sur la formation biologique du methane dans un sol riziere. *Ann. Inst. Pasteur, Paris 126A,* 247-258.

Rasmussen, R. A., Krasnec, J., and Pierotii, D. (1976). N_2O analysis in the atmosphere via electron capture-gas chromatography. *Geophys. Res. Lett. 3,* 615-618.

Redemann, C. T., Meikle, R. W., and Widofsky, J. G. (1964). Nutrient conserving agents. Loss of 2-chloro-6-(trichloromethyl)pyridine from soil. *J. Agric. Food Chem. 12,* 207-218.

Rice, W. A., and Paul, E. A. (1971). The acetylene reduction assay for measuring nitrogen fixation in waterlogged soil. *Can. J. Microbiol. 17,* 1049-1056.

Rittenberg, D. (1946). The preparation of gas samples for mass spectrographic analysis. *In* "Preparation and Measurement of Isotopic Tracers" (D. W. Wilson, ed.), pp. 31-42. Edwards Brothers, Ann Arbor, Michigan.

Salvas, P. L., and Taylor, B. F. (1980). Blockage of methanogenesis in marine sediments by the nitrification inhibitor 2-chloro-6-(trichloromethyl) pyridine (Nitrapyrin or N-Serve). *Curr. Microbiol. 4,* 305-308.

San Pietro, A. (1957). The measurement of stable isotopes. *In* "Methods in Enzymology" (S. P. Colowick and N. O. Kaplan, eds.), Vol. 4, pp. 473-488. Academic Press, New York.

Schell, D. M. (1978). Chemical and isotopic methods in nitrification studies. *In* "Microbiology--1978" (D. Schlessinger, ed.), pp. 292-295. Am. Soc. Microbiol., Washington, D. C.

Schmidt, E. L. (1973). Fluorescent antibody technique for the study of microbial ecology. *Bull. Ecol. Res. Commun. 17,* 67-76.

Schmidt, E. L. (1974). Quantitative autoecological study of microorganisms in soil by immunofluorescence. *Soil Sci. 118,* 141-149.

Schmidt, E. L. (1978). Nitrifying microorganisms and their methodology. *In* "Microbiology--1978" (D. Schlessinger, ed.), pp. 288-291. Am. Soc. Microbiol., Washington, D. C.

Schöllhorn, R., and Burris, R. H. (1966). Study of intermediates in nitrogen fixation. *Fed. Proc., Fed. Am. Soc. Exp. Biol. 25,* 710.

Schubert, K. R., and Evans, J. H. (1976). Hydrogen evolution: A major factor affecting the efficiency of nitrogen fixation in nodulated symbionts. *Proc. Natl. Acad. Sci. U. S. A. 73,* 1207-1211.

Seitzinger, S., Nixon, S., Pilson, M. E. Q., and Burke, S. (1980). Denitrification and N_2O production in near-shore marine sediments. *Geochim. Cosmochim. Acta 44,* 1853-1860.

Shattuck, G. E., and Alexander, M. (1963). A differential inhibitor of nitrifying microorganisms. *Soil Sci. Soc. Am. Proc. 27,* 600-601.

Sherr, B. F., and Payne, W. J. (1978). Effect of the *Spartina alterniflora* rootrhizome system on salt marsh soil denitrifying bacteria. *Appl. Environ. Microbiol. 35,* 724-729.

Simmonds, P. G. (1978). Direct determination of ambient carbon dioxide and nitrous oxide with a high-temperature ^{63}Ni electron-capture detector. *J. Chromatogr. 166,* 593-598.

Smith, D. W. (1980). An evaluation of marsh nitrogen fixation. *In* "Estuarine Perspectives" (V. S. Kennedy, ed.), pp. 135-142. Academic Press, New York.

Smith, M. S., Firestone, M. K., and Tiedje, J. M. (1978). The acetylene inhibition method for short-term measurement of soil denitrification and its evaluation using nitrogen-13. *Soil Sci. Soc. Am. J. 42,* 611-615.

Solorzano, L. (1969). Determination of ammonia in natural waters by the phenylhypochlorite method. *Limnol. Oceanogr. 14,* 799-801.

Somville, M. (1978). A method for the measurement of nitrification rates in water. *Water Res. 12,* 843-848.

Sørensen, J. (1978a). Capacity for denitrification and reduction of nitrate to ammonia in a coastal marine sediment. *Appl. Environ. Microbiol. 35,* 301-305.

Sørensen, J. (1978b). Denitrification rates in a marine sediment as measured by the acetylene inhibition technique. *Appl. Environ. Microbiol. 36,* 139-143.

Sørensen, J., Tiedje, J. M., and Firestone, M. (1980). Inhibition by sulfide of nitric and nitrous oxide reduction by denitrifying *Pseudomonas fluorescens. Appl. Environ. Microbiol. 39,* 105-108.

Stewart, W. D. P. (1971). Nitrogen fixation in the sea. *In* "Fertility of the Sea" (J. D. Costlow, ed.), Vol. 2, pp. 537-564. Gordon & Breach, New York.

Stewart, W. D. P., Fitzgerald, G. P., and Burris, R. H. (1967). *In situ* studies on N_2 fixation, using the acetylene reduction technique. *Proc. Natl. Acad. Sci. U. S. A. 58,* 2071-2078.

Strickland, J. D. H., and Parsons, T. R. (1972). A practical handbook of seawater analysis. *Bull. Fish. Res. Board Can. 167.*

Swinnerton, J. W., and Linnenbom, V. J. (1967). Determination of C_1-C_4 hydrocarbons in sea water by gas chromatography. *J. Chromatogr. 5,* 570-573.

Tam, T.-Y., and Knowles, R. (1979). Effects of sulfide and acetylene on nitrous oxide reduction by soil and by *Pseudomonas aeruginosa. Can. J. Microbiol. 25,* 1133-1138.

Thomas, J., Meeks, J. C., Wolk, C. P., Shaffer, P. W., Austin, S. M., and Chien, W.-S. (1977). Formation of glutamine from [^{13}N]ammonia, [^{13}N]dinitrogen and [^{14}C]glutamate by heterocysts isolated from *Anabaena cylindrica. J. Bacteriol. 129,* 1545-1555.

Tiedje, J. M. (1978). Denitrification in soil. *In* "Microbiology--1978" (D. Schlessinger, ed.), pp. 362-366. Am. Soc. Microbiol., Washington, D. C.

Tiedje, J. M., Firestone, R. B., Firestone, M. K., Betlach, M. R., Smith, M. S., and Caskey, W. H. (1979). Methods for the production and use of nitrogen-13 in studies of denitrification. *Soil Sci. Soc. Am. J. 43,* 709-716.

Topp, E., and Knowles, R., (1982). Nitrapyrin inhibits the obligate methylotrophs *Methylosinus trichosporium* and *Methylococcus capsulatus. FEMS Microbiol. Lett. 14,* 47-49.

Turner, G. L., and Gibson, A. H. (1980). Measurement of nitrogen fixation by indirect means. *In* "Methods for Evaluating Biological Nitrogen Fixation" (F. J. Bergersen, ed.), pp. 111-138. Wiley, New York.

van Berkum, P. (1980). Evaluation of acetylene reduction by excised roots for the determination of nitrogen fixation in grasses. *Soil Biol. Biochem. 12,* 141-145.

van Berkum, P., and Bohlool, B. B. (1980). Evaluation of nitrogen fixation by bacteria in association with roots of tropical grasses. *Microbiol. Rev. 44,* 491-517.

Van Raalte, C. D., and Patriquin, D. G. (1979). Use of the "Acetylene Blockage" technique for assaying denitrification in a salt marsh. *Mar. Biol. 52,* 315-320.

Van Raalte, C. D., Valiela, I., Carpenter, E. J., and Teal, J. M. (1974). Inhibition of nitrogen fixation in salt marshes measured by acetylene reduction. *Estuarine Coastal Mar. Sci. 2,* 301-305.

Wada, E., Tsuji, T., Saino, T., and Hattori, A. (1977). A simple procedure for mass spectrometric microanalysis of ^{15}N in particulate organic matter with special reference to ^{15}N-tracer experiments. *Anal. Biochem. 80,* 312-318.

Ward, B. B., and Perry, M. J. (1980). Immunofluorescent assay for the marine ammonium-oxidizing bacterium *Nitrosococcus oceanus*. *Appl. Environ. Microbiol. 39,* 913-918.

Ward, B. B., Olson, R. J., and Perry, M. J. (1982). Microbial nitrification rates in the primary nitrite maximum off Southern California. *Deep-Sea Res. 29,* 247-255.

Watanabe, I., and deGuzman, M. R. (1980). Effect of nitrate on acetylene disappearance from anaerobic soil. *Soil Biol. Biochem. 12,* 193-194.

Webb, K. L., and Wiebe, W. J. (1975). Nitrification on a coral reef. *Can. J. Microbiol. 21,* 1427-1431.

Wentworth, W. E., and Freeman, R. R. (1973). Measurement of atmospheric nitrous oxide using an electron capture detector in conjunction with gas chromatography. *J. Chromatogr. 79,* 322-324.

Wiebe, W. J., Johannes, R. E., and Webb, K. L. (1975). Nitrogen fixation in a coral reef community. *Science 188,* 257-259.

Witty, J. F. (1979). Acetylene reduction assay can overestimate nitrogen fixation in soil. *Soil Biol. Biochem. 11,* 209-210.

Wolk, C. P., Austin, S. M., Bortins, J., and Galonksky, A. (1974). Autoradiographic localization of ^{13}N after fixation of ^{13}N-labeled nitrogen gas by a heterocyst-forming blue-green alga. *J. Cell Biol. 61,* 440-453.

Yeomans, J. C., and Beauchamp, E. G. (1978). Limited inhibition of nitrous oxide reduction in soil in the presence of acetylene. *Soil Biol. Biochem. 10,* 517-519.

Yoshinari, T. (1976). Nitrous oxide in the sea. *Mar. Chem. 4,* 189-202.

Yoshinari, T. (1980). N_2O reduction by *Vibrio succinogenes*. *Appl. Environ. Microbiol. 39,* 81-84.

Yoshinari, T., and Knowles, R. (1976). Acetylene inhibition of nitrous oxide reduction by denitrifying bacteria. *Biochem. Biophys. Res. Commun. 69,* 705-710.

Zuberer, D. A., and Silver, W. S. (1978). Biological dinitrogen fixation (acetylene reduction) associated with Florida mangroves. *Appl. Environ. Microbiol. 35,* 567-575.

Chapter 23

ENZYMOLOGY OF NITROGEN ASSIMILATION

PAUL G. FALKOWSKI
Oceanographic Sciences Division
Department of Energy and Environment
Brookhaven National Laboratory
Upton, New York

I. INTRODUCTION

A. Perspective of Enzymological Research

The enzymology of the initial steps of inorganic nitrogen assimilation traditionally has been a subject of a research by plant biochemists and physiologists, microbiologists, molecular biologists, and geneticists. Because of the economic importance of nitrogen in agriculture, enzymological studies generally focus on higher plant species, particularly commercial crop plants. Relatively recently, however, ecological studies identifying inorganic nitrogen as a resource limiting primary production in the world's oceans have aroused interest in the enzymology of nitrogen assimilation in algae (e.g., Eppley et al., 1969; Eppley and Rogers, 1970; McCarthy and Eppley, 1972; Packard, 1979).

ISBN 0-12-160280-X

Much of the information on the biochemistry of nitrogen-assimilating enzymes in algae is available only for a few species. The old freshwater standbys *Chlorella* and *Chlamydomonas* were the first and remain the most extensively studied, closely followed by an assortment of nitrogen-fixing cyanobacteria such as *Nostoc* and *Anabaena*. Bringing up the rear are some marine species (of which one of the most popular is the neritic diatom *Skeletonema costatum*) on which a smattering of research has been done. Very few comparative studies have been done per se on the enzymology of nitrogen-assimilating enzymes (e.g., Packard, 1979), and it is therefore difficult to ascertain to what extent the few species studied are representative of natural phytoplankton communities.

Since biochemical processes are relatively conservative in an evolutionary sense (Baldwin, 1959) and because there is little information on the nitrogen enzymology of marine organisms, I have borrowed from nonmarine algal sources where necessary, and made comparisons with higher plant systems where appropriate (refer to P. Wheeler, Chapter 9; J. Paul, Chapter 8; J. Goldman and Glibert, Chapter 7, this volume).

B. Overview of Nitrogen Assimilation

While it is not the primary purpose of this chapter to review nitrogen metabolism in algae, some description of the processes involved and the position of the enzymes in the metabolic sequences are necessary for orientation. More detailed reviews of nitrogen assimilation are provided by Wheeler (Chapter 9, this volume) as well as Morris (1974), Falkowski (1978a), Hewitt and Cutting (1978), Collos and Slawyk (1980), McCarthy (1980), Miflin (1980), and Guerrero et al. (1981).

Five forms of inorganic nitrogen are available to phytoplankton in the marine environment: nitrate (NO_3^-), nitrite (NO_2^-), ammonium (NH_4^+), urea ($[NH_2]_2CO$), and dissolved gaseous dinitrogen (N_2). Although the last is assimilated only by prokaryotes such as cyano-

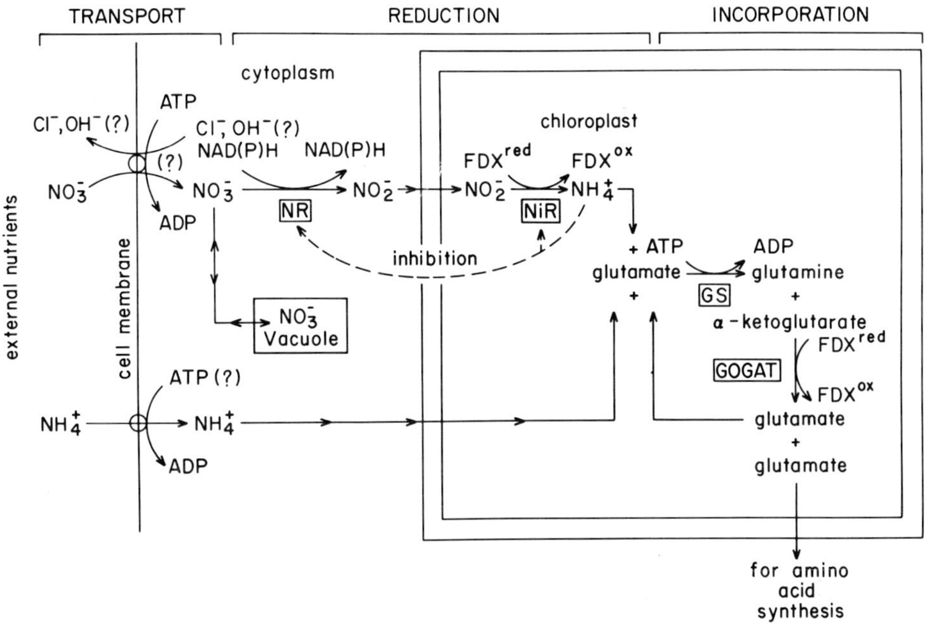

Fig. 1. Simplified schematic diagram showing the proposed pathway for the assimilation of nitrate, nitrite, and ammonium in algae. The inorganic nitrogen species is first transported into the cell via some active transport system(s) involving reversible ATP hydrolysis. Nitrate is reduced by nitrate reductase (NR) to nitrite in the cytoplasm. Nitrite is reduced to ammonium by nitrite reductase (NiR) in the chloroplast. Ammonium is incorporated into glutamate by glutamine synthetase (GS) to form glutamine. Finally, glutamine is reduced with α-ketoglutamate to form two molecules of glutamate by glutamate synthase (GOGAT). GS and GOGAT appear to be located in both the chloroplast and cytoplasm.

bacteria, it is important in that it represents a "new" nitrogen source, available from outside the hydrosphere, that is not significantly influenced by anthropogenic activities (Eppley, 1981). The enzymology of nitrogen fixation has been studied extensively and reviewed by Stewart (1973, 1980), Quispel (1974), Fogg (1974), Thorneley et al. (1978), and Orme-Johnson et al. (1977), and will be discussed only briefly here.

Whatever the form of nitrogen, assimilation consists of three basic processes (see Fig. 1). First, each form must be transported from the environment into the cell. Second, as only ammoniacal nitrogen is used in amino acid biosynthesis, each form must be converted to ammonium. In the case of the three oxidized forms (NO_3^-, NO_2^-, and N_2), one or more reduction steps are required; the complexed form, urea, must be hydrolyzed. Finally, ammonium is incorporated into carbon skeletons, derived ultimately from photosynthesis, to yield primary amino acids. This final process represents the actual synthesis of an organic nitrogen molecule from inorganic nitrogen.

Of the enzymes involved in the three basic processes, those that reduce oxidized nitrogen to ammonium are understood best; nitrogenase, nitrate reductase, and nitrite reductase have all been purified and characterized from algal sources. The enzymes responsible for ammonium incorporation have also been purified and characterized, though to a lesser extent; these include glutamate dehydrogenase (GDH), glutamine synthetase (GS), and glutamate synthase (GOGAT). Finally, the least is known about the enzymology and mechanisms of nitrogen uptake processes; there is no information about isolated, purified enzymes from either algal or higher plant sources that directly links a specific enzyme to a nitrogen transport reaction.

Some of the enzymological aspects of all the processes outlined will be reviewed in this chapter, but primary focus will be on: nitrate and nitrite reducases, glutamic dehydrogenase, glutamine synthetase, and glutamate synthase. I shall attempt to provide a brief description of the general physical characteristics of the proteins, the mechanisms of the reactions they catalyze, some aspects of their physiological regulation, and their potential use in understanding ecological processes.

II. UPTAKE

A. Nitrate and Nitrite

Although the kinetics of nitrate and nitrite uptake in marine phytoplankton have been studied extensively (e.g., Eppley and Coatsworth, 1968; Conway, 1977; MacIsaac and Dugdale, 1972; Serra et al., 1978), the actual mechanisms responsible for ion transport are largely unknown. As both nitrate and nitrite have a charge of -1 at pH 8 (i.e., the pH of seawater), are planar molecules, have similar molecular dimensions (ca. 2.4 Å total O-N-O length), and appear to interact competitively for the same uptake sites (Eppley and Coatsworth, 1968; Olson et al., 1980; but see Serra et al., 1978), for the purpose of the following discussion nitrite transport will be assumed to be similar if not identical to nitrate transport.

The mechanisms by which an ion can be transported against a negative chemical gradient at the expense of metabolic energy are of two major types: direct (primary) and indirect (secondary). These differ in whether energy is used directly for transport of the ion (active transport) or used to create a membrane potential that provides the driving force for ion translocation (symport or antiport processes; see Wheeler, Chapter 9, this volume). Both direct active transport and membrane potential gradient processes involve enzymes that reversibly hydrolyze ATP (ATPases).

Numerous studies have demonstrated clearly that nitrate transport is energy dependent. Simply summarized, the evidence is the following: (a) nitrate uptake is most frequently against the chemical potential gradient of the ion; (b) the uptake kinetics are saturable (i.e., the mechanism is probably not a diffusive process); (c) uptake and reduction are not always coupled and can be experimentally uncoupled, and the range of K_m values of nitrate reductase for NO_3^- rule it out as a possible means for promoting facilitated diffusion (see Packard, 1979); (d) nitrate uptake is inhibited by metabolic inhibitors such as 2,4 dinitrophenol (DNP) and carbonyl cyanide-*m*-chlorophenyl hydrozone (CCCP) (e.g., Ahmad and Morris, 1968; Falkowski and Stone, 1975; Kessler et al., 1970); (e) nitrate uptake and carbon fixation compete for metabolic energy and reducing equivalents (Falkowski and Stone, 1975; Collos and Slawyk, 1979); and (f) the Q_{10} for the uptake processes is ca. 2.0 or more.

Falkowski (1975) described an enzyme in membrane-enriched fractions of *Skeletonema costatum* that hydrolyzed ATP in the presence of NO_3^- and Cl^-. The enzyme had a bimodal pH profile with an optimum at pH 7.9 and a secondary optimum at pH 6.8-7.0. The energy of activation of the enzyme was ca. 16 kcal mol^{-1} and exhibited a break in the Arrhenius plot at 2.9°C, suggesting that the enzyme was membrane-bound. The half-saturation constant for activation of ATPase activity by NO_3^- was about 1 μM. The results of the studies generally suggested that nitrate transport could be directly coupled to ATP hydrolysis.

Thibaud ang Grignon (1981), examining the uptake of nitrate in corn roots, suggested that NO_3^- uptake was accomplished by OH^- antiport and the stoichiometry of the system appeared to be $2NO_3^-:1OH^-$. The study did not address specifically the question of whether a proton pump or an anion-activated ATPase (activated by NO_3^- and/or OH^- or otherwise) provided the energy for the transport process.

In a similar study of membrane potential and nitrate uptake in *Lemmna gibba*, Ullrich and Novacky (1981) inferred that NO_3^- uptake was accomplished by H^+ cotransport. This process is essentially the same as that described by Thibaud and Grignon (1981), since OH^- antiport and H^+ cotransport are indistinguishable; however, it is inferred that the transport of NO_3^- was directly coupled to ATP hydrolysis.

Butz and Jackson (1977) reviewed the then current theories of nitrate transport mechanisms and proposed that nitrate reductase was itself responsible for uptake of the ion. They hypothesized that some population of nitrate reductase might be membrane-bound, spanning the plasmalemma, and that it used ATP directly for nitrate transport while simultaneously reducing NO_3^- by means of NADH. This mechanism may be similar to that found in prokaryotes (cf. Beevers and Hageman, 1980).

It is obvious from the foregoing discussion that knowledge about nitrate uptake mechanisms in plants in general, let alone marine algae, is scanty. One difficulty in understanding the mechanisms and enzymology of the system is the lack of a good radioactive tracer for nitrogen (^{13}N has only a 10-min half-life). The use of $^{36}ClO_4^-$ (3×10^5-year half-life) as an analog of NO_3^- is promising (Tromballa and Broda, 1971), but has not been tested extensively in algal studies (cf. Beevers and Hageman, 1980). The mechanism of an enzyme transporting an ion is difficult to establish because the enzyme may not break or make any covalent bonds in the ion it transports. Although ATP may be hydrolyzed, there is no direct evidence that ion transport occurred because the assay systems are isotropic. One must compare the physiological kinetics of ion uptake *in vivo* with the biochemical properties of the enzyme implicated in the ion transport *in vitro*. In marine algae such a comparison is experimentally difficult to make because (a) the system is well buffered, so that small changes in pH are very hard to detect accurately; (b) measurement of membrane potentials and/or voltage-clamping techniques are not generally feasible; (c) only a small component of the ionic composition of seawater is nitrate or nitrite, so that ionic interference is a formidable problem; and (d) yields of membrane preparations from unicellular algae are very low and highly labile (cf. Falkowski, 1978a; Sullivan, 1978).

B. *Ammonium*

As with nitrate and nitrite uptake, the kinetics of ammonium uptake are much better understood than the mechanisms involved in transporting the ion across the plasmalemma (e.g., Healey, 1977). The recent use of radioactive analogs of ammonium, such as [^{14}C]-methylamine (Pelley and Bannister, 1979; Wheeler and Hellebust, 1981; Wheeler, 1980), in conjunction with short-term uptake experiments using [$^{15}NH_4^+$]ammonium (Glibert and Goldman, 1981) is increasing the potential for developing mechanistic transport models.

III. REDUCTION

Three enzymes are responsible for the reduction of nitrogen to ammonium in algae. The two most ubiquitous are nitrate and nitrite reductases, while the third, nitrogenase, occurs only in some prokaryotes (Yates, 1980). Nitrate reductase (EC 1.6.6.1, NADH nitrate oxidoreductase) is a cytoplasmic enzyme that catalyzes the reduction of nitrate to nitrite:

$$NO_3^- + 2e^- + 2H^+ \rightarrow NO_2^- + H_2O$$

(I)

Nitrite reductase (EC 1.7.7.1) is a chloroplastic enzyme catalyzing the reduction of nitrite to ammonium:

$$NO_2^- + 6e^- + 8H^+ \rightarrow NH_4^+ + 2H_2O$$

(II)

In vivo the reduction of nitrate (reaction I) is coupled to the oxidation of NAD(P)H and the reduction of nitrite (reaction II) is linked to ferredoxin, so that the overall stoichimetry for the reduction of nitrate to ammonium can be written as

$$NO_3^- + NAD(P)H + 6Fd\ (red) + 9H^+ \rightarrow NH_4^+ + NAD(P)^+ + 6Fd\ (ox) \quad (\Delta G = +69\ kcal\ mol^{-1} NH_4^+)$$

(III)

Nitrogenase, found in some marine cyanobacteria (e.g., *Trichodesmium* sp.), catalyzes the reduction of N_2 to ammonium:

$$N_2 + 3H_2O + 2H^+ \rightarrow 2NH_4^+ + 1.5\ O_2 \quad (\Delta G = +87\ kcal\ mol^{-1} NH_4^+)$$

(IV)

Thus, the fixation of N_2 requires about 25% more energy per mole of ammonium formed than the reduction of nitrate.

A. Nitrogenase

Nitrogenase (no assigned EC number) is a complex enzyme found in a few species of cyanobacteria and other prokaryotes. It consists of two oxygen-sensitive, iron-sulfur proteins, one of which contains Mo^{IV}. The Fe protein (nitrogenase reductase) binds Mg-ATP and transfers electrons from a donor to the MoFe protein (dinitrogen reductase), which reduces N_2. The total molecular weight of the dimeric Fe protein is ca. 60,000 to 65,000 and that of the MoFe protein, a tetramer, is ca. 225,000. Both proteins are oxygen sensitive, and both are required for N_2 fixation. In addition, the enzyme has an absolute requirement for Mg-ATP, which is thought, in cyanobacteria, to be derived from PS I activity *in vivo* (Stewart, 1973, 1980). Heterocysts, and presumably nonheterocystic cells containing nitrogenase, apparently lack PS II activity. The ATP:N stoichiometry is somewhat variable but under optimal conditions is about 12 ATP:N fixed (cf. Yates, 1980). No free intermediate of N_2 reduction, such as hydrazine (N_2H_4), has been isolated.

If *in vitro* no N_2 is provided as a substrate, H_2 is evolved. Hydrogenase activity is enhanced when reducing equivalents are produced but little or no N_2 is available as a sink for the reductant. Under such conditions H^+ may be reduced to H_2 (Burns and Hardy, 1975). In fact, even under optimal conditions, about 25% of the purified enzyme activity is "hydrogenase."

Nitrogenase is conveniently assayed by following the reduction of acetylene to ethylene with gas chromatography (Schöllhorn and Burris, 1966). The stoichiometric relationship between acetylene and N_2 reduction is not constant (see Capone, Chapter 4, this volume), however, and for quantitative studies must be measured by using $^{15}N_2$. An excellent review of the various assay methods for nitrogenase is presented by Burris (1974).

B. Nitrate Reductase

1. Purification

Nitrate reductase is easily purified from algal sources by affinity chromatography on blue Sepharose (such as commercial blue Sepharose CL6B, available from Pharmacia Fine Chemicals). The chromaphore Cibacron Blue F3G-A binds many NAD(P)H-requiring enzymes and was first suggested to bind nitrate reductase by Amy and Garrett (1974) and subsequently used in purification of the enzyme by Solomonson (1975).

Cells in log growth with nitrate as the sole nitrogen source are essential for the successful isolation and purification of the enzyme; nevertheless, such cells, especially diatoms, are not always rich in nitrate reductase activity. This activity can be enriched in batch cultures by resuspension of half the growth culture volume with half a volume of fresh nitrate-enriched medium 5-10 h prior to harvesting (typically overnight). This "shift-up" often results in a four- to tenfold increase in nitrate reductase activity, compared with that in regular batch cultures in log growth. For stability of nitrate reductase during storage, it is best to use large volumes of culture (10 liters or more), since purification and subsequent protein dilution results in loss of enzyme activity. Cells are most conveniently harvested by continuous flow centrifugation and disrupted by sonication (or passage through a French press) in cold potassium phosphate buffer (typically 100m*M*, pH 7.9) containing 1 m*M* dithiothreitol and 10 m*M* $MgSO_4$. Some investigators, following the procedures originally prescribed by higher-plant physiologists, add a small amount of polyvinylpyrrolidone to the extraction buffer to absorb phenols and tannins that might otherwise interfere with the extraction of the enzyme. Addition of detergents such as Triton X-100 (up to 1.0%) has no effect on the efficiency of extraction of nitrate reductase from the marine diatom *Skeletonema costatum*. The broken cell suspension is centrifuged at 15000 *g*, and the supernatant, containing nitrate reductase, can be used as a crude enzyme preparation (cf. Eppley et al., 1969).

For purification, the crude enzyme extract is usually precipitated with protamine or streptomycine sulfate followed by $(NH_4)_2SO_4$ fractionation. The protein fraction precipitating between 25 and 45% $(NH_4)_2SO_4$ is redissolved in 80 m*M* potassium phosphate buffer containing 2 m*M* $K_3Fe(CN)$ and a reduced sulfhydryl reagent. [The ferricyanide is added to "activate" the enzyme, i.e., to oxidize it (cf. Solomonson, 1975, 1978; Roldan and Butler, 1980).] This activated enzyme preparation is applied to a blue sepharose column and eluted first with a high molarity phosphate buffer (200-500 m*M*), which is followed by a linear gradient of 0-100 m*M* NADH in a low phosphate buffer (10-80 m*M*). Depending on the K_m of the enzyme with respect to NADH, nitrate reductase is eluted at some point in the NADH gradient.

The affinity chromatographic method allows for the purification of nitrate reductase to electrophoretic homogeneity in 4 h with an overall yield of 50-60%. Typical purification is 700- to 900-fold from the crude enzyme preparation. The method is far superior in efficiency and yield to the gel filtration/ion-exchange chromatographic methods formerly used (Amy and Garrett, 1974; Solomonson et al., 1975).

Purification of the cyanobacterial nitrate reductase (ferridoxin-dependent) has been achieved by the use of ferredoxin-Sepharose affinity chromatography (Manzano et al., 1978).

2. *Properties*

Assimilatory nitrate reductase is a large, soluble protein composed of three identical subunits (Solomonson et al., 1975). The native enzyme contains three prosthetic groups: flavin, heme (a *b*-type cytochrome), and Mo^{IV} in a proportion of 1:1:0.8 (Solomonson et al., 1975; Giri and Ramadoss, 1979). The reported molecular weights of the native enzyme vary, depending on the source and the method of determination, from 230,000 to 500,000 (Aparicio and Maldonado, 1978; Beavers and Hageman, 1980). The apparent molecular weight of the native enzyme is less on density gradient centrifugation than on gel filtration, suggesting that the protein is not spherical. Electron microscopy and laser diffraction studies suggest that the protein is globular, consisting of three C-shaped subunits oriented back to back at 120° (Giri and Ramadoss, 1979). With indirect methods used to determine the molecular weight and the partial specific volume of the purified enzyme, the "true" molecular weight of the protein is estimated to be between 280,000 and 356,000 (Giri and Ramadoss, 1979; Solomonson, 1975). Such a protein would contain approximately 900 amino acids per subunit. The amino acid composition of the enzyme is 90% neutral or acidic residues, with 20-21 half-cystine residues per subunit.

3. *Assay Methods*

At least three types of enzymatic activity are associated with fully active nitrate reductase (Amy and Garrett, 1974, distinguish four types). One activity is NADH oxidase, without necessarily the accompanying reduction of nitrate (diaphorase activity). A second activity is methyl viologen or FADH-dependent nitrate reduction (no NADH oxidation). A third activity is NADH-dependent nitrate reduction. A schematic representation of electron flow in the holoenzyme is presented in Fig. 2.

Diaphorase activity can be assayed by spectrophotometric measurement of NADH oxidation, leading to the reduction of FAD and cytochrome (Garrett and Nason, 1969). In the presence of nitrate, cytochrome undergoes rapid reoxidation, leading to the formation of nitrite. Nitrite production can be assayed colorimetrically by following the reaction with sulfanilamide and *N*-(1-naphtyl)ethylenediamine·2 HCℓ (Eppley, 1978). The measurement of nitrite produc-

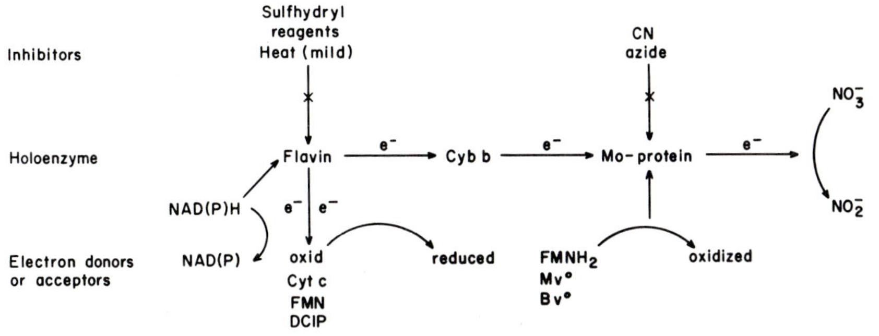

Fig. 2. Schematic diagram depicting the flow of electrons from NAD(P)H to NO_3^- in nitrate reductase (after Beevers and Hageman, 1980).

tion is the most common method of assaying nitrate reductase, but it is a poor method for kinetic or mechanistic studies because one cannot follow the reaction kinetics continuously. The method is, however, very well suited for field use and has been automated (Slawyk and Collos, 1976). Alternatively, one can follow continuously the nitrate-dependent oxidation of NADH spectrophotometrically or fluorometrically, or follow the utilization of NO_3^- by using ion-specific electrodes.

4. *Regulation of Enzyme Activity*

Thorough discussions of the regulation of nitrate reductase are presented by Hewitt et al. (1978), Beevers and Hageman (1980), and Guerrero et al. (1981). *In vivo* enzyme activity is repressed almost immediately on the addition of NH_4^+ (even in the presence of NO_3^-), but *in vitro* NH_4^+ has no effect. The enzyme is strongly inhibited *in vitro* by CN^-, hydroxylamine, and vanadate (Ramadoss, 1979), with a reported dissociation constant of 10^{-14} *M* for CN^- in purified *Chlorella* enzyme (Solomonson, 1975). Cyanide is a primary product of purine metabolism, and its role in the regulation of nitrate reductase *in vivo* is not obvious. Nevertheless, CN^- inhibition is completely reversible by oxidation with either ferricyanide or blue light (Roldan and Butler, 1980). Gewitz et al. (1978) presented data suggesting that CN^- was in fact an inhibitor of nitrate reductase *in vivo* in *C. fusca;* however, the inactivation was not related directly to ammonium repression of enzyme activity.

C. Nitrate Reductase

1. *Purification and Properties*

Like nitrogenase, nitrite reductase catalyzes a six-electron transfer, reducing NO_3^- to NH_4^+. In eukaryotic nitrite reduction, none of the hypothetical intermediates such as hyponitrite or hydroxylamine have been isolated *in vitro,* which suggests that the six-electron transfer is rapid and probably involves one active site. In bacteria and fungi, nitrite reductase is an iron-flavoprotein, requiring NADH and NADPH as reductants, respectively. In cyanobacteria, eukaryotic algae, and higher plants, ferredoxin appears to be the physiological electron donor.

The purification procedures described for nitrite reductase are relatively involved. Ho et al. (1976) describe a procedure for purification of the enzyme from *Porphyra yezoensis* involving three gel-filtration steps and two ion-exchange steps. Their overall yield was 3%, with 40-fold purification; however, the protein was electrophoretically homogeneous. More recently, the enzyme has been purified to homogeneity from spinach with high yields by affinity chromatography on ferredoxin-Sepharose (Ida, 1977).

Nitrite reductase is a single protein with a molecular weight of ca. 62,000 (in a diatom *Chaetoceros* sp., the MW is 120,000; Katagiri and Hattori, 1974, cited by Ho et al., 1976). The purified enzyme contains a seroheme and one Fe_2S_2 group per enzyme molecule. The enzyme has a pH optimum of ca. 7.5. The K_m of the enzyme with respect to ferredoxin is ca. 4×10^{-2} μM, and for NO_2^- it is ca.

800 μ*M*. Cyanide is a competitive inhibitor of NO_2^-, with a K_i of ca. 20 μ*M*.

2. *Assay Methods*

Nitrite reductase is usually assayed by measuring the disappearance of nitrite from the medium (Grant, 1970; Hattori and Uesugi, 1968; Eppley and Rogers, 1970). In practice one can provide the enzyme with an artificial electron source, such as methyl viologen, or a physiological electron source, ferredoxin. If methyl viologen is used, it must first be reduced with dithionite. The incubations are carried out *in vacuo*, and the reaction is terminated by shaking the tubes with air to oxidize the remaining methyl viologen. Nitrite is measured colorimetrically at 543 nm by the reaction with *N*-(1-naphthyl)ethylenediamine•2 HCl and sulfanilamide..

Nitrite reductase can also be assayed by following the production of ammonium, either by determining ammonium colorimetrically (Solorzano, 1969) or by using ion-specific electrodes.

IV. HYDROLYSIS

A. Urease

Urease (EC 3.5.1.5), the first enzyme to be crystallized and characterized as a protein (Sumner and Graham, 1925), is a metaloenzyme containing Ni (Dixon et al., 1975). It is inhibited by high concentrations of urea; the inhibition is reduced by the addition of glycine (cf. Baldwin, 1959). Urease is highly specific for urea and does not hydrolyze substituted ureas. It is relatively stable and is easily assayed by following the production of ammonium (e.g., Skokut and Filner, 1980).

B. ATP: Urea Amidolyase

In addition to hydrolysis by urease, there is another pathway based on ATP-dependent degradation of urea to ammonium and bicarbonate. This alternative pathway requires two enzymes that are found only in certain orders of the Chlorophyceae (Leftley and Syrett, 1973).

The first, urea carboxylase (EC 6.3.4.6), catalyzes the ATP-dependent carboxylation of urea to allophanate, and the second, allophanate hydrolase (EC 3.5.1.13), catalyzes the hydrolysis of allophanate to ammonium and bicarbonate. The overall reaction is

$$CO(NH_2)_2 + ATP + 3H_2O + HCO_3^- \rightarrow 2CO_2 + 2NH_4^+ + ADP + P_i \tag{V}$$

The assay of the enzyme depends on the CO_2 released from [^{14}C]urea (Thomas and Syrett, 1976). Because urea carboxylase (like many other carboxylases) requires biotin, avidin can be used as an inhibitor.

V. AMMONIUM INCORPORATION INTO AMINO ACIDS

There are at least three possible pathways for the assimilation of ammonium into amino acids, but careful kinetic studies by Folkes (1959) using $^{15}NH_4^+$ demonstrated that glutamic acid and glutamine were the primary initial products of ammonium incorporation. Until about 1974, the major pathway for ammonium assimilation was thought to be reductive amination of α-ketoglutarate, catalyzed by glutamic dehydrogenase (EC 1.4.1.4)(Lea and Miflin, 1974).

The discovery of glutamate synthase (GOGAT, EC 1.4.7.1, L-glutamate:NADP$^+$ oxidoreductase; deaminating, glutamate forming) in *Aerobacter aerogenes* grown under ammonium limitation (Tempest et al.,

1970), led to the hypothesis that NH_4^+ incorporation by glutamine synthease (GS, EC 6.3.1.2, L-glutamate:ammonia lyase, discovered by Krebs in 1935) was possible. This pathway was later supported by the demonstration that glutamine, not glutamate, is often the first labeled product of ammonium incorporation after N_2 fixation in *Anabaena cylindrica* using $^{13}N_2$ in 1- to 120 s incubations (Wolk et al., 1976). As comparative kinetic studies of the properties of glutamic dehydrogenase and glutamine synthetase were made, it became clear that the affinity of the former for ammonia was so low that it seemed improbable that this enzyme served in a primarily assimilatory role at physiological ammonium concentrations (Falkowski and Rivkin, 1976; Stewart and Rhodes, 1977). Many studies now suggest that ammonium incorporation is via the amide nitrogen of glutamine in the so-called GS/GOGAT pathway (e.g., Falkowski and Rivkin, 1976; Edge and Ricketts, 1978; Summons and Osmond, 1981), but there are only a few reports characterizing the enzymes in eukaryotic algae (cf. Miflin and Lea, 1980).

A. Glutamic Dehydrogenase

1. Properties

Glutamic dehydrogenase is a ubiquitous enzyme that has been studied in many algal species including *Chlorella vulgaris* (Morris and Syrett, 1965), *Ditylum brightwellii* (Eppley and Rogers, 1970), *Skeletonema costatum,* and others (Falkowski and Rivkin, 1976; Ahmed et al., 1977). The enzymology of glutamic dehydrogenases in plants has been reviewed recently by Stewart et al. (1980). Two isozymes are known, one of which occurs in the chloroplast (NADPH preference) and the other in mitochondria (NADH preference). The K_m of the algal chloroplast enzyme for NH_4^+ is relatively low compared with that of the higher plant enzyme, but is still too high to be physiologically important in ammonia assimilation. It varies from ca. 5 to 40 m*M*, depending on species.

Very little information is available about the structure of glutamic dehydrogenase in algae; most studies are from bovine liver and *Neurospora crassa* (cf. Smith et al., 1975). In *Chlorella sorokiniana*, the protein contains subunits with a molecular weight of ca. 60,000, and the holomeric enzyme is thought to be a hexamer (Gronostajski et al., 1978). In other studies, the enzyme has been reported to have molecular weights from 180,000 to ca. 250,000, and the molecular weights of the subunits suggest the holomeric protein is a tetramer. Amino acid analyses suggest the protein has six cysteine residues and no disulfide bonds and contains a metal, possibly Ca^{2+} or Mg^{2+}. The pH optimum of most glutamic dehydrogenases is between 7.6 and 8.2, but differs depending on the direction of the reaction (Gronostajski et al., 1978).

2. *Assay Methods*

Glutamic dehydrogenase can be assayed by following the oxidation of NAD(P)H spectrophotometrically (Morris and Syrett, 1965) or by following the appearance of $NAD(P)^+$ fluorometrically (Falkowski and Rivkin, 1976).

B. Glutamine Synthetase

1. *Purification and Properties*

To date, glutamine synthetase (GS) has not been purified or characterized from a marine eukaryote, so that all information is by analogy to other plant systems. Glutamine synthetase is known to occur in two isozymes: one found in the cytoplasm and the other in the chloroplast (Hirel and Godal, 1981; Kretovich et al., 1981). The enzyme has been purified to homogeneity from *Anabaena* by affinity chromatography on blue sepharose (Tuli et al., 1979). The yields are about 25% and purification is 90-fold. A similar level of purification can be achieved by density gradient centrifugation

(Emond et al., 1979). McMaster et al. (1980) purified the enzyme from *A. flosaquae* 380-fold using $(NH_4)_2SO_4$ precipitation, DEAE cellulose chromatography, and gel filtration. Purified GS from eukaryotic and higher plant sources consists of eight identical subunits, each having a molecular weight ranging from 45,000 to 58,000.

2. *Assay Methods*

Glutamine synthetase catalyzes two distinctly different reactions, referred to as "synthetase" and "transferase" activity. Assays of both types of activity are common (cf. Rhodes et al., 1975).

a. "Synthetase" Activity. In the formation of glutamine from glutamate and ammonia, GS requires *stoichiometric* amounts of ATP. The overall reaction

$$\text{glutamate} + NH_4^+ + \text{ATP} \xrightarrow[\text{Mg}_2^+]{\text{enzyme}} \text{glutamine} + \text{ADP} + P_i$$

is called a "synthetic" reaction and can be assayed in a number of ways.

i. One of the methods most frequently used takes advantage of the fact that hydroxylamine can be substituted for NH_4^+, and the product γ-glutamylhydroxamic acid yields a characteristic brown color with $FeC\ell_3$ (Elliot, 1953). This method is relatively simple and is preferred when ATPase activity is high (Shapiro and Stadtman, 1970); however, it is not often used for kinetic studies, especially when an accurate determination of the K_m for NH_4^+ is desired.

ii. A second method is to measure P_i formation colorimetrically (e.g., Fiske and Subbarow, 1925).

iii. A third method is to measure ATP hydrolysis by coupling the enzyme to excess pyruvate kinase and lactate dehydrogenase, and measuring the oxidation of NADH spectrophotometrically or fluorometrically (Shapiro and Stadtman, 1970; Falkowski and Rivkin, 1976). This has the advantage of allowing for the observation of the initial kinetics of the reaction, and, as long as excess phospho(enol) pyruvate is present, ADP is recycled back to ATP, maintaining a (near) constant high-energy nucleotide concentration.

iv. Finally, the production of gluamine can be measured either by paper chromatography or by following the production of [^{14}C]glutamine from [^{14}C]glutamate and separating the products by electrophoresis (cf. McKenzie et al., 1979).

b. "Transferase" Activity. GS also catalyzes the reaction

$$\text{glutamine} + \text{hydroxylamine} \xrightarrow[P_i]{ADP} \gamma\text{-glutamyl hydroxamate} + NH_4^+$$

The reaction proceeds much more slowly when ATP is substituted for ADP and requires only *catalytic* concentrations of nucleotide (cf. Meister, 1974). The product (γ-glutamyl hydroxamic acid) is measured colorimetrically after reaction with $FeC\ell_3$ (Shapiro and Stadtman, 1970).

It has been shown that the same enzyme catalyzes both reactions (Elliot, 1953). The rates and kinetics of the "synthetase" and "transferase" reactions are not identical but the ratio of activities may be constant (e.g., Harel et al., 1977).

3. Regulation

Glutamine synthetase appears to be a regulatory enzyme *in vivo* (Stewart et al., 1980; Miflin and Lea, 1980). Enzyme activity has been shown to vary with adenylate energy charge in higher plants (Weissman, 1976) and algae (Falkowski, 1978b; Sawhney and Nicholas, 1978). In addition, the enzyme is strongly inhibited *in vivo* by alanine and glycine (Rowell et al., 1977; McMaster et al., 1980). Glutamine synthetase has also been shown to be regulated in response to variations in light and nitrogen regimes (Tischner and Hüttermann, 1980). A useful metabolic inhibitor of GS is methionine sulfoxamine (e.g., Rigano et al., 1979). The enzyme appears to mistake the sulfoxin moiety of the antimetabolite for the γ-carboxylate of glutamate, forming an enzyme-inhibitor complex that essentially cannot be dissociated (cf. Walsh, 1979). Glutamine synthetase has also been implicated in controlling nitrogenase in *Klebsiella pneumoniae* (Shanmugam et al., 1978) and *Azobacter vinelandii* (Kleinschmidt and Kleiner, 1981). The role of GS appears to be that of regulating glutamine pools, which may repress or corepress (with NH_4^+) nitrogenase formation.

C. Glutamate Synthase

Glutamate synthase catalyzes the transfer of the amido group to α-ketoglutarate, leading to the formation of glutamate. The subsequent transamination of the amino nitrogen of glutamate to various α-keto acids is the primary pathway for the formation of most amino acids (Miflin and Lea, 1980; Givan, 1980).

1. Purification and Properties

Most reports on purified glutamate synthase to date deal with material from higher plants and cyanobacteria. The enzyme has been purified from a number of crops such as barley, maize, bean, and pea. A purification procedure has been described that uses $(NH_4)_2SO_4$ precipitation, followed by two DEAE-cellulose chromatography separations and sucrose density gradient centrifugation (Wallsgrove et al., 1977). This procedure resulted in a 170-fold purification with 24% yield. A procedure for partial purification from *Chlamydomonas reinhardii* based on $(NH_4)_2SO_4$ precipitation followed by gel filtration on Sephacryl S-200 was described by Guillimore and Sims (1981).

The enzyme is a multimeric, nonhem iron protein with a molecular weight of about 140,000 to 200,000. The K_m values for glutamine, α-ketoglutarate, and ferredoxin are ca. 330, 150, and 2 μ*M*, respectively (cf. Stewart et al., 1980). Cullimore and Sims (1981) and Matoh and Takahashi (1981) reported two isozymes, one utilizing ferredoxin and the other NADH. Their results suggest that the NADH isozyme is cold labile and sensitive to oxygen. The molecular weights were 165,000 for the ferredoxin enzyme and 240,000 for the NADH form. Both have pH optima at 7.5; however, the K_m for the NADH form with respect to glutamine was 900 μ*M* compared with 190 μ*M* for the ferredoxin form, and the K_m with respect to α-ketoglutarate was 7 μ*M* for the former and 170 μ*M* for the latter. The physiological role of the two enzymes has not been clearly established. As in the case of glutamine synthetase, one possibility is that the enzymes are spatially separated--one form (presumably the ferredoxin one) being located primarily in the chloroplast and the other in the cytosol. The spatial separation of the two isozymes, with differing cofactors, may allow for the incorporation of NH_4^+ in the dark.

2. *Assay Methods*

Glutamate synthase is unstable and requires very high concentrations of reduced sulfhydryls to maintain high activity. The enzyme can be assayed by (1) incubating with [^{14}C]glutamine and measuring the formation of [^{14}C]glutamate following ion-exchange chromatography (Cullimore and Sims, 1981) or electrophoresis (McKenzie et al., 1979); or (2) determining the formation of glutamate by means of paper chromatography (Wallsgrove et al., 1977). The reaction is inhibited by azaserine and to a lesser extent by albizziine (Wallsgrove et al., 1977; Miflin and Lea, 1980), which inhibit a wide range of glutamine-amide transfer reactions.

VI. THE APPLICATIONS OF ENZYME ASSAYS TO ECOLOGICAL STUDIES

Paradoxically, it seems that the potential uses of enzyme assays in ecological studies have been both overstated by ecologically oriented biochemists and underutilized by nonbiochemically oriented ecologists. One of the primary uses of enzymological studies in an environmental context is to investigate the physiological mechanisms by which an ecologically relevant process occurs. Such investigations are usually much more useful in providing qualitative rather than quantitative information. For example, Eppley et al. (1969) first proposed measuring nitrate reductase activity as a means of determining whether nitrate was used for growth. This idea evolved from physiological studies showing that nitrate reductase activity was greatly repressed by reduced nitrogen forms. The results of nitrate reductase assays could lead to a simple "yes" or "no" conclusion about nitrate utilization. Eppley's group subsequently incorporated nitrate reductase assays in field programs in an attempt to estimate how much nitrate was assimilated. These studies were based on the assumption that the reduction of nitrate was rate limiting (see McCarthy and Eppley, 1972). Although this may be true, it is often difficult to relate enzyme activity *in vitro* quantitatively to that *in vivo*, and therefore little confidence can be placed in quantitative projections of (for example) nitrate assimilation based on nitrate reductase activity. Beevers and Hageman (1980) discuss similar uncertainties with regard to the use of nitrate reductase assays in higher plant systems.

As a component of a larger applied research endeavor, enzymological research can be used to good advantage. For example, spatial and temporal distributions of nitrate reductase activity may be helpful in interpreting ^{15}N tracer studies in regions where ambient inorganic nitrogen concentrations are very low and are significantly changed by the addition of $^{15}NO_3^-$. Enzyme assays may also be useful in identifying rate-limiting steps or metabolic pathways.

Finally, enzymological studies are especially helpful in examining mechanisms of metabolic regulation, knowledge of which can, in turn, be valuable in interpreting ecological phenomena such as diel periodicity in nitrogen assimilation, effects of nitrogen enrichment on carbon fixation, the effects of light on nitrogen reduction, and many others.

ACKNOWLEDGMENT

This manuscript was prepared under the auspices of the United States Department of Energy under Contract No. DE-AC02-76H00016.

REFERENCES

Ahmad, J., and Morris, I. (1968). The effects of 2,4-dinitrophenol and other uncoupling agents on the assimilation of nitrate and nitrite by *Chlorella*. *Biochim. Biophys. Acta 162*, 32-38.

Ahmed, S. I., Kenner, R. A., and Packard, T. T. (1977). A comparative study of the glutamate dehydrogenase activity in several species of marine phytoplankton. *Mar. Biol. 39*, 93-101.

Amy, N. K., and Garrett, R. H. (1974). Purification and characterization of the nitrate reductase from the diatom *Thalassiosira pseudonana*. *Plant Physiol. 54*, 629-637.

Aparicio, P. J., and Maldonado, J. M. (1978). Regulation of nitrate assimilation in photosynthetic organisms. *In* "Nitrogen Assimilation of Plants" (E. J. Hewitt and C. V. Cutting, eds.), pp. 207-215. Academic Press, New York.

Baldwin, E. (1959). "Dynamic Aspects of Biochemistry," 3rd ed., Cambridge Univ. Press, London and New York.

Beevers, L., and Hageman, R. H. (1980). Nitrate and nitrite reduction. *In* "The Biochemistry of Plants" (B. J. Miflin, ed.) Vol. 5, pp. 115-168. Academic Press, New York.

Burns, R. C., and Hardy, R. W. F. (1975). "Nitrogen Fixation in Bacteria and Higher Plants." Springer-Verlag, Berlin and New York.

Burris, R. H. (1974). Methodology. *In* "The Biology of Nitrogen Fixation" (A. Quispel, ed.), pp. 9-33. Elsevier, New York.

Butz, R. G., and Jackson, W. A. (1977). A mechanism for nitrate transport and reduction. *Phytochemistry 16*, 409-417.

Collos, Y., and Slawyk, G. (1979). ^{13}C and ^{15}N uptake by marine phytoplankton. I. Influence of nitrogen source and concentration in laboratory cultures of diatoms. *J. Phycol. 15,* 186-190.

Collos, Y., and Slawyk, G. (1980). Nitrogen uptake and assimilation by marine phytoplankton. *In* "Primary Productivity in the Sea" (P. G. Falkowski, ed.), pp. 195-211. Plenum, New York.

Conway, H. L. (1977). Interactions of inorganic nitrogen in the uptake and assimilation by marine phytoplankton. *Mar. Biol. 39,* 221-232.

Cullimore, J. V., and Sims, A. P. (1981). Occurrence of two forms of glutamate synthase in *Chlamydomonas reinhardii. Phytochemistry 20,* 597-600.

Dixon, N. E., Gazzola, C., Blakeley, R. L., and Zerner, B. (1975). Jack bean urease (E.C. 3.5.1.5) a metalloenzyme. A simple biological role for nickel? *J. Am. Chem. Soc. 97,* 4131-4133.

Edge, P. A., and Ricketts, T. R. (1978). Studies on ammonium-assimilating enzymes of *Platymonas striata* Butcher (Prasinophyceae). *Planta 138,* 123-125.

Elliot, W. H. (1953). Isolation of glutamine synthetase and glutamotransferase from green peas. *J. Biol. Chem. 201,* 661-672.

Emond, D., Rondeau, N., and Cedergren, R. J. (1979). Distinctive properties of glutamine synthetase from the cyanobacterium *Anacystis nidulans. Can. J. Biochem. 57,* 843-851.

Eppley, R. W. (1978). Nitrate reductase in marine phytoplankton. *In* "Handbook of Phycological Methods" (J. A. Hellebust and J. S. Cragie, eds.), pp. 217-223. Cambridge Univ. Press, London and New York.

Eppley, R. W. (1981). Autotrophic production of particulate matter. *In* "Analysis of Marine Ecosystems" (A. R. Longhust, ed.), pp. 343-361. Academic Press, New York.

Eppley, R. W., and Coatsworth, J. L. (1968). Nitrate and nitrite uptake by *Ditylum brightwellii.* Kinetics and mechanisms. *J. Phycol. 4,* 151-156.

Eppley, R. W., and Rogers, J. N. (1970). Inorganic nitrogen assimilation of *Ditylum brightwelli,* a marine plankton diatom. *J. Phycol. 6,* 344-351.

Eppley, R. W., Coatsworth, J. L., and Solorzano, L. (1969). Studies of nitrate reductase in marine phytoplankton. *Limnol. Oceanogr. 14,* 194-205.

Falkowski, P. G. (1975). Nitrate uptake in marine phytoplankton: (Nitrate, chloride)-activated adenosine triphosphatase from *Skeletonema costatum* (Bacillariophycea). *J. Phycol. 12,* 448-450.

Falkowski, P. G. (1978a). The regulation of nitrate assimilation in lower plants. *In* "Nitrogen in the Environment" (D. R. Nielsen and J. G. MacDonald, eds.), Vol. 2, pp. 143-155. Academic Press, New York.

Falkowski, P. G. (1978b). Anion activated adenosine triphosphatases. *In* "Handbook of Phycological Methods" (J. A. Hellebust and J. S.

Cragie, eds.), pp. 225-261. Cambridge Univ. Press, London and New York.
Falkowski, P. G., and Rivkin, R. B. (1976). The role of glutamine synthetase in the incorporation of ammonium in *Skeletonema costatum* (Bacillariophyceae). *J. Phycol. 12,* 448-450.
Falkowski, P. G. and Stone, D. P. (1975). Nitrate uptake in marine phytoplankton: Energy sources and the interaction with carbon fixation. *Mar. Biol. 32,* 77-84.
Fiske, C. H. and Subbarow, Y. (1925). The colorometric determination of phosphorus. *J. Biol. Chem. 66,* 375-400.
Fogg, G. E. (1974). Nitrogen fixation. *In* "Algal Physiology and Biochemistry" (W.D.P. Stewart, ed.), pp. 560-582. Univ. California Press, Los Angeles.
Folkes, B. F. (1959). Position of amino acids in the assimilation of nitrogen and the synthesis of proteins in plants. *Symp. Soc. Exp. Biol. 13,* 126-147.
Garrett, R. H., and Nason, A. (1969). Further purification and properties of *Neurospora* nitrate reductase. *J. Biol. Chem. 244,* 2870-2882.
Gewitz, H. S., Piefke, J., and Vennesland, B. (1978). Nitrate reductase E.C. 1.6.6.1 of *Chlorella fusca.* Partial purification, cytochrome content, and presence of cyanide after *in vivo* inactivation. *Planta 141,* 323-328.
Giri, L., and Ramadoss, C. S. (1979). Physical studies on assimilatory nitrate reductase from *Chlorella vulgaris. J. Biol. Chem. 254,* 11703-11712.
Givan, C. V. (1980). Aminotransferases in higher plants. *In* "The Biochemistry of Plants" (B. J. Miflin, ed.), Vol. 5, pp. 329-357. Academic Press, New York.
Glibert, P. M., and Goldman, J. C. (1981). Rapid ammonium uptake by marine phytoplankton. *Mar. Biol. Lett. 2,* 25-32.
Grant, B. R. (1970). Nitrite reductase in *Dunaliella tertiolecta:* Isolation and properties. *Plant Cell Physiol. 11,* 55-64.
Gronostajski, R. M., Yeung, A. T., and Schmidt, R. R. (1978). Purification and properties of the inducible nicotinamide adenine dinucleotide phosphate-specific glutamic dehydrogenase from *Chlorella sorokiniana. J. Bacteriol. 134,* 621-628.
Guerrero, M. G., Vega, J. M., and Losada, M. (1981). The assimilatory nitrate-reducing system and its regulation. *Annu. Rev. Plant Physiol. 32,* 169-204.
Harel, E., Lea, P. J., and Miflin, B. J. (1977). The localization of enzymes of nitrogen assimilation in maize leaves and their activities during greening. *Planta 134,* 195-200.
Hattori, A., and Uesugi, I. (1968). Purification and properties of nitrite reductase from the blue-green alga *Anabaena cylindrica. Plant Cell Physiol. 9,* 689-699.
Healey, F. P. (1977). Ammonium and urea uptake by some freshwater algae. *Can. J. Bot. 55,* 61-69.
Hewitt, E. J., and Cutting, C. V., eds. (1979). "Nitrogen Assimilation of Plants." Academic Press, New York.

Hewitt, E. J., Hucklesby, D. P., Mann, A. F., Notton, B., and Rucklidge, G. J. (1978). Regulation of nitrate assimilation in plants. *In* "Nitrogen Assimilation of Plants" (E. J. Hewitt and C. V. Cutting, eds.) Academic Press, New York.

Hirel, B., and Godal, P. (1981). Glutamine synthetase isoforms in pea leaves: Intracellular location. *Z. Pflanzenphysiol. 102,* 315-319.

Ho, C.-H., Kawa, T., and Nisizawa, K. (1976). Purification and properties of a nitrite reductase from *Porphyra yezoensis* Ueda. *Plant Cell Physiol. 17,* 417-430.

Ida, S. (1977). Purification to homogeneity of spinach nitrite reductase by ferredoxin sepharose affinity chromatography. *J. Biochem. (Tokyo) 82,* 915-918.

Kessler, E., Hofmann, A., and Zemft, W. G. (1970). Inhibition of nitrite assimilation by uncouplers of phosphorylation. *Arch. Mikrobiol. 72,* 23-26.

Kleinschmidt, J. A., and Kleiner, D. (1981). Relationships between nitrogenase, glutamine synthetase, glutamine and energy charge in *Azobacter vinelandii*. *Arch. Microbiol. 128,* 412-415.

Kretovich, W. L., Eustigreiva, Z. G., Pushkin, A. V., and Dzhokharidze, (1981). Two forms of glutamine synthetase in leaves of *Cucurbita pepo*. *Phytochemistry 20,* 625-629.

Lea, P. J., and Miflin, B. J. (1974). Alternative route for nitrogen assimilation in higher plants. *Nature (London) 251,* 614-616.

Leftley, J. W., and Syrett, P. J. (1973). Urease and ATP:urea amidolyase activity in unicellular algae. *J. Gen. Microbiol. 77,* 109-115.

McCarthy, J. J. (1980). Nitrogen. *In* "The Physiological Ecology of Phytoplankton" (I. Morris, ed.), pp. 195-233. Blackwell, Oxford.

McCarthy, J. J., and Eppley, R. W. (1972). A comparison of chemical, isotopic, and enzymatic methods for measuring nitrogen assimilation of marine phytoplankton. *Limnol. Oceanogr. 17,* 371-382.

MacIsaac, J. J., and Dugdale, R. C. (1972). Interactions of light and inorganic nitrogen in controlling nitrogen uptake in the sea. *Deep-Sea Res. 19,* 209-232.

McKenzie, G. H., Ch'ng, A. L. and Gayler, K. R. (1979). Glutamine synthetase/glutamine: α-ketoglutarate aminotransferase in chloroplasts of the marine alga *Caulerpa simpliciuscual*. *Plant Physiol. 63,* 578-582.

McMaster, B. J., Danton, M. S., Storch, T. A., and Dunham, V. L. (1980). Regulation of glutamine synthetase E. C. 6.3.1.2 in the blue-green alga *Anabeana flos-aquae*. *Biochem. Biophys. Res. Commun. 96,* 975-983.

Manzano, C., Candau, P., and Guerrero, M. G. (1978). Affinity chromatography of *Anacystics nidulans* ferredoxin-nitrate reductase and NADP reductase on reduced ferredoxin-sepharose. *Anal. Biochem. 90,* 408-412.

Matoh, T., and Takahashi, E. (1981). Glutamate synthase in greening pea shoots. *Plant Cell Physiol. 22,* 727-731.

Meister, A. (1974). Glutamine synthetase of mammals. *In* "The Enzymes" (P. D. Boyer, ed.), 3rd ed. Vol. 10. pp. 699-754. Academic Press, New York.

Miflin, B. J., ed. (1980). "The Biochemistry of Plants," Vol. 5. Academic Press, New York.

Miflin, B. J., and Lea, P. J. (1980). Ammonia assimilation. *In* "The Biochemistry of Plants" (B. J. Miflin, ed.), Vol. 5, pp. 169-202. Academic Press, New York.

Morris, I. (1974). Nitrogen assimilation and protein synthesis. *In* "Algal Physiology and Biochemistry" (W.D.P. Stewart, ed.), pp. 583-609. Blackwell, Oxford.

Morris, I., and Syrett, P. J. (1965). The effect of nitrogen starvation on the activity of nitrate reductase and other enzymes in *Chlorella. J. Gen. Microbiol. 38,* 21-28.

Olson, R. J., Beeler, J., Soo Hoo, and Kiefer, D. A. (1980). Steady-state growth of the marine diatom *Thalassiosira pseudonana*. Uncoupled kinetics of nitrate uptake and nitrite production. *Plant Physiol. 66,* 383-389.

Orme-Johnson, W. H., Davis, L. C., Henzel, M. T., Averill, B. A., Orme-Johnson, N. R., Munck, E., and Zimmerman, R. (1977). *In* "Recent Developments in Nitrogen Fixation" W. Newton, J. R. Postgate, and C. Rodriguez-Barrueco, eds.), pp. 131-178. Academic Press, New York.

Packard, T. T. (1979). Half-saturation constants for nitrate reductase and nitrate translocation in marine phytoplankton. *Deep-Sea Res. 26A,* 321-326.

Pelley, J. L., and Bannister, T. T. (1979). Methylamine uptake in the green alga *Chlorella pyrenoidosa*. *J. Phycol. 15,* 110-112.

Quispel, A., ed. (1974). "The Biology of Nitrogen Fixation." Am. Elsevier, New York.

Ramadoss, C. S. (1979). The effect of vanadium on nitrate reductase of *Chlorella vulgaris*. *Planta 146,* 539-544.

Rhodes, D., Rendon, G. A., and Stewart, G. R. (1975). The control of glutamine synthetase level in *Lemna minor* L. *Planta 125,* 201-211.

Rigano, C., Rigano, V. D., Vona, U., and Fuggi, A. (1979). Glutamine synthetase activity ammonia assimilation, and control of nitrate reduction in the unicellular red alga *Cyanidium caldarium*. *Arch. Microbiol. 121,* 117-120.

Roldan, J. M., and Butler, W. L. (1980). Photoactivation of nitrate reductase from *Neurospora crassa*. *Photochem. Photobiol. 32,* 375-381.

Rowell, P., Enticott, S., and Stewart, W. D. P. (1977). Glutamine synthetase and nitrogenase activity in the blue-green alga *Anabaena cylindrica*. *New Phytol. 79,* 41-54.

Sawhney, S. K., and Nicholas, D. J. D. (1978). Effects of amino acids, adenine nucleotides and inorganic pyrophosphate on glutamine synthetase from *Anabaena cylindrica*. *Biochim. Biophys. Acta 527,* 485-496.

Schöllhorn, R., and Burris, R. H. (1966). Study of intermediates in nitrogen fixation. *Fed. Proc. Fed. Am. Soc. Exp. Biol. 25,* 710 (abstr.).

Serra, J. L., Llama, M. J., and Cadenas, E. (1978). Nitrate utilization by the diatom *Skeletonema costatum* I. Kinetics of nitrate uptake. *Plant Physiol. 62,* 987-990.

Shanmugam, K. T., O'Gara, F., Andersen, K., and Valentine, R. C. (1978). Biological nitrogen fixation. *Annu. Rev. Plant Physiol. 29,* 263-276.

Shapiro, B. M., and Stadtman, E. R. (1970). Glutamine synthetase *(Scherichia coli).* *In* "Methods in Enzymology" (H. Tabor and C. W. Tabor, eds.), Vol. 17A, pp. 910-922. Academic Press, New York.

Skokut, T. A., and Filner, P. (1980). Slow adaptive changes in urease levels of tobacco cells cultured on urea and other nitrogen sources. *Plant Physiol. 65,* 995-1003.

Slawyk, G., and Collos, Y. (1976). An automated assay for the determination of nitrate reductase in marine phytoplankton. *Mar. Biol. 34,* 23-26.

Smith, E. L., Austen, B. M., Blumenthals, K. M. and Nye, J. F. (1975). *In* "The Enzymes" (P. D. Boyer, ed.), 3rd ed., Vol. 11, pp. 293-367. Academic Press, New York.

Solomonson, L. P. (1975). Purification of NADH-nitrate reductase by affinity chromatography. *Plant Physiol. 56,* 553-855.

Solomonson, L. P. (1978). Structure of *Chlorella* nitrate reductase. *In* "Nitrogen Assimilation of Plants" (E. J. Hewitt and C. V. Cutting, eds.), pp. 199-205. Academic Press, New York.

Solomonson, T. P., Toumer, G. H., Hall, R. L., Borchers, R., and Bailey, J. L. (1975). Reduced nicotinamide adenine dinucleotide-nitrate reductase of *Chlorella vulgaris. J. Biol. Chem. 250,* 4120-4127.

Solorzano, L. (1969). Determination of ammonia in natural waters by the phenol-hypochlorite method. *Limnol. Oceanogr. 14,* 799-801.

Stewart, G. R., and Rhodes, D. (1977). A comparison of the characteristics of glutamine synthetase and glutamate dehydrogenase from *Lemna minor* L. *New Phytol. 79,* 257-268.

Stewart, G. R., Mann, A. F., and Fentem, P. A. (1980). Enzymes of glutamate formation: Glutamate dehydrogenase, glutamine synthetase, and glutamate synthase. *In* "The Biochemistry of Plants" (B. J. Miflin, ed.), Vol. 5, pp. 271-327. Academic Press, New York.

Stewart, W. D. P. (1973). Nitrogen fixation by photosynthetic microorganisms. *Annu. Rev. Microbiol. 27,* 283-316.

Stewart, W. D. P. (1980). Some aspects of structure and function in N_2-fixing cyanobacteria. *Annu. Rev. Microbiol. 34,* 497-536.

Sullivan, C. W. (1978). Ion transport and ion-stimulated adenosine triphophatases. *In* "Handbook of Phycological Methods" (J. A. Hellebust and J. S. Cragie, eds.), pp. 463-476. Cambridge Univ. Press, London and New York.

Summons, R. E., and Osmond, C. B. (1981). Nitrogen assimilation in the symbiotic marine alga *Gymnodinium microdriaticum:* Direct analysis of ^{15}N incorporation by GC-MS methods. *Phytochemistry 20,* 575-578.

Sumner, J. B., and Graham, V. A. (1925). The nature of insoluble urease. *Proc. Soc. Exp. Biol. Med. 22,* 504-506.

Tempest, D. W., Meers, J. L., and Brown, C. M. (1970). Synthesis of glutamate in *Aerobacter aerogenes* by a hitherto unknown route. *Biochem. J. 117,* 405-407.

Thibaud, J. B. and Grignon, C. (1981). Mechanisms of nitrate uptake in corn roots. *Plant Sci. Lett. 22,* 279-289.

Thomas, E. M., and Syrett, P. J. (1976). The assay of ATP:urea amidolyase in whole cells of *Chlorella. New Phytol. 76,* 409-413.

Thorneley, R. N. F., Eady, R. R., Smith, B. E., Lowe, D. J., Yates, M. G., O'Donnell, M. J., and Postgate, J. R. (1978). Biochemistry and mechanism of nitrogenase. *In* "Nitrogen Assimilation of Plants" (E. J. Hewitt and C. V. Cutting, eds.), pp. 27-43. Academic Press, New York.

Tischner, R., and Hüttermann, A. (1980). Regulation of glutamine synthetase E. C. 6.3.1.2 by light and during nitrogen deficiency in synchronous *Chlorella sorokiniana. Plant Physiol. 66,* 805-808.

Tromballa, H. W., and Broda, E. (1971). Das Verhalten von *Chlorella fusca* gegenuber Perchlorat und chlorat. *Arch. Mikrobiol. 78,* 214-223.

Tuli, R., Jarvali, N., and Thomas, J. (1979). A rapid method for the purification of glutamine synthetase E. C. 6.3.1.2 from the blue-green alga *Anabaena* L-31. *Indian J. Exp. Bot. 17,* 1239-1241.

Ullrich, W. R., and Novacky, A. (1981). Nitrate-dependent membrane potential changes and their induction in *Lemna gibba* G1. *Plant Sci. Lett. 22,* 211-217.

Wallgrove, R. M., Harel, E., Lea, P. J., and Miflin, B. J. (1977). Studies on glutamate synthase from the leaves of higher plants. *J. Exp. Bot. 28,* 588-596.

Walsh, C. (1979). "Enzymatic Reaction Mechanisms." Freeman, San Francisco, California.

Weissman, G. S. (1976). Glutamine synthetase regulation by energy charge in sunflower roots. *Plant Physiol. 57,* 339-343.

Wheeler, P. A. (1980). Use of methylammonium as an ammonium analogue in nitrogen transport and assimilation studies with *Cyclotella cryptica* (Bacilloriophyceae). *J. Phycol. 16,* 326-334.

Wheeler, P. A., and Hellebust, J. A. (1981). Uptake and concentration of alkylamines by a marine diatom. Effects of H^+ and K^+ and implications for the transport and accumulation of weak bases. *Plant Physiol. 67,* 367-372.

Wolk, C. P., Thomas, J., Shaffer, P. W., Austin, S. M., and Galonsky, A. (1976). Pathway of nitrogen metabolism after fixation of ^{13}N-labeled nitrogen gas by the cyanobacterium, *Anabaena cylindrica. J. Biol. Chem. 251,* 5027-5034.

Chapter 24

ECOSYSTEM SIMULATION MODELS: TOOLS FOR THE INVESTIGATION AND ANALYSIS OF NITROGEN DYNAMICS IN COASTAL AND MARINE ECOSYSTEMS

RICHARD L. WETZEL
Virginia Institute of Marine Science and
School of Marine Science
College of William and Mary
Gloucester Point, Virginia

RICHARD G. WIEGERT
Department of Zoology
University of Georgia
Athens, Georgia

I. INTRODUCTION

Simulation modeling of ecosystems has become a tool of the marine sciences only within the past 30 years. This is due primarily to the rapid technological advances made in the computer industry in the late 1950s and 1960s, but is also due to the development of multidisciplinary, holistic approaches in ecological research. Efforts to model coastal and marine ecosystems have attracted the interest and participation of not only biologists but also physical, chemical, and geological oceanographers and engineers Studies that involve a system modeling approach require a multidisciplinary effort; the interaction of the two benefits both the design and products of that research.

Models and the processes of modeling have become valuable research tools for a variety of reasons. They (a) provide a concep-

NITROGEN IN THE MARINE ENVIRONMENT

ISBN 0-12-160280-X

tual framework within which specific studies and experimental designs can be evaluated; (b) identify areas where information is lacking or inadequate; (c) determine sensitive or controlling parameters (processes); and (d) provide an experimental tool for analysis to aid future research and experimental designs. We have purposefully omitted model "predictiveness" from this list. Predictability, in the sense of a model being able to forecast changes in ecosystem structure or function as a result of anticipated perturbations, is not necessary to justify employing simulation models as research tools. We are of course speaking here of realistic explanatory models, not of statistical predictors, such as multiple regression models which have good predictive powers within the range of measured parameters, but which have no explanatory value.

We limit our discussion in this chapter to the philosophy and application of modeling as an investigative technique in basic research on nitrogen dynamics in coastal and open-ocean marine environments. We have omitted the vast and growing literature on water quality models that often includes various aspects of nutrient cycles. Recent reviews and symposia proceedings consider this area (e.g., Russel, 1975; Neilson and Cronin, 1981; O'Connor, 1981). Our approach is to discuss the development, construction, and analysis of ecosystem simulation models and to review specific studies that have incorporated nitrogen in the structure.

II. MODELS AND MODELING

Models are observer-defined abstractions of the real world. An interesting philosophical as well as applied discussion of general systems properties and modeling is presented in Weinberg (1975) and Weinberg and Weinberg (1979). Although we limit our discussion to mathematical constructions that are intended for digital computer simulation, models can be physical approaches, i.e., microcosms and macrocosms (e.g., Kitchens et al., 1979; Heinle et al., 1979) or some combination of the two (e.g., Nixon et al., 1979): theoretical treatments (e.g., Halfon, 1979); or statistical approaches (e.g., Ivakhnenko et al., 1979; Daniel and Wood, 1980).

In general, a modeling approach intended for computer simulation analysis follows procedures that include several phases not rigorously defined. These include: (1) conceptualization or compartmentalization, (2) mathematical formalization, (3) parameterization and documentation, (4) programming, (5) model simulation and analysis, and (6) validation-verification. An almost intangible aspect that pervades each phase is "redesign," which imparts a dynamic property to each step. In our own experience, it has been this characteristic that proved the most informative and exciting (as well as the one that ensures that no compartmental simulation model is every truly finished).

Conceptualization and compartmentalization of a model, together with assigning directed flows for the unit of exchange between compartments, define the structure of the modeled system. Many models evolve no further and serve to represent pictorially dynamic processes and potential controls in a static framework, analogous to an electronic circuit diagram. The basic purpose is to depict "state-of-the-art" (or our current paradigm) regarding functional relationships and, hopefully, explicit representation of feedback controls. Nixon (1981) and Webb (1981) employed conceptual models to illustrate and discuss nutrient dynamic processes in estuarine environments. Time-series compartmental diagrams illustrate quite nicely how conceptual structures change or evolve to reflect new information. For research in which several investigators are involved, model development to this stage serves principally as an organizational tool.

Dynamic models emerge from the objective implementation of the remaining steps to simulate compartmental behavior. Translating the directed arrows (fluxes) and their controls of the compartmental model into functional notation and ultimately into a simulation language accounts for the diversity in current models and the various modeling philosophies (Wiegert, 1975a, 1979). There is associated with all simulation models a large, implicit structure buried in the parameter values and flux equations. The mathematical structures should reflect explicitly the parameters and controls of interest (e.g., Wiegert, 1979; Wiegert and Wetzel, 1979).

The actual functions (equations) used to simulate a process or control that governs the dynamics of a compartment or state variable differ. Wiegert (1975b) and Shoemaker (1977) review the rationale for the more common mathematical relationships used in ecosystem simulation models. Basically, the problem or question at this development stage reduces to: What is an adequate (realistic) representation of flux for the process of interest that embodies both the assumptions of control and purposes of model simulation analyses? For example, Wetzel and Kowalski (1980) employed two functional forms for simulating the effect of light on the realized rate of seagrass photosynthesis:

$$Q_t = Q_o e^{-k_w z_t} \tag{1}$$

Equation (1) was used to predict light at the canopy top (Q_t) given incident rations (Q_o), the diffuse downwelling attenuation coefficient ($-k_w$) and water depth above the canopy (z_t). The equation was used to explore photosynthesis of the seagrasses under various light, water-column attenuations, and depth regimes, given an empirical derivation of net apparent photosynthesis and light. In a second series of investigations (actual field studies), studies of seagrass photosynthesis and submarine light were directed toward analysis of internal canopy structure, self-shading, and light attenuation within the canopy. Equation (1) was obviously inadequate to explore these effects in a fashion complementary of the field studies. The equation was reformulated to allow model simulation studies along these lines. Equation (2) was reformulated as;

$$Q_t = Q_o e^{-[k_w z_m + k_p (z_m - z_t)]} \tag{2}$$

Equation (2) incorporated measures of the level (height above the substrate) within the canopy structure of leaf area maximum (z_m) and light attenuation within the canopy (k_p)(Wetzel et al., 1981). Both equations served as adequate expressions for the simulation of light and photosynthesis in the seagrass community, but addressed quite different questions of potential control.

In a more general sense, flows in ecosystem simulation models can be characterized as either material or informational. Figure 1 is a simple box model of photosynthesis that illustrates both. In the example, the solid lines are material exchanges and the dashed lines are information flows and represent controls. To implement this model, mathematical relations are derived as functions of the compartments per se and the information flows (indicated by the control symbol on the photosynthesis flux). However, dynamic behavior of the model is associated with the two CO_2 compartments and the phytoplankton. In fact, most models to date that include nitrogen treat it in a similar manner, i.e., as a control on a material flux rather than a conserved exchange.

It is beyond the scope of this chapter to deal extensively with the programming of simulation models except to note that FORTRAN is probably the most common programming language used in simulation modeling, and that relatively simple numerical approximation methods are, at least initially, the methods of choice (Wiegert and Wetzel, 1974).

Analysis of simulation models range from rather straightforward sensitivity tests (Wiegert et al., 1975; Christian and Wetzel, 1979; Wiegert and Wetzel, 1979) to studies of stability characteristics (Kremer, 1979), to more mathematically sophisticated analyses of systems structure and control (e.g., Halfon, 1979). Levin (1975) reviews various methodologies and strategies for model analysis. For ecosystem-level simulation models, sensitivity analyses and stability characteristics are relatively simple model analysis tools that provide insight to functional relationships and controls. The

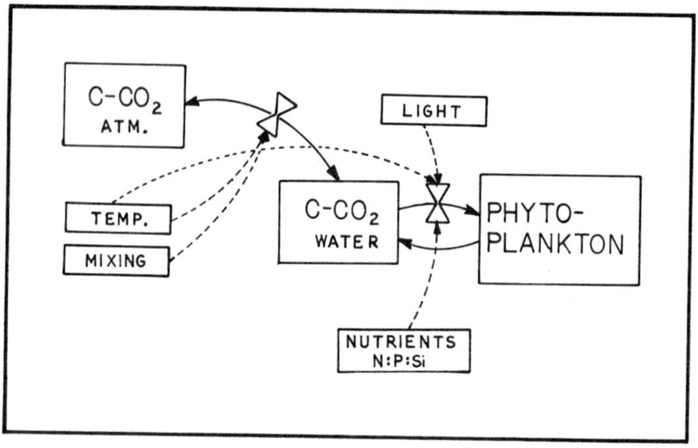

Fig. 1. Conceptual model of photosynthesis including exchange of mass (solid lines) and information controls (dashed lines).

results of these types of model analyses provide the greatest feedback between field/experimental research studies and the modeling effort.

Verification and validation of simulation models are terms generally used interchangeably in the modeling literature to describe or report a procedure for deciding the "correctness" of a model. Caswell (1976) reviews in detail the concept of validation. The procedure enters the modeling scenario in at least two steps: (1) mathematical structuring and programming and (2) simulation. Almost all discussions of the topic deal with the latter, i.e., establishing that the simulated behavior of the model, in some fashion, reasonably portrays our presumed measures of real-world dynamics. Comparing model output (i.e., compartment behavior) with time-series measures from the environment is the more common approach. If the two agree within certain bounds (mostly intuitive), the model is assumed "correct." However, the first and prerequisite step in model verification is establishing that the programmed structure is in fact behaving consistently with the assumed or derived mathematical structure. T. H. Probert (personal communication) suggests this can be a critical step (often ignored or at least not reported for ecosystem models), and thus should be incorporated in all modeling schemes. Based on our own experience, problems associated with model validation or verification are generally either a product of the original conceptual scheme or faulty assumptions in the structures of the model, i.e., our mathematical conceptualization and representation of the way we believe processes work and are controlled in nature are wrong. The intriguing possibility always exists that the model is correct and the measurements faulty.

Thus, simulation models of coastal and marine ecosystems can (1) provide an organizational framework for multidirected research efforts; (2) identify data requirements in terms of both "better" and new information; (3) test the conceptual and hypothetical structure of systems; and (4) using relatively simple techniques, evaluate principal controls and suggest new hypotheses.

III. NITROGEN AND SIMULATION MODELS OF COASTAL AND MARINE ECOSYSTEMS

Compared with terrestrial and freshwater ecosystems, simulation models of coastal and marine ecosystems are considerably fewer in number and tend to be more incomplete: incomplete in the sense of either the compartments considered or processes involved (Wiegert, 1975b, 1979). Models of nitrogen dynamics per se are even fewer. The majority of ecosystem simulation models that include nitrogen do so by treating nitrogen concentration (often without regard to specific compound) as a control on exchange between compartments. For most ecosystem simulation models, the unit of exchange is either energy (calories) or matter (biomass or carbon). Specific processes of the nitrogen cycle (e.g., nitrogen fixation, denitrification, nitrification, etc.) generally are not modeled explicitly. In particular, controls that govern the rates of these transformations are only implicit in the mathematical structure.

The scarcity of models of nitrogen dynamics is puzzling. The majority of ecosystem models simulate (in terms of mass balance) the behavior of state variables that generally are highly aggregated and the interactions are poorly understood (e.g., trophically designed models). But the conceptual model of nitrogen cycling was formulated many years ago. The specific transformations (processes) and their controls have been the subject of intense research for the past several decades and are relatively well understood, at least biochemically or physiologically. Also, a rather extensive data base exists from a variety of systems in terms of validation of model behavior, a situation not characteristic of marine trophically designed models. It may be that most basic research efforts in nitrogen cycling are process-oriented and tend to decouple the transformation measurement from the system as a whole. Holistic approaches are only now being more actively pursued.

In the following sections we discuss several specific examples of simulation models that include nitrogen either as a control or the unit of exchange. Based on our literature review for this, we note that many reports that address modeling and nitrogen dynamics are contained in the so-called "gray" literature and was largely unavailable to us. We have concentrated our review on publications from the refereed literature (journals) and texts.

A. Coastal and Open Ocean Systems

Conceptual models of coastal and open-ocean systems have largely adopted the classical, phytoplankton-herbivore-predator food chain analogy. Elaboration on this theme generally has followed the approach of increasing structural complexity (i.e., adding compartments) to include more species, life history stages, trophic levels, or incorporating more detailed treatments of environmental control (i.e., light, temperature, nutrients, physical properties, etc.) on specific processes. Model development and implementation generally have followed one or the other of these paths, but rarely both. Figure 2 illustrates the conceptual models as box and arrow diagrams for the discussion that follows. The convention followed in the illustration is: a solid line represents a matter (biomass, nutrient, carbon) exchange; a dotted or dashed line, an information or feedback control flow; a value symbol, a feedback-controlled flow.

Some of the first models developed were for coastal and open water systems. Riley, in a series of papers in the 1940s, developed a set of equations (the word "model" was not used in these earlier papers) to describe phytoplankton-zooplankton dynamics on Georges Bank (Riley, 1942; 1946, 1947a,b; Riley and Bumpus, 1946). The results of these studies and others are summarized in the classic monograph on the ecology of plankton in the western North Atlantic Ocean (Riley et al., 1949). Light, temperature, vertical mixing, and nutrients were treated as controls on specific rates of photosynthesis and growth of phytoplankton and zooplankton (Fig. 2a,b). Nutrient dynamics per se were not modeled, but empirical functions were derived to limit phytoplankton growth when concentration dropped below a specified level. Specifically, nitrogen was not considered limiting and phosphorus was. The main points addressed by the models were the role of light, vertical water-column stability, and zooplankton grazing as controls on productivity. Without considerable modification and increased complexity, Riley's models could not be used to investigate nitrogen dynamics. However, the basic conceptual structure is still prevalent in the more recent literature (Swartzman and Bentley, 1979; Webb, 1981).

Nearly two decades later, Riley (1967) developed a model to predict the distribution of nitrate and phosphate in coastal waters. The model was driven by a shoreward transport of nutrient-rich bottom water that mixes increasingly with surface waters in the onshore direction. A net, offshore, advective flow balanced the subsurface input. Remineralization of nitrogen and phosphorus was cal-

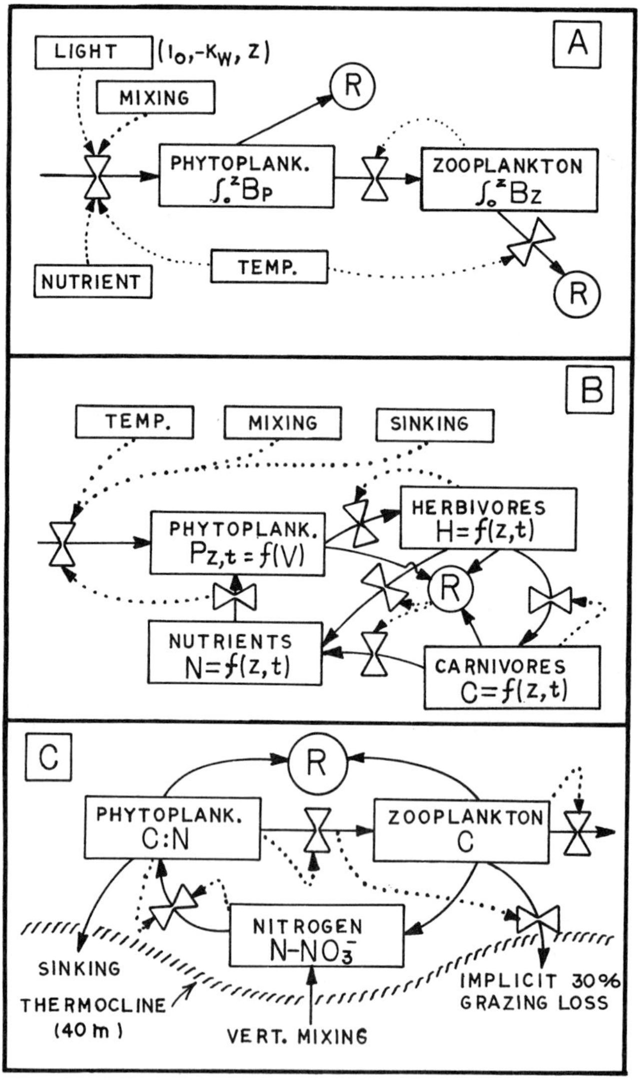

Fig. 2. Compartmental structure of selected simulation models of open-ocean and coastal marine phytoplankton-based systems. A and B after Riley (1946) and Riley et al. (1949), respectively; C after Steele (1974). Symbols are: I_o, surface light intensity; $-k_W$ vertical downwelling attenuation coefficient; Z, water depth, V, sinking rate; R, respiration. Symbols inside the boxes are compartment abbreviations.

culated as a percentage of utilization. Based on the results of studies using this model, Riley (1967) illustrated the adequacy of employing simple models to explain nutrient conditions and productivity in New England coastal waters. The model supported and perhaps enhanced the hypothesis of nitrogen being the principal limiting nutrient in coastal seas. Specifically, the modeling results offered an explanation for the generally low N:P ratios observed for nearshore waters. However, the actual nitrogen transformations were not explicity modeled.

Patten (1968) reviewed mathematical models of plankton production published through 1966. Twenty-four models were compared according to compartmental structure and interaction. Of the models that treated marine systems (excluding Riley's papers), only Cushing's (1959) included nutrients. We do not consider Davidson and Clymer's (1966) analysis a systems model since all equations were uncoupled. The majority of models, however, include light as a control. Thus, it appears no significant changes in either structural or functional characteristics of models relative to nitrogen dynamics occurred between Riley's original work and Patten's review.

Nearly 10 years after Riley's work, Steele (1974) developed and published a model of the North Sea planktonic ecosystem (Fig. 2c). In addition to incorporating light and mixing, as Riley did, Steele introduced nutrient kinetics (nitrogen uptake). As Steele (1974) points out, the kinetic relationship can most often be modeled with the Michaelis-Menten function

$$\mu = \mu_{max} \cdot \frac{[C]}{[K_c + C]} \cdot M_c \qquad (3)$$

where μ equals the realized uptake rate given the maximum specific rate μ_{max}, the half-saturation K_c (i.e., the concentration at which $\mu = 0.5\ \mu_{max}$), nitrogen concentration in the environment C, and the mass of the compartment M_c. This mathematical structure is and has continued to be the most common function used to describe nutrient kinetics of natural populations, although there are convincing physiological and mathematical arguments against it. For mixed, heterogeneous natural populations, μ_{max} and K_c have limited biological or ecological meaning, and significant errors can be introduced depending on the choice of parameter values (Williams, 1973). The function also contains no lower or upper bounds (thresholds) for uptake or population growth, an obvious biologically unrealistic assumption (Swartzman and Bentley, 1979; Wiegert, 1975b).

Steele (1974) modeled phytoplankton uptake of nitrogen [using Eq. (3); see Fig. 2c] from a dissolved inorganic pool that was controlled by the vertical mixing of nutrient-rich water below the thermocline (fixed at 40 M for the simulations) and by the regeneration of nitrogen by zooplankton excretion. Mortality of phytoplankton and zooplankton and fecal production represented particulate nitrogen states that were considered lost to the system by sinking below the mixed depth. Therefore, no nitrogen recycling within the mixed layer due to microbial processes or detrital sources are considered in the model.

Steele's (1974) analysis of nitrogen dynamics in the modeled system was limited to variation of the kinetic parameters, μ_{max} and K_c in the Michaelis-Menten function. A threefold variation in μ_{max} (0.2 or 0.3) produced different behavior in both phytoplankton and zooplankton compartments. A high μ_{max} (0.2 or 0.3) caused a rapid bloom response and low ambient nitrogen concentrations within the first 50 days of simulation. This was followed by a regular, limit-cycle response of lower-peak standing stocks and persistent low nitrogen concentration. A low μ_{max} (0.1) did not produce the typical bloom response. However, after an initial growth and decline period, a regular, limit-cycle response was evident that was similar in amplitude to the other simulation experiments but not in phase. Predicted nitrogen concentrations were also atypically high. Steele (1974) interpreted these model results, based on *in situ* observations, as evidence for the need to establish properly the relationship between nutrient concentration and biological uptake, although he concludes that the form of the mathematical expression is not important in terms of systems stability. Based on our previous discussion, we would argue that it is exactly consideration of this point that will ultimately limit the applicability of the simulation model as an analysis tool.

The most complex model structure for coastal or open-ocean systems recently introduced in the literature is the compartmental model of energy flux by Pomery (1979). Although the model is conceptual at present and does not include nitrogen flux, the major compartments involved in nitrogen transformations are present. The interesting aspect of the model is that it incorporates both detrital pathways and benthic processes. In order to develop realistic models of either energy flux or nutrient cycles in coastal seas, these processes and their controls must be considered together with greater hydrodynamic detail than has been characteristic.

B. *Upwelling Systems*

Boje and Tomczak (1978) discuss in general terms the concept of ecosystem analysis, models, and boundary conditions for upwelling ecosystems. The upwelling system off the Peruvian coast has been a site of intensive physical, chemical, and biological study since the late 1960s. Concomitant with the development of this research, there have been the design, construction, and simulation of models that couple carbon and nitrogen dynamics with the hydrodynamic properties of upwelling systems. The models have both served as analytical tools and provided feedback to specific research activities (Walsh, 1971).

Walsh and Dugdale (1971) report the results of the first-generation simulation model of nitrogen dynamics in the Peru upwelling system. The three-compartment nitrogen model [nutrients (NO_3^-), phytoplankton, and herbivores] was simulated in five spatial blocks that were depth stratified with an upper layer 11-km wide, 11-km long, and 10-m deep, and a lower layer extending from the upper 10-m boundary to the bottom. The spatial orientation of the blocks was located downstream in the upwelling plume. Transfer between blocks for each dompartment was driven by downstream, vertical, and lateral advection and diffusion. Biological transfers included nutrient uptake, grazing, excretion, and sinking for the various compartments. Nutrient uptake was modeled with a Michaelis-Menten function and modified to include a threshold response for grazing. Nutrient regeneration was implicitly incorporated in the model by setting the μ_{max} term [Eq. (3)] to reflect both NO_3^- and NH_3 uptake. Diel effects were also included to represent more realistically *(a priori)* nutrient uptake and grazing. By incorporating sine and cosine functions, Walsh and Dugdale (1971) restricted nutrient uptake to daylight hours and grazing to the night time. The herbivore compartment migrated between surface and bottom layers.

Walsh and Dugdale (1971) analyzed the model by adding terms sequentially (i.e., advection, diffusion, NO_3^- uptake, NO_3^- + NH_3 uptake, grazing, and sinking) to the model and comparing model output (compartment behavior) to field measurements. Based on the results of the first-generation model analyses, they suggested (1) a time lag was necessary between onset of primary production in newly upwelled water and grazing; (2) regenerated or recycled N (NH_3) was an important component of N uptake by phytoplankton, particularly in the downstream sections and, at least in the present version, was controlled by grazing; and (3) although the model behaved, in many respects, consistently with observed properties, model fidelity would be improved by increasing spatial detail (i.e., adding more blocks). The principal criticisms of the model were that the boundary condition set for the lower layer was unrealistic and probably was at some depth much less than the bottom (a point considered in later versions) and, as the authors indicate, many of the parameter values and interactions were not known. However, development and analysis of the model to this stage is an excellent example of the use of a model to generate hypotheses.

The Peru upwelling model structure and functional relations continued to evolve. Walsh (1975a) described a second-generation model that addressed many of the simulation results and questions generated by the first model version. Specifically, the second model version (1) increased spatial detail by fixing the lower boundary of the upper layer at the bottom of the surface Ekman layer (20 m) and fixing the lower-layer boundary at 80 m; (2) included the potential for multiple nutrients (NO_3^-, NH_3, SiO_4, and PO_4) and light limitation; (3) included grazing by both zooplankton and fish (anchoveta) and nutrient regeneration by zooplankton and anchoveta as a function of grazing intensity; and (4) added a detrital N compartment (however, remineralization and grazing of detrital N was not considered in the model). Also, hydrodynamic events (i.e., weak versus strong upwelling episodes) were included in the simulations by varying parameters in the functions controlling horizontal, lateral, and vertical advective and diffusive transport.

Model simulation analyses generally followed the procedure used for the first-generation model and addressed the potential controls of light and multiple nutrients, water circulation and mixing, food chain efficiency, and qualitative aspects of detrital N behavior. The increase in model structural and functional complexity allowed better resolution of compartmental behavior and controls. The results of the analyses are discussed in detail by Walsh (1975a) and go beyond the intent of this chapter.

In a second paper, Walsh (1975b) considers the role of feedback in governing carbon and nitrogen dynamics in the upwelling system using the same model. Overall, the agreement between predicted and observed concentrations for selected compartments was good. Walsh (1975a,b) points out what he considers the weaknesses of the model version, but perhaps more importantly he poses the more general question applicable to most if not all ecosystem models: How does one apply mesoscale simulation models to the actual system's scale (macrosacle)? A question that remains largely unresolved.

The modeling and research activities on upwelling systems conducted by Walsh and associates over the past decade are exemplary in terms of coupling basic research and systems analysis by ecosystem simulation modeling. For marine ecosystems, the nitrogen simulation models they report are probably the most complete in terms of process orientation. More recently, the models have been applied to various contemporary issues that include a preliminary assessment of potential yield (fish protein) to man (Walsh and Howe, 1976) and the effects of overfishing on the carbon-nitrogen cycling in upwelling systems (Walsh, 1981).

C. Marsh-Estuarine Systems

Simulation models of marsh or estuarine systems that include flows of nitrogen are scarce; models that simulate N as the conserved unit of mass balance are rare. One effort, which includes a discussion of the types of ecosystem models and a brief review of nitrogen models (to 1976), is that of Najarian and Harleman (1977). They discuss three categories of models (1) biodemographic models, emphasizing conservation of species or genetic information; (2) bioenergetic models organized around the conservation of energy; and (3)biogeochemical models, whose fluxes are the conserved mass of some element. The estuarine nitrogen model they develop and describe is the third type.

A seven-compartment, closed-matter loop model was developed to represent physical, chemical, and biological transformations of nitrogen compounds. As a first test of the conceptual model, Najarian and Harleman (1977) simulated the N dynamics of a chemostat system. The interesting result was the prediction of large, although damped, oscillations in several nitrogen compartments and a failure to reach steady state during the first 2000 hours (more than six residence times of 13 days).

Adding hydrodynamic components to the first model structure permitted coupling mass transport in an idealized advective estuary with the nitrogen transformation process.

Two models were investigated: A real-time model incorporated measured intratidal dispersion coefficients and transport processes. A second "flow-through" model assumed all transport due to freshwater input at the head of the estuary, and used a constant dispersion coefficient equal to the average of the spatially variable, real-time coefficients. Predicted ammonia concentrations using these two different models were as much as 100% in disagreement, even though the structure and parameter values of the biochemical transformations were identical. They concluded that the disparities of the models was a product of averaging the physical coefficients of the highly nonlinear, real-time model over a tidal cycle.

As a final demonstration of the interaction between biochemical transformation and physical transport processes, Najarian and Harleman (1977) adapted the model to the Potomac estuary. Again, large variation in nitrogen water-quality parameters were predicted intratidally, suggesting the utility of the real-time modeling approach as opposed to the flow-through averaging technique. They suggested that because more is known about physical transport processes than biochemical transformations, transport processes must be represented accurately in models to avoid obscuring potentially important dynamics caused by biochemical transformation in estuarine nitrogen models.

A second major effort involving estuarine modeling is that of Kremer and Nixon (1978). They describe a bioenergetic-type model of Narragansett Bay that incorporates a dynamic nutrient compartment as a control on certain energy flows. Inputs to the nitrogen component of this nutrient complex were by animal excretion, fluxes from the benthos (both dependent on the dynamics of the respective compartments), and from sewage entering the Bay, a forcing function. Nitrogen was removed as uptake by algae. Nitrogen excretion by animals was regulated by changes in temperature in accord with the C:N ratios of the food being consumed. Temperature was the sole factor regulating N regeneration by the benthos. Algal uptake varied with temperature and light but was also sensitive to the N composition of the algae. The model was unable to reproduce observed N changes in the Bay, possibly because of the failure to include a dissolved organic N component. Some simulation runs suggested N was the most limiting nutrient to production by phytoplankton and other components. Based on different half-saturation values for uptake of nitrogen and silicate, the authors suggested the latter was the most important nutrient causing cessation of a characteristic winter-spring bloom. Kremer and Nixon (1979) thus concluded that the limiting or controlling function of N versus Si was delicately balanced. Overall, they concluded that although the nutrient fluxes in the model simulations are reasonable, some additional factor is apparently missing.

The model of Barataria Bay developed by Hopkinson and Day (1977) is the biogeochemical type, but includes a detailed compartmentalization of both C and N. The authors state, however, that the field and literature data base on which the nitrogen model was built was extremely limited in comparison with the extensive base available for construction of the carbon model.

The N model included three atmospheric interactions: rain, N-fixation, and denitrification. In addition to the biotic compartments, both organic and inorganic soil nitrogen were included. Processes affecting these forms were mineralization, immobilization (incorporation by bacteria), nitrification, denitrification, and fixation.

Simulations of fluctuations in total N in the marsh soil and inorganic N in the water column agreed well with field measurements in Barataria Bay. However, soil ammonia was not simulated satisfactorily. Field results suggested that mineralization in the soil was balanced by uptake until autumn. The model simulations predicted only mineralization and thus loss throughout the year.

The Hopkinson-Day model was primarily a research hypothesis-generating tool. It was not considered an applied predictive model and simulations provided insights into further profitable research that might be conducted. With respect to the dynamics of nitrogen, suggestions were: (1) to study better the soil organic N component and its transformations, and (2) to investigate further temperature as a major control on productivity by *Spartina* (as opposed to light and N availability).

In general, simulation models of salt marsh and estuarine ecosystems have treated nitrogen in a cursory fashion, although recognizing its potential limiting influence on both phytoplankton and vascular plant production. Much of the work in these environments has been process-oriented and does not include simulation modeling as an analysis tool. Continued work with simulation models of bays and estuaries would do well to include a more detailed and complex representation of the major limiting nutrient transformations and compartmental transfers.

IV. SUMMARY AND CONCLUSIONS

We have attempted both to present the basic characteristics of a simulation modeling effort as an investigative tool and to review specific models of estuarine and marine ecosystems that included nitrogen. Our review is biased toward simulation models of entire ecosystems (or at least the principal components) and those published in the readily available literature. The chapter does not include terrestrial or freshwater models on nitrogen cycling or water quality models, all of which have a large literature.

In the context of the general discussion on models and modeling as investigative tools in basic and multidisciplinary research, there is an obvious paucity of detailed modeling efforts on nitrogen dynamics in estuarine and marine ecosystems. Two overall conclusions appear warranted: (1) The vast majority of models use matter (as carbon or biomass) or energy as the conserved unit and (2) nitrogen, primarily as NO_3^- and NH_3, is incorporated as control on specific flux processes and not dynamically simulated. The ecosystem models that actually conserve nitrogen in various compartmentalization schemes either are not validated (Hopkinson and Day, 1977) or accept simplifying assumptions regarding a complete simulation analysis of the nitrogen cycle (Walsh, 1975a). Over the past decade, various authors have implicated the potentially important role of dissolved organic nitrogen compounds in the marine nitrogen cycle. No model to our knowledge details the controls on or dynamics of this nitrogen state in simulation versions.

In shallow-water estuarine and coastal systems, i.e., where vertical mixing extends to the bottom, benthic nitrogen dynamics as influencing and potentially controlling water column properties are either implicit in the model structures or only grossly represented (Nixon, 1980, 1981; Kemp et al., 1983). In order to develop more realistic models of these systems, it seems paramount that this subsystem be explicit in the structural and functional model framework. Delaune and Patrich (1980) have recently demonstrated the importance of sedimentation in the nitrogen cycle of a Louisiana salt marsh, and Billen (1978) developed a nitrogen budget for North Sea sediments. Berner (1977) and Vanderborght et al. (1977) have developed models for nutrient regeneration and nitrogen diagenesis, respectively, for marine sediments. What is needed are models that realistically couple these important processes with the larger ecosystem.

These future models will probably not progress very far unless the chemical-biological phenomena are coupled with the physical-hydrodynamic properties of the modeled system. Also, unless realistic interaction equations that include the assumptions of control and feedback are derived for these models, the simulation model as an investigative and potential predictive tool will be compromised.

ACKNOWLEDGMENTS

We wish to thank our many colleagues, too numerous to name, who over the years have either directly or indirectly contributed to the general discussion and topics presented herein. Likewise, we apologize for anyone's "pet" model being excluded or overlooked. We also thank our anonymous reviewers for their comments which have largely been incorporated. Special thanks are due Ms. Carole Knox and Ms. Nancy White for their tireless editorial and secretarial assistance.

This chapter is a joint contribution of the Virginia Institute of Marine Science, Gloucester Point, Virginia (Contribution No. 1033) and the University of Georgia Marine Institute, Sapelo Island, Georgia (Contribution No. 445).

REFERENCES

Berner, R. A. (1977). Stoichiometric models for nutrient regeneration in anoxic sediments. *Limnol. Oceanogr. 22,* 781-786.

Billen, G. (1978). A budget of nitrogen recycling in North Sea sediments off the Belgina coast. *Estuarine Coastal Mar. Sci. 7,* 127-146.

Boje, J., and Tomczak, M. (1978). Ecosystem analysis and the definition of boundaries in upwelling regions. *In* "Upwelling Ecosystems" (R. Boje and M. Tomczak, eds.), pp. 3-11. Springer-Verlag, Berlin and New York.

Caswell, H. (1976). The validation problem. *In* "Systems Analysis and Simulation in Ecology" (B. C. Patten, ed.), Vol. 4, pp. 313-325. Academic Press, New York.

Christian, R. R., and Wetzel, R. L. (1979). Interaction between substrate, microbes, and consumers of *Spartina* detritus in estuaries. *In* "Estuarine Interactions" (M. L. Wiley, ed.), pp. 93-113. Academic Press, New York.

Cushing, D. H. (1959). On the nature of production in the sea. *Minist. Agric., Fish Food (G.B.), Fish. Invest. 22,* 1-40

Daniel, C., and Wood, F. S. (1980). "Fitting Equations to Data," 2nd ed. Wiley, New York.

Davidson, R. S., and Clymer, A. B. (1966). The desirability and applicability of simulating ecosystems. *Ann. N. Y. Acad. Sci. 128,* 790-794.

DeLaune, R. D., and Patrich, W. H., Jr. (1980). Rate of sedimentation and its role in nutrient cycling in a Louisiana salt marsh. *In* "Estuarine and Weland Processes" (P. Hamilton and K. B. Macdonald, eds.), pp. 401-412. Plenum, New York.

Halfon, E., ed. (1979). "Theoretical Systems Ecology." Academic Press, New York.

Heinle, D. R., Flemer, D. A., Huff, R. T., Sulkins, S. T., and Ulanowicz, R. E. (1979). Effects of perturbations on estuarine microcosms. *In* "Marsh-Estuarine Systems Simulation" (R. F. Dame, ed.), pp. 119-141. Univ. of South Carolina Press, Columbia.

Hopkinson, C. S., Jr., and Day, J. W., Jr. (1977). A model of the Barataria Bay salt marsh ecosystem. *In* "Ecosystem Modeling in Theory and Practice" (C. A. S. Hall,Jr. and J. W. Day, Jr., eds.), pp. 236-265. Wiley (Interscience), New York.

Ivakhnenko, A. G., Krotov, G. I., and Visotskg, V. N. (1979). Identification of the mathematical model of a complex system by the self-organization method. *In* "Theoretical Systems Ecology" (E. Halfon, ed.), pp. 325-352. Academic Press, New York.

Kemp, W. M., Wetzel, R. L., Boynton, W. R., E'Elia, C. F., and Stevenson, J. C. (1983). Nitrogen cycling at estuarine interfaces: Current concepts and research directions. *In* "Estuarine Comparisons" (V. S. Kennedy, ed.), pp. 209-230. Academic Press, New York.

Kitchens, W. M., Edwards, R. T., and Johnson, W. V. (1979). Development of a "living" salt marsh ecosystem model: A microecosystem approach. *In* "Marsh-Estuarine Systems Simulation" (R. F. Dame, ed.), pp. 107-117. Univ. of South Carolina Press, Columbia.

Kremer, J. N. (1979). An analysis of the stability characteristics of an estuarine ecosystem model. *In* "Marsh-Estuarine Systems Simulation" (R. F. Dame, ed.), pp. 189-206. Univ. of South Carolina Press, Columbia.

Kremer, J. N., and Nixon, S. W. (1978). "A Coastal Marine Ecosystem--Simulation and Analysis." Springer-Verlag, Berlin and New York.

Levin, S. A., ed. (1975). "Ecosystem Analysis and Prediction." Soc. Ind. Appl. Math., Philadelphia, Pennsylvania.

Najarian, T. O., and Harleman, D. R. F. (1977). A real time model of nitrogen-cycle dynamics in an estuarine system. *Prog. Water Technol. 84,* 323-345.

Neilson, B. J., and Cronin, L. E., eds. (1981). "Estuaries and Nutrients." Humana Press, Clifton, New Jersey.

Nixon, S. W. (1980). Between coastal marshes and coastal waters. *In* "Estuarine and Welands Processes" (R. Hamilton and K. B. MacDonald, eds.), pp. 437-525. Plenum, New York.

Nixon, S. W. (1981). Remineralization and nutrient cycling in coastal marine systems. *In* "Estuaries and Nutrients" (B. J. Neilson and L. E. Cronin, eds.), pp. 111-138. Humana Press, Clifton, New Jersey.

Nixon, S. W., Oviatt, C. A., Kremer, J. N., and Perez, K. (1979). The use of numerical models and laboratory microcosms in estuarine ecosystem analysis-simulations of winter phytoplankton bloom. *In* "Marsh-Estuarine Systems Simulation" (R. F. Dame, ed.), pp. 165-188. Univ. of South Carolina Press, Columbia.

O'Connor, D. J. (1981). Modeling of eutrophication in estuaries. *In* "Estuaries and Nutrients" (B. J. Neilson and L. E. Cronin, eds.), pp. 183-223. Humana Press, Clifton, New Jersey.

Patten, B. C. (1968). Mathematical models of plankton production. *Int. Rev. Gesamten. Hydrobiol. 53,* 357-408.

Pomeroy, L. R. (1979). Secondary production mechanisms of continental shelf communities. *In* "Ecological Processes in Coastal and Marine Systems" (R. J. Livingston, ed.), pp. 163-186. Plenum, New York.
Riley, G. A. (1942). The relationship of vertical turbulence and spring diatom flowering. *J. Mar. Res. 5,* 67-87.
Riley, G. A. (1946). Factors controlling phytoplankton population in Georges Bank. *J. Mar. Res. 6,* 54-73.
Riley, G. A. (1947a). A theoretical analysis of the zooplankton population of Georges Bank. *J. Mar. Res. 6,* 104-113.
Riley, G. A. (1947b). Seasonal fluctuations of the phytoplankton population in New England coastal waters. *J. Mar. Res. 6,* 114-125.
Riley, G. A. (1967). Mathematical model of nutrient conditions in coastal waters. *Bull. Bingham Oceanogr. Collect. 19,* 72-80.
Riley, G. A., and Bumpus, D. F. (1946). Phytoplankton-zooplankton relationships on Georges Bank. *J. Mar. Res. 6,* 33-47.
Riley, G. A., Stommel, H., and Bumpus, D. F. (1949). Quantitative ecology of the plankton of the Western North Atlantic. *Bull. Bingham Oceanogr. Collect. 12,* 1-169.
Russel, C. S., ed. (1975). "Ecological Modeling in a Resources Management Framework." Johns Hopkins Univ. Press, Baltimore, Maryland.
Shoemaker, C. A. (1977). Mathematical construction of ecological models. *In* "Ecosystem Modeling in Theory and Practice" (C. A. S. Hall, Jr. and J. W. Day, Jr., eds.), pp. 76-114. Wiley (Interscience), New York.
Steele, J. H. (1974). "The Structure of Marine Ecosystems." Harvard Univ. Press, Cambridge, Massachusetts.
Swartzman, G. L., and Bentley, R. (1979). A review and comparison of plankton simulation models. *J. Int. Soc. Ecol. Modeling 1,* 30-81.
Vanderborght, J. P., Wollast, R., and Billen, G. (1977). Kinetic models of diagensis in disturbed sediments. Part 2. Nitrogen diagensis. *Limnol. Oceanogr. 22,* 794-803.
Walsh, J. J. (1972). Implication of a systems approach to oceanography. *Science 176,* 969-975.
Walsh, J. J. (1975a). A spatial simulation model of the Peruvian upwelling ecosystem. *Deep-Sea Res. 22,* 201-236.
Walsh, J. J. (1975b). Utility of systems models: A consideration of some possible feedback loops of the Peruvian upwelling ecosystem. *In* "Estuarine Research" (L. E. Cronin, ed.), Vol. 1, pp. 617-633. Academic Press, New York.
Walsh, J. J. (1981). A carbon budget for overfishing off Peru. *Nature (London) 290,* 300-304.
Walsh, J. J., and Dugdale, R. C. (1971). A simulation model of the nitrogen flow in the Peruvian upwelling system. *Invest. Pesq. 35,* 309-330.
Walsh, J. J., and Howe, S. O. (1976). Protein from the sea: A comparison of the simulated nitrogen and carbon productivity of the

Peruvian upwelling ecosystem. *In* Systems Analysis and Simulation in Ecology" (B. C. Patten, ed.), Vol. 4, pp. 47-61. Academic Press, New York.

Webb, K. L. (1981). Conceptual models and processes of nutrient cycling in estuaries. *In* "Estuaries and Nutrients" (B. J. Neilson and L. E. Cronin, eds.), pp. 25-46. Humana Press, Clifton, New Jersey.

Weinberg, G. M. (1975). "An Introduction to General Systems Thinking." Wiley, New York.

Weinberg, G. M., and Weinberg, D. (1979). "On the Design of Stable Systems." Wiley, New York.

Wetzel, R. L., and Kowalski, M. S. (1980). *In situ* studies of light and photosynthesis in Chesapeake Bay seagrass communities. Paper presented at Atlantic Estuarine Res. Soc. Conf., Nov. 6-8, Virginia Beach, Va.

Wetzel, R. L., van Tine, R. F., and Penhale, P. A. (1981). "Light and Submerged Macrophyte Communities in the Chesapeak Bay: A Scientific Summary," SRAMSOE 260. *Va. Inst. Mar. Sci.*, Gloucester Point.

Wiegert, R. G. (1975a). Simulation models of ecosystems. *Annu. Rev. Ecol. Syst. 6,* 311-338.

Wiegert, R. G. (1975b). Mathematical representation of ecological interactions. *In* "Ecosystem Analysis and Prediction" (S. A. Levin, ed.), pp. 43-54. Soc. Ind. Appl. Math., Philadelphia, Pennsylvania.

Wiegert, R. G. (1979). Modeling coastal, estuarine, and marsh ecosystems state-of-the-art. *In* "Contemporary Quantitative Ecology and Related Ecometrics" (G. P. Patil and M. Rosenzweig, eds.), pp. 319-341. International Co-operative Publishing House, Fairland, Maryland.

Wiegert, R. G., and Wetzel, R. L. (1974). The effect of numerical integration technique on the simulation of carbon flow in a Georgia salt marsh. *Proc. Summer Comput. Simul. Conf.* Vol. 2, pp 275-277.

Wiegert, R. G., and Wetzel, R. L. (1979). Simulation experiments with a 14-compartment salt marsh model. *In* "Marsh-Estuarine Systems Simulation" (R. F. Dame, ed.), pp. 7-39. Univ. of South Carolina Press, Columbia.

Wiegert, R. G., Christian, R. R., Gallagher, J. L., Hall, J. R., Jones, R. D. H., and Wetzel, R. L. (1975). A preliminary ecosystem model of coastal Georgia *Spartina* marsh. *In* "Estuarine Research" (L. E. Cronin, ed.), Vol. 1, pp. 583-601. Academic Press, New York.

Williams, J. P. LeB. (1973). The validity of the application of simple kinetic analysis to heterogeneous microbial populations. *Limnol. Oceanogr. 18,* 159-164.

INDEX

A

B

C

H

I

K

M

N

O

T

U